# Schwarz Methods and Multilevel Preconditioners for Boundary Element Methods

Ernst P. Stephan · Thanh Tran

# Schwarz Methods and Multilevel Preconditioners for Boundary Element Methods

Springer

Ernst P. Stephan
Leibniz University Hannover
Hannover, Germany

Thanh Tran
The University of New South Wales
Sydney, NSW, Australia

ISBN 978-3-030-79282-4 ISBN 978-3-030-79283-1 (eBook)
https://doi.org/10.1007/978-3-030-79283-1

Mathematics Subject Classification: 65N30, 65N38, 65N55, 35J25, 45B05



This Springer imprint is published by the registered company Springer Nature Switzerland AG
The registered company address is: Gewerbestrasse 11, 6330 Cham, Switzerland

*Dedicated to our wives*
*Karin Sabine and Thanh-Binh.*

# Preface

Several textbooks and monographs on preconditioners for FEM (finite element methods) are available, but none exists for BEM (boundary element methods). For BEM, the topic is dealt with in research papers only and is focused on specific situations. A systematic approach and display of preconditioners for BEM is now done for the first time with this monograph.

BEM have become an important tool for the numerical solution of boundary value problems covering a rich area of applications in engineering and physics. Roughly speaking, BEM consist of the reformulation of the boundary value problems into boundary integral equations which are then solved by discretisation of the boundary into finite elements; see the books by Hsiao and Wendland [128], McLean [147], Nédélec [161], Sauter and Schwab [186], Steinbach [199], and the recent book by Gwinner and Stephan [92] which deals especially with contact problems.

While the reformulation of boundary value problems into boundary integral equations has a great advantage of reducing the domain dimension by one, the nature of integral operators is non-local whereas that of differential operators is local. As a consequence, boundary element discretisations of first-kind integral equations lead to systems of equations with dense and ill-conditioned matrices whose condition numbers grow with the degrees of freedom. (It should be noted that finite element discretisations of partial differential equations also result in ill-conditioned matrices, albeit sparse.)

Non-locality of boundary integral operators indicates that they can be analysed with the calculus of pseudodifferential operators, where the correct setting of the equations leads to the study of fractional-order Sobolev spaces (like $H^{1/2}$ and $H^{-1/2}$). This is a crucial difference to FEM where the underlying differential equations live in standard integral-order Sobolev spaces (like $H^1$). There are major differences between integral-order and fractional-order Sobolev spaces, which requires different analyses of the discretisations (by Galerkin, collocation or quadrature methods) of boundary integral equations. Moreover, as the convergence analysis for preconditioners requires sophisticated dealings with Sobolev norms, these differences mean new analyses are required when studying preconditioners for equations governed by non-local operators.

This book presents in detail the necessary analysis and focusses on additive and multiplicative Schwarz methods (domain-decomposition-type methods) and on multilevel methods to develop optimal preconditioners for the respective solvers. These solvers include the conjugate gradient method (for matrices resulting from the Galerkin boundary element approximation of the weakly singular and hypersingular integral equations governing Dirichlet and Neumann problems for the Laplacian) and GMRES (generalised minimal residual methods) for Helmholtz problems in acoustics.

The convergence theory for preconditioning of boundary integral equations has been developed over more than two decades, with results derived and presented in different ways and published in different papers for different equations. This monograph has the benefit of hindsight to put the whole theory in a systematic framework. It is our hope that the reader will benefit from this book to perform their own analysis for their specific problem.

Survey articles on subjects of the book are [204, 206]. The book is addressed to mathematicians and engineers as well as graduate students. There are other preconditioning techniques and fast solvers for boundary integral equations which are not investigated in this book, e.g., multigrid methods [35, 171, 172], the so-called operator preconditioning which uses operators of opposite orders to construct preconditioners [123, 149, 198], boundary element tearing and interconnecting (BETI) methods [132, 133, 134, 135], fast multipole methods (see, e.g., [9, 165, 194, 217]), panel clustering (see, e.g., [93, 99, 100]), and $H$-matrices (see, e.g., [95, 98, 97]).

The book has 16 chapters which are grouped into six parts.

Part I has two chapters and is the core for the analysis in subsequent chapters. Chapter 1 introduces the problems to be studied in the book. We are particularly interested in the screen and crack problems on open curves (in two dimensions) and open surfaces (in three dimensions). All three versions of the Galerkin boundary element method, namely the $h$-version, $p$-version, and $hp$-version, are considered. This first chapter sets the scene for the following chapters.

Chapter 2 provides with detail the framework of preconditioners by Schwarz methods. How to understand the mechanism of these preconditioning techniques for proper implementation? What is required in the analysis? What is the connection between the different forms of the preconditioners, namely the matrix form, variational form, and operator form? Each form has its own advantage and their connection and equivalence need to be clarified. All these issues are addressed in this chapter. Apart from the standard convergence theory for the preconditioned conjugate gradient method, we also present the convergence theory for preconditioners with GMRES in different settings, including problems with block matrices. To the best of our knowledge, this is the first systematic presentation on this topic. The chapter also presents the hybrid modified conjugate residual (HMCR) method for FEM-BEM coupling problems. The chapter ends with a section on the convergence theory for linear iterative methods which are used to solve preconditioned systems.

Part II covers preconditioners for two-dimensional problems. Chapter 3 presents the methods and analysis for additive and multiplicative Schwarz preconditioners. Both non-overlapping and overlapping methods are studied for the $h$-version Galerkin discretisation for the hypersingular and weakly singular integral equations. Chapter 4 treats the $p$-version Galerkin discretisation in the same manner as Chapter 3. Chapter 5 and Chapter 6 present multilevel methods for the $h$- and $hp$-versions of the hypersingular and weakly singular integral equations. A fully-discrete method is the topic of Chapter 7. Chapter 8 deals with indefinite problems when the boundary integral equations are the reformulation of boundary value problems with the Helmholtz equation. Non-overlapping additive Schwarz preconditioners are designed and analysed for the $h$- and $p$-version approximations of these boundary integral equations. Chapter 9 provides numerical results for the methods studied in Chapter 3–Chapter 6 and Chapter 8.

Part III deals with three-dimensional problems. As at this stage the reader should have been familiar with the required analysis, in this part we focus on the most general version of the Galerkin approximation, namely the $hp$-version. Results for the $h$-version and $p$-version, respectively, can be directly deduced by setting $p$ fixed, respectively, $h$ fixed, and disregard their corresponding treatment in the preconditioner. Two-level additive Schwarz preconditioners are considered in Chapter 10 for rectangular elements, while triangular elements are dealt with in Chapter 11. The last two chapters of this part study the effect of local mesh refinements to the condition number and propose strategies to overcome this side effect. Chapter 12 deals with the preconditioner by diagonal scaling, which effectively reduces the effect of local mesh refinements so that the condition number only grows as in the case of uniform refinements. Multilevel preconditioners with adaptive mesh refinements are the topic of Chapter 13.

Part IV is devoted to the treatment of preconditioners for the coupling of finite element and boundary element methods (Chapter 14). Both the symmetric and non-symmetric coupling methods with the GMRES solver are studied. A non-standard solver (HMCR) is also studied for the symmetric coupling. The unified theory developed in this chapter can be used to analyse different preconditioners.

Part V aims at showing robustness of additive Schwarz preconditioners. Departing from the standard boundary integral equations on open boundaries and the standard FEM and BEM considered in previous parts, we study in this part pseudo-differential equations on the sphere. Chapter 15 investigates preconditioners when spherical splines are used in the approximation. Chapter 16 deviates further from the boundary element context. In this chapter, meshless methods with radial basis functions are employed in the approximation. In both cases, it is shown that additive Schwarz preconditioners provide an efficient tool to solve the problem at a low cost. Even though spherical splines do not bear the same properties as finite and boundary elements, some modifications of the analysis are sufficient for this type of approximation. On the contrary, a completely different approach has to be employed for meshless methods.

Part VI contains three appendices where we collect important facts on interpolation and Sobolev spaces (Appendix A), boundary integral operators (Appendix B), and various technical lemmas (Appendix C) in order to make the book self-contained. In particular, in Appendix A we present new results on local and global properties of fractional-order Sobolev norms; see Subsection A.2.9. These new results help fill the gap in the analysis in current literature. This appendix also provides rigorous proofs of the equivalence of different types of Sobolev norms, which is required in the study of preconditioners.

It should be noted that the screen problem and crack problem (in linear elasticity) are the cases where the solution is most singular. The preconditioners developed in this book are also applicable to other problems defined on domains with closed boundaries, e.g., on polyhedral domains. Furthermore, even though we present preconditioners for the screen problem only, the same algorithms work for the crack problem (being a system of equations). It suffices to apply the preconditioners to the various components of the vector fields.

Altogether, this book showcases state-of-the-art preconditioning techniques for boundary integral equations with up-to-date research. Thorough treatments of interpolation spaces and Sobolev spaces of fractional orders familiarise the reader with necessary techniques in the analysis of preconditioners for first-kind boundary integral equations. The concise presentation of adaptive BEM, *hp*-version BEM, and coupling of FEM-BEM provides the reader with very efficient computational tools for solving practical problems with applications in mechanical engineering, acoustics, geodesy, and many other fields which involve boundary value problems in unbounded domains.

Hannover, Sydney,
April 2021

*Ernst P. Stephan*
*Thanh Tran*

# Acknowledgements

This book could not be completed without the tremendous support that we received from many different sources. First and foremost, the support of Mathematisches Forschungsinstitut Oberwolfach (through its Research-in-Pairs Grant, reference number 1815q) and the Australian Research Council (through DP160101755 and DP190101197) is gratefully acknowledged. A large part of the book was drafted during the authors' stay at MFO in April 2018 and during Stephan's visit to UNSW Sydney in November 2017. We also thank Springer Nature for its understanding of the significant delay in finalising the book due to the COVID-19 pandemic.

The authors thank their colleagues for their collaboration which has deeply influenced the contents of this book. They are (in alphabetical order): M. Ainsworth, M. Feischl, T. Führer, N. Heuer, T. Le Gia, F. Leydecker, M. Maischak, W. McLean, T.D. Pham, D. Praetorius, I.H. Sloan, and A. Wathen.

We take the opportunity to acknowledge our gratitude to our universities, the Leibniz University of Hannover, Germany, and the University of New South Wales, Sydney, Australia, for supporting our visits.

Last but not least, we thank our wives Karin Sabine and Thanh-Binh for their great understanding and support during the work-intensive time it took us to write this book.

# Contents

# Part I
# General Theory

# Chapter 1
# Introduction

This chapter sets the scene for the problems to be studied in the book. We first introduce in Section 1.1 typical problems from which boundary integral equations arise, namely, the scattering of time harmonic waves and cracks in elastic media. Single-layer and double-layer potentials with the appropriate fundamental solutions are given together with the first kind integral equations for Dirichlet and Neumann problems with weakly-singular and hypersingular boundary integral equations, respectively.

Section 1.2 briefly introduces the corresponding Galerkin schemes for solving approximately these boundary integral equations. It is well known that the solutions of the screen and crack problems in Subsection 1.1.1 and Subsection 1.1.2 are less regular near the screen or crack. The solutions of the boundary integral equations inherit this lack of smoothness. Therefore, to achieve desirable approximations, one uses besides the $h$-version (where better approximation is obtained by refining the mesh) the $p$-version (which yields better approximation by increasing the polynomial degree) and the $hp$-version as a combination of both previous versions.

Besides (locally) quasi-uniform meshes, one takes (algebraically or geometrically) graded meshes to obtain even more accurate approximate solutions of the integral equations. If the location where the solution becomes less regular is not known a priori, it pays off to apply adaptive schemes. These will also be mentioned in Section 1.2.

The Galerkin schemes described above result in systems of linear equations with dense and ill-conditioned matrices. As key solvers, we advocate the conjugate gradient method for symmetric positive definite matrices, and the GMRES for non-symmetric matrices. Both methods will be discussed in detail in Chapter 2. The ill conditioness of the matrices is discussed in Section 1.3.

As the matrices are ill-conditioned, the numbers of iterations of the solves blow up quickly with the degrees of freedom of the various Galerkin schemes. Therefore, preconditioning must be performed. As the matrices are dense, this is not a straightforward task. The following chapters describe and analyse domain decomposition and multilevel methods as a remedy for these defects.



E. P. Stephan and T. Tran, *Schwarz Methods and Multilevel Preconditioners for Boundary Element Methods*, https://doi.org/10.1007/978-3-030-79283-1_1

## 1.1 Model Problems

Let $\Gamma \subset \mathbb{R}^d$ be a (curved) slit when $d = 2$ or a screen with boundary $\gamma$ when $d = 3$, and let $\mathbf{n}$ be a normal vector to $\Gamma$. To fix the idea, we define $\Gamma$ as

$$\Gamma := \begin{cases} (-1,1) \times \{0\} & \text{if } d = 2, \\ (-1,1) \times (-1,1) \times \{0\} & \text{if } d = 3, \end{cases}$$

and denote by $\Gamma_1$ and $\Gamma_2$ the two sides of $\Gamma$ determined by $\mathbf{n}$; see Figure 1.1.

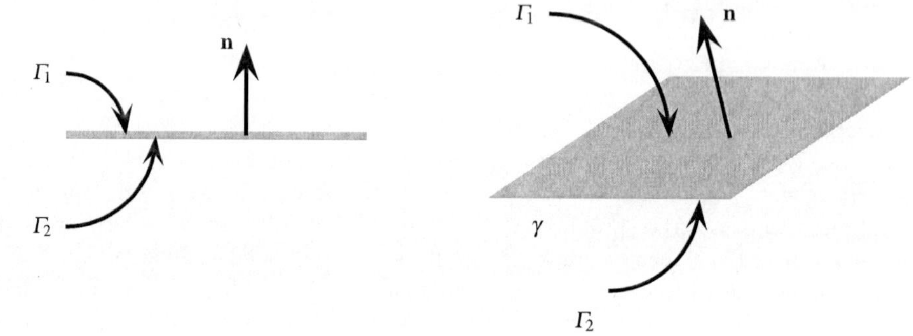


**Fig. 1.1** Screen $\Gamma \subset \mathbb{R}^d$ and sides $\Gamma_1, \Gamma_2$ (left: $d = 2$, right: $d = 3$)

### *1.1.1 Screen problems*

Denoting $\Omega_\Gamma := \mathbb{R}^d \setminus \overline{\Gamma}$ and $r := |x|$ (where $|x|$ is the Euclidean norm of $x \in \mathbb{R}^d$) we consider the Dirichlet screen problem

$$\text{(DSP)} \begin{cases} (\Delta + \kappa^2)u = 0 & \text{in } \Omega_\Gamma, \\ u = f_i & \text{on } \Gamma_i,\ i = 1,2, \end{cases}$$

and the Neumann screen problem

$$\text{(NSP)} \quad \begin{cases} (\Delta+\kappa^2)u = 0 & \text{in } \Omega_\Gamma, \\ \dfrac{\partial u}{\partial n} = g_i & \text{on } \Gamma_i,, \ i=1,2, \end{cases}$$

where $\mathrm{Im}(\kappa) \geq 0$ is the wave number, the data $f_i$ and $g_i$, $i = 1,2$, are given functions to be specified later. For each problem, the solution also satisfies the radiation condition at infinity, i.e., the condition when $r \to \infty$,

$$\text{(IC)} \quad \begin{cases} \text{For } \kappa = 0: & u(x) = \begin{cases} O(\log r) & \text{if } d=2, \\ O(1/r) & \text{if } d=3. \end{cases} \\ \text{For } \kappa \neq 0: & u(x) = \begin{cases} O(1/\sqrt{r}) & \text{if } d=2, \\ O(1/r) & \text{if } d=3, \end{cases} \\ \text{For } \kappa \neq 0: & \dfrac{\partial u}{\partial r} - i\kappa u = \begin{cases} o(1/\sqrt{r}) & \text{if } d=2, \\ o(1/r) & \text{if } d=3. \end{cases} \end{cases}$$

Let $G(x,y)$ denote the fundamental solution given by

$$G(x,y) = \begin{cases} -\frac{1}{2\pi}\log|x-y|, & d=2, \ \kappa=0, \\ \frac{i}{4}H_0^{(1)}(\kappa|x-y|), & d=2, \ \kappa \neq 0, \\ \frac{1}{4\pi}\dfrac{e^{i\kappa|x-y|}}{|x-y|}, & d=3, \end{cases}$$

where $H_0^1$ is the Hankel function of order zero and of the first kind. We recall the single-layer and double-layer potential operators

$$\mathfrak{S}v(x) := 2\int_\Gamma G(x,y)v(y)\,\mathrm{d}\sigma_y, \quad x \in \mathbb{R}^d,$$

$$\mathfrak{D}v(x) := 2\int_\Gamma \frac{\partial G(x,y)}{\partial n_y}v(y)\,\mathrm{d}\sigma_y, \quad x \in \mathbb{R}^d.$$

The representation formula states that the solution $u$ of (DSP) and (NSP) with the condition (IC) can be represented as

$$u(x) = \frac{1}{2}\mathfrak{D}[u]_\Gamma(x) - \frac{1}{2}\mathfrak{S}\left[\frac{\partial u}{\partial n}\right]_\Gamma(x), \quad x \in \Omega_\Gamma,$$

where, for any function $v$, the notation $[v]_\Gamma := \gamma_1 v - \gamma_2 v$ denotes the jump of $v$ across $\Gamma$. Here $\gamma_i$ is the trace operator onto $\Gamma_i$, $i = 1,2$. Hence, for the (DSP) problem it suffices to find $\left[\frac{\partial u}{\partial n}\right]_\Gamma$ whereas for the (NSP) problem $[u]_\Gamma$ is required.

Let $\mathfrak{V} : \widetilde{H}^{-1/2}(\Gamma) \to H^{1/2}(\Gamma)$ and $\mathfrak{W} : \widetilde{H}^{1/2}(\Gamma) \to H^{-1/2}(\Gamma)$ be defined by

$$\mathfrak{V}v(x) := 2\int_{\Gamma} G(x,y)v(y)\,\mathrm{d}s_y, \quad x \in \Gamma,$$

$$\mathfrak{W}v(x) := -2\frac{\partial}{\partial n_x}\int_{\Gamma} \frac{\partial G(x,y)}{\partial n_y}v(y)\,\mathrm{d}s_y, \quad x \in \Gamma.$$

It is known that $\mathfrak{V} : \widetilde{H}^{-1/2}(\Gamma) \to H^{1/2}(\Gamma)$ and $\mathfrak{W} : \widetilde{H}^{1/2}(\Gamma) \to H^{-1/2}(\Gamma)$ are bijective, provided that $\mathrm{Im}(\kappa) \geq 0$. It is noted that in the case that $\kappa = 0$ and $d = 2$, we further assume that $\mathrm{cap}(\Gamma) < 1$, where $\mathrm{cap}(\Gamma)$ denotes the logarithmic capacity of $\Gamma$. This assumption, which can always be obtained by scaling, ensures the positive definiteness of $\mathfrak{V}$; see e.g. [147]. For the definition of Sobolev spaces, please see Appendix A, and for further details on the properties of $\mathfrak{V}$ and $\mathfrak{W}$, please see Appendix B.

We assume that the Dirichlet and Neumann data in (DSP) and (NSP) satisfy $f_1, f_2 \in H^{1/2}(\Gamma)$ and $g_1, g_2 \in H^{-1/2}(\Gamma)$ such that $f_1 - f_2 \in \widetilde{H}^{1/2}(\Gamma)$, $g_1 - g_2 \in \widetilde{H}^{-1/2}(\Gamma)$ and $\int_\Gamma g_1 - g_2 = 0$. Letting $\varphi = \left[\frac{\partial u}{\partial n}\right]_\Gamma$ and $\psi = [u]_\Gamma$, then the (DSP)–(IC) problem is equivalent to the following weakly-singular integral equation

$$\mathfrak{V}\varphi(x) = f(x), \quad x \in \Gamma, \tag{1.1}$$

whereas the (NSP)–(IC) problem is equivalent to the hypersingular integral equation

$$\mathfrak{W}\psi(x) = -g(x), \quad x \in \Gamma. \tag{1.2}$$

The right-hand sides of (1.1) and (1.2) are defined by

$$f(x) = f_1(x) + f_2(x) + 2\int_{\Gamma} \big(f_1(y) - f_2(y)\big)\frac{\partial G(x,y)}{\partial n_y}\,\mathrm{d}\sigma_y$$

$$g(x) = g_1(x) + g_2(x) - 2\int_{\Gamma} \big(g_1(y) - g_2(y)\big)\frac{\partial G(x,y)}{\partial n_x}\,\mathrm{d}\sigma_y;$$

see [67] for the general case and [202] for the case when $f_1 = f_2$ and $g_1 = g_2$.

We define

$$a_{\mathfrak{V}}(\varphi,\psi) := \langle \mathfrak{V}\varphi, \psi\rangle \quad \forall \varphi, \psi \in \widetilde{H}^{-1/2}(\Gamma),$$
$$a_{\mathfrak{W}}(\varphi,\psi) := \langle \mathfrak{W}\varphi, \psi\rangle \quad \forall \varphi, \psi \in \widetilde{H}^{1/2}(\Gamma),$$

A weak formulation for (1.1) reads: Find $\varphi \in \widetilde{H}^{-1/2}(\Gamma)$ satisfying

$$a_{\mathfrak{V}}(\varphi,\phi) = \langle f,\phi\rangle, \quad \phi \in \widetilde{H}^{-1/2}(\Gamma). \tag{1.3}$$

Similarly, a weak formulation for (1.2) reads: Find $\psi \in \widetilde{H}^{1/2}(\Gamma)$ satisfying

$$a_{\mathfrak{W}}(\psi,\phi) = -\langle g,\phi\rangle, \quad \phi \in \widetilde{H}^{1/2}(\Gamma). \tag{1.4}$$

The existence and uniqueness of the solutions of (1.3) and (1.4) are studied in [68, 202, 212].

### *1.1.2 Crack problems in elasticity*

We also consider the following crack problems involving the Lamé operator. The Dirichlet crack problem is defined as follows. Given two functions $\mathbf{f}_1, \mathbf{f}_2 \in (H^{1/2}(\Gamma))^d$, $d = 2,3$, such that $\mathbf{f} := \mathbf{f}_1 - \mathbf{f}_2 \in (\widetilde{H}^{1/2}(\Gamma))^d$, find $\mathbf{u} \in (H^1_{\mathrm{loc}}(\Omega))^d$ satisfying

$$
\text{(DCP)} \quad \begin{cases} \mu\Delta\mathbf{u} + (\lambda+\mu)\,\mathrm{grad}\,\mathrm{div}\,\mathbf{u} = 0 & \text{in } \Omega_\Gamma, \\ \mathbf{u} = \mathbf{f}_i & \text{on } \Gamma_i,\ i = 1,2, \\ \mathbf{u}(x) = o(1) & \text{as } |x| \to \infty, \\ \nabla\mathbf{u}(x) = o(1/|x|) & \text{as } |x| \to \infty. \end{cases}
$$

The Neumann crack problem is defined as follows. Given two functions $\mathbf{g}_1, \mathbf{g}_2 \in (H^{-1/2}(\Gamma))^d$ such that $\mathbf{g} := \mathbf{g}_1 - \mathbf{g}_2 \in (\widetilde{H}^{-1/2}(\Gamma))^d$, find $\mathbf{u} \in (H^1_{\mathrm{loc}}(\Omega))^d$ satisfying

$$
\text{(NCP)} \quad \begin{cases} \mu\Delta\mathbf{u} + (\lambda+\mu)\,\mathrm{grad}\,\mathrm{div}\,\mathbf{u} = 0 & \text{in } \Omega_\Gamma, \\ T\mathbf{u} = \mathbf{g}_i & \text{on } \Gamma_i,\ i = 1,2, \\ \mathbf{u}(x) = o(1) & \text{as } |x| \to \infty, \\ \nabla\mathbf{u}(x) = o(1/|x|) & \text{as } |x| \to \infty, \end{cases}
$$

where the Lamé constants $\mu$ and $\lambda$ satisfy

$$
\mu > 0 \quad \text{and} \quad \begin{cases} \lambda + \mu > 0 & \text{for } d = 2, \\ \lambda + 2\mu > 0 & \text{for } d = 3, \end{cases}
$$

and where the traction $T\mathbf{u} \in \mathbb{R}^d$ is defined by

$$
T\mathbf{u} := \lambda(\mathrm{div}\,\mathbf{u})\mathbf{n} + 2\mu\frac{\partial\mathbf{u}}{\partial\mathbf{n}} + \mu\mathbf{n}\times\mathrm{curl}\,\mathbf{u}. \tag{1.5}
$$

It is shown in [201] that the solution of (DCP) admits the integral representation

$$
\mathbf{u}(x) = \int_\Gamma \mathbf{G}(x,y)[T\mathbf{u}](y)\,d\sigma_y, \quad x \in \Omega_\Gamma,
$$

where

$$
\mathbf{G}(x,y) := \frac{\lambda+3\mu}{4\pi(d-1)\mu(\lambda+2\mu)}\left\{\widetilde{G}(x,y)\mathbf{I}_{d\times d} + \frac{\lambda+\mu}{\lambda+3\mu}\frac{(x-y)(x-y)^\top}{|x-y|^d}\right\}
$$

with $\mathbf{I}_{d\times d}$ being the identity matrix of size $d\times d$ and

$$\widetilde{G}(x,y) := \begin{cases} -\log|x-y| & \text{for } d=2, \\ \dfrac{1}{|x-y|} & \text{for } d=3. \end{cases}$$

The jump of the traction $T\mathbf{u}$ across $\Gamma$, namely $[T\mathbf{u}]$, is the solution to the following weakly-singular integral equation

$$\mathfrak{V}\boldsymbol{\varphi} = \mathbf{f} \quad \text{on } \Gamma \tag{1.6}$$

where

$$\mathfrak{V}\boldsymbol{\varphi}(x) := 2\int_{\Gamma} \mathbf{G}(x,y)\boldsymbol{\varphi}(y)\, d\sigma_y, \quad x \in \Gamma,$$

and

$$\mathbf{f} := \mathbf{f}_1 + \mathbf{f}_2 + \Lambda(\mathbf{f}_1 - \mathbf{f}_2)$$

with

$$\Lambda\mathbf{v}(x) := 2\int_{\Gamma} \left[T_y\mathbf{G}(x,y)\right]^{\top} \mathbf{v}(y)\, d\sigma_y.$$

Here, $T_y$ is the traction $T$ defined in (1.5) acting in the $y$-variable.

It is also shown in [67, 201] that the solution of (NCP) admits the representation

$$\mathbf{u}(x) = \int_{\Gamma} [\mathbf{u}](y) \left[T_y\mathbf{G}(x,y)\right]^{\top} d\sigma_y$$

with the jump $[\mathbf{u}]$ being the solution of

$$\mathfrak{W}\boldsymbol{\psi} = -\mathbf{g} \quad \text{on } \Gamma \tag{1.7}$$

where

$$\mathfrak{W}\boldsymbol{\psi}(x) := -2T_x \int_{\Gamma} \left(T_y\mathbf{G}(x,y)\right)^{\top} \boldsymbol{\psi}(y)\, d\sigma_y$$

and

$$\mathbf{g}(x) := \left(\mathbf{g}_1(x) + \mathbf{g}_2(x)\right) + \Lambda'(\mathbf{g}_1 - \mathbf{g}_2)$$

where

$$\Lambda'\mathbf{v}(x) := 2T_x \int_{\Gamma} \mathbf{G}(x,y)\mathbf{v}(y)\, d\sigma_y.$$

The existence and uniqueness of weak solutions to (1.6) and (1.7) are studied in [67, 201].

It should be noted that corresponding settings can be formulated for problems on closed curves or closed surfaces. Moreover, even though we choose to present the settings for the screen problem, the preconditioners and analysis in the following chapters are applicable to the crack problem as well.

## 1.2 Galerkin Boundary Element Methods

The solutions of (1.3) and (1.4) are found numerically by solving these equations in finite dimensional spaces. More precisely, let $\widetilde{\mathscr{V}}$ and $\mathscr{V}$ be finite dimensional subspaces of $\widetilde{H}^{-1/2}(\Gamma)$ and $\widetilde{H}^{1/2}(\Gamma)$, respectively, to be defined later. The solution $\varphi$ of (1.3) is approximated by $\varphi_{\widetilde{\mathscr{V}}} \in \widetilde{\mathscr{V}}$ satisfying

$$a_{\mathfrak{V}}(\varphi_{\widetilde{\mathscr{V}}}, \phi) = \langle f, \phi \rangle, \quad \phi \in \widetilde{\mathscr{V}}. \tag{1.8}$$

Similarly, the solution of (1.4) is approximated by $\psi_{\mathscr{V}} \in \mathscr{V}$ satisfying

$$a_{\mathfrak{W}}(\psi_{\mathscr{V}}, \phi) = \langle g, \phi \rangle, \quad \phi \in \mathscr{V}. \tag{1.9}$$

The choice of $\widetilde{\mathscr{V}}$ and $\mathscr{V}$ determines different versions of the Galerkin method.

### 1.2.1 The $h$-version

Consider a partition $\mathscr{T}_h$ of $\Gamma$ into subintervals (when $d = 2$) or triangles/rectangles (when $d = 3$). On this mesh, we define $\widetilde{\mathscr{V}}$ to be the space of piecewise constant functions and $\mathscr{V}$ to be the space of continuous piecewise affine functions. In this version, the accuracy of the approximation is obtained by considering a family of partitions $\{\mathscr{T}_h\}_{h>0}$ with the space size $h$ tending to 0. The convergence of this approximation scheme is standard and goes back to the early works by Nédélec & Planchard [162], Hsiao & Wendland [127] for the weakly-singular integral equation, and Nédélec [160] for the hypersingular integral equation. For polygonal domains, see [65] and for cracks and screens [201, 202].

### 1.2.2 The $p$-version

In the $p$-version, we fix one partition $\mathscr{T}_h$ and define $\widetilde{\mathscr{V}}$ and $\mathscr{V}$ to be the space of piecewise polynomial functions and continuous piecewise polynomial functions of degree $p$, respectively. Accuracy is achieved by increasing $p$. The stability and convergence of the schemes is proved in [208]; see also [190]. More recent works include [30, 91].

### 1.2.3 The *hp*-version

This version uses the same finite element spaces as in the $p$-version and seeks to increase accuracy by both increasing $p$ and decreasing $h$. The stability and convergence of the schemes is proved in [209].

### 1.2.4 Graded meshes

On a geometric mesh, the $hp$-version yields convergence of exponential rates. Early works on curves include [17, 114, 203]. Other works on surfaces include [112, 124, 125, 142].

### 1.2.5 Adaptive schemes

If a certain accuracy of the solution $\varphi_{\widetilde{\mathscr{V}}}$ of (1.8) or $\psi_{\mathscr{V}}$ of (1.9) is required, adaptive mesh-refining algorithms of the type

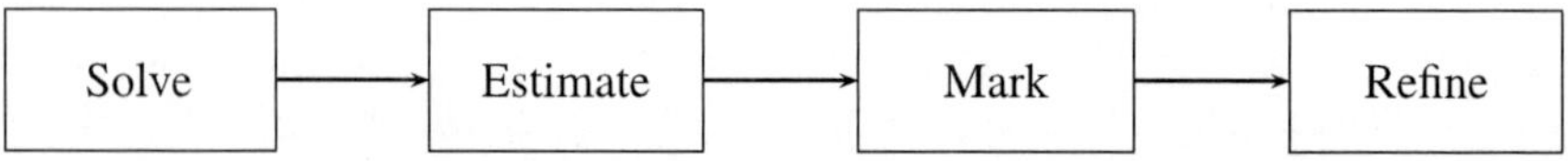


are used where, starting with a given initial triangulation $J_0$, a sequence of locally refined triangulations $J_l$ and corresponding Galerkin solutions $\varphi_l$ and $\psi_l$, respectively, are computed. Then the adaptive mesh refinement leads to a nested sequence of spaces $\widetilde{\mathscr{V}}(J_l) \subset \widetilde{\mathscr{V}}(J_{l+1})$ and $\mathscr{V}(J_l) \subset \mathscr{V}(J_{l+1})$, respectively, for $l = 0, 1, 2, \ldots$.

A posteriori error analysis for boundary integral equations with error estimation of residual type are derived in [53, 54, 55]. In recent years, convergence and optimality of adaptive BEM have been shown in [50, 74]; see also [92].

## 1.3 Condition Numbers

Equations (1.8) and (1.9) and corresponding equations for the $p$ and $hp$-versions result in matrix systems of the form

$$\mathbf{Au} = \mathbf{f}.$$

When the size of $\mathbf{A}$ is large, one resorts to the use of the conjugate gradient (CG) method if $\mathbf{A}$ is symmetric (which is the case when the wave number $\kappa$ in (DP) and (NP) is zero) or the generalised minimal residual (GMRES) method if $\mathbf{A}$ is unsymmetric (which is the case when $\kappa \neq 0$). These two methods will be discussed in detail in Chapter 2.

As will be seen in Chapter 2, if the matrix $\mathbf{A}$ is symmetric and positive definite, then the convergence of the CG method depends on the condition number

$$\kappa_2(\mathbf{A}) = \frac{\lambda_{\max}(\mathbf{A})}{\lambda_{\min}(\mathbf{A})},$$

where $\lambda_{\max}(\mathbf{A})$ and $\lambda_{\min}(\mathbf{A})$ are respectively the maximum and minimum eigenvalues of the matrix $\mathbf{A}$. On the other hand, if $\mathbf{A}$ is non-symmetric and positive definite, then the convergence of the GMRES method is dictated by the minimum eigenvalue of the symmetric part of $\mathbf{A}$ and the maximum eigenvalue of $\mathbf{A}^\top\mathbf{A}$. In the appendices, we will show how the condition numbers depend on the mesh size $h$ and the polynomial degree $p$.

As will be seen in Chapter 2, when the condition number is large, more iterations are required to obtain desirable accuracy of the approximation. In order to speed up the convergence, it is essential to apply preconditioning to the matrix. This is the subject of the next chapter.

# Chapter 2
# General Framework of Preconditioners

This chapter provides the general framework of preconditioning by additive and multiplicative Schwarz operators. The theory is developed for a general variational equation on a Hilbert space. Preconditioners are viewed from different angles: in the matrix form, variational form, and operator form. We discuss how to implement the preconditioner with the conjugate gradient method (CG), the generalised minimum residual method (GMRES), and with a linear iterative method. Most importantly, we present the theory of convergence which justifies what needs to be proved to show the efficiency of the designed preconditioner. A thorough investigation of bounds for the condition number and convergence rate is presented.

In Section 2.1 we introduce the general weak formulation together with the required assumptions. For convenience, we also present the CG and GMRES algorithms and error bounds for each algorithm.

The general concept of employing the CG and GMRES with preconditioners is presented in Section 2.2. Various versions of the preconditioned CG and GMRES algorithms are explained in this section. We also present in this section how a linear iterative method is used to solve the preconditioned system.

Domain decomposition methods are designed by using Schwarz operators, which is the subject of Section 2.3. This section is followed by a section on additive Schwarz preconditioners (Section 2.4) and a section for multiplicative Schwarz preconditioners (Section 2.5). In each section we present different forms of the preconditioners: the variational and operator forms (which are used in the analysis) and the matrix form (which is used in the implementation).

The last four sections are the core of the chapter and will often be referred to in the remainder of the book. Section 2.6 and Section 2.7 provide important ingredients for carrying out the analysis for preconditioners to be used with CG and GMRES, respectively. We present in these two sections the required estimates one has to show to be able to ensure convergence. Section 2.8 extends the commonly used CG and GMRES to other Krylov methods including ORTHODIR and the hybrid modified conjugate residual (HMCR) method which will be used for preconditioning of symmetric indefinite matrices (in FEM-BEM coupling problems). The chapter concludes with Section 2.9 which presents the convergence analysis for a linear it-

 
E. P. Stephan and T. Tran, *Schwarz Methods and Multilevel Preconditioners for Boundary Element Methods*, https://doi.org/10.1007/978-3-030-79283-1_2

erative method when it is used to solve a system preconditioned by the additive or multiplicative Schwarz preconditioner.

## 2.1 Finite Dimensional Problems

We present in this section the general form that covers all equations discussed in Section 1.2. The weak formulation has the following form.

Let $a(\cdot,\cdot) : \widetilde{H}^m(\Gamma) \times \widetilde{H}^m(\Gamma) \longrightarrow \mathbb{R}$ be a bilinear form, and $\ell \in H^{-m}(\Gamma)$, $m \in \mathbb{R}$, where $\Gamma$ is an open curve in $\mathbb{R}^2$ or an open surface in $\mathbb{R}^3$. Find $u \in \widetilde{H}^m(\Gamma)$ satisfying

$$a(u,v) = \langle \ell, v\rangle \quad \forall v \in \widetilde{H}^m(\Gamma). \tag{2.1}$$

When $m = 1/2$ equation (2.1) is the hypersingular integral equation arising from the Helmholtz or Laplace equation with a Neumann boundary condition. On the other hand, when $m = -1/2$ equation (2.1) is the weakly-singular integral equation arising from the Helmholtz or Laplace equation with a Dirichlet boundary condition.

The bilinear form $a(\cdot,\cdot)$ is assumed to satisfy one of the following assumptions:

**Assumption 2.1.**

(i) The bilinear form $a(\cdot,\cdot)$ is symmetric;
(ii) There exists a positive constant $C_{A_1}$ such that for any $u, v \in \widetilde{H}^m(\Gamma)$

$$|a(u,v)| \le C_{A_1} \|u\|_{\widetilde{H}^m(\Gamma)} \|v\|_{\widetilde{H}^m(\Gamma)};$$

(iii) There exists a positive constant $C_{A_2}$ such that for any $u \in \widetilde{H}^m(\Gamma)$

$$|a(u,u)| \ge C_{A_2} \|u\|^2_{\widetilde{H}^m(\Gamma)}.$$

**Assumption 2.2.**

(i) The bilinear form $a(\cdot,\cdot)$ can be decomposed as $a(\cdot,\cdot) = b(\cdot,\cdot) + k(\cdot,\cdot)$ where the bilinear form $b(\cdot,\cdot)$ is symmetric and $k(\cdot,\cdot)$ is skew-symmetric;
(ii) There exists a positive constant $C_{B_1}$ such that for any $u, v \in \widetilde{H}^m(\Gamma)$

$$|b(u,v)| \le C_{B_1} \|u\|_{\widetilde{H}^m(\Gamma)} \|v\|_{\widetilde{H}^m(\Gamma)}$$
$$|k(u,v)| \le C_{B_1} \|u\|_{\widetilde{H}^{m-1/2}(\Gamma)} \|v\|_{\widetilde{H}^m(\Gamma)}$$
$$|k(u,v)| \le C_{B_1} \|u\|_{\widetilde{H}^m(\Gamma)} \|v\|_{\widetilde{H}^{m-1/2}(\Gamma)}.$$

(iii) There exist positive constants $C_{B_2}$, $C_{B_3}$, and $C_{B_4}$ such that for any $u \in \widetilde{H}^m(\Gamma)$

$$b(u,u) \ge C_{B_2} \|u\|^2_{\widetilde{H}^m(\Gamma)},$$
$$\Re(a(u,u)) \ge C_{B_3} \|u\|^2_{\widetilde{H}^m(\Gamma)} - C_{B_4} \|u\|^2_{\widetilde{H}^{m-1/2}(\Gamma)}.$$

Returning to the model problems introduced in Chapter 1, it is noted that the weak formulations arising from the weakly-singular and hypersingular integral equations for the screen problem with $\kappa = 0$ satisfy Assumption 2.1, while the screen problem with $\kappa \neq 0$ and the crack problem satisfy Assumption 2.2; see [201, 202].

### 2.1.1 Galerkin equations and matrix systems

Let $\mathscr{V}$ be a finite-dimensional subspace of a Hilbert space $\mathscr{H}$ which has a basis $\{\Phi_1, \ldots, \Phi_N\}$, i.e.,

$$\mathscr{V} = \text{span}\{\Phi_1, \ldots, \Phi_N\}.$$

Then the Galerkin solution to (2.1) is given by: $u \in \mathscr{V}$ satisfying

$$a(u,v) = \langle f, v \rangle \quad \forall v \in \mathscr{V}. \tag{2.2}$$

Representing the solution $u$ by $u = \sum_{i=1}^N u_i \Phi_i$, we rewrite (2.2) as

$$\sum_{i=1}^N u_i a(\Phi_i, \Phi_j) = \langle f, \Phi_j \rangle, \quad j = 1, \ldots, N.$$

This set of equations can be represented in the matrix form as

$$\mathbf{A}\mathbf{u} = \mathbf{b} \tag{2.3}$$

where the matrix $\mathbf{A}$ and the vectors $\mathbf{u}$ and $\mathbf{b}$ are defined respectively by

$$\mathbf{A} = \left(a(\Phi_i, \Phi_j)\right)_{i,j=1}^N, \quad \mathbf{u} = (u_1, \ldots, u_N)^\top \quad \text{and} \quad \mathbf{b} = \left(\langle f, \Phi_1 \rangle, \ldots, \langle f, \Phi_N \rangle\right)^\top.$$

Here $\top$ denotes the transpose matrix.

In practice, the system (2.3) is large and therefore iterative methods are preferable to direct solvers. The next two subsections introduce the conjugate gradient method and generalised minimal residual method. The reader is referred to [16, 87, 96, 183] for the development of these methods and other iterative methods.

### 2.1.2 Solution by CG

When the matrix $\mathbf{A}$ in (2.3) is symmetric and positive definite, the conjugate gradient (CG) method can be used. The method, given in Algorithm 2.1, is a Krylov subspace method originally proposed by Hestenes & Stiefel [104]; see Section 2.8 for other Krylov subspace methods.

**Algorithm 2.1:** Conjugate Gradient Method

$k \leftarrow 0$; $\mathbf{u}_0 \leftarrow \mathbf{0}$; $\mathbf{r}_0 \leftarrow \mathbf{b}$
**while** $\mathbf{r}_k \neq \mathbf{0}$ **do**
  $k \leftarrow k+1$
  **if** $k = 1$ **then**
    $\mathbf{p}_1 = \mathbf{r}_0$
  **else**
    $\beta_k = \dfrac{\langle \mathbf{r}_{k-1}, \mathbf{r}_{k-1} \rangle_{\mathbb{R}^N}}{\langle \mathbf{r}_{k-2}, \mathbf{r}_{k-2} \rangle_{\mathbb{R}^N}}$
    $\mathbf{p}_k = \mathbf{r}_{k-1} + \beta_k \mathbf{p}_{k-1}$
  **end**
  $\alpha_k = \dfrac{\langle \mathbf{r}_{k-1}, \mathbf{r}_{k-1} \rangle_{\mathbb{R}^N}}{\langle \mathbf{A}\mathbf{p}_k, \mathbf{p}_k \rangle_{\mathbb{R}^N}}$
  $\mathbf{u}_k = \mathbf{u}_{k-1} + \alpha_k \mathbf{p}_k$
  $\mathbf{r}_k = \mathbf{r}_{k-1} - \alpha_k \mathbf{A}\mathbf{p}_k$
**end**
$\mathbf{u} = \mathbf{u}_k$

It is well known that (see e.g. [83]) after $k^{\text{th}}$ iterations, the error in the approximation is

$$\|\mathbf{u} - \mathbf{u}_k\|_{\mathbf{A}} \leq 2 \left( \frac{\sqrt{\kappa_2(\mathbf{A})} - 1}{\sqrt{\kappa_2(\mathbf{A})} + 1} \right)^k \|\mathbf{u} - \mathbf{u}_0\|_{\mathbf{A}}, \tag{2.4}$$

where the $\mathbf{A}$-norm is defined by $\|\mathbf{v}\|_{\mathbf{A}} = \sqrt{\mathbf{v}^\top \mathbf{A} \mathbf{v}}$ for all $\mathbf{v} \in \mathbb{R}^N$, and

$$\kappa_2(\mathbf{A}) = \lambda_{\max}(\mathbf{A}) / \lambda_{\min}(\mathbf{A})$$

is the condition number of $\mathbf{A}$ (with respect to the Euclidean norm).

It follows that a large value of $\kappa_2(\mathbf{A})$ results in a slow convergence of the CG method. To be precise, given a tolerance $\varepsilon > 0$, it takes approximately

$$\sqrt{\kappa_2(\mathbf{A})}\, |\log(\varepsilon/2)|/2$$

iterations to ensure that

$$\|\mathbf{u} - \mathbf{u}_k\|_{\mathbf{A}} \leq \varepsilon \|\mathbf{u} - \mathbf{u}_0\|_{\mathbf{A}}.$$

### *2.1.3 Solution by GMRES*

When the matrix $\mathbf{A}$ in (2.3) is non-symmetric and/or indefinite, the generalised minimal residual (GMRES) method can be used. The method, originally proposed by Saad & Schultz [184], is also a Krylov subspace method and is given in Algorithm 2.2.

**Algorithm 2.2:** Generalised Minimal Residual Method

**Start:**

1: Choose $\mathbf{u}_0$

**Arnoldi process:**

2: $\mathbf{r}_0 \leftarrow \mathbf{b} - \mathbf{A}\mathbf{u}_0,\ \beta \leftarrow \|\mathbf{r}_0\|_{\mathbb{R}^N},\ \mathbf{v}_1 \leftarrow \mathbf{r}_0/\beta$
3: **for** $j = 1, 2, \ldots, k$ **do**
4: $\quad \mathbf{p}_j \leftarrow \mathbf{A}\mathbf{v}_j$
5: $\quad$ **for** $i = 1, 2, \ldots, j$ **do**
6: $\qquad h_{i,j} \leftarrow \langle \mathbf{p}_j, \mathbf{v}_i \rangle_{\mathbb{R}^N}$
7: $\qquad \mathbf{p}_j \leftarrow \mathbf{p}_j - h_{i,j}\mathbf{v}_i$
8: $\quad$ **end for**
9: $\quad h_{j+1,j} \leftarrow \|\mathbf{p}_j\|_{\mathbb{R}^N}$
10: $\quad$ **if** $h_{j+1,j} = 0$ **then**
11: $\qquad j \leftarrow k$
12: $\quad$ **else**
13: $\qquad \mathbf{v}_{j+1} \leftarrow \mathbf{p}_j / h_{j+1,j}$
14: $\quad$ **end if**
15: **end for**
16: $\mathbf{V}_k \leftarrow [\mathbf{v}_1, \ldots, \mathbf{v}_k] \in \mathbb{R}^{N \times k}$
17: $\mathbf{H}_k \leftarrow (k+1) \times k$ Hessenberg matrix with nonzero entries $h_{i,j}$

**Approximate solution:**

18: $\mathbf{w}_k \leftarrow \arg\min_{\mathbf{w} \in \mathbb{R}^k} \|\beta \mathbf{e}_1 - \mathbf{H}_k \mathbf{w}\|_{\mathbb{R}^N}$ where $\mathbf{e}_1 = [1, 0, \ldots, 0]^\top \in \mathbb{R}^{k+1}$;
19: $\mathbf{u}_k \leftarrow \mathbf{u}_0 + \mathbf{V}_k \mathbf{w}_k$;

**Restart:**

20: **if** stopping criterion satisfied **then**
21: $\quad \mathbf{u} \leftarrow \mathbf{u}_k$; stop
22: **else**
23: $\quad \mathbf{u}_0 \leftarrow \mathbf{u}_k$ and go to Step 2.
24: **end if**

Let $\mathbf{A}_s$ be the symmetric part of $\mathbf{A}$, i.e., $\mathbf{A}_s := \dfrac{\mathbf{A} + \mathbf{A}^\top}{2}$. It is proved in [73] that if $\mathbf{A}_s$ is positive definite then

$$\|\mathbf{r}_k\|_{\mathbb{R}^N} \le \left[1 - \frac{\lambda_{\min}(\mathbf{A}_s)^2}{\lambda_{\max}(\mathbf{A}^\top \mathbf{A})}\right]^{k/2} \|\mathbf{r}_0\|_{\mathbb{R}^N}, \tag{2.5}$$

For the GMRES to converge, it is required that $\lambda_{\min}(\mathbf{A}_s)$ is bounded below away from zero, and $\lambda_{\max}(\mathbf{A}^\top \mathbf{A})$ is not too large. In Section 2.7 we will discuss how to estimate these eigenvalues.

## 2.2 Preconditioners

In this section we will explain how to expedite the convergence of the CG and GMRES algorithms by preconditioning. The reader is referred to [24, 87, 96, 183, 233] for more details.

### 2.2.1 *Preconditioned conjugate gradient method*

A preconditioner $\mathbf{C}$ is a symmetric matrix that satisfies $\mathbf{C}^{-1} \approx \mathbf{A}^{-1}$. Hence instead of solving $\mathbf{Au} = \mathbf{b}$ we solve a seemingly simpler equation $\mathbf{C}^{-1}\mathbf{Au} = \mathbf{C}^{-1}\mathbf{b}$. Note however that even though both $\mathbf{A}$ and $\mathbf{C}^{-1}$ are symmetric, in general $\mathbf{C}^{-1}\mathbf{A}$ is not symmetric. There are two approaches:

**Approach 1:** Since $\mathbf{C}^{-1/2}\mathbf{A}\mathbf{C}^{-1/2}$ is symmetric and positive definite, it is possible to obtain the solution $\mathbf{u}$ by successively solving with the CG method (Algorithm 2.1) the following two systems:

$$\begin{aligned} \widetilde{\mathbf{A}}\widetilde{\mathbf{u}} &= \widetilde{\mathbf{b}}, \\ \mathbf{C}^{1/2}\mathbf{u} &= \widetilde{\mathbf{u}}, \end{aligned} \tag{2.6}$$

with

$$\widetilde{\mathbf{A}} := \mathbf{C}^{-1/2}\mathbf{A}\mathbf{C}^{-1/2} \quad \text{and} \quad \widetilde{\mathbf{b}} := \mathbf{C}^{-1/2}\mathbf{b}.$$

It turns out that in practice we even do not have to solve the two systems in (2.6); see Algorithm 2.3.

**Approach 2:** Note that $\mathbf{C}^{-1}\mathbf{A}$ is symmetric with respect to the inner product $\langle\cdot,\cdot\rangle_C$ defined by $\langle\cdot,\cdot\rangle_C := \langle\mathbf{C}\cdot,\cdot\rangle_{\mathbb{R}^N}$, since for all $\mathbf{x},\mathbf{y} \in \mathbb{R}^N$

$$\begin{aligned} \langle\mathbf{C}^{-1}\mathbf{Ax},\mathbf{y}\rangle_{\mathbf{C}} &= \mathbf{y}^\top\mathbf{C}\mathbf{C}^{-1}\mathbf{Ax} \\ &= \mathbf{y}^\top\mathbf{Ax} \\ &= \mathbf{y}^\top\mathbf{A}^\top\mathbf{C}^{-\top}\mathbf{Cx} \\ &= \langle\mathbf{x},\mathbf{C}^{-1}\mathbf{Ay}\rangle_{\mathbf{C}}. \end{aligned}$$

Therefore we can solve

$$\widehat{\mathbf{A}}\mathbf{u} = \widehat{\mathbf{b}} \tag{2.7}$$

by the CG method (Algorithm 2.1) with the inner product $\langle\cdot,\cdot\rangle_{\mathbf{C}}$, where

$$\widehat{\mathbf{A}} := \mathbf{C}^{-1}\mathbf{A} \quad \text{and} \quad \widehat{\mathbf{b}} := \mathbf{C}^{-1}\mathbf{b}. \tag{2.8}$$

In the following we show that both approaches in fact entail the same algorithm. It should be noted that in the implementation neither $\mathbf{C}^{-1/2}$ (for Approach 1) nor $\mathbf{C}^{-1}$ (for Approach 2) need be computed. At each iteration step an equation of the form

$$\mathbf{C}\mathbf{z} = \mathbf{r}$$

is solved. Note however that the matrix $\mathbf{C}$ itself is not computed either. The preconditioner is designed so that $\mathbf{z}$ can be computed without an explicit form of $\mathbf{C}$; see Section 2.4.2.

Consider the first approach where we solve (2.6). Algorithm 2.1 applied to this equation reads:

$$\begin{aligned}
&k = 0;\ \widetilde{\mathbf{u}}_0 = \mathbf{0};\ \widetilde{\mathbf{r}}_0 = \widetilde{\mathbf{b}}\\
&\text{while } \widetilde{\mathbf{r}}_k \neq \mathbf{0}\\
&\qquad k = k+1\\
&\qquad \text{if } k = 1\\
&\qquad\qquad \widetilde{\mathbf{p}}_1 = \widetilde{\mathbf{r}}_0\\
&\qquad \text{else}\\
&\qquad\qquad \widetilde{\beta}_k = \frac{\langle \widetilde{\mathbf{r}}_{k-1}, \widetilde{\mathbf{r}}_{k-1}\rangle_{\mathbb{R}^N}}{\langle \widetilde{\mathbf{r}}_{k-2}, \widetilde{\mathbf{r}}_{k-2}\rangle_{\mathbb{R}^N}}\\
&\qquad\qquad \widetilde{\mathbf{p}}_k = \widetilde{\mathbf{r}}_{k-1} + \widetilde{\beta}_k \widetilde{\mathbf{p}}_{k-1}\\
&\qquad \text{end}\\
&\qquad \widetilde{\alpha}_k = \frac{\langle \widetilde{\mathbf{r}}_{k-1}, \widetilde{\mathbf{r}}_{k-1}\rangle_{\mathbb{R}^N}}{\left\langle \widetilde{\mathbf{A}}\widetilde{\mathbf{p}}_k, \widetilde{\mathbf{p}}_k\right\rangle_{\mathbb{R}^N}}\\
&\qquad \widetilde{\mathbf{u}}_k = \widetilde{\mathbf{u}}_{k-1} + \widetilde{\alpha}_k \widetilde{\mathbf{p}}_k\\
&\qquad \widetilde{\mathbf{r}}_k = \widetilde{\mathbf{r}}_{k-1} - \widetilde{\alpha}_k \widetilde{\mathbf{A}}\widetilde{\mathbf{p}}_k\\
&\text{end}\\
&\widetilde{\mathbf{u}} = \widetilde{\mathbf{u}}_k
\end{aligned} \tag{2.9}$$

It can be shown by induction that $\widetilde{\mathbf{r}}_k = \widetilde{\mathbf{b}} - \widetilde{\mathbf{A}}\widetilde{\mathbf{u}}_k$. If we define

$$\begin{aligned}
\mathbf{u}_k &:= \mathbf{C}^{-1/2}\widetilde{\mathbf{u}}_k, & \mathbf{r}_k &:= \mathbf{b} - \mathbf{A}\mathbf{u}_k,\\
\mathbf{z}_k &:= \mathbf{C}^{-1}\mathbf{r}_k, & \mathbf{p}_k &:= \mathbf{C}^{-1/2}\widetilde{\mathbf{p}}_k,\\
\alpha_k &:= \widetilde{\alpha}_k, & \beta_k &:= \widetilde{\beta}_k,
\end{aligned} \tag{2.10}$$

then

$$\begin{aligned}
\widetilde{\mathbf{r}}_k &= \mathbf{C}^{-1/2}\mathbf{b} - \mathbf{C}^{-1/2}\mathbf{A}\mathbf{C}^{-1/2}\mathbf{C}^{1/2}\mathbf{u}_k\\
&= \mathbf{C}^{-1/2}(\mathbf{b} - \mathbf{A}\mathbf{u}_k)\\
&= \mathbf{C}^{-1/2}\mathbf{r}_k,
\end{aligned}$$

so that

$$\langle \widetilde{\mathbf{r}}_k, \widetilde{\mathbf{r}}_k \rangle_{\mathbb{R}^N} = \langle \mathbf{z}_k, \mathbf{r}_k \rangle_{\mathbb{R}^N} \quad \text{and} \quad \left\langle \widetilde{\mathbf{A}}\widetilde{\mathbf{p}}_k, \widetilde{\mathbf{p}}_k \right\rangle_{\mathbb{R}^N} = \langle \mathbf{A}\mathbf{p}_k, \mathbf{p}_k \rangle_{\mathbb{R}^N},$$

implying

$$\beta_k = \frac{\langle \mathbf{z}_{k-1}, \mathbf{r}_{k-1} \rangle_{\mathbb{R}^N}}{\langle \mathbf{z}_{k-2}, \mathbf{r}_{k-2} \rangle_{\mathbb{R}^N}}, \qquad \mathbf{p}_k = \mathbf{z}_{k-1} + \beta_k \mathbf{p}_{k-1},$$

$$\alpha_k = \frac{\langle \mathbf{z}_{k-1}, \mathbf{r}_{k-1} \rangle_{\mathbb{R}^N}}{\langle \mathbf{A}\mathbf{p}_k, \mathbf{p}_k \rangle_{\mathbb{R}^N}}, \qquad \mathbf{u}_k = \mathbf{u}_{k-1} + \alpha_k \mathbf{p}_k.$$

It is also clear that if $\{\widetilde{\mathbf{u}}_k\}$ converges to $\widetilde{\mathbf{u}}$ then $\{\mathbf{u}_k\}$ converges to $\mathbf{u}$. Hence it follows from (2.9) that $\mathbf{u}_k$ can be computed by Algorithm 2.3. Steps 5 and 8 require the

**Algorithm 2.3:** Preconditioned Conjugate Gradient Method

1 $k = 0$; $\mathbf{u}_0 = \mathbf{0}$; $\mathbf{r}_0 = \mathbf{b}$
2 **while** $\mathbf{r}_k \neq \mathbf{0}$ **do**
3     $k = k + 1$
4     **if** $k = 1$ **then**
5         $\mathbf{z}_0 = \mathbf{C}^{-1}\mathbf{r}_0$
6         $\mathbf{p}_1 = \mathbf{z}_0$
7     **else**
8         $\mathbf{z}_{k-1} = \mathbf{C}^{-1}\mathbf{r}_{k-1}$
9         $\beta_k = \dfrac{\langle \mathbf{z}_{k-1}, \mathbf{r}_{k-1} \rangle_{\mathbb{R}^N}}{\langle \mathbf{z}_{k-2}, \mathbf{r}_{k-2} \rangle_{\mathbb{R}^N}}$
10         $\mathbf{p}_k = \mathbf{z}_{k-1} + \beta_k \mathbf{p}_{k-1}$
11     **end**
12     $\alpha_k = \dfrac{\langle \mathbf{z}_{k-1}, \mathbf{r}_{k-1} \rangle_{\mathbb{R}^N}}{\langle \mathbf{A}\mathbf{p}_k, \mathbf{p}_k \rangle_{\mathbb{R}^N}}$
13     $\mathbf{u}_k = \mathbf{u}_{k-1} + \alpha_k \mathbf{p}_k$
14     $\mathbf{r}_k = \mathbf{r}_{k-1} - \alpha_k \mathbf{A}\mathbf{p}_k$
15 **end**
16 $\mathbf{u} = \mathbf{u}_k$

solution of $\mathbf{C}\mathbf{z} = \mathbf{r}$ for some vector $\mathbf{r} \in \mathbb{R}^N$ and thus $\mathbf{C}$ should be designed such that this problem is easy (cheap) to solve; see Sections 2.4.2, 2.5.1, and 2.5.3.

As mentioned earlier, we can also solve (2.7) by Algorithm 2.1 using the inner product $\langle \cdot, \cdot \rangle_{\mathbf{C}}$ instead of the Euclidean inner product. This generates the following vectors and parameters

$$\widehat{\mathbf{p}}_1 = \widehat{\mathbf{r}}_0,$$

$$\widehat{\beta}_k = \frac{\langle \widehat{\mathbf{r}}_{k-1}, \widehat{\mathbf{r}}_{k-1} \rangle_{\mathbf{C}}}{\langle \widehat{\mathbf{r}}_{k-2}, \widehat{\mathbf{r}}_{k-2} \rangle_{\mathbf{C}}},$$

$$\widehat{\mathbf{p}}_k = \widehat{\mathbf{r}}_{k-1} + \widehat{\beta}_k \widehat{\mathbf{p}}_{k-1},$$

$$\widehat{\alpha}_k = \frac{\langle \widehat{\mathbf{r}}_{k-1}, \widehat{\mathbf{r}}_{k-1} \rangle_{\mathbf{C}}}{\left\langle \widehat{\mathbf{A}}\widehat{\mathbf{p}}_k, \widehat{\mathbf{p}}_k \right\rangle_{\mathbf{C}}}.$$

It can be shown that $\widehat{\mathbf{r}}_k = \widehat{\mathbf{b}} - \widehat{\mathbf{A}}\mathbf{u}_k$. Similarly to (2.10), we define

$$\mathbf{r}_k := \mathbf{b} - \mathbf{A}\mathbf{u}_k, \quad \mathbf{z}_k := \mathbf{C}^{-1}\mathbf{r}_k,$$
$$\mathbf{p}_k := \widehat{\mathbf{p}}_k,$$
$$\alpha_k := \widehat{\alpha}_k, \quad \beta_k := \widehat{\beta}_k.$$

Then

$$\widehat{\mathbf{r}}_k = \mathbf{C}^{-1}\mathbf{r}_k = \mathbf{z}_k$$

which implies

$$\langle \widehat{\mathbf{r}}_k, \widehat{\mathbf{r}}_k \rangle_{\mathbf{C}} = \langle \mathbf{z}_k, \mathbf{r}_k \rangle_{\mathbb{R}^N}$$

and

$$\left\langle \widehat{\mathbf{A}}\widehat{\mathbf{p}}_k, \widehat{\mathbf{p}}_k \right\rangle_{\mathbf{C}} = \langle \mathbf{A}\mathbf{p}_k, \mathbf{p}_k \rangle_{\mathbb{R}^N},$$

so that

$$\begin{aligned}
\beta_k &= \frac{\langle \mathbf{z}_{k-1}, \mathbf{r}_{k-1} \rangle_{\mathbb{R}^N}}{\langle \mathbf{z}_{k-2}, \mathbf{r}_{k-2} \rangle_{\mathbb{R}^N}}, \\
\alpha_k &= \frac{\langle \mathbf{z}_{k-1}, \mathbf{r}_{k-1} \rangle_{\mathbb{R}^N}}{\langle \mathbf{A}\mathbf{p}_k, \mathbf{p}_k \rangle_{\mathbb{R}^N}}, \\
\mathbf{p}_k &= \mathbf{z}_{k-1} + \beta_k \mathbf{p}_{k-1}, \\
\mathbf{u}_k &= \mathbf{u}_{k-1} + \alpha_k \mathbf{p}_k.
\end{aligned}$$

We proved that an application of Algorithm 2.1 to solve (2.7) using the inner product $\langle \cdot, \cdot \rangle_{\mathbf{C}}$ is indeed Algorithm 2.3.

In view of (2.4), the error in the approximation of $\mathbf{u}$ by $\mathbf{u}_k$ by Algorithm 2.3 is

$$\|\mathbf{u} - \mathbf{u}_k\|_{\mathbf{A}} \le 2 \left( \frac{\sqrt{\kappa_2(\widetilde{\mathbf{A}})} - 1}{\sqrt{\kappa_2(\widetilde{\mathbf{A}})} + 1} \right)^k \|\mathbf{u} - \mathbf{u}_0\|_{\mathbf{A}}, \tag{2.11}$$

### 2.2.2 Preconditioned GMRES

When equation (2.3) arises from (2.1) where the bilinear form satisfies Assumption 2.2, the matrix $\mathbf{A}$ is non-symmetric and indefinite. The preconditioner $\mathbf{C}$ in this case can be non-symmetric, but it has to be positive definite.

If we use GMRES method defined in Algorithm 2.2 to solve the preconditioned system

$$\widehat{\mathbf{A}}\widehat{\mathbf{u}} = \widehat{\mathbf{b}} \tag{2.12}$$

where $\widehat{\mathbf{A}} = \mathbf{C}^{-1}\mathbf{A}$, $\widehat{\mathbf{u}} = \mathbf{u}$, and $\widehat{\mathbf{b}} = \mathbf{C}^{-1}\mathbf{b}$, then we obtain Algorithm 2.4. This algorithm is often used in practice and is called the left preconditioned GMRES; see e.g. [183]. It is possible to design the right preconditioned GMRES to solve

$$\mathbf{A}\mathbf{C}^{-1}\overline{\mathbf{u}} = \mathbf{b} \quad \text{where} \quad \overline{\mathbf{u}} = \mathbf{C}\mathbf{u}.$$

In this book, we only focus on the left preconditioner. The reader is referred to [183] for more details on the right version.

It should be remarked that the symmetric part $\widehat{\mathbf{A}}_s$ of $\widehat{\mathbf{A}}$ is not necessarily positive definite, in particular with the preconditioners to be defined in this book (Chapter 8), as proved in [46]. Therefore the convergence given in (2.5) does not hold in general, and thus optimal results may not be achieved for these preconditioners. This loss of optimality can be avoided by using a more appropriate inner product than the Euclidean inner product, as suggested in [185]. Before presenting this inner product and its use to solve (2.12) we note that Step 18 in Algorithm 2.4 is the minimization of $\|\mathbf{r}_0\|_{\mathbb{R}^N}$ in the Krylov space (see [16, 87, 96, 183, 184])

$$\mathscr{K}_k := \operatorname{span}\{\mathbf{r}_0, \widehat{\mathbf{A}}\mathbf{r}_0, \ldots, \widehat{\mathbf{A}}^{k-1}\}.$$

Now let $\mathbf{B}$ be the stiffness matrix defined by the bilinear form $b(\cdot,\cdot)$ given in Assumption 2.2; namely

$$\mathbf{B} = \big(b(\Phi_i, \Phi_j)\big)_{i,j=1}^{N}.$$

By Assumption 2.2 this matrix is symmetric and positive definite. Therefore we can define another inner product and its associated norm

$$\langle \mathbf{x}, \mathbf{y} \rangle_{\mathbf{B}} := \langle \mathbf{B}\mathbf{x}, \mathbf{y} \rangle_{\mathbb{R}^N} \quad \text{and} \quad \|\mathbf{x}\|_{\mathbf{B}} := \langle \mathbf{x}, \mathbf{x} \rangle_{\mathbf{B}}^{1/2} = \langle \mathbf{B}\mathbf{x}, \mathbf{x} \rangle_{\mathbb{R}^N}^{1/2} \quad \forall \mathbf{x}, \mathbf{y} \in \mathbb{R}^N. \tag{2.13}$$

A new preconditioned GMRES algorithm is designed by replacing the Euclidean inner product and norm in Steps 2, 6, 9, and 18 in Algorithm 2.4 by the $\mathbf{B}$-inner product and the $\mathbf{B}$-norm. As shown in [185], the minimization of $\|\mathbf{r}_0\|_{\mathbf{B}}$ in the Krylov space $\mathscr{K}_k$ results in the same Step 18 as in Algorithm 2.4. Therefore we have the following algorithm whose convergence results will be shown in Section 2.7.1.

In the case the preconditioner $\mathbf{C}$ is symmetric and positive definite, it can be used to define an inner product used in Algorithm 2.5, namely instead of (2.13) we define

$$\langle \mathbf{x}, \mathbf{y} \rangle_{\mathbf{C}} := \langle \mathbf{C}\mathbf{x}, \mathbf{y} \rangle_{\mathbb{R}^N} \quad \text{and} \quad \|\mathbf{x}\|_{\mathbf{C}} := \langle \mathbf{x}, \mathbf{x} \rangle_{\mathbf{C}}^{1/2} = \langle \mathbf{C}\mathbf{x}, \mathbf{x} \rangle_{\mathbb{R}^N}^{1/2} \quad \forall \mathbf{x}, \mathbf{y} \in \mathbb{R}^N.$$

**Algorithm 2.4:** Preconditioned GMRES I

**Start:**

1: Choose $\mathbf{u}_0$

**Arnoldi process:**

2: $\mathbf{r}_0 \leftarrow \mathbf{C}^{-1}(\mathbf{b}-\mathbf{A}\mathbf{u}_0),\ \beta \leftarrow \|\mathbf{r}_0\|_{\mathbb{R}^N},\ \mathbf{v}_1 \leftarrow \mathbf{r}_0/\beta$
3: **for** $j = 1,2,\ldots,k$ **do**
4: $\quad \mathbf{p}_j \leftarrow \mathbf{C}^{-1}\mathbf{A}\mathbf{v}_j$
5: $\quad$ **for** $i = 1,2,\ldots,j$ **do**
6: $\qquad h_{i,j} \leftarrow \langle \mathbf{p}_j, \mathbf{v}_i \rangle_{\mathbb{R}^N}$
7: $\qquad \mathbf{p}_j \leftarrow \mathbf{p}_j - h_{i,j}\mathbf{v}_i$
8: $\quad$ **end for**
9: $\quad h_{j+1,j} \leftarrow \|\mathbf{p}_j\|_{\mathbb{R}^N}$
10: $\quad$ **if** $h_{j+1,j} = 0$ **then**
11: $\qquad j \leftarrow k$
12: $\quad$ **else**
13: $\qquad \mathbf{v}_{j+1} \leftarrow \mathbf{p}_j / h_{j+1,j}$
14: $\quad$ **end if**
15: **end for**
16: $\mathbf{V}_k \leftarrow [\mathbf{v}_1,\ldots,\mathbf{v}_k] \in \mathbb{R}^{N\times k}$
17: $\mathbf{H}_k \leftarrow (k+1)\times k$ Hessenberg matrix with nonzero entries $h_{i,j}$

**Approximate solution:**

18: $\mathbf{w}_k \leftarrow \arg\min_{\mathbf{w}\in\mathbb{R}^k} \|\beta \mathbf{e}_1 - \mathbf{H}_k \mathbf{w}\|_{\mathbb{R}^N}$ where $\mathbf{e}_1 = [1,0,\ldots,0]^\top \in \mathbb{R}^{k+1}$;
19: $\mathbf{u}_k \leftarrow \mathbf{u}_0 + \mathbf{V}_k \mathbf{w}_k$;

**Restart:**

20: **if** stopping criterion satisfied **then**
21: $\quad \mathbf{u} \leftarrow \mathbf{u}_k$; stop
22: **else**
23: $\quad \mathbf{u}_0 \leftarrow \mathbf{u}_k$ and go to Step 2.
24: **end if**

The $\mathbf{C}$-inner product and $\mathbf{C}$-norm are then used in Steps 2, 6, 9, and 18. Since in general $\mathbf{C}$ is not known explicitly, the above steps are not implementable by directly using $\mathbf{C}$. However, we can write $\beta$ and $\mathbf{v}_1$ in Step 2 as follows. Let $\mathbf{s}_0 = \mathbf{b} - \mathbf{A}\mathbf{u}_0$ so that $\mathbf{r}_0 = \mathbf{C}^{-1}\mathbf{s}_0$. Then

$$\beta = \|\mathbf{r}_0\|_{\mathbf{C}} = \langle \mathbf{C}\mathbf{r}_0, \mathbf{r}_0 \rangle_{\mathbb{R}^N}^{1/2} = \langle \mathbf{s}_0, \mathbf{C}^{-1}\mathbf{s}_0 \rangle_{\mathbb{R}^N}^{1/2} \quad \text{and} \quad \mathbf{v}_1 = \mathbf{C}^{-1}\mathbf{s}_0/\beta.$$

Similarly, letting $\mathbf{q}_j = \mathbf{A}\mathbf{v}_j = \mathbf{C}\mathbf{p}_j$ (see Step 4 in Algorithm 2.4 or Algorithm 2.5) we have

$$h_{i,j} = \langle \mathbf{p}_j, \mathbf{v}_i \rangle_{\mathbf{C}} = \langle \mathbf{q}_j, \mathbf{v}_i \rangle_{\mathbb{R}^N} \quad \text{and} \quad h_{j+1,j} = \|\mathbf{p}_j\|_{\mathbf{C}} = \langle \mathbf{C}\mathbf{p}_j, \mathbf{p}_j \rangle_{\mathbb{R}^N}^{1/2} = \langle \mathbf{q}_j, \mathbf{p}_j \rangle_{\mathbb{R}^N}^{1/2}.$$

Even though an explicit form of $\mathbf{C}^{-1}$ is not known, one can compute $\mathbf{C}^{-1}\mathbf{x}$ for any $\mathbf{x} \in \mathbb{R}^N$; see e.g. Section 2.4.2. Therefore, Algorithm 2.5 with the $\mathbf{C}$-inner prod-

**Algorithm 2.5:** Preconditioned GMRES II

**Start:**

1: Choose $\mathbf{u}_0$

**Arnoldi process:**

2: $\mathbf{r}_0 \leftarrow \mathbf{C}^{-1}(\mathbf{b} - \mathbf{A}\mathbf{u}_0)$, $\beta \leftarrow \|\mathbf{r}_0\|_{\mathbf{B}}$, $\mathbf{v}_1 \leftarrow \mathbf{r}_0/\beta$

3: **for** $j = 1, 2, \ldots, k$ **do**

4: $\quad\mathbf{p}_j \leftarrow \mathbf{C}^{-1}\mathbf{A}\mathbf{v}_j$

5: $\quad$**for** $i = 1, 2, \ldots, j$ **do**

6: $\quad\quad h_{i,j} \leftarrow \langle \mathbf{p}_j, \mathbf{v}_i \rangle_{\mathbf{B}}$

7: $\quad\quad \mathbf{p}_j \leftarrow \mathbf{p}_j - h_{i,j}\mathbf{v}_i$

8: $\quad$**end for**

9: $\quad h_{j+1,j} \leftarrow \|\mathbf{p}_j\|_{\mathbf{B}}$

10: $\quad$**if** $h_{j+1,j} = 0$ **then**

11: $\quad\quad j \leftarrow k$

12: $\quad$**else**

13: $\quad\quad \mathbf{v}_{j+1} \leftarrow \mathbf{p}_j / h_{j+1,j}$

14: $\quad$**end if**

15: **end for**

16: $\mathbf{V}_k \leftarrow [\mathbf{v}_1, \ldots, \mathbf{v}_k] \in \mathbb{R}^{N \times k}$

17: $\mathbf{H}_k \leftarrow (k+1) \times k$ Hessenberg matrix with nonzero entries $h_{i,j}$

**Approximate solution:**

18: $\mathbf{w}_k \leftarrow \arg\min_{\mathbf{w} \in \mathbb{R}^k} \|\beta \mathbf{e}_1 - \mathbf{H}_k \mathbf{w}\|_{\mathbb{R}^N}$ where $\mathbf{e}_1 = [1, 0, \ldots, 0]^\top \in \mathbb{R}^{k+1}$;

19: $\mathbf{u}_k \leftarrow \mathbf{u}_0 + \mathbf{V}_k \mathbf{w}_k$;

**Restart:**

20: **if** stopping criterion satisfied **then**

21: $\quad \mathbf{u} \leftarrow \mathbf{u}_k$; stop

22: **else**

23: $\quad \mathbf{u}_0 \leftarrow \mathbf{u}_k$ and go to Step 2.

24: **end if**

uct becomes Algorithm 2.6 below. We remark that one can repeat the proof in [185, Section 5] with the **C**-inner product to show that Step 18 is still the same.

## *2.2.3 Preconditioning and linear iteration*

A preconditioner **C** (not necessarily symmetric) can be used with a linear iterative method to find the approximate solution to $\mathbf{A}\mathbf{u} = \mathbf{b}$ as follows

$$\mathbf{u}_{k+1} = \mathbf{u}_k + \mathbf{C}^{-1}\left(\mathbf{b} - \mathbf{A}\mathbf{u}_k\right), \quad k = 0, 1, 2, \ldots. \tag{2.14}$$

We note here again that, as in the case of PCG or PGMRES, in order to find

$$\mathbf{C}^{-1}\left(\mathbf{b} - \mathbf{A}\mathbf{u}_k\right),$$

**Algorithm 2.6:** Preconditioned GMRES III – Symmetric preconditioner $\mathbf{C}$

**Start:**

1: Choose $\mathbf{u}_0$

**Arnoldi process:**

2: $\mathbf{s}_0 \leftarrow \mathbf{b} - \mathbf{A}\mathbf{u}_0,\ \beta \leftarrow \langle \mathbf{s}_0, \mathbf{C}^{-1}\mathbf{s}_0 \rangle_{\mathbb{R}^N}^{1/2},\ \mathbf{v}_1 \leftarrow \mathbf{C}^{-1}\mathbf{s}_0/\beta$

3: **for** $j = 1, 2, \ldots, k$ **do**

4: $\quad \mathbf{q}_j \leftarrow \mathbf{A}\mathbf{v}_j,\ \mathbf{p}_j \leftarrow \mathbf{C}^{-1}\mathbf{q}_j$

5: $\quad$ **for** $i = 1, 2, \ldots, j$ **do**

6: $\qquad h_{i,j} \leftarrow \langle \mathbf{q}_j, \mathbf{v}_i \rangle_{\mathbb{R}^N}$

7: $\qquad \mathbf{p}_j \leftarrow \mathbf{p}_j - h_{i,j}\mathbf{v}_i$

8: $\quad$ **end for**

9: $\quad h_{j+1,j} \leftarrow \langle \mathbf{q}_j, \mathbf{p}_j \rangle_{\mathbb{R}^N}^{1/2}$

10: $\quad$ **if** $h_{j+1,j} = 0$ **then**

11: $\qquad j \leftarrow k$

12: $\quad$ **else**

13: $\qquad \mathbf{v}_{j+1} \leftarrow \mathbf{p}_j / h_{j+1,j}$

14: $\quad$ **end if**

15: **end for**

16: $\mathbf{V}_k \leftarrow [\mathbf{v}_1, \ldots, \mathbf{v}_k] \in \mathbb{R}^{N \times k}$

17: $\mathbf{H}_k \leftarrow (k+1) \times k$ Hessenberg matrix with nonzero entries $h_{i,j}$

**Approximate solution:**

18: $\mathbf{w}_k \leftarrow \arg\min_{\mathbf{w} \in \mathbb{R}^k} \|\beta \mathbf{e}_1 - \mathbf{H}_k \mathbf{w}\|_{\mathbb{R}^N}$ where $\mathbf{e}_1 = [1, 0, \ldots, 0]^\top \in \mathbb{R}^{k+1}$

19: $\mathbf{u}_k \leftarrow \mathbf{u}_0 + \mathbf{V}_k \mathbf{w}_k$

**Restart:**

20: **if** stopping criterion satisfied **then**

21: $\quad \mathbf{u} \leftarrow \mathbf{u}_k$; stop

22: **else**

23: $\quad \mathbf{u}_0 \leftarrow \mathbf{u}_k$ and go to Step 2.

24: **end if**

we solve $\mathbf{C}\mathbf{d}_k = \mathbf{b} - \mathbf{A}\mathbf{u}_k$. The Jacobi and Gauss–Seidel methods are two examples of linear iterative methods with preconditioners.

We note that (2.14) can be written as

$$\mathbf{u}_{k+1} = \left(\mathbf{I} - \mathbf{C}^{-1}\mathbf{A}\right)\mathbf{u}_k + \mathbf{C}^{-1}\mathbf{b}, \quad k = 0, 1, 2, \ldots. \tag{2.15}$$

The matrix $\mathbf{I} - \mathbf{C}^{-1}\mathbf{A}$ is called the iteration matrix. See [96] for more details.

## 2.3 Schwarz Operators

### 2.3.1 *Required properties of a preconditioner*

Both the preconditioned conjugate gradient method and the preconditioned GMRES method involve the solution of $\mathbf{Cz} = \mathbf{r}$ for any residual vector $\mathbf{r}$; see Algorithms 2.3, 2.4, 2.5, and 2.6. So it is required that $\mathbf{C}$ be designed such that the above-mentioned system is easy to solve. There is no need to compute $\mathbf{C}$ or $\mathbf{C}^{-1}$. It is also required that the condition numbers of $\widetilde{\mathbf{A}}$ and $\widehat{\mathbf{A}}$ defined in (2.6) and (2.8) be small. It should be pointed out that there is a competition between these two aims. The extreme case of $\mathbf{C} = \mathbf{I}$ results in a quick solution $\mathbf{z}$ with no decrease in condition number. The other extreme is $\mathbf{C} = \mathbf{A}$ which results in a condition number equal 1 but the same difficulty in solving the original equation.

Before defining the Schwarz operators, we recall the definitions in Section 2.1.1 and observe that for all $u, v \in \mathscr{V}$ there hold

$$\langle u, v \rangle = \mathbf{v}^\top \mathbf{M} \mathbf{u} \quad \text{and} \quad a(u, v) = \mathbf{v}^\top \mathbf{A} \mathbf{u},$$

where $\mathbf{M}$ and $\mathbf{A}$ are $N \times N$ matrices with entries $\mathbf{M}_{ij} = \langle \Phi_i, \Phi_j \rangle$ and $\mathbf{A}_{ij} = a(\Phi_i, \Phi_j)$, respectively.

Let $M, A : \mathscr{V} \to \mathscr{V}$ be the operators having matrix representations $\mathbf{M}$ and $\mathbf{A}$ (with respect to the basis $\{\Phi_1, \ldots, \Phi_N\}$), respectively. Then

$$Mu = \sum_{m=1}^{N} \langle u, \Phi_m \rangle \Phi_m \quad \text{and} \quad \langle Au, v \rangle = a(u, Mv) \quad \forall u, v \in \mathscr{V}.$$

If we define $\mathbf{b}_i = \langle f, \Phi_i \rangle$ and $b = \sum_{i=1}^{N} \mathbf{b}_i \Phi_i$, then we have the following equivalences

$$\boxed{\mathbf{A}\mathbf{u} = \mathbf{b} \iff a(u, v) = \langle f, v \rangle \quad \forall v \in \mathscr{V} \iff Au = b.}$$

The first equation is used in the implementation, whereas last two equations are used for the analysis. They represent three different forms of the problem to be solved, namely, the matrix form, the variational form, and the operator form. In the sequel while designing preconditioners we develop the three corresponding forms.

### 2.3.2 *Ingredients of the Schwarz operators*

The Schwarz operators are defined from projections onto subspaces of $\mathscr{V}$. In the next definition, the spaces $\mathscr{V}_j$ are spaces with dimensions much less than that of $\mathscr{V}$.

#### 2.3.2.1 Subspace decomposition

For $j = 0,\dots,J$, let

$$\mathscr{V}_j = \operatorname{span}\{\Phi_1^{(j)},\dots,\Phi_{N_j}^{(j)}\}$$

be finite dimensional spaces with dimensions $N_j < N$. We denote by $R_j : \mathscr{V} \to \mathscr{V}_j$ the *restriction operator* from $\mathscr{V}$ onto $\mathscr{V}_j$ and by $R_j^t : \mathscr{V}_j \to \mathscr{V}$ the *prolongation operator* from $\mathscr{V}_j$ to $\mathscr{V}$. They are defined so that their matrix representations are the transpose matrices $\mathbf{R}_j$ and $\mathbf{R}_j^\top$, respectively. If $\mathscr{V}_j$ is a subspace of $\mathscr{V}$ then the entries of $\mathbf{R}_j$ and $\mathbf{R}_j^\top$ contain 0 and 1. We will denote by $\mathscr{B}$ and $\mathscr{B}_j$ the corresponding bases of $\mathscr{V}$ and $\mathscr{V}_j$, namely,

$$\mathscr{B} = \{\Phi_1,\dots,\Phi_N\} \quad \text{and} \quad \mathscr{B}_j = \{\Phi_1^{(j)},\dots,\Phi_{N_j}^{(j)}\}.$$

A representation

$$\mathscr{V} = R_0^t\mathscr{V}_0 + \cdots + R_J^t\mathscr{V}_J$$

is called a *subspace decomposition* of $\mathscr{V}$. In practice with FEM/BEM, the subspaces $\mathscr{V}_j$, $j = 1,\dots,J$, are defined from a decomposition of the spatial domain into subdomains. Hence the name *domain decomposition method*.

*Example 2.1.* Assume that $\mathscr{V} = \operatorname{span}\{\Phi_1,\Phi_2,\Phi_3\}$ and $\mathscr{V}_1 = \operatorname{span}\{\Phi_1,\Phi_3\}$. The restriction operator $R_1 : \mathscr{V} \to \mathscr{V}_1$ is defined by

$$R_1 u = b_1\Phi_1 + b_3\Phi_3 \quad \text{for any} \quad u = b_1\Phi_1 + b_2\Phi_2 + b_3\Phi_3,$$

and the prolongation operator $R_1^t : \mathscr{V}_1 \to \mathscr{V}$ is defined by

$$R_1^t v_1 = a_1\Phi_1 + 0\Phi_2 + a_3\Phi_3 \quad \text{for any} \quad v_1 = a_1\Phi_1 + a_3\Phi_3.$$

It can be seen that

$$[R_1]_{\mathscr{B},\mathscr{B}_1} = \mathbf{R}_1 = \begin{bmatrix} 1 & 0 & 0 \\ 0 & 0 & 1 \end{bmatrix} \quad \text{and} \quad [R_1^t]_{\mathscr{B}_1,\mathscr{B}} = \mathbf{R}_1^\top = \begin{bmatrix} 1 & 0 \\ 0 & 0 \\ 0 & 1 \end{bmatrix},$$

where $\mathscr{B} = \{\Phi_1,\Phi_2,\Phi_3\}$ and $\mathscr{B}_1 = \{\Phi_1^{(1)},\Phi_2^{(1)}\} := \{\Phi_1,\Phi_3\}$. The coordinate vector of $v_1$ as an element of $\mathscr{V}_1$ is $[v_1]_{\mathscr{B}_1} = \mathbf{v}_1 = (a_1,a_3)$, whereas the coordinate vector of $v_1$ as an element of $\mathscr{V}$ is $[v_1]_{\mathscr{B}} = \mathbf{R}_1^\top\mathbf{v}_1 = (a_1,0,a_3)$.

#### 2.3.2.2 Projections

Let $\mathscr{V}_j$, $R_j$ and $R_j^t$, $j = 0,\dots,J$, be the spaces and operators defined in Subsection 2.3.2.1. Let $a_j(\cdot,\cdot) : \mathscr{V}_j \times \mathscr{V}_j \to \mathbb{R}$ be a symmetric, positive definite bilinear form on $\mathscr{V}_j$. We define the following projection-like operators

$$\widetilde{P}_j : \mathscr{V} \to \mathscr{V}_j, \quad a_j(\widetilde{P}_j u, v_j) = a(u, R_j^t v_j) \quad \forall u \in \mathscr{V},\, v_j \in \mathscr{V}_j, \quad j = 0, \ldots, J. \tag{2.16}$$

The operators $\widetilde{P}_j$ are well defined because the bilinear forms $a_j(\cdot,\cdot)$ are coercive. The reason to call $\widetilde{P}_j$ a projection-like operator can be seen from the following lemma.

**Lemma 2.1.** *For $j = 0, \ldots, J$, if $a_j(\cdot,\cdot)$ is chosen to be the bilinear form inherited from that in $\mathscr{V}$, i.e.,*

$$a_j(u_j, v_j) = a(R_j^t u_j, R_j^t v_j) \quad \forall u_j, v_j \in \mathscr{V}_j, \tag{2.17}$$

*then*

$$\widetilde{P}_j R_j^t \widetilde{P}_j = \widetilde{P}_j. \tag{2.18}$$

*Proof.* The required result follows from (2.16) and (2.17) because

$$a_j(\widetilde{P}_j R_j^t \widetilde{P}_j u_j, v_j) = a(R_j^t \widetilde{P}_j u_j, R_j^t v_j) = a_j(\widetilde{P}_j u_j, v_j) \quad \forall u_j, v_j \in \mathscr{V}_j.$$ □

A projection can be defined from $\widetilde{P}_j$ as shown in the following lemma.

**Lemma 2.2.** *For $j = 0, \ldots, J$, let $P_j : \mathscr{V} \to \mathscr{V}$ be defined by*

$$P_j = R_j^t \widetilde{P}_j. \tag{2.19}$$

*Then the following statements hold:*

(i) *$P_j$ is symmetric with respect to the bilinear form $a(\cdot,\cdot)$, i.e., $a(P_j u, v) = a(u, P_j v)$ for all $u, v \in \mathscr{V}$.*
(ii) *$P_j$ is positive semidefinite, i.e., $a(P_j u, u) \geq 0$ for all $u \in \mathscr{V}$.*
(iii) *If $a_j(\cdot,\cdot)$ is chosen to satisfy (2.17), then $P_j$ is a projection, i.e., $P_j^2 = P_j$.*

*Proof.* (i) For any $u$, $v \in \mathscr{V}$ we obtain from the definitions of $P_j$ and $\widetilde{P}_j$ that

$$a(P_j u, v) = a_j(\widetilde{P}_j u, \widetilde{P}_j v) = a(u, R_j^t \widetilde{P}_j v) = a(u, P_j v).$$

Hence $P_j$ is symmetric.

(ii) For any $u \in \mathscr{V}$ we have

$$a(P_j u, u) = a(R_j^t \widetilde{P}_j u, u) = a_j(\widetilde{P}_j u, \widetilde{P}_j u) \geq 0$$

because $a_j(\cdot,\cdot)$ is positive definite.

(iii) With the choice (2.17), the identity $P_j^2 = P_j$ follows from (2.18) by applying $R_j^t$ to the left of both sides. This completes the proof of the lemma. □

### 2.3.3 Schwarz operators

**Definition 2.1.** *The additive Schwarz operator $P_{\mathrm{ad}} : \mathscr{V} \to \mathscr{V}$ is defined by*

$$P_{\text{ad}} = P_0 + \cdots + P_J. \tag{2.20}$$

*The non-symmetric multiplicative Schwarz operator* $P_{\text{mu}} : \mathscr{V} \to \mathscr{V}$ *is defined by*

$$P_{\text{mu}} = I - E_{\text{mu}}, \tag{2.21}$$

*where* $E_{\text{mu}} : \mathscr{V} \to \mathscr{V}$ *is the error propagation operator defined by*

$$E_{\text{mu}} = (I - P_J) \cdots (I - P_0). \tag{2.22}$$

*Finally, the symmetric multiplicative Schwarz operator* $P_{\text{smu}} : \mathscr{V} \to \mathscr{V}$ *is defined by*

$$P_{\text{smu}} = I - E_{\text{smu}}, \tag{2.23}$$

*where* $E_{\text{smu}} : \mathscr{V} \to \mathscr{V}$ *is defined by*

$$E_{\text{smu}} = E_{\text{mu}}^t E_{\text{mu}} = (I - P_0) \cdots (I - P_{J-1})(I - P_J)(I - P_J)(I - P_{J-1}) \cdots (I - P_0). \tag{2.24}$$

The following lemma is a consequence of Lemma 2.2.

**Lemma 2.3.** *The additive Schwarz operator* $P_{\text{ad}}$ *is symmetric with respect to the bilinear form* $a(\cdot,\cdot)$*, i.e.,*

$$a(P_{\text{ad}} u, v) = a(u, P_{\text{ad}} v) \quad \forall u, v \in \mathscr{V}.$$

*Moreover, if* (2.17) *holds then*

$$E_{\text{smu}} = (I - P_0) \cdots (I - P_{J-1})(I - P_J)(I - P_{J-1}) \cdots (I - P_0).$$

## 2.4 Additive Schwarz Preconditioners

We shall now describe how the Schwarz operators are used to define preconditioners. First we describe the matrix form of these operators. We then design the additive Schwarz preconditioner in the matrix form which helps to understand the implementation of the method. Finally, we show the variational and operator forms of this preconditioning technique. These forms are useful in the convergence analysis.

### *2.4.1 Matrix forms of* $P_j$ *and* $P_{\text{ad}}$

Let $\mathbf{A}_j := (a_j(\Phi_k^{(j)}, \Phi_l^{(j)}))_{k,l=1}^{N_j}$ for $j = 0, \ldots, J$. It can be easily seen that

$$a_j(u_j, v_j) = \mathbf{v}_j^\top \mathbf{A}_j \mathbf{u}_j \quad \forall u_j, v_j \in \mathscr{V}_j, \ j = 0, \ldots, J,$$

where $\mathbf{u}_j = [u_j]_{\mathscr{B}_j}$ and $\mathbf{v}_j = [v_j]_{\mathscr{B}_j}$. Here $\mathscr{B}_j$ is the basis of $\mathscr{V}_j$. If the bilinear form $a_j(\cdot,\cdot)$ in $\mathscr{V}_j$ is inherited from the bilinear form $a(\cdot,\cdot)$ in $\mathscr{V}$, then the matrices $\mathbf{A}_j$ and $\mathbf{A}$ are related as follows.

**Lemma 2.4.** *If* (2.17) *holds then the following relation holds*

$$\mathbf{A}_j = \mathbf{R}_j \mathbf{A} \mathbf{R}_j^\top. \tag{2.25}$$

*Proof.* For any $k, \ell = 1, \ldots, N_j$ we have

$$\begin{aligned} a_j(\Phi_k^{(j)}, \Phi_\ell^{(j)}) &= a(R_j^t \Phi_k^{(j)}, R_j^t \Phi_\ell^{(j)}) \\ &= \left(\mathbf{R}_j^\top \boldsymbol{\Phi}_\ell^{(j)}\right)^\top \mathbf{A} \mathbf{R}_j^\top \boldsymbol{\Phi}_k^{(j)} \\ &= \left(\boldsymbol{\Phi}_\ell^{(j)}\right)^\top \mathbf{R}_j \mathbf{A} \mathbf{R}_j^\top \boldsymbol{\Phi}_k^{(j)}. \end{aligned}$$

This implies (2.25). □

**Lemma 2.5.** *The matrix representations* $\widetilde{\mathbf{P}}_j$, $\mathbf{P}_j$, *and* $\mathbf{P}_{\mathrm{ad}}$ *of the operators* $\widetilde{P}_j$, $P_j$ *and* $P_{\mathrm{ad}}$, *respectively, are given by*

$$\widetilde{\mathbf{P}}_j = \mathbf{A}_j^{-1} \mathbf{R}_j \mathbf{A}, \quad \mathbf{P}_j = \mathbf{R}_j^\top \mathbf{A}_j^{-1} \mathbf{R}_j \mathbf{A} \quad \textit{and} \quad \mathbf{P}_{\mathrm{ad}} = \sum_{j=0}^{J} \mathbf{R}_j^\top \mathbf{A}_j^{-1} \mathbf{R}_j \mathbf{A}. \tag{2.26}$$

*Proof.* We have the following equivalences from the definition of $\widetilde{P}_j$:

$$\begin{aligned} a_j(\widetilde{P}_j u, v_j) = a(u, R_j^t v_j) &\iff \mathbf{v}_j^\top \mathbf{A}_j \widetilde{\mathbf{P}}_j \mathbf{u} = (\mathbf{R}_j^\top \mathbf{v}_j)^\top \mathbf{A} \mathbf{u} \\ &\iff \mathbf{v}_j^\top \mathbf{A}_j \widetilde{\mathbf{P}}_j \mathbf{u} = \mathbf{v}_j^\top \mathbf{R}_j \mathbf{A} \mathbf{u}. \end{aligned}$$

As this holds for all $u \in \mathscr{V}$ and $v_j \in \mathscr{V}_j$ we deduce $\mathbf{A}_j \widetilde{\mathbf{P}}_j = \mathbf{R}_j \mathbf{A}$ or $\widetilde{\mathbf{P}}_j = \mathbf{A}_j^{-1} \mathbf{R}_j \mathbf{A}$. The matrix representations of $P_j$ and $P_{\mathrm{ad}}$ follow from their definitions (2.19) and (2.20), concluding the proof of the lemma. □

Defining the *additive Schwarz preconditioner* by

$$\mathbf{C}_{\mathrm{ad}}^{-1} := \sum_{j=0}^{J} \mathbf{R}_j^\top \mathbf{A}_j^{-1} \mathbf{R}_j,$$

we deduce

$$\mathbf{P}_{\mathrm{ad}} = \mathbf{C}_{\mathrm{ad}}^{-1} \mathbf{A}. \tag{2.27}$$

### 2.4.2 Additive Schwarz preconditioner: the matrix form

Recall from Algorithm 2.3 that at each iteration step one needs to find $\mathbf{z} \in \mathbb{R}^N$ satisfying

$$\mathbf{C}_{\text{ad}}\mathbf{z} = \mathbf{r}$$

for some $\mathbf{r} \in \mathbb{R}^N$. This is done by the following algorithm, recalling that $N$ is the dimension of $\mathscr{V}$ whereas $N_j$ is the dimension of $\mathscr{V}_j$.

---

**Algorithm 2.7:** The matrix form of the additive Schwarz preconditioner

---

To compute $\mathbf{z} = \mathbf{C}_{\text{ad}}^{-1}\mathbf{r}$ for a given $\mathbf{r}$:
**for** $j = 0, \ldots, J$ **do**
  Compute $\mathbf{R}_j\mathbf{r} \in \mathbb{R}^{N_j}$
  Solve $\mathbf{A}_j\mathbf{y}_j = \mathbf{R}_j\mathbf{r}$ for $\mathbf{y}_j \in \mathbb{R}^{N_j}$
**end for**

$$\mathbf{z} \leftarrow \sum_{j=0}^{J} \mathbf{R}_j^\top \mathbf{y}_j = \sum_{j=0}^{J} \mathbf{R}_j^\top \mathbf{A}_j^{-1}\mathbf{R}_j\mathbf{r} \in \mathbb{R}^N$$

---

### 2.4.3 Additive Schwarz preconditioner: the variational form

Let $C_{\text{ad}}^{-1} : \mathscr{V} \to \mathscr{V}$ be the operator whose matrix representation is $\mathbf{C}_{\text{ad}}^{-1}$. For the analysis, it is sometimes more convenient to write Algorithm 2.7 in the following variational form.

---

**Algorithm 2.8:** The variational form of the additive Schwarz preconditioner

---

For any $r \in \mathscr{V}$
**for** $j = 0, \ldots, J$ **do**
  Compute $u_j \in \mathscr{V}_j$ satisfying

$$a_j(u_j, v_j) = \left\langle M^{-1}r, R_j^t v_j \right\rangle \quad \forall v_j \in \mathscr{V}_j$$

**end for**
Compute

$$C_{\text{ad}}^{-1} r = \sum_{j=0}^{J} R_j^t u_j$$

---

### 2.4.4 Additive Schwarz preconditioner: the operator form

It follows from (2.27) that $P_{\text{ad}} = C_{\text{ad}}^{-1}A$. The following equivalences then follow:

$$\boxed{Au = b \iff P_{\text{ad}}u = C_{\text{ad}}^{-1}b}$$

The right-hand side $C_{\text{ad}}^{-1}b$ has the following representation.

**Lemma 2.6.** *For all $j = 0, \ldots, J$ let $g_j \in \mathscr{V}_j$ satisfy*

$$a_j(g_j, v_j) = \langle f, R_j^t v_j \rangle \quad \forall v_j \in \mathscr{V}_j. \tag{2.28}$$

*Define $g \in \mathscr{V}$ by*

$$g = \sum_{j=0}^{J} R_j^t g_j.$$

*Then $g = C_{\text{ad}}^{-1}b$.*

*Proof.* Since any $v_j \in \mathscr{V}_j$ can be represented as $v_j = \sum_{k=1}^{N_j} (\mathbf{v}_j)_k \Phi_k^{(j)}$, we have

$$\langle f, R_j^t v_j \rangle = \sum_{k=1}^{N_j} (\mathbf{v}_j)_k \left\langle f, R_j^t \Phi_k^{(j)} \right\rangle = \sum_{k=1}^{N_j} (\mathbf{v}_j)_k (\mathbf{R}_j \mathbf{b})_k = \mathbf{v}_j^t \mathbf{R}_j \mathbf{b}.$$

Hence (2.28) implies

$$\mathbf{v}_j^\top \mathbf{A}_j \mathbf{g}_j = \mathbf{v}_j^\top \mathbf{R}_j \mathbf{b},$$

so that

$$\mathbf{g}_j = \mathbf{A}_j^{-1} \mathbf{R}_j \mathbf{b}.$$

The definition of $g$ gives

$$\mathbf{g} = \sum_{j=0}^{J} \mathbf{R}_j^\top \mathbf{A}_j^{-1} \mathbf{R}_j \mathbf{b} = \mathbf{C}_{\text{ad}}^{-1} \mathbf{b},$$

yielding the required identity. □

Thanks to Lemma 2.6, Algorithms 2.7 and 2.8 can be written in the operator form as in Algorithm 2.9 below.

## 2.5 Multiplicative Schwarz Preconditioners

In this section, we shall describe the matrix form and operator form for the non-symmetric and symmetric multiplicative Schwarz preconditioners.

**Algorithm 2.9:** The operator form of the additive Schwarz preconditioner

**for** $j = 0,\dots,J$ **do**
  Find $g_j \in \mathscr{V}_j$ satisfying $a_j(g_j,v_j) = \left\langle f, R_j^t v_j \right\rangle \quad \forall v_j \in \mathscr{V}_j$
**end for**
$g \leftarrow \sum_{j=0}^{J} R_j^t g_j \in \mathscr{V}$
Solve $P_{\mathrm{ad}} u = g$

### 2.5.1 *Multiplicative Schwarz preconditioner: the matrix form*

**Algorithm 2.10:** The matrix form of the multiplicative Schwarz preconditioner

To compute $\mathbf{z} = \mathbf{C}_{\mathrm{mu}}^{-1}\mathbf{r}$ for a given $\mathbf{r} \in \mathbb{R}^N$
$\mathbf{y}_0 \leftarrow \mathbf{R}_0^\top \mathbf{A}_0^{-1} \mathbf{R}_0 \mathbf{r}$;
**for** $j = 1,\dots,J$ **do**
  $\mathbf{y}_j \leftarrow \mathbf{y}_{j-1} + \mathbf{R}_j^\top \mathbf{A}_j^{-1} \mathbf{R}_j (\mathbf{r} - \mathbf{A}\mathbf{y}_{j-1})$
**end for**
$\mathbf{C}_{\mathrm{mu}}^{-1}\mathbf{r} \leftarrow \mathbf{y}_J$

### 2.5.2 *Multiplicative Schwarz preconditioner: the operator form*

We first derive a closed form for $\mathbf{C}_{\mathrm{mu}}^{-1}$.

**Lemma 2.7.** *If $\mathbf{E}_{\mathrm{mu}}$ is the matrix representation of the error propagation operator $E_{\mathrm{mu}}$ defined in* (2.22)*, then*

$$\mathbf{E}_{\mathrm{mu}} = (\mathbf{I} - \mathbf{P}_J)\cdots(\mathbf{I} - \mathbf{P}_0) \tag{2.29}$$

*and*

$$\mathbf{C}_{\mathrm{mu}}^{-1} = (\mathbf{I} - \mathbf{E}_{\mathrm{mu}})\mathbf{A}^{-1}. \tag{2.30}$$

*Proof.* The identity (2.29) follows directly from the definition (2.22). We prove (2.30). Let $\mathbf{C}_j = \mathbf{R}_j^\top \mathbf{A}_j^{-1} \mathbf{R}_j$ for $j = 0,\dots,J$. Then $\mathbf{y}_0 = \mathbf{C}_0 \mathbf{r}$ and

$$\begin{aligned}
\mathbf{y}_1 &= \mathbf{y}_0 + \mathbf{C}_1(\mathbf{r} - \mathbf{A}\mathbf{y}_0) \\
&= \mathbf{C}_0\mathbf{r} + \mathbf{C}_1\mathbf{r} - \mathbf{C}_1\mathbf{A}\mathbf{C}_0\mathbf{r} \\
&= \mathbf{A}^{-1}(\mathbf{A}\mathbf{C}_0\mathbf{r} + \mathbf{A}\mathbf{C}_1\mathbf{r} - \mathbf{A}\mathbf{C}_1\mathbf{A}\mathbf{C}_0\mathbf{r}) \\
&= \mathbf{A}^{-1}\big[\mathbf{I} - (\mathbf{I} - \mathbf{A}\mathbf{C}_1)(\mathbf{I} - \mathbf{A}\mathbf{C}_0)\big]\mathbf{r}.
\end{aligned}$$

By induction we have

$$\mathbf{y}_J = \mathbf{A}^{-1}\big[\mathbf{I} - (\mathbf{I} - \mathbf{AC}_J)(\mathbf{I} - \mathbf{AC}_{J-1})\cdots(\mathbf{I} - \mathbf{AC}_0)\big]\mathbf{r}.$$

Since $\mathbf{C}_{\mathrm{mu}}^{-1}\mathbf{r} = \mathbf{y}_J$ we infer

$$\mathbf{C}_{\mathrm{mu}}^{-1} = \mathbf{A}^{-1} - \mathbf{A}^{-1}(\mathbf{I} - \mathbf{AC}_J)(\mathbf{I} - \mathbf{AC}_{J-1})\cdots(\mathbf{I} - \mathbf{AC}_0).$$

Applying successively the identity

$$\mathbf{A}^{-1}(\mathbf{I} - \mathbf{AC}_j) = (\mathbf{I} - \mathbf{C}_j\mathbf{A})\mathbf{A}^{-1}$$

we obtain

$$\mathbf{A}^{-1}(\mathbf{I} - \mathbf{AC}_J)(\mathbf{I} - \mathbf{AC}_{J-1})\cdots(\mathbf{I} - \mathbf{AC}_0) = (\mathbf{I} - \mathbf{C}_J\mathbf{A})(\mathbf{I} - \mathbf{C}_{J-1}\mathbf{A})\cdots(\mathbf{I} - \mathbf{C}_0\mathbf{A})\mathbf{A}^{-1},$$

so that, noting (2.26) and (2.29),

$$\begin{aligned}\mathbf{C}_{\mathrm{mu}}^{-1} &= \big[\mathbf{I} - (\mathbf{I} - \mathbf{C}_J\mathbf{A})(\mathbf{I} - \mathbf{C}_{J-1}\mathbf{A})\cdots(\mathbf{I} - \mathbf{C}_0\mathbf{A})\big]\mathbf{A}^{-1}\\ &= \big[\mathbf{I} - (\mathbf{I} - \mathbf{P}_J)(\mathbf{I} - \mathbf{P}_{J-1})\cdots(\mathbf{I} - \mathbf{P}_0)\big]\mathbf{A}^{-1}\\ &= (\mathbf{I} - \mathbf{E}_{\mathrm{mu}})\mathbf{A}^{-1}.\end{aligned}$$

The lemma is proved. □

It follows from Lemma 2.7 that

$$C_{\mathrm{mu}}^{-1} = (I - E_{\mathrm{mu}})A^{-1}.$$

Reminding the definition of $P_{\mathrm{mu}}$, see (2.21), we deduce

$$P_{\mathrm{mu}} = C_{\mathrm{mu}}^{-1}A \quad \text{or} \quad \mathbf{P}_{\mathrm{mu}} = \mathbf{C}_{\mathrm{mu}}^{-1}\mathbf{A} \tag{2.31}$$

and the equivalence

$$\boxed{Au = b \iff P_{\mathrm{mu}}u = C_{\mathrm{mu}}^{-1}b}$$

It is noted that

$$\mathbf{C}_{\mathrm{mu}}^{-1}\mathbf{A} = \mathbf{I} - \mathbf{E}_{\mathrm{mu}} = \mathbf{I} - (\mathbf{I} - \mathbf{P}_J)\cdots(\mathbf{I} - \mathbf{P}_0)$$

is not symmetric, and therefore in the implementation the GMRES method or a linear iterative method is used instead of CG. A symmetric version of the multiplicative Schwarz preconditioner is designed in the next section.

### 2.5.3 *Symmetric multiplicative Schwarz*

The symmetric version of the symmetric multiplicative Schwarz preconditioner is given in Algorithm 2.11.

**Algorithm 2.11:** The matrix form of the symmetric multiplicative Schwarz preconditioner

To compute $\mathbf{z} = \mathbf{C}_{\mathrm{smu}}^{-1}\mathbf{r}$ for a given $\mathbf{r} \in \mathbb{R}^N$
1: $\mathbf{y}_0 \leftarrow \mathbf{R}_0^\top \mathbf{A}_0^{-1} \mathbf{R}_0 \mathbf{r}$
2: **for** $j = 1, \dots, J$ **do**
3: $\quad \mathbf{y}_j \leftarrow \mathbf{y}_{j-1} + \mathbf{R}_j^\top \mathbf{A}_j^{-1} \mathbf{R}_j (\mathbf{r} - \mathbf{A}\mathbf{y}_{j-1})$
4: **end for**
5: $\mathbf{y}_{J+1} \leftarrow \mathbf{y}_J$
6: **for** $j = J, \dots, 0$ **do**
7: $\quad \mathbf{y}_j \leftarrow \mathbf{y}_{j+1} + \mathbf{R}_j^\top \mathbf{A}_j^{-1} \mathbf{R}_j (\mathbf{r} - \mathbf{A}\mathbf{y}_{j+1})$
8: **end for**
9: $\mathbf{C}_{\mathrm{smu}}^{-1}\mathbf{r} \leftarrow \mathbf{y}_0$

*Remark 2.1.* If the local bilinear forms $a_j(\cdot,\cdot)$ satisfy (2.17) then Step 5 is not necessary and the loop in Step 6 starts from $J-1$ instead of $J$; cf. Lemma 2.3.

The matrix $\mathbf{C}_{\mathrm{smu}}^{-1}$ has the following closed form:

$$\mathbf{C}_{\mathrm{smu}}^{-1} = (\mathbf{I} - \mathbf{E}_{\mathrm{smu}})\mathbf{A}^{-1},$$

where

$$\mathbf{E}_{\mathrm{smu}} = (\mathbf{I} - \mathbf{P}_0)\cdots(\mathbf{I} - \mathbf{P}_{J-1})(\mathbf{I} - \mathbf{P}_J)(\mathbf{I} - \mathbf{P}_J)(\mathbf{I} - \mathbf{P}_{J-1})\cdots(\mathbf{I} - \mathbf{P}_0).$$

Therefore,

$$\mathbf{P}_{\mathrm{smu}} = \mathbf{C}_{\mathrm{smu}}^{-1}\mathbf{A}. \tag{2.32}$$

## 2.6 Convergence Theory for Preconditioners with PCG

This section forms the background for the convergence analysis of the Schwarz methods. In view of (2.11), we shall explain what is to be proved to justify that a preconditioner is efficient. We follow the analysis in [219].

In the sequel, by $P_{\mathrm{pre}}$ we mean either $P_{\mathrm{ad}}$ or $P_{\mathrm{smu}}$. Similarly for $\mathbf{C}_{\mathrm{pre}}^{-1}$.

### *2.6.1 Extremal eigenvalues of preconditioned matrices*

Recall from Section 2.2.1 that solving the system $\mathbf{A}\mathbf{u} = \mathbf{b}$ by the PCG method is equivalent to solving the system $\widetilde{\mathbf{A}}\widetilde{\mathbf{u}} = \widetilde{\mathbf{b}}$ with the Euclidean inner product $\langle\cdot,\cdot\rangle_{\mathbb{R}^N}$ or the system $\widehat{\mathbf{A}}\mathbf{u} = \widehat{\mathbf{b}}$ with the $\mathbf{C}$-inner product $\langle\cdot,\cdot\rangle_{\mathbf{C}} := \langle \mathbf{C}_{\mathrm{pre}}\cdot,\cdot\rangle_{\mathbb{R}^N}$, where

$$\widetilde{\mathbf{A}} = \mathbf{C}_{\mathrm{pre}}^{-1/2}\mathbf{A}\mathbf{C}_{\mathrm{pre}}^{-1/2} \quad \text{and} \quad \widehat{\mathbf{A}} = \mathbf{C}_{\mathrm{pre}}^{-1}\mathbf{A}. \tag{2.33}$$

According to (2.4), it is necessary to estimate the condition number $\kappa_2(\widetilde{\mathbf{A}})$ and $\kappa_{\mathbf{C}}(\widehat{\mathbf{A}})$ where

$$\kappa_2(\widetilde{\mathbf{A}}) := \lambda_{\max}(\widetilde{\mathbf{A}})/\lambda_{\min}(\widetilde{\mathbf{A}}) \quad \text{and} \quad \kappa_{\mathbf{C}}(\widehat{\mathbf{A}}) := \|\widehat{\mathbf{A}}\|_{\mathbf{C}}\|\widehat{\mathbf{A}}^{-1}\|_{\mathbf{C}}.$$

The following lemma shows that $\kappa_{\mathbf{C}}(\widehat{\mathbf{A}})$ can be estimated by estimating the extremal eigenvalues of $\widehat{\mathbf{A}}$.

**Lemma 2.8.** *The following statements hold.*

(i) *The matrices $\widetilde{\mathbf{A}}$ and $\widehat{\mathbf{A}}$ have the same eigenvalues.*
(ii) *The extremal eigenvalues of $\widehat{\mathbf{A}}$ and the condition number $\kappa_{\mathbf{C}}(\widehat{\mathbf{A}})$ satisfy*

$$\lambda_{\max}(\widehat{\mathbf{A}}) = \max_{\mathbf{x}\in\mathbb{R}^N} \frac{\left\langle \widehat{\mathbf{A}}\mathbf{x},\mathbf{x}\right\rangle_{\mathbf{C}}}{\langle \mathbf{x},\mathbf{x}\rangle_{\mathbf{C}}} = \|\widehat{\mathbf{A}}\|_{\mathbf{C}},$$

$$\lambda_{\min}(\widehat{\mathbf{A}}) = \min_{\mathbf{x}\in\mathbb{R}^N} \frac{\left\langle \widehat{\mathbf{A}}\mathbf{x},\mathbf{x}\right\rangle_{\mathbf{C}}}{\langle \mathbf{x},\mathbf{x}\rangle_{\mathbf{C}}} = \frac{1}{\|\widehat{\mathbf{A}}^{-1}\|_{\mathbf{C}}},$$

$$\kappa_{\mathbf{C}}(\widehat{\mathbf{A}}) = \frac{\lambda_{\max}(\widehat{\mathbf{A}})}{\lambda_{\min}(\widehat{\mathbf{A}})} = \frac{\lambda_{\max}(\widetilde{\mathbf{A}})}{\lambda_{\min}(\widetilde{\mathbf{A}})}.$$

*Proof.* In this proof we write $\mathbf{C}$ for $\mathbf{C}_{\mathrm{pre}}$.

(i) The first statement is obvious from

$$\widehat{\mathbf{A}}\mathbf{u} = \lambda\mathbf{u} \iff \mathbf{C}^{-1/2}\widetilde{\mathbf{A}}\mathbf{C}^{1/2}\mathbf{u} = \lambda\mathbf{u} \iff \widetilde{\mathbf{A}}\mathbf{C}^{1/2}\mathbf{u} = \lambda\mathbf{C}^{1/2}\mathbf{u} \iff \widetilde{\mathbf{A}}\widetilde{\mathbf{u}} = \lambda\widetilde{\mathbf{u}}$$

where $\widetilde{\mathbf{u}} = \mathbf{C}^{1/2}\mathbf{u}$.

(ii) For the second statement, we first consider $\lambda_{\max}(\widehat{\mathbf{A}})$. Due to part (i), the symmetry of $\widetilde{\mathbf{A}}$, and the symmetry and positive definiteness of $\mathbf{C}$ we have, with $\widehat{\mathbf{x}} = \mathbf{C}^{-1/2}\mathbf{x}$,

$$\begin{aligned}
\lambda_{\max}(\widehat{\mathbf{A}}) = \lambda_{\max}(\widetilde{\mathbf{A}}) &= \max_{\mathbf{x}\in\mathbb{R}^N} \frac{\left\langle \mathbf{C}^{-1/2}\mathbf{A}\mathbf{C}^{-1/2}\mathbf{x},\mathbf{x}\right\rangle_{\mathbb{R}^N}}{\langle \mathbf{x},\mathbf{x}\rangle_{\mathbb{R}^N}} \\
&= \max_{\widehat{\mathbf{x}}\in\mathbb{R}^N} \frac{\left\langle \mathbf{C}^{-1/2}\mathbf{A}\widehat{\mathbf{x}},\mathbf{C}^{1/2}\widehat{\mathbf{x}}\right\rangle_{\mathbb{R}^N}}{\left\langle \mathbf{C}^{1/2}\widehat{\mathbf{x}},\mathbf{C}^{1/2}\widehat{\mathbf{x}}\right\rangle_{\mathbb{R}^N}} \\
&= \max_{\widehat{\mathbf{x}}\in\mathbb{R}^N} \frac{\langle \mathbf{A}\widehat{\mathbf{x}},\widehat{\mathbf{x}}\rangle_{\mathbb{R}^N}}{\langle \mathbf{C}\widehat{\mathbf{x}},\widehat{\mathbf{x}}\rangle_{\mathbb{R}^N}} = \max_{\widehat{\mathbf{x}}\in\mathbb{R}^N} \frac{\left\langle \widehat{\mathbf{A}}\widehat{\mathbf{x}},\widehat{\mathbf{x}}\right\rangle_{\mathbf{C}}}{\langle \widehat{\mathbf{x}},\widehat{\mathbf{x}}\rangle_{\mathbf{C}}}.
\end{aligned}$$

On the other hand, we have, with $\mathbf{w} = \mathbf{C}^{1/2}\mathbf{v}$,

$$\|\widehat{\mathbf{A}}\|_{\mathbf{C}}^2 = \max_{\mathbf{v}\in\mathbb{R}^N} \frac{\left\langle \widehat{\mathbf{A}}\mathbf{v},\widehat{\mathbf{A}}\mathbf{v}\right\rangle_{\mathbf{C}}}{\langle \mathbf{v},\mathbf{v}\rangle_{\mathbf{C}}} = \max_{\mathbf{v}\in\mathbb{R}^N} \frac{(\widehat{\mathbf{A}}\mathbf{v})^\top\mathbf{C}\widehat{\mathbf{A}}\mathbf{v}}{\mathbf{v}^\top\mathbf{C}\mathbf{v}} = \max_{\mathbf{v}\in\mathbb{R}^N} \frac{\mathbf{v}^\top\mathbf{A}\mathbf{C}^{-1}\mathbf{A}\mathbf{v}}{\mathbf{v}^\top\mathbf{C}\mathbf{v}}$$

$$= \max_{\mathbf{v}\in\mathbb{R}^N} \frac{\mathbf{w}^\top \widetilde{\mathbf{A}}^\top \widetilde{\mathbf{A}}\mathbf{w}}{\mathbf{w}^\top \mathbf{w}} = \max_{\mathbf{w}\in\mathbb{R}^N} \frac{\mathbf{w}^\top \widetilde{\mathbf{A}}^2 \mathbf{w}}{\mathbf{w}^\top \mathbf{w}} = \lambda_{\max}(\widetilde{\mathbf{A}})^2 = \lambda_{\max}(\widehat{\mathbf{A}})^2,$$

where we used the fact that $\widetilde{\mathbf{A}}$ and thus $\widetilde{\mathbf{A}}^2$ are symmetric. Therefore, the results for $\lambda_{\max}(\widehat{\mathbf{A}})$ hold. Similarly, the results for $\lambda_{\min}(\widehat{\mathbf{A}})$ hold. The last two identities follow from the definition of $\kappa_{\mathbf{C}}(\widehat{\mathbf{A}})$ and part (i). □

The above lemma tells us that regardless of which approach we use, for convergence analysis it suffices to estimate the extremal eigenvalues of $\widetilde{\mathbf{A}}$. Lemma C.6 in Appendix C shows that instead of estimating the eigenvalues of $\widetilde{\mathbf{A}}$, we can estimate the extremal eigenvalues of $\mathbf{A}^{1/2}\mathbf{C}_{\text{pre}}^{-1}\mathbf{A}^{1/2}$. The next lemma shows how the extremal eigenvalues of $\mathbf{A}^{1/2}\mathbf{C}_{\text{pre}}^{-1}\mathbf{A}^{1/2}$ can be expressed in terms of the bilinear form $a(\cdot,\cdot)$ and the preconditioner operator $P_{\text{pre}}$. It is this form of expression that will be used in the analysis.

**Lemma 2.9.** *The following identities hold*

$$\lambda_{\min}(\mathbf{C}_{\text{pre}}^{-1/2}\mathbf{A}\mathbf{C}_{\text{pre}}^{-1/2}) = \inf_{u\in\mathscr{V}} \frac{a(P_{\text{pre}}u,u)}{a(u,u)}$$

*and*

$$\lambda_{\max}(\mathbf{C}_{\text{pre}}^{-1/2}\mathbf{A}\mathbf{C}_{\text{pre}}^{-1/2}) = \sup_{u\in\mathscr{V}} \frac{a(P_{\text{pre}}u,u)}{a(u,u)}.$$

*Proof.* It follows from Lemma C.6 in Appendix C and the Rayleigh quotient that

$$\lambda_{\min}(\mathbf{C}_{\text{pre}}^{-1/2}\mathbf{A}\mathbf{C}_{\text{pre}}^{-1/2}) = \lambda_{\min}(\mathbf{A}^{1/2}\mathbf{C}_{\text{pre}}^{-1}\mathbf{A}^{1/2}) = \inf_{\mathbf{v}\in\mathbb{R}^N} \frac{\mathbf{v}^\top\mathbf{A}^{1/2}\mathbf{C}_{\text{pre}}^{-1}\mathbf{A}^{1/2}\mathbf{v}}{\mathbf{v}^\top\mathbf{v}}.$$

Recalling that $\mathbf{C}_{\text{pre}}^{-1}\mathbf{A} = \mathbf{P}_{\text{pre}}$, see (2.27), (2.31), and (2.32), and letting $\mathbf{u} = \mathbf{A}^{-1/2}\mathbf{v}$, we deduce

$$\lambda_{\min}(\mathbf{C}_{\text{pre}}^{-1/2}\mathbf{A}\mathbf{C}_{\text{pre}}^{-1/2}) = \inf_{\mathbf{u}\in\mathbb{R}^N} \frac{\mathbf{u}^\top\mathbf{A}\mathbf{C}_{\text{pre}}^{-1}\mathbf{A}\mathbf{u}}{\mathbf{u}^\top\mathbf{A}\mathbf{u}} = \inf_{\mathbf{u}\in\mathbb{R}^N} \frac{\mathbf{u}^\top\mathbf{A}\mathbf{P}_{\text{pre}}\mathbf{u}}{\mathbf{u}^\top\mathbf{A}\mathbf{u}} = \inf_{u\in\mathscr{V}} \frac{a(P_{\text{pre}}u,u)}{a(u,u)}.$$

Similar arguments give the result for $\lambda_{\max}$. □

The above lemma prompts us to define the extremal eigenvalues and condition number of the Schwarz operator $P_{\text{pre}}$ as in the following subsection.

### 2.6.2 Condition numbers of Schwarz operators

**Definition 2.2.** *The maximum eigenvalue, minimum eigenvalue, and condition number of the Schwarz operator $P_{\text{pre}}$ are defined by*

$$\lambda_{\max}(P_{\text{pre}}) := \sup_{u\in\mathscr{V}} \frac{a(P_{\text{pre}}u,u)}{a(u,u)},$$
$$\lambda_{\min}(P_{\text{pre}}) := \inf_{u\in\mathscr{V}} \frac{a(P_{\text{pre}}u,u)}{a(u,u)},$$
$$\kappa(P_{\text{pre}}) := \frac{\lambda_{\max}(P_{\text{pre}})}{\lambda_{\min}(P_{\text{pre}})}.$$

The following lemma is a direct consequence of Lemma 2.9.

**Lemma 2.10.** *If*

$$C_0^{-2}a(u,u) \le a(P_{\text{pre}}u,u) \le C_1^2 a(u,u) \quad \forall u \in \mathscr{V}, \tag{2.34}$$

*then*

$$\lambda_{\max}(P_{\text{pre}}) \le C_1^2 \quad \textit{and} \quad \lambda_{\min}(P_{\text{pre}}) \ge C_0^{-2}$$

*so that*

$$\kappa_2(\mathbf{C}_{\text{pre}}^{-1/2}\mathbf{A}\mathbf{C}_{\text{pre}}^{-1/2}) = \kappa(P_{\text{pre}}) \le C_0^2 C_1^2.$$

Recalling that the condition number $\kappa_2(\mathbf{A})$ grows significantly with the degrees of freedom $N$ (see Section 1.3) the goal is to design preconditioners $\mathbf{C}^{-1}$ so that (2.34) hold with $C_0$ and $C_1$ independent of (or depending moderately on) the variables in concern, i.e., $J$ (number of subspaces) and $N$.

### 2.6.3 Stability of decomposition – A lower bound for $\lambda_{\min}(P_{\text{ad}})$

**Assumption 2.3 (Stability of decomposition).** There exists a constant $C_0$ such that every $u \in \mathscr{V}$ admits a decomposition

$$u = \sum_{j=0}^{J} R_j^t u_j, \quad u_j \in \mathscr{V}_j, \tag{2.35}$$

which satisfies

$$\sum_{j=0}^{J} a_j(u_j,u_j) \le C_0^2 a(u,u). \tag{2.36}$$

**Proposition 2.1.** *Let Assumption 2.3 be satisfied. Then the following statements hold.*

(i) *For any* $u \in \mathscr{V}$

$$a(P_{\text{ad}}u,u) \ge C_0^{-2}a(u,u). \tag{2.37}$$

*Hence*

$$\lambda_{\min}(P_{\text{ad}}) \ge C_0^{-2}.$$

(ii) $P_{\text{ad}}$ *is positive-definite.*

(iii) *For any* $u \in \mathscr{V}$

$$a(P_{\mathrm{ad}}^{-1}u,u) = \min_{\substack{u_j \in \mathscr{V}_j \\ u=\sum_{j=0}^J R_j^t u_j}} \sum_{j=0}^J a_j(u_j,u_j). \tag{2.38}$$

*Proof.* We first prove (2.37). The decomposition (2.35), the Cauchy-Schwarz inequality, (2.16), and (2.36) imply

$$\begin{aligned}
a(u,u) &= \sum_{j=0}^J a(u,R_j^t u_j) = \sum_{j=0}^J a_j(\widetilde{P}_j u,u_j) \\
&\le \left(\sum_{j=0}^J a_j(\widetilde{P}_j u,\widetilde{P}_j u)\right)^{1/2} \left(\sum_{j=0}^J a_j(u_j,u_j)\right)^{1/2} \\
&\le C_0 \left(\sum_{j=0}^J a_j(\widetilde{P}_j u,\widetilde{P}_j u)\right)^{1/2} a(u,u)^{1/2}.
\end{aligned}$$

Hence, using again (2.16), (2.19) and (2.20) we deduce

$$a(u,u) \le C_0^2 \sum_{j=0}^J a_j(\widetilde{P}_j u,\widetilde{P}_j u) = C_0^2 \sum_{j=0}^J a(R_j^t \widetilde{P}_j u,u) = C_0^2 \sum_{j=0}^J a(P_j u,u) = C_0^2 a(P_{\mathrm{ad}} u,u).$$

This gives a lower bound for $\lambda_{\min}(P_{\mathrm{ad}})$ if one invokes Lemma 2.10. It also implies that $P_{\mathrm{ad}}$ is positive definite.

It remains to show (2.38). By using again (2.35), (2.16), the Cauchy–Schwarz inequality, and (2.20) we obtain

$$\begin{aligned}
a(P_{\mathrm{ad}}^{-1}u,u) &= \sum_{j=0}^J a(P_{\mathrm{ad}}^{-1}u,R_j^t u_j) = \sum_{j=0}^J a_j(\widetilde{P}_j P_{\mathrm{ad}}^{-1}u,u_j) \\
&\le \left(\sum_{j=0}^J a_j(\widetilde{P}_j P_{\mathrm{ad}}^{-1}u,\widetilde{P}_j P_{\mathrm{ad}}^{-1}u)\right)^{1/2} \left(\sum_{j=0}^J a_j(u_j,u_j)\right)^{1/2} \\
&= \left(\sum_{j=0}^J a(P_{\mathrm{ad}}^{-1}u,P_j P_{\mathrm{ad}}^{-1}u)\right)^{1/2} \left(\sum_{j=0}^J a_j(u_j,u_j)\right)^{1/2} \\
&= a(P_{\mathrm{ad}}^{-1}u,u)^{1/2} \left(\sum_{j=0}^J a_j(u_j,u_j)\right)^{1/2},
\end{aligned}$$

which implies

$$a(P_{\mathrm{ad}}^{-1}u,u) \le \min_{\substack{u_j \in \mathscr{V}_j \\ u=\sum_{j=0}^J R_j^t u_j}} \sum_{j=0}^J a_j(u_j,u_j).$$

Equality occurs when $u_j = \widetilde{P}_j P_{\mathrm{ad}}^{-1}u$. □

### 2.6.4 Coercivity of decomposition – An upper bound for $\lambda_{\max}(P_{\text{ad}})$

**Assumption 2.4 (Coercivity of decomposition).** There exists a constant $C_1$ such that for every $u \in \mathscr{V}$ if

$$u = \sum_{j=0}^{J} R_j^t u_j, \quad u_j \in \mathscr{V}_j,$$

then

$$a(u,u) \le C_1^2 \sum_{j=0}^{J} a_j(u_j,u_j).$$

**Proposition 2.2.** *Let Assumption 2.4 and* (2.17) *be satisfied. Assume further that $P_{\text{ad}}$ is positive definite. Then for any $u \in \mathscr{V}$ the following statement holds*

$$a(P_{\text{ad}}u,u) \le C_1^2 a(u,u)$$

*and thus $\lambda_{\max}(P_{\text{ad}}) \le C_1^2$.*

*Proof.* Since $P_{\text{ad}}$ is positive definite, for any $u \in \mathscr{V}$, we can define $v = P_{\text{ad}}^{-1/2} u$. Then the symmetry of $P_{\text{ad}}$ yields

$$a(u,u) = a(P_{\text{ad}}^{1/2} v, P_{\text{ad}}^{1/2} v) = a(P_{\text{ad}}v,v). \tag{2.39}$$

On the other hand, for any $w \in \mathscr{V}$, since $P_{\text{ad}}w = \sum_{j=0}^{J} R_j^t \widetilde{P}_j w$, Assumption 2.4 and (2.17) imply

$$a(P_{\text{ad}}w, P_{\text{ad}}w) \le C_1^2 \sum_{j=0}^{J} a_j(\widetilde{P}_j w, \widetilde{P}_j w) = C_1^2 \sum_{j=0}^{J} a(P_j w, P_j w).$$

Hence,

$$a(P_{\text{ad}}u,u) = a(P_{\text{ad}}^{1/2} u, P_{\text{ad}}^{1/2} u) = a(P_{\text{ad}}v, P_{\text{ad}}v) \le C_1^2 \sum_{j=0}^{J} a(P_j v, P_j v)$$

Since $P_j$ is a projection, this implies

$$a(P_{\text{ad}}u,u) \le C_1^2 \sum_{j=0}^{J} a(P_j v, v) = C_1^2 a(P_{\text{ad}}v,v).$$

This inequality and (2.39) give the required estimate. □

### 2.6.5 Strengthened Cauchy–Schwarz inequalities and local stability

**Assumption 2.5 (Strengthened Cauchy–Schwarz Inequalities).** There exist constants $\varepsilon_{ij} \in [0,1)$, $i, j = 1, \ldots, J$, such that

$$\left|a(R_i^t u_i, R_j^t u_j)\right| \leq \varepsilon_{ij} a(R_i^t u_i, R_i^t u_i)^{1/2} a(R_j^t u_j, R_j^t u_j)^{1/2} \quad \forall u_i \in \mathscr{V}_i,\ u_j \in \mathscr{V}_j.$$

We denote by $\rho(\mathscr{E})$ the spectral radius of the matrix $\mathscr{E} = (\varepsilon_{ij})$. The inequality is trivial with $\varepsilon_{ij} = 1$; however, in that case $\rho(\mathscr{E}) = J$, resulting in a poor upper bound for $\lambda_{\max}(P_{\text{ad}})$; see Lemma 2.11 and Proposition 2.2.

**Assumption 2.6 (Local stability).** There exists $\omega > 0$ such that

$$a(R_j^t u_j, R_j^t u_j) \leq \omega a_j(u_j, u_j) \quad \forall u_j \in \mathscr{V}_j,\ j = 0, \ldots, J.$$

This local stability obviously holds with $\omega = 1$ if $a_j(\cdot,\cdot)$ is chosen by (2.17).

**Lemma 2.11.** *Assume that Assumptions 2.5 and 2.6 hold. Then Assumption 2.4 holds.*

*Proof.* For any $u = \sum_{j=0}^J R_j^t u_j \in \mathscr{V}$ with $u_j \in \mathscr{V}_j$, denoting $v_0 = R_0^t u_0$ and $v_1 = \sum_{j=1}^J R_j^t u_j$ so that $u = v_0 + v_1$, we have (with the help of the Cauchy–Schwarz inequality)

$$\begin{aligned}
a(u,u) &= a(v_0,v_0) + a(v_1,v_1) + 2a(v_0,v_1) \\
&\leq a(v_0,v_0) + a(v_1,v_1) + 2a(v_0,v_0)^{1/2} a(v_1,v_1)^{1/2} \\
&\leq 2a(v_0,v_0) + 2a(v_1,v_1) \\
&= 2a(R_0^t u_0, R_0^t u_0) + 2 \sum_{i,j=1}^J a(R_i^t u_i, R_j^t u_j).
\end{aligned}$$

The Strengthened Cauchy–Schwarz inequalities then imply

$$\begin{aligned}
a(u,u) &= 2a(R_0^t u_0, R_0^t u_0) + 2 \sum_{i,j=1}^J \varepsilon_{ij} a(R_i^t u_i, R_i^t u_i)^{1/2} a(R_j^t u_j, R_j^t u_j)^{1/2} \\
&\leq 2a(R_0^t u_0, R_0^t u_0) + 2\rho(\mathscr{E}) \sum_{j=1}^J a(R_j^t u_j, R_j^t u_j) \\
&\leq 2(\rho(\mathscr{E}) + 1) \sum_{j=0}^J a(R_j^t u_j, R_j^t u_j).
\end{aligned}$$

The local stability assumption yields

$$a(u,u) \leq 2\omega(\rho(\mathscr{E})+1)\sum_{j=0}^{J} a_j(u_j,u_j).$$

Hence Assumption 2.4 holds with $C_1^2 = 2\omega(\rho(\mathscr{E})+1)$. □

Therefore, due to Proposition 2.2, Assumptions 2.3, 2.5, and 2.6 provide the following upper bound for the maximum eigenvalue of $P_{\mathrm{ad}}$:

$$\lambda_{\max}(P_{\mathrm{ad}}) \leq 2\omega(\rho(\mathscr{E})+1).$$

The following two lemmas give a better upper bound.

**Lemma 2.12.** *Under Assumption 2.6, the following statement holds, for all $u \in \mathscr{V}$ and $j = 0,\ldots,J$,*

$$a(P_j u, P_j u) \leq \omega^2 a(u,u) \quad \text{and} \quad a(P_j u, u) \leq \omega a(u,u).$$

*Proof.* For $j = 0,\ldots,J$ and $u \in \mathscr{V}$, we have from the definitions of $P_j$ and $\widetilde{P}_j$, and Assumption 2.6

$$\begin{aligned} a(P_j u, P_j u) &= a(R_j^t \widetilde{P}_j u, R_j^t \widetilde{P}_j u) \leq \omega a_j(\widetilde{P}_j u, \widetilde{P}_j u) = \omega a(u, P_j u) \\ &\leq \omega a(u,u)^{1/2} a(P_j u, P_j u)^{1/2} \end{aligned}$$

which implies the first inequality. The second inequality is obtained by using the Cauchy–Schwarz inequality

$$a(P_j u, u) \leq a(P_j u, P_j u)^{1/2} a(u,u)^{1/2} \leq \omega a(u,u).$$ □

**Lemma 2.13.** *Under the Assumptions 2.5 and 2.6, the following estimate holds, for all $u \in \mathscr{V}$,*

$$a(P_{\mathrm{ad}} u, u) \leq \omega(\rho(\mathscr{E})+1) a(u,u).$$

*Therefore, $\lambda_{\max}(P_{\mathrm{ad}}) \leq \omega(\rho(\mathscr{E})+1)$.*

*Proof.* Let $\widehat{P}$ be defined as in Lemma 2.15, i.e., $\widehat{P} = \sum_{j=1}^{J} P_j$. Then

$$a(P_{\mathrm{ad}} u, u) \leq a(P_0 u, u) + a(\widehat{P} u, u).$$

Lemmas 2.15 and 2.12 give

$$a(P_{\mathrm{ad}} u, u) \leq \omega a(u,u) + \omega\rho(\mathscr{E}) a(u,u) = \omega(1+\rho(\mathscr{E})) a(u,u).$$ □

### 2.6.6 An upper bound for $\lambda_{\max}(P_{\mathrm{smu}})$

**Proposition 2.3.** *For any $u \in \mathscr{V}$ the following inequality holds*

$$a(P_{\text{smu}}u,u) \leq a(u,u).$$

*Therefore,* $\lambda_{\max}(P_{\text{smu}}) \leq 1$.

*Proof.* It follows from (2.23) and (2.24) that

$$\begin{aligned} a(P_{\text{smu}}u,u) &= a(u,u) - a(E_{\text{mu}}^t E_{\text{mu}}u,u) \\ &= a(u,u) - a(E_{\text{mu}}u, E_{\text{mu}}u) \\ &\leq a(u,u). \end{aligned}$$

Hence $\lambda_{\max}(P_{\text{smu}}) \leq 1$. □

### 2.6.7 A lower bound for $\lambda_{\min}(P_{\text{smu}})$

**Lemma 2.14.** *Assume that Assumptions 2.5 and 2.6 are satisfied. For any* $j,k = 1,\ldots,J$ *and any* $u,v \in \mathscr{V}$ *the following inequalities hold*

$$\begin{aligned} a(P_j u, v) &\leq a(P_j u,u)^{1/2} a(P_j v, v)^{1/2}, \\ a(P_j u, P_k v) &\leq \omega \varepsilon_{jk} a(P_j u,u)^{1/2} a(P_k v,v)^{1/2}. \end{aligned}$$

*Proof.* Definitions (2.16) and (2.19), and the Cauchy–Schwarz inequality yield

$$\begin{aligned} a(P_j u, v) = a(u, P_j v) = a(u, R_j^t \widetilde{P}_j v) &= a_j(\widetilde{P}_j u, \widetilde{P}_j v) \\ &\leq a_j(\widetilde{P}_j u, \widetilde{P}_j u)^{1/2} a_j(\widetilde{P}_j v, \widetilde{P}_j v)^{1/2} \\ &= a(u, P_j u)^{1/2} a(v, P_j v)^{1/2} = a(P_j u, u)^{1/2} a(P_j v, v)^{1/2} \end{aligned}$$

and

$$\begin{aligned} a(P_j u, P_k v) &\leq \varepsilon_{jk} a(P_j u, P_j u)^{1/2} a(P_k P v, P_k v)^{1/2} \\ &\leq \omega \varepsilon_{jk} a_j(\widetilde{P}_j u, \widetilde{P}_j u)^{1/2} a_k(\widetilde{P}_k v, \widetilde{P}_k v)^{1/2} \\ &= \omega \varepsilon_{jk} a(P_j u, u)^{1/2} a(P_k v, v)^{1/2}, \end{aligned}$$

proving the lemma. □

**Lemma 2.15.** *Under Assumptions 2.5 and 2.6, if* $\widehat{P} = \sum_{j=1}^{J} P_j$ *then*

$$a(\widehat{P}u,u) \leq \omega \rho(\mathscr{E}) a(u,u).$$

*Proof.* It follows from Lemma 2.14 that

$$a(\widehat{P}u, \widehat{P}u) = \sum_{j,k=1}^{J} a(P_j u, P_k u) \leq \omega \sum_{j,k=1}^{J} \varepsilon_{jk} a(P_j u, u)^{1/2} a(P_k u, u)^{1/2}.$$

Since $\mathscr{E}$ is symmetric we have

$$\mathbf{v}^\top \mathscr{E} \mathbf{v} \le \rho(\mathscr{E}) \mathbf{v}^\top \mathbf{v} \quad \forall \mathbf{v} \in \mathbb{R}^J.$$

In particular, with $\mathbf{v} = (a(u, P_j u)^{1/2})_{j=1}^J$ we obtain

$$\begin{aligned} a(\widehat{P}u, \widehat{P}u) &\le \omega \rho(\mathscr{E}) \sum_{j=1}^J a(u, P_i u) = \omega \rho(\mathscr{E}) a(u, \widehat{P}u) \\ &\le \omega \rho(\mathscr{E}) a(u,u)^{1/2} a(\widehat{P}u, \widehat{P}u)^{1/2}, \end{aligned}$$

so that

$$a(\widehat{P}u, \widehat{P}u)^{1/2} \le \omega \rho(\mathscr{E}) a(u,u)^{1/2}.$$

Hence

$$a(\widehat{P}u, u) \le a(\widehat{P}u, \widehat{P}u)^{1/2} a(u,u)^{1/2} \le \omega \rho(\mathscr{E}) a(u,u),$$

proving the lemma. □

**Proposition 2.4.** *Let Assumptions 2.3, 2.5 and 2.6 be satisfied, and assume that* $\omega \in (0,2)$. *Then for any* $u \in \mathscr{V}$ *there holds*

$$a(P_{\text{smu}} u, u) \ge \frac{2-\omega}{(1 + 2\widehat{\omega}^2 \rho(\mathscr{E})^2) C_0^2} a(u,u), \tag{2.40}$$

*where* $\widehat{\omega} = \max\{1, \omega\}$. *Therefore,*

$$\lambda_{\min}(P_{\text{smu}}) \ge \frac{2-\omega}{(1 + 2\widehat{\omega}^2 \rho(\mathscr{E})^2) C_0^2}. \tag{2.41}$$

*Proof.* Let

$$E_{-1} = I \quad \text{and} \quad E_j = (I - P_j) \cdots (I - P_0), \quad j = 0, \ldots, J.$$

We recall from the proof of Proposition 2.3 that

$$a(P_{\text{smu}} u, u) = a(u,u) - a(E_J u, E_J u). \tag{2.42}$$

Inequality (2.40) is proved if, for any $u \in \mathscr{V}$,

$$(2-\omega) \sum_{j=0}^J a(P_j E_{j-1} u, E_{j-1} u) \le a(u,u) - a(E_J u, E_J u) \tag{2.43}$$

and

$$a(P_{\text{ad}} u, u) \le \left(1 + 2\widehat{\omega}^2 \rho(\mathscr{E})^2\right) \sum_{j=0}^J a(E_{j-1} u, P_j E_{j-1} u). \tag{2.44}$$

Indeed, it follows from (2.42)–(2.44) that

$$a(P_{\mathrm{smu}}u,u) \geq \frac{(2-\omega)a(P_{\mathrm{ad}}u,u)}{(1+2\widehat{\omega}^2\rho(\mathscr{E})^2)}.$$

The required result now follows from (2.37).

It remains to show (2.43) and (2.44). Noting the property of telescoping sums we have

$$a(u,u)-a(E_Ju,E_Ju) = \sum_{j=0}^{J}\Big(a(E_{j-1}u,E_{j-1}u)-a(E_ju,E_ju)\Big).$$

Since $E_j = (I-P_j)E_{j-1}$ implying $E_ju = E_{j-1}u - P_jE_{j-1}u$, it follows that

$$\begin{aligned}
&a(u,u)-a(E_Ju,E_Ju)\\
&\quad= \sum_{j=0}^{J}\Big(a(E_{j-1}u,E_{j-1}u)-a(E_{j-1}u-P_jE_{j-1}u,E_{j-1}u-P_jE_{j-1}u)\Big)\\
&\quad= \sum_{j=0}^{J}\Big(2a(E_{j-1}u,P_jE_{j-1}u)-a(P_jE_{j-1}u,P_jE_{j-1}u)\Big).
\end{aligned}$$

Assumption 2.6, (2.16), and (2.19) imply

$$\begin{aligned}
a(u,u)-a(E_Ju,E_Ju) &\geq \sum_{j=0}^{J}\Big(2a(E_{j-1}u,P_jE_{j-1}u)-\omega a_j(\widetilde{P}_jE_{j-1}u,\widetilde{P}_jE_{j-1}u)\Big)\\
&= (2-\omega)\sum_{j=0}^{J}a(E_{j-1}u,P_jE_{j-1}u).
\end{aligned}$$

Finally, to prove (2.44) we first note that $E_k = (I-P_k)E_{k-1}$ so that

$$E_{k-1} = E_k + P_kE_{k-1}, \quad k=1,2,\ldots,J.$$

Taking the sum for $k$ from 1 to $j-1$ we deduce

$$I = E_{j-1} + P_0 + \sum_{k=1}^{j-1}P_kE_{k-1}.$$

Therefore,

$$a(P_ju,u) = a(P_ju,E_{j-1}u) + a(P_ju,P_0u) + \sum_{k=1}^{j-1}a(P_ju,P_kE_{k-1}u).$$

Applying Lemma 2.14 to the first two terms and applying Lemma 2.14 to each term in the sum on the right-hand side, we infer

$$a(P_ju,u) \leq a(P_ju,u)^{1/2}\Big(a(P_jE_{j-1}u,E_{j-1}u)^{1/2} + a(P_jP_0u,P_0u)^{1/2}\Big)$$

$$+\omega\sum_{k=1}^{j}\varepsilon_{jk}a(P_ju,u)^{1/2}a(P_kE_{k-1}u,E_{k-1}u)^{1/2},$$

so that

$$\begin{aligned}a(P_ju,u)^{1/2} &\le a(P_jE_{j-1}u,E_{j-1}u)^{1/2}+a(P_jP_0u,P_0u)^{1/2}\\&\quad+\omega\sum_{k=1}^{j-1}\varepsilon_{jk}a(P_kE_{k-1}u,E_{k-1}u)^{1/2}\\&\le a(P_jP_0u,P_0u)^{1/2}+\widehat{\omega}\sum_{k=1}^{j}\varepsilon_{jk}a(P_kE_{k-1}u,E_{k-1}u)^{1/2}\Big).\end{aligned}$$

Hence by squaring both sides and using $(a+b)^2\le 2a^2+2b^2$ we deduce

$$a(P_ju,u)\le 2a(P_jP_0u,P_0u)+2\widehat{\omega}^2\left(\sum_{k=1}^{J}\varepsilon_{jk}a(P_kE_{k-1}u,E_{k-1}u)^{1/2}\right)^2$$

Summing over $j=1,\dots,J$ we obtain

$$\begin{aligned}a(\widehat{P}u,u)&\le 2a(\widehat{P}P_0u,P_0u)+2\widehat{\omega}^2\sum_{j=1}^{J}\left(\sum_{k=1}^{J}\varepsilon_{jk}a(P_kE_{k-1}u,E_{k-1}u)^{1/2}\right)^2\\&=: T_1+T_2.\end{aligned}$$

Using successively Lemma 2.15, Assumption 2.6, and the definition of $\widetilde{P}_0$ we infer

$$T_1\le 2\omega\rho(\mathscr{E})a(P_0u,P_0u)\le 2\omega^2\rho(\mathscr{E})a_0(\widetilde{P}_0u,\widetilde{P}_0u)=2\omega^2\rho(\mathscr{E})a(P_0u,u).$$

To estimate $T_2$ we recall that

$$\|\mathscr{E}\mathbf{v}\|_{\ell_2}^2\le\rho(\mathscr{E})^2\|\mathbf{v}\|_{\ell_2}^2,\quad \mathbf{v}\in\mathbb{R}^J,$$

and thus with $\mathbf{v}=\big(a(P_kE_{k-1}u,E_{k-1}u)^{1/2}\big)_{k=1}^J$ we deduce

$$\sum_{j=1}^{J}\left(\sum_{k=1}^{J}\varepsilon_{jk}a(P_kE_{k-1}u,E_{k-1}u)^{1/2}\right)^2\le\rho(\mathscr{E})^2\sum_{j=1}^{J}a(P_jE_{j-1}u,E_{j-1}u).$$

Hence

$$T_2=\le 2\widehat{\omega}^2\rho(\mathscr{E})^2\sum_{j=1}^{J}a(P_jE_{j-1}u,E_{j-1}u).$$

Therefore, assuming $\rho(\mathscr{E})\ge 1$ we obtain

$$a(\widehat{P}u,u) \le 2\widehat{\omega}^2\rho(\mathscr{E})^2 \sum_{j=0}^{J} a(P_jE_{j-1}u, E_{j-1}u).$$

Finally,

$$a(P_{\text{ad}}u,u) = a(P_0u,u) + a(\widehat{P}u,u) \le \left(1+2\widehat{\omega}^2\rho(\mathscr{E})^2\right) \sum_{j=0}^{J} a(P_jE_{j-1}u, E_{j-1}u),$$

proving (2.44), and thus completing the proof of (2.40). Inequality (2.41) is a consequence of (2.40). □

### 2.6.8 Summary

We summarise the above analysis in this section. Table 2.1 shows bounds for the maximum and minimum eigenvalues obtained under various assumptions.

| Assumptions | | | | Extremum eigenvalues |
|---|---|---|---|---|
| 2.3 | 2.4 (∗) | 2.5 | 2.6 | |
| Stability | Coercivity | Strengthened Cauchy-Schwarz | Local stability | |
| X | | | | $\lambda_{\min}(P_{\text{ad}}) \ge C_0^{-2}$ |
| | X | | | $\lambda_{\max}(P_{\text{ad}}) \le C_1^2$ |
| | | X | X | $\lambda_{\max}(P_{\text{ad}}) \le \omega(\rho(\mathscr{E})+1)$ |
| X | | X | X | $\lambda_{\min}(P_{\text{smu}}) \ge \dfrac{2-\omega}{(1+2\widehat{\omega}^2\rho(\mathscr{E})^2)C_0^2}$ |

**Table 2.1** Bounds for extremal eigenvalues from assumptions. (∗) Further assumption: $P_{\text{ad}}$ is positive definite.

The following theorems are direct consequences of the above results.

**Theorem 2.1.** *If Assumptions 2.3 and 2.4 hold, then*

$$\kappa(P_{\text{ad}}) \le C_0^2C_1^1.$$

*If Assumptions 2.3, 2.5, and 2.6 hold, then*

$$\kappa(P_{\text{ad}}) \le C_0^2\omega(\rho(\mathscr{E})+1).$$

*In particular, if the local bilinear forms $a_j(\cdot,\cdot)$, $j=0,\ldots,J$ are chosen by* (2.17), *then Assumption 2.6 holds with $\omega=1$ and therefore*

$$\kappa(P_{\text{ad}}) \le C_0^2(\rho(\mathscr{E})+1).$$

**Theorem 2.2.** *If Assumptions 2.3, 2.5, and 2.6 hold, then*

$$\kappa(P_{\text{smu}}) \leq \frac{(1+2\widehat{\omega}^2\rho(\mathscr{E})^2)C_0^2}{2-\omega}.$$

*In particular, with the choice* (2.17) *for the bilinear forms* $a_j(\cdot,\cdot)$ *we have*

$$\kappa(P_{\text{smu}}) \leq (1+2\rho(\mathscr{E})^2)C_0^2.$$

## 2.7 Convergence Theory for Preconditioners with GMRES

### *2.7.1 Preconditioned GMRES with the* B*-inner product*

In this section we derive the convergence results for Algorithm 2.5. Recalling the convergence result (2.5) proved in [73, Theorem 3.3], we define the corresponding transpose and extremal eigenvalues with respect to the **B**-inner product (2.13) as follows:

$$\begin{aligned}
\left\langle \widehat{\mathbf{A}}^{\top_{\mathrm{B}}}\mathbf{v},\mathbf{w}\right\rangle_{\mathbf{B}} &:= \left\langle \mathbf{v},\widehat{\mathbf{A}}\mathbf{w}\right\rangle_{\mathbf{B}} \qquad \forall \mathbf{v},\mathbf{w}\in\mathbb{R}^N\\
\widehat{\mathbf{A}}_s^{\mathrm{B}} &:= \frac{\widehat{\mathbf{A}}+\widehat{\mathbf{A}}^{\top_{\mathrm{B}}}}{2}\\
\lambda_{\min}^{\mathbf{B}}(\widehat{\mathbf{A}}_s^{\mathrm{B}}) &:= \inf_{\mathbf{v}\in\mathbb{R}^N}\frac{\left\langle \widehat{\mathbf{A}}_s^{\mathrm{B}}\mathbf{v},\mathbf{v}\right\rangle_{\mathbf{B}}}{\langle \mathbf{v},\mathbf{v}\rangle_{\mathbf{B}}}\\
\lambda_{\max}^{\mathbf{B}}(\widehat{\mathbf{A}}^{\top_{\mathrm{B}}}\widehat{\mathbf{A}}) &:= \sup_{\mathbf{v}\in\mathbb{R}^N}\frac{\left\langle \widehat{\mathbf{A}}^{\top_{\mathrm{B}}}\widehat{\mathbf{A}}\mathbf{v},\mathbf{v}\right\rangle_{\mathbf{B}}}{\langle \mathbf{v},\mathbf{v}\rangle_{\mathbf{B}}}.
\end{aligned} \tag{2.45}$$

We note that

$$\begin{aligned}
\lambda_{\min}^{\mathbf{B}}(\widehat{\mathbf{A}}_s^{\mathrm{B}}) &= \inf_{\mathbf{v}\in\mathbb{R}^N}\frac{\left\langle \widehat{\mathbf{A}}\mathbf{v},\mathbf{v}\right\rangle_{\mathbf{B}}}{\langle \mathbf{v},\mathbf{v}\rangle_{\mathbf{B}}}\\
\lambda_{\max}^{\mathbf{B}}(\widehat{\mathbf{A}}^{\top_{\mathrm{B}}}\widehat{\mathbf{A}}) &= \sup_{\mathbf{v}\in\mathbb{R}^N}\frac{\left\langle \widehat{\mathbf{A}}\mathbf{v},\widehat{\mathbf{A}}\mathbf{v}\right\rangle_{\mathbf{B}}}{\langle \mathbf{v},\mathbf{v}\rangle_{\mathbf{B}}} = ||\widehat{\mathbf{A}}||_{\mathbf{B}}^2.
\end{aligned} \tag{2.46}$$

If the preconditioner **C** is designed such that $\widehat{\mathbf{A}}_s^{\mathrm{B}}$ is positive definite with respect to the **B**-inner product, then $\lambda_{\min}^{\mathbf{B}}(\widehat{\mathbf{A}}_s^{\mathrm{B}}) > 0$. By repeating the proof of [73, Theorem 3.3] we can prove that

$$\|\widehat{\mathbf{r}}_k\|_{\mathbf{B}} \leq \left[1 - \frac{\left(\lambda_{\min}^{\mathbf{B}}(\widehat{\mathbf{A}}_s^{\mathbf{B}})\right)^2}{\lambda_{\max}^{\mathbf{B}}(\widehat{\mathbf{A}}^{\mathsf{T}_c}\widehat{\mathbf{A}})}\right]^{k/2} \|\widehat{\mathbf{r}}_0\|_{\mathbf{B}}. \tag{2.47}$$

A similar convergence result can also be obtained by applying Algorithm 2.2 to the system

$$\widetilde{\mathbf{A}}\widetilde{\mathbf{u}} = \widetilde{\mathbf{b}} \tag{2.48}$$

where

$$\widetilde{\mathbf{A}} := \mathbf{B}^{1/2}\widehat{\mathbf{A}}\mathbf{B}^{-1/2}, \quad \widetilde{\mathbf{u}} := \mathbf{B}^{1/2}\widehat{\mathbf{u}}, \quad \widetilde{\mathbf{b}} := \mathbf{B}^{1/2}\widehat{\mathbf{b}}.$$

First we show that solving (2.48) by Algorithm (2.2) (with the Euclidean inner product) yield the same iterates as solving (2.12) by Algorithm 2.4; see [156, Lemma 3.1].

**Lemma 2.16.** *For $k \in \{1,\ldots,N\}$, let $\widetilde{\mathbf{u}}_k$ be the approximate solution of* (2.48) *generated by GMRES with the Euclidean inner product (Algorithm 2.2) and initial guess $\widetilde{\mathbf{u}}_0$, and let $\widehat{\mathbf{u}}_k$ be the approximate solution of* (2.12) *generated by PGMRES with the $\langle\cdot,\cdot\rangle_{\mathbf{B}}$-inner product (Algorithm 2.4) and initial guess $\widehat{\mathbf{u}}_0 = \mathbf{B}^{-1/2}\widetilde{\mathbf{u}}_0$. Then*

$$\widehat{\mathbf{u}}_k = \mathbf{B}^{-1/2}\widetilde{\mathbf{u}}_k.$$

*Moreover, for both approaches the corresponding residuals $\widetilde{\mathbf{r}}_i$, $\widehat{\mathbf{r}}_i$, and search directions $\widetilde{\mathbf{p}}_i$, $\widehat{\mathbf{p}}_i$, $i = 1,\ldots,k+1$, satisfy*

$$\widehat{\mathbf{r}}_i = \mathbf{B}^{-1/2}\widetilde{\mathbf{r}}_i, \quad \|\widehat{\mathbf{r}}_i\|_{\mathbf{B}} = \|\widetilde{\mathbf{r}}_i\|_{\mathbb{R}^N}$$
$$\widehat{\mathbf{p}}_i = \mathbf{B}^{-1/2}\widetilde{\mathbf{p}}_i, \quad \|\widehat{\mathbf{p}}_i\|_{\mathbf{B}} = \|\widetilde{\mathbf{p}}_i\|_{\mathbb{R}^N}.$$

*Proof.* For simplicity of notation we will denote $\mathbf{B}^{1/2}$ by $\mathbf{D}$. Then

$$\widehat{\mathbf{A}} = \mathbf{D}^{-1}\widetilde{\mathbf{A}}\mathbf{D}, \quad \widehat{\mathbf{b}} = \mathbf{D}^{-1}\widetilde{\mathbf{b}}, \quad \widehat{\mathbf{u}}_0 = \mathbf{D}^{-1}\widetilde{\mathbf{u}}_0.$$

With the notations of Algorithms 2.2 and 2.5, we have

$$\begin{aligned} \widehat{\mathbf{r}}_0 &= \widehat{\mathbf{b}} - \widehat{\mathbf{A}}\widehat{\mathbf{u}}_0 = \mathbf{D}^{-1}\left(\widetilde{\mathbf{b}} - \widetilde{\mathbf{A}}\widetilde{\mathbf{u}}_0\right) = \mathbf{D}^{-1}\widetilde{\mathbf{r}}_0 \\ \|\widehat{\mathbf{r}}_0\|_{\mathbf{B}}^2 &= \langle \mathbf{D}\widehat{\mathbf{r}}_0, \mathbf{D}\widehat{\mathbf{r}}_0\rangle_{\mathbb{R}^N} = \|\widetilde{\mathbf{r}}_0\|_{\mathbb{R}^N}^2 \\ \widehat{\mathbf{v}}_1 &= \widehat{\mathbf{r}}_0/\|\widehat{\mathbf{r}}_0\|_{\mathbf{B}} = \mathbf{D}^{-1}\widetilde{\mathbf{v}}_1. \end{aligned}$$

For $j = 1,\ldots,k$ and $i = 1,\ldots,j$ we also have

$$\begin{aligned} \widehat{h}_{i,j} &= \left\langle \widehat{\mathbf{A}}\widehat{\mathbf{v}}_j, \widehat{\mathbf{v}}_i \right\rangle_{\mathbf{B}} = \left\langle \mathbf{D}\widehat{\mathbf{A}}\widehat{\mathbf{v}}_j, \mathbf{D}\widehat{\mathbf{v}}_i \right\rangle_{\mathbb{R}^N} = \left\langle \widetilde{\mathbf{A}}\widetilde{\mathbf{v}}_j, \widetilde{\mathbf{v}}_i \right\rangle_{\mathbb{R}^N} = \widetilde{h}_{i,j} \\ \widehat{\mathbf{p}}_j &= \widehat{\mathbf{A}}\widehat{\mathbf{v}}_j - \sum_{i=1}^{j} \widehat{h}_{i,j}\widehat{\mathbf{v}}_j = \mathbf{D}^{-1}\left(\widetilde{\mathbf{A}}\widetilde{\mathbf{v}}_j - \sum_{i=1}^{j} \widetilde{h}_{i,j}\widetilde{\mathbf{v}}_j\right) = \mathbf{D}^{-1}\widetilde{\mathbf{p}}_j \\ \widehat{h}_{j+1,j} &= \|\widehat{\mathbf{p}}_j\|_{\mathbf{B}} = \langle \mathbf{D}\widehat{\mathbf{p}}_j, \mathbf{D}\widehat{\mathbf{p}}_i\rangle_{\mathbb{R}^N} = \|\widetilde{\mathbf{p}}_j\|_{\mathbb{R}^N} = \widetilde{h}_{j+1,j} \end{aligned}$$

$$\widehat{\mathbf{v}}_{j+1} = \frac{\widehat{\mathbf{p}}_j}{\widehat{h}_{j+1,j}} = \frac{\mathbf{D}^{-1}\widetilde{\mathbf{p}}_j}{\widetilde{h}_{j+1,j}} = \mathbf{D}^{-1}\widetilde{\mathbf{v}}_{j+1}.$$

Obviously,

$$\langle \widehat{\mathbf{p}}_i, \widehat{\mathbf{p}}_j \rangle_{\mathbf{B}} = \langle \mathbf{D}\widehat{\mathbf{p}}_i, \mathbf{D}\widehat{\mathbf{p}}_j \rangle_{\mathbb{R}^N} = \langle \widetilde{\mathbf{p}}_i, \widetilde{\mathbf{p}}_j \rangle_{\mathbb{R}^N} = \delta_{i,j} \quad \text{(Kronecker delta)}.$$

Furthermore,

$$\widehat{\mathscr{K}_k} := \{\widehat{\mathbf{r}}_0, \widehat{\mathbf{A}}\widehat{\mathbf{r}}_0, \ldots, \widehat{\mathbf{A}}^{k-1}\widehat{\mathbf{r}}_0\} = \mathbf{D}^{-1}\{\widetilde{\mathbf{r}}_0, \widetilde{\mathbf{A}}\widetilde{\mathbf{r}}_0, \ldots, \widetilde{\mathbf{A}}^{k-1}\widetilde{\mathbf{r}}_0\} =: \mathbf{D}^{-1}\widetilde{\mathscr{K}_k}.$$

From

$$\widehat{\mathbf{w}}_k = \underset{\widehat{\mathbf{z}} \in \widehat{\mathscr{K}_k}}{\arg\min} \|\widehat{\mathbf{b}} - \widehat{\mathbf{A}}(\widehat{\mathbf{r}}_0 + \widehat{\mathbf{z}})\|_{\mathbf{B}} = \underset{\widetilde{\mathbf{z}} \in \widetilde{\mathscr{K}_k}}{\arg\min} \|\widetilde{\mathbf{b}} - \widetilde{\mathbf{A}}(\widetilde{\mathbf{r}}_0 + \widetilde{\mathbf{z}})\|_{\mathbb{R}^N} = \widetilde{\mathbf{w}}_k$$

it follows that $\widehat{\mathbf{u}}_k = \mathbf{D}^{-1}\widetilde{\mathbf{u}}_k$, completing the proof. □

If $\widetilde{\mathbf{A}}_s$ is positive definite, a convergence result similar to (2.47) can be derived by applying (2.5) to $\widetilde{\mathbf{r}}_k$:

$$\|\widetilde{\mathbf{r}}_k\|_{\mathbb{R}^N} \le \left[1 - \frac{\left(\lambda_{\min}(\widetilde{\mathbf{A}}_s)\right)^2}{\lambda_{\max}(\widetilde{\mathbf{A}}^\top\widetilde{\mathbf{A}})}\right]^{k/2} \|\widetilde{\mathbf{r}}_0\|_{\mathbb{R}^N}. \tag{2.49}$$

In fact, one can show the following lemma.

**Lemma 2.17.** *The following statement holds:*

$$\lambda_{\min}^{\mathbf{B}}(\widehat{\mathbf{A}}_s^{\mathrm{B}}) = \lambda_{\min}(\widetilde{\mathbf{A}}_s) \quad \textit{and} \quad \lambda_{\max}^{\mathbf{B}}(\widehat{\mathbf{A}}^{\top_{\mathrm{B}}}\widehat{\mathbf{A}}) = \lambda_{\max}(\widetilde{\mathbf{A}}^\top\widetilde{\mathbf{A}}).$$

*Proof.* First we note from (2.45) and (2.13) that, with $\mathbf{w} = \mathbf{B}^{1/2}\mathbf{v}$,

$$\begin{aligned}
\lambda_{\min}^{\mathbf{B}}(\widehat{\mathbf{A}}_s^{\mathrm{B}}) &= \inf_{\mathbf{v}\in\mathbb{R}^N} \frac{\left\langle \widehat{\mathbf{A}}\mathbf{v}, \mathbf{v} \right\rangle_{\mathbf{B}}}{\langle \mathbf{v}, \mathbf{v} \rangle_{\mathbf{B}}} = \inf_{\mathbf{v}\in\mathbb{R}^N} \frac{\left\langle \mathbf{B}^{1/2}\widehat{\mathbf{A}}\mathbf{v}, \mathbf{B}^{1/2}\mathbf{v} \right\rangle_{\mathbb{R}^N}}{\left\langle \mathbf{B}^{1/2}\mathbf{v}, \mathbf{B}^{1/2}\mathbf{v} \right\rangle_{\mathbb{R}^N}} \\
&= \inf_{\mathbf{w}\in\mathbb{R}^N} \frac{\left\langle \widetilde{\mathbf{A}}\mathbf{w}, \mathbf{w} \right\rangle_{\mathbb{R}^N}}{\langle \mathbf{w}, \mathbf{w} \rangle_{\mathbb{R}^N}} = \inf_{\mathbf{w}\in\mathbb{R}^N} \frac{\left\langle \widetilde{\mathbf{A}}_s\mathbf{w}, \mathbf{w} \right\rangle_{\mathbb{R}^N}}{\langle \mathbf{w}, \mathbf{w} \rangle_{\mathbb{R}^N}} = \lambda_{\min}(\widetilde{\mathbf{A}}_s).
\end{aligned}$$

Similarly, for the maximum eigenvalue we have

$$\begin{aligned}
\lambda_{\max}^{\mathbf{B}}(\widehat{\mathbf{A}}^{\top_{\mathrm{B}}}\widehat{\mathbf{A}}) &= \sup_{\mathbf{v}\in\mathbb{R}^N} \frac{\left\langle \widehat{\mathbf{A}}\mathbf{v}, \widehat{\mathbf{A}}\mathbf{v} \right\rangle_{\mathbf{B}}}{\langle \mathbf{v}, \mathbf{v} \rangle_{\mathbf{B}}} = \sup_{\mathbf{v}\in\mathbb{R}^N} \frac{\left\langle \mathbf{B}^{1/2}\widehat{\mathbf{A}}\mathbf{v}, \mathbf{B}^{1/2}\widehat{\mathbf{A}}\mathbf{v} \right\rangle_{\mathbb{R}^N}}{\left\langle \mathbf{B}^{1/2}\mathbf{v}, \mathbf{B}^{1/2}\mathbf{v} \right\rangle_{\mathbb{R}^N}} \\
&= \sup_{\mathbf{w}\in\mathbb{R}^N} \frac{\left\langle \widetilde{\mathbf{A}}\mathbf{w}, \widetilde{\mathbf{A}}\mathbf{w} \right\rangle_{\mathbb{R}^N}}{\langle \mathbf{w}, \mathbf{w} \rangle_{\mathbb{R}^N}} = \sup_{\mathbf{w}\in\mathbb{R}^N} \frac{\left\langle \widetilde{\mathbf{A}}^\top\widetilde{\mathbf{A}}\mathbf{w}, \mathbf{w} \right\rangle_{\mathbb{R}^N}}{\langle \mathbf{w}, \mathbf{w} \rangle_{\mathbb{R}^N}} = \lambda_{\max}(\widetilde{\mathbf{A}}^\top\widetilde{\mathbf{A}}).
\end{aligned}$$ □

As with the symmetric case using PCG, we now want to relate these eigenvalues with the bilinear forms defining the matrix system. Let $Q$ be the Schwarz operator designed for an indefinite problem, whose matrix representation is $\mathbf{Q} = \mathbf{C}^{-1}\mathbf{A}$; see (2.27). (A precise definition of $Q$ will be given in Chapter 8.) Note that in general $\mathbf{Q}$ is non-symmetric. Similarly to Lemma 2.9 for the case of symmetric $\mathbf{A}$, we now have the following result for non-symmetric $\mathbf{A}$.

**Lemma 2.18.** *The following statements hold:*

$$\sup_{v \in \mathscr{V}} \frac{b(Qv, Qv)}{b(v,v)} = \lambda_{\max}(\widetilde{\mathbf{A}}^{\top}\widetilde{\mathbf{A}}) = \lambda_{\max}^{\mathbf{B}}(\widehat{\mathbf{A}}^{\top_{\mathrm{B}}}\widehat{\mathbf{A}}),$$
$$\inf_{v \in \mathscr{V}} \frac{b(Qv, v)}{b(v,v)} = \lambda_{\min}(\widetilde{\mathbf{A}}_s) = \lambda_{\min}^{\mathbf{B}}(\widehat{\mathbf{A}}_s^{\mathrm{B}}).$$

*Proof.* The required identities are a result of (2.46) and Lemma 2.17, noting that $\mathbf{Q} = \widehat{\mathbf{A}}$ and that

$$\frac{b(Qv, Qv)}{b(v,v)} = \frac{\left\langle \widehat{\mathbf{A}}\mathbf{v}, \widehat{\mathbf{A}}\mathbf{v} \right\rangle_{\mathbf{B}}}{\langle \mathbf{v}, \mathbf{v} \rangle_{\mathbf{B}}} \quad \text{and} \quad \frac{b(Qv, v)}{b(v,v)} = \frac{\left\langle \widehat{\mathbf{A}}\mathbf{v}, \mathbf{v} \right\rangle_{\mathbf{B}}}{\langle \mathbf{v}, \mathbf{v} \rangle_{\mathbf{B}}}.$$

□

As a consequence, we have the following lemma; cf. [185].

**Lemma 2.19.** *Assume that the Schwarz operator $Q$ is designed such that*

$$0 < c_Q \leq \inf_{v \in \mathscr{V}} \frac{b(Qv, v)}{b(v,v)} \quad \textit{and} \quad \sup_{v \in \mathscr{V}} \frac{b(Qv, Qv)}{b(v,v)} \leq C_Q^2. \tag{2.50}$$

*Let $\widehat{\mathbf{r}}_k$ and $\widetilde{\mathbf{r}}_k$ be the residuals obtained by solving* (2.12) *and* (2.48) *by Algorithms 2.5 and 2.2, respectively. Then*

$$\|\widehat{\mathbf{r}}_k\|_{\mathbf{B}} = \|\widetilde{\mathbf{r}}_k\|_{\mathbb{R}^N} \leq \left(1 - \frac{c_Q^2}{C_Q^2}\right)^{k/2} \|\widetilde{\mathbf{r}}_0\|_{\mathbb{R}^N} = \left(1 - \frac{c_Q^2}{C_Q^2}\right)^{k/2} \|\widehat{\mathbf{r}}_0\|_{\mathbf{B}}. \tag{2.51}$$

*Furthermore, if $\overline{\mathbf{r}}_k$ is the residual obtained by solving* (2.12) *with Algorithm 2.4 then*

$$\|\overline{\mathbf{r}}_k\|_{\mathbb{R}^N} \leq \sqrt{\kappa_2(\mathbf{B})} \left(1 - \frac{c_Q^2}{C_Q^2}\right)^{k/2} \|\overline{\mathbf{r}}_0\|_{\mathbb{R}^N} \tag{2.52}$$

*where $\kappa_2(\mathbf{B}) = \lambda_{\max}(\mathbf{B})/\lambda_{\min}(\mathbf{B})$.*

*Proof.* Estimate (2.51) is a result of Lemma 2.16, Lemma 2.18, (2.49) and (2.50). Estimate (2.52) is proved in [185]. □

### 2.7.2 Preconditioned GMRES with the C-inner product

When the stiffness matrix $\mathbf{A}$ is non-symmetric but the preconditioner $\mathbf{C}$ is symmetric (and positive definite), one can use Algorithm 2.6 to solve $\mathbf{Au} = \mathbf{b}$; see [78]. Defining $\lambda_{\min}^{\mathbf{C}}(\widehat{\mathbf{A}}_s^{\mathbf{C}})$ and $\lambda_{\max}^{\mathbf{C}}(\widehat{\mathbf{A}}^{\mathsf{T}_\mathbf{C}}\widehat{\mathbf{A}})$ in the same manner as in (2.45) but with the $\mathbf{B}$-inner product replaced by the $\mathbf{C}$-inner product, and assuming that $\mathbf{C}$ is designed such that $\lambda_{\min}^{\mathbf{C}}(\widehat{\mathbf{A}}_s^{\mathbf{C}}) > 0$ we obtain similarly to (2.47)

$$\|\widehat{\mathbf{r}}_k\|_\mathbf{C} \leq \left[1 - \frac{\lambda_{\min}^{\mathbf{C}}(\widehat{\mathbf{A}}_s^{\mathbf{C}})^2}{\lambda_{\max}^{\mathbf{C}}(\widehat{\mathbf{A}}^{\mathsf{T}_\mathbf{C}}\widehat{\mathbf{A}})}\right]^{k/2} \|\widehat{\mathbf{r}}_0\|_\mathbf{C}. \tag{2.53}$$

Similarly to Lemma 2.18 we have the following lemma.

**Lemma 2.20.** *Assume that the preconditioner* $\mathbf{C}$ *is defined such that*

$$0 < \alpha \leq \inf_{\substack{\mathbf{v}\in\mathbb{R}^N\\ \mathbf{v}\neq\mathbf{0}}} \frac{\langle \mathbf{Av}, \mathbf{v}\rangle_{\mathbb{R}^N}}{\langle \mathbf{Cv}, \mathbf{v}\rangle_{\mathbb{R}^N}} \quad \text{and} \quad \sup_{\substack{\mathbf{v}\in\mathbb{R}^N\\ \mathbf{v}\neq\mathbf{0}}} \frac{\langle \mathbf{C}^{-1}\mathbf{Av}, \mathbf{Av}\rangle_{\mathbb{R}^N}}{\langle \mathbf{Cv}, \mathbf{v}\rangle_{\mathbb{R}^N}} \leq \beta^2. \tag{2.54}$$

*Then*

$$\|\widehat{\mathbf{r}}_k\|_\mathbf{C} \leq \left(1 - \frac{\alpha^2}{\beta^2}\right)^{k/2} \|\widehat{\mathbf{r}}_0\|_\mathbf{C}.$$

*Proof.* The result is a direct consequence of (2.53) and the following identities

$$\lambda_{\min}^{\mathbf{C}}(\widehat{\mathbf{A}}_s^{\mathbf{C}}) = \inf_{\substack{\mathbf{v}\in\mathbb{R}^N\\ \mathbf{v}\neq\mathbf{0}}} \frac{\langle \mathbf{C}^{-1}\mathbf{Av}, \mathbf{v}\rangle_\mathbf{C}}{\langle \mathbf{v}, \mathbf{v}\rangle_\mathbf{C}} = \inf_{\substack{\mathbf{v}\in\mathbb{R}^N\\ \mathbf{v}\neq\mathbf{0}}} \frac{\langle \mathbf{Av}, \mathbf{v}\rangle_{\mathbb{R}^N}}{\langle \mathbf{Cv}, \mathbf{v}\rangle_{\mathbb{R}^N}}$$

$$\lambda_{\max}^{\mathbf{C}}(\widehat{\mathbf{A}}^{\mathsf{T}_\mathbf{C}}\widehat{\mathbf{A}}) = \sup_{\substack{\mathbf{v}\in\mathbb{R}^N\\ \mathbf{v}\neq\mathbf{0}}} \frac{\langle \mathbf{C}^{-1}\mathbf{Av}, \mathbf{C}^{-1}\mathbf{Av}\rangle_\mathbf{C}}{\langle \mathbf{v}, \mathbf{v}\rangle_\mathbf{C}} = \sup_{\substack{\mathbf{v}\in\mathbb{R}^N\\ \mathbf{v}\neq\mathbf{0}}} \frac{\langle \mathbf{C}^{-1}\mathbf{Av}, \mathbf{Av}\rangle_{\mathbb{R}^N}}{\langle \mathbf{Cv}, \mathbf{v}\rangle_{\mathbb{R}^N}}. \qquad \square$$

### 2.7.3 Preconditioned GMRES with block matrices

In this subsection, we are concerned with convergence results when GMRES is used to solve the matrix equation

$$\mathbf{Au} = \mathbf{b} \tag{2.55}$$

where

$$\mathbf{A} := \begin{pmatrix} \mathbf{A}_{11} & \mathbf{A}_{12} \\ \mathbf{A}_{21} & \mathbf{A}_{22} \end{pmatrix} \in \mathbb{R}^{N\times N}, \quad \mathbf{u} := \begin{pmatrix} \mathbf{u}_1 \\ \mathbf{u}_2 \end{pmatrix} \in \mathbb{R}^N \quad \text{and} \quad \mathbf{b} := \begin{pmatrix} \mathbf{b}_1 \\ \mathbf{b}_2 \end{pmatrix} \in \mathbb{R}^N.$$

Here $\mathbf{A}_{ii} \in \mathbb{R}^{n_i \times n_i}$, $\mathbf{A}_{ij} \in \mathbb{R}^{n_i \times n_j}$, and $\mathbf{u}_i, \mathbf{b}_i \in \mathbb{R}^{n_i}$, $i, j = 1, 2$, with $n_1 + n_2 = N$. We assume that $\mathbf{A}_{ii}$, $i = 1, 2$, is symmetric and positive definite, and that $\mathbf{A}$ is positive definite but can be non-symmetric. Letting

$$\mathbf{A}_{\mathrm{diag}} := \begin{pmatrix} \mathbf{A}_{11} & \mathbf{0} \\ \mathbf{0} & \mathbf{A}_{22} \end{pmatrix} \quad \text{and} \quad \mathbf{A}_{\mathrm{as}} := \begin{pmatrix} \mathbf{0} & \mathbf{A}_{12} \\ \mathbf{A}_{21} & \mathbf{0} \end{pmatrix},$$

then $\mathbf{A} = \mathbf{A}_{\mathrm{diag}} + \mathbf{A}_{\mathrm{as}}$ with $\mathbf{A}_{\mathrm{diag}}$ being symmetric and positive definite.

This matrix $\mathbf{A}$ arises, for example, from the use of FEM-BEM coupling in solving interface problems; see Chapter 14. Readers who are not familiar to or not interested in these problems can skip this section in the first reading.

Assume that $\mathbf{C}_{ii}^{-1} \in \mathbb{R}^{n_i \times n_i}$ is a preconditioner of $\mathbf{A}_{ii}$, $i = 1, 2$, which satisfies

$$d_i \langle \mathbf{C}_{ii}\mathbf{v}_i, \mathbf{v}_i \rangle_{\mathbb{R}^{n_i}} \le \langle \mathbf{A}_{ii}\mathbf{v}_i, \mathbf{v}_i \rangle_{\mathbb{R}^{n_i}} \le D_i \langle \mathbf{C}_{ii}\mathbf{v}_i, \mathbf{v}_i \rangle_{\mathbb{R}^{n_i}}, \quad d_i > 0,\ D_i > 0, \tag{2.56}$$

for all $\mathbf{v}_i \in \mathbb{R}^{n_i}$, $i = 1, 2$. Let

$$\mathbf{C} := \begin{pmatrix} \mathbf{C}_{11} & \mathbf{0} \\ \mathbf{0} & \mathbf{C}_{22} \end{pmatrix}.$$

Then

$$d \langle \mathbf{C}\mathbf{v}, \mathbf{v} \rangle_{\mathbb{R}^N} \le \left\langle \mathbf{A}_{\mathrm{diag}}\mathbf{v}, \mathbf{v} \right\rangle_{\mathbb{R}^N} \le D \langle \mathbf{C}\mathbf{v}, \mathbf{v} \rangle_{\mathbb{R}^N} \quad \forall \mathbf{v} \in \mathbb{R}^N \tag{2.57}$$

where $d := \min\{d_1, d_2\}$, $D := \max\{D_1, D_2\}$. Note that (2.57) is equivalent to

$$d \left\langle \mathbf{A}_{\mathrm{diag}}^{-1}\mathbf{w}, \mathbf{w} \right\rangle_{\mathbb{R}^N} \le \langle \mathbf{C}^{-1}\mathbf{w}, \mathbf{w} \rangle_{\mathbb{R}^N} \le D \left\langle \mathbf{A}_{\mathrm{diag}}^{-1}\mathbf{w}, \mathbf{w} \right\rangle_{\mathbb{R}^N} \quad \forall \mathbf{w} \in \mathbb{R}^N. \tag{2.58}$$

This equivalence follows from the choice $\mathbf{w} = \mathbf{A}_{\mathrm{diag}}^{1/2} \big( \mathbf{A}_{\mathrm{diag}}^{-1/2} \mathbf{C} \mathbf{A}_{\mathrm{diag}}^{-1/2} \big)^{1/2} \mathbf{A}_{\mathrm{diag}}^{1/2} \mathbf{v}$ and elementary calculations.

We will precondition equation (2.55) with $\mathbf{C}$. The following convergence result holds.

**Lemma 2.21.** *Assume that* (2.57) *holds. Let*

$$c_{\mathbf{A}} := \inf_{\substack{\mathbf{v} \in \mathbb{R}^N \\ \mathbf{v} \ne \mathbf{0}}} \frac{\langle \mathbf{A}\mathbf{v}, \mathbf{v} \rangle_{\mathbb{R}^N}}{\left\langle \mathbf{A}_{\mathrm{diag}}\mathbf{v}, \mathbf{v} \right\rangle_{\mathbb{R}^N}}, \qquad C_{\beta_1} := \sup_{\substack{\mathbf{v} \in \mathbb{R}^{n_1} \\ \mathbf{v} \ne \mathbf{0}}} \frac{\left\langle \mathbf{A}_{21}^{\top} \mathbf{A}_{22}^{-1} \mathbf{A}_{21}\mathbf{v}, \mathbf{v} \right\rangle_{\mathbb{R}^{n_1}}}{\langle \mathbf{A}_{11}\mathbf{v}, \mathbf{v} \rangle_{\mathbb{R}^{n_1}}},$$

$$C_{\beta_2} := \sup_{\substack{\mathbf{v} \in \mathbb{R}^{n_2} \\ \mathbf{v} \ne \mathbf{0}}} \frac{\left\langle \mathbf{A}_{12}^{\top} \mathbf{A}_{11}^{-1} \mathbf{A}_{12}\mathbf{v}, \mathbf{v} \right\rangle_{\mathbb{R}^{n_2}}}{\langle \mathbf{A}_{22}\mathbf{v}, \mathbf{v} \rangle_{\mathbb{R}^{n_2}}}, \qquad \widetilde{C} := \max\{C_{\beta_1}, C_{\beta_2}\},$$

$$C_{\beta_3} := \sup_{\substack{\mathbf{v} = (\mathbf{v}_1, \mathbf{v}_2) \in \mathbb{R}^{n_1} \times \mathbb{R}^{n_2} \\ \mathbf{v} \ne \mathbf{0}}} \frac{2 \left| \langle \mathbf{A}_{12}\mathbf{v}_2, \mathbf{v}_1 \rangle_{\mathbb{R}^{n_1}} + \langle \mathbf{A}_{21}\mathbf{v}_1, \mathbf{v}_2 \rangle_{\mathbb{R}^{n_2}} \right|}{\left\langle \mathbf{A}_{\mathrm{diag}}\mathbf{v}, \mathbf{v} \right\rangle_{\mathbb{R}^N}},$$

*and let*

$$\alpha := \begin{cases} d & \text{if } \mathbf{A}_{12}^\top = -\mathbf{A}_{21}, \\ c_{\mathbf{A}} d, & \text{otherwise}, \end{cases}$$

$$\beta^2 := \begin{cases} D^2(\widetilde{C}+1) & \text{if } \mathbf{A}_{12}^\top = -\mathbf{A}_{21}, \\ D^2(\widetilde{C}+C_{\beta_3}+1) & \text{otherwise}. \end{cases}$$

*Then*

$$\|\widehat{\mathbf{r}}_k\|_{\mathbf{C}} \le \left(1 - \frac{\alpha^2}{\beta^2}\right)^{k/2} \|\widehat{\mathbf{r}}_0\|_{\mathbf{C}}. \tag{2.59}$$

*Proof.* We note that if $\mathbf{A}_{12}^\top = -\mathbf{A}_{21}$, then $C_{\beta_3} = 0$. Due to Lemma 2.20, it suffices to prove that $\alpha$ and $\beta$ satisfy (2.54). We first consider $\alpha$.

- If $\mathbf{A}_{12}^\top = -\mathbf{A}_{21}$, then since $\mathbf{A} = \mathbf{A}_{\text{diag}} + \mathbf{A}_{\text{as}}$ we have, for $\mathbf{v} = (\mathbf{v}_1, \mathbf{v}_2) \in \mathbb{R}^{n_1} \times \mathbb{R}^{n_2}$,

$$\begin{aligned} \langle \mathbf{A}\mathbf{v}, \mathbf{v}\rangle_{\mathbb{R}^N} &= \left\langle \mathbf{A}_{\text{diag}}\mathbf{v}, \mathbf{v}\right\rangle_{\mathbb{R}^N} + \langle \mathbf{A}_{\text{as}}\mathbf{v}, \mathbf{v}\rangle_{\mathbb{R}^N} \\ &= \left\langle \mathbf{A}_{\text{diag}}\mathbf{v}, \mathbf{v}\right\rangle_{\mathbb{R}^N} + \langle \mathbf{A}_{12}\mathbf{v}_2, \mathbf{v}_1\rangle_{\mathbb{R}^{n_1}} + \langle \mathbf{A}_{21}\mathbf{v}_1, \mathbf{v}_2\rangle_{\mathbb{R}^{n_2}} \\ &= \left\langle \mathbf{A}_{\text{diag}}\mathbf{v}, \mathbf{v}\right\rangle_{\mathbb{R}^N} + \left\langle \mathbf{v}_2, \mathbf{A}_{12}^\top\mathbf{v}_1\right\rangle_{\mathbb{R}^{n_2}} + \langle \mathbf{A}_{21}\mathbf{v}_1, \mathbf{v}_2\rangle_{\mathbb{R}^{n_2}} \\ &= \left\langle \mathbf{A}_{\text{diag}}\mathbf{v}, \mathbf{v}\right\rangle_{\mathbb{R}^N}. \end{aligned}$$

Hence, assumption (2.57) implies

$$\alpha = d \le \inf_{\substack{\mathbf{v}\in\mathbb{R}^N \\ \mathbf{v}\neq\mathbf{0}}} \frac{\left\langle \mathbf{A}_{\text{diag}}\mathbf{v}, \mathbf{v}\right\rangle_{\mathbb{R}^N}}{\langle \mathbf{C}\mathbf{v}, \mathbf{v}\rangle_{\mathbb{R}^N}} = \inf_{\substack{\mathbf{v}\in\mathbb{R}^N \\ \mathbf{v}\neq\mathbf{0}}} \frac{\langle \mathbf{A}\mathbf{v}, \mathbf{v}\rangle_{\mathbb{R}^N}}{\langle \mathbf{C}\mathbf{v}, \mathbf{v}\rangle_{\mathbb{R}^N}}.$$

- If $\mathbf{A}_{12}^\top \neq -\mathbf{A}_{21}$, then it follows from the definition of $c_{\mathbf{A}}$ and from (2.57) that

$$\langle \mathbf{A}\mathbf{v}, \mathbf{v}\rangle_{\mathbb{R}^N} \ge c_{\mathbf{A}} \left\langle \mathbf{A}_{\text{diag}}\mathbf{v}, \mathbf{v}\right\rangle_{\mathbb{R}^N} \ge c_{\mathbf{A}} d \,\langle \mathbf{C}\mathbf{v}, \mathbf{v}\rangle_{\mathbb{R}^N},$$

which implies

$$\alpha = c_{\mathbf{A}} d \le \inf_{\substack{\mathbf{v}\in\mathbb{R}^N \\ \mathbf{v}\neq\mathbf{0}}} \frac{\langle \mathbf{A}\mathbf{v}, \mathbf{v}\rangle_{\mathbb{R}^N}}{\langle \mathbf{C}\mathbf{v}, \mathbf{v}\rangle_{\mathbb{R}^N}}.$$

Next we prove the estimate for $\beta^2$. For any $\mathbf{v} = (\mathbf{v}_1, \mathbf{v}_2) \in \mathbb{R}^{n_1} \times \mathbb{R}^{n_2}$ we have by using (2.58) which is equivalent to assumption (2.57)

$$\left\langle \mathbf{C}^{-1}\mathbf{A}\mathbf{v}, \mathbf{A}\mathbf{v}\right\rangle_{\mathbb{R}^N} \le D\left\langle \mathbf{A}_{\text{diag}}^{-1}\mathbf{A}\mathbf{v}, \mathbf{A}\mathbf{v}\right\rangle_{\mathbb{R}^N} = D\left\langle \mathbf{A}^\top\mathbf{A}_{\text{diag}}^{-1}\mathbf{A}\mathbf{v}, \mathbf{v}\right\rangle_{\mathbb{R}^N}.$$

Noting that

$$\mathbf{A}^\top\mathbf{A}_{\text{diag}}^{-1}\mathbf{A} = \begin{pmatrix} \mathbf{A}_{11} & \mathbf{A}_{21}^\top \\ \mathbf{A}_{12}^\top & \mathbf{A}_{22} \end{pmatrix} \begin{pmatrix} \mathbf{A}_{11}^{-1} & \mathbf{0} \\ \mathbf{0} & \mathbf{A}_{22}^{-1} \end{pmatrix} \begin{pmatrix} \mathbf{A}_{11} & \mathbf{A}_{12} \\ \mathbf{A}_{21} & \mathbf{A}_{22} \end{pmatrix}$$

$$= \mathbf{A}_{\mathrm{diag}} + \begin{pmatrix} \mathbf{A}_{21}^\top \mathbf{A}_{22}^{-1} \mathbf{A}_{21} & \mathbf{0} \\ \mathbf{0} & \mathbf{A}_{12}^\top \mathbf{A}_{11}^{-1} \mathbf{A}_{12} \end{pmatrix} + \begin{pmatrix} \mathbf{0} & \mathbf{A}_{12} + \mathbf{A}_{21}^\top \\ \mathbf{A}_{12}^\top + \mathbf{A}_{21} & \mathbf{0} \end{pmatrix}$$

we deduce

$$\begin{aligned}
\left\langle \mathbf{A}^\top \mathbf{A}_{\mathrm{diag}}^{-1} \mathbf{A}\mathbf{v}, \mathbf{v} \right\rangle_{\mathbb{R}^N} &= \left\langle \mathbf{A}_{\mathrm{diag}} \mathbf{v}, \mathbf{v} \right\rangle_{\mathbb{R}^N} \\
&\quad + \left\langle \mathbf{A}_{21}^\top \mathbf{A}_{22}^{-1} \mathbf{A}_{21} \mathbf{v}_1, \mathbf{v}_1 \right\rangle_{\mathbb{R}^{n_1}} + \left\langle \mathbf{A}_{12}^\top \mathbf{A}_{11}^{-1} \mathbf{A}_{12} \mathbf{v}_2, \mathbf{v}_2 \right\rangle_{\mathbb{R}^{n_2}} \\
&\quad + \left\langle (\mathbf{A}_{12} + \mathbf{A}_{21}^\top) \mathbf{v}_2, \mathbf{v}_1 \right\rangle_{\mathbb{R}^{n_1}} + \left\langle (\mathbf{A}_{12}^\top + \mathbf{A}_{21}) \mathbf{v}_1, \mathbf{v}_2 \right\rangle_{\mathbb{R}^{n_2}} \\
&= \left\langle \mathbf{A}_{\mathrm{diag}} \mathbf{v}, \mathbf{v} \right\rangle_{\mathbb{R}^N} \\
&\quad + \left\langle \mathbf{A}_{21}^\top \mathbf{A}_{22}^{-1} \mathbf{A}_{21} \mathbf{v}_1, \mathbf{v}_1 \right\rangle_{\mathbb{R}^{n_1}} + \left\langle \mathbf{A}_{12}^\top \mathbf{A}_{11}^{-1} \mathbf{A}_{12} \mathbf{v}_2, \mathbf{v}_2 \right\rangle_{\mathbb{R}^{n_2}} \\
&\quad + 2 \left\langle \mathbf{A}_{12} \mathbf{v}_2, \mathbf{v}_1 \right\rangle_{\mathbb{R}^{n_1}} + 2 \left\langle \mathbf{A}_{21} \mathbf{v}_1, \mathbf{v}_2 \right\rangle_{\mathbb{R}^{n_2}} \\
&\le \left\langle \mathbf{A}_{\mathrm{diag}} \mathbf{v}, \mathbf{v} \right\rangle_{\mathbb{R}^N} + C_{\beta_1} \left\langle \mathbf{A}_{11} \mathbf{v}_1, \mathbf{v}_1 \right\rangle_{\mathbb{R}^{n_1}} + C_{\beta_2} \left\langle \mathbf{A}_{22} \mathbf{v}_2, \mathbf{v}_2 \right\rangle_{\mathbb{R}^{n_2}} \\
&\quad + C_{\beta_3} \left\langle \mathbf{A}_{\mathrm{diag}} \mathbf{v}, \mathbf{v} \right\rangle_{\mathbb{R}^N} \\
&\le (\widetilde{C} + C_{\beta_3} + 1) \left\langle \mathbf{A}_{\mathrm{diag}} \mathbf{v}, \mathbf{v} \right\rangle_{\mathbb{R}^N}.
\end{aligned}$$

Altogether we have, using again (2.57),

$$\begin{aligned}
\left\langle \mathbf{C}^{-1} \mathbf{A}\mathbf{v}, \mathbf{A}\mathbf{v} \right\rangle_{\mathbb{R}^N} &\le D(\widetilde{C} + C_{\beta_3} + 1) \left\langle \mathbf{A}_{\mathrm{diag}} \mathbf{v}, \mathbf{v} \right\rangle_{\mathbb{R}^N} \\
&\le D^2(\widetilde{C} + C_{\beta_3} + 1) \left\langle \mathbf{C}\mathbf{v}, \mathbf{v} \right\rangle_{\mathbb{R}^N} \\
&= \beta^2 \left\langle \mathbf{C}\mathbf{v}, \mathbf{v} \right\rangle_{\mathbb{R}^N},
\end{aligned}$$

proving the lemma. □

## 2.8 Other Krylov Subspace Methods

In this section we introduce some other Krylov subspace methods for symmetric indefinite matrices. These methods, which will be useful for solving FEM-BEM coupling problems in Chapter 14, are Generalised Conjugate Residual methods, which are of Krylov's subspace type. More details can be found in [16, 87, 183]. The reader who is not interested in FEM-BEM coupling problems can skip this section.

### 2.8.1 General Krylov subspace methods

First we follow [183] to explain briefly how the Krylov subspace methods are developed. They belong to a larger class of methods, the projection methods. Recall equation (2.3)

$$\mathbf{A}\mathbf{u} = \mathbf{b}$$

where $\mathbf{A} \in \mathbb{R}^{N\times N}$ is a nonsingular matrix. Suppose that $\mathbf{u}_0$ is given and a sequence of iteration vectors $\mathbf{u}_j$, $j = 1,\dots,m-1$, has been computed. Let $\mathbf{r}_j := \mathbf{b} - \mathbf{A}\mathbf{u}_j$, $j = 0,\dots,m-1$, denote the residual vectors. In a projection method, the next iterate $\mathbf{u}_m$ is defined by

$$\mathbf{u}_m \in \mathbf{u}_0 + \mathscr{K}_m \quad \text{such that} \quad \langle \mathbf{r}_m, \mathbf{v} \rangle_{\mathbb{R}^N} = 0 \quad \forall \mathbf{v} \in \mathscr{L}_m \tag{2.60}$$

where $\mathscr{K}_m$ and $\mathscr{L}_m$ are two $m$-dimensional subspaces of $\mathbb{R}^N$. This condition is called the Petrov–Galerkin condition. Assuming exact arithmetic, due to this condition, a projection method gives the exact solution $\mathbf{u}$ after at most $N$ iterations.

For any $\mathbf{s} \in \mathbb{R}^N$, $\mathbf{B} \in \mathbb{R}^{N\times N}$, and $m = 1,2,\dots$, the Krylov subspace of dimension $m = 1,2,\dots$ generated by $\mathbf{s}$ and $\mathbf{B}$ is defined by

$$\mathscr{K}_m(\mathbf{s},\mathbf{B}) := \operatorname{span}\{\mathbf{s}, \mathbf{B}\mathbf{s}, \dots, \mathbf{B}^{m-1}\mathbf{s}\}. \tag{2.61}$$

If the subspace $\mathscr{K}_m$ in (2.60) is chosen to be $\mathscr{K}_m(\mathbf{r}_0,\mathbf{A})$, then the projection method is called a Krylov subspace method. Different choices of the space $\mathscr{L}_m$ give rise to different Krylov subspace methods. Three commonly seen choices are $\mathscr{L}_m = \mathscr{K}_m$, $\mathscr{L}_m = \mathbf{A}\mathscr{K}_m$, and $\mathscr{L}_m = \mathscr{K}_m(\mathbf{r}_0,\mathbf{A}^\top)$.

It can be shown, see [183, Proposition 5.2], that if $\mathbf{A}$ is symmetric and positive definite, and if $\mathscr{L}_m = \mathscr{K}_m$, then $\mathbf{u}_m$ satisfies (2.60) if and only if it minimises the error $\mathbf{e} := \mathbf{u} - \widetilde{\mathbf{u}}$ over the space $\mathbf{u}_0 + \mathscr{K}_m$ in the $\mathbf{A}$-norm, i.e.,

$$\mathbf{u}_m = \operatorname*{arg\,min}_{\widetilde{\mathbf{u}} \in \mathbf{u}_0 + \mathscr{K}_m} \|\mathbf{u} - \widetilde{\mathbf{u}}\|_{\mathbf{A}}. \tag{2.62}$$

Here we recall that $\|\mathbf{x}\|_{\mathbf{A}} = \langle \mathbf{x},\mathbf{x}\rangle_{\mathbf{A}}^{1/2} = \langle \mathbf{A}\mathbf{x},\mathbf{x}\rangle_{\mathbb{R}^N}$. On the other hand, it can be shown, see [183, Proposition 5.3], that if $\mathbf{A}$ is an arbitrary square matrix and if $\mathscr{L}_m = \mathbf{A}\mathscr{K}_m$ then $\mathbf{u}_m$ satisfies (2.60) if and only if it minimises the residual $\|\mathbf{b} - \mathbf{A}\widetilde{\mathbf{u}}\|_{\mathbb{R}^N}$, i.e.,

$$\mathbf{u}_m = \operatorname*{arg\,min}_{\widetilde{\mathbf{u}} \in \mathbf{u}_0 + \mathscr{K}_m} \|\mathbf{b} - \mathbf{A}\widetilde{\mathbf{u}}\|_{\mathbb{R}^N}. \tag{2.63}$$

The CG method (Algorithm 2.1) for a symmetric and positive definite matrix $\mathbf{A}$ is a Krylov subspace method with $\mathscr{K}_m = \mathscr{L}_m = \mathscr{K}_m(\mathbf{r}_0,\mathbf{A})$. The new iterate $\mathbf{u}_m$ satisfies (2.62). The GMRES method (Algorithm 2.2) for an arbitrary square matrix $\mathbf{A}$ is a Krylov method with $\mathscr{K}_m = \mathscr{K}_m(\mathbf{v}_1,\mathbf{A})$ and $\mathscr{L}_m = \mathbf{A}\mathscr{K}_m$, where $\mathbf{v}_1 := \mathbf{r}_0/\|\mathbf{r}_0\|_{\mathbb{R}^N}$. The new iterate $\mathbf{u}_m$ satisfies (2.63).

In the next subsection we generalise the CG and GMRES methods.

### 2.8.2 The generalised three-term CG methods

In this subsection we consider the equation

$$\mathbf{G}\mathbf{u} = \mathbf{g} \tag{2.64}$$

where $\mathbf{G}$ is a non-singular $N \times N$ matrix. We generalise the methods discussed in Section 2.8.1.

Let $\mathbf{H}$ be a symmetric positive definite matrix that defines a norm, called the $\mathbf{H}$-norm, $\|\mathbf{x}\|_{\mathbf{H}} := \langle \mathbf{x}, \mathbf{x} \rangle_{\mathbf{H}}^{1/2} = \langle \mathbf{H}\mathbf{x}, \mathbf{x} \rangle_{\mathbb{R}^N}^{1/2}$. Suppose that $\mathbf{u}_0$ is given and that a sequence of iteration vectors $\mathbf{u}_j$, $j = 1, \ldots, m+1$, has been computed. Let $\mathbf{r}_j := \mathbf{g} - \mathbf{G}\mathbf{u}_j$ and $\mathbf{e}_j := \mathbf{u} - \mathbf{u}_j$ be the residual vectors and error vectors, respectively. The iterates $\mathbf{u}_j$ are assumed to be computed by minimising the $\mathbf{H}$-norm of the error $\mathbf{e}_j$ over the Krylov subspace $\mathscr{K}_j(\mathbf{r}_0, \mathbf{G})$, namely

$$\mathbf{u}_{j+1} := \operatorname*{arg\,min}_{\widetilde{\mathbf{u}} \in \mathbf{u}_0 + \mathscr{K}_{j+1}(\mathbf{r}_0, \mathbf{G})} \|\mathbf{u} - \widetilde{\mathbf{u}}\|_{\mathbf{H}}, \quad j = 0, \ldots, m; \tag{2.65}$$

c.f., (2.62). This is equivalent to

$$\langle \mathbf{e}_{j+1}, \mathbf{v} \rangle_{\mathbf{H}} = 0 \quad \forall \mathbf{v} \in \mathscr{K}_{j+1}(\mathbf{r}_0, \mathbf{G}), \quad j = 0, \ldots, m. \tag{2.66}$$

Noting that $\mathbf{u}_m \in \mathbf{u}_0 + \mathscr{K}_m \subset \mathbf{u}_0 + \mathscr{K}_{m+1}$, and letting $\mathbf{d}_m := \mathbf{u}_{m+1} - \mathbf{u}_m \in \mathscr{K}_{m+1}(\mathbf{r}_0, \mathbf{G})$ and $\mathbf{d} := \widetilde{\mathbf{u}} - \mathbf{u}_m \in \mathscr{K}_{m+1}(\mathbf{r}_0, \mathbf{G})$, we have $\mathbf{e}_{m+1} = \mathbf{e}_m + \mathbf{d}_m$ so that

$$\langle \mathbf{d}_m, \mathbf{v} \rangle_{\mathbf{H}} = \langle \mathbf{e}_m, \mathbf{v} \rangle_{\mathbf{H}} \quad \forall \mathbf{v} \in \mathscr{K}_{m+1}(\mathbf{r}_0, \mathbf{G}).$$

Since

$$\langle \mathbf{e}_m, \mathbf{v} \rangle_{\mathbf{H}} = 0 \quad \forall \mathbf{v} \in \mathscr{K}_m(\mathbf{r}_0, \mathbf{G})$$

it follows that

$$\langle \mathbf{d}_m, \mathbf{v} \rangle_{\mathbf{H}} = 0 \quad \forall \mathbf{v} \in \mathscr{K}_m(\mathbf{r}_0, \mathbf{G}).$$

This condition means that an $\mathbf{H}$-orthogonal basis of $\mathscr{K}_{m+1}(\mathbf{r}_0, \mathbf{G})$ will deliver the set of conjugate search directions $\{\mathbf{p}_j\}_{j=0}^m$. This set can be calculated by means of a standard Gram–Schmidt orthogonalisation process:

$$\mathbf{p}_0 := \mathbf{r}_0, \quad \mathbf{p}_{j+1} := \mathbf{G}\mathbf{p}_j - \sum_{i=0}^{j} \sigma_{ji} \mathbf{p}_i, \quad \sigma_{ji} := \frac{\langle \mathbf{G}\mathbf{p}_j, \mathbf{p}_i \rangle_{\mathbf{H}}}{\langle \mathbf{p}_i, \mathbf{p}_i \rangle_{\mathbf{H}}}, \quad j = 0, \ldots, m. \tag{2.67}$$

Then

$$\mathbf{u}_{m+1} = \mathbf{u}_m + \alpha_m \mathbf{p}_m$$

where $\alpha_m$ mus be chosen such that (2.66) holds. This yields

$$\alpha_m = \frac{\langle \mathbf{e}_m, \mathbf{p}_m \rangle_{\mathbf{H}}}{\langle \mathbf{p}_m, \mathbf{p}_m \rangle_{\mathbf{H}}}.$$

Under conditions to be addressed now, one can avoid working out the expensive recurrence formula (2.67) and turn to a three-term recursion instead:

$$\mathbf{p}_{k+1} = \mathbf{G}\mathbf{p}_k - \sigma_{k,k-1}\mathbf{p}_{k-1} - \sigma_{kk}\mathbf{p}_k. \tag{2.68}$$

We follow [11] to call this algorithm *three-term* CG($\mathbf{H},\mathbf{G}$) for the solution of (2.64). We note that Algorithm 2.1 is three-term CG($\mathbf{A},\mathbf{A}$) while Algorithm 2.3 is three-term CG($\mathbf{A},\mathbf{C}^{-1}\mathbf{A}$).

Let $\mathbf{G}^{\top_\mathbf{H}}$ be the $\mathbf{H}$-adjoint of $\mathbf{G}$, i.e., the matrix which satisfies

$$\left\langle \mathbf{G}^{\top_\mathbf{H}}\mathbf{x},\mathbf{y}\right\rangle_\mathbf{H} = \langle \mathbf{x},\mathbf{G}\mathbf{y}\rangle_\mathbf{H} \quad \forall \mathbf{x},\mathbf{y}\in\mathbb{R}^N.$$

Then

$$\mathbf{G}^{\top_\mathbf{H}} = (\mathbf{H}\mathbf{G}\mathbf{H}^{-1})^\top = \mathbf{H}^{-1}\mathbf{G}^\top\mathbf{H}. \tag{2.69}$$

If $\mathbf{G}^{\top_\mathbf{H}} = \mathbf{G}$, we say that $\mathbf{G}$ is $\mathbf{H}$-symmetric. As a special case of [11, Theorem 2.1], we have

**Theorem 2.3.** *If* $\mathbf{G}$ *is* $\mathbf{H}$*-symmetric, then three-term CG(*$\mathbf{H},\mathbf{G}$*) gives the exact solution of* (2.64) *within a finite number of steps.*

Algorithm 2.12, called Orthodir($\mathbf{H},\mathbf{G}$), implements CG($\mathbf{H},\mathbf{G}$) for $\mathbf{H}$-symmetric matrices.

**Algorithm 2.12:** Orthodir($\mathbf{H},\mathbf{G}$)

1 $\mathbf{p}_0 \leftarrow \mathbf{r}_0$
2 $\alpha_i \leftarrow \dfrac{\langle \mathbf{e}_i,\mathbf{p}_i\rangle_\mathbf{H}}{\langle \mathbf{p}_i,\mathbf{p}_i\rangle_\mathbf{H}}$
3 $\mathbf{u}_{i+1} \leftarrow \mathbf{u}_i + \alpha_i\mathbf{p}_i$
4 $\mathbf{r}_{i+1} \leftarrow \mathbf{r}_i - \alpha_i\mathbf{G}\mathbf{p}_i$
5 $\gamma_i \leftarrow \dfrac{\langle \mathbf{G}\mathbf{p}_i,\mathbf{p}_i\rangle_\mathbf{H}}{\langle \mathbf{p}_i,\mathbf{p}_i\rangle_\mathbf{H}}$
6 $\sigma_i \leftarrow \dfrac{\langle \mathbf{G}\mathbf{p}_i,\mathbf{p}_{i-1}\rangle_\mathbf{H}}{\langle \mathbf{p}_{i-1},\mathbf{p}_{i-1}\rangle_\mathbf{H}}$
7 $\mathbf{p}_{i+1} \leftarrow \mathbf{G}\mathbf{p}_i - \gamma_i\mathbf{p}_i - \sigma_i\mathbf{p}_{i-1}$

The computation of $\alpha_i$ in Step 2 involves the unknown quantity $\mathbf{e}_i$. Thus, $\mathbf{H}$ must be chosen so that $\alpha_i$ can be expressed in terms of $\mathbf{G}$ and $\mathbf{g}$; see [11].

### 2.8.3 *The hybrid modified conjugate residual (HMCR) method*

We now apply Orthodir($\mathbf{H},\mathbf{G}$) (Algorithm 2.12) to the preconditioned system

$$\mathbf{C}^{-1}\mathbf{A}\mathbf{u} = \mathbf{C}^{-1}\mathbf{b} \tag{2.70}$$

where $\mathbf{A}$ is symmetric and indefinite. We restrict ourselves to the case when $\mathbf{C}$ is symmetric and positive definite even though generalisations to more general $\mathbf{C}$ are available; see [11].

Let

$$\mathbf{G} := \mathbf{C}^{-1}\mathbf{A} \quad \text{and} \quad \mathbf{H} := \mathbf{A}\mathbf{C}^{-1}\mathbf{A}. \tag{2.71}$$

Then $\mathbf{G}$ is $\mathbf{H}$-symmetric. Indeed, it follows from (2.69), (2.71), and the symmetry of $\mathbf{A}$ and $\mathbf{C}$ that

$$\mathbf{G}^{\top_{\mathbf{H}}} = (\mathbf{A}\mathbf{C}^{-1}\mathbf{A})^{-1}(\mathbf{C}^{-1}\mathbf{A})^{\top}(\mathbf{A}\mathbf{C}^{-1}\mathbf{A}) = \underbrace{\mathbf{A}^{-1}\mathbf{C}\mathbf{A}^{-1}\mathbf{A}\mathbf{C}^{-1}\mathbf{A}}_{=\mathbf{I}}\mathbf{C}^{-1}\mathbf{A} = \mathbf{G}.$$

Hence Orthodir$(\mathbf{H},\mathbf{G})$ is applicable and gives the exact solution of (2.70) after a finite number of steps; see Theorem 2.3. The choice (2.71) also guarantees that $\alpha_i$ in Step 2 of Algorithm 2.12 is computable

$$\alpha_i = \frac{\langle \mathbf{e}_i, \mathbf{p}_i\rangle_{\mathbf{H}}}{\langle \mathbf{p}_i, \mathbf{p}_i\rangle_{\mathbf{H}}} = \frac{\langle \mathbf{A}\mathbf{C}^{-1}\mathbf{A}(\mathbf{u}-\mathbf{u}_i), \mathbf{p}_i\rangle}{\langle \mathbf{A}\mathbf{C}^{-1}\mathbf{A}\mathbf{p}_i, \mathbf{p}_i\rangle} = \frac{\langle \mathbf{r}_i, \mathbf{C}^{-1}\mathbf{A}\mathbf{p}_i\rangle}{\langle \mathbf{A}\mathbf{p}_i, \mathbf{C}^{-1}\mathbf{A}\mathbf{p}_i\rangle}.$$

Therefore, Orthodir$(\mathbf{A}\mathbf{C}^{-1}\mathbf{A}, \mathbf{C}^{-1}\mathbf{A})$ yields Algorithm 2.13 for solving (2.70).

**Algorithm 2.13:** MCR-ODIR$(\mathbf{C}^{-1}\mathbf{A})$

Given $\mathbf{u}_0$, $\mathbf{r}_0 \leftarrow \mathbf{b} - \mathbf{A}\mathbf{u}_0$ and $\mathbf{p}_0 \leftarrow \mathbf{C}^{-1}\mathbf{r}_0$
**for** $i = 0, 1, 2, \dots$ **do**
  $\alpha_i \leftarrow \dfrac{\langle \mathbf{C}^{-1}\mathbf{A}\mathbf{p}_i, \mathbf{r}_i\rangle}{\langle \mathbf{C}^{-1}\mathbf{A}\mathbf{p}_i, \mathbf{A}\mathbf{p}_i\rangle}$
  $\mathbf{u}_{i+1} \leftarrow \mathbf{u}_i + \alpha_i \mathbf{p}_i$
  $\mathbf{r}_{i+1} \leftarrow \mathbf{r}_i - \alpha_i \mathbf{A}\mathbf{p}_i$
  **if** *stopping criterion* **then**
    stop
  **else**
    $\gamma_i \leftarrow \dfrac{\langle \mathbf{A}\mathbf{C}^{-1}\mathbf{A}\mathbf{p}_i, \mathbf{C}^{-1}\mathbf{A}\mathbf{p}_i\rangle}{\langle \mathbf{C}^{-1}\mathbf{A}\mathbf{p}_i, \mathbf{A}\mathbf{p}_i\rangle}$
    $\sigma_i \leftarrow \dfrac{\langle \mathbf{C}^{-1}\mathbf{A}\mathbf{p}_i, \mathbf{A}\mathbf{p}_{i-1}\rangle}{\langle \mathbf{C}^{-1}\mathbf{A}\mathbf{p}_{i-1}, \mathbf{A}\mathbf{p}_{i-1}\rangle}$
    $\mathbf{p}_{i+1} \leftarrow \mathbf{C}^{-1}\mathbf{A}\mathbf{p}_i - \gamma_i \mathbf{p}_i - \sigma_i \mathbf{p}_{i-1}$
  **end**
  Next $i$
**end**

One can define a sequence of Krylov subspaces which is different from the subspace $\mathscr{K}_j(\mathbf{r}_0, \mathbf{G})$; see Subsection 2.8.2. Let

$$\mathbf{s}_j = \mathbf{C}^{-1}\mathbf{r}_j, \quad j = 0, 1, \dots,$$

be the preconditioned residuals. We replace $\mathscr{K}_{j+1}(\mathbf{r}_0, \mathbf{G})$ in (2.65) by $\mathscr{K}_{j+1}(\mathbf{s}_0, \mathbf{G})$. Under certain conditions to be given in Theorem 2.4, the three-term recursion (2.68) can be replaced by a two-term recursion so that the resulting algorithm converges within a finite number of steps; see Algorithm 2.14.

**Algorithm 2.14:** MCR-OMIN($\mathbf{C}^{-1}\mathbf{A}$)

Given $\mathbf{u}_0$, $\mathbf{r}_0 \leftarrow \mathbf{b} - \mathbf{A}\mathbf{u}_0$, $\mathbf{s}_0 \leftarrow \mathbf{C}^{-1}\mathbf{r}_0$, $\mathbf{p}_0 \leftarrow \mathbf{C}^{-1}\mathbf{s}_0$
**for** $i = 0, 1, 2, \ldots$ **do**
  $\alpha_i \leftarrow \dfrac{\langle \mathbf{C}^{-1}\mathbf{A}\mathbf{p}_i, \mathbf{r}_i \rangle}{\langle \mathbf{C}^{-1}\mathbf{A}\mathbf{p}_i, \mathbf{A}\mathbf{p}_i \rangle}$
  $\mathbf{u}_{i+1} \leftarrow \mathbf{u}_i + \alpha_i \mathbf{p}_i$
  $\mathbf{r}_{i+1} \leftarrow \mathbf{r}_i - \alpha_i \mathbf{A}\mathbf{p}_i$
  $\mathbf{s}_{i+1} \leftarrow \mathbf{s}_i - \alpha_i \mathbf{C}^{-1}\mathbf{A}\mathbf{p}_i$
  **if** *(stopping criterion)* **then**
    stop
  **else**
    $\beta_i \leftarrow \dfrac{\langle \mathbf{C}^{-1}\mathbf{A}\mathbf{p}_i, \mathbf{A}\mathbf{s}_{i+1} \rangle}{\langle \mathbf{C}^{-1}\mathbf{A}\mathbf{p}_i, \mathbf{A}\mathbf{p}_i \rangle}$
    $\mathbf{p}_{i+1} \leftarrow \mathbf{s}_{i+1} - \beta_i \mathbf{p}_i$
  **end**
  Next $i$
**end**

We now give a theorem on the relation between MCR-ODIR and MCR-OMIN; see [11].

**Theorem 2.4.** *The following statements are equivalent.*

(i) MCR-OMIN($\mathbf{C}^{-1}\mathbf{A}$) *and* MCR-ODIR($\mathbf{C}^{-1}\mathbf{A}$) *are equivalent, namely they generate the same sequence of iterates* $\{\mathbf{u}_k\}$, *given the same initial vector* $\mathbf{u}_0$*;*
(ii) MCR-OMIN($\mathbf{C}^{-1}\mathbf{A}$) *converges for each starting vector* $\mathbf{u}_0$ *to the exact solution of* (2.70)*;*
(iii) *The matrix* $\mathbf{H}\mathbf{C}^{-1}\mathbf{A} = \mathbf{A}\mathbf{C}^{-1}\mathbf{A}\mathbf{C}^{-1}\mathbf{A}$ *is definite;*
(iv) *In* MCR-OMIN($\mathbf{C}^{-1}\mathbf{A}$) *the scalar* $\alpha_j$ *is non-zero for all* $j \geq 0$*;*
(v) $\mathscr{K}_{j+1}(\mathbf{s}_0, \mathbf{C}^{-1}\mathbf{A}) = \mathrm{span}\{\mathbf{s}_0, \ldots, \mathbf{s}_j\}$.

MCR-OMIN($\mathbf{C}^{-1}\mathbf{A}$) takes one scalar product less than MCR-ODIR($\mathbf{C}^{-1}\mathbf{A}$) and requires less storage. However, the difficulty with applying MCR-OMIN to indefinite linear system is that $\beta_i$ may become zero; in which case the iteration is tracked in the current Krylov subspace. This effect, which is named "stalling", can be avoided by switching from MCR-OMIN to the robust MCR-ODIR each time when $\alpha_i$ becomes smaller than a certain prescribed tolerance. This yields the hybrid modified conjugate residual (HMCR) algorithm [58], which is given in Algorithm 2.15.

We note that MCR-ODIR($\mathbf{C}^{-1}\mathbf{A}$), MCR-OMIN($\mathbf{C}^{-1}\mathbf{A}$), and HMCR($\mathbf{C}^{-1}\mathbf{A}$) are MINRES methods for symmetric indefinite matrices. Since $\mathbf{A}$ is symmetric and indefinite, the matrix $\mathbf{H}\mathbf{C}^{-1}\mathbf{A}$ is also indefinite, and thus the MCR-OMIN($\mathbf{C}^{-1}\mathbf{A}$)

**Algorithm 2.15:** HMCR($\mathbf{C}^{-1}\mathbf{A}$)

Given $\mathbf{u}_0$ and a tolerance $\overline{\alpha}$, set $\mathbf{r}_0 \leftarrow \mathbf{b} - \mathbf{A}\mathbf{u}_0,\ \mathbf{p}_0 \leftarrow \mathbf{C}^{-1}\mathbf{r}_0$
**for** $i = 0, 1, 2, \ldots$ **do**
  $\alpha_i \leftarrow \dfrac{\langle \mathbf{C}^{-1}\mathbf{A}\mathbf{p}_i, \mathbf{r}_i \rangle}{\langle \mathbf{C}^{-1}\mathbf{A}\mathbf{p}_i, \mathbf{A}\mathbf{p}_i \rangle}$
  $\mathbf{u}_{i+1} \leftarrow \mathbf{u}_i + \alpha_i \mathbf{p}_i$
  $\mathbf{r}_{i+1} \leftarrow \mathbf{r}_i - \alpha_i \mathbf{A}\mathbf{p}_i$
  $\mathbf{s}_{i+1} \leftarrow \mathbf{s}_i - \alpha_i \mathbf{C}^{-1}\mathbf{A}\mathbf{p}_i$
  **if** *(stopping criterion)* **then**
    stop
  **end**
  **if** $|\alpha_i| < \overline{\alpha}$ **then**
    *** do ODIR ***
    **if** $|\alpha_{i-1}| < \overline{\alpha}$ **then**
      $c \leftarrow 1$
    **else**
      $c \leftarrow -1/\alpha_{i-1}$
    **end**
    $\gamma_i \leftarrow \dfrac{\langle \mathbf{A}\mathbf{C}^{-1}\mathbf{A}\mathbf{p}_i, \mathbf{C}^{-1}\mathbf{A}\mathbf{p}_i \rangle}{\langle \mathbf{C}^{-1}\mathbf{A}\mathbf{p}_i, \mathbf{A}\mathbf{p}_i \rangle}$
    $\sigma_i \leftarrow \dfrac{\langle \mathbf{C}^{-1}\mathbf{A}\mathbf{p}_i, \mathbf{A}\mathbf{p}_{i-1} \rangle}{\langle \mathbf{C}^{-1}\mathbf{A}\mathbf{p}_{i-1}, \mathbf{A}\mathbf{p}_{i-1} \rangle}$
    $\mathbf{p}_{i+1} \leftarrow \mathbf{C}^{-1}\mathbf{A}\mathbf{p}_i - \gamma_i \mathbf{p}_i - \sigma_i \mathbf{p}_{i-1}$
  **else**
    *** do OMIN ***
    $\beta_i \leftarrow \dfrac{\langle \mathbf{C}^{-1}\mathbf{A}\mathbf{p}_i, \mathbf{A}\mathbf{s}_{i+1} \rangle}{\langle \mathbf{C}^{-1}\mathbf{A}\mathbf{p}_i, \mathbf{A}\mathbf{p}_i \rangle}$
    $\mathbf{p}_{i+1} \leftarrow \mathbf{s}_{i+1} - \beta_i \mathbf{p}_i$
  **end**
  Next $i$
**end**

part of the foregoing algorithm does not necessarily converge. Hence the method of choice is HMCR($\mathbf{C}^{-1}\mathbf{A}$).

The error vector has to be minimised with respect to the $\mathbf{H}$-norm. This implies

$$\begin{aligned} \|\mathbf{e}_{k+1}\|_{\mathbf{H}} &= \langle \mathbf{C}^{-1}\mathbf{A}(\mathbf{u} - \mathbf{u}_{k+1}), \mathbf{A}(\mathbf{u} - \mathbf{u}_{k+1}) \rangle_{\mathbb{R}^N}^{1/2} = \|\mathbf{b} - \mathbf{A}\mathbf{u}_{k+1}\|_{\mathbf{C}^{-1}} \\ &= \min \left\{ \|\mathbf{b} - \mathbf{A}\widetilde{\mathbf{u}}\|_{\mathbf{C}^{-1}} : \widetilde{\mathbf{u}} \in \mathbf{u}_0 + \mathscr{K}_{k+1}(\mathbf{C}^{-1}\mathbf{r}_0, \mathbf{C}^{-1}\mathbf{A}) \right\}. \end{aligned}$$

Here $\mathscr{K}_{k+1}(\mathbf{C}^{-1}\mathbf{r}_0, \mathbf{C}^{-1}\mathbf{A})$ is defined in (2.61). Thus

$$\frac{\|\mathbf{b} - \mathbf{A}\mathbf{u}_{k+1}\|_{\mathbf{C}^{-1}}}{\|\mathbf{b} - \mathbf{A}\mathbf{u}_0\|_{\mathbf{C}^{-1}}} = \min_{p \in \Pi_k} \frac{\|(\mathbf{I} - \mathbf{A}(p(\mathbf{C}^{-1}\mathbf{A}))\mathbf{C}^{-1})\mathbf{r}_0\|_{\mathbf{C}^{-1}}}{\|\mathbf{r}_0\|_{\mathbf{C}^{-1}}}$$

$$= \min_{\substack{p \in \Pi_{k+1} \\ p(0)=1}} \frac{\|p(\mathbf{A}\mathbf{C}^{-1})\mathbf{r}_0\|_{\mathbf{C}^{-1}}}{\|\mathbf{r}_0\|_{\mathbf{C}^{-1}}} = \min_{\substack{p \in \Pi_{k+1} \\ p(0)=1}} \max_{\lambda_j \in \Lambda(\mathbf{C}^{-1}\mathbf{A})} |p(\lambda_j)|$$

where $\Pi_k$ is the space of real polynomials of degree at most $k$ and $\Lambda(\mathbf{C}^{-1}\mathbf{A})$ is the set of eigenvalues of $\mathbf{C}^{-1}\mathbf{A}$. Hence in order to be able to find a good preconditioner $\mathbf{C}$, one has to investigate the spectra $\Lambda$ of $\mathbf{A}$ and of its submatrices.

If $E$ is an inclusion set for the eigenvalues (which are real but may be positive or negative), then

$$\frac{\|\mathbf{b} - \mathbf{A}\mathbf{u}_k\|_{\mathbf{C}^{-1}}}{\|\mathbf{b} - \mathbf{A}\mathbf{u}_0\|_{\mathbf{C}^{-1}}} \leq \min_{\substack{p \in \Pi_k \\ p(0)=1}} \max_{x \in E} |p(x)|.$$

Let

$$E = E(\alpha) := [-a, -b\alpha] \cup [c\alpha^2, d]$$

where $a$, $b$, $c$, $d$, and $\alpha$ are positive and $\alpha$ is asymptotically small, and let

$$e_k := \min_{\substack{p \in \Pi_k \\ p(0)=1}} \max_{x \in E} |p(x)|.$$

Then it is proved in [232, Theorem 4.1] that

$$\lim_{k \to \infty} e_k^{1/k} \leq 1 - \alpha^{3/2} \sqrt{\frac{bc}{ad}} + O(\alpha^{5/2}).$$

Hence we have the following lemma.

**Lemma 2.22.** *Assume that the eigenvalues of* $\mathbf{C}^{-1}\mathbf{A}$ *belong to*

$$[-a, -b\alpha] \cup [c\alpha^2, d]$$

*where $a$, $b$, $c$, $d$, and $\alpha$ are positive and $\alpha$ is asymptotically small. Then*

$$\lim_{k \to \infty} \left( \frac{\|\mathbf{r}_k\|_{\mathbf{C}^{-1}}}{\|\mathbf{r}_0\|_{\mathbf{C}^{-1}}} \right)^{1/k} \leq 1 - \alpha^{3/2} \sqrt{\frac{bc}{ad}} + O(\alpha^{5/2})$$

*where* $\mathbf{r}_k := \mathbf{b} - \mathbf{A}\mathbf{u}_k$.

*Remark 2.2.* The above result is a generalisation of [58, Theorem 3.2] which assumes that the eigenvalues of $\mathbf{C}^{-1}\mathbf{A}$ belong to symmetric inclusion sets $[-a, -b] \cup [c, d]$ with $-a < -b < 0 < c < d$ and $a - b = d - c$. This theorem states that

$$\left( \frac{\|\mathbf{r}_k\|_{\mathbb{R}^N}}{\|\mathbf{r}_0\|_{\mathbb{R}^N}} \right)^{1/k} \leq 2^{1/2k} \left( \frac{1 - \sqrt{bc/ad}}{1 + \sqrt{bc/ad}} \right)^{1/2}.$$

Therefore, the numbers of iterations of the preconditioned minimum residual method which are required to solve $\mathbf{A}\mathbf{u} = \mathbf{b}$ (with $\mathbf{A}$ being symmetric and indefinite) up to a given accuracy is bounded by $O(1/\sqrt{bc/ad})$.

## 2.9 Convergence Theory for Linear Iterative Methods with Preconditioners

As mentioned in Subsection 2.2.3, a preconditioner can be used to accelerate the convergence of a linear iterative method. We briefly present here the convergence analysis of this preconditioning technique; see [96] for more details.

Recall the definition of $\mathbf{u}_k$ and $\mathbf{u}$ in Subsection 2.2.3. Defining $\mathbf{e}_k := \mathbf{u}_k - \mathbf{u}$ and using the identity

$$\mathbf{u} = \mathbf{u} + \mathbf{C}^{-1}\big(\mathbf{b} - \mathbf{A}\mathbf{u}\big) = \big(\mathbf{I} - \mathbf{C}^{-1}\mathbf{A}\big)\mathbf{u} + \mathbf{C}^{-1}\mathbf{b},$$

we deduce from (2.15) that

$$\mathbf{e}_{k+1} = \big(\mathbf{I} - \mathbf{C}^{-1}\mathbf{A}\big)\mathbf{e}_k, \quad k = 0, 1, 2, \ldots.$$

It follows that if there exists a norm such that

$$\|\mathbf{I} - \mathbf{C}^{-1}\mathbf{A}\| < 1,$$

then the preconditioned iterative method converges. We call $\|\mathbf{I} - \mathbf{C}^{-1}\mathbf{A}\|$ the convergence rate of the method.

We show in the following subsections that under some assumptions,

$$\|\mathbf{I} - \mathbf{C}^{-1}\mathbf{A}\|_{\mathbf{A}} < 1 \quad \text{and} \quad \|\mathbf{I} - \mathbf{C}^{-1}\mathbf{A}\|_{\mathbf{C}} < 1$$

where we recall that the $\|\cdot\|_{\mathbf{E}}$-norm of a matrix $\mathbf{D}$ is defined by

$$\|\mathbf{D}\|_{\mathbf{E}} = \sup_{\mathbf{x}\in\mathbb{R}^N} \frac{\langle \mathbf{D}\mathbf{x}, \mathbf{D}\mathbf{x}\rangle_{\mathbf{E}}}{\langle \mathbf{x}, \mathbf{x}\rangle_{\mathbf{E}}} = \sup_{\mathbf{x}\in\mathbb{R}^N} \frac{\langle \mathbf{E}\mathbf{D}\mathbf{x}, \mathbf{D}\mathbf{x}\rangle_{\mathbb{R}^N}}{\langle \mathbf{E}\mathbf{x}, \mathbf{x}\rangle_{\mathbb{R}^N}}.$$

Here, $N$ is the size of the matrix $\mathbf{E}$ and $\mathbf{D}$. We consider two different cases: the case when $\mathbf{C}$ is symmetric (which arises from the additive Schwarz method or symmetric multiplicative Schwarz method), and the case when $\mathbf{C}$ is non-symmetric (which arises from the non-symmetric multiplicative Schwarz method).

### *2.9.1 Symmetric preconditioner*

In this subsection, the preconditioner $\mathbf{C}$ is defined from the additive Schwarz operator, see (2.27), or the multiplicative Schwarz operator, see (2.32), and thus $\mathbf{C}$ is symmetric. Recalling the definition of $\widetilde{\mathbf{A}}$ and $\widehat{\mathbf{A}}$ in (2.33), we first prove the following lemma.

**Lemma 2.23.** *We have*

$$\|\mathbf{I} - \widehat{\mathbf{A}}\|_{\mathbf{C}} = \big|1 - \lambda_{\min}(\widetilde{\mathbf{A}})\big| \tag{2.72}$$

*and*

$$\|\mathbf{I}-\widehat{\mathbf{A}}\|_{\mathbf{A}} \le \sqrt{1-2\lambda_{\min}(\widetilde{\mathbf{A}})+\lambda_{\max}^2(\widetilde{\mathbf{A}})}. \tag{2.73}$$

*Proof.* First we note that, for any $\mathbf{v} \in \mathbb{R}^N$,

$$\begin{aligned}
\|(\mathbf{I}-\widehat{\mathbf{A}})\mathbf{v}\|_{\mathbf{C}}^2 &= \mathbf{v}^\top(\mathbf{I}-\widehat{\mathbf{A}})^\top\mathbf{C}(\mathbf{I}-\widehat{\mathbf{A}})\mathbf{v} \\
&= \mathbf{v}^\top(\mathbf{C}-\mathbf{A})(\mathbf{I}-\mathbf{C}^{-1}\mathbf{A})\mathbf{v} \\
&= \mathbf{v}^\top\mathbf{C}^{1/2}(\mathbf{I}-\mathbf{C}^{-1/2}\mathbf{A}\mathbf{C}^{-1/2})\mathbf{C}^{1/2}\mathbf{C}^{-1/2}(\mathbf{I}-\mathbf{C}^{-1/2}\mathbf{A}\mathbf{C}^{-1/2})\mathbf{C}^{1/2}\mathbf{v} \\
&= \mathbf{w}^\top(\mathbf{I}-\widetilde{\mathbf{A}})^2\mathbf{w}
\end{aligned}$$

where $\mathbf{w} = \mathbf{C}^{1/2}\mathbf{v}$. Hence

$$\begin{aligned}
\|\mathbf{I}-\widehat{\mathbf{A}}\|_{\mathbf{C}}^2 &= \max_{\mathbf{v}\in\mathbb{R}^N} \frac{\|(\mathbf{I}-\widehat{\mathbf{A}})\mathbf{v}\|_{\mathbf{C}}^2}{\|\mathbf{v}\|_{\mathbf{C}}^2} = \max_{\mathbf{v}\in\mathbb{R}^N} \frac{\|(\mathbf{I}-\widehat{\mathbf{A}})\mathbf{v}\|_{\mathbf{C}}^2}{\mathbf{v}^\top\mathbf{C}\mathbf{v}} \\
&= \max_{\mathbf{w}\in\mathbb{R}^N} \frac{\mathbf{w}^\top(\mathbf{I}-\widetilde{\mathbf{A}})^2\mathbf{w}}{\mathbf{w}^\top\mathbf{w}} = \lambda_{\max}^2(\mathbf{I}-\widetilde{\mathbf{A}}).
\end{aligned}$$

On the other hand,

$$\lambda_{\max}(\mathbf{I}-\widetilde{\mathbf{A}}) = 1-\lambda_{\min}(\widetilde{\mathbf{A}}).$$

This proves (2.72). To prove (2.73), we write

$$\begin{aligned}
\|(\mathbf{I}-\widehat{\mathbf{A}})\mathbf{v}\|_{\mathbf{A}}^2 &= \big\langle \mathbf{A}(\mathbf{v}-\widehat{\mathbf{A}}\mathbf{v}), \mathbf{v}-\widehat{\mathbf{A}}\mathbf{v}\big\rangle_{\mathbb{R}^N} \\
&= \langle \mathbf{A}\mathbf{v}, \mathbf{v}\rangle_{\mathbb{R}^N} - 2\langle \mathbf{A}\mathbf{v}, \widehat{\mathbf{A}}\mathbf{v}\rangle_{\mathbb{R}^N} + \langle \mathbf{A}\widehat{\mathbf{A}}\mathbf{v}, \widehat{\mathbf{A}}\mathbf{v}\rangle_{\mathbb{R}^N} \\
&= \|\mathbf{v}\|_{\mathbf{A}}^2 - 2\langle \mathbf{A}\mathbf{v}, \widehat{\mathbf{A}}\mathbf{v}\rangle_{\mathbb{R}^N} + \|\widehat{\mathbf{A}}\mathbf{v}\|_{\mathbf{A}}^2
\end{aligned}$$

so that

$$\begin{aligned}
\|\mathbf{I}-\widehat{\mathbf{A}}\|_{\mathbf{A}}^2 &= \max_{\mathbf{v}\in\mathbb{R}^N} \frac{\|(\mathbf{I}-\widehat{\mathbf{A}})\mathbf{v}\|_{\mathbf{A}}^2}{\|\mathbf{v}\|_{\mathbf{A}}^2} \\
&\le 1 - 2\min_{\mathbf{v}\in\mathbb{R}^N} \frac{\langle \mathbf{A}\mathbf{v}, \widehat{\mathbf{A}}\mathbf{v}\rangle_{\mathbb{R}^N}}{\langle \mathbf{A}\mathbf{v}, \mathbf{v}\rangle_{\mathbb{R}^N}} + \|\widehat{\mathbf{A}}\|_{\mathbf{A}}^2.
\end{aligned}$$

Lemma 2.8 gives $\|\widehat{\mathbf{A}}\|_{\mathbf{A}} = \lambda_{\max}(\widetilde{\mathbf{A}})$. Moreover, with $\mathbf{w} = \mathbf{A}^{1/2}\mathbf{v}$ we have

$$\begin{aligned}
\min_{\mathbf{v}\in\mathbb{R}^N} \frac{\langle \mathbf{A}\mathbf{v}, \widehat{\mathbf{A}}\mathbf{v}\rangle_{\mathbb{R}^N}}{\langle \mathbf{A}\mathbf{v}, \mathbf{v}\rangle_{\mathbb{R}^N}} &= \min_{\mathbf{v}\in\mathbb{R}^N} \frac{\mathbf{v}^\top\mathbf{A}\mathbf{C}^{-1}\mathbf{A}\mathbf{v}}{\mathbf{A}\mathbf{v}^\top\mathbf{v}} \\
&= \min_{\mathbf{w}\in\mathbb{R}^N} \frac{\mathbf{w}^\top\mathbf{A}^{1/2}\mathbf{C}^{-1}\mathbf{A}^{1/2}\mathbf{w}}{\mathbf{w}^\top\mathbf{w}} \\
&= \lambda_{\min}(\mathbf{A}^{1/2}\mathbf{C}^{-1}\mathbf{A}^{1/2}).
\end{aligned}$$

Lemma C.6 implies

$$\min_{\mathbf{v}\in\mathbb{R}^N} \frac{\langle \mathbf{A}\mathbf{v}, \widehat{\mathbf{A}}\mathbf{v}\rangle_{\mathbb{R}^N}}{\langle \mathbf{A}\mathbf{v}, \mathbf{v}\rangle_{\mathbb{R}^N}} = \lambda_{\min}\big(\widetilde{\mathbf{A}}\big).$$

Altogether we obtain (2.73). □

Using the above lemma, we can prove the following convergence result for the additive and symmetric multiplicative Schwarz methods.

**Theorem 2.5.** *If* $\mathbf{C}$ *is a symmetric positive definite preconditioner, then the linear iteration*

$$\mathbf{u}_{k+1} = \mathbf{u}_k + \mathbf{C}^{-1}\big(\mathbf{b} - \mathbf{A}\mathbf{u}_k\big), \quad k = 0, 1, 2, \ldots,$$

*converges if*

$$\big|1 - \lambda_{\min}(\widetilde{\mathbf{A}})\big| < 1 \quad \text{or} \quad \sqrt{1 - 2\lambda_{\min}(\widetilde{\mathbf{A}}) + \lambda_{\max}^2(\widetilde{\mathbf{A}})} < 1.$$

*In particular, the following statements hold:*

*(i) Let* $\mathbf{C}$ *be defined by the additive Schwarz algorithm,* $\mathbf{C} = \mathbf{C}_{\mathrm{ad}}$. *If*

$$\lambda_{\min}(\mathbf{C}_{\mathrm{ad}}^{-1/2}\mathbf{A}\mathbf{C}_{\mathrm{ad}}^{-1/2}) < 1$$

*and if Assumption 2.3 (stability of the decomposition) holds, then the convergence rate* $\|\mathbf{I} - \mathbf{C}_{\mathrm{ad}}^{-1}\mathbf{A}\|_{\mathbf{C}_{\mathrm{ad}}}$ *satisfies*

$$\|\mathbf{I} - \mathbf{C}_{\mathrm{ad}}^{-1}\mathbf{A}\|_{\mathbf{C}_{\mathrm{ad}}} \le 1 - C_0^2$$

*where* $C_0$ *is the constant given by Assumption 2.3.*

*(ii) Let* $\mathbf{C}$ *is defined by the symmetric multiplicative Schwarz algorithm,* $\mathbf{C} = \mathbf{C}_{\mathrm{smu}}$. *If*

$$\lambda_{\min}(\mathbf{C}_{\mathrm{smu}}^{-1/2}\mathbf{A}\mathbf{C}_{\mathrm{smu}}^{-1/2}) < 1$$

*and if Assumptions 2.3, 2.5, and 2.6 hold, then the convergence rate satisfies*

$$\|\mathbf{I} - \mathbf{C}_{\mathrm{smu}}^{-1}\mathbf{A}\|_{\mathbf{C}_{\mathrm{smu}}} \le 1 - \frac{2-\omega}{(1 + 2\widehat{\omega}^2\rho(\mathscr{E})^2)C_0^2}$$

*where the constants* $C_0$, $\omega$, $\rho(\mathscr{E})$ *are defined in these assumptions, and* $\widehat{\omega} = \max\{1, \omega\}$.

*Proof.* The convergence of the iterative method is an obvious consequence of Lemma 2.23. Part (i) follows from Proposition 2.1 and Lemma 2.9, while part (ii) follows from Proposition 2.4 and Lemma 2.9. □

The condition

$$\sqrt{1 - 2\lambda_{\min}(\widetilde{\mathbf{A}}) + \lambda_{\max}^2(\widetilde{\mathbf{A}})} < 1$$

is not possible if $\lambda_{\min}(\widetilde{\mathbf{A}}) \to 0$ and $\lambda_{\max}(\widetilde{\mathbf{A}}) > 1$. A damping factor $\Theta$ is needed to guarantee convergence. The iteration takes the form

$$\mathbf{u}_{k+1} = \mathbf{u}_k + \Theta \mathbf{C}^{-1}\big(\mathbf{b} - \mathbf{A}\mathbf{u}_k\big), \quad k = 0,1,2,\ldots. \tag{2.74}$$

The following theorem is taken from [96].

**Theorem 2.6.** *The iterative method* (2.74) *converges if and only if*

$$0 < \Theta < \frac{2}{\lambda_{\max}(\mathbf{C}^{-1}\mathbf{A})}.$$

*The convergence rate is given by*

$$\|\mathbf{I} - \mathbf{C}^{-1}\mathbf{A}\|_{\mathbf{C}} = \max\{\big|1 - \Theta\lambda_{\min}(\mathbf{C}^{-1}\mathbf{A})\big|, \big|1 - \Theta\lambda_{\max}(\mathbf{C}^{-1}\mathbf{A})\big|\}.$$

*The optimal convergence rate is attained when*

$$\Theta_{\text{opt}} = \frac{2}{\lambda_{\max}(\mathbf{C}^{-1}\mathbf{A}) + \lambda_{\min}(\mathbf{C}^{-1}\mathbf{A})}$$

*and in that case*

$$\|\mathbf{I} - \mathbf{C}^{-1}\mathbf{A}\|_{\mathbf{C}} = \frac{\kappa(\mathbf{C}^{-1}\mathbf{A}) - 1}{\kappa(\mathbf{C}^{-1}\mathbf{A}) + 1}.$$

*Proof.* See [96, Theorems 4.4.2 and 4.4.3]. □

It is noted that this convergence rate is slower than that of the PCG method, which is

$$\frac{\sqrt{\kappa(\mathbf{C}^{-1}\mathbf{A})} - 1}{\sqrt{\kappa(\mathbf{C}^{-1}\mathbf{A})} + 1}.$$

### 2.9.2 Non-symmetric preconditioner

To show the convergence of the non-symmetric multiplicative Schwarz preconditioner, we need the following result. Recalling the definition of $P_j$ in Definition 2.1, we define

$$E_{-1} := I \quad \text{and} \quad E_j := (I - P_j)E_{j-1}, \quad j = 0,\ldots,J. \tag{2.75}$$

We note that $E_J = E_{\text{mu}}$ which is defined in Definition 2.1. We first prove the following two lemmas.

**Lemma 2.24.** *For all $v \in \mathscr{V}$ we have*

$$\sum_{j=0}^{J} a(P_j E_{j-1}v, E_{j-1}v) = a(v,v) - a(E_J v, E_J v).$$

*Proof.* This lemma is first proved in [39, Lemma 2.1]. We present it here for completeness. For any $j = 0,\ldots,J$ and $v \in \mathscr{V}$ we deduce from (2.75) and the fact that $P_j$

is a projection onto $\mathscr{V}_j$

$$\begin{aligned}a(E_j v, E_j v) &= a\big((I-P_j)E_{j-1}v, (I-P_j)E_{j-1}v\big)\\ &= a(E_{j-1}v, E_{j-1}v) - 2a(P_jE_{j-1}v, E_{j-1}v) + a(P_jE_{j-1}v, P_jE_{j-1}v)\\ &= a(E_{j-1}v, E_{j-1}v) - a(P_jE_{j-1}v, E_{j-1}v).\end{aligned}$$

Hence

$$a(P_jE_{j-1}v, E_{j-1}v) = a(E_{j-1}v, E_{j-1}v) - a(E_jv, E_jv).$$

By summing over $j$ from 0 to $J$ and noting that $E_{-1} = I$ we deduce the required result. □

**Lemma 2.25.** *Assume that Assumptions 2.4 and 2.5 are satisfied. We have*

$$a(P_{\mathrm{ad}}v, v) \le C_2^2 \sum_{j=0}^{J} a(P_jE_{j-1}v, E_{j-1}v) \quad \forall v \in \mathscr{V}$$

*where*

$$C_2^2 := 2\max\{C_1^2, \rho^2(\mathscr{E})\}$$

*with $C_1$ and $\rho(\mathscr{E})$ being the constants defined in these assumptions.*

*Proof.* It follows from the definition of $E_j$ that

$$I - E_k = \sum_{j=0}^{k} P_jE_{j-1}, \quad k = 0, \ldots, J.$$

Consequently, we have

$$\begin{aligned}a(P_kv, v) &= a(P_kv, E_{k-1}v) + a(P_kv, (I-E_{k-1})v)\\ &= a(P_kv, P_kE_{k-1}v) + \sum_{j=0}^{k-1} a(P_kv, P_jE_{j-1}v)\\ &= \sum_{j=0}^{k} a(P_kv, P_jE_{j-1}v)\end{aligned}$$

which implies

$$a(P_{\mathrm{ad}}v, v) = \sum_{k=1}^{J} a(P_kv, v) = \sum_{k=1}^{J} a(P_kv, P_0v) + \sum_{k=1}^{J}\sum_{j=1}^{k} a(P_kv, P_jE_{j-1}v).$$

The Cauchy–Schwarz and Strengthened Cauchy–Schwarz inequalities (Assumption 2.5) yield

$$a(P_{\mathrm{ad}}v, v) \le \Big(\sum_{k=1}^{J} a(P_kv, v)\Big)^{1/2}\Big(\sum_{k=1}^{J} a(P_kP_0v, P_0v)\Big)^{1/2}$$

$$
\begin{aligned}
&+\sum_{k=1}^{J}\sum_{j=1}^{J}\varepsilon_{kj}a(P_kv,P_kv)^{1/2}a(P_jE_{j-1}v,P_jE_{j-1}v)^{1/2}\\
&=\Big(\sum_{k=1}^{J}a(P_kv,v)\Big)^{1/2}\Big(\sum_{k=1}^{J}a(P_kP_0v,P_0v)\Big)^{1/2}\\
&+\sum_{k=1}^{J}\sum_{j=1}^{J}\varepsilon_{kj}a(P_kv,v)^{1/2}a(P_jE_{j-1}v,E_{j-1}v)^{1/2}
\end{aligned}
$$

where in the last step we used the projection property of $P_j$. Note that, for any symmetric matrix $\mathbf{B}=(B_{kj})\in\mathbb{R}^{J\times J}$ and any vectors $\mathbf{x}=(x_k)\in\mathbb{R}^J,\mathbf{y}=(y_k)\in\mathbb{R}^J$,

$$
\sum_{k=1}^{J}\sum_{j=1}^{J}B_{kj}x_ky_j=\langle\mathbf{Bx},\mathbf{y}\rangle_{\mathbb{R}^J}\le\|\mathbf{B}\|_2\|\mathbf{x}\|_2\|\mathbf{y}\|_2=\rho(\mathbf{B})\|\mathbf{x}\|_2\|\mathbf{y}\|_2,
$$

where $\|\cdot\|_2$ denotes the Euclidean norm. Hence,

$$
\begin{aligned}
a(P_{\text{ad}}v,v)\le&\Big(\sum_{k=1}^{J}a(P_kv,v)\Big)^{1/2}\Big(\sum_{k=1}^{J}a(P_kP_0v,P_0v)\Big)^{1/2}\\
&+\rho(\mathscr{E})\Big(\sum_{k=1}^{J}a(P_kv,v)\Big)^{1/2}\Big(\sum_{k=1}^{J}a(P_kE_{k-1}v,E_{k-1}v)\Big)^{1/2},
\end{aligned}
$$

which implies

$$
\begin{aligned}
a(P_{\text{ad}}v,v)&\le2\sum_{k=1}^{J}a(P_kP_0v,P_0v)+2\rho^2(\mathscr{E})\sum_{k=1}^{J}a(P_kE_{k-1}v,E_{k-1}v)\\
&\le2a(P_{\text{ad}}P_0v,P_0v)+2\rho^2(\mathscr{E})\sum_{k=1}^{J}a(P_kE_{k-1}v,E_{k-1}v).
\end{aligned}
$$

Assumption 2.4 and Proposition 2.2 give

$$
\begin{aligned}
a(P_{\text{ad}}v,v)&\le2C_1^2a(P_0v,P_0v)+2\rho^2(\mathscr{E})\sum_{k=1}^{J}a(P_kE_{k-1}v,E_{k-1}v)\\
&\le C_2^2\sum_{k=0}^{J}a(P_kE_{k-1}v,E_{k-1}v).
\end{aligned}
\qquad\square
$$

We are now ready to prove the convergence result for the linear iterative method with a non-symmetric multiplicative Schwarz preconditioner.

**Theorem 2.7.** *Assume that $\mathbf{C}=\mathbf{C}_{\text{mu}}$ is a preconditioner defined from the (non-symmetric) multiplicative Schwarz method. Under Assumptions 2.3, 2.4, and* (2.5)*, we have*

$$
\|\mathbf{I}-\mathbf{C}_{\text{mu}}^{-1}\mathbf{A}\|_{\mathbf{A}}\le1-\frac{1}{C_0^2C_2^2}
$$

*where $C_0$ is given by Assumption 2.3 and $C_2$ is given by Lemma 2.25.*

*Proof.* Lemma 2.7 gives $\mathbf{C}_{\rm mu}^{-1} = (\mathbf{I}-\mathbf{E}_{\rm mu})\mathbf{A}^{-1}$ so that

$$\mathbf{I}-\mathbf{C}_{\rm mu}^{-1}\mathbf{A} = \mathbf{E}_{\rm mu}.$$

It also follows from this lemma that $\mathbf{E}_{\rm mu}$ is the matrix representation of the operator $E_J$ defined by (2.75). Therefore, the theorem is proved if we prove

$$\|E_J\|_a \le 1-\frac{1}{C_0^2C_2^2}$$

where

$$\|E_J\|_a = \max_{v\in\mathscr{V}}\frac{a(E_Jv,E_Jv)}{a(v,v)}.$$

It follows from Proposition 2.1 Lemma 2.25 that

$$a(v,v) \le C_0^2 a(P_{\rm ad}v,v) \le C_0^2C_2^2\sum_{j=0}^{J} a(P_jE_{j-1}v,E_{j-1}v) \quad \forall v\in\mathscr{V}.$$

This together with Lemma 2.24 gives

$$a(v,v) \le C_0^2C_2^2a(v,v) - C_0^2C_2^2a(E_Jv,E_Jv) \quad \forall v\in\mathscr{V},$$

which implies

$$a(E_Jv,E_Jv) \le \left(1-\frac{1}{C_0^2C_2^2}\right)a(v,v) \quad \forall v\in\mathscr{V}.$$

This proves the theorem. □

# Part II
# Two-Dimensional Problems

# Chapter 3
# Two-Level Methods: the $h$-Version

In this chapter, we define different two-level methods for the $h$-version of the Galerkin approximation for both the hypersingular and weakly singular integral equations on a curve in $\mathbb{R}^2$. These are non-overlapping and overlapping additive and multiplicative Schwarz methods . As discussed in Chapter 2, it suffices to define a subspace decomposition for each preconditioner. The contents of this chapter are based on the papers [144, 220, 222, 227] by the authors. Numerical results which underline the theory will be presented in Chapter 9.

The novelty of this chapter is that in order to prove the coercivity of the decomposition, we use an inequality involving the global and local $\widetilde{H}^{1/2}$-norm which has recently been proved in [224]; see Theorem A.11. This is in place of Theorem A.10 as ubiquitously seen in the literature, including the above mentioned references. The use of Theorem A.10 yields a larger upper bound for the maximum eigenvalue of the Schwarz operators; see Lemma 3.2.

For convenience, we recall here the hypersingular integral equation (1.2) and the weakly singular integral equation (1.1) for $d = 2$, i.e., $\Gamma = (-1, 1)$, and for $\kappa = 0$:

$$\mathfrak{W}u(x) = g(x) \quad \text{and} \quad \mathfrak{V}u(x) = f(x), \quad x \in \Gamma,$$

where

$$\mathfrak{W}v(x) := -\frac{1}{\pi}\frac{\partial}{\partial n_x}\int_\Gamma \frac{\partial}{\partial n_y}\log|x-y|v(y)\,d\sigma_y = -\frac{1}{\pi}\,\text{f.p.}\int_\Gamma \frac{v(y)}{|x-y|^2}\,d\sigma_y, \quad x \in \Gamma,$$

$$\mathfrak{V}v(x) := -\frac{1}{\pi}\int_\Gamma \log|x-y|v(y)\,d\sigma_y, \quad x \in \Gamma.$$

Here, f.p. denotes a finite part integral in the sense of Hadamard; see e.g. [92, 94, 128, 147, 186].

The corresponding Galerkin equations are as follows. Find $\psi_{\mathscr{V}} \in \mathscr{V}$ and $\varphi_{\widetilde{\mathscr{V}}} \in \widetilde{\mathscr{V}}$ satisfying

$$a_{\mathfrak{W}}(\psi_{\mathscr{V}}, \phi) = \langle g, \phi \rangle, \quad \phi \in \mathscr{V}, \tag{3.1}$$



E. P. Stephan and T. Tran, *Schwarz Methods and Multilevel Preconditioners for Boundary Element Methods*, https://doi.org/10.1007/978-3-030-79283-1_3

and

$$a_{\mathfrak{V}}(\varphi_{\widetilde{\mathscr{V}}},\phi) = \langle f,\phi\rangle, \quad \phi \in \widetilde{\mathscr{V}}, \tag{3.2}$$

where $\mathscr{V} \subset \widetilde{H}^{1/2}(\Gamma)$ and $\widetilde{\mathscr{V}} \subset \widetilde{H}^{-1/2}(\Gamma)$ are two finite-dimensional spaces to be defined later. Here, the bilinear forms $a_{\mathfrak{W}}(\cdot,\cdot)$ and $a_{\mathfrak{V}}(\cdot,\cdot)$ are defined by

$$\begin{aligned} a_{\mathfrak{W}}(\varphi,\psi) &:= \langle \mathfrak{W}\varphi,\psi\rangle \quad \forall\varphi,\psi \in \widetilde{H}^{1/2}(\Gamma), \\ a_{\mathfrak{V}}(\varphi,\psi) &:= \langle \mathfrak{V}\varphi,\psi\rangle \quad \forall\varphi,\psi \in \widetilde{H}^{-1/2}(\Gamma). \end{aligned}$$

## 3.1 Additive Schwarz Methods

We first define the two-level mesh and the finite-dimensional spaces $\mathscr{V}$ and $\widetilde{\mathscr{V}}$.

**Definition 3.1 (Two-level mesh and boundary-element spaces).**
*The coarse mesh: We first divide $\Gamma$ into disjoint subdomains $\Gamma_i$ with length $H_i$, $i = 1,\dots,J$, so that $\overline{\Gamma} = \cup_{i=1}^{J}\overline{\Gamma}_i$, and denote by $H$ the maximum value of $H_i$.*

*The fine mesh: Each $\Gamma_i$ is further divided into disjoint subintervals $\Gamma_{ij}$, $j = 1,\dots,N_i$, so that $\overline{\Gamma}_i = \cup_{j=1}^{N_i}\overline{\Gamma}_{ij}$. The maximum length of the subintervals $\Gamma_{ij}$ in $\Gamma_i$ is denoted by $h_i$, and the maximum value of $h_i$ is denoted by $h$. We require that the fine mesh is locally quasi-uniform, i.e., it is quasi-uniform in each subdomain $\Gamma_i$.*

*Boundary element spaces: On the coarse mesh we define*

$$\begin{aligned} \mathscr{V}_0 &:= \{v \in C(\Gamma) : v(\pm 1) = 0,\ v|_{\Gamma_i} \text{ is affine, } i = 1,\dots,J\} \\ \widetilde{\mathscr{V}}_0 &:= \{v : \Gamma \to \mathbb{R} : v|_{\Gamma_i} \text{ is constant for } i = 1,\dots,J \text{ and } \langle v,1\rangle_{L^2(\Gamma)} = 0\}. \end{aligned}$$

*On the fine mesh we define*

$$\begin{aligned} \mathscr{V} &:= \{v \in C(\Gamma) : v(\pm 1) = 0,\ v|_{\Gamma_{ij}} \text{ is affine, } j = 1,\dots,N_i,\ i = 1,\dots,J\} \\ \widetilde{\mathscr{V}} &:= \{v : \Gamma \to \mathbb{R} : v|_{\Gamma_{ij}} \text{ is constant, } j = 1,\dots,N_i,\ i = 1,\dots,J\}. \end{aligned}$$

We now design non-overlapping additive Schwarz preconditioners for the two equations (3.2) and (3.1).

### 3.1.1 Non-overlapping methods

#### 3.1.1.1 The hypersingular integral equation

As explained in Chapter 2, in order to design an additive Schwarz preconditioner to (3.1) it suffices to define a subspace decomposition of $\mathscr{V}$.

**Definition 3.2 (Subspace decomposition 1).** *On the two-level mesh defined in Definition 3.1 we define*

$$\mathscr{V}_i := \left\{v \in \mathscr{V} \,:\, \operatorname{supp} v \subset \overline{\Gamma}_i\right\} = \mathscr{V} \cap \widetilde{H}^{1/2}(\Gamma_i), \quad i = 1,\dots,J, \tag{3.3}$$

*and define the decomposition* $\mathscr{V} = \mathscr{V}_0 + \mathscr{V}_1 + \cdots + \mathscr{V}_J$.

The additive Schwarz preconditioner is defined as in Section 2.4 using this subspace decomposition. Note that the bilinear form $a_{\mathfrak{W}}(\cdot,\cdot)$ satisfies (2.17). As given by Theorem 2.1, in order to estimate the condition number of $\kappa(P_{\text{add}})$ it suffices to show the stability and coercivity of the decomposition, namely Assumption 2.3 and Assumption 2.4 in Section 2.6. This is the purpose of the next two lemmas.

**Lemma 3.1 (Stability of decomposition).** *Any* $v \in \mathscr{V}$ *admits a decomposition*

$$v = \sum_{i=0}^{J} v_i, \quad v_i \in \mathscr{V}_i, \tag{3.4}$$

*satisfying*

$$\sum_{i=0}^{J} a_{\mathfrak{W}}(v_i, v_i) \lesssim \max_{1\le i\le J}\left(1 + \log\frac{H_i}{h_i}\right) a_{\mathfrak{W}}(v, v)$$

*where the constant is independent of* $v$, $H$ *and* $h$.

*Proof.* Noting Lemma B.3 in Appendix B, we will find a decomposition of the form (3.4) that satisfies

$$\sum_{i=0}^{J} \|v_i\|^2_{\widetilde{H}_{\mathrm{I}}^{1/2}(\Gamma)} \lesssim \max_{1\le i\le J}\left(1 + \log\frac{H_i}{h_i}\right) \|v\|^2_{\widetilde{H}_{\mathrm{I}}^{1/2}(\Gamma)}.$$

Due to Lemma A.13 it suffices to show that there exists a decomposition satisfying

$$\|v_0\|^2_{\widetilde{H}_{\mathrm{I}}^{1/2}(\Gamma)} + \sum_{i=1}^{J} \|v_i\|^2_{\widetilde{H}_{\mathrm{I}}^{1/2}(\Gamma_i)} \lesssim \max_{1\le i\le J}\left(1 + \log\frac{H_i}{h_i}\right) \|v\|^2_{\widetilde{H}_{\mathrm{I}}^{1/2}(\Gamma)}. \tag{3.5}$$

The decomposition (3.4) is defined by first defining $v_0 := \Pi_H v$, the linear interpolant of $v$ at the coarse mesh points, and define $v_i := (v - v_0)|_{\Gamma_i}$ for $i = 1,\dots,J$. Note that $v_i \in \widetilde{H}^{1/2}(\Gamma_i)$ (because $v_i$ vanishes at the endpoints of $\Gamma_i$) and that the decomposition $v = v_0 + \cdots + v_J$ is unique. It is given by Lemma C.38 (with $p = 1$) that

$$\|v_0\|^2_{\widetilde{H}_{\mathrm{I}}^{1/2}(\Gamma)} \lesssim \max_{1\le i\le J}\left(1 + \log\frac{H_i}{h_i}\right) \|v\|^2_{\widetilde{H}_{\mathrm{I}}^{1/2}(\Gamma)}. \tag{3.6}$$

On the other hand, Lemma A.8 and Lemma C.20 yield, for any $i = 1,\dots,J$,

$$\|v_i\|^2_{\widetilde{H}_{\mathrm{I}}^{1/2}(\Gamma_i)} \simeq \|v_i\|^2_{\widetilde{H}_{\mathrm{w}}^{1/2}(\Gamma_i)} \lesssim |v_i|^2_{H^{1/2}(\Gamma_i)} + \left(1 + \log\frac{H_i}{h_i}\right) \|v_i\|^2_{L^\infty(\Gamma_i)}. \tag{3.7}$$

We note that local quasi-uniformity is used here to obtain (3.7). For any $\alpha \in \mathbb{R}$, noting that $v_0 - \alpha = \Pi_H(v - \alpha)$ so that $\|v_0 - \alpha\|_{L^\infty(\Gamma_i)} \le \|v - \alpha\|_{L^\infty(\Gamma_i)}$, we have

$$\begin{aligned}\|v_i\|^2_{L^\infty(\Gamma_i)} &\le 2\left(\|v - \alpha\|^2_{L^\infty(\Gamma_i)} + \|v_0 - \alpha\|^2_{L^\infty(\Gamma_i)}\right)\\ &\le 4\|v - \alpha\|^2_{L^\infty(\Gamma_i)} = 4\|v - \alpha\|^2_{L^\infty(\Gamma_{ij})},\end{aligned} \tag{3.8}$$

for some $j \in \{1, \ldots, N_i\}$. Denoting $\Gamma_{ij} = (a,b)$, since $v$ is a linear function on $(a,b)$ we can assume, without loss of generality, that

$$\|v - \alpha\|^2_{L^\infty(\Gamma_{ij})} = |v(a) - \alpha|^2. \tag{3.9}$$

Recalling that

$$|v|^2_{H^{1/2}(a,b)} := \int_a^b \int_a^b \frac{|v(x) - v(y)|^2}{|x-y|^2}\, dx\, dy \tag{3.10}$$

and choosing $\alpha = v(b)$ we obtain from (3.8) and (3.9) that

$$\begin{aligned}\|v_i\|^2_{L^\infty(\Gamma_i)} &\le 4|v(a) - v(b)|^2 = 4\int_a^b \int_a^b \frac{|v(a) - v(b)|^2}{|a-b|^2}\, dx\, dy\\ &= 4\int_a^b \int_a^b \frac{|v(x) - v(y)|^2}{|x-y|^2}\, dx\, dy\\ &= 4|v|^2_{H^{1/2}(\Gamma_{ij})} \le 4|v|^2_{H^{1/2}(\Gamma_i)}.\end{aligned}$$

This together with (3.7) yields

$$\|v_i\|^2_{\widetilde{H}_{\mathrm{I}}^{1/2}(\Gamma_i)} \lesssim |v_i|^2_{H^{1/2}(\Gamma_i)} + \left(1 + \log\frac{H_i}{h_i}\right)|v|^2_{H^{1/2}(\Gamma_i)}. \tag{3.11}$$

Summing over $i$, using (3.10) and (3.6), we obtain

$$\begin{aligned}\sum_{i=1}^J \|v_i\|^2_{\widetilde{H}_{\mathrm{I}}^{1/2}(\Gamma_i)} &\lesssim \left|\sum_{i=1}^J v_i\right|^2_{H^{1/2}(\Gamma)} + \max_{1\le i\le J}\left(1 + \log\frac{H_i}{h_i}\right)|v|^2_{H^{1/2}(\Gamma)}\\ &= |v - v_0|^2_{H^{1/2}(\Gamma)} + \max_{1\le i\le J}\left(1 + \log\frac{H_i}{h_i}\right)|v|^2_{H^{1/2}(\Gamma)}\\ &\lesssim \|v_0\|^2_{\widetilde{H}_{\mathrm{S}}^{1/2}(\Gamma)} + \|v\|^2_{\widetilde{H}_{\mathrm{S}}^{1/2}(\Gamma)} + \max_{1\le i\le J}\left(1 + \log\frac{H_i}{h_i}\right)\|v\|^2_{\widetilde{H}_{\mathrm{S}}^{1/2}(\Gamma)}\\ &\lesssim \max_{1\le i\le J}\left(1 + \log\frac{H_i}{h_i}\right)\|v\|^2_{\widetilde{H}_{\mathrm{I}}^{1/2}(\Gamma)}.\end{aligned}$$

This inequality and (3.6) imply (3.5), completing the proof of the lemma. □

The next lemma gives the coercivity property of the decomposition.

**Lemma 3.2 (Coercivity of decomposition).** *Any $v = \sum_{i=0}^{J} v_i$ with $v_i \in \mathscr{V}_i$ satisfies*

$$a_{\mathfrak{W}}(v,v) \lesssim \sum_{i=0}^{J} a_{\mathfrak{W}}(v_i,v_i).$$

*The constant is independent of $v$, $H$ and $h$.*

*Proof.* Due to Lemma B.3 it suffices to prove

$$\|v\|^2_{\widetilde{H}_{\mathrm{I}}^{1/2}(\Gamma)} \lesssim \sum_{i=0}^{J} \|v_i\|^2_{\widetilde{H}_{\mathrm{I}}^{1/2}(\Gamma)}. \tag{3.12}$$

First we note that

$$\|v\|^2_{\widetilde{H}_{\mathrm{I}}^{1/2}(\Gamma)} \le 2\Big(\|v_0\|^2_{\widetilde{H}_{\mathrm{I}}^{1/2}(\Gamma)} + \Big\|\sum_{i=1}^{J} v_i\Big\|^2_{\widetilde{H}_{\mathrm{I}}^{1/2}(\Gamma)}\Big).$$

Theorem A.11 then gives

$$\|v\|^2_{\widetilde{H}_{\mathrm{I}}^{1/2}(\Gamma)} \lesssim 2\Big(\|v_0\|^2_{\widetilde{H}_{\mathrm{I}}^{1/2}(\Gamma)} + \sum_{i=1}^{J} \|v_i\|^2_{\widetilde{H}_{\mathrm{I}}^{1/2}(\Gamma)}\Big)$$

proving the lemma. □

The following theorem provides an upper bound for the condition number of the additive Schwarz operator.

**Theorem 3.1.** *The condition number of the additive Schwarz operator $P_{\mathrm{ad}}$ associated with the decomposition given in Definition 3.2 satisfies*

$$\kappa(P_{\mathrm{ad}}) \lesssim \max_{1\le i\le J}\left(1+\log\frac{H_i}{h_i}\right)$$

*where the constant is independent of $H$ and $h$.*

*Proof.* This is a result of Theorem 2.1, Lemma 3.1, and Lemma 3.2. □

#### 3.1.1.2 The weakly-singular integral equation

We now consider the weakly-singular integral equation (3.2). As for the hypersingular integral equation, it suffices to define a subspace decomposition and prove the coercivity and stability of that decomposition. We note that the feasibility of a subspace decomposition (in particular in the overlapping case) is implicitly given in the stability condition; see Assumption 2.3.

Briefly speaking, the strategy is as follows. For any subspace decomposition of the space of piecewise constant functions $\widetilde{\mathscr{V}}$ (defined in Definition 3.1), in order to prove the coercivity and stability of the decomposition, we will use corresponding results obtained for the corresponding decomposition of the space $\mathscr{V}$ of piecewise linear functions. This can be achieved by resorting to Lemma A.19 in the Appendix, which requires that the functions in $\widetilde{\mathscr{V}}$ have zero integral means. Therefore the decomposition of a function $v \in \widetilde{\mathscr{V}}$ has to consist of functions with zero integral mean and a global constant.

For a function $v$ defined on a set $D$, we adopt this notation

$$\mu_D(v) := \frac{1}{|D|} \int_D v \, \mathrm{d}x \tag{3.13}$$

where $|D|$ is the Lebesgue measure of $D$. We also define

$$\widetilde{\mathscr{V}}_{-1} = \operatorname{span}\{1\}. \tag{3.14}$$

We first prove the following generic result which will be used for both the non-overlapping and overlapping methods for the weakly-singular integral equation.

**Lemma 3.3.** *Let $\Gamma$ be partitioned into*

$$\overline{\Gamma} = \bigcup_{i=0}^{J} \overline{G}_i,$$

*where $\{G_i\}$ can be disjoint or overlap subdomains. We define*

$$\begin{gathered} \mathscr{S}_0 := \mathscr{V}_0, \quad \mathscr{S}_i := \mathscr{V} \cap \widetilde{H}^{1/2}(G_i), \quad i = 1, \dots, J, \\ \widetilde{\mathscr{S}}_{-1} := \widetilde{\mathscr{V}}_{-1}, \quad \widetilde{\mathscr{S}}_0 := \widetilde{\mathscr{V}}_0, \\ \widetilde{\mathscr{S}}_i := \{ v \in \widetilde{\mathscr{V}} \cap \widetilde{H}^{-1/2}(G_i) : \langle v, 1 \rangle_{L^2(G_i)} = 0 \}. \quad i = 1, \dots, J. \end{gathered}$$

*Assume that $\mathscr{V}$ is decomposed by*

$$\mathscr{V} = \mathscr{S}_0 + \mathscr{S}_1 + \cdots + \mathscr{S}_J \tag{3.15}$$

*and $\widetilde{\mathscr{V}}$ by*

$$\widetilde{\mathscr{V}} = \widetilde{\mathscr{S}}_{-1} + \widetilde{\mathscr{S}}_0 + \cdots + \widetilde{\mathscr{S}}_J. \tag{3.16}$$

(i) *Assume the following property of the decomposition* (3.15). *Any $w \in \mathscr{V}$ having the form*

$$w = w_0 + \cdots + w_J \quad \textit{with} \quad w_j \in \mathscr{S}_j \tag{3.17}$$

*satisfies*

$$\|w\|^2_{\widetilde{H}^{1/2}(\Gamma)} \le C_{\mathrm{coe}} \sum_{i=0}^{J} \|w_i\|^2_{\widetilde{H}^{1/2}(\Gamma)}. \tag{3.18}$$

*Then any $v \in \widetilde{\mathscr{V}}$ having the form*

$$v = v_{-1} + v_0 + \cdots + v_J \quad \textit{with} \quad v_j \in \widetilde{\mathscr{S}}_j, \ j = -1, 0, \dots, J, \tag{3.19}$$

*satisfies*

$$\|v\|^2_{\widetilde{H}_{\mathrm{I}}^{-1/2}(\Gamma)} \le C C_{\mathrm{coe}} \sum_{i=-1}^{J} \|v_i\|^2_{\widetilde{H}_{\mathrm{I}}^{-1/2}(\Gamma)}. \tag{3.20}$$

(ii) *Assume the following property of the decomposition* (3.15). *Any* $w \in \mathscr{V}$ *admits a decomposition* (3.17) *which satisfies*

$$\sum_{i=0}^{J} \|w_i\|^2_{\widetilde{H}^{1/2}(\Gamma)} \le C_{\mathrm{sta}} \|w\|^2_{\widetilde{H}^{1/2}(\Gamma)}. \tag{3.21}$$

*Then any* $v \in \widetilde{\mathscr{V}}$ *admits a decomposition* (3.19) *which satisfies*

$$\sum_{i=-1}^{J} \|v_i\|^2_{\widetilde{H}_{\mathrm{I}}^{-1/2}(\Gamma)} \le C C_{\mathrm{sta}} \|v\|^2_{\widetilde{H}_{\mathrm{I}}^{-1/2}(\Gamma)}. \tag{3.22}$$

*This means the decomposition* (3.16) *has the same properties as* (3.15).

*Proof.* Assume that the assumption in Part (i) holds, i.e., if $w \in \mathscr{V}$ satisfies (3.17), then (3.18) holds. Consider $v \in \widetilde{\mathscr{S}}$ having the form (3.19) and define

$$\widetilde{v} := v_0 + v_1 + \cdots + v_J.$$

Since $\widetilde{v}$ and $v_i$, $i = 0, \dots, J$, have zero integral means, we can invoke Lemma A.19 to define

$$w := I_{1/2}\widetilde{v} \in \mathscr{V}, \quad \text{and} \quad w_i := I_{1/2} v_i \in \mathscr{S}_i, \quad i = 0, \dots, J,$$

which satisfy

$$w = w_0 + \cdots + w_J,$$
$$\widetilde{v} = D_{1/2} w, \quad v_i = D_{1/2} w_i, \; i = 0, \dots, J,$$

and

$$\|w\|_{\widetilde{H}_{\mathrm{I}}^{1/2}(\Gamma)} \simeq \|\widetilde{v}\|_{\widetilde{H}_{\mathrm{I}}^{-1/2}(\Gamma)}, \quad \|w_i\|_{\widetilde{H}_{\mathrm{I}}^{1/2}(\Gamma)} \simeq \|v_i\|_{\widetilde{H}_{\mathrm{I}}^{-1/2}(\Gamma)}, \; i = 0, \dots, J. \tag{3.23}$$

It follows from (3.18) and (3.23) that

$$\|\widetilde{v}\|^2_{\widetilde{H}_{\mathrm{I}}^{-1/2}(\Gamma)} \le C_{\mathrm{coe}} C_{ID} \sum_{i=0}^{J} \|v_i\|^2_{\widetilde{H}_{\mathrm{I}}^{-1/2}(\Gamma)}$$

where $C_{ID}$ is the constant defined from the constants in (3.23). Since $v = v_{-1} + \widetilde{v}$, the triangle inequality, Cauchy–Schwarz inequality, and the above inequality imply

$$\|v\|^2_{\widetilde{H}_{\mathrm{I}}^{-1/2}(\Gamma)} \le 2\|v_{-1}\|^2_{\widetilde{H}_{\mathrm{I}}^{-1/2}(\Gamma)} + 2\|\widetilde{v}\|^2_{\widetilde{H}_{\mathrm{I}}^{-1/2}(\Gamma)}$$

$$\le \max\{2, 2C_{\rm coe}C_{ID}\} \sum_{i=-1}^{J} \|v_i\|^2_{\widetilde{H}_{\rm I}^{-1/2}(\Gamma)}.$$

This proves (3.20).

Now assume that the assumption of Part (ii) holds, i.e., for any $w \in \mathscr{V}$, there exists a decomposition of $w$ of the form (3.17) such that (3.21) is true. Consider $v \in \widetilde{\mathscr{V}}$. We define

$$v_{-1} := \mu_\Gamma(v) \in \mathscr{S}_{-1} \quad \text{and} \quad \widetilde{v} := v - v_{-1}.$$

Then $\widetilde{v}$ belongs to $\widetilde{\mathscr{V}}$ with integral mean zero. Let $w := I_{1/2}\widetilde{v}$. Then $w$ belongs to $\mathscr{V}$ which by assumption admits a decomposition (3.17) satisfying

$$\sum_{i=0}^{J} \|w_i\|^2_{\widetilde{H}_{\rm I}^{1/2}(\Gamma)} \le C_{\rm sta}\|w\|^2_{\widetilde{H}_{\rm I}^{1/2}(\Gamma)}.$$

Defining

$$v_i := D_{1/2}w_i \in \mathscr{S}_i, \quad i = 0, \ldots, J,$$

we have $\widetilde{v} = v_0 + \cdots + v_J$, which implies that $v$ admits a decomposition (3.19). Using the same argument as in the proof of Part (i), we obtain

$$\sum_{i=0}^{J} \|v_i\|^2_{\widetilde{H}_{\rm I}^{-1/2}(\Gamma)} \le C_{\rm sta}C_{ID}\|\widetilde{v}\|^2_{\widetilde{H}_{\rm I}^{-1/2}(\Gamma)}.$$

It remains to prove

$$\|v_{-1}\|^2_{\widetilde{H}_{\rm I}^{-1/2}(\Gamma)} \lesssim \|v\|^2_{\widetilde{H}_{\rm I}^{-1/2}(\Gamma)}.$$

Since $v_{-1} = \mu_\Gamma(v) = \langle v, 1\rangle_\Gamma / |\Gamma|$ we have, noting that $\|1\|_{\widetilde{H}_{\rm I}^{-1/2}(\Gamma)}\|1\|_{H_{\rm I}^{1/2}(\Gamma)}$ is constant,

$$\begin{aligned}\|v_{-1}\|_{\widetilde{H}_{\rm I}^{-1/2}(\Gamma)} &= \frac{|\langle v, 1\rangle|}{|\Gamma|}\|1\|_{\widetilde{H}_{\rm I}^{-1/2}(\Gamma)} \le \frac{1}{|\Gamma|}\|1\|_{\widetilde{H}_{\rm I}^{-1/2}(\Gamma)}\|1\|_{H_{\rm I}^{1/2}(\Gamma)}\|v\|_{\widetilde{H}_{\rm I}^{-1/2}(\Gamma)} \\ &\lesssim \|v\|_{\widetilde{H}_{\rm I}^{-1/2}(\Gamma)}.\end{aligned}$$

□

We will now define a subspace decomposition for a non-overlapping method and apply Lemma 3.3 to this decomposition. Let $v$ be a function in $\widetilde{\mathscr{V}}$ and let $v_i := v|_{\Gamma_i} - \mu_{\Gamma_i}(v|_{\Gamma_i})$. Then $v_i$ is a piecewise constant function on the fine mesh, has support in $\Gamma_i$, and satisfies $\langle v_i, 1\rangle_{L^2(\Gamma_i)} = 0$. Moreover,

$$v = \sum_{i=1}^{J} v|_{\Gamma_i} = v_H + \sum_{i=1}^{J} v_i$$

where $v_H$ is the piecewise constant function on the coarse mesh which satisfies $v_H\big|_{\Gamma_i} = \mu_{\Gamma_i}(v|_{\Gamma_i})$. Defining $v_{-1} := \mu_\Gamma(v_H)$ and $v_0 := v_H - v_{-1}$ we then have

$$v = \sum_{i=-1}^{J} v_i. \tag{3.24}$$

We note that $v_{-1} \in \widetilde{\mathcal{V}}_{-1}$ and $v_0 \in \widetilde{\mathcal{V}}_0$ defined in (3.14) and Definition 3.1. The above decomposition of $v$ assures us that we can define the following subspace decomposition for $\widetilde{\mathcal{V}}$.

**Definition 3.3 (Subspace decomposition 2).** *Define on the two-level mesh given in Definition 3.1 the subspaces*

$$\widetilde{\mathcal{V}}_i := \left\{ v \in \widetilde{\mathcal{V}} : \operatorname{supp} v \subset \overline{\Gamma}_i, \quad \langle v, 1 \rangle_{L^2(\Gamma_i)} = 0 \right\}, \quad i = 1, \dots, J,$$

*and the subspace decomposition*

$$\widetilde{\mathcal{V}} = \widetilde{\mathcal{V}}_{-1} \oplus \widetilde{\mathcal{V}}_0 \oplus \cdots \oplus \widetilde{\mathcal{V}}_J.$$

Before stating the coercivity and stability of the above decomposition, we comment on the basis for $\widetilde{\mathcal{V}}$. We note that $\dim \widetilde{\mathcal{V}} = N_1 + \cdots + N_J$; see Definition 3.1. An obvious basis for $\widetilde{\mathcal{V}}$ is the set of all brick functions which equal one in one subinterval $\Gamma_{ij}$ and zero elsewhere. However, for both the analysis and implementation we use Haar basis functions to define bases for the above subspaces and thus for $\widetilde{\mathcal{V}}$.

Firstly, the basis functions for the global subspace $\widetilde{\mathcal{V}}_0$ consists of

$$\chi_0^h(x) \equiv 1 \quad \text{and} \quad \chi_i^h(x) := \begin{cases} \dfrac{1}{H_i}, & x \in \Gamma_i, \\ -\dfrac{1}{H_{i+1}}, & x \in \Gamma_{i+1}, \\ 0, & \text{otherwise}, \end{cases} \qquad i = 1, \dots, J-1. \tag{3.25}$$

The functions $\chi_i^h$, $i = 1, \dots, J-1$, are derivatives of the hat functions forming a basis for the space $\mathcal{V}_0$ defined in Definition 3.1.

Next, each local subspace $\widetilde{\mathcal{V}}_i$, $i = 1, \dots, J$, has the following basis functions

$$\chi_{ij}^h(x) = \begin{cases} \dfrac{1}{h_{i,j}}, & x \in \Gamma_{ij}, \\ -\dfrac{1}{h_{i,j+1}}, & x \in \Gamma_{i,j+1}, \\ 0, & \text{otherwise}, \end{cases} \qquad j = 1, \dots, N_i - 1. \tag{3.26}$$

These functions are derivatives of the hat functions forming a basis for the subspace $\mathcal{V}_i$ defined in Definition 3.2. From the decomposition of $\widetilde{\mathcal{V}}$ defined in Definition 3.3 we have $\widetilde{\mathcal{V}} = \operatorname{span}\left\{ \chi_0^h, \chi_i^h, \chi_{ij}^h, \ i = 1, \dots, J, \ j = 1, \dots, N_i - 1 \right\}$.

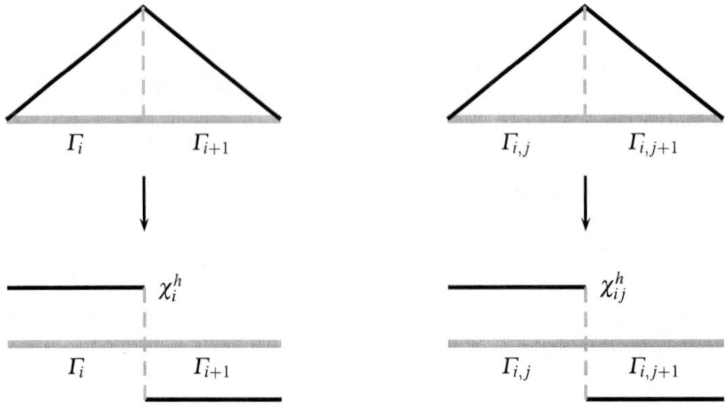


**Fig. 3.1** Haar basis functions defined in (3.25) and (3.26)

Invoking Lemma 3.3 with $\mathscr{S}_i = \mathscr{V}_i$, $i = 0,\dots,J$, and $\widetilde{\mathscr{S}_j} = \widetilde{\mathscr{V}_j}$, $j = -1,\dots,J$, we obtain the following lemmas.

**Lemma 3.4 (Stability of decomposition).** *Any $v \in \widetilde{\mathscr{V}}$ admits a decomposition*

$$v = \sum_{i=-1}^{J} v_i, \quad v_i \in \widetilde{\mathscr{V}_i},\ i = -1,\dots,J,$$

*satisfying (with a constant independent of $v$, $H$, and $h$)*

$$\sum_{i=-1}^{J} a_{\mathfrak{V}}(v_i, v_i) \lesssim \max_{1\le i\le J}\left(1 + \log\frac{H_i}{h_i}\right) a_{\mathfrak{V}}(v, v).$$

*Proof.* The lemma is a consequence of Lemma 3.1 and Lemma 3.3 with

$$C_{\mathrm{sta}} \simeq \max_{1\le i\le J}\left(1 + \log\frac{H_i}{h_i}\right). \qquad \square$$

**Lemma 3.5 (Coercivity of decomposition).** *Any $v \in \widetilde{\mathscr{V}}$ having the form $v = \sum_{i=-1}^{J} v_i$ with $v_i \in \widetilde{\mathscr{V}_i}$ satisfies*

$$a_{\mathfrak{V}}(v, v) \lesssim \sum_{i=-1}^{J} a_{\mathfrak{V}}(v_i, v_i).$$

*The constant is independent of $v$, $H$, and $h$.*

*Proof.* The lemma is a consequence of Lemma 3.2 and Lemma 3.3 with $C_{\mathrm{coe}} \simeq 1$. $\square$

Combining Lemmas 3.4 and 3.5, and invoking Theorem 2.1 we then have

**Theorem 3.2.** *The condition number of the additive Schwarz operator $P_{\rm ad}$ satisfies*

$$\kappa(P_{\rm ad}) \lesssim \max_{1\le i\le J}\left(1+\log\frac{H_i}{h_i}\right).$$

*The constant is independent of $H$ and $h$.*

In the next subsection, we design overlapping additive Schwarz preconditioners for equations (3.1) and (3.2).

## *3.1.2 Overlapping methods*

### 3.1.2.1 The hypersingular integral equation

We consider again the two-level mesh defined in Definition 3.1. It is assumed that the meshes are locally quasi-uniform. To simplify our arguments, we assume that the lengths of the subdomains and subintervals are in the order of $H$ and $h$, respectively.

**Definition 3.4 (Subspace decomposition 3).** *We now extend each subdomain $\Gamma_i$ on both sides (except for the left side of $\Gamma_1$ and the right side of $\Gamma_J$ which are the endpoints of $\Gamma$) so that each element lies entirely in one extended subdomain and that the length of the overlap between two extended subdomains $\Gamma_i'$ and $\Gamma_{i+1}'$ is on the order of $\delta$ for some $\delta\in(0,H]$. The smallest possible overlap is of length $2h$ whereas the largest is of length $2H$, which implies*

$$|\Gamma_i'|\sim H. \tag{3.27}$$

*We note that the endpoints of $\Gamma_i'$ coincide with fine-mesh points; see Figure 3.2.*

*We define $\mathscr{V}_0$ to be the space of continuous piecewise-linear functions on the coarse mesh, vanishing at the endpoints of $\Gamma$, and $\mathscr{V}_i=\mathscr{V}\cap\widetilde{H}^{1/2}(\Gamma_i')$, $i=1,\dots,J$. The finite-dimensional space $\mathscr{V}$ defined in Definition 3.1 is now decomposed as $\mathscr{V}=\mathscr{V}_0+\mathscr{V}_1+\cdots+\mathscr{V}_J$.*

As in the case of the non-overlapping preconditioner, it is necessary to prove the coercivity and stability of the decomposition defined in Definition 3.4.

**Lemma 3.6 (Coercivity of decomposition).** *For any $v\in\mathscr{V}$, if $v=\sum_{i=0}^J v_i$ for some $v_i\in\mathscr{V}_i$, then*

$$a_{\mathfrak{W}}(v,v)\lesssim\sum_{i=0}^J a_{\mathfrak{W}}(v_i,v_i).$$

*Proof.* First we note that the Cauchy–Schwarz inequality implies

$$\|v\|^2_{\widetilde{H}_{\rm I}^{1/2}(\Gamma)}\lesssim\|v_0\|^2_{\widetilde{H}_{\rm I}^{1/2}(\Gamma)}+\Big\|\sum_{i=1}^J v_i\Big\|^2_{\widetilde{H}_{\rm I}^{1/2}(\Gamma)}.$$

By construction there are at most 3 subdomains $\overline{\Gamma}'_i$ to which any $x \in \Gamma$ can belong. (The utmost case happens when $\delta = H$.) In other words, the index set $\mathscr{I} = \{1,\dots,J\}$ can be partitioned into

$$\mathscr{I} = \mathscr{I}_1 \cup \mathscr{I}_2 \cup \mathscr{I}_3$$

where each $\mathscr{I}_k$, $k = 1,2,3$, has the property

$$\Gamma'_i \cap \Gamma'_j = \varnothing \quad \text{if } i,j \in \mathscr{I}_k,\ i \neq j.$$

Hence using successively the Cauchy–Schwarz inequality and Theorem A.11 we obtain

$$\begin{aligned}\|\sum_{i=1}^{J} v_i\|^2_{\widetilde{H}_{\rm I}^{1/2}(\Gamma)} &= \|\sum_{k=1}^{3}\sum_{i\in\mathscr{I}_k} v_i\|^2_{\widetilde{H}_{\rm I}^{1/2}(\Gamma)} \le 3\sum_{k=1}^{3}\|\sum_{i\in\mathscr{I}_k} v_i\|^2_{\widetilde{H}_{\rm I}^{1/2}(\Gamma)}\\ &\le 3\sum_{k=1}^{3}\sum_{i\in\mathscr{I}_k}\|v_i\|^2_{\widetilde{H}_{\rm I}^{1/2}(\Gamma)} = 3\sum_{i=1}^{J}\|v_i\|^2_{\widetilde{H}_{\rm I}^{1/2}(\Gamma)}.\end{aligned}$$

The above argument is called the colouring argument; see *e.g.* [34, 105]. Altogether, we have

$$\|v\|^2_{\widetilde{H}_{\rm I}^{1/2}(\Gamma)} \lesssim \sum_{i=0}^{J}\|v_i\|^2_{\widetilde{H}_{\rm I}^{1/2}(\Gamma)}.$$

The desired result now follows from Lemma B.3. □

The next lemma is essential to prove stability of the decomposition.

**Lemma 3.7.** *Let $\{\theta_l : l = 1,\dots,L\}$ be a partition of unity on $\Gamma := (a,b)$ such that*

$$0 \le \theta_l \le 1, \quad \sum_{l=1}^{L}\theta_l = 1, \quad \left|\frac{d\theta_l}{dx}\right| \le \frac{C}{\delta}, \tag{3.28}$$

*for some constants $C$ and $\delta$, and let $I_l := [a_l,b_l] = \operatorname{supp}\theta_l$.*

1. *For any $w \in \widetilde{H}^{1/2}(\Gamma)$ there holds*

$$\begin{aligned}\sum_{l=1}^{L}\|\theta_l w\|^2_{\widetilde{H}_{\rm I}^{1/2}(I_l)} &\le 4C^2\sum_{l=1}^{L}\frac{b_l-a_l}{\delta^2}\|w\|^2_{L^2(I_l)} + 2\sum_{l=1}^{L}|w|^2_{H^{1/2}(I_l)}\\ &\quad + \int_\Gamma \frac{|w(x)|^2}{b-x}\,dx + \int_\Gamma\frac{|w(x)|^2}{x-a}\,dx.\end{aligned} \tag{3.29}$$

2. *There exist a positive constant $\tilde{C}$ and a partition of unity $\{\theta_l : l = 1,\dots,L\}$ composed of piecewise-linear functions on a mesh of size $\delta \le \min_l(b_l-a_l)/2$ such that for any $w \in \widetilde{H}^{1/2}(\Gamma)$ there holds*

$$\sum_{l=1}^{L} \|\theta_l w\|^2_{\widetilde{H}_1^{1/2}(I_l)} \le \tilde{C} \sum_{l=1}^{L} \left(1+\log\frac{b_l-a_l}{\delta}\right)^2 \left(\frac{1}{b_l-a_l}\|w\|^2_{L^2(I_l)} + |w|^2_{H^{1/2}(I_l)}\right)$$
$$+ \int_\Gamma \frac{|w(x)|^2}{b-x}\,dx + \int_\Gamma \frac{|w(x)|^2}{x-a}\,dx. \tag{3.30}$$

*Proof.* In view of the definition of the $\widetilde{H}^{1/2}$-norm, we will estimate

$$T_1 := \int_{I_l}\int_{I_l} \frac{|(\theta_l w)(x) - (\theta_l w)(y)|^2}{|x-y|^2}\,dx\,dy,$$

$$T_2 := \int_{I_l} \frac{|(\theta_l w)(x)|^2}{b_l - x}\,dx \quad \text{and} \quad T_3 := \int_{I_l} \frac{|(\theta_l w)(x)|^2}{x-a_l}\,dx.$$

By using (3.28) we have

$$T_1 \le 2\int_{I_l}\int_{I_l} \frac{|\theta_l(x)-\theta_l(y)|^2}{|x-y|^2}|w(x)|^2\,dx\,dy + 2\int_{I_l}\int_{I_l} \frac{|w(x)-w(y)|^2}{|x-y|^2}|\theta_l(y)|^2\,dx\,dy$$
$$\le 2C^2\frac{b_l-a_l}{\delta^2}\|w\|^2_{L^2(I_l)} + 2|w|^2_{H^{1/2}(I_l)}.$$

For the term $T_2$ we note that $\theta(b_l)=0$ when $l=1,\dots,L-1$ and use (3.28) again to obtain

$$T_2 = \int_{I_l} \frac{|\theta_l(b_l)-\theta_l(x)|^2}{b_l-x}|w(x)|^2\,dx \le C^2\frac{b_l-a_l}{\delta^2}\|w\|^2_{L^2(I_l)}.$$

When $l=L$ (*i.e.* , $b_l=b$) there may not hold $\theta_L(b)=0$, but we can estimate $T_2$ as

$$T_2 = \int_{I_L} \frac{|(\theta_L w)(x)|^2}{b-x}\,dx \le \int_\Gamma \frac{|w(x)|^2}{b-x}\,dx.$$

Similarly, for $T_3$ we have

$$T_3 \le C^2\frac{b_l-a_l}{\delta^2}\|w\|^2_{L^2(I_l)} \quad \text{when } 2\le l\le L$$

and

$$T_3 \le \int_\Gamma \frac{|w(x)|^2}{x-a}\,dx \quad \text{when } l=1.$$

Summing over $l=1,\dots,L$ we obtain (3.29).

The proof of (3.30) mainly follows [72]. For simplicity of notations we assume, without loss of generality, $I_l=[0,\beta]$ and write $I$ and $\theta$ instead of $I_l$ and $\theta_l$. We will prove that each of $T_1$, $T_2$ (when $1\le l\le L-1$), and $T_3$ (when $2\le l\le L$) is bounded by

$$\left(1+\log\frac{\beta}{\delta}\right)^2\left(\frac{1}{\beta}\|w\|^2_{L^2(I)}+|w|^2_{H^{1/2}(I)}\right). \tag{3.31}$$

We define $\theta$ to be linear in $[0,\delta]$ and $[\beta-\delta,\beta]$, and equal to 1 in $[\delta,\beta-\delta]$, *i.e.* ,

$$\theta(x) := \begin{cases} \frac{x}{\delta}, & 0\le x\le\delta,\\ 1, & \delta<x<\beta-\delta,\\ \frac{\beta-x}{\delta}, & \beta-\delta\le x\le\beta. \end{cases}$$

Splitting the integral over $I$ into three integrals over $I_1 := [0,\delta]$, $I_2 := [\delta,\beta-\delta]$, and $I_3 := [\beta-\delta,\beta]$, and denoting, for any $v$,

$$A_{ij}(v) := \iint\limits_{I_i\times I_j} \frac{|v(x)-v(y)|^2}{|x-y|^2}\,dx\,dy,$$

we observe that, by symmetry, in order to estimate $T_1$ it suffices to consider $A_{11}(\theta w)$, $A_{12}(\theta w)$, $A_{13}(\theta w)$, $A_{22}(\theta w)$, $A_{23}(\theta w)$, and $A_{33}(\theta w)$. There is no need to consider $A_{22}(\theta w)$ because $A_{22}(\theta w) = A_{22}(w)$. The symmetric shape of $\theta$ implies the similarity of $A_{11}(\theta w)$ and $A_{33}(\theta w)$, and of $A_{12}(\theta w)$ and $A_{23}(\theta w)$. There remains three terms to be considered: $A_{11}(\theta w)$, $A_{12}(\theta w)$, and $A_{13}(\theta w)$.

From the definition of $\theta$ it is easy to see that

$$A_{11}(\theta w) \le \frac{2}{\delta}\|w\|^2_{L^2(0,\delta)} + 2A_{11}(w).$$

It was proved in [72, Lemma 3.4] that

$$\frac{1}{\delta}\|w\|^2_{L^2(0,\delta)} \le \tilde{c}\left(1+\log\frac{\beta}{\delta}\right)\left(\frac{1}{\beta}\|w\|^2_{L^2(I)}+|w|^2_{H^{1/2}(I)}\right), \tag{3.32}$$

where $\tilde{c}$ is independent of $w$, $\beta$, and $\delta$. Thus $A_{11}(\theta w)$ is bounded by (3.31). For the term $A_{12}(\theta w)$ we have

$$\begin{aligned} A_{12}(\theta w) &= \frac{1}{\delta^2}\int_0^{\delta}\int_{\delta}^{\beta-\delta} \frac{|xw(x)-\delta w(y)|^2}{|x-y|^2}\,dy\,dx \\ &\le \frac{2}{\delta^2}\int_0^{\delta}\left(\int_{\delta}^{\beta-\delta}\frac{dy}{|x-y|^2}\right)|x-\delta|^2|w(x)|^2\,dx + 2A_{12}(w) \\ &\le \frac{2}{\delta}\|w\|^2_{L^2(0,\delta)} + 2A_{12}(w). \end{aligned}$$

Using again (3.32) we obtain the estimate for $A_{12}(\theta w)$. Finally, for $A_{13}(\theta w)$, since the assumption $\delta\le\beta/2$ implies that $\frac{\beta}{2}-x\ge 0$ for $x\in I_1$ and $y-\frac{\beta}{2}\ge 0$ for $y\in I_3$, we find

$$
\begin{aligned}
A_{13}(\theta w) &= \frac{1}{\delta^2}\int_0^{\delta}\int_{\beta-\delta}^{\beta} \frac{|xw(x)-(\beta-y)w(y)|^2}{|x-y|^2}\,\mathrm{d}y\,\mathrm{d}x \\
&\le \frac{2}{\delta^2}\int_0^{\delta}\int_{\beta-\delta}^{\beta} \frac{|x+y-\beta|^2}{|x-y|^2}|w(x)|^2\,\mathrm{d}y\,\mathrm{d}x \\
&\quad + \frac{2}{\delta^2}\int_0^{\delta}\int_{\beta-\delta}^{\beta} |\beta-y|^2\frac{|w(x)-w(y)|^2}{|x-y|^2}\,\mathrm{d}y\,\mathrm{d}x \\
&\le \frac{2}{\delta^2}\int_0^{\delta}\int_{\beta-\delta}^{\beta} \frac{|(y-\frac{\beta}{2})-(\frac{\beta}{2}-x)|^2}{|(y-\frac{\beta}{2})+(\frac{\beta}{2}-x)|^2}|w(x)|^2\,\mathrm{d}y\,\mathrm{d}x + 2A_{13}(w) \\
&\le \frac{2}{\delta}\|w\|^2_{L^2(0,\delta)} + 2A_{13}(w).
\end{aligned}
$$

Inequality (3.32) yields the estimate for $A_{13}(\theta w)$, implying $T_1$ is bounded by (3.31). The proof for $T_2$ follows by using again the definition of $\theta$:

$$
\begin{aligned}
T_2 &\le \int_0^{\beta-\delta} \frac{|w(x)|^2}{\beta-x}\,\mathrm{d}x + \frac{1}{\delta^2}\int_{\beta-\delta}^{\beta}(\beta-x)|w(x)|^2\,\mathrm{d}x \\
&\le \int_0^{\beta-\delta} \frac{|w(x)|^2}{\beta-x}\,\mathrm{d}x + \frac{1}{\delta}\|w\|^2_{L^2(\beta-\delta,\beta)}.
\end{aligned}
$$

It was proved in [72, Lemma 3.5] that

$$
\int_0^{\beta-\delta} \frac{|w(x)|^2}{\beta-x}\,\mathrm{d}x \le \bar{c}\left(1+\log\frac{\beta}{\delta}\right)^2\left(\frac{1}{\beta}\|w\|^2_{L^2(I)} + |w|_{H^{1/2}(I)}\right).
$$

This inequality and an analogue to (3.32) yield the bound (3.31) for $T_2$. Similar arguments hold for $T_3$ which completes the proof. □

We need the following approximation property for the proof of stability. Since the operator $-d^2/dx^2$ with domain of definition $\widetilde{H}^1(\Gamma) = H^1_0(\Gamma)$ is positive definite and self-adjoint, we can define

$$
\Lambda := \sqrt{-\frac{d^2}{dx^2}} : \widetilde{H}^{1/2}(\Gamma) \to H^{-1/2}(\Gamma), \tag{3.33}
$$

which in turn is self-adjoint as an operator. Moreover, see [36],

$$
\langle \Lambda\xi, \xi\rangle \simeq \|\xi\|^2_{\widetilde{H}^{1/2}_{\mathrm{I}}(\Gamma)} \qquad \forall \xi \in \widetilde{H}^{1/2}(\Gamma), \tag{3.34}
$$

where $\langle \cdot, \cdot \rangle$ denotes the duality product between $\widetilde{H}^{1/2}(\Gamma)$ and $H^{-1/2}(\Gamma)$ which extends the $L^2$-inner product in $\Gamma$. Let $P_H : \widetilde{H}^{1/2}(\Gamma) \longrightarrow \mathscr{V}_0$ be the projection defined by the inner product $\langle \Lambda \cdot, \cdot \rangle$, *i.e.*

$$\langle \Lambda P_H v, w \rangle = \langle \Lambda v, w \rangle \quad \forall v \in \widetilde{H}^{1/2}(\Gamma),\, w \in \mathscr{V}_0. \tag{3.35}$$

Using standard arguments one can prove that for any $v \in \widetilde{H}^{1/2}(\Gamma)$

$$\|P_H v\|_{\widetilde{H}_I^{1/2}(\Gamma)} \lesssim \|v\|_{\widetilde{H}_I^{1/2}(\Gamma)} \tag{3.36}$$

and

$$\|P_H v - v\|_{L^2(\Gamma)} \lesssim H^{1/2} \|v\|_{\widetilde{H}_I^{1/2}(\Gamma)}. \tag{3.37}$$

**Lemma 3.8 (Stability of decomposition).** *Any $v \in \mathscr{V}$ admits a decomposition*

$$v = \sum_{i=0}^{J} v_i, \quad v_i \in \mathscr{V}_i,$$

*satisfying*

$$\sum_{i=0}^{J} a_{\mathfrak{W}}(v_i, v_i) \lesssim (1 + \log^2 \frac{H}{\delta}) a_{\mathfrak{W}}(v, v),$$

*where the constant involved is independent of not only $v$, $H$ and $h$ but also $\delta$.*

*Proof.* Consider a partition of unity on $\Gamma$ composed of piecewise-linear functions $\theta_i$ which are defined as in the proof of Lemma 3.7 so that supp $\theta_i = \overline{\Gamma}_i'$; see Figure 3.2. The step-size of the mesh on which $\theta_i$ is defined is $2\delta$, the size of the overlap.

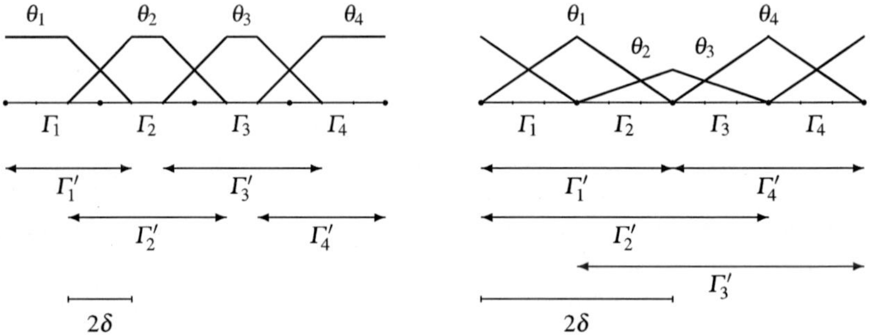


**Fig. 3.2** Partitions of unity with smallest (left) and largest (right) possible overlaps. Note $\theta_2 = \theta_3$ on $\overline{\Gamma}_2' \cap \overline{\Gamma}_3' = \overline{\Gamma}_2 \cup \overline{\Gamma}_3$ on the right picture.

For any $v \in \mathscr{V}$ we define a decomposition of $v$ as a sum of functions in $\mathscr{V}_i$ as follows. Let $v_0 = P_H v$ and $v_i = \Pi_h(\theta_i w)$, $i = 1, \ldots, J$, where $w = v - v_0$ and $\Pi_h$ denotes the linear interpolation operator which interpolates a continuous function at the fine-mesh points. Then

$$\sum_{i=0}^{J} v_i = v_0 + \sum_{i=1}^{J} \Pi_h(\theta_i w) = v_0 + \Pi_h\left(\sum_{i=1}^{J} \theta_i w\right) = v_0 + \Pi_h w = v.$$

Moreover, since $\theta_i w$ is a piecewise-quadratic function on $\Gamma$, Lemma C.38 gives

$$\|v_i\|_{\widetilde{H}_{\rm I}^{1/2}(\Gamma)} \lesssim \|\theta_i w\|_{\widetilde{H}_{\rm I}^{1/2}(\Gamma)}.$$

Hence by using (3.36), (3.30), and (3.37) we obtain, noting (3.27),

$$\begin{aligned}
\sum_{i=0}^{J} \|v_i\|^2_{\widetilde{H}_{\rm I}^{1/2}(\Gamma)} &\lesssim \|v\|^2_{\widetilde{H}_{\rm I}^{1/2}(\Gamma)} + \sum_{i=1}^{J} \|\theta_i w\|^2_{\widetilde{H}_{\rm I}^{1/2}(\Gamma_i')} \\
&\lesssim \|v\|^2_{\widetilde{H}_{\rm I}^{1/2}(\Gamma)} + (1 + \log^2 \frac{H}{\delta}) \left(\frac{1}{H} \|w\|^2_{L^2(\Gamma)} + \|w\|^2_{\widetilde{H}_{\rm I}^{1/2}(\Gamma)}\right) \\
&\lesssim (1 + \log^2 \frac{H}{\delta}) \|v\|^2_{\widetilde{H}_{\rm I}^{1/2}(\Gamma)}.
\end{aligned}$$

The required result is then a consequence of Lemma B.3. □

Combining Lemmas 3.6 and 3.8 we obtain:

**Theorem 3.3.** *The condition number of the additive Schwarz operator P is bounded by*

$$\kappa(P_{\rm ad}) \lesssim (1 + \log^2 \frac{H}{\delta})$$

*where the constant involved is independent of $H$, $h$, and $\delta$.*

#### 3.1.2.2 The weakly-singular integral equation

We now consider an overlapping preconditioner for the weakly-singular integral equation (3.2). As in the case of the non-overlapping preconditioner, it is crucial to define the subspace decomposition.

**Definition 3.5 (Subspace decomposition 4).** *The subspace decomposition is now defined as in Definition 3.3 with*

$$\widetilde{\mathscr{V}}_i := \left\{v \in \widetilde{\mathscr{V}} : \operatorname{supp} v \subset \overline{\Gamma}'_i, \quad \langle v, 1\rangle_{L^2(\Gamma_i')} = 0\right\}, \quad i = 1, \ldots, J,$$

*where the subdomains $\Gamma_i'$, $i = 1, \ldots, J$, are defined in Definition 3.4.*

We note that $\widetilde{\mathscr{V}}_i \subset \widetilde{H}^{-1/2}(\Gamma_i')$ for $i = 1, \ldots, J$. As in the case of the non-overlapping preconditioner, the following lemmas on coercivity and stability of the decomposition defined in Definition 3.5 are consequences of Lemma 3.6, Lemma 3.8, and Lemma 3.3.

**Lemma 3.9 (Coercivity of decomposition).** *For any $v \in \widetilde{\mathscr{V}}$ if*

$$v = \sum_{i=-1}^{J} v_i, \quad v_i \in \widetilde{\mathscr{V}}_i,$$

*where $\widetilde{\mathscr{V}}_i$, $i = -1, 0, \ldots, J$, are defined in Definition 3.5, then*

$$a_{\mathfrak{Y}}(v,v) \lesssim \sum_{i=-1}^{J} a_{\mathfrak{Y}}(v_i, v_i)$$

*where the constant is independent of $v$, $H$, $h$, and $\delta$.*

**Lemma 3.10 (Stability of decomposition).** *Any $v \in \widetilde{\mathscr{V}}$ admits a decomposition*

$$v = \sum_{i=-1}^{J} v_i, \quad v_i \in \widetilde{\mathscr{V}}_i,$$

*satisfying*

$$\sum_{i=-1}^{J} a_{\mathfrak{Y}}(v_i, v_i) \lesssim (1 + \log^2 \frac{H}{\delta}) a_{\mathfrak{Y}}(v,v),$$

*where the constant is independent of $v$, $H$, $h$, and $\delta$.*

Combining Lemmas 3.10 and 3.9 we obtain

**Theorem 3.4.** *The condition number of the additive Schwarz operator $P_{\text{ad}}$ satisfies*

$$\kappa(P_{\text{ad}}) \lesssim \left(1 + \log^2 \frac{H}{\delta}\right)$$

*where the constant involved is independent of $H$, $h$, and $\delta$.*

## 3.2 Multiplicative Schwarz Methods

On the same subspace decompositions defined in Section 3.1 we can design corresponding multiplicative Schwarz preconditioners, both nonoverlapping and overlapping methods.

## 3.2.1 Non-overlapping methods

### 3.2.1.1 The hypersingular integral equation

On the subspace decomposition defined in Definition 3.2, a symmetric multiplicative Schwarz operator $P_{\text{smu}}$ is designed as in Chapter 2. In view of Theorem 2.2, in order to estimate the condition number of $P_{\text{smu}}$ it is necessary to prove the validity of Assumptions 2.3, 2.5, and 2.6.

Assumption 2.3 (Stability of decomposition) is proved in Lemma 3.1. Assumption 2.6 (Local stability) is obvious with $\omega = 1$. In order to prove the validity of Assumption 2.5 (Strengthened Cauchy–Schwarz), one needs to find $\varepsilon_{\mathfrak{W}}^{ij} \in (0,1)$, $i, j = 1, \ldots, J$, satisfying

$$\varepsilon_{\mathfrak{W}}^{ij} = \sup_{\substack{v_i \in \mathscr{V}_i \\ v_j \in \mathscr{V}_j}} \frac{a_{\mathfrak{W}}(v_i, v_j)}{a_{\mathfrak{W}}(v_i, v_i)^{1/2} a_{\mathfrak{W}}(v_j, v_j)^{1/2}}.$$

The above inequality obviously holds for $\varepsilon_{\mathfrak{W}}^{ij} = 1$. The aim is to estimate the spectral radius $\rho(\mathscr{E}_{\mathfrak{W}})$ of the matrix $\mathscr{E}_{\mathfrak{W}} = (\varepsilon_{\mathfrak{W}}^{ij})_{i,j=1}^{J}$; see Theorem 2.2. In the case that the bilinear form $a_{\mathfrak{W}}(\cdot,\cdot)$ is defined from a differential operator, $\varepsilon_{\mathfrak{W}}^{ij} = \delta_{ij}$ (the Kronecker delta) because $\mathscr{V}_i \cap \mathscr{V}_j = \varnothing$ if $i \neq j$. Hence $\rho(\mathscr{E}_{\mathfrak{W}}) = 1$.

Since $\mathscr{E}_{\mathfrak{W}}$ is a dense matrix for the hypersingular integral equation, this optimal result cannot be achieved. We obtain this rough estimate for $\rho(\mathscr{E}_{\mathfrak{W}})$

$$\rho(\mathscr{E}_{\mathfrak{W}}) \leq \|\mathscr{E}_{\mathfrak{W}}\|_{\infty} \leq J$$

where $J$ is the number of subdomains. Theorem 2.2 yields

$$\kappa(P_{\text{smu}}) \lesssim J^2 \max_{1 \leq i \leq J} (1 + \log \frac{H_i}{h_i}). \tag{3.38}$$

Similarly, if the non-symmetric multiplicative Schwarz preconditioner $\mathbf{C}_{\text{mu}}^{-1}$ is used together with the linear iterative method

$$\mathbf{u}_{k+1} = \mathbf{u}_k + \mathbf{C}_{\text{mu}}^{-1}(\mathbf{b} - \mathbf{A}\mathbf{u}_k), \quad k = 0, 1, 2, \ldots,$$

then Theorem 2.7 gives the following estimate for the convergence rate

$$\|\mathbf{I} - \mathbf{C}_{\text{mu}}^{-1}\mathbf{A}\|_{\mathbf{A}} \leq 1 - \frac{c}{J^2 \max\limits_{1 \leq i \leq J} \left(1 + \log \dfrac{H_i}{h_i}\right)} \tag{3.39}$$

where $c$ is a constant.

#### 3.2.1.2 The weakly-singular integral equation

Similarly to the hypersingular integral equation, if the symmetric and non-symmetric multiplicative Schwarz methods are defined on the non-overlapping decomposition given in Definition 3.3, then estimates similar to (3.38) and (3.39) hold.

### 3.2.2 Overlapping methods

#### 3.2.2.1 The hypersingular integral equation

In the same manner, we can define on the subspace decomposition defined in Definition 3.4, symmetric and non-symmetric multiplicative Schwarz preconditioners and obtain

$$\kappa(P_{\text{smu}}) \lesssim J^2(1+\log^2\frac{H}{\delta})$$

and

$$\|\mathbf{I}-\mathbf{C}_{\text{mu}}^{-1}\mathbf{A}\|_{\mathbf{A}} \le 1-\frac{c}{J^2(1+\log^2\frac{H}{\delta})}.$$

#### 3.2.2.2 The weakly-singular integral equation

Similar results hold for the overlapping Schwarz preconditioner defined with the overlapping decomposition Definition 3.5.

### 3.2.3 A special case

A bound for $\kappa(P_{\text{smu}})$ that is independent of $J$ can be obtained in the special case when $h_i = h$ and $H_i = H = 2h$ for all $i$. For ease of notation, we assume that $\Gamma = (-1,1)$ and let the partition of $\Gamma$ be denoted by

$$x_j = -1+jh, \quad h = \frac{2}{N}, \quad j = 0,\dots,N. \tag{3.40}$$

The boundary element spaces $\mathscr{V}$ and $\widetilde{\mathscr{V}}$ defined in Definition 3.1 are now defined in the mesh with size $h$, while the coarse spaces $\mathscr{V}_0$ and $\widetilde{\mathscr{V}}_0$ are defined on the coarse mesh with mesh size $H$. The subdomains are $\Gamma_i := (x_{i-1}, x_{i+1})$, $i = 1,\dots,N-1$, which are overlapping. Let $\phi_i$ denote the hat function that satisfies $\phi_i(x_j) = \delta_{ij}$, and let $\widetilde{\phi}_i = d\phi_i/dx$, $i = 1,\dots,N-1$. Then the local subspaces are

$$\mathscr{V}_i = \text{span}\{\phi_i\}, \quad i = 1,\dots,N-1. \tag{3.41}$$

We also define, for $i = 1, \ldots, N$,

$$\bar{\phi}_i(x) = \begin{cases} 1 & \text{if } x \in (x_{i-1}, x_i), \\ 0 & \text{otherwise.} \end{cases}$$

#### 3.2.3.1 The hypersingular integral equation

The multiplicative Schwarz operator is associated with the subspace decomposition

$$\mathscr{V} = \mathscr{V}_0 + \mathscr{V}_1 + \cdots + \mathscr{V}_{N-1} \tag{3.42}$$

where $\mathscr{V}_1, \ldots, \mathscr{V}_{N-1}$ are defined in (3.41). We prove the following Strengthened Cauchy–Schwarz Lemma.

**Lemma 3.11.** *Let $\mathscr{E}_{\mathfrak{W}}$ be the matrix with entries*

$$\varepsilon_{\mathfrak{W}}^{ij} = \sup_{\substack{v_i \in \mathscr{V}_i \\ v_j \in \mathscr{V}_j}} \frac{a_{\mathfrak{W}}(v_i, v_j)}{a_{\mathfrak{W}}(v_i, v_i)^{1/2} a_{\mathfrak{W}}(v_j, v_j)^{1/2}}, \quad i, j = 1, \ldots, N-1.$$

*Then the spectral radius $\rho(\mathscr{E}_{\mathfrak{W}})$ is bounded independent of $N$, i.e.,*

$$\rho(\mathscr{E}_{\mathfrak{W}}) \lesssim 1.$$

*Proof.* Since

$$\rho(\mathscr{E}_{\mathfrak{W}}) \le \|\mathscr{E}_{\mathfrak{W}}\|_\infty := \max_{1 \le i \le N-1} \sum_{j=1}^{N-1} |\varepsilon_{\mathfrak{W}}^{ij}|, \tag{3.43}$$

it suffices to prove

$$\|\mathscr{E}_{\mathfrak{W}}\|_\infty \lesssim 1. \tag{3.44}$$

First we note that (3.41) gives

$$\varepsilon_{\mathfrak{W}}^{ij} = \frac{|a_{\mathfrak{W}}(\phi_i, \phi_j)|}{a_{\mathfrak{W}}(\phi_i, \phi_i)^{1/2} a_{\mathfrak{W}}(\phi_j, \phi_j)^{1/2}}, \quad i, j = 1, \ldots, N-1. \tag{3.45}$$

Next we recall a well-known result

$$a_{\mathfrak{W}}(u, v) = a_{\mathfrak{V}}(\dot{u}, \dot{v}) \quad \forall u, v \in \widetilde{H}^{1/2}(\Gamma) \tag{3.46}$$

where $\dot{u}$ (respectively, $\dot{v}$) denotes the derivative of $u$ (respectively, $v$) with respect to the arc-length, which is the usual derivative $du/dx$ (respectively, $dv/dx$) in this special case. Since

$$\widetilde{\phi}_i = \frac{1}{h}\left(\bar{\phi}_i - \bar{\phi}_{i+1}\right), \tag{3.47}$$

we have

$$a_{\mathfrak{W}}(\phi_i,\phi_j) = a_{\mathfrak{V}}(\widetilde{\phi}_i,\widetilde{\phi}_j) = \frac{1}{h^2} a_{\mathfrak{V}}(\bar{\phi}_i - \bar{\phi}_{i+1}, \bar{\phi}_j - \bar{\phi}_{j+1})$$
$$= \frac{1}{h^2}\Big(a_{\mathfrak{V}}(\bar{\phi}_i,\bar{\phi}_j) - a_{\mathfrak{V}}(\bar{\phi}_{i+1},\bar{\phi}_j) - a_{\mathfrak{V}}(\bar{\phi}_i,\bar{\phi}_{j+1}) + a_{\mathfrak{V}}(\bar{\phi}_{i+1},\bar{\phi}_{j+1})\Big). \tag{3.48}$$

Let

$$F(z) = \begin{cases} z^2 \log|z| & \text{for } z \neq 0, \\ 0 & \text{for } z = 0. \end{cases} \tag{3.49}$$

Then by using the formula

$$\int_c^d \int_a^b \log|x-y|\,dx\,dy = \frac{1}{2}\Big(F(c-b) - F(d-b) - F(c-a) + F(d-a)\Big) - \frac{3}{2}(b-a)(d-c), \tag{3.50}$$

we deduce

$$a_{\mathfrak{V}}(\bar{\phi}_i,\bar{\phi}_j) = -\frac{1}{2\pi}\Big(F(x_{i-1}-x_j) - F(x_i - x_j) - F(x_{i-1} - x_{j-1}) + F(x_i - x_{j-1})\Big) + \frac{3h^2}{2\pi}.$$

Noting that

$$F(x_\ell - x_k) = h^2\Big(F(\ell - k) + |\ell - k|^2 \log h\Big) \tag{3.51}$$

we have

$$F(x_{i-1}-x_j) - F(x_i - x_j) - F(x_{i-1} - x_{j-1}) + F(x_i - x_{j-1})$$
$$= h^2\Big(F(i-j-1) - 2F(i-j) + F(i-j+1) + 2\log h\Big)$$

so that

$$a_{\mathfrak{V}}(\bar{\phi}_i,\bar{\phi}_j) = -\frac{h^2}{2\pi}\Big(F(i-j-1) - 2F(i-j) + F(i-j+1) + 2\log h - 3\Big). \tag{3.52}$$

By letting, for $m = 1, 2, \ldots,$

$$G(m) := -2F(m-1) + 3F(m) - 2F(m+1) + F(m-2)/2 + F(m+2)/2,$$

we deduce from (3.48) and (3.52)

$$a_{\mathfrak{W}}(\phi_i,\phi_j) = \frac{1}{\pi} G(i-j).$$

In particular, noting that $F$ is an even function, we have

$$a_{\mathfrak{W}}(\phi_i,\phi_i) = \frac{1}{\pi}G(0) = \frac{1}{\pi}\Big(3F(0) - 4F(1) + F(2)\Big) = \frac{4\log 2}{\pi}. \tag{3.53}$$

Hence, it follows from (3.45) that

$$\varepsilon_{\mathfrak{W}}^{ij} = \frac{1}{4\log 2}|G(i-j)|.$$

In view of (3.43) and (3.44), it suffices to prove

$$\sum_{m=0}^{\infty} |G(m)| \lesssim 1. \tag{3.54}$$

To do so, we define

$$f(x) := F(x+1) + F(x-1) - 2F(x) \quad \text{for any } x \in \mathbb{R} \tag{3.55}$$

to obtain

$$G(m) = \frac{1}{2}\Big(f(m+1) + f(m-1) - 2f(m)\Big).$$

Hence (3.54) is a consequence of Lemma C.50 in the Appendix. This completes the proof of the lemma. □

Consequently, we have the following theorem.

**Theorem 3.5.** *(i) If $P_{\text{smu}}$ is the symmetric multiplicative Schwarz operator associated with the subspace decomposition* (3.42)*, then the condition number of $P_{\text{smu}}$ satisfies*

$$\kappa(P_{\text{smu}}) \lesssim 1.$$

*(ii) If $\mathbf{C}_{\text{mu}}$ is the non-symmetric multiplicative Schwarz preconditioner associated with the subspace decomposition* (3.42)*, then the convergence rate of the linear iterative method satisfies*

$$\|\mathbf{I} - \mathbf{C}_{\text{mu}}^{-1}\mathbf{A}\|_{\mathbf{A}} \le C < 1.$$

*Proof.* Part (i) is a consequence of Lemma 3.11, Lemma 3.8 (noting that $H/\delta$ is constant in this case), and Theorem 2.2. Part (ii) is a consequence of Lemma 3.11, Lemma 3.8, Lemma 3.6, and Theorem 2.7. □

*Remark 3.1.* The same analysis as in Lemma 3.11 can be carried out for the nonoverlapping method. In this case the subspace $\mathcal{V}_i$ is defined by (3.41) at odd nodal points $x_i$ only. The supports of functions in two different subspaces $V_i$ are disjoint.

#### 3.2.3.2 The weakly-singular integral equation

The special case when $H = 2h$ with the fine mesh defined by (3.40) gives the same result for the weakly-singular integral equation. The space $\widetilde{\mathcal{V}}$ is decomposed as

$$\widetilde{\mathcal{V}} = \widetilde{\mathcal{V}}_0 + \widetilde{\mathcal{V}}_1 + \cdots + \widetilde{\mathcal{V}}_{N-1} \tag{3.56}$$

where

$$\widetilde{\mathcal{V}}_i := \operatorname{span}\{\widetilde{\phi}_i\}, \quad i = 1, \ldots, N-1, \quad \text{and} \quad \widetilde{\mathcal{V}}_0 := \operatorname{span}\{\widetilde{\phi}_N\} \tag{3.57}$$

with $\widetilde{\phi}_N \equiv 1$.

**Lemma 3.12.** *Let $\mathscr{E}_{\mathfrak{V}}$ be the matrix with entries*

$$\varepsilon_{\mathfrak{V}}^{ij} = \sup_{\substack{v_i \in \widetilde{\mathcal{V}}_i \\ v_j \in \widetilde{\mathcal{V}}_j}} \frac{a_{\mathfrak{V}}(v_i, v_j)}{a_{\mathfrak{V}}(v_i, v_i)^{1/2} a_{\mathfrak{V}}(v_j, v_j)^{1/2}}, \quad i, j = 1, \ldots, N.$$

*Then the spectral radius $\rho(\mathscr{E}_{\mathfrak{V}})$ is bounded independent of $N$, i.e.,*

$$\rho(\mathscr{E}_{\mathfrak{V}}) \lesssim 1.$$

*Proof.* As in the proof of Lemma 3.11 it suffices to prove

$$\max_{1 \le i \le N} \sum_{j=1}^{N} |\varepsilon_{\mathfrak{V}}^{ij}| \lesssim 1. \tag{3.58}$$

The definition (3.57) of $\widetilde{\mathcal{V}}_i$ gives

$$\varepsilon_{\mathfrak{V}}^{ij} = \frac{|a_{\mathfrak{V}}(\widetilde{\phi}_i, \widetilde{\phi}_j)|}{a_{\mathfrak{V}}(\widetilde{\phi}_i, \widetilde{\phi}_i)^{1/2} a_{\mathfrak{V}}(\widetilde{\phi}_j, \widetilde{\phi}_j)^{1/2}}, \quad i, j = 1, \ldots, N. \tag{3.59}$$

It follows from (3.46) and (3.45) that

$$\varepsilon_{\mathfrak{V}}^{ij} = \varepsilon_{\mathfrak{W}}^{ij}, \quad i, j = 1, \ldots, N-1.$$

This and Lemma 3.11 yield

$$\max_{1 \le i \le N-1} \sum_{j=1}^{N-1} \varepsilon_{\mathfrak{V}}^{ij} \lesssim 1.$$

Hence, (3.58) is proved if we prove

$$\sum_{j=1}^{N} \varepsilon_{\mathfrak{V}}^{Nj} \lesssim 1. \tag{3.60}$$

In view of (3.59), to compute $\varepsilon_{\mathfrak{V}}^{Nj}$ we need to compute $a_{\mathfrak{V}}(\widetilde{\phi}_j, \widetilde{\phi}_j)$ and $a_{\mathfrak{V}}(\widetilde{\phi}_N, \widetilde{\phi}_j)$ for $j = 1, \ldots, N$. For $j = 1, \ldots, N-1$, (3.46) and (3.53) give

$$a_{\mathfrak{V}}(\widetilde{\phi}_j, \widetilde{\phi}_j) = \frac{4 \log 2}{\pi}.$$

For $j = N$ we have, recalling that $\phi_N \equiv 1$ on $\Gamma = [-1,1] = [x_0, x_N]$,

$$a_{\mathfrak{V}}(\widetilde{\phi}_N, \widetilde{\phi}_N) = -\frac{1}{\pi}\int_\Gamma\int_\Gamma \log|x-y|\,dx\,dy = \frac{6-4\log 2}{\pi} > 0.$$

Moreover, it follows from (3.47), (3.50), and (3.51) that

$$\begin{aligned}
a_{\mathfrak{V}}(\widetilde{\phi}_N, \widetilde{\phi}_j) &= \frac{1}{h} a_{\mathfrak{V}}(\widetilde{\phi}_N, \bar{\phi}_j - \bar{\phi}_{j+1}) \\
&= -\frac{1}{\pi h}\int_{x_0}^{x_N}\int_{x_{j-1}}^{x_j} \log|x-y|\,\mathrm{d}y\,\mathrm{d}x + \frac{1}{\pi h}\int_{x_0}^{x_N}\int_{x_j}^{x_{j+1}} \log|x-y|\,\mathrm{d}y\,\mathrm{d}x \\
&= -\frac{1}{2\pi h}\Big(F(x_0 - x_j) - F(x_N - x_j) - F(x_0 - x_{j-1}) + F(x_N - x_{j-1}) \\
&\qquad - F(x_0 - x_{j+1}) + F(x_N - x_{j+1}) + F(x_0 - x_j) - F(x_N - x_j)\Big) \\
&= \frac{h}{2\pi}\Big(f(j) - f(N-j)\Big)
\end{aligned}$$

where $F$ and $f$ are defined in (3.49) and (3.55), respectively. Letting

$$\widetilde{G}(x) := f(x) - f(N-x), \quad x \in \mathbb{R},$$

we deduce

$$\varepsilon_{\mathfrak{V}}^{Nj} = \frac{h}{2}\frac{1}{\sqrt{6-4\log 2}}\frac{1}{\sqrt{4\log 2}}|\tilde{G}(j)|.$$

Therefore, to prove (3.60) it suffices to show

$$\frac{h}{2}\sum_{j=1}^{N-1}|\tilde{G}(j)| = \frac{1}{N}\sum_{j=1}^{N-1}|\tilde{G}(j)| \lesssim 1.$$

This follows from Lemma C.51 in the Appendix, completing the proof of the lemma. □

Similarly to the case of the hypersingular integral equation, we deduce for the following theorem for the weakly-singular integral equation.

**Theorem 3.6.** *(i) If $P_{\text{smu}}$ is the symmetric multiplicative Schwarz operator associated with the subspace decomposition* (3.56), *then the condition number of $P_{\text{smu}}$ satisfies*

$$\kappa(P_{\text{smu}}) \lesssim 1.$$

*(ii) If $\mathbf{C}_{\text{mu}}$ is the non-symmetric multiplicative Schwarz preconditioner associated with the subspace decomposition* (3.56), *then the convergence rate of the linear iterative method satisfies*

$$\|\mathbf{I} - \mathbf{C}_{\text{mu}}^{-1}\mathbf{A}\|_{\mathbf{A}} \le C < 1.$$

*Proof.* Part (i) is a consequence of Lemma 3.12, Lemma 3.10, and Theorem 2.2. Part (ii) is a consequence of Lemma 3.11, Lemma 3.10, Lemma 3.9, and Theorem 2.7. □

## 3.3 Numerical Results

The theoretical results in this chapter are underlined by numerical results reported in Chapter 9. The following table relates the theorems in this chapter with the tables and figures in Chapter 9 which corroborate them.

| Equation | Preconditioner | | Theorem | Table | Figure |
|---|---|---|---|---|---|
| Hypersingular | Additive Schwarz | PCG–Non-overlapping | 3.1 | 9.2, 9.3 | 9.3, 9.5 |
| | | PCG–Overlapping | 3.3 | 9.3, 9.4 | 9.4, 9.5 |
| Weakly-singular | Additive Schwarz | PCG–Non-overlapping | 3.2 | 9.1 | 9.2 |

**Table 3.1** Theoretical results in Chapter 3 and relevant numerical results in Chapter 9

Numerical results for the non-symmetric multiplicative Schwarz preconditioner implemented with the linear iterative solver can be found in [101, 144].

# Chapter 4
# Two-Level Methods: the $p$-Version

In this chapter, we define two-level additive and multiplicative Schwarz methods for the $p$-version of the Galerkin approximation for both the hypersingular and weakly-singular integral equations on $\Gamma = [-1, 1]$. Similarly to the case of the $h$-version, it suffices to define the subspace decomposition and prove its stability and coercivity. The contents of this chapter are based on publications [144, 222, 228] by the authors. However, differently from the above references, we use Theorem A.11 instead of Theorem A.10 to prove the coercivity of the decomposition. This is to ensure a constant bound independent of the mesh size. Numerical results which underline the theory will be presented in Chapter 9.

For convenience, we recall the Galerkin equations to be solved: Find $\psi_{\mathscr{V}} \in \mathscr{V}$ and $\varphi_{\widetilde{\mathscr{V}}} \in \widetilde{\mathscr{V}}$ satisfying

$$a_{\mathfrak{W}}(\psi_{\mathscr{V}}, \phi) = \langle g, \phi \rangle, \quad \phi \in \mathscr{V}, \tag{4.1}$$

and

$$a_{\mathfrak{V}}(\varphi_{\widetilde{\mathscr{V}}}, \phi) = \langle f, \phi \rangle, \quad \phi \in \widetilde{\mathscr{V}}, \tag{4.2}$$

where $\mathscr{V} \subset \widetilde{H}^{1/2}(\Gamma)$ and $\widetilde{\mathscr{V}} \subset \widetilde{H}^{-1/2}(\Gamma)$ are two finite-dimensional spaces to be defined later, while the bilinear forms are defined by

$$\begin{aligned} a_{\mathfrak{W}}(\varphi, \psi) &:= \langle \mathfrak{W}\varphi, \psi \rangle \quad \forall \varphi, \psi \in \widetilde{H}^{1/2}(\Gamma), \\ a_{\mathfrak{V}}(\varphi, \psi) &:= \langle \mathfrak{V}\varphi, \psi \rangle \quad \forall \varphi, \psi \in \widetilde{H}^{-1/2}(\Gamma), \end{aligned}$$

with

$$\mathfrak{W}v(x) := -\frac{1}{\pi} \frac{\partial}{\partial n_x} \int_\Gamma \frac{\partial}{\partial n_y} \log|x-y| v(y) \, d\sigma_y = -\frac{1}{\pi} \, \text{f.p.} \int_\Gamma \frac{v(y)}{|x-y|^2} \, d\sigma_y, \quad x \in \Gamma,$$

$$\mathfrak{V}v(x) := -\frac{1}{\pi} \int_\Gamma \log|x-y| v(y) \, d\sigma_y, \quad x \in \Gamma.$$



E. P. Stephan and T. Tran, *Schwarz Methods and Multilevel Preconditioners for Boundary Element Methods*, https://doi.org/10.1007/978-3-030-79283-1_4

In the first section of this chapter, we design and analyse non-overlapping and overlapping additive Schwarz methods for both equations (4.1) and (4.2). The second section is devoted to the corresponding multiplicative Schwarz versions.

The finite-dimensional spaces $\mathscr{V}$ and $\widetilde{\mathscr{V}}$ are defined in the following definition.

**Definition 4.1.** *For some positive integer $N$, we fix on $\Gamma$ a regular mesh of size $h = 1/N$ defined by*

$$x_i := -1 + ih, \ i = 0, \dots, N.$$

*Let $\Gamma_j := (x_{j-1}, x_j)$, $j = 1, \dots, N$, we define the spaces $\mathscr{V}$ and $\widetilde{\mathscr{V}}$ by*

$$\begin{aligned} \mathscr{V} &:= \{v \in C(\Gamma) : v(\pm 1) = 0, \ v|_{\Gamma_j} \in \mathbb{P}_p, \quad j = 1, \dots, N\}, \\ \widetilde{\mathscr{V}} &:= \{v : \Gamma \to \mathbb{R} : v|_{\Gamma_j} \in \mathbb{P}_p, \quad j = 1, \dots, N\}, \end{aligned} \tag{4.3}$$

*where $\mathbb{P}_p$ is the space of polynomials of degree at most $p$, $p \ge 1$.*

We note that

$$\mathscr{V} \subset \widetilde{H}^{1/2}(\Gamma), \quad \dim \mathscr{V} = pN - 1, \quad \widetilde{\mathscr{V}} \subset \widetilde{H}^{-1/2}(\Gamma), \quad \dim \widetilde{\mathscr{V}} = N(p+1).$$

## 4.1 Additive Schwarz Methods

### 4.1.1 Non-overlapping methods

#### 4.1.1.1 The hypersingular integral equation

Let $\phi_k \in \mathscr{V}$, $k = 1, \dots, N-1$, be hat functions satisfying

$$\phi_k(x_l) := \begin{cases} 1 & \text{if } k = l, \\ 0 & \text{if } k \ne l. \end{cases}$$

For $q = 2, \dots, p$ and $j = 1, \dots, N$ we define $\mathscr{L}_{q,j}$ as the affine image of $\mathscr{L}_q$ onto $\Gamma_j$, see Definition 4.1, where

$$\mathscr{L}_q(x) := \int_{-1}^{x} L_{q-1}(s)\, ds,$$

with $L_{q-1}$ being the Legendre polynomial of degree $q-1$. We extend $\mathscr{L}_{q,j}$ by 0 outside $\Gamma_j$. It is clear that

$$\mathscr{B} := \{\phi_1, \dots, \phi_{N-1}, \mathscr{L}_{2,1}, \dots, \mathscr{L}_{2,N}, \dots, \mathscr{L}_{p,1}, \dots, \mathscr{L}_{p,N}\}$$

is a basis for $\mathscr{V}$.

**Definition 4.2 (Subspace decomposition 1).** *Let $\mathcal{V}_0$ be the space of continuous piecewise-linear functions on $\Gamma$ vanishing at $\pm 1$, and $\mathcal{V}_j$ be the space of polynomials of degree at most $p$ on $\Gamma_j$ which vanish at the end points of $\overline{\Gamma}_j$, $j = 1,\dots,N$. More precisely,*

$$\mathcal{V}_0 := \operatorname{span}\{\phi_1,\dots,\phi_{N-1}\},$$
$$\mathcal{V}_j := \operatorname{span}\{\mathcal{L}_{2,j},\dots,\mathcal{L}_{p,j}\}, \quad j = 1,\dots,N.$$

*A non-overlapping decomposition of $\mathcal{V}$ can be defined as a direct sum:*

$$\mathcal{V} = \mathcal{V}_0 \oplus \mathcal{V}_1 \oplus \cdots \oplus \mathcal{V}_N. \tag{4.4}$$

The space $\mathcal{V}_0$ serves the same purpose as the coarse space in the $h$-version. Since this is a direct sum, any function $u \in \mathcal{V}$ admits a unique decomposition

$$u = u_0 + \cdots + u_N, \quad u_i \in \mathcal{V}_i,\ i = 0,\dots,N. \tag{4.5}$$

We also denote

$$\widetilde{u} := u_1 + \cdots + u_N. \tag{4.6}$$

As in the case of the $h$-version, we prove the coercivity and stability of this decomposition.

**Lemma 4.1 (Coercivity of decomposition).** *For any $u \in \mathcal{V}$, the unique decomposition* (4.5) *satisfies*

$$a_{\mathfrak{W}}(u,u) \lesssim \sum_{i=0}^{N} a_{\mathfrak{W}}(u_i,u_i)$$

*where the constant is independent of $u$, $p$, and $N$.*

*Proof.* With $\widetilde{u}$ defined in (4.6) we have $\widetilde{u}|_{\Gamma_i} = u_i$. By using Lemma B.3 and Theorem A.11, and by noting that $\operatorname{supp} u_j \subset \overline{\Gamma}_j$ for $j = 1,\dots,N$ we obtain

$$\begin{aligned}
a_{\mathfrak{W}}(u,u) &\simeq \|u\|^2_{\widetilde{H}_{\mathrm{I}}^{1/2}(\Gamma)} \\
&\lesssim \|u_0\|^2_{\widetilde{H}_{\mathrm{I}}^{1/2}(\Gamma)} + \|\widetilde{u}\|^2_{\widetilde{H}_{\mathrm{I}}^{1/2}(\Gamma)} \\
&= \|u_0\|^2_{\widetilde{H}_{\mathrm{I}}^{1/2}(\Gamma)} + \sum_{i=1}^{N} \|u_i\|^2_{\widetilde{H}_{\mathrm{I}}^{1/2}(\Gamma)} \\
&\lesssim \sum_{i=0}^{N} a_{\mathfrak{W}}(u_i,u_i),
\end{aligned}$$

proving the lemma. □

**Lemma 4.2 (Stability of decomposition).** *For any $u \in \mathcal{V}$, the unique decomposition* (4.5) *satisfies*

$$\sum_{i=0}^{N} a_{\mathfrak{W}}(u_i,u_i) \lesssim (1+\log p)^2 a_{\mathfrak{W}}(u,u)$$

*where the constant is independent of u, p, and N.*

*Proof.* Due to Lemma B.3 and Lemma A.13, the required result is proved if we prove

$$\|u_0\|^2_{\widetilde{H}^{1/2}_{\mathrm{I}}(\Gamma)} + \sum_{i=1}^{N} \|u_i\|^2_{\widetilde{H}^{1/2}_{\mathrm{I}}(\Gamma_i)} \lesssim (1+\log p)^2 \|u\|^2_{\widetilde{H}^{1/2}_{\mathrm{I}}(\Gamma)}. \tag{4.7}$$

First we use Lemma A.6 and (C.50) in Theorem C.1 (noting that $u_i$ vanishes at the endpoints of $\Gamma_i$) to obtain

$$\|u_i\|^2_{\widetilde{H}^{1/2}_{\mathrm{I}}(\Gamma_i)} = \|\widehat{u}_i\|^2_{\widetilde{H}^{1/2}_{\mathrm{I}}(\widehat{\Gamma})} \lesssim |\widehat{u}_i|^2_{H^{1/2}(\widehat{\Gamma})} + (1+\log p)\|\widehat{u}_i\|^2_{L^\infty(\widehat{\Gamma})}$$

where $\widehat{u}_i$ is defined from $u_i$ as usual when $\Gamma_i$ is rescaled into the reference element $\widehat{\Gamma} = (-1,1)$. That means $\widehat{u}_i(\hat{x}) = u_i(x)$ for $x \in \Gamma_i$ and $\hat{x} \in \widehat{\Gamma}$ where $\hat{x}$ is the affine image of $x$. Since $\|\widehat{u}_i\|^2_{L^\infty(\widehat{\Gamma})} = \|u_i\|^2_{L^\infty(\Gamma_i)}$, we deduce with the help of Lemma A.4

$$\|u_i\|^2_{\widetilde{H}^{1/2}_{\mathrm{I}}(\Gamma_i)} \lesssim |u_i|^2_{H^{1/2}(\Gamma_i)} + (1+\log p)\|u_i\|^2_{L^\infty(\Gamma_i)}.$$

Since $u_i = (u-u_0)|_{\Gamma_i}$, for any $\alpha \in \mathbb{R}$ we have

$$\|u_i\|_{L^\infty(\Gamma_i)} \le \|u-\alpha\|_{L^\infty(\Gamma_i)} + \|u_0-\alpha\|_{L^\infty(\Gamma_i)}.$$

As $\widetilde{u}$ vanishes at the mesh points, $u_0$ is the piecewise linear interpolant of $u$ on this mesh. Therefore $\|u_0-\alpha\|_{L^\infty(\Gamma_i)} \le \|u-\alpha\|_{L^\infty(\Gamma_i)}$, implying

$$\|u_i\|_{L^\infty(\Gamma_i)} \le 2\|u-\alpha\|_{L^\infty(\Gamma_i)}.$$

Choosing $\alpha = u(x_i)$, we can invoke Lemma C.26 to obtain

$$\|u_i\|^2_{L^\infty(\Gamma_i)} \lesssim (1+\log p)|u|^2_{H^{1/2}(\Gamma_i)}.$$

Altogether, we deduce noting that $u_i = (u-u_0)|_{\Gamma_i}$

$$\begin{aligned}
\|u_i\|^2_{\widetilde{H}^{1/2}_{\mathrm{I}}(\Gamma_i)} &\lesssim |u_i|^2_{H^{1/2}(\Gamma_i)} + (1+\log p)^2|u|^2_{H^{1/2}(\Gamma_i)} \\
&\lesssim |u_0|^2_{H^{1/2}(\Gamma_i)} + |u|^2_{H^{1/2}(\Gamma_i)} + (1+\log p)^2|u|^2_{H^{1/2}(\Gamma_i)} \\
&\lesssim |u_0|^2_{H^{1/2}(\Gamma_i)} + (1+\log p)^2|u|^2_{H^{1/2}(\Gamma_i)}.
\end{aligned}$$

Summing over $i$ from 1 to $N$ and invoking Lemma C.39, we infer

$$\sum_{i=1}^{N} \|u_i\|^2_{\widetilde{H}^{1/2}_{\mathrm{I}}(\Gamma_i)} \lesssim (1+\log p)^2 \sum_{i=1}^{N} |u|^2_{H^{1/2}(\Gamma_i)} \le (1+\log p)^2 |u|^2_{H^{1/2}(\Gamma)}$$

where in the last step we used the definition of the $H^{1/2}$-seminorm in (A.25). Noting the definition of the $\|\cdot\|_{\widetilde{H}^{1/2}_{\mathrm{w}}(\Gamma)}$ norm defined in Section A.2.6.1, and noting that this norm is equivalent to the $\|\cdot\|_{\widetilde{H}_{\mathrm{I}}^{1/2}(\Gamma)}$ norm with constant depending only on the measure of $\Gamma$, we deduce

$$\sum_{i=1}^{N} \|u_i\|^2_{\widetilde{H}_{\mathrm{I}}^{1/2}(\Gamma_i)} \lesssim (1+\log p)^2 \|u\|^2_{\widetilde{H}_{\mathrm{w}}^{1/2}(\Gamma)} \simeq (1+\log p)^2 \|u\|^2_{\widetilde{H}_{\mathrm{I}}^{1/2}(\Gamma)}.$$

The triangle inequality, definition (4.6), Theorem A.10, and the above inequality imply

$$\begin{aligned}
\|u_0\|^2_{\widetilde{H}_{\mathrm{I}}^{1/2}(\Gamma)} &\lesssim \|u\|^2_{\widetilde{H}_{\mathrm{I}}^{1/2}(\Gamma)} + \|\widetilde{u}\|^2_{\widetilde{H}_{\mathrm{I}}^{1/2}(\Gamma)} \\
&\leq \|u\|^2_{\widetilde{H}_{\mathrm{I}}^{1/2}(\Gamma)} + \sum_{i=1}^{N} \|u_i\|^2_{\widetilde{H}_{\mathrm{I}}^{1/2}(\Gamma_i)} \\
&\lesssim (1+\log p)^2 \|\widetilde{u}\|^2_{\widetilde{H}_{\mathrm{I}}^{1/2}(\Gamma)}.
\end{aligned}$$

This proves (4.7) with a constant independent of $p$ and $N$. The lemma is proved. □

Combining Lemmas 4.1 and 4.2, and using the general theory in Chapter 2, we obtain

**Theorem 4.1.** *The condition number* $\kappa(P_{\mathrm{ad}})$ *of the additive Schwarz operator* $P_{\mathrm{ad}}$ *associated with the direct sum decomposition* (4.4) *satisfies*

$$\kappa(P_{\mathrm{ad}}) \lesssim (1+\log p)^2$$

*where the constant is independent of* $p$ *and* $N$.

#### 4.1.1.2 The weakly-singular integral equation

As with the hypersingular integral equation, we decompose $\widetilde{\mathscr{V}}$ (defined in (4.3)) as a direct sum.

**Definition 4.3 (Subspace decomposition 2).** *For* $p = 0,1,2,\ldots,$ *and* $j = 1,\ldots,N$, *let* $L_{p,j}$ *be the affine image of the Legendre polynomial* $L_p$ *of degree* $p$ *onto the subinterval* $\Gamma_j$. *We define*

$$\begin{aligned}
\widetilde{\mathscr{V}}_0 &:= \operatorname{span}\{L_{0,1},\ldots,L_{0,N}\}, \\
\widetilde{\mathscr{V}}_j &:= \operatorname{span}\{L_{1,j},\ldots,L_{p,j}\}, \quad j = 1,\ldots,N,
\end{aligned}$$

*and decompose* $\widetilde{\mathscr{V}}$ *as a direct sum*

$$\widetilde{\mathscr{V}} = \widetilde{\mathscr{V}}_0 \oplus \cdots \oplus \widetilde{\mathscr{V}}_N. \tag{4.8}$$

Note that $\widetilde{\mathscr{V}}_0$ is the space of piecewise-constant functions on $\Gamma$ and $\widetilde{\mathscr{V}}_j$ is the space of polynomials of degree at most $p$ on $\Gamma_j$ which have zero integral means (property of Legendre polynomials of degree $q \geq 1$). The space $\widetilde{\mathcal{V}}_0$ serves the same purpose as the coarse space in the $h$ version. The subspace decomposition (4.8) results in a unique decomposition for each $u \in \widetilde{\mathcal{V}}$, namely

$$u = u_0 + \cdots + u_N, \quad u_j \in \widetilde{\mathcal{V}}_j, \ j = 0, \ldots, N. \tag{4.9}$$

The coercivity and stability of the decomposition (4.8) can be obtained by invoking Lemma 3.3 as in the $h$-version studied in Chapter 3.

**Lemma 4.3 (Coercivity of decomposition).** *For any $u \in \widetilde{\mathcal{V}}$, the unique decomposition* (4.9) *satisfies*

$$a_{\mathfrak{W}}(u,u) \lesssim \sum_{i=0}^{N} a_{\mathfrak{W}}(u_i,u_i)$$

*where the constant is independent of $u$, $p$, and $N$.*

*Proof.* The proof is similar to that of Lemma 3.5 and is omitted. □

**Lemma 4.4 (Stability of decomposition).** *For each $u \in \widetilde{\mathcal{V}}$, the unique decomposition* (4.9) *satisfies*

$$\sum_{i=0}^{N} a_{\mathfrak{W}}(u_i,u_i) \lesssim (1+\log(p+1))^2 a_{\mathfrak{W}}(u,u)$$

*where the constant is independent of $u$, $p$, and $N$.*

*Proof.* The proof follows that of Lemma 3.4 and is omitted. □

Combining Lemmas 2.10, 4.3 and 4.4 we obtain

**Theorem 4.2.** *The condition number $\kappa(P_{\mathrm{ad}})$ of the additive Schwarz operator $P_{\mathrm{ad}}$ associated with the decomposition* (4.8) *satisfies*

$$\kappa(P_{\mathrm{ad}}) \lesssim (1+\log(p+1))^2$$

*where the constant is independent of $p$ and $N$.*

## 4.1.2 Overlapping methods

### 4.1.2.1 The hypersingular integral equation

**Definition 4.4 (Subspace decomposition 3).** *Considering overlapping subdomains*

$$\Gamma_i' = \Gamma_i \cup \{x_i\} \cup \Gamma_{i+1}, \quad i = 1, \ldots, N-1,$$

*we now decompose $\mathcal{V}$ by*

$$\mathcal{V} = \mathcal{V}_0 + \cdots + \mathcal{V}_{N-1}$$

*where $\mathcal{V}_0$ is the space of piecewise-linear functions vanishing at the endpoints of $\Gamma$, and*

$$\mathcal{V}_i := (\mathcal{V} \setminus \mathcal{V}_0) \cap \widetilde{H}^{1/2}(\Gamma_i'), \quad i = 1, \ldots, N-1.$$

The coercivity of the above decomposition of $\mathcal{V}$ is obvious; [228, Lemma 5.1].

**Lemma 4.5 (Coercivity of decomposition).** *For any $u \in \mathcal{V}$, if $u = \sum_{i=0}^{N-1} u_i$ for some $u_i \in \mathcal{V}_i$ then*

$$a_{\mathfrak{W}}(u,u) \lesssim \sum_{i=0}^{N-1} a_{\mathfrak{W}}(u_i,u_i)$$

*where the constant is independent of $u$, $p$ and $N$.*

*Proof.* The proof is similar to that of Lemma 3.6 and is omitted. □

To prove the stability of the decomposition in Definition 4.4, we need to find for each $u \in \mathcal{V}$ a representation of the form $u = u_0 + \cdots + u_{N-1}$ with $u_i \in \mathcal{V}_i$. Firstly, we define, similarly to the projection $P_H$ defined in (3.35), the projection $P_h : \widetilde{H}^{1/2}(\Gamma) \to \mathcal{V}_0$ by

$$\langle \Lambda P_h v, w \rangle = \langle \Lambda v, w \rangle \quad \forall v \in \widetilde{H}^{1/2}(\Gamma), \, w \in \mathcal{V}_0,$$

where $\Lambda$ is defined in (3.33). Note that the mesh size in the present case is $h$. For each $u \in \mathcal{V}$, let

$$u_0 := P_h u \in \mathcal{V}_0. \tag{4.10}$$

Then, similarly to (3.36) and (3.37), we have

$$\|u_0\|_{\widetilde{H}_{\mathrm{I}}^{1/2}(\Gamma)} \lesssim \|u\|_{\widetilde{H}_{\mathrm{I}}^{1/2}(\Gamma)} \quad \text{and} \quad \|u - u_0\|_{L^2(\Gamma)} \lesssim h^{1/2} \|u\|_{\widetilde{H}_{\mathrm{I}}^{1/2}(\Gamma)}. \tag{4.11}$$

Next we consider a partition of unity $\{\theta_i : i = 1, \ldots, N-1\}$ composed of piecewise-linear functions $\theta_i$ satisfying

$$\operatorname{supp} \theta_i \subset \overline{\Gamma}_i', \quad 0 \le \theta_i \le 1, \quad \sum_{i=1}^{N-1} \theta_i = 1, \quad \left| \frac{d\theta_i}{dx} \right| \le \frac{1}{h}, \tag{4.12}$$

where $\Gamma_i'$ is defined in Definition 4.4. Figure 4.1 plots these functions for $N = 4$.

Let $w := u - u_0$. Then $\theta_i w$ (having support in $\overline{\Gamma_i'}$) comprises two parts, $\theta_i w|_{\Gamma_i}$ and $\theta_i w|_{\Gamma_{i+i}}$, each being a polynomial of degree at most $p+1$. We define

$$u_i = A^{-1} T_p^* A(\theta_i w), \quad i = 1, \ldots, N-1, \tag{4.13}$$

where $T_p^*$ and $A$ are defined in Lemma C.43 and Lemma C.44, respectively. These functions $u_0, \ldots, u_{N-1}$ define a decomposition for $u$, as shown in the following lemma.

**Lemma 4.6.** *For each $i = 1, \ldots, N-1$ the function $u_i$ defined by* (4.13) *belongs to $\mathcal{V}_i$. Moreover,*

$$u = u_0 + \cdots + u_{N-1}.$$

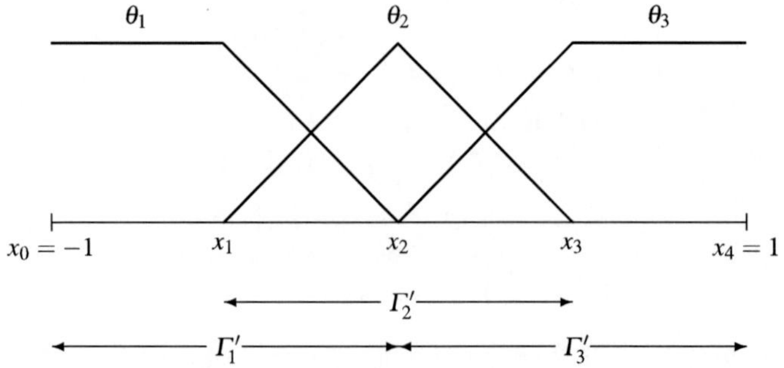


**Fig. 4.1** Partition of unity $\{\theta_1, \theta_2, \theta_3\}$

*Proof.* First we prove $u_i \in \mathscr{V}_i$ for $i = 1,\dots,N-1$. For each $i \in \{1,\dots,N-1\}$, the function $\theta_i w$ has support in $\overline{\Gamma_i'}$. Moreover, $(\theta_i w)|_{\Gamma_i}$ and $(\theta_i w)|_{\Gamma_{i+1}}$ are polynomials of degree at most $p+1$. Due to the definition of $A$ and $T_p^*$, the function $u_i$ also has support in $\overline{\Gamma_i'}$ and its restrictions on $\Gamma_i$ and $\Gamma_{i+1}$ are polynomials of degree at most $p$. Since $x_{i-1}$, $x_i$, and $x_{i+1}$ become interpolation points $z_1$ and $z_{p+1}$ after the affine mappings, see Lemma C.41, we have

$$
\begin{aligned}
u_i|_{\Gamma_{i+1}}(x_{i+1}) &= \theta_i w(x_{i+1}) = 0,\\
u_i|_{\Gamma_i}(x_{i-1}) &= \theta_i w(x_{i-1}) = 0,\\
u_i|_{\Gamma_{i+1}}(x_i) &= u_i|_{\Gamma_i}(x_i).
\end{aligned}
$$

Hence $u_i$ is a continuous function on $\Gamma_i'$ vanishing at $x_{i-1}$ and $x_{i+1}$, whose restrictions on $\Gamma_i$ and $\Gamma_{i+1}$ are polynomials of degrees at most $p$. Therefore $u_i \in \mathscr{V}_i$.

In order to guarantee that the functions $u_0, \dots, u_{N-1}$ form a decomposition for $u$ we need to check that

$$
w = \sum_{i=1}^{N-1} u_i = \sum_{i=1}^{N-1} A^{-1} T_p^* A(\theta_i w).
$$

But this comes from the linearity of $A$ and $T_p^*$, the fact that $\sum_{i=1}^{N-1} \theta_i = 1$ and that $T_p^* f = f$ if $f$ is a polynomial of degree at most $p$. $\square$

**Lemma 4.7 (Stability of decomposition).** *Any $u \in \mathscr{V}$ admits a decomposition $u = u_0 + \cdots + u_{N-1}$ defined by* (4.10) *and* (4.13) *that satisfies*

$$
\sum_{i=0}^{N-1} a_{\mathfrak{W}}(u_i, u_i) \lesssim a_{\mathfrak{W}}(u, u)
$$

*where the constant is independent of u and p, but may depend on* $\mu := h_{\max}/h_{\min}$ *which can be chosen to be constant.*

*Proof.* As usual, we will prove

$$\sum_{i=0}^{N-1} \|u_i\|^2_{\widetilde{H}_{\mathrm{I}}^{1/2}(\Gamma_i')} \lesssim \|u\|^2_{\widetilde{H}_{\mathrm{I}}^{1/2}(\Gamma)}.$$

Lemmas A.6 and A.5 imply

$$\|u_i\|^2_{\widetilde{H}_{\mathrm{I}}^{1/2}(\Gamma_i')} \simeq \|u_i\|^2_{\widetilde{H}_{\mathrm{w}}^{1/2}(\Gamma_i')}$$

where the constant is independent of $u$, $p$, and $h$. Invoking successively Lemma C.44 and Lemma C.43 we obtain

$$\begin{aligned}
\|u_i\|^2_{\widetilde{H}_{\mathrm{w}}^{1/2}(\Gamma_i')} &\le \mu \|T_p^* A(\theta_i w)\|^2_{\widetilde{H}_{\mathrm{w}}^{1/2}(-1,1)} \\
&\simeq \mu \|T_p^* A(\theta_i w)\|^2_{\widetilde{H}_{\mathrm{I}}^{1/2}(-1,1)} \\
&\lesssim \mu \|A(\theta_i w)\|^2_{\widetilde{H}_{\mathrm{I}}^{1/2}(-1,1)} \\
&\simeq \mu \|A(\theta_i w)\|^2_{\widetilde{H}_{\mathrm{w}}^{1/2}(-1,1)} \\
&\le \mu^2 \|\theta_i w\|^2_{\widetilde{H}_{\mathrm{w}}^{1/2}(\Gamma_i')} \\
&\simeq \mu^2 \|\theta_i w\|^2_{\widetilde{H}_{\mathrm{I}}^{1/2}(\Gamma_i')}
\end{aligned}$$

where in the last step we used again Lemmas A.6 and A.5. Here $\mu = h_{\max}/h_{\min}$ and the hidden constant is independent of $u$, $p$, and $h$. By using (3.29) and (4.11), noting that $\delta \sim h$ and $|\Gamma_i'| \sim 2h$, we obtain

$$\begin{aligned}
\sum_{i=0}^{N-1} \|u_i\|^2_{\widetilde{H}_{\mathrm{I}}^{1/2}(\Gamma_i')} &\lesssim \|u\|^2_{\widetilde{H}_{\mathrm{I}}^{1/2}(\Gamma)} + \frac{1}{h}\|w\|^2_{L^2(\Gamma)} + \|w\|^2_{\widetilde{H}_{\mathrm{I}}^{1/2}(\Gamma)} \\
&\lesssim \|u\|^2_{\widetilde{H}_{\mathrm{I}}^{1/2}(\Gamma)},
\end{aligned}$$

completing the proof of the lemma. □

Theorem 2.1, Lemma 4.5, and Lemma 4.7 yield the following theorem.

**Theorem 4.3.** *The condition number of the overlapping additive Schwarz operator* $P_{\mathrm{ad}}$ *defined from the subspace decomposition defined in Definition 4.4 is bounded independently of p and N.*

#### 4.1.2.2 The weakly-singular integral equation

Similarly to the case of the $h$-version, we are able to design and analyse an overlapping method for the weakly-singular integral equation.

**Definition 4.5 (Subspace decomposition 4).** *The subspace decomposition $\widetilde{\mathscr{V}} = \widetilde{\mathscr{V}}_0 + \cdots + \widetilde{\mathscr{V}}_{N-1}$ is defined with*

$$\widetilde{\mathscr{V}}_j := (\widetilde{\mathscr{V}} \setminus \widetilde{\mathscr{V}}_0) \cap \widetilde{H}^{-1/2}(\Gamma_j'), \quad j = 1, \ldots, N-1,$$

*where $\Gamma_j'$ is defined in Definition 4.4.*

We have the following theorem.

**Theorem 4.4.** *The condition number of the additive Schwarz preconditioner associated with the decomposition defined in Definition 4.5 is bounded independently of p and N.*

*Proof.* The proof is carried out in the same manner as in the non-overlapping case and is omitted. □

## 4.2 Multiplicative Schwarz Methods

We now define on the same subspace decompositions defined in Section 4.1 the corresponding multiplicative Schwarz preconditioners. The analysis is carried out in the following subsections.

### *4.2.1 Non-overlapping methods*

#### 4.2.1.1 The hypersingular integral equation

We shall in this subsection design a multiplicative method for the $p$-version of the Galerkin boundary element method applied to the hypersingular integral equation.

We consider again the subspace decomposition defined in Definition 4.2. Due to Theorem 2.2, it suffices to show a Strengthened Cauchy–Schwarz inequality.

**Lemma 4.8.** *Let $\mathscr{V}_i$ be defined as in Definition 4.2 and let*

$$\theta_{ij} := \sup_{u \in \mathscr{V}_i, v \in \mathscr{V}_j} \frac{a_{\mathfrak{W}}(u,v)}{a_{\mathfrak{W}}(u,u)^{1/2} a_{\mathfrak{W}}(v,v)^{1/2}}, \quad i,j = 1,\ldots,N.$$

*Then, with a constant independent of i, j, and p, we have*

$$\theta_{ij} \lesssim \frac{1}{(|i-j|+1)^2}.$$

*If $\Theta$ is the $N \times N$ matrix defined by $\Theta = \{\theta_{ij}\}_{1 \le i,j \le N}$, then with a constant independent of p,*

$$\|\Theta\|_2 \lesssim 1.$$

*Proof.* Let $u = \sum_{k=2}^{p} u_k \mathscr{L}_{k,i}$ and $v = \sum_{\ell=2}^{p} v_\ell \mathscr{L}_{\ell,j}$. Then

$$\begin{aligned}
a_{\mathfrak{W}}(u,v) &= \sum_{k=2}^{p} \sum_{\ell=2}^{p} u_k v_\ell a_{\mathfrak{W}}(\mathscr{L}_{k,i}, \mathscr{L}_{\ell,j}) \\
&= \sum_{k=2}^{p} \sum_{\ell=2}^{p} u_k v_\ell \frac{2}{|\Gamma_i|} \frac{2}{|\Gamma_j|} a_{\mathfrak{V}}(L_{k-1,i}, L_{\ell-1,j}).
\end{aligned} \tag{4.14}$$

Noting that $\operatorname{dist}(\Gamma_i, \Gamma_j) = h(|i-j|-1)$, $i \neq j$, due to Lemma C.17 (compare with [106, Lemma 1]) we have, with $|\Gamma_i| = |\Gamma_j| = h$ and $|i-j| \ge 2$,

$$\begin{aligned}
a_{\mathfrak{V}}(L_{k-1,i}, L_{\ell-1,j}) &\lesssim \frac{|\Gamma_i|\,|\Gamma_j| \max(|\Gamma_i|/2, |\Gamma_j|/2)^{k+\ell-2}}{\operatorname{dist}(\Gamma_i,\Gamma_j)^{k+\ell-2}\sqrt{(2k-1)(2\ell-1)}} \\
&= \frac{h^2 (1/2)^{k+\ell-2}}{(|i-j|-1)^{k+\ell-2}\sqrt{(2k-1)(2\ell-1)}} \\
&\lesssim \frac{h^2}{(k-1)(\ell-1)(|i-j|-1)^{k+\ell-2}\sqrt{(2k-1)(2\ell-1)}}.
\end{aligned} \tag{4.15}$$

From (4.14) and (4.15) we obtain

$$\begin{aligned}
|a_{\mathfrak{W}}(u,v)| &\lesssim \sum_{k=2}^{p} \sum_{\ell=2}^{p} \frac{|u_k|\,|v_\ell|}{(k-1)(\ell-1)(|i-j|-1)^{k+\ell-2}\sqrt{(2k-1)(2\ell-1)}} \\
&\simeq \sum_{k=2}^{p} \sum_{\ell=2}^{p} \frac{|u_k|}{(k-1)\sqrt{2k-1}} \frac{|v_\ell|}{(\ell-1)\sqrt{2\ell-1}} \frac{1}{(|i-j|-1)^{k+\ell-2}} \\
&\lesssim \left( \sum_{k=2}^{p} \left( \frac{|u_k|}{(k-1)\sqrt{2k-1}} \right)^2 \right)^{1/2} \left( \sum_{\ell=2}^{p} \left( \frac{|v_\ell|}{(\ell-1)\sqrt{2\ell-1}} \right)^2 \right)^{1/2} \\
&\quad \times \left\| \left\{ \frac{1}{(|i-j|-1)^{k+\ell-2}} \right\}_{k,\ell=2,\dots,p} \right\|_2.
\end{aligned} \tag{4.16}$$

For $|i-j| \ge 3$ we have

$$\begin{aligned}
\left\| \left\{ \frac{1}{(|i-j|-1)^{k+\ell-2}} \right\}_{k,\ell=2,\dots,p} \right\|_2 &\le \left\| \left\{ \frac{1}{(|i-j|-1)^{k+\ell-2}} \right\}_{k,\ell=2,\dots,p} \right\|_\infty \\
&= \max_{2 \le k \le p} \sum_{\ell=2}^{p} \frac{1}{(|i-j|-1)^{k+\ell-2}} \le \sum_{\ell=2}^{p} \frac{1}{(|i-j|-1)^{\ell}} \\
&\le \frac{1}{(|i-j|-1)^2} \sum_{\ell=2}^{\infty} \frac{1}{(|i-j|-1)^{\ell-2}}
\end{aligned}$$

$$\lesssim \frac{1}{(|i-j|-1)^2}. \tag{4.17}$$

Due to Lemma C.16 we have (note the scaling)

$$\left(\sum_{k=2}^{p}\left(\frac{|u_k|}{(k-1)\sqrt{2k-1}}\right)^2\right)^{1/2} \lesssim \|u\|_{\tilde{H}_I^{1/2}(\Gamma_i)} \lesssim \|u\|_{\tilde{H}_I^{1/2}(\Gamma)} \simeq a_{\mathfrak{W}}(u,u)^{1/2} \tag{4.18}$$

where the constant in the second inequality and thus the constant in the equivalence depend on $h$, which is fixed. Using (4.16), (4.17) and (4.18) we obtain

$$\theta_{ij} \lesssim \frac{1}{(|i-j|-1)^2} \quad \text{for } |i-j| \geq 3. \tag{4.19}$$

Due to the Cauchy–Schwarz inequality we have $\theta_{ij} \leq 1$ for $|i-j| \leq 2$ independent of $p$ and $N$. Therefore, for any $i, j = 1, \ldots, N$ we have

$$\theta_{ij} \lesssim \frac{1}{(|i-j|+1)^2}.$$

Finally we have

$$\|\Theta\|_2 \leq \|\Theta\|_\infty \lesssim \max_{1\leq i\leq N} \sum_{j=1}^{N} \frac{1}{(|i-j|+1)^2} \lesssim \sum_{j=1}^{N} \frac{1}{j^2} \lesssim 1.$$

This completes the proof of the lemma. □

Theorem 2.2, Theorem 2.7, Lemma 4.1, Lemma 4.2, and Lemma 4.8 yield

**Theorem 4.5.** *(i) The condition number of the symmetric multiplicative Schwarz operator $P_{\text{smu}}$ defined with the subspace decomposition* (4.4) *satisfies*

$$\kappa(P_{\text{smu}}) \lesssim (1+\log p)^2.$$

*(ii) The convergence rate of the linear iterative method associated with the non-symmetric multiplicative Schwarz preconditioner defined with the subspace decomposition* (4.4) *satisfies*

$$\|\mathbf{I} - \mathbf{C}_{\text{mu}}^{-1}\mathbf{A}\|_{\mathbf{A}} \leq 1 - \frac{c}{(1+\log p)^2}$$

*where $c$ is a constant.*

#### 4.2.1.2 The weakly-singular integral equation

Consider the subspace decomposition defined in Definition 4.3. Again our task is to prove a Strengthened Cauchy–Schwarz inequality.

**Lemma 4.9.** *Let $\widetilde{\mathscr{V}}_i$ be defined as in Definition 4.3, and let*

$$\theta_{ij} := \sup_{u\in\widetilde{\mathscr{V}}_i, v\in\widetilde{\mathscr{V}}_j} \frac{a(u,v)}{a(u,u)^{1/2}a(v,v)^{1/2}}.$$

*If $0 < h \le 1$, then*

$$\theta_{ij} \lesssim \frac{1}{(|i-j|+1)^2}.$$

*The constant is independent of $i$, $j$, and $p$. Moreover, if $\Theta = \{\theta_{ij}\}_{1\le i,j\le N}$, then*

$$\|\Theta\|_2 \lesssim 1$$

*where the constant is independent of $p$.*

*Proof.* Let $u = \sum_{k=1}^{p} u_k L_{k,i}$ and $v = \sum_{\ell=1}^{p} v_k L_{\ell,j}$. Then

$$a_{\mathfrak{V}}(u,v) = \sum_{k=1}^{p}\sum_{\ell=1}^{p} u_k v_\ell \langle \mathfrak{V} L_{k,i}, L_{\ell,j}\rangle_{L^2(\Gamma_j)}. \tag{4.20}$$

Due to Lemma C.17 (compare with [106, Lemma 1]) we have with $|\Gamma_i| = |\Gamma_j| = h$ and $\operatorname{dist}(\Gamma_i,\Gamma_j) = h(|i-j|-1)$

$$\begin{aligned}
\langle \mathfrak{V} L_{k,i}, L_{\ell,j}\rangle_{L^2(\Gamma_j)} &\lesssim \frac{|\Gamma_i|\,|\Gamma_j|\max(|\Gamma_i|/2,|\Gamma_j|/2)^{k+\ell}}{\operatorname{dist}(\Gamma_i,\Gamma_j)^{k+\ell}\sqrt{(2k+1)(2\ell+1)}}\\
&= \frac{h^2(1/2)^{k+\ell}}{(|i-j|-1)^{k+\ell}\sqrt{(2k+1)(2\ell+1)}}\\
&\lesssim \frac{h^2}{k\ell(|i-j|-1)^{k+\ell}\sqrt{(2k+1)(2\ell+1)}}.
\end{aligned} \tag{4.21}$$

From (4.20) and (4.21) we obtain

$$\begin{aligned}
|a_{\mathfrak{V}}(u,v)| &\lesssim h^2 \sum_{k=1}^{p}\sum_{\ell=1}^{p} \frac{|u_k|\,|v_\ell|}{k\ell(|i-j|-1)^{k+\ell}\sqrt{(2k+1)(2\ell+1)}}\\
&= h^2 \sum_{k=1}^{p}\sum_{\ell=1}^{p} \frac{|u_k|}{k\sqrt{2k+1}}\,\frac{|v_\ell|}{\ell\sqrt{2\ell+1}}\,\frac{1}{(|i-j|-1)^{k+\ell}}\\
&\lesssim h^2 \left(\sum_{k=1}^{p}\left(\frac{|u_k|}{k\sqrt{2k+1}}\right)^2\right)^{1/2} \left(\sum_{\ell=1}^{p}\left(\frac{|v_\ell|}{\ell\sqrt{2\ell+1}}\right)^2\right)^{1/2} \times\\
&\qquad \left\| \left\{ \frac{1}{(|i-j|-1)^{k+\ell}} \right\}_{k,\ell=1,\dots,p} \right\|_2.
\end{aligned} \tag{4.22}$$

For $|i-j| \ge 3$ we have

$$\begin{aligned}\left\|\left\{\frac{1}{(|i-j|-1)^{k+\ell}}\right\}_{k,\ell=1,\dots,p}\right\|_2 &\le \left\|\left\{\frac{1}{(|i-j|-1)^{k+\ell}}\right\}_{k,\ell=1,\dots,p}\right\|_\infty\\ &= \max_{1\le k\le p}\sum_{\ell=1}^{p}\frac{1}{(|i-j|-1)^{k+\ell}} \le \sum_{\ell=1}^{p}\frac{1}{(|i-j|-1)^{\ell+1}}\\ &\le \frac{1}{(|i-j|-1)^2}\sum_{\ell=0}^{\infty}\frac{1}{(|i-j|-1)^{\ell}}\\ &\lesssim \frac{1}{(|i-j|-1)^2}. \end{aligned} \tag{4.23}$$

Due to Lemma C.14 we have for $0<h\le 1$ (note the scaling)

$$h\left(\sum_{k=1}^{p}\left(\frac{|u_k|}{k\sqrt{2k+1}}\right)^2\right)^{1/2} \lesssim \|u\|_{\widetilde{H}_{\mathrm{I}}^{-1/2}(\Gamma_i)} \lesssim \|u\|_{\widetilde{H}_{\mathrm{I}}^{-1/2}(\Gamma)} \lesssim a_{\mathfrak{V}}(u,u)^{1/2}. \tag{4.24}$$

The constant depends on $h$ which is fixed. Using (4.22), (4.23) and (4.24) we obtain for $|i-j|\ge 3$

$$\theta_{ij} \le \frac{C}{(|i-j|-1)^2}.$$

Due to the Cauchy–Schwarz inequality we have $\theta_{ij}\le 1$ for $|i-j|\le 2$ independent of $p$ and $N$. Therefore, for any $i,j=1,\dots,N$ there holds

$$\theta_{ij} \le \frac{C}{(|i-j|+1)^2}.$$

Finally we have

$$\|\Theta\|_2 \le \|\Theta\|_\infty \lesssim \max_{1\le i\le N}\sum_{j=1}^{N}\frac{1}{(|i-j|+1)^2} \lesssim \sum_{j=1}^{N}\frac{1}{j^2} \lesssim 1.$$

This completes the proof of the lemma. □

Theorem 2.2, Theorem 2.7, Lemma 4.3, Lemma 4.4, and Lemma 4.9 yield

**Theorem 4.6.** *(i) The condition number of the symmetric multiplicative Schwarz operator $P_{\mathrm{smu}}$ defined with the subspace decomposition* (4.8) *satisfies*

$$\kappa(P_{\mathrm{smu}}) \lesssim (1+\log p)^2.$$

*(ii) The convergence rate of the linear iterative method associated with the nonsymmetric multiplicative Schwarz preconditioner defined with the subspace decomposition* (4.8) *satisfies*

$$\|\mathbf{I}-\mathbf{C}_{\mathrm{mu}}^{-1}\mathbf{A}\|_{\mathbf{A}} \le 1-\frac{c}{(1+\log p)^2}$$

*where $c$ is a constant.*

## 4.2.2 Overlapping methods

### 4.2.2.1 The hypersingular integral equation

We will in this subsection design an overlapping method for the $p$-version of the Galerkin boundary element method applied to the hypersingular integral equation. The subspace decomposition is the same as for the additive Schwarz method, namely the subspace decomposition 3 defined in Definition 4.4.

Similarly to the case of non-overlapping methods, it suffices to show a Strengthened Cauchy–Schwarz inequality. We first notice that for $i = 1,\dots,N-1$, a basis for $\mathscr{V}_i$ consists of $\mathscr{L}_{q,i}$ and $\mathscr{L}_{q,i+1}$, where $\mathscr{L}_{q,i}$ is defined in Subsection 4.1.1.1. Then a function $u \in \mathscr{V}_i$ can be represented by

$$u = \sum_{k=2}^{p} u_{k,i}\mathscr{L}_{k,i} + \sum_{k=2}^{p} u_{k,i+1}\mathscr{L}_{k,i+1}. \tag{4.25}$$

**Lemma 4.10.** *Let $\mathscr{V}_i$ be defined as in Definition 4.4 and let*

$$\theta_{ij} := \sup_{u\in\mathscr{V}_i, v\in\mathscr{V}_j} \frac{a_{\mathfrak{W}}(u,v)}{a_{\mathfrak{W}}(u,u)^{1/2}a_{\mathfrak{W}}(v,v)^{1/2}}, \quad i,j = 1,\dots,N-1.$$

*If $\Theta = \{\theta_{ij}\}_{1\le i,j\le N-1}$, then*

$$\|\Theta\|_2 \lesssim 1.$$

*The constant is independent of p.*

*Proof.* The proof is similar to that of Lemma 4.8. Instead of (4.14) we now have, due to (4.25),

$$a_{\mathfrak{W}}(u,v) = \sum_{m=0}^{1}\sum_{n=0}^{1}\sum_{k=2}^{p}\sum_{\ell=2}^{p} u_{k,i+m}v_{\ell,j+n}\left\langle \mathfrak{W}\mathscr{L}_{k,i+m}, \mathscr{L}_{\ell,j+n}\right\rangle_{L^2(\Gamma_{j+n})}.$$

Note that $\operatorname{dist}(\Gamma_i', \Gamma_j') = h(|i-j|-2)$, $i \neq j$. Similarly to (4.15), for $|i-j| \geq 3$, we have

$$\left\langle \mathfrak{W}L_{k-1,i+m}, L_{\ell-1,j+n}\right\rangle_{L^2(\Gamma_{j+n})} \lesssim \frac{h^2}{(k-1)(\ell-1)(|i-j|-2)^{k+\ell-2}\sqrt{(2k-1)(2\ell-1)}}.$$

Repeating the argument as in the proof of Lemma 4.8 we obtain, similarly to (4.19),

$$\theta_{ij} \lesssim \frac{1}{(|i-j|-2)^2} \quad \text{for } |i-j| \geq 4.$$

The remainder of the proof goes as before and is omitted. □

As a consequence of Theorem 2.2, Theorem 2.7, Lemmas 4.5, 4.7 and 4.10, we have the following theorem.

**Theorem 4.7.** *(i) The condition number of the overlapping symmetric multiplicative Schwarz operator $P_{\text{smu}}$ defined from the subspace decomposition given in Definition 4.4 is bounded independent of p.*

*(ii) The convergence rate of the linear iterative method associated with the nonsymmetric multiplicative Schwarz preconditioner defined with the subspace decompositiongiven in Definition 4.4 is bounded by a positive constant less than* 1.

#### 4.2.2.2 The weakly-singular integral equation

The overlapping method for this equation can be carried out in exactly the same manner. We use again the subspace decomposition 4 defined in Definition 4.4. Then a function $u \in \widetilde{\mathscr{V}}_j$, $j = 1, \ldots, N-1$, can be represented by

$$u = \sum_{k=1}^{p} u_{k,j} L_{k,j} + \sum_{k=1}^{p} u_{k,j+1} L_{k,j+1}.$$

We state here the Strengthened Cauchy–Schwarz inequality and leave the proof to the reader.

**Lemma 4.11.** *Let $\widetilde{\mathscr{V}}_i$ be defined as in Definition 4.5 and let*

$$\theta_{ij} := \sup_{u \in \widetilde{\mathscr{V}}_i, v \in \widetilde{\mathscr{V}}_j} \frac{a_{\mathfrak{V}}(u,v)}{a_{\mathfrak{V}}(u,u)^{1/2} a_{\mathfrak{V}}(v,v)^{1/2}}, \quad i,j = 1, \ldots, N-1.$$

*If $\Theta = \{\theta_{ij}\}_{1 \le i,j \le N-1}$, then*

$$\|\Theta\|_2 \lesssim 1$$

*where the constant is independent of p.*

As a consequence of Theorem 2.2, Theorem 2.7, Lemma 4.11, and the stability and coercivity of the decomposition (the details of which are left for the reader to verify), we have the following result.

**Theorem 4.8.** *(i) The condition number of the overlapping symmetric multiplicative Schwarz operator $P_{\text{smu}}$ defined from the subspace decomposition given in Definition 4.5 is bounded independent of p.*

*(ii) The convergence rate of the linear iterative method associated with the nonsymmetric multiplicative Schwarz preconditioner defined with the subspace decomposition given in Definition 4.5 is bounded by a positive constant less than* 1.

## 4.3 Numerical Results

The following table relates the theoretical results obtained in this chapter with the numerical results obtained in Chapter 9. These numerical results corroborate the theoretical results.

| Equation | Preconditioner | | Theorem | Table | Figure |
|---|---|---|---|---|---|
| Hypersingular | Additive Schwarz | PCG–Non-overlapping | 4.1 | 9.13–9.18 | 9.7 |
| | | PCG–Overlapping | 4.3 | 9.14–9.17 | |
| | Multiplicative Schwarz | Linear iterative method | 4.5 | | 9.10–9.11 |
| Weakly-singular | Additive Schwarz | PCG–Non-overlapping | 4.2 | 9.7–9.12 | 9.6 |
| | Multiplicative Schwarz | Linear iterative method | 4.6 | | 9.8–9.9 |

**Table 4.1** Theoretical results in Chapter 4 and relevant numerical results in Chapter 9

# Chapter 5
# Multilevel Methods: the $h$-Version

When applying the two-level methods for the $h$-version (see Chapter 3) we have to solve a global problem on the coarse space $\mathscr{V}_0$ or $\widetilde{\mathscr{V}}_0$; cf. Algorithms 2.7 and 2.11. The inclusion of this global problem is necessary; see e.g. [71, 237]. However, the dimensions of $\mathscr{V}_0$ and $\widetilde{\mathscr{V}}_0$, determined by the coarse mesh size $H$, are large when $H$ is small; see Definition 3.1. Therefore it is natural to carry out the same algorithm for this global problem, which leads to a multilevel method.

In this chapter, we analyse multilevel additive and multiplicative Schwarz preconditioners when the $h$-version Galerkin approximation is used to solve the hypersingular and weakly-singular integral equations. The results for the additive Schwarz preconditioner in this chapter improve the initial results proved in our paper [227], while those for the non-symmetric multiplicative method were first obtained in [144], and those for the symmetric multiplicative method are new.

## 5.1 Additive Schwarz Methods

### 5.1.1 *The hypersingular integral equation*

The multilevel method is defined as follows. Starting with a coarse mesh

$$\mathscr{N}^1 : -1 = x_0^1 < x_1^1 = 0 < x_2^1 = 1,$$

we successively divide each subinterval into two equal intervals to obtain the meshes $\mathscr{N}^\ell$ with mesh points $x_i^\ell$, $i = 0, \dots, N_\ell + 1$, $\ell = 1, \dots, L$, where $N_\ell = 2^\ell - 1$. Hence, if $h_\ell$ is the mesh size of $\mathscr{N}^\ell$, $\ell = 1, \dots, L-1$, then $h_\ell = 2h_{\ell+1}$. For $\ell = 1, \dots, L$, let $\mathscr{V}^\ell$ be the finite-dimensional space associated with $\mathscr{N}^\ell$ which contains continuous and piecewise-linear functions vanishing at the endpoints $\pm 1$. We have a nested sequence of spaces

$$\mathscr{V}^1 \subset \mathscr{V}^2 \subset \cdots \subset \mathscr{V}^L.$$



E. P. Stephan and T. Tran, *Schwarz Methods and Multilevel Preconditioners for Boundary Element Methods*, https://doi.org/10.1007/978-3-030-79283-1_5

Let $\left\{\phi_1^\ell,\dots,\phi_{N_\ell}^\ell\right\}$ be the nodal basis for $\mathscr{V}^\ell$. Then $\operatorname{supp}\phi_i^\ell = [x_{i-1}^\ell, x_{i+1}^\ell]$, $i = 1,\dots,N_\ell$. For $\ell = 1,\dots,L$, we decompose $\mathscr{V}^\ell$ as

$$\mathscr{V}^\ell = \sum_{i=1}^{N_\ell} \mathscr{V}_i^\ell, \tag{5.1}$$

with $\mathscr{V}_i^\ell = \operatorname{span}\{\phi_i^\ell\}$ for $i = 1,\dots,N_\ell$. Eventually, $\mathscr{V}^L$ is decomposed as

$$\mathscr{V}^L = \sum_{\ell=1}^{L}\sum_{i=1}^{N_\ell} \mathscr{V}_i^\ell. \tag{5.2}$$

Let $P_{\mathscr{V}_i^\ell} : \mathscr{V}^L \to \mathscr{V}_i^\ell$ be defined for any $v \in \mathscr{V}^L$ by

$$a_{\mathfrak{W}}(P_{\mathscr{V}_i^\ell} v, w) = a_{\mathfrak{W}}(v, w) \quad \text{for all } w \in \mathscr{V}_i^\ell \tag{5.3}$$

and let

$$T_\ell := \sum_{i=1}^{N_\ell} P_{\mathscr{V}_i^\ell}. \tag{5.4}$$

The multilevel additive Schwarz operator is defined as

$$P_{MAS} = \sum_{\ell=1}^{L}\sum_{i=1}^{N_\ell} P_{\mathscr{V}_i^\ell} = \sum_{\ell=1}^{L} T_\ell. \tag{5.5}$$

This multilevel method was originally designed in [71] for finite element discretisations of elliptic differential equations. An alternative discussion was given in [40] with the so-called BPX preconditioner. We shall prove in this section that this method is also a good preconditioner for the hypersingular integral equation.

We will first prove the stability of the decomposition (5.2).

**Lemma 5.1 (Stability of decomposition).** *Any $v \in \mathscr{V}$ admits a decomposition*

$$v = \sum_{\ell=1}^{L}\sum_{i=1}^{N_\ell} v_i^\ell, \quad v_i^\ell \in \mathscr{V}_i^\ell,$$

*that satisfies*

$$\sum_{\ell=1}^{L}\sum_{i=1}^{N_\ell} a_{\mathfrak{W}}(v_i^\ell, v_i^\ell) \lesssim a_{\mathfrak{W}}(v, v).$$

*The constant is independent of $L$ and $N_\ell$, $\ell = 1,\dots,L$.*

*Proof.* Similarly to the case of the two-level method, we define for each $\ell = 1,\dots,L$ a projector $\widetilde{P}_{\mathscr{V}^\ell} : \widetilde{H}^{1/2}(\Gamma) \to \mathscr{V}^\ell$, $v \mapsto \widetilde{P}_{\mathscr{V}^\ell}$ by

$$\widetilde{a}(\widetilde{P}_{\mathscr{V}^\ell} v, \phi) = \widetilde{a}(v, \phi) \quad \text{for any } \phi \in \mathscr{V}^\ell$$

where the bilinear form $\widetilde{a}(\cdot,\cdot)$ is defined by

$$\widetilde{a}(\phi,\psi) := \langle \Lambda\phi, \psi\rangle, \quad \phi,\ \psi \in \widetilde{H}^{1/2}(\Gamma).$$

Here the operator $\Lambda$ is defined in (3.33). Similarly to (3.37) we now have

$$\|\widetilde{P}_{\mathcal{V}^\ell} v - v\|_{L^2(\Gamma)} \le c h_\ell^{1/2} \|v\|_{\widetilde{H}^{1/2}(\Gamma)}, \tag{5.6}$$

for any $v \in \widetilde{H}^{1/2}(\Gamma)$. With the introduction of $\widetilde{P}_{\mathcal{V}^\ell}$, we can first decompose $v \in \mathcal{V}^L$ as $v = \sum_{\ell=1}^L v^\ell$ by letting

$$v^\ell := \left(\widetilde{P}_{\mathcal{V}^\ell} - \widetilde{P}_{\mathcal{V}^{\ell-1}}\right) v \quad \text{for any } \ell = 1,\dots,L, \quad \text{with } \widetilde{P}_{V^0} \equiv 0. \tag{5.7}$$

We then further decompose $v^\ell$ as

$$v^\ell = \sum_{i=1}^{N_\ell} v_i^\ell, \quad \text{with } v_i^\ell := \Pi^\ell(\theta_i^\ell v^\ell) \in \mathcal{V}_i^\ell.$$

Here $\Pi^\ell$ is the standard nodal value interpolation operator from $C(\Gamma)$ onto $\mathcal{V}^\ell$ and, for each $\ell = 1,\dots,L$, $\left\{\theta_i^\ell\right\}_{i=1}^{N_\ell}$ is a partition of unity defined as in (4.12). We note that the definition (5.7) of $v^\ell$ yields

$$\widetilde{a}(v^\ell, v^k) = 0 \quad \text{for } 1 \le \ell \ne k \le L. \tag{5.8}$$

In fact, assume that $\ell < k$. Then $v^\ell \in \mathcal{V}^\ell \subset \mathcal{V}^{k-1} \subset \mathcal{V}^k$, and therefore

$$\begin{aligned}
\widetilde{a}(v^k, v^\ell) &= \widetilde{a}(\widetilde{P}_{\mathcal{V}^k} v, v^\ell) - \widetilde{a}(\widetilde{P}_{\mathcal{V}^{k-1}} v, v^\ell) \\
&= \widetilde{a}(v, v^\ell) - \widetilde{a}(v, v^\ell) = 0.
\end{aligned}$$

There also holds from the definition (5.7) of $v^\ell$ that

$$\widetilde{P}_{\mathcal{V}^{\ell-1}} v^\ell = 0 \quad \text{for any } \ell = 1,\dots,L. \tag{5.9}$$

With the above decomposition of $v \in \mathcal{V}^L$, $v = \sum_{\ell=1}^L \sum_{j=1}^{N_\ell} v_j^\ell$, we can now prove the required estimate. We have

$$\begin{aligned}
\|v_i^\ell\|^2_{\widetilde{H}_{\mathrm{I}}^{1/2}(\Gamma_i^\ell)} &\lesssim \|\theta_i^\ell v^\ell\|^2_{\widetilde{H}_{\mathrm{I}}^{1/2}(\Gamma_i^\ell)} \lesssim \|\theta_i^\ell v^\ell\|_{\widetilde{H}_{\mathrm{I}}^0(\Gamma_i^\ell)} \|\theta_i^\ell v^\ell\|_{\widetilde{H}_{\mathrm{I}}^1(\Gamma_i^\ell)} \\
&\lesssim \|v^\ell\|_{L^2(\Gamma_i^\ell)} \left(h_\ell^{-1} \|v^\ell\|_{L^2(\Gamma_i^\ell)} + \|v^\ell\|_{H^1(\Gamma_i^\ell)}\right) \\
&\lesssim \left(h_\ell^{-1} \|v^\ell\|^2_{L^2(\Gamma_i^\ell)} + h_\ell^{-1/2} \|v^\ell\|_{L^2(\Gamma_i^\ell)} h_\ell^{1/2} \|v^\ell\|_{H^1(\Gamma_i^\ell)}\right) \\
&\lesssim \left(h_\ell^{-1} \|v^\ell\|^2_{L^2(\Gamma_i^\ell)} + h_\ell \|v^\ell\|^2_{H^1(\Gamma_i^\ell)}\right).
\end{aligned}$$

Summing over $i = 1,\dots,N_\ell$ we achieve

$$\sum_{i=1}^{N_\ell} \|v_i^\ell\|^2_{\widetilde{H}_{\mathrm{I}}^{1/2}(\Gamma_i^\ell)} \le c\left(h_\ell^{-1}\|v^\ell\|^2_{L^2(\Gamma)} + h_\ell\|v^\ell\|^2_{\widetilde{H}_{\mathrm{I}}^{1}(\Gamma)}\right). \tag{5.10}$$

It is now necessary to estimate $\|v^\ell\|^2_{L^2(\Gamma)}$. By writing, see (5.9),

$$v^\ell = (I - \widetilde{P}_{\mathscr{V}^{\ell-1}})v^\ell,$$

and by using (5.6), we obtain

$$\|v^\ell\|_{L^2(\Gamma)} = \|(I - \widetilde{P}_{\mathscr{V}^{\ell-1}})v^\ell\|_{L^2(\Gamma)} \lesssim h_{\ell-1}^{1/2}\|v^\ell\|_{\widetilde{H}_{\mathrm{I}}^{1/2}(\Gamma)}. \tag{5.11}$$

Inequalities (5.10), (5.11) and the inverse property of the boundary element functions now yield

$$\sum_{i=1}^{N_\ell} \|v_i^\ell\|^2_{\widetilde{H}_{\mathrm{I}}^{1/2}(\Gamma_i^\ell)} \lesssim \|v^\ell\|^2_{\widetilde{H}_{\mathrm{I}}^{1/2}(\Gamma)}. \tag{5.12}$$

By using successively Lemma B.3, Lemma A.13, (5.12), (3.34), and (5.8), we obtain

$$\begin{aligned}
\sum_{\ell=1}^{L}\sum_{i=1}^{N_\ell} a_{\mathfrak{W}}(v_i^\ell, v_i^\ell) &\simeq \sum_{\ell=1}^{L}\sum_{i=1}^{N_\ell} \|v_i^\ell\|^2_{\widetilde{H}_{\mathrm{I}}^{1/2}(\Gamma)} \lesssim \sum_{\ell=1}^{L}\sum_{i=1}^{N_\ell} \|v_i^\ell\|^2_{\widetilde{H}_{\mathrm{I}}^{1/2}(\Gamma_i^\ell)} \\
&\lesssim \sum_{\ell=1}^{L} \|v^\ell\|^2_{\widetilde{H}_{\mathrm{I}}^{1/2}(\Gamma)} \simeq \sum_{\ell=1}^{L} \widetilde{a}(v^\ell, v^\ell) \\
&= \widetilde{a}(v,v) \simeq \|v\|^2_{\widetilde{H}_{\mathrm{I}}^{1/2}(\Gamma)} \simeq a_{\mathfrak{W}}(v,v).
\end{aligned}$$

The lemma is proved. □

The operator $T_\ell$ defined by (5.4) has the following property.

**Lemma 5.2 (Strengthened Cauchy–Schwarz inequality).** *There exist constants $C > 0$ and $\gamma \in (0,1)$ such that, for any $v \in \mathscr{V}^k$ where $1 \le k \le \ell \le L$, the following inequality holds*

$$a_{\mathfrak{W}}(T_\ell v, v) \le C\gamma^{2(\ell-k)} a_{\mathfrak{W}}(v,v).$$

*Proof.* Since $\mathscr{V}_i^\ell = \operatorname{span}\,\{\phi_i^\ell\}$, for any $v \in \mathscr{V}^L$ we have

$$P_{\mathscr{V}_i^\ell} v = \frac{a_{\mathfrak{W}}(v, \phi_i^\ell)}{a_{\mathfrak{W}}(\phi_i^\ell, \phi_i^\ell)}\phi_i^\ell,$$

so that

$$T_\ell v = \sum_{i=1}^{N_\ell} \frac{a_{\mathfrak{W}}(v, \phi_i^\ell)}{a_{\mathfrak{W}}(\phi_i^\ell, \phi_i^\ell)}\phi_i^\ell,$$

which implies

$$a_{\mathfrak{W}}(T_\ell v, v) = \sum_{i=1}^{N_\ell} \frac{a_{\mathfrak{W}}(v, \phi_i^\ell)^2}{a_{\mathfrak{W}}(\phi_i^\ell, \phi_i^\ell)}. \tag{5.13}$$

It is necessary to estimate $a_{\mathfrak{W}}(v, \phi_i^\ell)$. We define, for each $\ell = 1, \dots, L$, the operator $\mathfrak{W}_\ell : \mathscr{V}^\ell \to \mathscr{V}^\ell$ by

$$\langle \mathfrak{W}_\ell \phi, \psi \rangle = a_{\mathfrak{W}}(\phi, \psi) \quad \text{for any } \phi, \psi \in \mathscr{V}^\ell.$$

Then, by noting that $\operatorname{supp} \phi_i^\ell = \overline{\Gamma}'_{\ell,i} := [x_{i-1}^\ell, x_{i+1}^\ell]$, we have

$$a_{\mathfrak{W}}(v, \phi_i^\ell)^2 = \langle \mathfrak{W}_\ell v, \phi_i^\ell \rangle^2 \le \|\mathfrak{W}_\ell v\|^2_{L^2(\Gamma'_{\ell,i})} \|\phi_i^\ell\|^2_{L^2(\Gamma'_{\ell,i})}. \tag{5.14}$$

It follows from Lemma C.12 that

$$\|\phi_i^\ell\|^2_{L^2(\Gamma'_{\ell,i})} \simeq h_\ell \simeq h_\ell \|\phi_i^\ell\|^2_{\widetilde{H}_{\mathrm{I}}^{1/2}(\Gamma'_{\ell,i})}. \tag{5.15}$$

Inequalities (5.13), (5.14), (5.15), and the equivalence $a_{\mathfrak{W}}(\phi_i^\ell, \phi_i^\ell) \simeq \|\phi_i^\ell\|^2_{\widetilde{H}_{\mathrm{I}}^{1/2}(\Gamma'_{\ell,i})}$ imply

$$a_{\mathfrak{W}}(T_\ell v, v) \lesssim h_\ell \sum_{i=1}^{N_\ell} \|\mathfrak{W}_\ell v\|^2_{L^2(\Gamma'_{\ell,i})} \lesssim h_\ell \|\mathfrak{W}_\ell v\|^2_{L^2(\Gamma)}. \tag{5.16}$$

By using the definition of $\mathfrak{W}_\ell$, the mapping property of the hypersingular integral operator $\mathfrak{W}$, and the inverse property we obtain

$$\begin{aligned}
\|\mathfrak{W}_\ell v\|^2_{L^2(\Gamma)} &= \langle \mathfrak{W}_\ell v, \mathfrak{W}_\ell v \rangle \\
&= \langle \mathfrak{W} v, \mathfrak{W}_\ell v \rangle \\
&\le \|\mathfrak{W}_\ell v\|_{\widetilde{H}^\varepsilon(\Gamma)} \|\mathfrak{W} v\|_{H^{-\varepsilon}(\Gamma)} \\
&\lesssim \|\mathfrak{W}_\ell v\|_{\widetilde{H}^\varepsilon(\Gamma)} \|v\|_{\widetilde{H}^{1-\varepsilon}(\Gamma)} \\
&\lesssim h_\ell^{-\varepsilon} h_k^{-1/2+\varepsilon} \|\mathfrak{W}_\ell v\|_{L^2(\Gamma)} \|v\|_{\widetilde{H}^{1/2}(\Gamma)}
\end{aligned}$$

for any $\varepsilon > 0$. Hence

$$\|\mathfrak{W}_\ell v\|^2_{L^2(\Gamma)} \le C h_\ell^{-2\varepsilon} h_k^{-1+2\varepsilon} \|v\|^2_{\widetilde{H}^{1/2}(\Gamma)}. \tag{5.17}$$

Inequalities (5.16) and (5.17) then give

$$a_{\mathfrak{W}}(T_\ell v, v) \le C h_\ell^{1-2\varepsilon} h_k^{-1+2\varepsilon} a_{\mathfrak{W}}(v, v) = C \gamma^{2(\ell-k)} a_{\mathfrak{W}}(v, v),$$

where $\gamma = (1/2)^{(1-2\varepsilon)/2}$, proving the lemma. □

Using the above Strengthened Cauchy–Schwarz inequality we prove the following result for estimating the maximum eigenvalue of the operator $P_{MAS}$.

**Lemma 5.3.** *The multilevel additive Schwarz operator $P_{MAS}$ satisfies*

$$a_{\mathfrak{W}}(P_{MAS}v,v) \lesssim a_{\mathfrak{W}}(v,v) \quad \textit{for any } v \in \mathscr{V}^L \tag{5.18}$$

*where the constant is independent of the number of levels and the number of mesh points.*

*Proof.* The proof is similar to [38, Theorem 3.1] and is included here for completeness. For $\ell = 1,\dots,L$ let $P_\ell : \mathscr{V}^L \to \mathscr{V}^\ell$ be defined for any $v \in \mathscr{V}^L$ as

$$a_{\mathfrak{W}}(P_\ell v,w) = a_{\mathfrak{W}}(v,w) \quad \text{for any } w \in \mathscr{V}^\ell.$$

Then for any $v \in \mathscr{V}^L$ we can write

$$P_\ell v = \sum_{k=1}^{\ell} (P_k - P_{k-1})v, \tag{5.19}$$

where $P_0 \equiv 0$. Hence

$$\begin{aligned} a_{\mathfrak{W}}(P_{MAS}v,v) &= \sum_{\ell=1}^{L} a_{\mathfrak{W}}(T_\ell v,v) = \sum_{\ell=1}^{L} a_{\mathfrak{W}}(T_\ell v,P_\ell v) \\ &= \sum_{\ell=1}^{L}\sum_{k=1}^{\ell} a_{\mathfrak{W}}\big(T_\ell v,(P_k - P_{k-1})v\big). \end{aligned}$$

Since $a_{\mathfrak{W}}(T_\ell\cdot,\cdot)$ is symmetric and positive definite, we can use Cauchy–Schwarz's inequality with respect to that quadratic form to obtain

$$a_{\mathfrak{W}}(P_{MAS}v,v) \le \sum_{\ell=1}^{L}\sum_{k=1}^{\ell} a_{\mathfrak{W}}(T_\ell v,v)^{1/2}\, a_{\mathfrak{W}}\big(T_\ell(P_k - P_{k-1})v,(P_k - P_{k-1})v\big)^{1/2}.$$

Since $\mathscr{V}^{k-1} \subset \mathscr{V}^k \subset \mathscr{V}^\ell$, we can invoke Lemma 5.2 to deduce

$$\begin{aligned} a_{\mathfrak{W}}\big(T_\ell(P_k - P_{k-1})v,(P_k - P_{k-1})v\big)^{1/2} &\le C^{1/2}\gamma^{\ell-k} a_{\mathfrak{W}}\big((P_k - P_{k-1})v,(P_k - P_{k-1})v\big)^{1/2} \\ &= C^{1/2}\gamma^{\ell-k} a_{\mathfrak{W}}\big((P_k - P_{k-1})v,v\big)^{1/2} \end{aligned}$$

where the above equality is obtained by a simple calculation. Consequently, with $S := \sum_{i=0}^{\infty} \gamma^i$,

$$\begin{aligned} a_{\mathfrak{W}}(P_{MAS}v,v) &\le \sum_{\ell=1}^{L}\sum_{k=1}^{\ell} \gamma^{(\ell-k)/2} a_{\mathfrak{W}}(T_\ell v,v)^{1/2}\, C^{1/2}\gamma^{(\ell-k)/2}\, a_{\mathfrak{W}}\big((P_k - P_{k-1})v,v\big)^{1/2} \\ &\le \sum_{\ell=1}^{L}\sum_{k=1}^{\ell} \Big(\frac{1}{2S}\gamma^{\ell-k} a_{\mathfrak{W}}(T_\ell v,v) + \frac{CS}{2}\gamma^{\ell-k} a_{\mathfrak{W}}\big((P_k - P_{k-1})v,v\big)\Big) \\ &\le \frac{1}{2}\sum_{\ell=1}^{L} a_{\mathfrak{W}}(T_\ell v,v) + \frac{CS^2}{2}\sum_{k=1}^{L} a_{\mathfrak{W}}\big((P_k - P_{k-1})v,v\big) \end{aligned}$$

$$= \frac{1}{2} a_{\mathfrak{W}}(P_{MAS}v, v) + \frac{CS^2}{2} a_{\mathfrak{W}}(P_L v, v)$$
$$= \frac{1}{2} a_{\mathfrak{W}}(P_{MAS}v, v) + \frac{CS^2}{2} a_{\mathfrak{W}}(v, v),$$

where in the second last step we used (5.19). Inequality (5.18) then follows. □

Lemma 5.1 and Proposition 2.1 yield $\lambda_{\min}(P_{MAS}) \gtrsim 1$. Moreover, Lemma 5.3 yields $\lambda_{\max}(P_{MAS}) \lesssim 1$. Therefore we have the following result.

**Theorem 5.1.** *The condition number of the multilevel additive Schwarz method defined by (5.5) has a bounded condition number*

$$\kappa(P_{MAS}) \lesssim 1.$$

### *5.1.2 The weakly-singular integral equation*

We proceed as in the case of the two-level method in Subsection 3.1.1.2. Let $\mathscr{N}_\ell$, $\ell = 1, 2, \ldots, L$ be the meshes defined in Subsection 5.1.1. We define (compare with (3.25) and (3.26))

$$\chi_0^1(x) \equiv 1 \quad \text{and} \quad \chi_i^\ell(x) := \begin{cases} \dfrac{1}{h_\ell}, & x \in (x_{i-1}^\ell, x_i^\ell), \\ -\dfrac{1}{h_\ell}, & x \in (x_i^\ell, x_{i+1}^\ell), \\ 0, & \text{otherwise}, \end{cases}$$

where $i = 1, \ldots, N_\ell$, and $\ell = 1, \ldots, L$. (We note that by the way the meshes are defined, $N_\ell$ is an odd number.) The subspaces on level $\ell$ are defined by

$$\widetilde{\mathscr{V}}_0^1 = \operatorname{span}\{\chi_0^1\}, \quad \widetilde{\mathscr{V}}_i^\ell = \operatorname{span}\{\chi_i^\ell\}, \quad i = 1, \ldots, N_\ell.$$

By using the same argument as in Subsection 3.1.1.2, see (3.24), we can show that any $v \in \widetilde{\mathscr{V}}$ can be decomposed as

$$v = v_0 + \sum_{\ell=1}^{L} \sum_{\substack{i=1 \\ i \text{ odd}}}^{N_\ell} v_i^\ell \tag{5.20}$$

where $v_0 \in \widetilde{\mathscr{V}}_0^1$ and $v_i^\ell \in \widetilde{\mathscr{V}}_i^\ell$. Indeed, for $i = 1, 3, 5, \ldots, N_L$, we can define $v_i^L \in \widetilde{\mathscr{V}}_i^L$ by

$$v_i^L := v|_{\widetilde{\Gamma}_i^L} - \mu_{\widetilde{\Gamma}_i^L}(v)$$

where $\widetilde{\Gamma}_i^L := (x_{i-1}^L, x_{i+1}^L)$ is a subinterval on level $L-1$. Hence

$$v = v_{L-1} + \sum_{\substack{i=1 \\ i \text{ odd}}}^{N_L} v_i^L$$

where $v_{L-1}$ is a piecewise constant function on the mesh $\mathscr{N}^{L-1}$ which satisfies $(v_{L-1})\big|_{\widetilde{\Gamma}_i^L} := \mu_{\widetilde{\Gamma}_i^L}(v)$. Repeating the same argument, we arrive at (5.20).

Consequently the boundary element space $\widetilde{\mathscr{V}}$ consisting of piecewise constant functions on the mesh $\mathscr{N}_L$ can be decomposed by

$$\widetilde{\mathscr{V}} = \widetilde{\mathscr{V}}_0^1 + \sum_{\ell=1}^{L} \sum_{i=1}^{N_\ell} \widetilde{\mathscr{V}}_i^\ell. \tag{5.21}$$

We note that the functions $\chi_0^1, \chi_i^\ell, i = 1,3,5,\ldots,N_\ell, \ell = 1,\ldots,L$, form a hierarchical basis for $\widetilde{\mathscr{V}}$. However, a decomposition with these basis functions does not yield an optimal preconditioner; see [231].

The multilevel additive Schwarz operator is now defined as

$$P_{MAS} = P_{\widetilde{\mathscr{V}}_0^1} + \sum_{\ell=1}^{L} \sum_{i=1}^{N_\ell} P_{\widetilde{\mathscr{V}}_i^\ell}, \tag{5.22}$$

where $P_{\widetilde{\mathscr{V}}_i^\ell} : \widetilde{\mathscr{V}}^L \to \widetilde{\mathscr{V}}_i^\ell$ is defined for any $v \in \widetilde{\mathscr{V}}^L$ by

$$a_{\mathfrak{V}}(P_{\widetilde{\mathscr{V}}_i^\ell} v, w) = a_{\mathfrak{V}}(v, w) \quad \text{for all } w \in \widetilde{\mathscr{V}}_i^\ell. \tag{5.23}$$

By using Lemma A.19 and the same argument as in Subsection 3.1.1.2, we can derive from Lemma 5.1 and Lemma 5.2 the following results, the proofs of which are left as exercises.

**Lemma 5.4.** *Any $v \in \widetilde{\mathscr{V}}$ admits a decomposition*

$$v = v_0^1 + \sum_{\ell=1}^{L} \sum_{i=1}^{N_\ell} v_i^\ell, \quad v_i^\ell \in \widetilde{\mathscr{V}}_i^\ell,$$

*that satisfies*

$$a_{\mathfrak{V}}(v_0^1, v_0^1) + \sum_{\ell=1}^{L} \sum_{i=1}^{N_\ell} a_{\mathfrak{V}}(v_i^\ell, v_i^\ell) \lesssim a_{\mathfrak{V}}(v, v).$$

*The constant is independent of $L$ and $N_\ell$, $\ell = 1, \ldots, L$.*

**Lemma 5.5 (Strengthened Cauchy–Schwarz inequality).** *Let*

$$\widetilde{T}_\ell = \sum_{i=1}^{N_\ell} P_{\widetilde{\mathscr{V}}_i^\ell}.$$

*There exist constants $C > 0$ and $\gamma \in (0,1)$ such that for any $v \in \widetilde{\mathscr{V}}^k$ where $1 \le k \le \ell \le L$ the following inequality holds*

$$a_{\mathfrak{W}}(\widetilde{T}_\ell v, v) \le C\gamma^{2(\ell-k)} a_{\mathfrak{W}}(v,v).$$

As a consequence, we have the following result.

**Theorem 5.2.** *The condition number of the multilevel additive Schwarz method defined by (5.22) has a bounded condition number*

$$\kappa(P_{MAS}) \lesssim 1.$$

## 5.2 Multiplicative Schwarz methods

### 5.2.1 The hypersingular integral equation

For the multiplicative Schwarz preconditioners for this equation, we use the same subspace decomposition defined by (5.2) and projections $P_{\mathcal{V}_i^\ell}$, $i = 1,\dots,N_\ell$, $\ell = 1,\dots,L$, defined by (5.3). The multilevel symmetric multiplicative Schwarz operator is defined by, see (2.23),

$$P_{MSMS} = I - E_{MSMS}. \tag{5.24}$$

Here, see (2.24),

$$E_{MSMS} = E_{MMS}^t\, E_{MMS}$$

with

$$E_{MMS} = \prod_{\ell=1}^{L} T_\ell$$

where $T_\ell$ is defined by (5.4).

The convergence of the multilevel symmetric multiplicative Schwarz is given by (2.11). As in previous chapters, Lemma 2.10 implies that it suffices to estimate $\kappa(P_{MSMS})$.

The analysis in Subsection 5.1.1 yields the following theorem.

**Theorem 5.3.** *The condition number of the multilevel symmetric multiplicative Schwarz $P_{MSMS}$ defined by* (5.24) *satisfies*

$$\kappa(P_{MSMS}) \lesssim 1.$$

*Proof.* The theorem is a consequence of Lemma 5.1 (Stability of decomposition), Lemma 5.2 (Strengthened Cauchy–Schwarz inequality), and Theorem 2.2, noting that Assumption 2.6 (local stability) is obvious due to the choice of the bilinear forms in subspaces to be $a_{\mathfrak{W}}(\cdot,\cdot)$. □

### *5.2.2 The weakly-singular integral equation*

For the symmetric multiplicative Schwarz preconditioner for this equation, we use the same subspace decomposition defined by (5.21). The multilevel multiplicative Schwarz operator is defined by (5.24) where $E_{MMS}$ is defined by (2.24) with the projections $P_{\widetilde{\mathscr{V}}_0^1}$ and $P_{\widetilde{\mathscr{V}}_i^\ell}$, $i = 1,\ldots,N_\ell$, $\ell = 1,\ldots,L$, defined by (5.23).

As a consequence of the analysis in Subsection 5.1.2 we obtain the following theorem.

**Theorem 5.4.** *The condition number of the multilevel symmetric multiplicative Schwarz $P_{MMS}$ defined by* (5.24) *satisfies*

$$\kappa(P_{MSMS}) \lesssim 1.$$

*Proof.* The theorem is a consequence of Lemma 5.4 (Stability of decomposition), Lemma 5.5 (Strengthened Cauchy–Schwarz inequality), and Theorem 2.2, noting that Assumption 2.6 (local stability) is obvious due to the choice of the bilinear forms in subspaces to be $a_{\mathfrak{V}}(\cdot,\cdot)$. □

## 5.3 Numerical Results

The following table shows tables and figures in Chapter 9 which support the theoretical results obtained in this chapter.

| Equation | Preconditioner | Theorem | Table | Figure |
|---|---|---|---|---|
| Hypersingular | Additive Schwarz (PCG) | 5.1 | 9.2, 9.3 | 9.3 |
| Weakly-singular | Additive Schwarz (PCG) | 5.2 | 9.1 | 9.2 |

**Table 5.1** Theoretical results in Chapter 5 and relevant numerical results in Chapter 9

# Chapter 6
# Additive Schwarz Methods for the $hp$-Version

In this chapter, we report on additive Schwarz preconditioners for the $hp$-version of the hypersingular integral equation on a slit or a polygon in $\mathbb{R}^2$. Section 6.1 develops preconditioners when quasi-uniform meshes are used whereas Section 6.2 deals with geometrically graded meshes. These results first appeared in [120] and [229].

We recall here the Galerkin equation to be considered:

$$a_{\mathfrak{W}}(\psi_{\mathscr{V}}, \phi) = \langle g, \phi \rangle, \quad \phi \in \mathscr{V} \tag{6.1}$$

where the boundary element space $\mathscr{V}$ is to be defined later. It is noted that if $\Gamma$ is a polygon, then the energy space $\widetilde{H}^{1/2}(\Gamma)$ for the hypersingular integral operator coincides with the space $H^{1/2}(\Gamma)$. In this case, in the definition of the boundary element space the vanishing assumption at the endpoints of $\Gamma$ is invalid. However, continuity of these functions is still required. In Section 6.1 we only present the results for the case when $\Gamma$ is a slit. Extension to polygons is obvious. Section 6.2 presents results for both the slit and polygon. We will not elaborate on the different boundary element spaces as it is clear from the context.

In Section 6.3 we will briefly mention how the results for the weakly-singular integral equation can be obtained in a similar manner.

## 6.1 Preconditioners with Quasi-uniform Meshes

We will develop in this section two-level methods (non-overlapping and overlapping) for the hypersingular integral equation (6.1). A similar approach as in Chapter 3 and Chapter 4 can be employed to obtain similar results for the weakly-singular integral equation. We will not repeat that discussion here. Subsection 6.1.1 presents a non-overlapping method whereas an overlapping preconditioner is presented in Subsection 6.1.2.

The two-level preconditioners designed in this subsection are based on the following two-level mesh and boundary-element spaces.



E. P. Stephan and T. Tran, *Schwarz Methods and Multilevel Preconditioners for Boundary Element Methods*, https://doi.org/10.1007/978-3-030-79283-1_6

**Definition 6.1 (Two-level mesh and boundary-element spaces).**
*The coarse mesh: We first divide $\Gamma$ into disjoint subdomains $\Gamma_i$ with length $H_i$, $i = 1,\dots,J$, so that $\overline{\Gamma} = \cup_{i=1}^{J}\overline{\Gamma}_i$, and denote by $H$ the maximum value of $H_i$.*

*The fine mesh: Each $\Gamma_i$ is further divided into disjoint subintervals $\Gamma_{ij}$, $j = 1,\dots,N_i$, so that $\overline{\Gamma}_i = \cup_{j=1}^{N_i}\overline{\Gamma}_{ij}$. The maximum length of the subintervals $\Gamma_{ij}$ in $\Gamma_i$ is denoted by $h_i$, and the maximum value of $h_i$ is denoted by $h$. We require that the fine mesh is locally quasi-uniform, i.e., it is quasi-uniform in each subdomain $\Gamma_i$.*

*Boundary element spaces:*

$$
\begin{aligned}
\mathscr{V} &:= \{v \in C(\Gamma) : v|_{\Gamma_{ij}} \in \mathbb{P}_{p_{ij}}(\Gamma_{ij});\ i = 1,\dots,J;\ j = 1,\dots,N_i;\ v(\pm 1) = 0\},\\
\mathscr{V}_h &:= \{v \in C(\Gamma) : v|_{\Gamma_{ij}} \in \mathbb{P}_1(\Gamma_{ij});\ i = 1,\dots,J;\ j = 1,\dots,N_i;\ v(\pm 1) = 0\},\\
\mathscr{V}_H &:= \{v \in C(\Gamma) : v|_{\Gamma_i} \in \mathbb{P}_1(\Gamma_i)\ \mathit{for}\ i = 1,\dots,J;\ v(\pm 1) = 0\},\\
\mathscr{V}_h^0(\Gamma_i) &:= \{v \in \mathscr{V}_h : \operatorname{supp} v \subset \overline{\Gamma}_i\},\\
\mathscr{V}_p^0(\Gamma_{ij}) &:= \{v \in \mathscr{V} : \operatorname{supp} v \subset \overline{\Gamma}_{ij}\},\\
\mathscr{V}_p &:= \bigoplus_{i=1}^{J}\bigoplus_{j=1}^{N_i}\mathscr{V}_p^0(\Gamma_{ij}).
\end{aligned}
$$

*Here $\mathbb{P}_q(G)$ denotes the set of polynomials of degree at most $q$ which are defined on the set $G$.*

For simplicity of notations, in this section we will assume

$$p_{ij} = p > 0 \quad \forall j = 1,\dots,N_i,\ i = 1,\dots,J.$$

Recalling the definition in Section 4.1.1.1 of affine images of antiderivatives of Legendre polynomials, we have

$$\mathscr{V}_p^0(\Gamma_{ij}) = \operatorname{span}\{\mathscr{L}_{q,ij} : 2 \le q \le p\} \tag{6.2}$$

where $\mathscr{L}_{q,ij}$ is the affine image of $\mathscr{L}_q$ onto $\Gamma_{ij}$.

It is clear from Definition 6.1 that $\mathscr{V}$ can be decomposed into

$$\mathscr{V} = \mathscr{V}_h \oplus \mathscr{V}_p. \tag{6.3}$$

If we further decompose $\mathscr{V}_h$ and $\mathscr{V}_p$ we will obtain different additive Schwarz methods for equation (6.1).

### 6.1.1 A two-level non-overlapping method

Due to (6.3) we can define the non-overlapping method via a two-level decomposition of $\mathscr{V}_h$ as follows. The two blocks $\mathscr{V}_h$ and $\mathscr{V}_p$ can be decomposed in an obvious way by

$$\mathscr{V}_h = \mathscr{V}_H \oplus \left( \bigoplus_{i=1}^{J} \mathscr{V}_h^0(\Gamma_i) \right) \tag{6.4}$$

and

$$\mathscr{V}_p = \oplus \left( \bigoplus_{i=1}^{J} \bigoplus_{j=1}^{N_i} \mathscr{V}_p^0(\Gamma_{ij}) \right). \tag{6.5}$$

The above decompositions result in the following two-level decomposition for $\mathscr{V}$.

**Definition 6.2.** *The boundary element space $\mathscr{V}$ is decomposed by the following direct sum:*

$$\mathscr{V} = \mathscr{V}_H \oplus \left( \bigoplus_{i=1}^{J} \mathscr{V}_h^0(\Gamma_i) \right) \oplus \left( \bigoplus_{i=1}^{J} \bigoplus_{j=1}^{N_i} \mathscr{V}_p^0(\Gamma_{ij}) \right).$$

The above decomposition is a direct decomposition because the intersection of any two subspaces contains only the zero function.

Let $\Pi_H : C(\Gamma) \to \mathscr{V}_H$ and $\Pi_h : C(\Gamma) \to \mathscr{V}_h$ be the usual interpolation operators. A function $v \in \mathscr{V}$ can be uniquely decomposed by

$$v = v^H + \sum_{i=1}^{J} v_i^h + \sum_{i=1}^{J} \sum_{j=1}^{N_i} v_{ij}^p \tag{6.6}$$

where

$$v^H = \Pi_H v, \quad \sum_{i=1}^{J} v_i^h = \Pi_h v - v^H, \quad \text{and} \quad \sum_{i=1}^{J} \sum_{j=1}^{N_i} v_{ij}^p = v - \Pi_h v. \tag{6.7}$$

As in the previous chapters, we prove the coercivity and stability of this decomposition. The proof for coercivity is simple.

**Lemma 6.1 (Coercivity of decomposition).** *For any $v \in \mathscr{V}$*

$$a_{\mathfrak{W}}(v,v) \lesssim a_{\mathfrak{W}}(v^H, v^H) + \sum_{i=1}^{J} a_{\mathfrak{W}}(v_i^h, v_i^h) + \sum_{i=1}^{J} \sum_{j=1}^{N_i} a_{\mathfrak{W}}(v_{ij}^p, v_{ij}^p).$$

*The constant is independent of $v$, $H$, $h$, and $p$.*

*Proof.* Noting the disjoint supports of $v_i^h$, $i = 1, \dots, J$, and of $v_{ij}^p$, $j = 1, \dots, N_i$ and $i = 1, \dots, J$, we obtain by recalling (6.6) and using the Cauchy–Schwarz inequality and Theorem A.11

$$\begin{aligned}
\|v\|^2_{\widetilde{H}_{\mathrm{I}}^{1/2}(\Gamma)} &\lesssim \|v^H\|^2_{\widetilde{H}_{\mathrm{I}}^{1/2}(\Gamma)} + \Big\|\sum_{i=1}^{J} v_j^h\Big\|^2_{\widetilde{H}_{\mathrm{I}}^{1/2}(\Gamma)} + \Big\|\sum_{i=1}^{J} \sum_{j=1}^{N_i} v_{ij}^p\Big\|^2_{\widetilde{H}_{\mathrm{I}}^{1/2}(\Gamma)} \\
&\lesssim \|v^H\|^2_{\widetilde{H}_{\mathrm{I}}^{1/2}(\Gamma)} + \sum_{i=1}^{J} \|v_j^h\|^2_{\widetilde{H}_{\mathrm{I}}^{1/2}(\Gamma)} + \sum_{i=1}^{J} \sum_{j=1}^{N_i} \|v_{ij}^p\|^2_{\widetilde{H}_{\mathrm{I}}^{1/2}(\Gamma)}.
\end{aligned}$$

This proves the lemma. □

Next, we prove the stability of the decomposition.

**Lemma 6.2 (Stability of decomposition).** *For any* $v \in \mathscr{V}$

$$a_{\mathfrak{W}}(v^H, v^H) + \sum_{i=1}^{J} a_{\mathfrak{W}}(v_i^h, v_i^h) + \sum_{i=1}^{J}\sum_{j=1}^{N_i} a_{\mathfrak{W}}(v_{ij}^p, v_{ij}^p) \lesssim (1+\log p)^2 \max_{1\le i\le J}\left(1+\log\frac{H_i}{h_i}\right) a_{\mathfrak{W}}(v,v).$$

*The constant is independent of v, H, h, and p.*

*Proof.* As before, we need to prove

$$\|v^H\|^2_{\widetilde{H}_{\mathrm{I}}^{1/2}(\Gamma)} + \sum_{i=1}^{J} \|v_i^h\|^2_{\widetilde{H}_{\mathrm{I}}^{1/2}(\Gamma_i)} + \sum_{i=1}^{J}\sum_{j=1}^{N_i} \|v_{ij}^p\|^2_{\widetilde{H}_{\mathrm{I}}^{1/2}(\Gamma_{ij})} \lesssim (1+\log p)^2 \max_{1\le i\le J}\left(1+\log\frac{H_i}{h_i}\right) \|v\|^2_{\widetilde{H}_{\mathrm{I}}^{1/2}(\Gamma)}. \tag{6.8}$$

The first and third terms on the left-hand side are bounded by the right-hand side due to (6.7) and Lemma C.38. To show the boundedness of the second term, we note that $v_i^h$ can be rewritten as $v_i^h = (\Pi_h v - \Pi_H(\Pi_h v))|_{\Gamma_i}$. Invoking Lemma C.38 with $v$ in that lemma replaced by $\Pi_h v$, a piecewise polynomial of degree 1, we deduce

$$\sum_{i=1}^{J} \|v_i^h\|^2_{\widetilde{H}_{\mathrm{I}}^{1/2}(\Gamma_i)} \lesssim \max_{1\le i\le J}\left(1+\log\frac{H_i}{h_i}\right) \|\Pi_h v\|^2_{\widetilde{H}_{\mathrm{I}}^{1/2}(\Gamma)}.$$

Invoking the same lemma again yields the required bound. □

Lemma 6.1, Lemma 6.2, and Theorem 2.1 give the following result.

**Theorem 6.1.** *The condition number of the two-level additive Schwarz operator* $P_{\mathrm{ad}}$ *associated with the direct sum decomposition defined in Definition 6.2 satisfies*

$$\kappa(P_{\mathrm{ad}}) \lesssim (1+\log p)^2 \max_{1\le i\le J}\left(1+\log\frac{H_i}{h_i}\right).$$

### *6.1.2 A two-level overlapping method*

The overlapping subdomains $\Gamma_i'$ are defined as follows. We extend each subdomain $\Gamma_i$ on both sides (except for the left side of $\Gamma_1$ and the right side of $\Gamma_J$ which are the endpoints of $\Gamma$) so that the length of the overlap between two adjacent extended subdomains $\Gamma_i'$ and $\Gamma_{i+1}'$ is proportional to $\delta$ for some $\delta \in (0,H)$. We assume that the endpoints of $\Gamma_i'$ coincide with the fine mesh points. This means the overlap is a union of subintervals $\Gamma_{ij}$.

The overlapping method is defined via the following subspace decomposition.

**Definition 6.3.** *The boundary element space $\mathscr{V}$ is decomposed as follows:*

$$\mathscr{V} = \mathscr{V}_H \cup \left( \bigcup_{i=1}^{J} \mathscr{V}_h^0(\Gamma_i') \right) \cup \left( \bigcup_{i=1}^{J} \bigcup_{j=1}^{N_i} \mathscr{V}_p^0(\Gamma_{ij}) \right)$$

The above decomposition is coercive and stable, as given in the next two lemmas.

**Lemma 6.3 (Coercivity of decomposition).** *If $v \in \mathscr{V}$ satisfies $v = v^H + \sum_{i=1}^J v_i^h + \sum_{i=1}^J \sum_{j=1}^{N_i} v_{ij}^p$ where $v^H \in \mathscr{V}_H$, $v_i^h \in \mathscr{V}_h^0(\Gamma_i')$ and $v_{ij}^p \in \mathscr{V}_p^0(\Gamma_{ij})$, then*

$$a_{\mathfrak{W}}(v,v) \lesssim a_{\mathfrak{W}}(v^H,v^H) + \sum_{i=1}^{J} a_{\mathfrak{W}}(v_i^h,v_i^h) + \sum_{i=1}^{J} \sum_{j=1}^{N_i} a_{\mathfrak{W}}(v_{ij}^p,v_{ij}^p).$$

*The constant is independent of $v$, $h$, $H$, and $p$.*

*Proof.* The proof is similar to that of Lemma 3.6 by using the colouring argument, and is omitted. □

**Lemma 6.4 (Stability of decomposition).** *For any $v \in \mathscr{V}$, there exists a decomposition $v = v^H + \sum_{i=1}^J v_i^h + \sum_{i=1}^J \sum_{j=1}^{N_i} v_{ij}^p$ with $v^H \in \mathscr{V}_H$, $v_i^h \in \mathscr{V}_h^0(\Gamma_i')$ and $v_{ij}^p \in \mathscr{V}_p^0(\Gamma_{ij})$, then*

$$\begin{aligned} a_{\mathfrak{W}}(v^H,v^H) &+ \sum_{i=1}^{J} a_{\mathfrak{W}}(v_i^h,v_i^h) + \sum_{i=1}^{J} \sum_{j=1}^{N_i} a_{\mathfrak{W}}(v_{ij}^p,v_{ij}^p) \\ &\lesssim (1+\log p)^2 \left(1+\log \frac{H}{\delta}\right)^2 a_{\mathfrak{W}}(v,v). \end{aligned}$$

*The constant is independent of $v$, $H$, $h$, $\delta$, and $p$.*

*Proof.* Consider a partition of unity which consists of piecewise-linear functions $\theta_i$, $i = 1,\dots,J$, supported in $\overline{\Gamma}_i'$ as defined in Lemma 3.7. Here $\delta$ is the overlap size. Let $P_H : \mathscr{V} \to \mathscr{V}_H$ be defined as in (3.35), recall (3.36) and (3.37). We define

$$\begin{aligned} v^H &:= P_H v, \\ w &:= v - v^H, \\ v_i^h &:= \Pi_h(\theta_i w), \\ v_{ij}^p &:= (w - \Pi_h w)|_{\Gamma_{ij}}. \end{aligned}$$

Then since $\theta_i w$ has support in $\Gamma_i'$, so does $v_i^h$ and thus $v_i^h \in \mathscr{V}_h^0(\Gamma_i')$. It is also easy to see that $\operatorname{supp} v_{ij}^p \in \overline{\Gamma}_{ij}$, implying $v_{ij}^p \in \mathscr{V}_p^0(\Gamma_{ij})$. Moreover,

$$\sum_{i=1}^{J} v_i^h = \Pi_h \left( \sum_{i=1}^{J} \theta_i w \right) = \Pi_h w = \Pi_h v - v^H.$$

Hence

$$\begin{aligned} v^H + \sum_{i=1}^{J} v_i^h + \sum_{i=1}^{J} \sum_{j=1}^{N_i} v_{ij}^p &= \Pi_h v + \sum_{i=1}^{J} \sum_{j=1}^{N_i} (w - \Pi_h w)|_{\Gamma_{ij}} \\ &= \Pi_h v + w - \Pi_h (v - v^H) \\ &= \Pi_h v + w - \Pi_h v + v^H = v. \end{aligned}$$

Therefore, the above defined is a valid decomposition of $v$. Next we prove the inequality

$$\begin{aligned} \|v^H\|^2_{\widetilde{H}_{\mathrm{I}}^{1/2}(\Gamma)} + \sum_{i=1}^{J} \|v_i^h\|^2_{\widetilde{H}_{\mathrm{I}}^{1/2}(\Gamma_i)} + \sum_{i=1}^{J} \sum_{j=1}^{N_i} \|v_{ij}^p\|^2_{\widetilde{H}_{\mathrm{I}}^{1/2}(\Gamma_{ij})} \\ \lesssim (1 + \log p)^2 \left( 1 + \log \frac{H}{\delta} \right)^2 \|v\|^2_{\widetilde{H}_{\mathrm{I}}^{1/2}(\Gamma)}. \end{aligned} \tag{6.9}$$

The first term on the left-hand side is bounded by the right-hand side due to (3.36). Inequality (3.37) gives

$$\frac{1}{H} \|w\|^2_{L^2(\Gamma)} \lesssim \|v\|^2_{\widetilde{H}_{\mathrm{I}}^{1/2}(\Gamma)} \quad \text{and} \quad \|w\|^2_{\widetilde{H}_{\mathrm{I}}^{1/2}(\Gamma)} \lesssim \|v\|^2_{\widetilde{H}_{\mathrm{I}}^{1/2}(\Gamma)}.$$

To show that the second term on the left-hand side of (6.9) is bounded by the right-hand side, we invoke Lemma C.38 and Lemma 3.7 to obtain

$$\begin{aligned} \sum_{i=1}^{J} \|v_i^h\|^2_{\widetilde{H}_{\mathrm{I}}^{1/2}(\Gamma_i')} &\lesssim (1 + \log p)^2 \sum_{i=1}^{J} \|\theta_i w\|^2_{\widetilde{H}_{\mathrm{I}}^{1/2}(\Gamma_i')} \\ &\lesssim (1 + \log p)^2 \sum_{i=1}^{J} \left( 1 + \log \frac{|\Gamma_i'|}{\delta} \right)^2 \left( \frac{1}{|\Gamma_i'|} \|w\|^2_{L^2(\Gamma_i')} + |w|^2_{H^{1/2}(\Gamma_i')} \right) \\ &\quad + \int_\Gamma \frac{|w(x)|^2}{1-x} \, \mathrm{d}x + \int_\Gamma \frac{|w(x)|^2}{x+1} \, \mathrm{d}x \\ &\lesssim (1 + \log p)^2 \left( 1 + \log \frac{H}{\delta} \right)^2 \left( \frac{1}{H} \|w\|^2_{L^2(\Gamma)} + |w|^2_{H^{1/2}(\Gamma)} \right) \\ &\quad + \int_\Gamma \frac{|w(x)|^2}{1-x} \, \mathrm{d}x + \int_\Gamma \frac{|w(x)|^2}{x+1} \, \mathrm{d}x \\ &\lesssim (1 + \log p)^2 \left( 1 + \log \frac{H}{\delta} \right)^2 \|v\|^2_{\widetilde{H}_{\mathrm{w}}^{1/2}(\Gamma)} \end{aligned}$$

$$\simeq (1+\log p)^2 \left(1+\log \frac{H}{\delta}\right)^2 \|v\|^2_{\widetilde{H}_{\rm I}^{1/2}(\Gamma)}.$$

Finally, to prove that the third term on the left-hand side of (6.9) is bounded by the right-hand side, we note that

$$v_{ij}^p = (w - \Pi_h w)|_{\Gamma_{ij}} = (v - \Pi_h v)|_{\Gamma_{ij}}$$

and invoke Lemma C.38 again. □

Combining Lemma 6.3, Lemma 6.4, and Theorem 2.1 we obtain the following theorem.

**Theorem 6.2.** *The condition number of the two-level additive Schwarz operator $P_{\rm ad}$ associated with the overlapping decomposition defined in Definition 6.3 satisfies*

$$\kappa(P_{\rm ad}) \lesssim (1+\log p)^2 \left(1+\log \frac{H}{\delta}\right)^2.$$

### 6.1.3 Multilevel methods

We first consider a multilevel method in $p$, which is initially studied in [120]. Noting (6.2) and (6.5), we can decompose $\mathscr{V}_p$ by

$$\mathscr{V}_p = \bigoplus_{i=1}^{J} \bigoplus_{j=1}^{N_i} \bigoplus_{q=2}^{p} \mathscr{W}_{ij}^q \tag{6.10}$$

where

$$\mathscr{W}_{ij}^q := \operatorname{span}\{\mathscr{L}_{q,ij}\}.$$

We then obtain the following subspace decomposition of $\mathscr{V}$:

$$\mathscr{V} = \mathscr{V}_h \oplus \left( \bigoplus_{i=1}^{J} \bigoplus_{j=1}^{N_i} \bigoplus_{q=2}^{p} \mathscr{W}_{ij}^q \right). \tag{6.11}$$

Any $u \in \mathscr{V}$ admits a unique decomposition

$$u = u_0 + \sum_{i=1}^{J} \sum_{j=1}^{N_i} \sum_{q=2}^{p} u_{ij}^q, \quad u_0 \in \mathscr{V}_h, \quad u_{ij}^q \in \mathscr{W}_{ij}^q. \tag{6.12}$$

We denote

$$u_{ij} := \sum_{q=2}^{p} u_{ij}^q \quad \text{and} \quad w_p := \sum_{j=1}^{J} \sum_{i=1}^{N_i} u_{ij}. \tag{6.13}$$

As usual, we prove the stability and coercivity of this decomposition.

**Lemma 6.5 (Coercivity of decomposition).** *The decomposition* (6.12) *satisfies*

$$a_{\mathfrak{W}}(u,u) \lesssim a_{\mathfrak{W}}(u_0,u_0) + \sum_{i=1}^{J}\sum_{j=1}^{N_i}\sum_{q=2}^{p} a_{\mathfrak{W}}(u_{ij}^q,u_{ij}^q).$$

*The constant is independent of u, J, and p.*

*Proof.* It follows from (6.12) and (6.13) that $u = u_0 + w_p = u_0 + \sum_{j=1}^{J}\sum_{i=1}^{N_i} u_{ij}$. Noting that $u_{ij} \in \widetilde{H}^{1/2}(\Gamma_{ij})$ is the restriction of $w_p$ onto $\Gamma_{ij}$ and using Lemma B.3, Theorem A.11, together with standard arguments, we obtain

$$a_{\mathfrak{W}}(u,u) \lesssim a_{\mathfrak{W}}(u_0,u_0) + \sum_{i=1}^{J}\sum_{j=1}^{N_i} a_{\mathfrak{W}}(u_{ij},u_{ij}). \tag{6.14}$$

It remains to show that for $i = 1,\ldots,J$, $j = 1,\ldots,N_i$

$$a_{\mathfrak{W}}(u_{ij},u_{ij}) \lesssim \sum_{q=2}^{p} a_{\mathfrak{W}}(u_{ij}^q,u_{ij}^q)$$

or equivalently

$$\|u_{ij}\|^2_{\widetilde{H}_{\mathrm{I}}^{1/2}(\Gamma)} \lesssim \sum_{q=2}^{p} \|u_{ij}^q\|^2_{\widetilde{H}_{\mathrm{I}}^{1/2}(\Gamma)}$$

where the constant is independent of all parameters involved. This is a direct consequence of Lemma C.15 by mapping the interval $\Gamma_j$ onto $I = (-1,1)$ and using Lemma A.6. □

**Lemma 6.6 (Stability of decomposition).** *Any $u \in \mathcal{V}$ admits a unique decomposition* (6.12) *which satisfies*

$$a_{\mathfrak{W}}(u_0,u_0) + \sum_{i=1}^{J}\sum_{j=1}^{N_i}\sum_{q=2}^{p} a_{\mathfrak{W}}(u_{ij}^q,u_{ij}^q) \lesssim p(1+\log p)^2 a_{\mathfrak{W}}(u,u).$$

*The constant is independent of u, J, and p.*

*Proof.* A verbatim proof along the lines of the proof of Lemma 4.2 gives

$$a_{\mathfrak{W}}(u_0,u_0) + \sum_{i=1}^{J}\sum_{j=1}^{N_i} a_{\mathfrak{W}}(u_{ij},u_{ij}) \lesssim (1+\log p)^2 a_{\mathfrak{W}}(u,u).$$

It remains to show that for each $j = 1,\ldots,N_i$ and $i = 1,\ldots,J$

$$\sum_{q=2}^{p} a_{\mathfrak{W}}(u_{ij}^q,u_{ij}^q) \lesssim p\,a_{\mathfrak{W}}(u_{ij},u_{ij})$$

or equivalently

$$\sum_{q=2}^{p} \|u_{ij}^{q}\|^2_{\widetilde{H}_{\rm I}^{1/2}(\Gamma_{ij})} \lesssim p\|u_{ij}\|^2_{\widetilde{H}_{\rm I}^{1/2}(\Gamma_{ij})}$$

where the constant is independent of all involving parameters. Again this is a direct consequence of Lemma C.15 and Lemma A.6. □

It follows from Lemma 6.5, Lemma 6.6, and Theorem 2.1 the following theorem concerning a multilevel decomposition of the $p$-block $\mathscr{V}_p$.

**Theorem 6.3.** *The additive Schwarz operator* $P_{\rm ad}$ *associated with the multilevel decomposition* (6.11) *for the hp-version with quasi-uniform meshes has condition number bounded as*

$$\kappa(P_{\rm ad}) \lesssim p\big(1+\log p\big)^2.$$

*The positive constant is independent of all parameters involved.*

If we further decompose the $h$-block $\mathscr{V}_h$ by (6.4) then

$$\mathscr{V} = \mathscr{V}_H \oplus \left(\bigoplus_{i=1}^{J} \mathscr{V}_h^0(\Gamma_i)\right) \oplus \left(\bigoplus_{i=1}^{J}\bigoplus_{j=1}^{N_i}\bigoplus_{q=2}^{p} \mathscr{W}_{ij}^q\right). \tag{6.15}$$

Theorem 6.1 and Theorem 6.3 yield

**Theorem 6.4.** *The condition number of the additive Schwarz operator associated with the decomposition* (6.15) *satisfies*

$$\kappa(P_{\rm ad}) \lesssim p(1+\log p)^2 \max_{1\le i\le J}\left(1+\log\frac{H_i}{h_i}\right).$$

A full multilevel method can be defined with the $h$-block decomposed as in Section 5.1.1, see (5.2) and the $p$-block decomposed as (6.10). We then obtain from Theorem 5.1 and Theorem 6.3 the following theorem.

**Theorem 6.5.** *The additive Schwarz operator* $P_{\rm ad}$ *associated with the multilevel decomposition* (5.2) *and* (6.11) *for the hp-version with quasi-uniform meshes has condition number bounded as*

$$\kappa(P_{\rm ad}) \lesssim p\big(1+\log p\big)^2.$$

*The positive constant is independent of all parameters involved.*

We note that even though the condition number of the $p$-multilevel additive Schwarz operator grows faster than that of the two-level method, the most efficient method is the full multilevel method in Theorem 6.5.

## 6.2 Preconditioners with Geometric Meshes

In this section, we design a two-level non-overlapping preconditioner (Subsection 6.2.1) and a multilevel preconditioner (Subsection 6.2.2) for the case when a geometric mesh is used to accelerate the convergence rate of the Galerkin scheme.

It is known that this type of meshes yields an exponentially fast convergence (in the energy norm) whereas an $hp$-version with a quasi-uniform mesh converges only with an algebraic rate. Concerning the convergence theory we refer to [17, 114, 115, 208, 209]. Hence a preconditioner for the $hp$-version with geometric meshes is of practical importance.

We first define the ansatz space for the $hp$-version with geometric meshes on $(-1,0)$. On $(-1,1)$ the geometric meshes are defined by symmetry. The space on the whole polygon $\Gamma$ can then be obtained by affine mappings of $(-1,1)$ onto the edges of $\Gamma$.

The geometric mesh is designed so that it is geometrically graded towards the endpoints $\pm 1$ of $\Gamma$ where singularities of the solutions occur. This mesh is defined by levels. We will define the mesh points on the half interval $[-1,0]$ and use symmetry to obtain the other half.

Let $K$ be a positive integer, which is usually referred to as the level of the geometric mesh, and let $\sigma$ be a constant such that $0<\sigma<1$. The mesh points are defined as

$$\begin{cases} x_0^{(K)} &= -1, \\ x_j^{(K)} &= -1+\sigma^{K-j}, \quad j=1,\dots,K. \end{cases} \tag{6.16}$$

We note that the mesh points of level $K$ are obtained from those of level $K-1$ by keeping all the points of level $K-1$, adding one more point near the endpoint $-1$, namely $-1+\sigma^{K-1}$, and then reordering the index.

The ansatz space $\mathscr{V}_\sigma^K$ (on $(-1,0)$) is now spanned by piecewise polynomial functions on $(-1,0)$ whose restrictions on $(x_{j-1}^{(K)},x_j^{(K)})$ are polynomials of degree at most $p_j=j$, $j=1,\dots,K$. On $\Gamma$ the space $\mathscr{V}_\sigma^K$ is obtained by affine mappings as mentioned above. For the hypersingular operator it is also required that the trial functions are continuous. We note that the dimension of $\mathscr{V}_\sigma^K$ is

$$M := \dim(\mathscr{V}_\sigma^K) = K^2+K-1. \tag{6.17}$$

In order to decompose $\mathscr{V}_\sigma^K$ as a direct sum as in Definition 6.2, we need to define a coarse mesh in association with the fine mesh (6.16). For our analysis, it is essential to maintain local quasi-uniformity, i.e., quasi-uniformity in each subdomain. Thus it is appropriate to group a fixed number of subintervals $(x_{j-1}^{(K)},x_j^{(K)})$ to form a subdomain.

Let $k_0<K$ be a fixed positive integer. There exists an integer $m$ such that

$$x_1^{(K-mk_0)} \le 0 = x_K^{(K)} < x_1^{(K-(m+1)k_0)}.$$

The coarse mesh points in $(-1,0)$ are the first points of the fine meshes of different levels $K, K-k_0, K-2k_0, \ldots, K-mk_0$ which are close to $-1$, namely,

$$-1 = x_0^{(K)} < x_1^{(K)} < x_1^{(K-k_0)} < x_1^{(K-2k_0)} < x_1^{(K-3k_0)} < \cdots < x_1^{(K-mk_0)} \le x_K^{(K)} = 0. \tag{6.18}$$

Depending on whether $x_1^{(K-mk_0)} < x_K^{(K)}$ or $x_1^{(K-mk_0)} = x_K^{(K)}$ there will be $m+3$ or $m+2$ points, i.e. $m+2$ or $m+1$ subdomains, in the half interval $[-1,0]$. Since

$$x_1^{(K-ik_0)} = x_{ik_0+1}^K \quad \text{for } i = 0, 1, \ldots, m,$$

each subdomain

$$\Gamma_i := (x_1^{(K-(i-1)k_0)}, x_1^{(K-ik_0)}) = (x_{(i-1)k_0+1}^{(K)}, x_{ik_0+1}^{(K)}) \tag{6.19}$$

consists of $k_0$ subintervals $(x_{j-1}^{(K)}, x_j^{(K)})$. Since $k_0$ is fixed with respect to $K$, this procedure ensures quasi-uniformity in each $\Gamma_i$. Moreover, there exists a constant $c$ (depending on $k_0$) such that

$$\frac{H_i}{h_i} \le c, \tag{6.20}$$

where, we recall that, $H_i$ is the length of $\Gamma_i$ and $h_i$ is the maximum length of the subintervals in $\Gamma_i$.

We can now define the coarse-mesh space $\mathscr{V}_H$ and the local spaces $\mathscr{V}_h^0(\Gamma_i)$ and $\mathscr{V}_p^0(\Gamma_{ij})$ as in Definition 6.1. We note that the dimensions of the spaces $\mathscr{V}_H$ and $\mathscr{V}_h^0(\Gamma_i)$ defined on these coarse and fine meshes are

$$\dim \mathscr{V}_H = \begin{cases} 2m+3 & \text{if } x_1^{(K-mk_0)} < x_K^{(K)}, \\ 2m+1 & \text{if } x_1^{(K-mk_0)} = x_K^{(K)}, \end{cases}$$

and

$$\dim \mathscr{V}_h^0(\Gamma_i) = k_0 - 1.$$

Note that $m$ is proportional to $k_0^{-1}$. Therefore if we choose $k_0$ to be small so that dim $\mathscr{V}_h^0(\Gamma_i)$ (and hence the size of the subproblem in this subspace) is small, then dim $\mathscr{V}_H$ (and hence the size of the subproblem in this subspace) is large, and vice-versa.

### 6.2.1 A two-level preconditioner

Let $\mathscr{V}_{\sigma,h}$, $\mathscr{V}_{\sigma,H}$, $\mathscr{V}_{\sigma,h}^0(\Gamma_i)$, $\mathscr{V}_p^0(\Gamma_{ij})$, and $\mathscr{V}_p$ be defined in the same manner as $\mathscr{V}_h$, $\mathscr{V}_H$, $\mathscr{V}_h^0(\Gamma_i)$, $\mathscr{V}_p^0(\Gamma_{ij})$, and $\mathscr{V}_p$, respectively; see Definition 6.1. Here $\Gamma_i$ and $\Gamma_{ij}$ are defined by (6.16), (6.18), and (6.19). Then we can define the following two-level subspace decomposition for $\mathscr{V}_{\sigma,h}$ and $\mathscr{V}_\sigma^K$:

$$\mathscr{V}_{\sigma,h} = \mathscr{V}_{\sigma,H} \oplus \left( \bigoplus_{i=1}^{J} \mathscr{V}_{\sigma,h}^{0}(\Gamma_i) \right), \tag{6.21}$$

$$\mathscr{V}_{\sigma}^{K} = \mathscr{V}_{\sigma,h} \oplus \mathscr{V}_{p}, \tag{6.22}$$

and thus

$$\mathscr{V}_{\sigma}^{K} = \mathscr{V}_{\sigma,H} \oplus \left( \bigoplus_{i=1}^{J} \mathscr{V}_{\sigma,h}^{0}(\Gamma_i) \right) \oplus \left( \bigoplus_{i=1}^{J} \bigoplus_{j=1}^{N_i} \mathscr{V}_{p}^{0}(\Gamma_{ij}) \right). \tag{6.23}$$

We deduce from Theorem 6.1 the following result.

**Theorem 6.6.** *The condition number of the additive Schwarz operator $P_{\rm ad}$ associated with the decomposition* (6.23) *satisfies*

$$\kappa(P_{\rm ad}) \lesssim \log^2 M$$

*where $M$ is the degrees of freedom defined by* (6.17).

*Proof.* It is noted that Theorem 6.1 requires local quasi-uniformity of the two-level mesh, and that the way the two-level geometric mesh is designed ensures this property. Therefore, this theorem still applies in this case. Let $J = \dim \mathscr{V}_H + 1$ be the number of subdomains. Then it follows from Theorem 6.1 and (6.20) that

$$\begin{aligned} \kappa(P_{\rm ad}) &\lesssim (1+\log p)^2 \max_{1\le i\le J} (1+\log \frac{H_i}{h_i}) \\ &\lesssim (1+\log p)^2. \end{aligned}$$

By the definition of $\mathscr{V}_{\sigma}^{K}$ we have

$$p = \max_{1\le j\le K} p_j = \max_{1\le j\le K} j = K. \tag{6.24}$$

Hence

$$\kappa(P_{\rm ad}) \lesssim \log^2 K.$$

Since

$$M = \dim \mathscr{V}_{\sigma}^{K} = K^2 + K - 1, \tag{6.25}$$

we have

$$\kappa(P_{\rm ad}) \lesssim \log^2 M.$$

□

### 6.2.2 A multilevel preconditioner

The $p$-block can be decomposed as in (6.10). However, a multilevel decomposition of the $h$-block by (5.2) is impossible because of the structure of the geometric mesh. Nevertheless,

$$\dim(\mathscr{V}_h) = 2K \simeq \sqrt{M}.$$

Hence, the size of the subproblem on this $h$-block is small compared to the size of the $p$-block. Thus, a two-level preconditioner for the $h$-block, as designed in Subsection 6.2.1, in conjunction with the multilevel method for the $p$-block gives an efficient preconditioner for the $hp$-version with geometric meshes.

We state the main theorem of this section.

**Theorem 6.7.** *With the two-level geometric mesh defined by* (6.16) *and* (6.18)*, the condition number of the additive Schwarz operator* $P_{\mathrm{ad}}$ *associated with the decomposition* (6.22)*,* (6.21)*, and* (6.10) *satisfies*

$$\kappa(P_{\mathrm{ad}}) \lesssim \sqrt{M} \log^2 M,$$

*where $M$ is the degrees of freedom defined by* (6.17).

*Proof.* Noting (6.24) and (6.25), we derive this theorem directly from Theorem 6.4 and Theorem 6.6. □

## 6.3 Results for the Weakly-Singular Integral Equation

We recall the Galerkin equation for the weakly-singular integral equation

$$a_{\mathfrak{V}}(\psi_{\widetilde{\mathscr{V}}}, \phi) = \langle f, \phi \rangle, \quad \phi \in \widetilde{\mathscr{V}}. \tag{6.26}$$

Results for both types of meshes, namely quasi-uniform and geometric meshes, can be derived. In the case of quasi-uniform meshes (see Definition 6.1) the space $\widetilde{\mathscr{V}}$ is the space of polynomials of degree $p \geq 0$ on each subinterval $\Gamma_{ij}$ for $j = 1, \dots, N_i$ and $i = 1, \dots, J$. In the case of geometric meshes defined by (6.16) and (6.18), the space $\widetilde{\mathscr{V}} = \widetilde{\mathscr{V}}_\sigma^K$ consists of functions which are not necessarily continuous whose restriction on $(x_{j-1}^{(K)}, x_j^{(K)})$ is a polynomial of degree $p_j = j-1$, $j = 1, \dots, K$.

Let $\widetilde{\mathscr{V}}_h$ and $\widetilde{\mathscr{V}}_{\sigma,h}$ be the space of piecewise constant functions on the quasi-uniform mesh and geometric mesh, respectively. We also define

$$\widetilde{\mathscr{V}}_p^0 = \operatorname{span}\{L_{q,ij} : 1 \leq q \leq p\} \quad \text{and} \quad \widetilde{\mathscr{V}}_p = \bigoplus_{i=1}^{J} \bigoplus_{j=1}^{N_i} \widetilde{\mathscr{V}}_p^0(\Gamma_{ij})$$

where $L_{q,ij}$ denotes the affine image of the Legendre polynomial of degree $q$ onto $\Gamma_{ij}$. Corresponding to the decompositions (6.3) and (6.22) we now have

$$\widetilde{\mathscr{V}} = \widetilde{\mathscr{V}}_h \oplus \widetilde{\mathscr{V}}_p \quad \text{and} \quad \widetilde{\mathscr{V}}_\sigma^K = \widetilde{\mathscr{V}}_{\sigma,h} \oplus \widetilde{\mathscr{V}}_p.$$

Different decompositions of $\widetilde{\mathscr{V}}_h$, $\widetilde{\mathscr{V}}_{\sigma,h}$, and $\widetilde{\mathscr{V}}_p$ result in different preconditioners for (6.26). We obtain the following theorems, the proofs of which follow along the lines of the proofs in Chapter 3, Chapter 4, and Chapter 5 by using the corresponding results for (6.1). We omit the proofs and refer interested readers to [120].

In the case of quasi-uniform meshes, we obtain the following results.

**Theorem 6.8.** *The condition number of the additive Schwarz operator $P_{\mathrm{ad}}$ associated with the two-level decomposition*

$$\widetilde{\mathscr{V}} = \widetilde{\mathscr{V}}_h \oplus \left( \bigoplus_{i=1}^{J} \bigoplus_{j=1}^{N_i} \widetilde{\mathscr{V}}_p^0(\Gamma_{ij}) \right)$$

*where $\widetilde{\mathscr{V}}_h$ is decomposed as in Definition 3.3 satisfies*

$$\kappa(P_{\mathrm{ad}}) \lesssim (1+\log p)^2 \max_{1\le i\le J} \left(1+\log \frac{H_i}{h_i}\right).$$

*The constant is independent of N and p.*

If we apply multilevel decompositions for both the $h$-block $\widetilde{\mathscr{V}}_h$ and the $p$-block $\widetilde{\mathscr{V}}_p$, we obtain the following theorem.

**Theorem 6.9.** *Let $\widetilde{\mathscr{W}_{ij}^q} = \mathrm{span}\{L_{q,ij}\}$. If $P_{\mathrm{ad}}$ is the multilevel additive Schwarz operator associated with the decomposition*

$$\widetilde{\mathscr{V}} = \widetilde{\mathscr{V}}_h \oplus \left( \bigoplus_{i=1}^{J} \bigoplus_{j=1}^{N_i} \bigoplus_{q=1}^{p} \widetilde{\mathscr{W}_{ij}^q} \right)$$

*where the h-block $\widetilde{\mathscr{V}}_h$ is decomposed by (5.21), then its condition number $\kappa(P_{\mathrm{ad}})$ satisfies*

$$\kappa(P_{\mathrm{ad}}) \lesssim (p+1)\big(1+\log(p+1)\big)^2.$$

*The constant is independent of N and p.*

In the case of geometric meshes, we obtain the following results, recalling property (6.20) of this two-level mesh.

**Theorem 6.10.** *If $P_{\mathrm{ad}}$ is the additive Schwarz operator associated with the two-level decomposition*

$$\widetilde{\mathscr{V}}_\sigma^K = \widetilde{\mathscr{V}}_{\sigma,h} \oplus \left( \bigoplus_{i=1}^{J} \bigoplus_{j=1}^{N_i} \widetilde{\mathscr{V}}_p^0(\Gamma_{ij}) \right)$$

*where $\widetilde{\mathscr{V}}_{\sigma,h}$ is decomposed as in Definition 3.3, then the condition number $\kappa(P_{\mathrm{ad}})$ satisfies*

$$\kappa(P_{\mathrm{ad}}) \lesssim \log^2 K \simeq \log^2 M.$$

*The constants are independent of K and M.*

**Theorem 6.11.** *The additive Schwarz operator $P_{\mathrm{ad}}$ associated with the multilevel decomposition*

$$\widetilde{\mathscr{V}}^K_\sigma = \widetilde{\mathscr{V}}_{\sigma,h} \oplus \left(\bigoplus_{i=1}^{J} \bigoplus_{j=1}^{N_i} \bigoplus_{q=1}^{p} \widetilde{\mathscr{W}}^q_{ij}\right)$$

*where* $\widetilde{\mathscr{V}}_{\sigma,h}$ *is decomposed as in Definition 3.3, then the condition number* $\kappa(P_{\mathrm{ad}})$ *satisfies*

$$\kappa(P_{\mathrm{ad}}) \lesssim \sqrt{M}\log^2 M.$$

*The constants are independent of* $M$.

## 6.4 Numerical Results

The theoretical results in this chapter are corroborated by the numerical results presented in Chapter 9. The following table shows the relevant theorems and figures.

| Equation | Preconditioner | | Theorem | Figure |
|---|---|---|---|---|
| Hypersingular | Two-level non-overlapping | | 6.1 | 9.12–9.15 |
| | Two-level overlapping | | 6.2 | 9.12–9.14, 9.16 |
| Weakly-singular | Quasi-uniform meshes | Two-level | 6.8 | 9.17, |
| | | Multilevel | 6.9 | 9.17 |
| | Geometric meshes | Two-level | 6.10 | 9.18 |
| | | Multilevel | 6.11 | 9.18 |

**Table 6.1** Theoretical results in Chapter 6 and relevant numerical results in Chapter 9

# Chapter 7
# A Fully Discrete Method

A fully-discrete and symmetric method for the weakly-singular boundary integral equation is studied in [148]. In this chapter we discuss the use of additive Schwarz methods as preconditioners for the linear system arising from this discretization. Non-overlapping and overlapping two-level methods, and a multilevel method are discussed. We prove that the condition numbers of the preconditioned systems grow at most logarithmically with the degrees of freedom. The results in this chapter were first reported in [221].

We would like to note that for the problem considered in this chapter, due to the smoothness of the solution, a direct method may be sufficient in solving the linear system. However, this study is a step along the way to developing preconditioners for more demanding problems, where they are truly needed. Furthermore, the results in this paper apply not only to the logarithmic boundary integral equation but also to any equation of the form (7.5) in which the bilinear form $\ell(\cdot,\cdot)$ is equivalent to the $H^{-1/2}$ norm.

In the next section, we reintroduce the Dirichlet boundary value problem on a domain with smooth boundary and present a parametrization for the smooth boundary to derive the boundary integral equation. This derivation is similar to that in Chapter 1.

## 7.1 The Boundary Integral Equation and a Fully Discrete Method

Let $\Gamma$ be a smooth closed curve in $\mathbb{R}^2$. The equation to be considered in this chapter has the form

$$\frac{1}{2\pi}\int_{\Gamma} \phi(\mathbf{y}) \log\frac{\alpha}{|\mathbf{x}-\mathbf{y}|}\, d\sigma_{\mathbf{y}} = F(\mathbf{x}), \quad \mathbf{x}\in\Gamma, \tag{7.1}$$

where $\alpha$ is a positive parameter to be specified later. Let $\gamma$ be a 1-periodic parametric representation of $\Gamma$ such that $\gamma \in C^\infty$ and $|\gamma'(x)| \neq 0$ for $x \in \mathbb{R}$ and let



E. P. Stephan and T. Tran, *Schwarz Methods and Multilevel Preconditioners for Boundary Element Methods*, https://doi.org/10.1007/978-3-030-79283-1_7

$$u(x) := \frac{1}{2\pi}\phi[\gamma(x)]|\gamma'(x)| \quad \text{and} \quad g(x) := F[\gamma(x)].$$

Then we can recast (7.1) as

$$\int_0^1 K(x,y)u(y)\,dy = g(x), \quad x \in [0,1], \tag{7.2}$$

where

$$K(x,y) := \log\frac{\alpha}{|\gamma(x)-\gamma(y)|} \quad \forall x,y \in [0,1]. \tag{7.3}$$

For all $s \in \mathbb{R}$, let $H^s$ denote the usual 1-periodic Sobolev space of order $s$ with norm denoted by $\|\cdot\|_s$. With the bilinear form $\ell(\cdot,\cdot) : H^{-1/2} \times H^{-1/2} \to \mathbb{R}$ defined by

$$\ell(v,w) := \int_0^1\int_0^1 K(x,y)v(x)w(y)\,dx\,dy, \quad v,w \in H^{-1/2}, \tag{7.4}$$

the weak form of (7.2) has the form: find $u \in H^{-1/2}$ such that

$$\ell(u,v) = (g,v) \quad \forall v \in H^{-1/2}. \tag{7.5}$$

Here, $(\cdot,\cdot)$ denotes the inner product in $L^2(0,1)$.

The parameter $\alpha$ in the definition (7.3) of $K$ is chosen to satisfy

$$\alpha \geq \text{logarithmic capacity of } \Gamma, \tag{7.6}$$

so that there exists a positive constant $\lambda$ satisfying

$$\ell(v,v) \geq \lambda\|v\|_{-1/2}^2 \quad \forall v \in H^{-1/2}. \tag{7.7}$$

For an explanation of the logarithmic capacity, see *e.g.* [121]. In practice, a simple way of guaranteeing that (7.6) holds is to choose $\alpha$ to be greater than the diameter of $\Gamma$.

## 7.2 The Fully-Discrete and Symmetric Method

The approximate solution $u_h^*$ to $u$ satisfies

$$\ell(u_h^*,v) = (g,v) \quad \forall v \in \mathscr{S}_h, \tag{7.8}$$

where $\mathscr{S}_h$ is a finite-dimensional subspace of $H^{-1/2}$. The stiffness matrix and the right-hand side of the matrix system arising from (7.8) involve complicated integrals that usually cannot be evaluated analytically. Hence, we are led to consider a fully-discrete Galerkin method in which the numerical solution $u_h \in \mathscr{S}_h$ satisfies

$$\ell_h(u_h, v) = \langle g, v \rangle \quad \forall v \in \mathscr{S}_h. \tag{7.9}$$

Here $\ell_h$ and $\langle\cdot,\cdot\rangle$ are certain discrete approximations to $\ell$ and $(\cdot,\cdot)$. The forms $\ell_h$ and $\langle\cdot,\cdot\rangle$ are constructed in such a way that $u_h$ achieves the same rates of convergence as $u_h^*$ in the appropriate Sobolev norms. In [148] $\ell_h$ is designed to preserve the symmetry of the exact problem so that

$$\ell_h(v, w) = \ell_h(w, v) \quad \forall v, w \in \mathscr{S}_h.$$

We choose $\mathscr{S}_h$ to be the space of piecewise-constant functions on a 1-periodic uniform mesh with $N$ subintervals between 0 and 1 defined by

$$t_j := jh, \quad h := \frac{1}{N}, \quad j = 0, \ldots, N.$$

The discrete bilinear form $\ell_h(\cdot,\cdot)$ and the discrete inner product $\langle\cdot,\cdot\rangle$ are defined as follows; see [148, Section 5]. Let

$$a = 1/2, \quad b = 0.1262814358793956, \quad c = 0.6841827226000933, \quad d = 1/6,$$

or

$$a = 1/2, \quad b = 0.4992036265530720, \quad c = 0.0398933386986113, \quad d = 1/6,$$

and let

$$\begin{aligned}
s_1 &= s_6 = a, & \sigma_1 &= \sigma_6 = b, & w_1 &= w_6 = d,\\
s_2 &= s_5 = 0, & \sigma_2 &= \sigma_5 = c, & w_2 &= w_5 = 1/2 - 2d,\\
s_3 &= s_4 = -a, & \sigma_3 &= \sigma_4 = b, & w_3 &= w_4 = d.
\end{aligned}$$

Let the quadrature weights be $\{w_1, \ldots, w_6\}$ and the quadrature points be

$$(\xi_p, \eta_p) := \frac{1}{2}(1 + s_p - \sigma_p, 1 + s_p + \sigma_p), \quad p = 1, \ldots, 6.$$

We define

$$\ell_h(v, w) := h^2 \sum_{j=0}^{N-1} \sum_{k=0}^{N-1} \sum_{p=1}^{6} w_p\, K(x_{jp}, y_{kp}) v(y_{kp}) w(x_{jp}),$$

where

$$x_{jp} := t_j + \xi_p h \quad \text{and} \quad y_{kp} := t_k + \eta_p h.$$

We note that the quadrature points $(\xi_p, \eta_p)$, $p = 1, \ldots, 6$, are symmetric about both diagonals of the unit square.

We next define the discrete inner product $\langle\cdot,\cdot\rangle$ as a two-point Gauss–Legendre quadrature rule of degree of precision 2 with quadrature points $\xi_j'$ and weights $w_j'$, $j = 1, 2$, given by

$$\xi'_1 = \frac{1}{2}\left(1 - \frac{1}{\sqrt{3}}\right), \quad \xi'_2 = \frac{1}{2}\left(1 + \frac{1}{\sqrt{3}}\right), \quad w'_1 = w'_2 = 1/2.$$

Thus the discrete inner product is defined by

$$\langle v, w \rangle := h \sum_{j=0}^{N-1} \sum_{p=1}^{2} w'_p v(x'_{jp}) w(x'_{jp}),$$

where $x'_{jp} := t_j + \xi'_p h$.

With $\ell_h$ and $\langle \cdot, \cdot \rangle$ so defined, Theorem 5.1 in [148] yields, for $h$ sufficiently small, the unique existence of the solution $u_h \in \mathscr{S}_h$ of (7.9) which satisfies

$$\|u_h - u\|_s \le c h^{t-s} \|u\|_{t+\max(-1-s,0)},$$

for $s < 1/2$, $-1/2 < t$, and $-1 \le s \le t \le 1$. It is also shown in [148] that the discrete Galerkin matrix is positive definite.

It is noted that the method designed above is of translation-invariant order 3 and that the orders of precision of the left-hand side and right-hand side are both at least 3; see [148, Definition 4.2 and page 337].

Before describing the preconditioners, we prove the following lemma which yields the equivalence of the two bilinear forms $\ell$ and $\ell_h$ on $\mathscr{S}_h \times \mathscr{S}_h$.

**Lemma 7.1.** *The following statement holds*

$$\ell(v, v) \simeq \ell_h(v, v) \quad \forall v \in \mathscr{S}_h.$$

*The constants are independent of h.*

*Proof.* The boundedness of the form $\ell(\cdot, \cdot)$ together with Theorem 2.1 (i) and Corollary 3.9 in [148] immediately gives $\ell(v, v) \lesssim \ell_h(v, v)$. To prove $\ell_h(v, v) \lesssim \ell(v, v)$, we first note that Lemmas 4.4 and 3.5 in [148] yield, respectively (noting Theorem 2.1 (v) in the same paper),

$$|a_h(v, v)| \lesssim |a(v, v)| + \|v\|^2_{-1/2} \lesssim \|v\|^2_{-1/2} \quad \forall v \in \mathscr{S}_h$$

and

$$|b_h(v, v)| \lesssim |b(v, v)| + \|v\|^2_{-1/2} \quad \forall v \in \mathscr{S}_h,$$

where $a_h(\cdot, \cdot)$ and $b_h(\cdot, \cdot)$ are defined as in [148] so that

$$\ell_h(v, w) = a_h(v, w) + b_h(v, w) \quad \forall v, w \in \mathscr{S}_h.$$

Therefore by using (7.7) we obtain, for all $v \in \mathscr{S}_h$,

$$\begin{aligned} \ell_h(v, v) &\le |a_h(v, v)| + |b_h(v, v)| \\ &\lesssim \|v\|^2_{-1/2} + |b(v, v)| \\ &\lesssim \|v\|^2_{-1/2} + \ell(v, v) \lesssim \ell(v, v). \end{aligned}$$

□

## 7.3 Two-level Methods

For simplicity, we assume in this chapter a two-level uniform mesh defined as follows.

*The coarse mesh:* We first divide $I$ into disjoint subdomains $I_i$ with length $H$, for $i = 1,\ldots,J$, so that $\overline{I} = \cup_{i=1}^{J}\overline{I}_i$.

*The fine mesh:* Each $I_i$ is further divided into disjoint subintervals $I_{ij}$ with length $h$, for $j = 1,\ldots,N$, so that $\overline{I}_i = \cup_{j=1}^{N}\overline{I}_{ij}$.

Recall that the finite-dimensional space $\mathscr{S} = \mathscr{S}_h$ is defined as the space of 1-periodic piecewise-constant functions on the fine mesh. We define, for both the non-overlapping and overlapping methods, the space $\mathscr{S}_0$ as the space of 1-periodic piecewise-constant functions on the coarse mesh.

As in Chapter 3, the subspaces $\mathscr{S}_i$, $i = 1,\ldots,J$, will be defined as spaces of functions which are derivatives of piecewise-linear functions. On the fine mesh we also define the space $\mathscr{V}$ of continuous piecewise-linear functions.

### 7.3.1 A non-overlapping method

Let

$$\mathscr{V}_i := \left\{ v \in \mathscr{V} \,|\, \operatorname{supp} v \subset \overline{I}_i \right\}, \quad i = 1,\ldots,J,$$

and let

$$\mathscr{S}_i := \left\{ w \,|\, w = v' \text{ for some } v \in \mathscr{V}_i \right\}, \quad i = 1,\ldots,J. \tag{7.10}$$

Bases for $\mathscr{S}_i$ can be formed by the Haar basis functions; see Chapter 3.

The decomposition of $\mathscr{S}$ by

$$\mathscr{S} = \mathscr{S}_0 + \cdots + \mathscr{S}_J, \tag{7.11}$$

with $\mathscr{S}_i$ defined as above allows us to define the projections $P_j : \mathscr{S} \to \mathscr{S}_j$ by

$$L(P_j v, w) = L(v, w) \quad \forall v \in \mathscr{S}, w \in \mathscr{S}_j,\ j = 0,\ldots,J. \tag{7.12}$$

As usual, the additive Schwarz operator is defined by

$$P := \sum_{j=0}^{J} P_j. \tag{7.13}$$

When $L = \ell$ (Galerkin approximation), we denote $P$ by $P_G$. When $L = \ell_h$ (fully-discrete method), we denote $P$ by $P_F$.

**Theorem 7.1.** *The condition number of the non-overlapping additive Schwarz operator $P_F$ is bounded as*

$$\kappa(P_F) \lesssim 1 + \log \frac{H}{h}.$$

*Proof.* Lemma 3.4 and Lemma 3.5 in Chapter 3 are proved for the bilinear form defined with $L = \ell$. Due to Lemma 7.1 the same results hold for $L = \ell_h$. Invoking Theorem 2.1 we obtain the desired result. □

### 7.3.2 An overlapping method

We now extend each subdomain $I_i$ on each side by a fixed number of subintervals so that the length of the overlap between two extended subdomains $I_i'$ and $I_{i+1}'$ is $2\delta$ for some $\delta \in (0, H]$ which is a multiple of $h$. This implies

$$|I_i'| \sim H.$$

Let $\mathscr{V}_i$, $i = 1, \ldots, J$, be defined by

$$\mathscr{V}_i := \left\{ v \in \mathscr{V} \,|\, \operatorname{supp} v \subset \overline{I}_i' \right\}, \quad i = 1, \ldots, J.$$

Then the subspaces $\mathscr{S}_i$, $i = 1, \ldots, J$, are defined as in (7.10).

The subspace decomposition (7.11) using these subspaces yields an overlapping decomposition which together with (7.12) and (7.13) completely defines the additive Schwarz operators $P_G$ and $P_F$.

**Theorem 7.2.** *The condition number of the overlapping additive Schwarz operator $P_F$ is bounded as*

$$\kappa(P_F) \lesssim 1 + \log^2 \frac{H}{\delta}.$$

*Proof.* Similarly to the proof of Theorem 7.1, this is a result of Lemma 3.6, Lemma 3.8, Lemma 7.1, and Theorem 2.1. □

## 7.4 A Multilevel Method

The multilevel method is defined in exactly the same manner as in Chapter 5. The following theorem is then a result of the results obtained in that chapter and Lemma 7.1.

**Theorem 7.3.** *The condition number of the multilevel additive Schwarz operator $P_F$ is bounded independently of the number of level $L$ and the number of mesh points.*

## 7.5 Numerical Experiments

### 7.5.1 Implementation issues

We briefly discuss in this section how two-level methods are implemented. The implementation of the multilevel method is discussed in [227, 244].

Let $\phi_j$ be the brick function in the interval $(t_{j-1}, t_j)$ and let $\mathbf{A}_{kl}$, $k, l = 1, \ldots, J$, be sub-blocks of the stiffness matrix $\mathbf{A}$ arising from (7.9) defined by $\mathbf{A}_{kl} := (a_{ij}^{kl})$ with $a_{ij}^{kl} = \ell_h(\phi_i, \phi_j)$. Here $\phi_i$ and $\phi_j$ are brick functions such that

$$\operatorname{supp}\phi_i \subset \overline{I}_k \quad \text{and} \quad \operatorname{supp}\phi_j \subset \overline{I}_l$$

for the non-overlapping method, and

$$\operatorname{supp}\phi_i \subset \overline{I}_k' \quad \text{and} \quad \operatorname{supp}\phi_j \subset \overline{I}_l'$$

for the overlapping method. Then $\mathbf{A}$ has the form

$$\begin{bmatrix} \mathbf{A}_{11} & \mathbf{A}_{12} & \ldots & \mathbf{A}_{1J} \\ & \mathbf{A}_{22} & \ldots & \mathbf{A}_{2J} \\ & & \ddots & \vdots \\ & & & \mathbf{A}_{JJ} \end{bmatrix}$$

The matrix representation $\mathbf{B}$ of the non-overlapping preconditioner will now have the form

$$\begin{bmatrix} \mathbf{A}_{11}^{-1} & \mathbf{O} & \ldots & \mathbf{O} \\ & \mathbf{A}_{22}^{-1} & \ldots & \mathbf{O} \\ & & \ddots & \vdots \\ & & & \mathbf{A}_{JJ}^{-1} \end{bmatrix} + \mathbf{R}_0\mathbf{A}_0^{-1}\mathbf{R}_0^\top,$$

where $\mathbf{A}_0$ is the stiffness matrix on the coarse mesh level, $\mathbf{R}_0$ is the matrix representation of the interpolation operator from the coarse mesh level to the fine mesh level, and $\mathbf{R}_0^\top$ is its transpose; see Subsection 2.3.2.1. Let $\mathbf{v}$ be the vector representation of a function $v = \sum_{i=0}^J v_i \in \mathscr{S}$, $v_i \in \mathscr{S}_i$.

In the implementation, there is no need to compute the matrix $\mathbf{B}$ explicitly. To compute $\mathbf{B}\mathbf{v}$, it suffices to solve the local problems $\mathbf{A}_{ii}\mathbf{w}_i = \mathbf{v}_i$, $i = 1, \ldots, J$, and the global problem $\mathbf{A}_0\mathbf{w}_0 = \mathbf{R}_0^\top\mathbf{v}_0$, and compute $\mathbf{B}\mathbf{v}$ by

$$\mathbf{B}\mathbf{v} = \mathbf{R}_0\mathbf{w}_0 + \sum_{i=1}^J \mathbf{w}_i.$$

However, before doing so one has to transform the representation of the function $v$ in the usual basis to the Haar basis; see Section 7.3. After obtaining the resulting function from $\mathbf{B}\mathbf{v}$, a reverse transform to the usual basis is necessary.

### 7.5.2 Numerical results

Equation (7.1) is the reformulation (by the boundary integral equation method) of the boundary value problem with the Laplace equation $\Delta U = 0$ on $\Omega \subset \mathbb{R}^2$, subject to a Dirichlet boundary condition $U = f$ on $\Gamma = \partial\Omega$; see Chapter 1. In our numerical experiment we take $\Gamma$ to be the ellipse with semi-axes $a_1 = 4$ and $a_2 = 2$. The boundary data is chosen to be $f(\mathbf{x}) = \sqrt{|x_1 + x_2^2|}$, where $\mathbf{x} = (x_1, x_2)$. The right-hand side $F(\mathbf{x})$ of (7.1) is given by

$$F(\mathbf{x}) := \frac{1}{2} f(\mathbf{x}) + \frac{1}{2\pi} \int_{\Gamma} f(\mathbf{y}) \frac{\partial}{\partial \nu_{\mathbf{y}}} \log \frac{1}{|\mathbf{x} - \mathbf{y}|} \, d\sigma_{\mathbf{y}}, \quad \mathbf{x} \in \Gamma.$$

The logarithmic capacity of $\Gamma$ in this case is $(a_1 + a_2)/2 = 3$. We chose $\alpha = 3.5$.

The iterations were stopped when the $l_2$ norm of the residual was less than $10^{-9}$, and all calculations were performed in double precision. The Lanczos algorithm was used to compute the condition numbers.

Table 7.1 shows the condition numbers and numbers of iterations for the unpreconditioned systems and the systems preconditioned by the non-overlapping, overlapping, and multilevel methods designed in Section 7.3 and Section 7.4. Table 7.2 shows the cpu times for each of these methods.

| | Condition numbers | | | | Iterations | | | |
|---|---|---|---|---|---|---|---|---|
| DoF | (1) | (2) | (3) | (4) | (1) | (2) | (3) | (4) |
| 4 | 0.366E+01 | 0.366E+01 | 0.366E+01 | 0.234E+01 | 2 | 2 | 2 | 2 |
| 8 | 0.474E+01 | 0.474E+01 | 0.474E+01 | 0.457E+01 | 4 | 4 | 4 | 4 |
| 16 | 0.116E+02 | 0.193E+02 | 0.474E+01 | 0.665E+01 | 7 | 8 | 8 | 7 |
| 32 | 0.242E+02 | 0.253E+02 | 0.315E+01 | 0.290E+01 | 15 | 12 | 11 | 10 |
| 64 | 0.489E+02 | 0.319E+02 | 0.281E+01 | 0.105E+02 | 24 | 14 | 10 | 13 |
| 128 | 0.977E+02 | 0.389E+02 | 0.105E+02 | 0.122E+02 | 34 | 15 | 12 | 13 |
| 256 | 0.195E+03 | 0.467E+02 | 0.115E+02 | 0.144E+02 | 47 | 16 | 13 | 14 |
| 512 | 0.390E+03 | 0.550E+02 | 0.128E+02 | 0.164E+02 | 64 | 17 | 13 | 14 |
| 1024 | 0.779E+03 | 0.641E+02 | 0.145E+02 | 0.183E+02 | 87 | 18 | 14 | 14 |

**Table 7.1** (1): unpreconditioned, (2): preconditioned by non-overlapping method, (3): preconditioned by overlapping method, (4): preconditioned by multilevel method

| | CPU times | | | |
|---|---|---|---|---|
| DoF | (1) | (2) | (3) | (4) |
| 4 | 0.635E-01 | 0.704E-03 | 0.669E-03 | 0.648E-01 |
| 8 | 0.106E-02 | 0.922E-03 | 0.980E-03 | 0.468E-02 |
| 16 | 0.153E-02 | 0.201E-01 | 0.219E-01 | 0.106E-01 |
| 32 | 0.410E-02 | 0.343E-01 | 0.371E-01 | 0.206E-01 |
| 64 | 0.142E-01 | 0.615E-01 | 0.536E-01 | 0.396E-01 |
| 128 | 0.572E-01 | 0.149E+00 | 0.169E+00 | 0.695E-01 |
| 256 | 0.594E+00 | 0.608E+00 | 0.533E+00 | 0.264E+00 |
| 512 | 0.719E+01 | 0.359E+01 | 0.281E+01 | 0.165E+01 |
| 1024 | 0.520E+02 | 0.188E+02 | 0.150E+02 | 0.878E+01 |

**Table 7.2** (1): unpreconditioned, (2): preconditioned by non-overlapping method, (3): preconditioned by overlapping method, (4): preconditioned by multilevel method

# Chapter 8
# Indefinite Problems

In this chapter we design additive Schwarz methods for boundary integral equations arising from the Helmholtz equation in 2D; see Section 1.1.1. The integral operators take the forms

$$\mathfrak{V}_\kappa \varphi(x) = \frac{i}{2} \int_\Gamma H_0^{(1)}(\kappa|x-y|)\varphi(y)\,\mathrm{d}s_y, \qquad x \in \Gamma,$$

$$\mathfrak{W}_\kappa \varphi(x) = -\frac{i}{2}\frac{\partial}{\partial n_x} \int_\Gamma \frac{\partial H_0^{(1)}}{\partial n_y}(\kappa|x-y|)\varphi(y)\,\mathrm{d}s_y, \quad x \in \Gamma,$$

with $\kappa$ being a nonzero real number and $H_0^{(1)}$ the first-kind Hankel function of order zero. The corresponding equations (1.1) and (1.2) admit the following weak formulation: Given $\ell \in H^{-m}(\Gamma)$, find $\varphi \in \widetilde{H}^m(\Gamma)$ satisfying

$$a(\varphi, \psi) = \langle \ell, \psi \rangle \quad \forall \psi \in \widetilde{H}^m(\Gamma)$$

where

$$a(\varphi, \psi) = a_{\mathfrak{V}}(\varphi, \psi) = \langle \mathfrak{V}_\kappa \varphi, \psi \rangle, \quad \varphi, \psi \in \widetilde{H}^{-1/2}(\Gamma), \quad m = -1/2, \tag{8.1}$$

or

$$a(\varphi, \psi) = a_{\mathfrak{W}}(\varphi, \psi) = \langle \mathfrak{W}_\kappa \varphi, \psi \rangle, \quad \varphi, \psi \in \widetilde{H}^{1/2}(\Gamma), \quad m = 1/2. \tag{8.2}$$

In the next section we will present the general theory for indefinite problems satisfying some required assumptions. In subsequent sections we will show that the hypersigular equation (8.2) and the weakly-singular equation (8.1) satisfy these assumptions and thus we will apply the proposed preconditioners to these equations.

 
E. P. Stephan and T. Tran, *Schwarz Methods and Multilevel Preconditioners for Boundary Element Methods*, https://doi.org/10.1007/978-3-030-79283-1_8

## 8.1 General Theory for Indefinite Problems

In this section we consider the following general problem:

Given $\ell \in H^{-m}(\Gamma)$, $m \in \mathbb{R}$, find $\varphi \in \widetilde{H}^m(\Gamma)$ satisfying

$$a(\varphi, \psi) = \langle \ell, \psi \rangle \quad \forall \psi \in \widetilde{H}^m(\Gamma) \tag{8.3}$$

where the bilinear form $a(\cdot,\cdot) : \widetilde{H}^m(\Gamma) \times \widetilde{H}^m(\Gamma) \to \mathbf{C}$ satisfies several assumptions as stated below. Let $\mathscr{V}$ be a finite dimensional subspace of $\widetilde{H}^m(\Gamma)$. The solution $\varphi$ is approximated by $u \in \mathscr{V}$ satisfying

$$a(u, v) = \langle \ell, v \rangle \quad \forall v \in \mathscr{V}. \tag{8.4}$$

### 8.1.1 Assumptions

First we assume that the bilinear form $a(\cdot,\cdot)$ satisfies Assumption 2.2 (page 14), which we recall here for the reader's convenience.

**Assumption 8.1.**

(i) The bilinear form $a(\cdot,\cdot)$ can be decomposed as $a(\cdot,\cdot) = b(\cdot,\cdot) + k(\cdot,\cdot)$ where the bilinear form $b(\cdot,\cdot) : \widetilde{H}^m(\Gamma) \times \widetilde{H}^m(\Gamma) \to \mathbb{R}$ is symmetric and the bilinear form $k(\cdot,\cdot) : \widetilde{H}^m(\Gamma) \times \widetilde{H}^m(\Gamma) \to \mathbf{C}$ is skew-symmetric;

(ii) There exists a positive constant $C_{B_1}$ such that for any $u, v \in \widetilde{H}^m(\Gamma)$

$$|b(u,v)| \le C_{B_1} \|u\|_{\widetilde{H}^m(\Gamma)} \|v\|_{\widetilde{H}^m(\Gamma)},$$
$$|k(u,v)| \le C_{B_1} \|u\|_{\widetilde{H}^{m-1/2}(\Gamma)} \|v\|_{\widetilde{H}^m(\Gamma)},$$
$$|k(u,v)| \le C_{B_1} \|u\|_{\widetilde{H}^m(\Gamma)} \|v\|_{\widetilde{H}^{m-1/2}(\Gamma)}.$$

(iii) There exist positive constants $C_{B_2}$, $C_{B_3}$, and $C_{B_4}$ such that for any $u \in \widetilde{H}^m(\Gamma)$

$$b(u,u) \ge C_{B_2} \|u\|^2_{\widetilde{H}^m(\Gamma)},$$
$$\Re(a(u,u)) \ge C_{B_3} \|u\|^2_{\widetilde{H}^m(\Gamma)} - C_{B_4} \|u\|^2_{\widetilde{H}^{m-1/2}(\Gamma)}. \quad \text{(Gårding's inequality)}$$

Assumption 8.1 ensures the existence and uniqueness of (8.3) and (8.4).

As usual, domain decomposition methods for (8.4) are designed based on subspace decompositions of $\mathscr{V}$. This space is assumed to have the following properties.

**Assumption 8.2.** The space $\mathscr{V}$ is decomposed by

$$\mathscr{V} = \mathscr{V}_0 + \cdots + \mathscr{V}_J$$

where $\mathscr{V}_j$, $j = 0, \ldots, J$, are subspaces of $\mathscr{V}$ such that

$$\mathscr{V}_j = \{v \in \mathscr{V} : \operatorname{supp} v \subset \overline{\Gamma_j}\}, \quad j = 1, \ldots, J.$$

Here $\Gamma_j$, $j = 1, \ldots, J$, are subdomains satisfying $\overline{\Gamma} = \overline{\Gamma}_1 \cup \cdots \cup \overline{\Gamma}_J$. These subdomains may or may not overlap. Furthermore we assume that

(i) there exists a positive constant $C_{\mathscr{V}}$ such that for all $v_j \in \mathscr{V}_j$, $j = 1, \ldots, J$, the following statement holds with $v := v_1 + \cdots + v_J$

$$b(v, v) \le C_{\mathscr{V}} \sum_{j=1}^{J} b(v_j, v_j); \tag{8.5}$$

(ii) each $v \in \mathscr{V}$ admits a decomposition $v = v_0 + \cdots + v_J$ with $v_j \in \mathscr{V}_j$ satisfying

$$\sum_{j=0}^{J} b(v_j, v_j) \le \widetilde{C}_{\mathscr{V}} b(v, v). \tag{8.6}$$

The constants $C_{\mathscr{V}}$ and $\widetilde{C}_{\mathscr{V}}$ are independent of $v$ and $v_j$ but may depend on other parameters in the problem.

Finally, we make the following assumption on the relation between the symmetric and positive definite bilinear form $b(\cdot,\cdot)$ and the non-symmetric and non-positive definite bilinear form $a(\cdot,\cdot)$.

**Assumption 8.3.** There exists a positive constant $C_b$ such that for any $v_j \in \mathscr{V}_j$ with $j = 1, \ldots, J$, the following inequality holds

$$b(v_j, v_j) \le C_b \Re(a(v_j, v_j)).$$

In Chapter 2, Section 2.7, we presented the convergence theory for preconditioners to be implemented with GMRES. In the next subsection, we design preconditioners that yield the upper and lower bounds in (2.50). This in turn gives the residual estimates (2.51) and (2.52).

## *8.1.2 Additive Schwarz operators*

**Lemma 8.1.** *Under Assumption 8.1, the following projections are well defined:*

$$\begin{aligned} Q_j &: \mathscr{V} \longrightarrow \mathscr{V}_j, & a(Q_j w, v_j) &= a(w, v_j) & &\forall w \in \mathscr{V},\ v_j \in \mathscr{V}_j,\ j = 0, \ldots, J, \\ P_j &: \mathscr{V} \longrightarrow \mathscr{V}_j, & b(P_j w, v_j) &= \Re\big(a(w, v_j)\big) & &\forall w \in \mathscr{V},\ v_j \in \mathscr{V}_j,\ j = 1, \ldots, J. \end{aligned}$$

*Moreover,*

$$\|Q_0 v\|_{\widetilde{H}^m(\Gamma)} \le C_{Q_0} \|v\|_{\widetilde{H}^m(\Gamma)} \quad \forall v \in \widetilde{H}^m(\Gamma).$$

*Proof.* The results are direct consequences of Gårding's inequality (Assumption 8.1 (iii)). □

We make the following assumption on the projection $Q_0$.

**Assumption 8.4.** There exists a positive function $w(\alpha)$ satisfying

$$\lim_{\alpha \to 0^+} w(\alpha) = 0$$

such that for any $v \in \mathscr{V}$ the following inequality holds

$$\|v - Q_0 v\|_{\widetilde{H}^{m-1/2}(\Gamma)} \leq w^{1/2}(\alpha) \|v\|_{\widetilde{H}^m(\Gamma)}.$$

(In the next section, $\alpha$ is the coarse mesh size $H$ for the $h$-version and the fine mesh size $h$ for the $p$-version.)

Since $a(\cdot,\cdot)$ is non-symmetric, so are $Q_j$ and $P_j$, resulting in the non-symmetry of the following Schwarz operators.

**Definition 8.1.** *The additive Schwarz operators are defined by*

$$Q^{(1)} = Q_0 + Q_1 + \cdots + Q_J \quad \textit{and} \quad Q^{(2)} = Q_0 + P_1 + \cdots + P_J.$$

We can now design two algorithms from the operators $Q^{(1)}$ and $Q^{(2)}$.

---

**Algorithm 8.1:** Preconditioner $Q^{(1)}$

---

Find $u \in \mathscr{V}$ satisfying

$$Q^{(1)} u = g^{(1)} \tag{8.7}$$

where $g^{(1)} \in \mathscr{V}$ is defined by

$$g^{(1)} := \sum_{j=0}^{J} g_j^{(1)} \quad \text{with} \quad a(g_j^{(1)}, v_j) = \langle \ell, v_j \rangle \quad \forall v_j \in \mathscr{V}_j,\ j = 0, \ldots, J.$$

---

**Algorithm 8.2:** Preconditioner $Q^{(2)}$

---

Find $u \in \mathscr{V}$ satisfying

$$Q^{(2)} u = g^{(2)} \tag{8.8}$$

where $g^{(2)} \in \mathscr{V}$ is defined

$$g^{(2)} := g_0^{(1)} + \sum_{j=1}^{J} g_j^{(2)} \quad \text{with} \quad b(g_j^{(2)}, v_j) = \langle \ell, v_j \rangle, \quad v_j \in \mathscr{V}_j,\ j = 1, \ldots, J.$$

---

**Lemma 8.2.** *Under Assumptions 8.1–8.3, there exist* $\widetilde{c}_Q > 0$ *and* $\widetilde{C}_Q > 0$ *satisfying*

$$\widetilde{c}_Q b(v,v) \leq \sum_{j=0}^{J} b(Q_j v, Q_j v) \leq \widetilde{C}_Q b(v,v) \quad \forall v \in \mathscr{V} \tag{8.9}$$

*and*

$$\widetilde{c}_P b(v,v) \le b(Q_0 v, Q_0 v) + \sum_{j=1}^{J} b(P_j v, P_j v) \le \widetilde{C}_P b(v,v) \quad \forall v \in \mathscr{V}. \tag{8.10}$$

*Proof.* The following inequality obtained from Assumption 8.1 (i)–(ii) will be frequently used in the proof:

$$|a(u,v)| \le 2C_{B_1} \|u\|_{\widetilde{H}^m(\Gamma)} \|v\|_{\widetilde{H}^m(\Gamma)} \quad \forall u,v \in \widetilde{H}^m(\Gamma). \tag{8.11}$$

We first estimate $b(Q_0 v, Q_0 v)$ which is required in both (8.9) and (8.10). Assumption 8.1 (ii)–(iii) and Lemma 8.1 give

$$b(Q_0 v, Q_0 v) \le C_{Q_0}^2 C_{B_1} C_{B_2}^{-1} b(v,v). \tag{8.12}$$

To prove the right inequality in (8.9), we define $\widetilde{Q} := Q_1 + \cdots + Q_J$ and use successively Assumption 8.3, the definition of $Q_j$, (8.11), Assumption 8.1 (iii) and Assumption 8.2 (i) to obtain (with $c$ being a generic constant)

$$\begin{aligned}
\sum_{j=1}^{J} b(Q_j v, Q_j v) &\lesssim \sum_{j=1}^{J} \Re\big(a(Q_j v, Q_j v)\big) = \sum_{j=1}^{J} \Re\big(a(v, Q_j v)\big) = \Re\big(a(v, \widetilde{Q}v)\big) \\
&\lesssim \big|a(v,\widetilde{Q}v)\big| \lesssim \|v\|_{\widetilde{H}^m(\Gamma)} \|\widetilde{Q}v\|_{\widetilde{H}^m(\Gamma)} \\
&\lesssim b(v,v)^{1/2}\, b(\widetilde{Q}v, \widetilde{Q}v)^{1/2} \lesssim b(v,v)^{1/2} \Big(\sum_{j=1}^{J} b(Q_j v, Q_j v)\Big)^{1/2}
\end{aligned}$$

so that

$$\sum_{j=1}^{J} b(Q_j v, Q_j v) \lesssim b(v,v)$$

where the constant depends only on $C_b$, $C_{B_1}$, $C_{B_2}$, and $C_{\mathscr{V}}$. Combining the above estimate with (8.12) we obtain the right-hand side of (8.9) with

$$\widetilde{C}_Q = \widetilde{C}_Q(C_{Q_0}, C_{B_1}, C_{B_2}, C_b, C_{\mathscr{V}}). \tag{8.13}$$

Similarly, if $\widetilde{P} := P_1 + \cdots + P_J$ then we have from the definition of $P_j$, (8.11), and Assumption 8.1 (iii), and Assumption 8.2 (i) that

$$\begin{aligned}
\sum_{j=1}^{J} b(P_j v, P_j v) &= \sum_{j=1}^{J} \Re\big(a(v, P_j v)\big) = \Re\big(a(v, \widetilde{P}v)\big) \le \big|a(v, \widetilde{P}v)\big| \\
&\lesssim \|v\|_{\widetilde{H}^m(\Gamma)} \|\widetilde{P}v\|_{\widetilde{H}^m(\Gamma)} \lesssim b(v,v)^{1/2} b(\widetilde{P}v, \widetilde{P}v)^{1/2} \\
&\lesssim b(v,v)^{1/2} \Big(\sum_{j=1}^{J} b(P_j v, P_j v)\Big)^{1/2},
\end{aligned}$$

implying

$$\sum_{j=1}^{J} b(P_j v, P_j v) \lesssim b(v,v).$$

This inequality and (8.12) yield the right inequality in (8.10) with

$$\widetilde{C}_P = \widetilde{C}_P(C_{Q_0}, C_{B_1}, C_{B_2}, C_b, C_{\mathscr{V}}).$$

To prove the left inequality in (8.9) we employ Gårding's inequality and the boundedness of $b(\cdot,\cdot)$ (see Assumption 8.1 (ii)) to obtain

$$b(v,v) \le C_{B_5}\left(\Re\big(a(v,v)\big) + \|v\|^2_{\widetilde{H}^{m-1/2}(\Gamma)}\right) \tag{8.14}$$

where $C_{B_5}$ depends on $C_{B_1}$, $C_{B_3}$, and $C_{B_4}$. Let $v = v_0 + \cdots + v_J$ be any decomposition of $v \in \mathscr{V}$ where $v_j \in \mathscr{V}_j$, $j = 0, \ldots, J$. Then the definition of $Q_j$, (8.11), the Cauchy–Schwarz inequality, and Assumption 8.1 (iii) give

$$\begin{aligned}
\Re\big(a(v,v)\big) &\le \sum_{j=0}^{J} \big|a(v,v_j)\big| \\
&= \sum_{j=0}^{J} \big|a(Q_j v, v_j)\big| \\
&\le 2C_{B_1} \sum_{j=0}^{J} \|Q_j v\|_{\widetilde{H}^m(\Gamma)} \|v_j\|_{\widetilde{H}^m(\Gamma)} \\
&\le 2C_{B_1} \Big(\sum_{j=0}^{J} \|Q_j v\|^2_{\widetilde{H}^m(\Gamma)}\Big)^{1/2} \Big(\sum_{j=0}^{J} \|v_j\|^2_{\widetilde{H}^m(\Gamma)}\Big)^{1/2} \\
&\le 2C_{B_1} C_{B_2}^{-1} \Big(\sum_{j=0}^{J} b(Q_j v, Q_j v)\Big)^{1/2} \Big(\sum_{j=0}^{J} b(v_j, v_j)\Big)^{1/2}.
\end{aligned}$$

In particular, if the decomposition $v = v_0 + \cdots + v_J$ is the one given by Assumption 8.2 (ii), then

$$\Re\big(a(v,v)\big) \le 2C_{B_1} C_{B_2}^{-1} \widetilde{C}_{\mathscr{V}}^{1/2} \Big(\sum_{j=0}^{J} b(Q_j v, Q_j v)\Big)^{1/2} b(v,v)^{1/2}. \tag{8.15}$$

The final term in (8.14) can be estimated by using the triangle inequality, Lemma 8.1 and Assumption 8.4 as follows:

$$\begin{aligned}
\|v\|^2_{\widetilde{H}^{m-1/2}(\Gamma)} &\le 2\|v - Q_0 v\|^2_{\widetilde{H}^{m-1/2}(\Gamma)} + 2\|Q_0 v\|^2_{\widetilde{H}^{m-1/2}(\Gamma)} \\
&\le 2C_{B_2}^{-1}\, w(\alpha) b(v,v) + 2C_{B_2}^{-1} b(Q_0 v, Q_0 v)^{1/2} b(v,v)^{1/2}.
\end{aligned} \tag{8.16}$$

Altogether, (8.14), (8.15), and (8.16) yield

$$
\begin{aligned}
&\left(1-2C_{B_2}^{-1}C_{B_5}w(\alpha)\right)b(v,v)\\
&\leq 2C_{B_2}^{-1}C_{B_5}\left(C_{B_1}\widetilde{C}_{\mathscr{V}}^{1/2}\Big(\sum_{j=0}^{J}b(Q_jv,Q_jv)\Big)^{1/2}+b(Q_0v,Q_0v)^{1/2}\right)b(v,v)^{1/2}\\
&\leq 2C_{B_2}^{-1}C_{B_5}\left(C_{B_1}\widetilde{C}_{\mathscr{V}}^{1/2}+1\right)\Big(\sum_{j=0}^{J}b(Q_jv,Q_jv)\Big)^{1/2}b(v,v)^{1/2}.
\end{aligned}
$$

Since $w(\alpha)\to 0$ when $\alpha\to 0$, the above inequality implies, for $\alpha$ sufficiently small,

$$
\left(1-2C_{B_2}^{-1}C_{B_5}w(\alpha)\right)^2 b(v,v)\leq 4C_{B_2}^{-2}C_{B_5}^2\left(C_{B_1}\widetilde{C}_{\mathscr{V}}^{1/2}+1\right)^2\sum_{j=0}^{J}b(Q_jv,Q_jv)
$$

giving the left inequality in (8.9) with

$$
\widetilde{c_Q}^{-1}\simeq 4C_{B_2}^{-2}C_{B_5}^2\left(C_{B_1}\widetilde{C}_{\mathscr{V}}^{1/2}+1\right)^2. \tag{8.17}
$$

To prove the left inequality in (8.10) we estimate the first term on the right-hand side of (8.14) by

$$
\begin{aligned}
\Re\big(a(v,v)\big) &= \sum_{j=0}^{J}\Re\big(a(v,v_j)\big)\\
&= \Re\big(a(Q_0v,v_0)\big)+\sum_{j=1}^{J}b(P_jv,v_j)\\
&\leq 2C_{B_1}C_{B_2}^{-1}b(Q_0v,Q_0v)^{1/2}b(v_0,v_0)^{1/2}+\sum_{j=1}^{J}b(P_jv,P_jv)^{1/2}b(v_j,v_j)^{1/2}
\end{aligned}
$$

where in the last step we used (8.11) and Assumption 8.1 (iii). The proof can be proceeded as above and is omitted. □

**Assumption 8.5.** For all $v\in\mathscr{V}$

$$
\begin{aligned}
\Big|\sum_{j=0}^{J}k(Q_jv,v-Q_jv)\Big| &\leq w(\alpha)b(v,v),\\
|k(v-Q_0v,Q_0v)|+\Big|\sum_{j=1}^{J}k(v,P_jv)\Big| &\leq w(\alpha)b(v,v),
\end{aligned}
$$

where $w(\alpha)$ is the function given in Assumption 8.4.

As the main result of this section, we prove that under Assumptions 8.1–8.5, estimates (2.50) hold and thus we can invoke Lemma 2.19 in Chapter 2 for the convergence of the preconditioned GMRES applied to (8.4).

**Theorem 8.1.** *Under Assumptions 8.1–8.5, there exist $c_Q>0$ and $C_Q>0$ such that for any $v\in\mathscr{V}$ and $\ell=1,2$ the following estimates hold*

$$b(Q^{(\ell)}v, Q^{(\ell)}v) \le C_Q^2 b(v,v) \tag{8.18}$$

*and*

$$c_Q b(v,v) \le b(v, Q^{(\ell)}v). \tag{8.19}$$

*Proof.* We first prove (8.18) for $\ell = 1$. By using successively Assumption 8.1 (ii), Assumption 8.2 (i), and (8.9) we obtain

$$\begin{aligned}
b(Q^{(1)}v, Q^{(1)}v) &\le C_{B_1} \|Q^{(1)}v\|^2_{\widetilde{H}^m(\Gamma)} \\
&\le 2C_{B_1} \left( \|Q_0 v\|^2_{\widetilde{H}^m(\Gamma)} + \Big\|\sum_{j=1}^{J} Q_j v\Big\|^2_{\widetilde{H}^m(\Gamma)} \right) \\
&\le 2C_{B_1} C_{B_2}^{-1} \left( b(Q_0 v, Q_0 v) + b\Big(\sum_{j=1}^{J} Q_j v, \sum_{j=1}^{J} Q_j v\Big) \right) \\
&\le 2C_{B_1} C_{B_2}^{-1} \left( b(Q_0 v, Q_0 v) + C_{\mathscr{V}} \sum_{j=1}^{J} b(Q_j v, Q_j v) \right) \\
&\le 2C_{B_1} C_{B_2}^{-1} (1 + C_{\mathscr{V}}) \sum_{j=0}^{J} b(Q_j v, Q_j v) \\
&\le 2C_{B_1} C_{B_2}^{-1} (1 + C_{\mathscr{V}})\, \widetilde{C}_Q b(v,v)
\end{aligned}$$

where $\widetilde{C}_Q$ is given in (8.13). This means (8.18) holds with

$$C_Q^2 = 2C_{B_1} C_{B_2}^{-1} (1 + C_{\mathscr{V}}) \widetilde{C}_Q. \tag{8.20}$$

Similar result holds for $\ell = 2$ by using (8.10) instead of (8.9).

Next we prove (8.19) for $\ell = 1$. By using the definition of $Q_j$, (8.9), and Assumption 8.5 we obtain

$$\begin{aligned}
b(v, Q^{(1)}v) &= \sum_{j=0}^{J} b(v, Q_j v) \\
&\ge \sum_{j=0}^{J} b(Q_j v, Q_j v) - \sum_{j=0}^{J} |b(Q_j v, v - Q_j v)| \\
&= \sum_{j=0}^{J} b(Q_j v, Q_j v) - \sum_{j=0}^{J} |k(Q_j v, v - Q_j v)| \\
&\ge \widetilde{c}_Q b(v,v) - w(\alpha) b(v,v).
\end{aligned}$$

The required result is proved with

$$c_Q = \widetilde{c}_Q - w(\alpha) \tag{8.21}$$

which is positive if we can choose $\alpha > 0$ such that $w(\alpha)$ is sufficiently smaller than $\widetilde{c}_Q$.

Finally, for $\ell = 2$ we have by using the definition of $P_j$, $j = 1, \ldots, J$,

$$b(P_j v, P_j v) = \Re\big(a(v, P_j v)\big) = b(v, P_j v) + \Re\big(k(v, P_j v)\big)$$

so that

$$b(v, P_j v) = b(P_j v, P_j v) - \Re\big(k(v, P_j v)\big), \quad j = 1, \ldots, J.$$

Hence

$$\begin{aligned} b(v, Q^{(2)} v) &= b(v, Q_0 v) + \sum_{j=1}^{J} b(v, P_j v) \\ &= b(Q_0 v, Q_0 v) + b(v - Q_0 v, Q_0 v) + \sum_{j=1}^{J} b(P_j v, P_j v) - \sum_{j=1}^{J} \Re\big(k(v, P_j v)\big). \end{aligned}$$

By using the definition of $Q_0$ we deduce $b(v - Q_0 v, Q_0 v) = -k(v - Q_0 v, Q_0 v)$ and therefore we obtain, by rearranging the equation, using (8.10) and Assumption 8.5,

$$\begin{aligned} b(v, Q^{(2)} v) &= b(Q_0 v, Q_0 v) + \sum_{j=1}^{J} b(P_j v, P_j v) - k(v - Q_0 v, Q_0 v) - \sum_{j=1}^{J} \Re\big(k(v, P_j v)\big) \\ &\geq \widetilde{c}_P b(v, v) - \big|k(v - Q_0 v, Q_0 v)\big| - \sum_{j=1}^{J} \big|k(v, P_j v)\big| \\ &\geq \widetilde{c}_P b(v, v) - w(\alpha) b(v, v). \end{aligned}$$

Hence (8.19) is proved for $\ell = 2$ by choosing $\alpha > 0$ sufficiently small. □

As a direct consequence of Theorem 8.1, we prove that (8.7) and (8.8) have a unique solution and this solution is also the solution of (8.4).

**Theorem 8.2.** *Equations* (8.7) *and* (8.8) *have unique solutions. Moreover, equations* (8.4), (8.7), *and* (8.8) *are equivalent.*

*Proof.* It is clear from (8.19) and Assumption 8.1 (iii) that, for $\ell = 1, 2$, if $Q^{(\ell)} u = 0$ then $u = 0$. Hence (8.7) and (8.8) have at most one solution. Since these equations are in a finite dimensional space, uniqueness also implies existence.

Recall that Assumption 8.1 ensures the existence and uniqueness of the solution of (8.4). Let $u \in \mathscr{V}$ be the solution of (8.4). Then it follows from Definition 8.1 that, for all $v_j \in \mathscr{V}_j$, $j = 0, \ldots, J$,

$$a(Q_j u, v_j) = a(u, v_j) = \langle f, v_j \rangle = a(g_j^{(1)}, v_j).$$

Hence the above identity implies $Q_j u = g_j^{(1)}$ so that $Q^{(1)} u = g^{(1)}$, i.e., $u$ is also the solution of (8.7).

Similarly, for $j = 1, \ldots, J$

$$b(P_j u, v_j) = \Re\big(a(u, v_j)\big) = \langle f, v_j \rangle = b(g_j^{(2)}, v_j),$$

implying $b(P_j u - g_j^{(2)}, v_j) = 0$ for all $v_j \in \mathscr{V}_j$. The positive definiteness of $b(\cdot,\cdot)$ yields $P_j u = g_j^{(2)}$ so that $Q^{(2)}u = g^{(2)}$, i.e., $u$ is the solution of (8.8).

The converse can be seen by uniqueness. In fact, if $u$ is the solution of (8.7) which is not equal to the solution of (8.4), then the proof above shows that (8.7) has two solutions, which is a contradiction. Similarly for (8.8). □

## 8.2 Hypersingular Integral Equation

In this section we apply the theory developed in Section 8.1 to the hypersingular integral equation (8.2). We consider different preconditioners for both the $h$-version and $p$-version. In each case, it suffices to check on the assumptions posed in Section 8.1, namely Assumptions 8.1–8.5.

First we consider Assumption 8.1 which is universal for both the $h$-version and $p$-version.

*Checking of Assumption 8.1:* The bilinear forms are defined, for any $v$, $w \in \widetilde{H}^{1/2}(\Gamma)$, by

$$\begin{aligned} a_{\mathfrak{W}}(v,w) &:= \langle \mathfrak{W}_\kappa v, w\rangle\,, \\ b_{\mathfrak{W}}(v,w) &:= \langle \mathfrak{W}_0 v, w\rangle\,, \\ k_{\mathfrak{W}}(v,w) &:= \langle \mathfrak{K}_{\mathfrak{W}} v, w\rangle\,, \end{aligned} \tag{8.22}$$

where, for $x \in \Gamma$,

$$\begin{aligned} \mathfrak{W}_\kappa v(x) &:= -\frac{i}{2}\frac{\partial}{\partial n_x}\int_\Gamma \frac{\partial H_0^1(\kappa|x-y|)}{\partial n_y} v(y)\,d\sigma_y, \\ \mathfrak{W}_0 v(x) &:= -\frac{1}{\pi}\,\text{f.p.}\int_\Gamma \frac{v(y)}{|x-y|^2}\,d\sigma_y = \frac{1}{\pi}\frac{\partial}{\partial n_x}\int_\Gamma \frac{\partial}{\partial n_y}\log\frac{1}{|x-y|}\,d\sigma_y, \\ \mathfrak{K}_{\mathfrak{W}} v(x) &:= \mathfrak{W}_\kappa v(x) - \mathfrak{W}_0 v(x). \end{aligned}$$

Here, $\kappa$ is the wave number in the Helmholtz equation; see Subsection 1.1.1.

It is noted that the Hankel function $H_0^1$ has the following well known expansion

$$\begin{aligned} &\frac{i}{2}H_0^1(\kappa|x-y|) \\ &= \sum_{j=0}^{\infty} c_j|\kappa(x-y)|^{2j} - \frac{1}{\pi}\log|\kappa(x-y)|\Big(1+\sum_{j=1}^{\infty} d_j|\kappa(x-y)|^{2j}\Big), \end{aligned} \tag{8.23}$$

which converges for $|\kappa(x-y)| < R_0$ for some $R_0 > 0$.

It is shown in [235] that Assumption 8.1 holds true with $m = 1/2$; see also [65, 92].

Next we reconsider the subspace decompositions developed in Chapter 3 and Chapter 4, and show that Assumptions 8.2–8.5 hold true for these considerations. It should be noted that the bilinear form $a_{\mathfrak{W}}(\cdot,\cdot)$ in those chapters is the symmetric positive definite bilinear form $b_{\mathfrak{W}}(\cdot,\cdot)$ considered in this chapter.

### 8.2.1 The $h$-version

Consider the subspace decomposition of $\mathscr{V}$ defined in Definitions 3.1 and 3.2:

$$\mathscr{V} = \mathscr{V}_0 + \mathscr{V}_1 + \cdots + \mathscr{V}_J$$

where

$$\begin{aligned}
\mathscr{V} &:= \big\{v \in C(\Gamma) : v(\pm 1) = 0,\ v|_{\Gamma_{ij}} \text{ is affine},\ j = 1,\dots,N_i,\ i = 1,\dots,J\big\},\\
\mathscr{V}_0 &:= \big\{v \in C(\Gamma) : v(\pm 1) = 0,\ v|_{\Gamma_i} \text{ is affine},\ i = 1,\dots,J\big\},\\
\mathscr{V}_i &:= \big\{v \in \mathscr{V} : \operatorname{supp} v \subset \overline{\Gamma}_i\big\} = \mathscr{V} \cap \widetilde{H}^{1/2}(\Gamma_i), \quad i = 1,\dots,J,
\end{aligned}$$

with $\Gamma_i$ and $\Gamma_{ij}$ being disjoint subintervals such that

$$\overline{\Gamma} = \bigcup_{i=1}^{J} \overline{\Gamma}_i \quad \text{and} \quad \overline{\Gamma}_i = \bigcup_{j=1}^{N_i} \overline{\Gamma}_{ij},\ i = 1,\dots,J.$$

Recalling that $H_i = |\Gamma_i|$ and $h_i = \max_{1\le j\le N_i} |\Gamma_{ij}|$ we denote

$$H := \max_{1\le i\le J} H_i \quad \text{and} \quad 1 + \log\frac{H}{h} := \max_{1\le i\le J}\Big(1 + \log\frac{H_i}{h_i}\Big).$$

We assume that there exists a positive constant $C$ independent of $J$ such that

$$H^{1/4}\Big(1 + \log\frac{H}{h}\Big) \le C. \tag{8.24}$$

The above assumption dictates how fast $\log(H/h)$ could go to infinity. In fact, if $\varepsilon$ is the positive constant in (8.28), then it suffices to assume $H^{1-3\varepsilon}\big(1+\log(H/h)\big) \le C$; see the proof of Theorem 8.3.

*Checking of Assumption 8.2:* As were shown in Lemma 3.1 and Lemma 3.2, inequalities (8.5) and (8.6) hold true with

$$\widetilde{C}_{\mathscr{V}} = C_0\Big(1 + \log\frac{H}{h}\Big) \quad \text{and} \quad C_{\mathscr{V}} = C_1 \tag{8.25}$$

where $C_0$ and $C_1$ are constants independent of $H$ and $h$, $j = 1,\dots,J$.

*Checking of Assumption 8.3:* We show in the following lemma that this assumption is satisfied.

**Lemma 8.3.** *Under assumption* (8.24), *there exist positive constants $C_b$ and $H_0$ independent of $J$ such that if $0 < H \le H_0$ then for any $v_j \in \mathscr{V}_j$, the following inequality holds*

$$b_{\mathfrak{W}}(v_j, v_j) \le C_b \Re\big(a_{\mathfrak{W}}(v_j, v_j)\big), \quad j = 1, \ldots, J.$$

*Proof.* First we note that if $v_j \in \mathscr{V}_j$, $j = 1, \ldots, J$, then since $\operatorname{supp} v_j \subset \overline{\Gamma}_j$ and $|\Gamma_j| = H_j$ we have by using Lemma A.6

$$\|v_j\|^2_{L^2(\Gamma)} = \|v_j\|^2_{\widetilde{H}^0_{\rm I}(\Gamma_j)} = H_j \|\widehat{v}_j\|^2_{\widetilde{H}^0_{\rm I}(\widehat{\Gamma})} \le H_j \|\widehat{v}_j\|^2_{\widetilde{H}^{1/2}_{\rm I}(\widehat{\Gamma})} = H_j \|v_j\|^2_{\widetilde{H}^{1/2}_{\rm I}(\Gamma_j)}. \tag{8.26}$$

Invoking Lemma A.8 and Lemma C.21 and noting that $v_j$ vanishes at the endpoints of $\Gamma_j$, we obtain

$$\|v_j\|^2_{\widetilde{H}^{1/2}_{\rm I}(\Gamma_j)} \simeq \|v_j\|^2_{\widetilde{H}^{1/2}_{\rm w}(\Gamma_j)} \lesssim \Big(1 + \log \frac{H_j}{h_j}\Big)^2 |v_j|^2_{H^{1/2}(\Gamma_j)}.$$

Since $|v_j|_{H^{1/2}(\Gamma_j)} \le |v_j|_{H^{1/2}(\Gamma)}$, it follows that

$$\begin{aligned}\|v_j\|^2_{\widetilde{H}^{1/2}_{\rm I}(\Gamma_j)} &\le \Big(1 + \log \frac{H_j}{h_j}\Big)^2 |v_j|^2_{H^{1/2}(\Gamma)} \le \Big(1 + \log \frac{H_j}{h_j}\Big)^2 \|v_j\|^2_{\widetilde{H}^{1/2}_{\rm w}(\Gamma)} \\ &\simeq \Big(1 + \log \frac{H_j}{h_j}\Big)^2 \|v_j\|^2_{\widetilde{H}^{1/2}_{\rm I}(\Gamma)} \simeq \Big(1 + \log \frac{H_j}{h_j}\Big)^2 b_{\mathfrak{W}}(v_j, v_j)\end{aligned}$$

where we used again Lemma A.8 and used Assumption 8.1 (iii). This inequality together with (8.26) and the assumption (8.24) implies

$$\|v_j\|^2_{L^2(\Gamma)} \lesssim H^{1/2} b_{\mathfrak{W}}(v_j, v_j). \tag{8.27}$$

Therefore, Assumption 8.1 (iii) (Gårding's inequality) yields

$$\begin{aligned}b_{\mathfrak{W}}(v_j, v_j) &\lesssim \Re\big(a_{\mathfrak{W}}(v_j, v_j)\big) + \|v_j\|^2_{L^2(\Gamma)} \\ &\lesssim \Re\big(a_{\mathfrak{W}}(v_j, v_j)\big) + H^{1/2} b_{\mathfrak{W}}(v_j, v_j)\end{aligned}$$

implying

$$(1 - H^{1/2}) b_{\mathfrak{W}}(v_j, v_j) \lesssim \Re\big(a_{\mathfrak{W}}(v_j, v_j)\big).$$

This proves the lemma with $H_0$ chosen to be smaller than 1. □

*Checking of Assumption 8.4:* This assumption is satisfied with $m = 1/2$ and

$$w(H) = cH^{1-2\varepsilon} \tag{8.28}$$

for some $\varepsilon > 0$, as shown in the following lemma.

**Lemma 8.4.** *For every $\varepsilon > 0$ there exist positive constants $c$ and $H_0$ such that for any $H \in (0, H_0]$ and $v \in \mathscr{V}$ the following estimates hold*

$$\|Q_0 v - v\|_{L^2(\Gamma)} \le c H^{1/2-\varepsilon} \|v\|_{\widetilde{H}^{1/2}(\Gamma)}.$$

*Proof.* The results are consequences of Gårding's inequality and Nitsche's trick. The detailed proof is omitted. We note that the presence of $\varepsilon$ in the estimate is due to the fact that the projection $Q_0$ is defined with the bilinear form $a_{\mathfrak{W}}(\cdot,\cdot)$ which inherits the singularity due to $\Gamma$ being an open curve (or a screen in 3D). When $\Gamma$ is a closed curve (or closed surface in 3D) the estimate is optimal in the order of $O(H^{1/2})$; see [235]. □

In order to prove that Assumption 8.5 is satisfied, we first prove the following lemma.

**Lemma 8.5.** *Under assumption* (8.24), *for any $v \in \mathscr{V}$, the following inequality holds*

$$\|Q_j v\|^2_{L^2(\Gamma)} \le H_j^{1/2} \|Q_j v\|^2_{\widetilde{H}_I^{1/2}(\Gamma)}, \quad j = 1, \ldots, J.$$

*Proof.* The proof follows in the same manner as the proof of (8.27). □

*Checking of Assumption 8.5:* Finally we show the assumption on $k_{\mathfrak{W}}(\cdot,\cdot)$.

**Lemma 8.6.** *Under assumption* (8.24), *there exists $H_0 > 0$ such that for $0 < H \le H_0$, the bilinear form $k_{\mathfrak{W}}(\cdot,\cdot)$ defined in* (8.22) *satisfies for all $v \in \mathscr{V}$*

$$\left| \sum_{j=0}^{J} k_{\mathfrak{W}}(v - Q_j v, Q_j v) \right| \lesssim H^{1/4} b_{\mathfrak{W}}(v,v),$$

$$|k_{\mathfrak{W}}(v - Q_0 v, Q_0 v)| + \left| \sum_{j=1}^{J} k_{\mathfrak{W}}(v, P_j v) \right| \lesssim H^{1/4} b_{\mathfrak{W}}(v,v).$$

*Proof.* Firstly, for any $v \in \mathscr{V}$ we have by using Assumption 8.1 and Lemma 8.4

$$\begin{aligned} |k_{\mathfrak{W}}(v - Q_0 v, Q_0 v)| &\lesssim \|v - Q_0 v\|_{L^2(\Gamma)} \|Q_0 v\|_{\widetilde{H}^{1/2}(\Gamma)} \\ &\lesssim H^{1/2-\varepsilon} \|v\|^2_{\widetilde{H}^{1/2}(\Gamma)} \\ &\lesssim H^{1/2-\varepsilon} b_{\mathfrak{W}}(v,v). \end{aligned} \tag{8.29}$$

Next we have by using again Assumption 8.1

$$\begin{aligned} \left| \sum_{j=1}^{J} k_{\mathfrak{W}}(v - Q_j v, Q_j v) \right| &\le \left| \sum_{j=1}^{J} k_{\mathfrak{W}}(v, Q_j v) \right| + \left| \sum_{j=1}^{J} k_{\mathfrak{W}}(Q_j v, Q_j v) \right| \\ &= \left| k_{\mathfrak{W}}\Big(v, \sum_{j=1}^{J} Q_j v\Big) \right| + \left| \sum_{j=1}^{J} k_{\mathfrak{W}}(Q_j v, Q_j v) \right| \end{aligned}$$

$$\lesssim \|v\|_{\widetilde{H}_{\mathrm{I}}^{1/2}(\Gamma)} \Big\|\sum_{j=1}^{J} Q_j v\Big\|_{L^2(\Gamma)} + \sum_{j=1}^{J} \|Q_j v\|_{\widetilde{H}_{\mathrm{I}}^{1/2}(\Gamma)} \|Q_j v\|_{L^2(\Gamma)}.$$

Due to the fact that $\operatorname{supp} Q_j v \subset \overline{\Gamma}_j$ for $j = 1, \ldots, J$, we have

$$\begin{aligned}
\Big\|\sum_{j=1}^{J} Q_j v\Big\|_{L^2(\Gamma)} &= \Big(\sum_{i=1}^{J} \int_{\Gamma_i} \Big|\sum_{j=1}^{J} Q_j v(x)\Big|^2 \, dx\Big)^{1/2} \\
&= \Big(\sum_{i=1}^{J} \int_{\Gamma_i} \big|Q_i v(x)\big|^2 \, dx\Big)^{1/2} = \Big(\sum_{i=1}^{J} \|Q_i v\|_{L^2(\Gamma_i)}^2\Big)^{1/2}. \qquad (8.30)
\end{aligned}$$

Therefore, by invoking Lemma 8.5 we infer

$$\begin{aligned}
\Big|\sum_{j=1}^{J} k_{\mathfrak{W}}(v - Q_j v, Q_j v)\Big| &\lesssim \|v\|_{\widetilde{H}_{\mathrm{I}}^{1/2}(\Gamma)} \Big(\sum_{j=1}^{J} \|Q_j v\|_{L^2(\Gamma)}^2\Big)^{1/2} \\
&\quad + \sum_{j=1}^{J} \|Q_j v\|_{\widetilde{H}_{\mathrm{I}}^{1/2}(\Gamma)} \|Q_j v\|_{L^2(\Gamma)} \\
&\lesssim H^{1/4} \|v\|_{\widetilde{H}_{\mathrm{I}}^{1/2}(\Gamma)} \Big(\sum_{j=1}^{J} \|Q_j v\|_{\widetilde{H}_{\mathrm{I}}^{1/2}(\Gamma)}^2\Big)^{1/2} \\
&\quad + H^{1/4} \sum_{j=1}^{J} \|Q_j v\|_{\widetilde{H}_{\mathrm{I}}^{1/2}(\Gamma)}^2 \\
&\lesssim H^{1/4} b_{\mathfrak{W}}(v,v)^{1/2} \Big(\sum_{j=1}^{J} b_{\mathfrak{W}}(Q_j v, Q_j v)\Big)^{1/2} \\
&\quad + H^{1/4} \sum_{j=1}^{J} b_{\mathfrak{W}}(Q_j v, Q_j v).
\end{aligned}$$

Invoking Lemma 8.2 we deduce

$$\Big|\sum_{j=1}^{J} k_{\mathfrak{W}}(v - Q_j v, Q_j v)\Big| \lesssim H^{1/4} \, b_{\mathfrak{W}}(v,v).$$

The above inequality and (8.29) give the first inequality in the lemma.

The second inequality in the lemma can be shown by estimating $|\sum_{j=1}^{J} k_{\mathfrak{W}}(v, P_j v)|$ in the same manner as $|\sum_{j=1}^{J} k_{\mathfrak{W}}(v, Q_j v)|$ above. We omit the details. □

Having proved Assumptions 8.1–8.5, we can now invoke Theorem 8.1 and Lemma 2.19 to have the convergence result.

**Theorem 8.3.** *Under assumption* (8.24), *there exists* $H_0 > 0$ *such that for any* $H \in (0, H_0]$ *the following residual estimate holds*

$$\|\mathbf{r}_k\|_{\mathbb{R}^N} \lesssim \left(1 - \frac{1}{\big(1+\log(H/h)\big)^2}\right)^{k/2} \|\mathbf{r}_0\|_{\mathbb{R}^N}$$

*where* $\mathbf{r}_k$ *is the residual defined in Subsection 2.2.2.*

*Proof.* We recall from Lemma 2.19 that

$$\|\mathbf{r}_k\|_{\mathbb{R}^N} \lesssim \left(1 - \frac{c_Q^2}{C_Q^2}\right)^{k/2} \|\mathbf{r}_0\|_{\mathbb{R}^N} \tag{8.31}$$

where $c_Q$ and $C_Q$ are given by Theorem 8.1. Due to (8.13), (8.20), and (8.25), we have

$$C_Q^2 \simeq C_{\mathscr{V}}\,(1+C_{\mathscr{V}}) \simeq 1.$$

On the other hand, due to (8.17), (8.21), (8.25), and (8.28), we have

$$c_Q \simeq \widetilde{C}_{\mathscr{V}}^{-1} - w(H) \simeq \left(1+\log\frac{H}{h}\right)^{-1} - H^{1-2\varepsilon} \gtrsim \frac{1}{1+\log(H/h)} - H^{1/2}$$
$$\geq \frac{1}{1+\log(H/h)} - \frac{C^2}{\big(1+\log(H/h)\big)^2}$$

where $C$ is the constant in (8.24). We are only concerned with the case when $1 + \log(H/h) \to \infty$, because in this case $c_Q \to 0$ so that the convergence rate goes to 1, which implies a slow convergence of the preconditioned GMRES. However, in this case

$$\frac{1}{1+\log(H/h)} - \frac{C^2}{\big(1+\log(H/h)\big)^2} \geq \frac{1}{2\big(1+\log(H/h)\big)}$$

when $H$ and $h$ are sufficiently small. Consequently

$$1 - \frac{c_Q^2}{C_Q^2} \lesssim 1 - \frac{1}{\big(1+\log(H/h)\big)^2}$$

which together with (8.31) proves the theorem. □

## 8.2.2 The *p*-version

For this version, we consider the subspace decomposition defined in Definition 4.1 and Definition 4.2, namely

$$\mathscr{V} = \mathscr{V}_0 \oplus \mathscr{V}_1 \oplus \cdots \oplus \mathscr{V}_N$$

where

$$\mathscr{V} := \big\{v \in C(\Gamma) : v(\pm 1) = 0,\ v|_{\Gamma_j} \in \mathbb{P}_p, \quad j = 1,\dots,N\big\}$$

$$\mathscr{V}_0 := \{v \in C(\Gamma) : v(\pm 1) = 0,\ v|_{\Gamma_j} \in \mathbb{P}_1, \quad j = 1,\dots,N\}$$
$$\mathscr{V}_j := \mathscr{V} \cap \widetilde{H}^{1/2}(\Gamma_j), \quad j = 1,\dots,N.$$

Here, $\mathbb{P}_p$ is the set of polynomials of degree $p$, and

$$\Gamma_j := (x_{j-1}, x_j), \quad x_j = -1 + 2j/N, \quad j = 1,\dots,N.$$

Analogously to assumption (8.24) we assume the following property on the mesh size $h$ and polynomial degree $p$. For any $p_0 > 0$, there exists $h_0 > 0$ such that

$$h_0^{1/4}(1 + \log p_0) \lesssim 1. \tag{8.32}$$

*Checking of Assumption 8.2:* It was shown in Lemma 4.1 and Lemma 4.2 that inequalities (8.5) and (8.6) hold with

$$\widetilde{C}_V \simeq (1 + \log p)^2 \quad \text{and} \quad C_V = \text{constant}.$$

*Checking of Assumption 8.3:* Similarly to the $h$-version, we have the following lemma.

**Lemma 8.7.** *Under the assumption* (8.32), *there exists positive constants $C_b$ and $h_0$ such that for $j = 1,\dots,N$ and $0 < h \le h_0$ the following inequality holds*

$$b_{\mathfrak{W}}(v_j, v_j) \le C_b \Re\big(a_{\mathfrak{W}}(v_j, v_j)\big), \quad v_j \in \mathscr{V}_j.$$

*Proof.* The proof is very similar to that of Lemma 8.3, except that in this version $|\Gamma_j| = h$ and thus $H_j$ is replaced by $h$. Moreover, we invoke Lemma C.26 instead of Lemma C.21, and use assumption (8.32) in lieu of (8.24). The details are omitted. □

*Checking of Assumption 8.4:* Without giving further detail we state that this assumption is satisfied with $m = 1/2$ and $w(h) = ch^{1-2\varepsilon}$ for some $\varepsilon > 0$.

*Checking of Assumption 8.5:* Similarly to Lemma 8.6, we have exactly the same result for the $p$-version, with $H_0$ and $H$ replaced by $h_0$ and $h$. The detailed proof is omitted.

Collecting all the estimates in the above assumption checks, we obtain the following convergence result.

**Theorem 8.4.** *Under the assumption* (8.32), *there exists $h_0 > 0$ such that for any $h$ satisfying $0 < h \le h_0$ and $p = 1,\dots,p_0$ the following residual estimate holds*

$$\|\mathbf{r}_k\|_{\mathbb{R}^N} \lesssim \left(1 - \frac{1}{(1 + \log p)^4}\right)^{k/2} \|\mathbf{r}_0\|_{\mathbb{R}^N}.$$

*Proof.* Similarly to the $h$-version, we have

$$C_Q \simeq 1 \quad \text{and} \quad c_Q \simeq (1+\log p)^{-2} - h^{1-2\varepsilon}.$$

The proof follows along the lines of that of Theorem 8.3. □

## 8.3 Weakly-Singular Integral Equation

We now consider the application of the general theory in Section 8.1 to the weakly-singular integral equation (8.1). Similarly to the case of the hypersingular integral equation, it suffices to show that Assumptions 8.1–8.5 are satisfied.

*Checking of Assumption 8.1:* The bilinear forms are defined, for any $v, w \in \widetilde{H}^{1/2}(\Gamma)$, by

$$\begin{aligned} a_{\mathfrak{V}}(v,w) &:= \langle \mathfrak{V}_\kappa v, w\rangle, \\ b_{\mathfrak{V}}(v,w) &:= \langle \mathfrak{V}_0 v, w\rangle, \\ k_{\mathfrak{V}}(v,w) &:= \langle \mathfrak{K}_{\mathfrak{V}} v, w\rangle, \end{aligned} \tag{8.33}$$

where, for $x \in \Gamma$,

$$\begin{aligned} \mathfrak{V}_\kappa v(x) &:= \frac{i}{2}\int_\Gamma H_0^1(\kappa|x-y|)v(y)\,d\sigma_y, \\ \mathfrak{V}_0 v(x) &:= -\frac{1}{\pi}\int_\Gamma \log|x-y| v(y)\,d\sigma_y, \\ \mathfrak{K}_{\mathfrak{V}} v(x) &:= \mathfrak{V}_\kappa v(x) - \mathfrak{V}_0 v(x); \end{aligned}$$

see (8.23). It is shown in [212] that Assumption 8.1 holds true with $m = -1/2$.

In the following we will reconsider the subspace decompositions developed in Chapter 3 and Chapter 4, and show that Assumptions 8.2–8.5 hold true for these considerations. As in the case of the hypersingular integral equation, the bilinear form $a_{\mathfrak{V}}(\cdot,\cdot)$ in those chapters is the symmetric positive definite bilinear form $b_{\mathfrak{V}}(\cdot,\cdot)$ considered in this chapter.

### 8.3.1 The *h*-version

Recall the subspace decomposition of

$$\widetilde{\mathscr{V}} := \{v : \Gamma \to \mathbb{R} : v|_{\Gamma_{ij}} \text{ is constant, } j = 1,\dots,N_i,\ i = 1,\dots,J\}$$

defined in Definitions 3.1 and 3.3:

$$\widetilde{\mathscr{V}} = \widetilde{\mathscr{V}}_{-1} + \widetilde{\mathscr{V}}_0 + \cdots + \widetilde{\mathscr{V}}_J.$$

As in the case of the hypersingular integral equation, it suffices to check that Assumptions 8.1–8.5 are satisfied.

*Checking of Assumption 8.2:* As were shown in Lemma 3.4 and Lemma 3.5, inequalities (8.5) and (8.6) hold true with

$$\widetilde{C}_{\mathscr{V}} = C_0\left(1+\log\frac{H}{h}\right) \quad \text{and} \quad C_{\mathscr{V}} = C_2$$

where $C_0$ and $C_2$ are constants independent of $H$ and $h$.

*Checking of Assumption 8.3:* This assumption can be shown in the same manner as in the proof of Lemma 8.3 by using Gårding's inequality, except that instead of (8.26) we now use the following relation of the $\widetilde{H}^{-1}$-norm and $\widetilde{H}^{-1/2}$-norm. Noting that $\langle v_j, 1\rangle_{L^2(\Gamma)} = 0$, we can invoke Lemma A.19 and use (8.27) to have

$$\|v_j\|^2_{\widetilde{H}_{\mathrm{I}}^{-1}(\Gamma)} \simeq \|I_0 v_j\|^2_{\widetilde{H}_{\mathrm{I}}^{0}(\Gamma)} \lesssim H^{1/2}\|I_0 v_j\|^2_{\widetilde{H}_{\mathrm{I}}^{1/2}(\Gamma)} \simeq H^{1/2}\|v_j\|^2_{\widetilde{H}_{\mathrm{I}}^{-1/2}(\Gamma)}. \tag{8.34}$$

*Checking of Assumption 8.4:* This assumption is satisfied with $m = -1/2$ and $w(H) = cH^{1-2\varepsilon}$ for some $\varepsilon > 0$, namely

$$\|Q_0 v - v\|_{\widetilde{H}^{-1}(\Gamma)} \lesssim H^{1/2-\varepsilon}\|v\|_{\widetilde{H}^{-1/2}(\Gamma)}, \quad 0 < H \le H_0. \tag{8.35}$$

The proof is exactly the same as that of Lemma 8.4 and is omitted.

*Checking of Assumption 8.5:* Finally we check on the assumption of the bilinear form $k_{\mathfrak{V}}(\cdot,\cdot)$.

**Lemma 8.8.** *For any $\varepsilon > 0$ there exists $H_0 > 0$ such that for all $H \in (0, H_0]$, the bilinear form $k_{\mathfrak{V}}(\cdot,\cdot)$ defined in (8.33) satisfies for all $v \in \mathscr{V}$*

$$\left|\sum_{j=0}^{J} k_{\mathfrak{V}}(v - Q_j v, Q_j v)\right| \lesssim H^{1/4} b_{\mathfrak{V}}(v,v),$$

$$|k_{\mathfrak{V}}(v - Q_0 v, Q_0 v)| + \left|\sum_{j=1}^{J} k_{\mathfrak{V}}(v, P_j v)\right| \lesssim H^{1/4} b_{\mathfrak{V}}(v,v).$$

*Proof.* Firstly, for any $v \in \mathscr{V}$ we have by using Assumption 8.1, Lemma 8.1 (with $m = -1/2$), and (8.35)

$$\begin{aligned}|k_{\mathfrak{V}}(v - Q_0 v, Q_0 v)| &\lesssim \|v - Q_0 v\|_{\widetilde{H}^{-1}(\Gamma)}\|Q_0 v\|_{\widetilde{H}^{-1/2}(\Gamma)} \\ &\lesssim H^{1/2-\varepsilon}\|v\|^2_{\widetilde{H}^{-1/2}(\Gamma)} \lesssim H^{1/2-\varepsilon} b_{\mathfrak{V}}(v,v). \end{aligned} \tag{8.36}$$

Next we have by using again Assumption 8.1

$$\begin{aligned}\Big|\sum_{j=1}^{J} k_{\mathfrak{V}}(v - Q_j v, Q_j v)\Big| &\le \Big|\sum_{j=1}^{J} k_{\mathfrak{V}}(v, Q_j v)\Big| + \Big|\sum_{j=1}^{J} k_{\mathfrak{V}}(Q_j v, Q_j v)\Big| \\ &= \Big|k_{\mathfrak{V}}\Big(v, \sum_{j=1}^{J} Q_j v\Big)\Big| + \Big|\sum_{j=1}^{J} k_{\mathfrak{V}}(Q_j v, Q_j v)\Big| \\ &\lesssim \|v\|_{\widetilde{H}^{-1/2}(\Gamma)} \Big\|\sum_{j=1}^{J} Q_j v\Big\|_{\widetilde{H}^{-1}(\Gamma)} \\ &\quad + \sum_{j=1}^{J} \|Q_j v\|_{\widetilde{H}^{-1/2}(\Gamma)} \|Q_j v\|_{\widetilde{H}^{-1}(\Gamma)} \\ &\lesssim \|v\|_{\widetilde{H}^{-1/2}(\Gamma)} \Big\|\sum_{j=1}^{J} Q_j v\Big\|_{\widetilde{H}^{-1}(\Gamma)} + H^{1/4} \sum_{j=1}^{J} \|Q_j v\|^2_{\widetilde{H}^{-1/2}(\Gamma)}\end{aligned}$$

where in the last step we used (8.34). For the term $\|\sum_{j=1}^{J} Q_j v\|_{\widetilde{H}^{-1}(\Gamma)}$ we use again Lemma A.19 (noting that $\sum_{j=1}^{J} Q_j v$ has mean zero) and use the same argument as in (8.30) to obtain

$$\Big\|\sum_{j=1}^{J} Q_j v\Big\|_{\widetilde{H}^{-1}(\Gamma)} \simeq \Big\|\sum_{j=1}^{J} I_0 Q_j v\Big\|_{\widetilde{H}^{0}(\Gamma)} = \Big(\sum_{j=1}^{J} \|I_0 Q_j v\|^2_{\widetilde{H}^{0}(\Gamma)}\Big)^{1/2}.$$

Using (8.27) and Lemma A.19 we deduce

$$\Big\|\sum_{j=1}^{J} Q_j v\Big\|_{\widetilde{H}^{-1}(\Gamma)} \lesssim H^{1/4}\Big(\sum_{j=1}^{J} \|I_0 Q_j v\|^2_{\widetilde{H}_I^{1/2}(\Gamma)}\Big)^{1/2} \simeq H^{1/4}\Big(\sum_{j=1}^{J} \|Q_j v\|^2_{\widetilde{H}^{-1/2}(\Gamma)}\Big)^{1/2}$$

so that

$$\begin{aligned}\Big|\sum_{j=1}^{J} k_{\mathfrak{V}}(v - Q_j v, Q_j v)\Big| &\lesssim H^{1/4} \|v\|_{\widetilde{H}^{-1/2}(\Gamma)} \Big(\sum_{j=1}^{J} \|Q_j v\|^2_{\widetilde{H}^{-1/2}(\Gamma)}\Big)^{1/2} \\ &\quad + H^{1/4} \sum_{j=1}^{J} \|Q_j v\|^2_{\widetilde{H}^{-1/2}(\Gamma)}.\end{aligned}$$

Hence, by using Assumption 8.1 and Lemma 8.2 we deduce

$$\Big|\sum_{j=1}^{J} k_{\mathfrak{V}}(v - Q_j v, Q_j v)\Big| \lesssim H^{1/4}\, b_{\mathfrak{V}}(v, v).$$

The above inequality and (8.36) give the first inequality in the lemma.

The second inequality in the lemma can be shown by estimating $|\sum_{j=1}^{J} k_{\mathfrak{V}}(v, P_j v)|$ in the same manner as $|\sum_{j=1}^{J} k_{\mathfrak{V}}(v, Q_j v)|$ above. □

Having proved Assumptions 8.1–8.5, we can now invoke Theorem 8.1 and Lemma 2.19 to have the following convergence result.

**Theorem 8.5.** *Under the assumption* (8.24)*, there exists* $H_0 > 0$ *such that for any* $H \in (0, H_0]$ *the following residual estimate holds*

$$\|\mathbf{r}_k\|_{\mathbb{R}^N} \lesssim \left(1 - \frac{1}{(1+\log(H/h))^2}\right)^{k/2} \|\mathbf{r}_0\|_{\mathbb{R}^N}.$$

*Proof.* The proof is similar to that of Theorem 8.3 and is omitted. □

### 8.3.2 The *p*-version

For this version, we consider the subspace decomposition defined in Definition 4.1 and Definition 4.3, namely

$$\widetilde{\mathscr{V}} = \widetilde{\mathscr{V}}_0 \oplus \widetilde{\mathscr{V}}_1 \oplus \cdots \oplus \widetilde{\mathscr{V}}_N$$

where

$$\begin{aligned}
\widetilde{\mathscr{V}} &:= \left\{v : \Gamma \to \mathbb{R} : v|_{\Gamma_{ij}} \in \mathbb{P}_p, \quad i = 1,\dots,N,\ j = 1,\dots,N_i\right\} \\
\widetilde{\mathscr{V}}_0 &:= \left\{v : \Gamma \to \mathbb{R} : v|_{\Gamma_i} \in \mathbb{P}_0, \quad i = 1,\dots,N\right\} \\
\widetilde{\mathscr{V}}_j &:= \widetilde{\mathscr{V}} \cap \widetilde{H}^{-1/2}(\Gamma_j), \quad j = 1,\dots,N.
\end{aligned}$$

Here, $\mathbb{P}_p$ is the set of polynomials of degree $p$, and

$$\Gamma_j := (x_{j-1}, x_j), \quad x_j = -1 + 2j/N, \quad j = 1,\dots,N.$$

*Checking of Assumption 8.2:* It was shown in Lemma 4.3 and Lemma 4.4 that inequalities (8.5) and (8.6) hold with

$$\widetilde{C}_{\mathscr{V}} \simeq (1+\log p)^2 \quad \text{and} \quad C_{\mathscr{V}} \simeq 1.$$

*Checking of Assumption 8.3:* Similarly to the $h$-version, we have the following lemma.

**Lemma 8.9.** *Under the assumption* (8.32)*, there exists positive constants* $C_b$ *and* $h_0$ *such that for* $j = 1,\dots,N$ *and* $0 < h \le h_0$ *the following inequality holds*

$$b(v_j, v_j) \le C_b \Re\big(a(v_j, v_j)\big), \quad v_j \in \widetilde{\mathscr{V}}_j.$$

*Proof.* The proof is exactly the same to that of Lemma 8.3, except that in this version $|\Gamma_j| = h$ and thus $H_j$ is replaced by $h$. □

*Checking of Assumption 8.4:* Without giving further detail we state that this assumption is satisfied with $m = -1/2$ and $w(h) = ch^{1-2\varepsilon}$ for some $\varepsilon > 0$.

*Checking of Assumption 8.5:* Similarly to Lemma 8.6, we have exactly the same result for the $p$-version, with $H_0$ and $H$ replaced by $h_0$ and $h$. The detailed proof is omitted.

Collecting all the estimates in the above assumption checks, we obtain the following convergence result.

**Theorem 8.6.** *Under the assumption* (8.32), *there exists* $h_0 > 0$ *such that for any* $h$ *satisfying* $0 < h \leq h_0$ *and* $p = 0, 1, \ldots, p_0$, *the following residual estimate holds*

$$\|\mathbf{r}_k\|_{\mathbb{R}^N} \lesssim \Big(1 - \frac{1}{(1+\log(p+1))^4}\Big)^{k/2} \|\mathbf{r}_0\|_{\mathbb{R}^N}.$$

*Proof.* The proof follows along the lines of that of Theorem 8.4 and is omitted. □

## 8.4 Numerical Results

The theoretical results obtained in this chapter are supported by numerical results in Chapter 9. The connection is shown in the following table.

| Equation | Preconditioner | Theorem | Table |
|---|---|---|---|
| Hypersingular | $h$-version | 8.3 | 9.6 |
| | $p$-version | 8.4 | 9.20 |
| Weakly-singular | $h$-version | 8.5 | 9.5 |
| | $p$-version | 8.6 | 9.19 |

**Table 8.1** Theoretical results in Chapter 8 and relevant numerical results in Chapter 9

## 8.5 Indefinite Problems in Three-Dimensions

Since the general theory developed in Section 8.1 is independent of the dimension of the problem, similar results as those derived in Section 8.2 can be obtained in the case of three dimensions. We omit the detailed analysis, which can be found in [207], and only present here some numerical results.

We solve equation (1.2) with $\Gamma = (-1,1) \times (-1,1) \times \{0\}$, $\kappa = 5$, and $g(x) \equiv 1$ on a uniform triangular mesh by using the non-overlapping and overlapping preconditioner; see Chapter 10. The results for the non-overlapping preconditioner are reported in Table 8.2 while those for the overlapping preconditioner are reported in Table 8.3. In both tables, the number of iterations and CPU times (in seconds), with various values of $H/h$, are listed.

| DoF | WP | Non-overlapping | | | |
|---|---|---|---|---|---|
| | | $H/h=2$ | $H/h=4$ | $H/h=8$ | $H/h=16$ |
| 9 | 6 (0.01) | 6 (0.01) | | | |
| 49 | 17 (0.02) | 17 (0.01) | 17 (0.02) | | |
| 225 | 23 (0.02) | 20 (0.03) | 20 (0.02) | 21 (0.04) | |
| 961 | 31 (0.15) | 21 (0.24) | 21 (0.12) | 23 (0.20) | 23 (0.62) |
| 3969 | 44 (3.02) | 21 (4.39) | 21 (1.62) | 21 (1.75) | 26 (4.16) |
| 16129 | 63 (84.94) | 21 (93.72) | 21 (32.18) | 21 (29.86) | 24 (41.06) |

**Table 8.2** Number of iterations and CPU times (in parentheses). WP: without preconditioner

| DoF | Overlapping | | | | | | | |
|---|---|---|---|---|---|---|---|---|
| | $\delta=h$ | | | | $\delta=2h$ | | | |
| | $H/h=2$ | $H/h=4$ | $H/h=8$ | $H/h=16$ | $H/h=2$ | $H/h=4$ | $H/h=8$ | $H/h=16$ |
| 9 | 6 (0.02) | | | | 6 (0.02) | | | |
| 49 | 19 (0.01) | 18 (0.02) | | | 17 (0.03) | 20 (0.02) | | |
| 225 | 28 (0.04) | 27 (0.03) | 22 (0.05) | | 22 (0.09) | 26 (0.08) | 26 (0.09) | |
| 961 | 30 (0.39) | 27 (0.23) | 26 (0.34) | 25 (0.86) | 26 (0.65) | 29 (0.48) | 27 (0.57) | 27 (1.17) |
| 3969 | 30 (6.53) | 28 (2.50) | 27 (2.84) | 28 (6.12) | 31 (8.34) | 31 (3.92) | 28 (4.13) | 28 (7.99) |
| 16129 | 30 (135.47) | 28 (43.76) | 27 (41.14) | 29 (58.46) | 35 (166.31) | 31 (53.10) | 29 (49.80) | 29 (69.04) |

**Table 8.3** Number of iterations and CPU times (in parentheses)

# Chapter 9
# Implementation Issues and Numerical Experiments

In this chapter we discuss issues with implementation for the two-dimensional problems discussed in this part of the book. First we mention some important issues in implementation of the methods and then present the different numerical results for the different methods studied in Chapters 3–6 and Chapter 8. The stiffness matrix arising from the fully-discrete method studied in Chapter 7 is computed slightly differently and the numerical results for this study have been presented in that chapter.

The equations to be solved in this chapter take the form (1.1) and (1.2) on the interval $\Gamma = (-1, 1)$. The Galerkin boundary element method solves these equations in the finite dimensional space $\mathscr{S}$ where $\mathscr{S} = \widetilde{\mathscr{V}}$ is the space of piecewise polynomials of degree $p \geq 0$ for (1.1), and $\mathscr{S} = \mathscr{V}$ is the space of continuous piecewise polynomials of degree $p \geq 1$ vanishing at $\pm 1$ for (1.2); see (1.8) and (1.9).

Many of the numerical results reported in this chapter are obtained by using `maiprogs` written by Matthias Maischak. See the online documents Book of Numerical Experiments and User Guide Technical Manual at
`http://people.brunel.ac.uk/ mastmmm/`.

## 9.1 Implementation Issues

There are two different ways to benefit from the Schwarz preconditioners. They can be used in conjunction with the preconditioned conjugate gradient method (Algorithm 2.3) if the matrix is symmetric and positive definite, or with the preconditioned GMRES method (Algorithm 2.4 or Algorithm 2.5) if the matrix is indefinite or nonsymmetric. They can also be used together with a linear iterative method (Richardson, Jacobi, or Gauss–Seidel method). The analysis for the first approach requires estimation of the condition number (as done in Section 2.6), while the analysis for the second approach requires estimation of the convergence rate; see Section 2.9. It is noted that estimates for condition numbers give estimates for convergence rates, and vice versa; see [241].



E. P. Stephan and T. Tran, *Schwarz Methods and Multilevel Preconditioners for Boundary Element Methods*, https://doi.org/10.1007/978-3-030-79283-1_9

In this chapter, for the additive Schwarz methods we present numerical results using the first approach (PCG or PGMRES) while for the multiplicative Schwarz methods we present results using the second approach (linear iterative methods). The implementation of the additive Schwarz preconditioners is explained in Algorithm 2.7 and the implementation of the multiplicative Schwarz preconditioner, which is like that of multigrid methods, is explained in Algorithm 2.10 or Algorithm 2.11.

For both the additive and multiplicative versions, it is necessary to explain how the restriction and prolongation matrices $\mathbf{R}_j$ and $\mathbf{R}_j^\top$ are computed. It suffices to explain the additive Schwarz case. We have to distinguish two different cases: two-level and multilevel preconditioners.

(i) In a two-level method, the space $\mathscr{S}$ is decomposed by

$$\mathscr{S} = \mathscr{S}_0 + \mathscr{S}_1 + \cdots + \mathscr{S}_J$$

where $\mathscr{S}_0$ is the boundary element space defined on a coarse mesh. We recall the following matrix representation of the additive Schwarz operator which has been developed in Chapter 2, Section 2.3. Let $\mathbf{A}_j$ denote the Galerkin matrix corresponding to $\mathscr{S}_j$, $j = 0, \ldots, J$ and $\mathbf{A}$ the matrix corresponding to $\mathscr{S}$. It is shown in (2.27) that the matrix representation of the additive Schwarz operator $P_{\mathrm{ad}}$ is

$$\mathbf{P}_{\mathrm{ad}} = \mathbf{C}_{\mathrm{ad}}^{-1}\mathbf{A}$$

where

$$\mathbf{C}_{\mathrm{ad}}^{-1} = \sum_{j=0}^{J} \mathbf{R}_j^\top \mathbf{A}_j^{-1} \mathbf{R}_j$$

with $\mathbf{R}_j$ being the matrix representation of the restriction operator from the space $\mathscr{S}$ onto $\mathscr{S}_j$ and $\mathbf{R}_j^\top$ its transpose. Assume that $\{\phi_i : i = 1, \ldots, M\}$ is a basis for $\mathscr{S}$, and $\{\phi_i : i = N_{j-1}+1, \ldots, N_j\}$ is a basis for $\mathscr{S}_j$, $j = 1, \ldots, J$. Then, for $j = 1, \ldots, J$, the $(\ell, k)$-th entry of $\mathbf{R}_j$, $c_{\ell,k}^{(j)}$, is defined by

$$c_{\ell,k}^{(j)} := \begin{cases} 1 & \text{if } \ell = k \in \{N_{j-1}+1, \ldots, N_j\}, \\ 0 & \text{otherwise.} \end{cases}$$

The simple form of the projection matrices $\mathbf{R}_j$ is due to the fact that the same basis functions are used in $\mathscr{S}$ and in the subspaces $\mathscr{S}_j$, $j = 1, \ldots, J$.

In the case of the two-level non-overlapping method, the matrices $\mathbf{A}_j$ are simply the diagonal blocks of $\mathbf{A}$, the preconditioner $\mathbf{C}_{\mathrm{ad}}^{-1}$ can be written as

$$\mathbf{C}_{\mathrm{ad}}^{-1} = \mathbf{R}_0^\top \mathbf{A}_0^{-1} \mathbf{R}_0 + \begin{pmatrix} \mathbf{A}_1 & & & \\ & \mathbf{A}_2 & & \\ & & \ddots & \\ & & & \mathbf{A}_J \end{pmatrix}^{-1}.$$

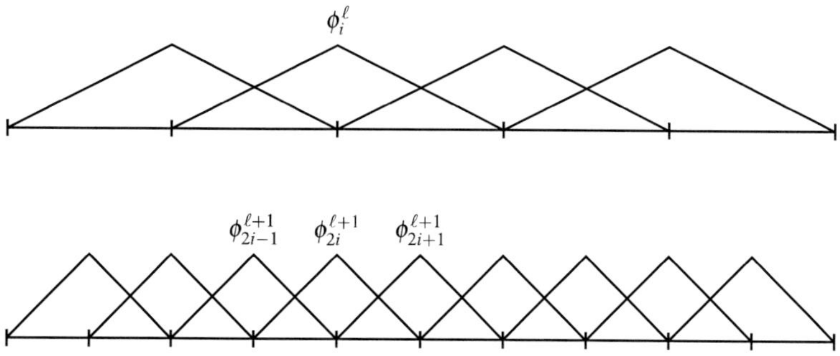


**Fig. 9.1** Hat basis functions on levels $\ell$ and $\ell+1$

In the case of the two-level overlapping method, the subblocks $\mathbf{A}_j$, $j = 1,\ldots,J$, overlap and we only have

$$\mathbf{C}_{\text{ad}}^{-1} = \mathbf{R}_0^\top \mathbf{A}_0^{-1} \mathbf{R}_0 + \begin{pmatrix} \mathbf{A}_1^{-1} & & & \\ & \mathbf{A}_2^{-1} & & \\ & & \ddots & \\ & & & \mathbf{A}_J^{-1} \end{pmatrix}.$$

In the implementation, the matrix $\mathbf{C}_{\text{ad}}^{-1}$ is not calculated; however, Algorithm 2.7 is used.

(ii) In a multilevel method, the subspaces form a nested sequence

$$\mathscr{S}^0 \subset \mathscr{S}^1 \subset \cdots \mathscr{S}^L$$

where, for each $\ell = 0,\ldots,L$, the space $\mathscr{S}^\ell$ has as basis functions $\phi_i^\ell$, $i = 1,\ldots,N_\ell = \dim(\mathscr{S}^\ell)$. Here the restriction matrix $\mathbf{R}_\ell$ from $\mathscr{S}^{\ell+1}$ to $\mathscr{S}^\ell$ is an $N_\ell \times N_{\ell+1}$ matrix whose entries $c_{ij}^\ell$, $i = 1,\ldots,N_\ell$ and $j = 1,\ldots,N_{\ell+1}$, are defined via the following relation between the basis functions on level $\ell$ and level $\ell+1$:

$$\phi_i^\ell = \sum_{j=1}^{N_{\ell+1}} c_{ij}^\ell \phi_j^{\ell+1}.$$

In the case of hat functions on a uniform mesh, as described in Section 5.1.1, the above formula becomes (see Figure 9.1)

$$\phi_i^\ell = \frac{1}{2}\phi_{2i-1}^{\ell+1} + \phi_{2i}^{\ell+1} + \frac{1}{2}\phi_{2i+1}^{\ell+1}, \quad i = 1,\ldots,N_\ell.$$

In the case of the Haar basis a similar relation holds

$$\psi_0^\ell = \psi_0^{\ell+1} \equiv 1, \quad \psi_i^\ell = \frac{1}{2}\psi_{2i-1}^{\ell+1} + \psi_{2j}^{\ell+1} + \frac{1}{2}\psi_{2i+1}^{\ell+1}, \quad i = 1,\dots,N_\ell - 1.$$

These relations of basis functions on different levels define the restriction matrix $\mathbf{R}_\ell$.

Each subspace $\mathscr{S}^\ell$ is furthered decomposed as $\mathscr{S}^\ell = \text{span}\{\phi_1^\ell\} \oplus \cdots \oplus \text{span}\{\phi_{N_\ell}^\ell\}$. Consequently, the multilevel additive Schwarz preconditioner becomes a multilevel diagonal preconditioner, and Algorithm 2.7 becomes Algorithm 9.1.

**Algorithm 9.1:** The matrix form of the multilevel additive Schwarz preconditioner

To compute $\mathbf{z} = \mathbf{C}_{\text{ad}}^{-1}\mathbf{r}$ for a given $\mathbf{r}$:
**for** $j = 0,\dots,L$ **do**
  Compute $\mathbf{R}_j\mathbf{r} \in \mathbb{R}^{N_j}$
  Compute $\mathbf{y}_j = \mathbf{D}_j^{-1}\mathbf{R}_j\mathbf{r}$ for $\mathbf{y}_j \in \mathbb{R}^{N_j}$ where $\mathbf{D}_j$ is the diagonal matrix defined from the diagonal of $\mathbf{A}_j$
**end for**
$\mathbf{z} \leftarrow \sum_{j=0}^{L} \mathbf{R}_j^\top \mathbf{y}_j \in \mathbb{R}^N$

Note that $\mathbf{R}_j$ are sparse matrices. Therefore, in both Algorithm 2.7 and Algorithm 9.1, the multiplication of $\mathbf{R}_j$ and $\mathbf{R}_j^\top$ to a vector costs only $O(N)$ operations in contrary to $O(N^2)$ in case of a full matrix.

The following numerical results have been reported in [120, 143, 210, 211, 222, 228, 229]. They are reprinted here with copyright permission.

## 9.2 Numerical Results for the $h$-Version

In this section, we report on the numerical experiments with the weakly-singular and hypersingular integral equations studied in Chapter 3, Chapter 5, and Chapter 8, when the $h$-version Galerkin boundary element method is used. The right-hand sides of both equations (1.1) and (1.2) are chosen to be $f(x) \equiv 1$ and $g(x) \equiv 1$. For each equation, we solve with different preconditioners, namely, two-level non-overlapping additive Schwarz, multilevel methods, and the BPX preconditioner which is introduced in [40] for partial differential equations. In [80] the BPX preconditioner is studied for the weakly-singular integral equation.

We note that the condition numbers are computed by using the Lanczos algorithm incorporated in the iterative solvers (CG, PCG, GMRES, and PGMRES); see [83].

The results for Chapter 3 and Chapter 5 involving the boundary integral equations arising from the Laplace equation are reported in Tables 9.1–9.4 and Figures 9.2–9.5. The conjugate gradient method (Algorithm 2.1) is used to solve the unpreconditioned system, while the preconditioned conjugate gradient method (Algorithm 2.3 is used to solve the preconditioned system.

The results for Chapter 8 involving the boundary integral equations arising from the Helmholtz equation are reported in Table 9.5, and Table 9.6. For these equations, the GMRES and preconditioned GMRES (Algorithms 2.2 and 2.4) are used.

| $N$ | Condition number | | | | Number of iterations | | | |
|---|---|---|---|---|---|---|---|---|
| | | | | | CG | PCG | | |
| | **A** | two-level | multilevel | BPX | | two-level | multilevel | BPX |
| 16 | 15.7545 | 2.1310 | 5.1005 | 4.0543 | 8 | 8 | 8 | 8 |
| 32 | 32.6847 | 2.6175 | 6.0171 | 4.8520 | 20 | 12 | 11 | 12 |
| 64 | 65.1292 | 2.9767 | 6.9188 | 8.8454 | 33 | 16 | 14 | 17 |
| 128 | 129.6566 | 3.1926 | 7.8182 | 10.4593 | 45 | 19 | 16 | 19 |
| 256 | 259.8891 | 3.3153 | 8.7199 | 12.1835 | 66 | 21 | 17 | 22 |
| 512 | 517.1460 | 3.3777 | 9.6253 | 14.0330 | 85 | 21 | 19 | 23 |
| 1024 | 1036.1733 | 3.4108 | 10.5344 | 16.0186 | 125 | 22 | 19 | 24 |
| 2048 | 2066.6248 | 3.4255 | 11.4468 | 18.1483 | 168 | 22 | 19 | 26 |
| 4096 | 4119.1740 | 3.4338 | 12.3619 | 20.4271 | 226 | 23 | 19 | 29 |
| 8192 | | 3.4361 | 13.2791 | 22.8595 | | 23 | 19 | 30 |
| 16384 | | 3.4370 | 14.1980 | 25.4481 | | 23 | 19 | 31 |
| 32768 | | 3.4379 | 15.1182 | 28.1890 | | 23 | 19 | 32 |
| 65536 | | 3.4406 | 16.0395 | 31.1033 | | 24 | 19 | 33 |

**Table 9.1** Weakly-singular integral equation (3.2) with $f(x) \equiv 1$. [Theory: Chapter 3, Section 3.1.1.2 (two-level); Chapter 5, Section 5.1.2 (multilevel); and References [40, 80] (BPX).]

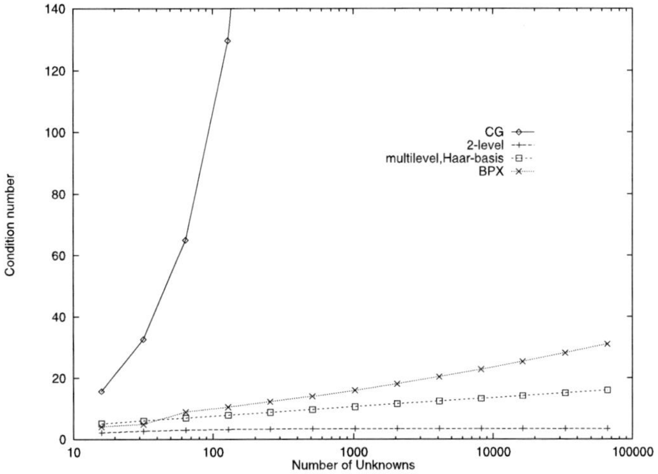


**Fig. 9.2** Weakly-singular integral equation (3.2) with $f(x) \equiv 1$. [Theory: Chapter 3, Section 3.1.1.2 (two-level); Chapter 5, Section 5.1.2 (multilevel); and Reference [40, 80] (BPX).]

| $N$ | Condition number | | | Number of iterations | | |
|---|---|---|---|---|---|---|
| | **A** | two-level | multilevel | CG | PCG | |
| | | | | | two-level | multilevel |
| 15 | 7.7629 | 2.1475 | 3.0353 | 8 | 7 | 8 |
| 31 | 15.5445 | 2.2072 | 3.4613 | 11 | 11 | 11 |
| 63 | 31.1092 | 2.2162 | 3.7561 | 17 | 12 | 14 |
| 127 | 62.4163 | 2.2276 | 3.9714 | 26 | 13 | 16 |
| 255 | 125.0924 | 2.2299 | 4.1335 | 38 | 13 | 17 |
| 511 | 250.4733 | 2.2262 | 4.2578 | 55 | 12 | 17 |
| 1023 | 501.2394 | 2.2250 | 4.3545 | 78 | 12 | 17 |
| 2047 | 1002.7757 | 2.2236 | 4.4308 | 109 | 12 | 17 |
| 4095 | 2005.8634 | 2.2225 | 4.4917 | 154 | 12 | 18 |
| 8191 | | 2.2160 | 4.5408 | | 11 | 18 |
| 16383 | | 2.2153 | 4.5808 | | 11 | 18 |
| 32767 | | 2.2150 | 4.6138 | | 11 | 18 |
| 65535 | | 2.2067 | 4.6413 | | 10 | 18 |

**Table 9.2** Hypersingular integral equation (3.1) with $g(x) \equiv 1$. [Theory: Chapter 3, Section 3.1.1.1 (two-level); Chapter 5, Section 5.1.1 (multilevel).]

| DoF | Condition number | | | | Iterations | | | |
|---|---|---|---|---|---|---|---|---|
| | CG | PCG | | | CG | PCG | | |
| | | MAS | NP | OP | | MAS | NP | OP |
| 3 | 0.2012E+01 | 1.644 | | | 2 | 2 | | |
| 7 | 0.3865E+01 (0.77) | 2.407 | 2.608 | 3.026 | 4 | 4 | 4 | 4 |
| 15 | 0.7578E+01 (0.88) | 3.035 | 3.597 | 2.839 | 5 | 6 | 6 | 6 |
| 31 | 0.1535E+02 (0.97) | 3.461 | 4.667 | 2.561 | 9 | 8 | 7 | 7 |
| 63 | 0.3089E+02 (0.99) | 3.755 | 5.826 | 2.468 | 13 | 9 | 8 | 7 |
| 127 | 0.6220E+02 (1.00) | 3.971 | 7.115 | 2.550 | 19 | 10 | 9 | 7 |
| 255 | 0.1249E+03 (1.00) | 4.131 | 8.564 | 2.786 | 27 | 10 | 9 | 8 |
| 511 | 0.2503E+03 (1.00) | 4.255 | 10.190 | 3.120 | 39 | 11 | 10 | 8 |
| 1023 | 0.5010E+03 (1.00) | 4.350 | 12.010 | 3.505 | 55 | 11 | 11 | 8 |
| 2047 | 0.1003E+04 (1.00) | 4.425 | 14.010 | 3.933 | 79 | 11 | 11 | 8 |

**Table 9.3** Hypersingular integral equation: Condition number and number of iterations of different methods. CG: without preconditioner (number in brackets: order of increase); MAS: multilevel additive Schwarz method; NP: nonoverlapping preconditioner with $H = 1/2$; OP: overlapping preconditioner with $H = 1/2$ and $\delta = h = 2/\text{DoF}$. [Theory: Chapter 3, Section 3.1.1.1 (NP), Section 3.1.2.1 (OP); Chapter 5, Section 5.1.1 (MAS).]

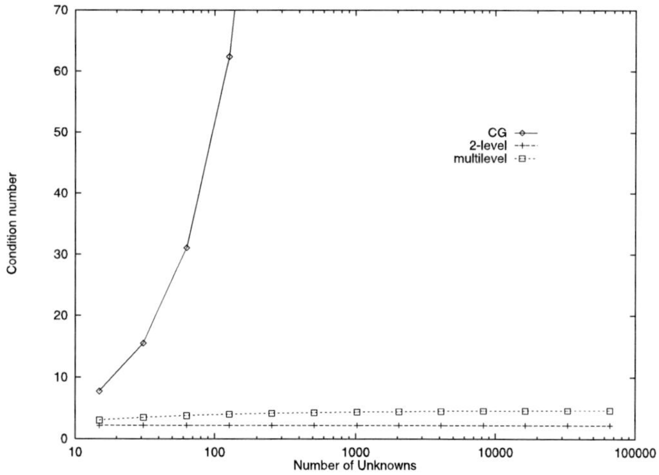


**Fig. 9.3** Hypersingular integral equation (3.1) with $g(x) \equiv 1$. [Theory: Chapter 3, Section 3.1.1.1 (two-level); Chapter 5, Section 5.1.1 (multilevel).]

| DoF | Overlapping additive Schwarz: condition number | | | | | |
|---|---|---|---|---|---|---|
| | $\delta = h$ | $\delta = 2h$ | $\delta = 4h$ | $\delta = h$ | $\delta = 2h$ | $\delta = 4h$ |
| | $H/\delta = 2$ | | | $H/\delta = 4$ | | |
| 7 | 3.026 | 2.838 | | 2.515 | | |
| 15 | 3.035 | 3.262 | 2.876 | 2.839 | 2.554 | |
| 31 | 3.087 | 3.266 | 3.328 | 2.930 | 2.900 | 2.577 |
| 63 | 3.103 | 3.294 | 3.342 | 2.896 | 2.986 | 2.932 |
| 127 | 3.117 | 3.302 | 3.367 | 2.900 | 2.955 | 3.015 |
| 255 | 3.123 | 3.303 | 3.371 | 2.895 | 2.957 | 2.985 |
| 511 | 3.125 | 3.302 | 3.371 | 2.890 | 2.952 | 2.987 |
| 1023 | 3.124 | 3.299 | 3.369 | 2.885 | 2.947 | 2.982 |
| 2047 | 3.122 | 3.298 | 3.367 | 2.882 | 2.943 | 2.976 |
| | $H/\delta = 8$ | | | $H/\delta = 16$ | | |
| 15 | 2.298 | | | | | |
| 31 | 2.561 | 2.320 | | 2.238 | | |
| 63 | 2.597 | 2.594 | 2.333 | 2.468 | 2.219 | |
| 127 | 2.621 | 2.630 | 2.613 | 2.480 | 2.472 | 2.208 |
| 255 | 2.614 | 2.654 | 2.648 | 2.482 | 2.495 | 2.478 |
| 511 | 2.605 | 2.648 | 2.673 | 2.460 | 2.490 | 2.501 |
| 1023 | 2.597 | 2.639 | 2.666 | 2.442 | 2.467 | 2.496 |
| 2047 | 2.592 | 2.632 | 2.657 | 2.429 | 2.449 | 2.474 |
| | $H/\delta = 32$ | | | $H/\delta = 64$ | | |
| 63 | 2.388 | | | | | |
| 127 | 2.550 | 2.272 | | 2.631 | | |
| 255 | 2.575 | 2.567 | 2.259 | 2.786 | 2.609 | |
| 511 | 2.534 | 2.568 | 2.564 | 2.797 | 2.764 | 2.600 |
| 1023 | 2.499 | 2.529 | 2.566 | 2.752 | 2.775 | 2.755 |
| 2047 | 2.475 | 2.495 | 2.529 | 2.716 | 2.731 | 2.766 |
| | $H/\delta = 128$ | | | $H/\delta = 256$ | | |
| 255 | 2.954 | | | | | |
| 511 | 3.120 | 2.927 | | 3.338 | | |
| 1023 | 3.116 | 3.092 | 2.915 | 3.505 | 3.306 | |
| 2047 | 3.056 | 3.090 | 3.079 | 3.499 | 3.473 | 3.292 |
| | $H/\delta = 512$ | | | $H/\delta = 1024$ | | |
| 1023 | 3.773 | | | | | |
| 2047 | 3.933 | 3.737 | | 4.253 | | |

**Table 9.4** Hypersingular integral equation: overlapping additive Schwarz, $h = 2/\text{DoF}$ fine mesh size, $\delta$: overlap size, $H$: coarse mesh size. [Theory: Chapter 3, Section 3.1.2.1.]

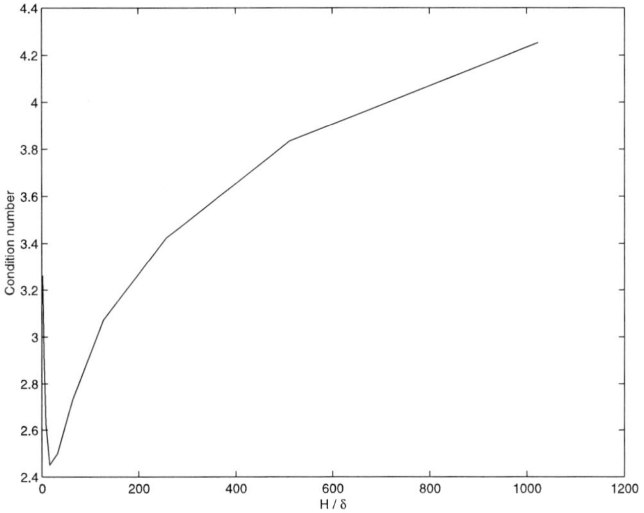


**Fig. 9.4** Hypersingular integral equation: Condition number of the overlapping preconditioned system as a function of $H/\delta$ when DoF = 2047. $H$: coarse mesh size, $\delta$: overlap size. [Theory: Chapter 3, Section 3.1.2.1.]

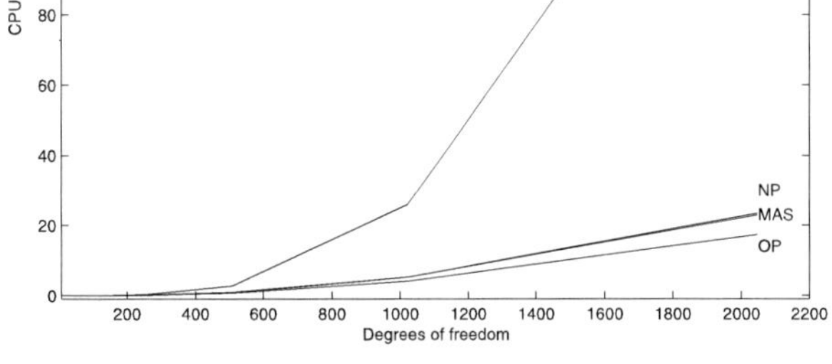


| DoF | H | |
|---|---|---|
| | NP | OP |
| 7 | 1/2 | 1/2 |
| 15 | 1/2 | 1/2 |
| 31 | 1/2 | 1/2 |
| 63 | 1/2 | 1/2 |
| 127 | 1/4 | 1/4 |
| 255 | 1/8 | 1/8 |
| 511 | 1/32 | 1/16 |
| 1023 | 1/64 | 1/32 |
| 2047 | 1/128 | 1/64 |

**Fig. 9.5** Hypersingular integral equation: CPU time of different methods. CG: without preconditioner, MAS: multilevel additive Schwarz, NP: non-overlapping, OP: overlapping, with $H$ chosen as in the side table and $\delta = h = 2/\text{DoF}$ for OP. [Theory: Chapter 3 and Chapter 5.]

| Number of iterations | | | | | | | | | | |
|---|---|---|---|---|---|---|---|---|---|---|
| | | | PGMRES | | | | | | | |
| $\kappa$ | $N$ | GMRES | $Q^{(1)}$ | | | | $Q^{(2)}$ | | | |
| | | | $H/h=$ | | | | | | | |
| | | | 2 | 4 | 8 | 16 | 2 | 4 | 8 | 16 |
| 2 | 32 | 18 | 16 | 14 | 10 | 7 | 16 | 14 | 10 | 8 |
| | 64 | 26 | 21 | 22 | 16 | 11 | 21 | 22 | 16 | 11 |
| | 128 | 34 | 21 | 24 | 24 | 18 | 21 | 24 | 24 | 18 |
| | 256 | 44 | 22 | 24 | 27 | 27 | 22 | 24 | 27 | 27 |
| | 512 | 55 | 22 | 25 | 27 | 29 | 22 | 25 | 27 | 29 |
| | 1024 | 69 | 23 | 25 | 27 | 30 | 23 | 25 | 27 | 30 |
| 10 | 32 | 17 | 16 | 15 | 11 | 7 | 16 | 16 | 14 | 13 |
| | 64 | 28 | 25 | 23 | 17 | 12 | 25 | 23 | 19 | 15 |
| | 128 | 37 | 24 | 29 | 26 | 20 | 24 | 29 | 26 | 21 |
| | 256 | 46 | 23 | 27 | 32 | 28 | 23 | 27 | 32 | 29 |
| | 512 | 58 | 23 | 27 | 30 | 35 | 23 | 27 | 30 | 35 |
| | 1024 | 72 | 24 | 27 | 29 | 33 | 24 | 27 | 30 | 33 |

| CPU times | | | | | | | | | | |
|---|---|---|---|---|---|---|---|---|---|---|
| | | | PGMRES | | | | | | | |
| $\kappa$ | $N$ | GMRES | $Q^{(1)}$ | | | | $Q^{(2)}$ | | | |
| | | | $H/h=$ | | | | | | | |
| | | | 2 | 4 | 8 | 16 | 2 | 4 | 8 | 16 |
| 2 | 32 | 0.00 | 0.01 | 0.01 | 0.00 | 0.00 | 0.01 | 0.01 | 0.00 | 0.01 |
| | 64 | 0.01 | 0.02 | 0.02 | 0.01 | 0.01 | 0.02 | 0.02 | 0.01 | 0.01 |
| | 128 | 0.04 | 0.04 | 0.04 | 0.04 | 0.03 | 0.05 | 0.05 | 0.05 | 0.04 |
| | 256 | 0.23 | 0.18 | 0.15 | 0.16 | 0.17 | 0.20 | 0.18 | 0.20 | 0.21 |
| | 512 | 1.77 | 1.10 | 0.88 | 0.89 | 0.96 | 1.19 | 1.01 | 1.03 | 1.11 |
| | 1024 | 13.14 | 6.97 | 5.08 | 5.15 | 5.67 | 7.01 | 5.17 | 5.24 | 5.77 |
| 10 | 32 | 0.00 | 0.01 | 0.01 | 0.00 | 0.00 | 0.01 | 0.01 | 0.01 | 0.01 |
| | 64 | 0.01 | 0.02 | 0.02 | 0.01 | 0.01 | 0.02 | 0.02 | 0.02 | 0.01 |
| | 128 | 0.04 | 0.05 | 0.05 | 0.05 | 0.04 | 0.06 | 0.06 | 0.05 | 0.05 |
| | 256 | 0.24 | 0.20 | 0.19 | 0.21 | 0.19 | 0.25 | 0.23 | 0.28 | 0.24 |
| | 512 | 1.85 | 1.15 | 1.02 | 1.06 | 1.24 | 1.33 | 1.16 | 1.25 | 1.43 |
| | 1024 | 13.69 | 7.03 | 5.31 | 5.38 | 6.06 | 7.30 | 5.86 | 6.06 | 6.60 |

**Table 9.5** Weakly-singular integral equation (8.1) for the Helmholtz equation with $f(x) \equiv 1$ and wave number $\kappa = 2$ and $\kappa = 10$. The two preconditioners $Q^{(1)}$ and $Q^{(2)}$ are defined in Definition 8.1. [Theory: Chapter 8, Section 8.3.1.]

| $N$ | Condition number | | Number of iterations | |
|---|---|---|---|---|
| | **A** | $Q^{(1)}$ | GMRES with **A** | PGMRES with $Q^{(1)}$ |
| 16 | 9.04 | 2.17 | 8 | 8 |
| 32 | 12.46 | 2.22 | 12 | 11 |
| 64 | 24.90 | 2.24 | 19 | 14 |
| 128 | 49.77 | 2.26 | 29 | 15 |
| 256 | 99.50 | 2.26 | 43 | 15 |
| 512 | 198.96 | 2.26 | 62 | 15 |

**Table 9.6** Hypersingular integral equation (8.2) with $g(x) \equiv 1$ and wave number $\kappa = 2$. The preconditioner $Q^{(1)}$ defined in Definition 8.1. [Theory: Chapter 8, Section 8.2.1.]

## 9.3 Numerical Results for the $p$-Version

In this section we report on the numerical results when the $p$-version Galerkin boundary element method is used to solve the boundary integral equations in concern. These numerical results underline the theoretical results obtained in Chapter 4, Chapter 6, and Chapter 8, Sections 8.2.2 and 8.3.2. For additive Schwarz preconditioners, see Tables 9.7–9.20 and Figures 9.6–9.7. For multiplicative Schwarz preconditioners, see Figures 9.8–9.11.

| $p$ | Minimum eigenvalue | | | Maximum eigenvalue | | | Condition number | | |
|---|---|---|---|---|---|---|---|---|---|
| | CG | PCG | | CG | PCG | | CG | PCG | |
| | | two-level | multilevel | | two-level | multilevel | | two-level | multilevel |
| 1 | 5.41e-2 | 0.55 | 0.53 | 0.52 | 1.43 | 1.47 | 9.67 | 2.59 | 2.75 |
| 2 | 1.87e-2 | 0.39 | 0.38 | 0.52 | 1.43 | 1.47 | 27.90 | 3.65 | 3.88 |
| 3 | 9.10e-3 | 0.33 | 0.29 | 0.52 | 1.49 | 1.52 | 57.57 | 4.46 | 5.27 |
| 4 | 5.16e-3 | 0.29 | 0.23 | 0.52 | 1.49 | 1.52 | 101.53 | 5.15 | 6.50 |
| 5 | 3.22e-3 | 0.26 | 0.20 | 0.52 | 1.52 | 1.54 | 162.54 | 5.77 | 7.82 |
| 6 | 2.15e-3 | 0.24 | 0.17 | 0.52 | 1.52 | 1.54 | 243.82 | 6.32 | 9.06 |
| 7 | 1.50e-3 | 0.23 | 0.15 | 0.52 | 1.54 | 1.54 | 348.35 | 6.83 | 10.37 |
| 8 | 1.09e-3 | 0.21 | 0.13 | 0.52 | 1.55 | 1.55 | 479.38 | 7.29 | 11.62 |
| 9 | 8.19e-4 | 0.20 | 0.12 | 0.52 | 1.56 | 1.55 | 639.93 | 7.73 | 12.91 |
| 10 | 6.29e-4 | 0.19 | 0.11 | 0.52 | 1.56 | 1.55 | 833.20 | 8.14 | 14.16 |

**Table 9.7** Weakly-singular integral equation (4.2) with $f(x) \equiv 1$: $p$-version with two subintervals. [Theory: Chapter 4, Section 4.1.1.2 (two-level) and Chapter 6, Section 6.2 (multilevel).]

| $N = 16$ | | | | | | | | |
|---|---|---|---|---|---|---|---|---|
| $p$ | DoF | Condition number | | Iterations | | CPU (sec) | | |
| | | CG | PCG | CG | PCG | CG | PCG |
| 1 | 32 | 114.78 | 3.75 | 23 | 17 | 0.00 | 0.00 |
| 2 | 48 | 292.68 (1.35) | 4.30 | 44 | 18 | 0.01 | 0.01 |
| 3 | 64 | 590.10 (1.73) | 5.60 | 62 | 22 | 0.01 | 0.01 |
| 4 | 80 | 1013.59 (1.88) | 6.02 | 80 | 23 | 0.02 | 0.01 |
| 5 | 96 | 1604.98 (2.06) | 6.97 | 103 | 24 | 0.03 | 0.01 |
| 6 | 112 | 2374.78 (2.15) | 7.32 | 113 | 25 | 0.04 | 0.02 |
| 7 | 128 | 3365.43 (2.26) | 8.08 | 133 | 26 | 0.06 | 0.02 |
| 8 | 144 | 4587.78 (2.32) | 8.39 | 153 | 26 | 0.08 | 0.02 |
| 9 | 160 | 6084.37 (2.40) | 9.03 | 169 | 26 | 0.11 | 0.03 |
| 10 | 176 | 7865.87 (2.44) | 9.31 | 195 | 27 | 0.15 | 0.03 |
| 11 | 192 | 9974.92 (2.49) | 9.86 | 210 | 28 | 0.19 | 0.04 |
| 12 | 208 | 12422.04 (2.52) | 10.12 | 231 | 28 | 0.26 | 0.05 |
| 13 | 224 | 15249.94 (2.56) | 10.61 | 246 | 28 | 0.32 | 0.05 |
| 14 | 240 | 18469.06 (2.58) | 10.84 | 272 | 28 | 0.40 | 0.06 |
| 15 | 256 | 22122.19 (2.62) | 11.28 | 291 | 29 | 0.49 | 0.07 |
| 16 | 272 | 26219.49 (2.63) | 11.50 | 327 | 29 | 0.63 | 0.08 |
| 17 | 288 | 30804.54 (2.66) | 11.90 | 354 | 30 | 0.76 | 0.09 |
| 18 | 304 | 35881.72 (2.67) | 12.10 | 363 | 30 | 0.91 | 0.10 |
| 19 | 320 | 41507.71 (2.69) | 12.47 | 393 | 30 | 1.11 | 0.11 |
| 20 | 336 | 47966.68 (2.82) | 12.65 | 415 | 30 | 1.26 | 0.12 |

**Table 9.8** Weakly-singular integral equation with $f(x) \equiv 1$, $p$-version. $N$: number of subintervals, CG: without preconditioner, PCG: non-overlapping additive Schwarz preconditioner. [Theory: Chapter 4, Section 4.1.1.2.]

| | | $N = 32$ | | | | | | |
|---|---|---|---|---|---|---|---|---|
| $p$ | DoF | Condition number | | | Iterations | | CPU (sec) | |
| | | CG | | PCG | CG | PCG | CG | PCG |
| 1 | 64 | 231.20 | | 3.79 | 48 | 22 | 0.01 | 0.01 |
| 2 | 96 | 588.05 | (1.35) | 4.32 | 76 | 20 | 0.02 | 0.01 |
| 3 | 128 | 1184.97 | (1.73) | 5.65 | 105 | 24 | 0.05 | 0.02 |
| 4 | 160 | 2034.33 | (1.88) | 6.06 | 132 | 23 | 0.09 | 0.03 |
| 5 | 192 | 3220.56 | (2.06) | 7.01 | 156 | 26 | 0.14 | 0.04 |
| 6 | 224 | 4764.08 | (2.15) | 7.36 | 181 | 25 | 0.20 | 0.04 |
| 7 | 256 | 6750.46 | (2.26) | 8.13 | 220 | 28 | 0.31 | 0.06 |
| 8 | 288 | 9200.85 | (2.32) | 8.44 | 247 | 26 | 0.43 | 0.07 |
| 9 | 320 | 12200.98 | (2.40) | 9.08 | 277 | 27 | 0.59 | 0.08 |
| 10 | 352 | 15771.70 | (2.44) | 9.36 | 316 | 27 | 0.80 | 0.10 |
| 11 | 384 | 19998.87 | (2.49) | 9.92 | 347 | 28 | 1.27 | 0.13 |
| 12 | 416 | 24903.08 | (2.52) | 10.17 | 378 | 28 | 1.58 | 0.15 |
| 13 | 448 | 30570.33 | (2.56) | 10.66 | 393 | 29 | 1.88 | 0.18 |
| 14 | 480 | 37021.07 | (2.58) | 10.89 | 446 | 29 | 2.43 | 0.20 |
| 15 | 512 | 44341.31 | (2.62) | 11.34 | 485 | 30 | 2.96 | 0.23 |
| 16 | 544 | 52553.41 | (2.63) | 11.55 | 522 | 30 | 3.56 | 0.25 |
| 17 | 576 | 61736.76 | (2.66) | 11.96 | 573 | 30 | 4.32 | 0.28 |
| 18 | 608 | 72016.98 | (2.69) | 12.15 | 614 | 30 | 5.18 | 0.32 |
| 19 | 640 | 83129.70 | (2.65) | 12.53 | 638 | 31 | 5.95 | 0.36 |
| 20 | 672 | 99704.00 | (3.54) | 12.71 | 657 | 31 | 6.71 | 0.39 |

**Table 9.9** Weakly-singular integral equation with $f(x) \equiv 1$, $p$-version. $N$: number of subintervals, CG: without preconditioner, PCG: non-overlapping additive Schwarz preconditioner. [Theory: Chapter 4, Section 4.1.1.2.]

| $N = 64$ | | | | | | | | |
|---|---|---|---|---|---|---|---|---|
| $p$ | DoF | Condition number | | | Iterations | | CPU (sec) | |
| | | CG | | PCG | CG | PCG | CG | PCG |
| 1 | 128 | 463.25 | | 3.81 | 72 | 22 | 0.03 | 0.02 |
| 2 | 192 | 1177.47 | (1.35) | 4.33 | 109 | 20 | 0.09 | 0.03 |
| 3 | 256 | 2372.38 | (1.73) | 5.66 | 159 | 24 | 0.22 | 0.06 |
| 4 | 320 | 4072.30 | (1.88) | 6.07 | 198 | 23 | 0.42 | 0.07 |
| 5 | 384 | 6446.51 | (2.06) | 7.03 | 251 | 26 | 0.75 | 0.11 |
| 6 | 448 | 9535.54 | (2.15) | 7.38 | 286 | 25 | 1.15 | 0.14 |
| 7 | 512 | 13510.86 | (2.26) | 8.14 | 330 | 26 | 1.73 | 0.18 |
| 8 | 576 | 18414.54 | (2.32) | 8.45 | 385 | 26 | 2.53 | 0.22 |
| 9 | 640 | 24418.31 | (2.40) | 9.10 | 429 | 27 | 3.49 | 0.27 |
| 10 | 704 | 31563.67 | (2.44) | 9.37 | 470 | 28 | 4.64 | 0.34 |
| 11 | 768 | 40022.60 | (2.49) | 9.93 | 534 | 28 | 7.72 | 0.48 |
| 12 | 832 | 49836.07 | (2.52) | 10.18 | 559 | 30 | 10.29 | 0.63 |
| 13 | 896 | 61176.35 | (2.56) | 10.68 | 658 | 29 | 15.11 | 0.75 |
| 14 | 960 | 74084.03 | (2.58) | 10.91 | 696 | 29 | 19.51 | 0.90 |
| 15 | 1024 | 88731.70 | (2.62) | 11.35 | 717 | 30 | 24.10 | 1.10 |
| 16 | 1088 | 105137.80 | (2.63) | 11.56 | 828 | 31 | 31.41 | 1.28 |
| 17 | 1152 | 123573.94 | (2.67) | 11.97 | 819 | 31 | 35.12 | 1.45 |
| 18 | 1216 | 143949.95 | (2.67) | 12.17 | 936 | 32 | 45.32 | 1.81 |
| 19 | 1280 | 166169.50 | (2.65) | 12.54 | 1016 | 31 | 55.22 | 1.82 |
| 20 | 1344 | 197207.23 | (3.34) | 12.71 | 1106 | 31 | 80.17 | 2.03 |

**Table 9.10** Weakly-singular integral equation with $f(x) \equiv 1$, $p$-version. $N$: number of subintervals, CG: without preconditioner, PCG: non-overlapping additive Schwarz preconditioner. [Theory: Chapter 4, Section 4.1.1.2.]

| $N = 128$ | | | | | | | | |
|---|---|---|---|---|---|---|---|---|
| $p$ | DoF | Condition number | | | Iterations | | CPU (sec) | |
| | | CG | | PCG | CG | PCG | CG | PCG |
| 1 | 256 | 926.93 | | 3.81 | 106 | 22 | 0.15 | 0.07 |
| 2 | 384 | 2355.64 | (1.35) | 4.33 | 163 | 20 | 0.48 | 0.10 |
| 3 | 512 | 4745.99 | (1.73) | 5.66 | 233 | 24 | 1.22 | 0.19 |
| 4 | 640 | 8146.45 | (1.88) | 6.07 | 268 | 23 | 2.17 | 0.25 |
| 5 | 768 | 12895.73 | (2.06) | 7.03 | 348 | 26 | 4.70 | 0.45 |
| 6 | 896 | 19074.80 | (2.15) | 7.38 | 421 | 27 | 9.60 | 0.72 |
| 7 | 1024 | 27026.73 | (2.26) | 8.15 | 495 | 26 | 16.58 | 0.97 |
| 8 | 1152 | 36835.55 | (2.32) | 8.45 | 579 | 28 | 25.01 | 1.33 |
| 9 | 1280 | 48844.85 | (2.40) | 9.10 | 637 | 28 | 34.78 | 1.65 |
| 10 | 1408 | 63137.53 | (2.44) | 9.37 | 686 | 29 | 46.38 | 2.11 |
| 11 | 1536 | 80057.70 | (2.49) | 9.93 | 750 | 28 | 62.21 | 2.47 |
| 12 | 1664 | 99687.17 | (2.52) | 10.18 | 875 | 30 | 86.89 | 3.15 |
| 13 | 1792 | 122370.66 | (2.56) | 10.68 | 932 | 29 | 163.47 | 3.60 |
| 14 | 1920 | 148189.27 | (2.58) | 10.91 | 1012 | 31 | 139.20 | 4.48 |
| 15 | 2048 | 177488.10 | (2.61) | 11.36 | 1135 | 31 | 198.52 | 5.63 |
| 16 | 2176 | 210342.04 | (2.63) | 11.57 | 1198 | 32 | 212.50 | 5.91 |
| 17 | 2304 | 247110.42 | (2.66) | 11.98 | 1265 | 31 | 248.65 | 6.36 |
| 18 | 2432 | 287901.31 | (2.67) | 12.17 | 1369 | 32 | 299.08 | 7.25 |
| 19 | 2560 | 332784.60 | (2.68) | 12.55 | 1366 | 31 | 329.19 | 7.75 |
| 20 | 2688 | 384781.09 | (2.83) | 12.74 | 1498 | 33 | 397.32 | 9.06 |

**Table 9.11** Weakly-singular integral equation with $f(x) \equiv 1$, $p$-version. $N$: number of subintervals, CG: without preconditioner, PCG: non-overlapping additive Schwarz preconditioner. [Theory: Chapter 4, Section 4.1.1.2.]

| $p$ | Number of iterations | | | | | |
|---|---|---|---|---|---|---|
| | $N = 2$ | | | $N = 4$ | | |
| | CG | PCG | | CG | PCG | |
| | | multilevel | two-level | | multilevel | two-level |
| 1 | 4 | 2 | 2 | 8 | 4 | 4 |
| 2 | 6 | 3 | 3 | 15 | 6 | 6 |
| 3 | 9 | 4 | 4 | 20 | 8 | 8 |
| 4 | 12 | 5 | 5 | 27 | 10 | 10 |
| 5 | 15 | 6 | 6 | 36 | 12 | 10 |
| 6 | 17 | 7 | 6 | 43 | 14 | 11 |
| 7 | 21 | 8 | 6 | 52 | 15 | 11 |
| 8 | 24 | 9 | 6 | 56 | 17 | 11 |
| 9 | 27 | 10 | 6 | 65 | 19 | 12 |
| 10 | 33 | 11 | 6 | 78 | 22 | 13 |

**Table 9.12** Weakly-singular integral equation (4.2) with $f(x) \equiv 1$: $p$-version. $N$: number of subintervals. [Theory: Chapter 4, Section 4.1.1.2 (two-level); Chapter 6, Section 6.2 (multilevel).]

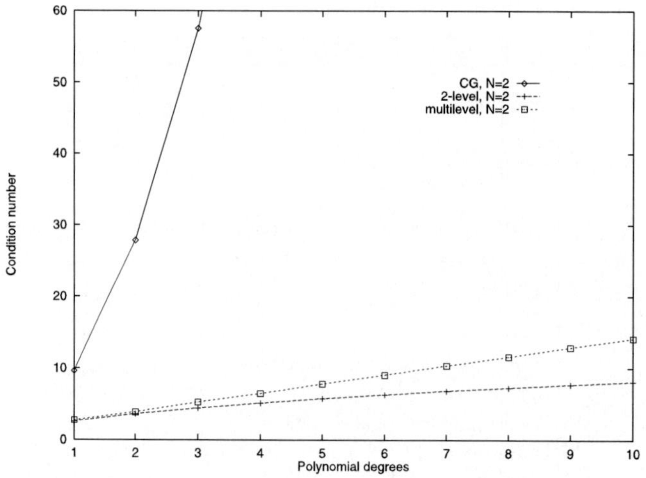


**Fig. 9.6** Weakly-singular integral equation (4.2) with $f(x) \equiv 1$: $p$-version with two subintervals, using CG and PCG with two-level and multilevel additive Schwarz preconditioners. [Theory: Chapter 4, Section 4.1.1.2 (two-level); Chapter 6, Section 6.2 (multilevel).]

| $p$ | Minimum eigenvalue | | | Maximum eigenvalue | | | Condition number | | |
|---|---|---|---|---|---|---|---|---|---|
| | CG | PCG | | CG | PCG | | CG | PCG | |
| | | two-level | multilevel | | two-level | multilevel | | two-level | multilevel |
| 2 | 0.21 | 0.55 | 0.55 | 0.93 | 1.27 | 1.27 | 4.44 | 2.31 | 2.31 |
| 3 | 7.40e-2 | 0.39 | 0.39 | 0.94 | 1.39 | 1.39 | 12.75 | 3.55 | 3.55 |
| 4 | 3.62e-2 | 0.33 | 0.29 | 0.94 | 1.39 | 1.43 | 26.04 | 4.16 | 4.85 |
| 5 | 2.06e-2 | 0.29 | 0.24 | 0.94 | 1.40 | 1.47 | 45.88 | 4.84 | 6.14 |
| 6 | 1.29e-2 | 0.26 | 0.20 | 0.94 | 1.40 | 1.48 | 73.32 | 5.31 | 7.41 |
| 7 | 8.58e-2 | 0.24 | 0.17 | 0.94 | 1.41 | 1.50 | 109.94 | 5.84 | 8.69 |
| 8 | 6.01e-2 | 0.23 | 0.15 | 0.94 | 1.40 | 1.50 | 156.98 | 6.21 | 9.95 |
| 9 | 4.37e-3 | 0.21 | 0.13 | 0.94 | 1.40 | 1.51 | 215.97 | 6.61 | 11.24 |
| 10 | 3.27e-3 | 0.20 | 0.12 | 0.94 | 1.40 | 1.52 | 288.23 | 6.95 | 12.50 |

**Table 9.13** Hypersingular integral equation (4.1) with $f(x) \equiv 1$: $p$-version with two subintervals. [Theory: Chapter 8, Section 8.2.2.]

| $N = 16$ | | | | | | | | | | |
|---|---|---|---|---|---|---|---|---|---|---|
| | | Condition number | | | Iterations | | | CPU $\times 10$ secs | | |
| $p$ | DoF | CG | PCG | | CG | PCG | | CG | PCG | |
| | | | NP | OP | | NP | OP | | NP | OP |
| 2 | 31 | 11.09 | 3.75 | 2.94 | 17 | 17 | 15 | 0.02 | 0.04 | 0.05 |
| 3 | 47 | 19.98 (1.45) | 4.30 | 3.01 | 26 | 17 | 17 | 0.04 | 0.05 | 0.07 |
| 4 | 63 | 41.00 (2.50) | 5.60 | 3.09 | 42 | 22 | 18 | 0.08 | 0.09 | 0.10 |
| 5 | 79 | 69.18 (2.34) | 6.02 | 3.10 | 52 | 22 | 17 | 0.13 | 0.11 | 0.12 |
| 6 | 95 | 110.14 (2.55) | 6.97 | 3.13 | 62 | 23 | 19 | 0.20 | 0.13 | 0.16 |
| 7 | 111 | 162.09 (2.51) | 7.32 | 3.13 | 73 | 24 | 18 | 0.28 | 0.16 | 0.18 |
| 8 | 127 | 230.26 (2.63) | 8.08 | 3.15 | 91 | 25 | 18 | 0.42 | 0.20 | 0.21 |
| 9 | 143 | 313.13 (2.61) | 8.39 | 3.15 | 99 | 25 | 18 | 0.54 | 0.23 | 0.25 |
| 10 | 159 | 415.83 (2.69) | 9.03 | 3.16 | 113 | 26 | 19 | 0.74 | 0.28 | 0.30 |
| 11 | 175 | 536.86 (2.68) | 9.31 | 3.16 | 124 | 26 | 19 | 0.99 | 0.34 | 0.36 |
| 12 | 191 | 681.36 (2.74) | 9.86 | 3.17 | 136 | 27 | 19 | 1.25 | 0.39 | 0.41 |
| 13 | 207 | 847.83 (2.73) | 10.12 | 3.17 | 158 | 27 | 19 | 1.77 | 0.46 | 0.48 |
| 14 | 223 | 1041.39 (2.77) | 10.61 | 3.18 | 164 | 27 | 19 | 2.11 | 0.52 | 0.55 |
| 15 | 239 | 1260.55 (2.77) | 10.84 | 3.17 | 184 | 28 | 19 | 2.69 | 0.60 | 0.59 |
| 16 | 255 | 1510.44 (2.80) | 11.28 | 3.18 | 202 | 28 | 19 | 3.36 | 0.67 | 0.67 |
| 17 | 271 | 1789.53 (2.80) | 11.50 | 3.18 | 206 | 28 | 19 | 3.99 | 0.77 | 0.75 |
| 18 | 287 | 2103.01 (2.82) | 11.90 | 3.18 | 221 | 29 | 19 | 4.72 | 0.87 | 0.83 |
| 19 | 303 | 2449.24 (2.82) | 12.10 | 3.18 | 229 | 29 | 19 | 5.71 | 1.04 | 0.92 |
| 20 | 319 | 2833.71 (2.84) | 12.47 | 3.19 | 261 | 29 | 19 | 7.38 | 1.06 | 0.99 |

**Table 9.14** Hypersingular integral equation with $g(x) \equiv 1$, $p$-version. CG: without preconditioner; PCG with non-overlapping preconditioner (NP) and with overlapping preconditioner (OP); $N$: number of subintervals. Number in brackets: order of increase. [Theory: Chapter 4, Section 4.1.1.1 and Section 4.1.2.1.]

| $N = 32$ | | | | | | | | | | | |
|---|---|---|---|---|---|---|---|---|---|---|---|
| $p$ | DoF | Condition number | | | | Iterations | | | CPU (sec) | | |
| | | CG | | PCG | | CG | PCG | | CG | PCG | |
| | | | | NP | OP | | NP | OP | | NP | OP |
| 2 | 63 | 18.35 | | 3.79 | 2.97 | 29 | 20 | 16 | 0.01 | 0.01 | 0.01 |
| 3 | 95 | 20.14 | (0.23) | 4.32 | 3.03 | 37 | 19 | 17 | 0.01 | 0.01 | 0.01 |
| 4 | 127 | 41.58 | (2.52) | 5.65 | 3.11 | 49 | 23 | 18 | 0.02 | 0.02 | 0.02 |
| 5 | 159 | 69.66 | (2.31) | 6.06 | 3.12 | 63 | 22 | 17 | 0.04 | 0.02 | 0.03 |
| 6 | 191 | 110.93 | (2.55) | 7.01 | 3.15 | 75 | 23 | 18 | 0.06 | 0.03 | 0.03 |
| 7 | 223 | 163.12 | (2.50) | 7.36 | 3.16 | 88 | 24 | 18 | 0.10 | 0.04 | 0.04 |
| 8 | 255 | 231.73 | (2.63) | 8.13 | 3.17 | 106 | 25 | 19 | 0.15 | 0.05 | 0.06 |
| 9 | 287 | 315.03 | (2.61) | 8.44 | 3.17 | 125 | 25 | 18 | 0.22 | 0.06 | 0.06 |
| 10 | 319 | 418.34 | (2.69) | 9.08 | 3.19 | 139 | 26 | 19 | 0.29 | 0.08 | 0.08 |
| 11 | 351 | 540.01 | (2.68) | 9.36 | 3.18 | 156 | 26 | 19 | 0.42 | 0.10 | 0.10 |
| 12 | 383 | 685.33 | (2.74) | 9.92 | 3.19 | 178 | 27 | 19 | 0.58 | 0.13 | 0.12 |
| 13 | 415 | 852.67 | (2.73) | 10.17 | 3.19 | 193 | 27 | 19 | 0.79 | 0.15 | 0.13 |
| 14 | 447 | 1047.28 | (2.77) | 10.66 | 3.20 | 206 | 28 | 19 | 1.00 | 0.17 | 0.15 |
| 15 | 479 | 1267.58 | (2.77) | 10.89 | 3.20 | 218 | 28 | 19 | 1.21 | 0.19 | 0.17 |
| 16 | 511 | 1518.79 | (2.80) | 11.34 | 3.20 | 240 | 28 | 19 | 1.49 | 0.21 | 0.19 |
| 17 | 543 | 1799.42 | (2.80) | 11.55 | 3.20 | 257 | 28 | 19 | 1.82 | 0.24 | 0.21 |
| 18 | 575 | 2114.40 | (2.82) | 11.96 | 3.21 | 274 | 29 | 19 | 2.21 | 0.27 | 0.24 |
| 19 | 607 | 2465.71 | (2.84) | 12.15 | 3.21 | 300 | 29 | 19 | 2.73 | 0.31 | 0.27 |
| 20 | 639 | 2846.90 | (2.80) | 12.53 | 3.21 | 324 | 30 | 19 | 3.29 | 0.35 | 0.30 |

**Table 9.15** Hypersingular integral equation with $g(x) \equiv 1$, $p$-version. CG: without preconditioner; PCG with non-overlapping preconditioner (NP) and with overlapping preconditioner (OP); $N$: number of subintervals. Number in brackets: order of increase. [Theory: Chapter 4, Section 4.1.1.1 and Section 4.1.2.1.]

| $N = 64$ | | | | | | | | | | | |
|---|---|---|---|---|---|---|---|---|---|---|---|
| | | Condition number | | | Iterations | | | CPU (sec) | | | |
| $p$ | DoF | CG | PCG | | CG | PCG | | CG | PCG | | |
| | | | NP | OP | | NP | OP | | NP | OP | |
| 2 | 127 | 34.90 | 3.81 | 2.97 | 32 | 19 | 16 | 0.01 | 0.02 | 0.02 | |
| 3 | 191 | 35.18 (0.02) | 4.33 | 3.04 | 40 | 19 | 17 | 0.03 | 0.03 | 0.04 | |
| 4 | 255 | 43.57 (0.74) | 5.66 | 3.12 | 56 | 21 | 18 | 0.08 | 0.05 | 0.05 | |
| 5 | 319 | 69.78 (2.11) | 6.07 | 3.13 | 72 | 22 | 18 | 0.15 | 0.07 | 0.07 | |
| 6 | 383 | 111.28 (2.56) | 7.03 | 3.16 | 87 | 23 | 18 | 0.26 | 0.10 | 0.10 | |
| 7 | 447 | 163.40 (2.49) | 7.38 | 3.16 | 102 | 23 | 18 | 0.41 | 0.13 | 0.12 | |
| 8 | 511 | 232.18 (2.63) | 8.14 | 3.18 | 118 | 25 | 18 | 0.62 | 0.17 | 0.15 | |
| 9 | 575 | 315.56 (2.60) | 8.45 | 3.18 | 134 | 25 | 18 | 0.88 | 0.21 | 0.19 | |
| 10 | 639 | 419.05 (2.69) | 9.10 | 3.19 | 155 | 26 | 18 | 1.27 | 0.27 | 0.22 | |
| 11 | 703 | 540.88 (2.68) | 9.37 | 3.19 | 175 | 26 | 18 | 2.22 | 0.36 | 0.31 | |
| 12 | 767 | 686.43 (2.74) | 9.93 | 3.20 | 197 | 27 | 19 | 3.17 | 0.47 | 0.40 | |
| 13 | 831 | 854.01 (2.73) | 10.18 | 3.20 | 213 | 27 | 18 | 4.19 | 0.58 | 0.46 | |
| 14 | 895 | 1048.92 (2.77) | 10.68 | 3.20 | 223 | 28 | 19 | 5.29 | 0.74 | 0.58 | |
| 15 | 959 | 1269.53 (2.77) | 10.91 | 3.20 | 246 | 28 | 19 | 6.83 | 0.88 | 0.69 | |
| 16 | 1023 | 1521.11 (2.80) | 11.35 | 3.21 | 275 | 28 | 19 | 8.94 | 1.04 | 0.80 | |
| 17 | 1087 | 1802.05 (2.80) | 11.56 | 3.21 | 292 | 28 | 19 | 10.90 | 1.19 | 0.91 | |
| 18 | 1151 | 2117.51 (2.82) | 11.97 | 3.21 | 321 | 29 | 19 | 13.68 | 1.37 | 1.02 | |
| 19 | 1215 | 2467.33 (2.83) | 12.17 | 3.21 | 344 | 29 | 19 | 16.72 | 1.67 | 1.15 | |
| 20 | 1279 | 2849.45 (2.81) | 12.54 | 3.21 | 369 | 29 | 19 | 20.28 | 1.72 | 1.27 | |

**Table 9.16** Hypersingular integral equation with $g(x) \equiv 1$, $p$-version. CG: without preconditioner; PCG with non-overlapping preconditioner (NP) and with overlapping preconditioner (OP); $N$: number of subintervals. Number in brackets: order of increase. [Theory: Chapter 4, Section 4.1.1.1 and Section 4.1.2.1.]

| N = 128 | | | | | | | | | | | |
|---|---|---|---|---|---|---|---|---|---|---|---|
| | | Condition number | | | Iterations | | | CPU (sec) | | |
| $p$ | DoF | CG | PCG | | CG | PCG | | CG | PCG | |
| | | | NP | OP | | NP | OP | | NP | OP |
| 2 | 255 | 68.59 | 3.81 | 2.97 | 40 | 19 | 16 | 0.06 | 0.06 | 0.07 |
| 3 | 383 | 69.09 (0.02) | 4.33 | 3.04 | 46 | 19 | 16 | 0.14 | 0.10 | 0.10 |
| 4 | 511 | 70.26 (0.06) | 5.66 | 3.11 | 62 | 21 | 17 | 0.32 | 0.17 | 0.16 |
| 5 | 639 | 70.29 (0.00) | 6.07 | 3.12 | 75 | 21 | 17 | 0.61 | 0.23 | 0.22 |
| 6 | 767 | 111.98 (2.55) | 7.03 | 3.15 | 91 | 23 | 17 | 1.25 | 0.40 | 0.34 |
| 7 | 895 | 163.48 (2.45) | 7.38 | 3.16 | 109 | 23 | 17 | 2.51 | 0.62 | 0.52 |
| 8 | 1023 | 232.43 (2.64) | 8.15 | 3.17 | 127 | 24 | 18 | 4.29 | 0.91 | 0.75 |
| 9 | 1151 | 315.70 (2.60) | 8.45 | 3.18 | 148 | 24 | 18 | 6.47 | 1.16 | 0.95 |
| 10 | 1279 | 419.30 (2.69) | 9.10 | 3.19 | 170 | 25 | 18 | 9.40 | 1.50 | 1.18 |
| 11 | 1407 | 541.12 (2.68) | 9.37 | 3.19 | 189 | 25 | 18 | 12.98 | 1.85 | 1.45 |
| 12 | 1535 | 686.76 (2.74) | 9.93 | 3.19 | 205 | 26 | 18 | 17.23 | 2.35 | 1.74 |
| 13 | 1663 | 854.37 (2.73) | 10.18 | 3.19 | 220 | 26 | 18 | 22.27 | 2.80 | 2.11 |
| 14 | 1791 | 1049.39 (2.77) | 10.68 | 3.20 | 242 | 27 | 18 | 29.11 | 3.53 | 2.44 |
| 15 | 1919 | 1270.06 (2.77) | 10.91 | 3.20 | 266 | 27 | 18 | 37.53 | 3.99 | 2.89 |
| 16 | 2047 | 1521.76 (2.80) | 11.36 | 3.21 | 291 | 28 | 18 | 111.9 | 4.78 | 3.26 |
| 17 | 2175 | 1802.71 (2.79) | 11.57 | 3.21 | 305 | 28 | 18 | 55.41 | 5.30 | 3.69 |
| 18 | 2303 | 2118.45 (2.82) | 11.98 | 3.21 | 341 | 28 | 19 | 68.92 | 5.87 | 4.21 |
| 19 | 2431 | 2468.17 (2.83) | 12.17 | 3.21 | 360 | 28 | 19 | 80.34 | 6.52 | 4.75 |
| 20 | 2559 | 2852.32 (2.82) | 12.55 | 3.21 | 394 | 29 | 19 | 97.14 | 7.41 | 5.15 |

**Table 9.17** Hypersingular integral equation with $g(x) \equiv 1$, $p$-version. CG: without preconditioner; PCG with non-overlapping preconditioner (NP) and with overlapping preconditioner (OP); $N$: number of subintervals. Number in brackets: order of increase. [Theory: Chapter 4, Section 4.1.1.1 and Section 4.1.2.1.]

| $p$ | Number of iterations | | |
|---|---|---|---|
| | CG | PCG | |
| | | multilevel | two-level |
| 2 | 2 | 2 | 2 |
| 3 | 3 | 3 | 3 |
| 4 | 4 | 4 | 4 |
| 5 | 5 | 5 | 5 |
| 6 | 7 | 6 | 6 |
| 7 | 8 | 7 | 7 |
| 8 | 9 | 8 | 7 |
| 9 | 12 | 9 | 7 |
| 10 | 13 | 10 | 7 |

**Table 9.18** Hypersingular integral equation (4.1) with $g(x) \equiv 1$: $p$-version. Number of subintervals = 2. [Theory: Chapter 4, Section 4.1.1.1 (two-level) and Chapter 6, Section 6.2 (multilevel).]

| Number of iterations | | | | | | | | | |
|---|---|---|---|---|---|---|---|---|---|
| $\kappa$ | $p$ | DoF | GMRES | PGMRES | | DoF | GMRES | PGMRES | |
| | | | | $Q^{(1)}$ | $Q^{(2)}$ | | | $Q^{(1)}$ | $Q^{(2)}$ |
| $\kappa = 2$ | | $N = 16$ | | | | $N = 32$ | | | |
| | 4 | 80 | 44 | 23 | 23 | 160 | 64 | 25 | 25 |
| | 6 | 112 | 55 | 25 | 25 | 224 | 77 | 27 | 27 |
| | 8 | 144 | 64 | 26 | 27 | 288 | 91 | 29 | 29 |
| | 10 | 176 | 72 | 28 | 28 | 352 | 105 | 31 | 31 |
| | 12 | 208 | 82 | 29 | 29 | 416 | 114 | 32 | 32 |
| | 14 | 240 | 89 | 29 | 29 | 480 | 124 | 33 | 33 |
| | 16 | 272 | 95 | 30 | 30 | 544 | 135 | 33 | 33 |
| | 18 | 304 | 103 | 30 | 30 | 608 | 145 | 34 | 34 |
| | 20 | 336 | 110 | 31 | 31 | 672 | 155 | 35 | 35 |
| | | $N = 64$ | | | | $N = 128$ | | | |
| | 4 | 320 | 90 | 26 | 26 | 640 | 119 | 28 | 28 |
| | 6 | 448 | 112 | 28 | 28 | 896 | 149 | 30 | 30 |
| | 8 | 576 | 127 | 30 | 30 | 1152 | 179 | 32 | 32 |
| | 10 | 704 | 144 | 32 | 32 | 1408 | 204 | 33 | 33 |
| | 12 | 832 | 159 | 33 | 33 | 1664 | 222 | 34 | 34 |
| | 14 | 960 | 175 | 34 | 34 | 1920 | 239 | 35 | 35 |
| | 16 | 1088 | 190 | 34 | 34 | 2176 | 258 | 36 | 36 |
| | 18 | 1216 | 203 | 35 | 35 | 2432 | 279 | 37 | 37 |
| | 20 | 1344 | 215 | 36 | 36 | 2688 | 297 | 38 | 38 |
| $\kappa = 10$ | | $N = 16$ | | | | $N = 32$ | | | |
| | 4 | 80 | 45 | 26 | 27 | 160 | 65 | 31 | 31 |
| | 6 | 112 | 57 | 28 | 29 | 224 | 80 | 33 | 34 |
| | 8 | 144 | 65 | 30 | 31 | 288 | 94 | 35 | 35 |
| | 10 | 176 | 74 | 32 | 32 | 352 | 108 | 37 | 37 |
| | 12 | 208 | 84 | 33 | 33 | 416 | 118 | 38 | 38 |
| | 14 | 240 | 91 | 34 | 34 | 480 | 128 | 39 | 39 |
| | 16 | 272 | 99 | 35 | 35 | 544 | 138 | 40 | 40 |
| | 18 | 304 | 105 | 35 | 36 | 608 | 148 | 40 | 40 |
| | 20 | 336 | 113 | 36 | 37 | 672 | 159 | 40 | 40 |
| | | $N = 64$ | | | | $N = 128$ | | | |
| | 4 | 320 | 94 | 29 | 29 | 640 | 124 | 29 | 29 |
| | 6 | 448 | 115 | 32 | 32 | 896 | 154 | 31 | 31 |
| | 8 | 576 | 130 | 34 | 33 | 1152 | 185 | 33 | 33 |
| | 10 | 704 | 149 | 35 | 35 | 1408 | 211 | 35 | 35 |
| | 12 | 832 | 164 | 36 | 36 | 1664 | 226 | 36 | 36 |
| | 14 | 960 | 179 | 37 | 37 | 1920 | 247 | 37 | 37 |
| | 16 | 1088 | 195 | 38 | 38 | 2176 | 268 | 38 | 38 |
| | 18 | 1216 | 210 | 39 | 39 | 2432 | 287 | 39 | 39 |
| | 20 | 1344 | 220 | 40 | 39 | 2688 | 306 | 39 | 39 |

**Table 9.19** Weakly-singular integral equation (8.1), $p$-version with $f(x) \equiv 1$, wave number $\kappa$, number of subintervals $N$. The preconditioners $Q^{(1)}$ and $Q^{(2)}$ are defined in Definition 8.1. [Theory: Chapter 8, Section 8.3.2.]

| $p$ | Condition number | | Number of iterations | |
|---|---|---|---|---|
| | GMRES | PGMRES $Q^{(1)}$ | GMRES | PGMRES $Q^{(1)}$ |
| 2 | 3.99 | 2.96 | 2 | 2 |
| 3 | 14.58 | 3.46 | 3 | 3 |
| 4 | 29.12 | 4.09 | 4 | 4 |
| 5 | 51.20 | 4.58 | 5 | 5 |
| 6 | 81.54 | 5.06 | 6 | 6 |
| 7 | 122.18 | 5.48 | 7 | 7 |
| 8 | 174.29 | 5.87 | 8 | 7 |
| 9 | 239.73 | 6.23 | 9 | 7 |
| 10 | 319.81 | 6.56 | 11 | 7 |

**Table 9.20** Hypersingular integral equation (8.2) with $g(x) \equiv 1$ and wave number $\kappa = 2$: $p$-version with 2 subintervals. [Theory: Chapter 8, Section 8.2.2.]

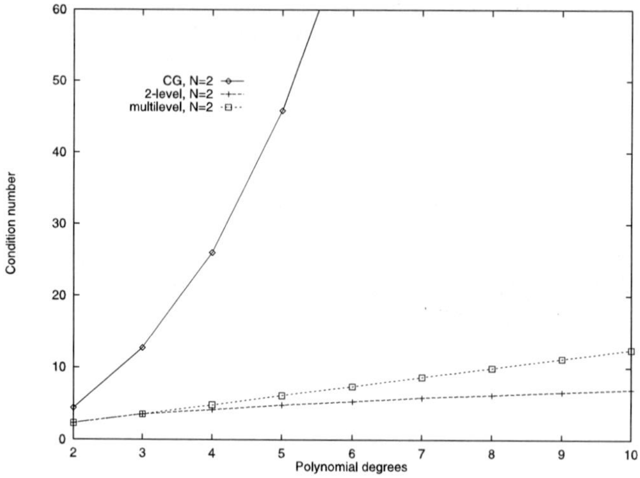


**Fig. 9.7** Hypersingular integral equation (4.1) with $f(x) \equiv 1$: $p$-version with two subintervals, using CG and PCG with two-level and multilevel additive Schwarz preconditioners. [Theory: Chapter 4, Section 4.1.1.1 (two-level); Chapter 6, Section 6.2 (multilevel).]

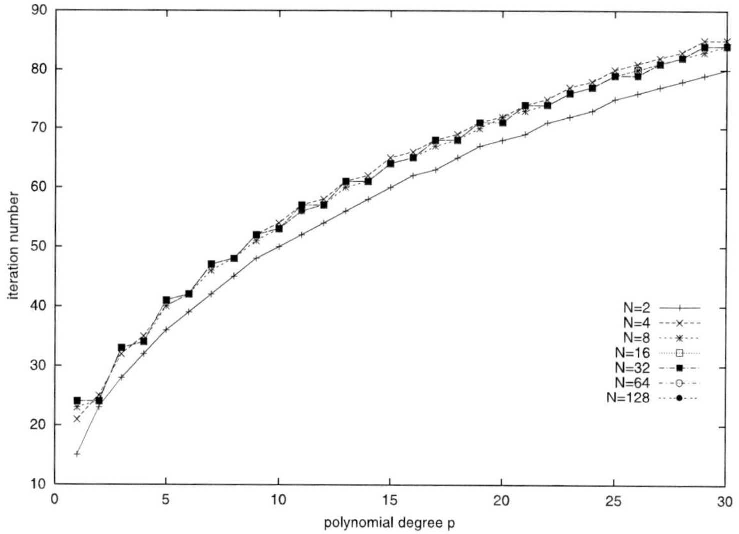


**Fig. 9.8** Weakly-singular integral equation, $p$-version, iteration numbers, multiplicative Schwarz preconditioner. [Theory: Chapter 4, Section 4.2.1.2.]

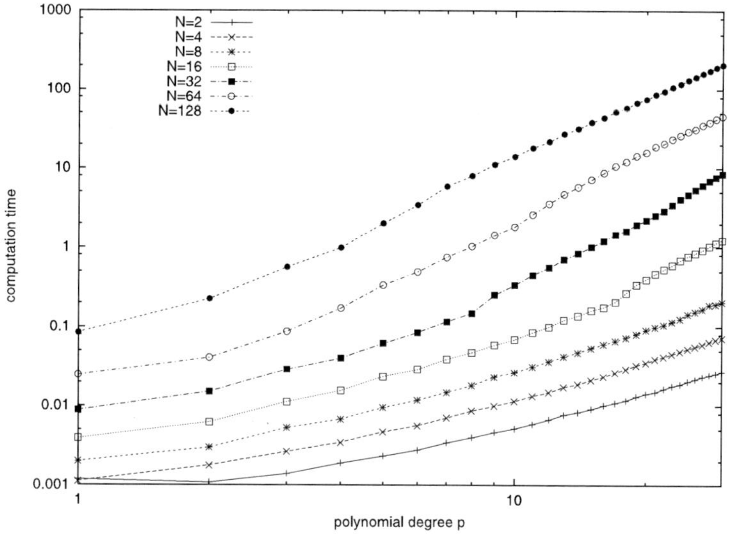


**Fig. 9.9** Weakly-singular integral equation, $p$-version, computational time, multiplicative Schwarz preconditioner. [Theory: Chapter 4, Section 4.2.1.2.]

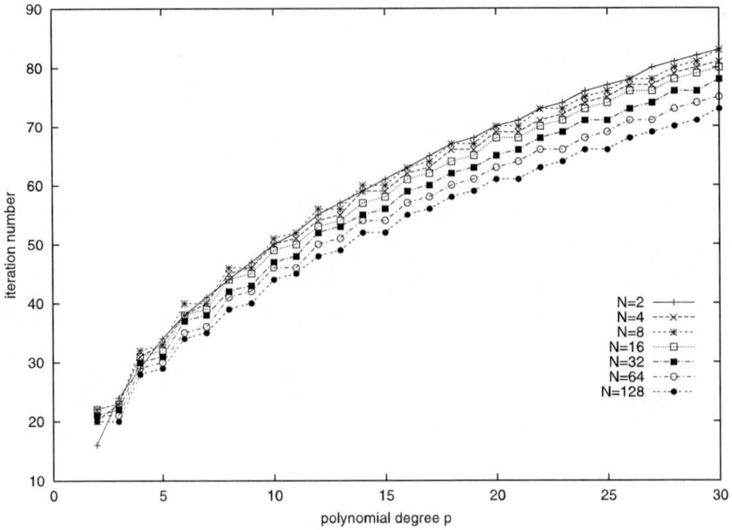


**Fig. 9.10** Hypersingular integral equation, $p$-version, iteration numbers, multiplicative Schwarz preconditioner. [Theory: Chapter 4, Section 4.2.1.1.]

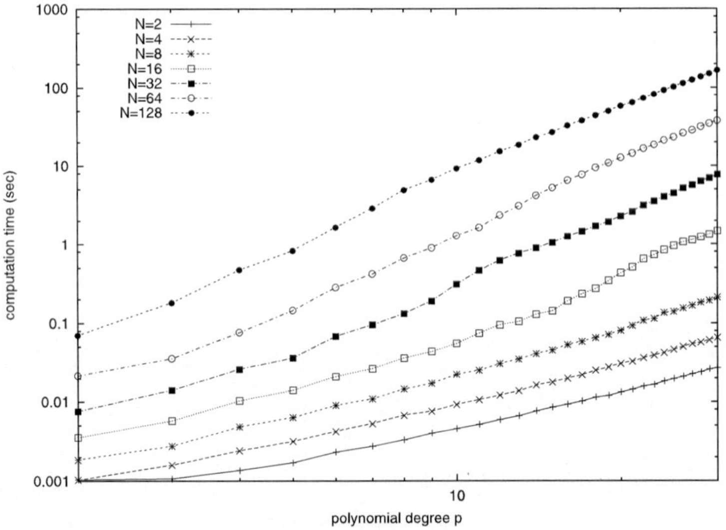


**Fig. 9.11** Hypersingular integral equation, $p$-version, computational time, multiplicative Schwarz preconditioner. [Theory: Chapter 4, Section 4.2.1.1.]

## 9.4 Numerical Results for the *hp*-Version

The numerical results in this section corroborate the theoretical results developed in Chapter 6. Figures 9.12–9.17 show the results for quasi-uniform meshes, while Figure 9.18 shows the results for geometric meshes.

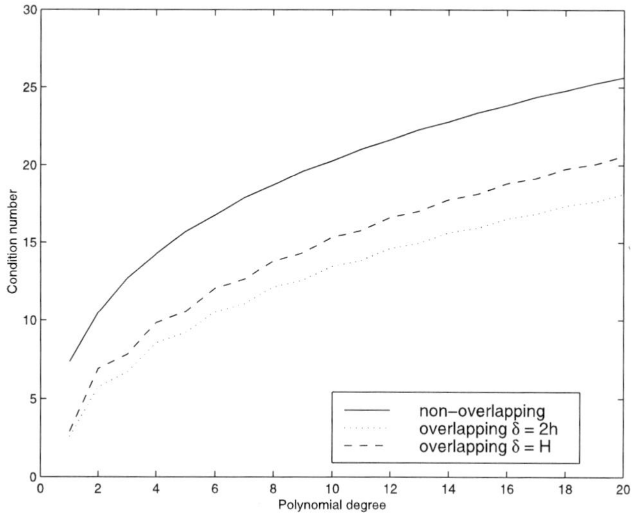


**Fig. 9.12** Hypersingular integral equation: Condition numbers when $N = 128$ and $H = 1/2$. [Theory: Chapter 6.]

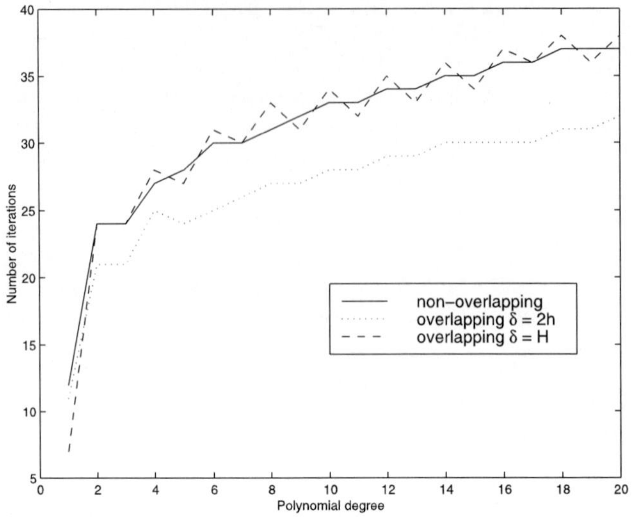


**Fig. 9.13** Hypersingular integral equation: Numbers of iterations when $N = 128$ and $H = 1/2$. [Theory: Chapter 6.]

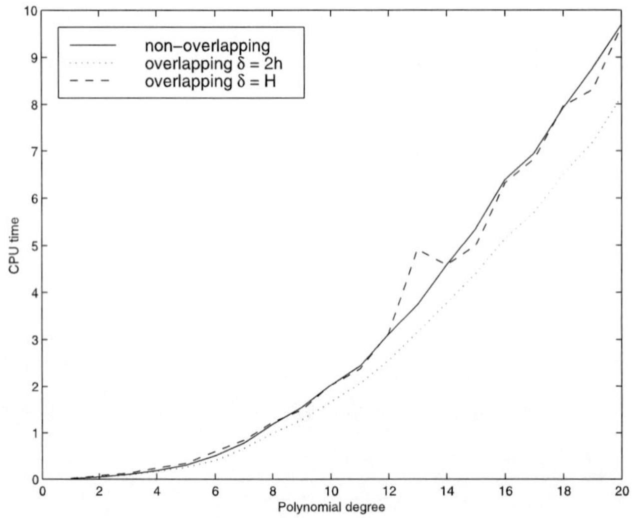


**Fig. 9.14** Hypersingular integral equation: CPU times when $N = 128$ and $H = 1/2$. [Theory: Chapter 6.]

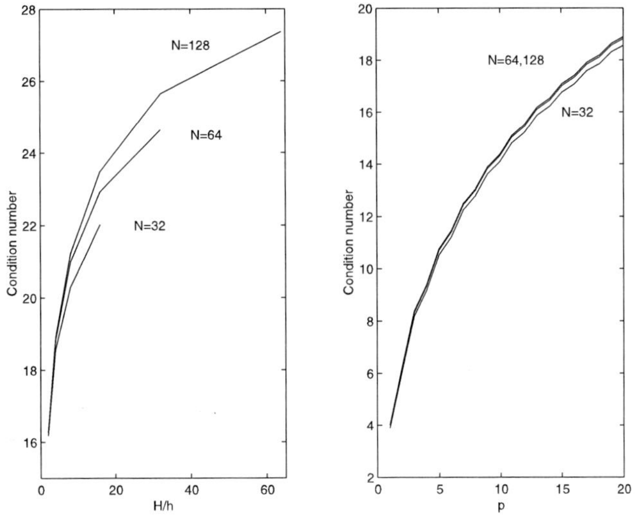


**Fig. 9.15** Hypersingular integral equation: non-overlapping additive Schwarz. Condition numbers as a function of $H/h$ when $p = 20$ (left figure) and as a function of $p$ when $H/h = 4$ (right figure). [Theory: Chapter 6.]

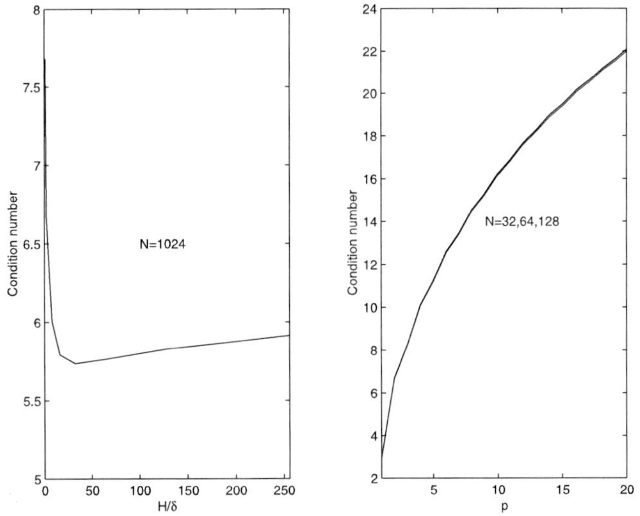


**Fig. 9.16** Hypersingular integral equation: overlapping additive Schwarz. Condition numbers as a function of $H/\delta$ when $p = 2$ (left figure) and as a function of $p$ when $H/h = 4$ (right figure). [Theory: Chapter 6.]

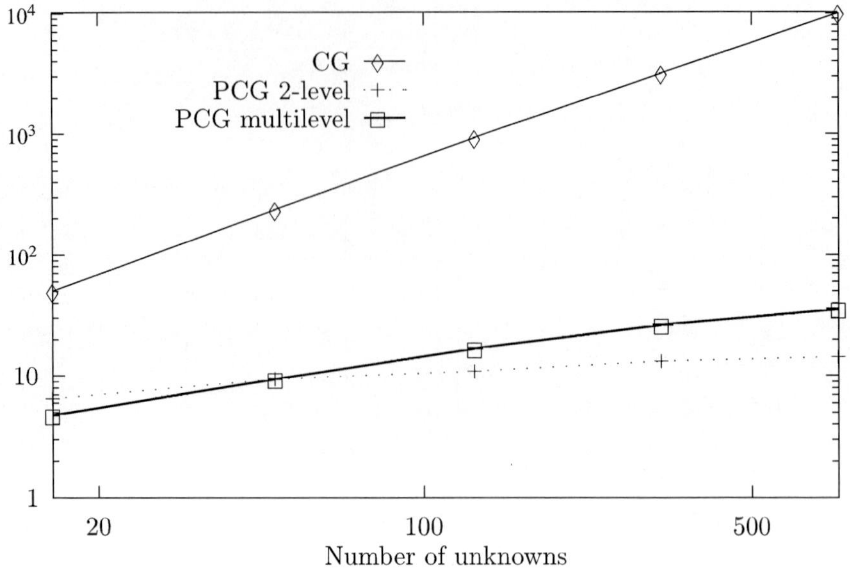


**Fig. 9.17** Weakly-singular integral equation: condition numbers for the *hp*-version with quasi-uniform meshes. [Theory: Chapter 6, Section 6.2.]

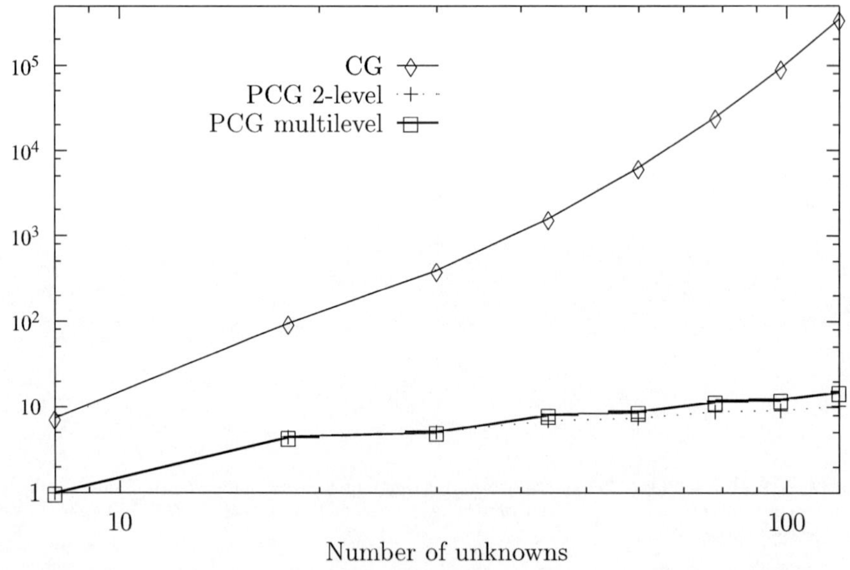


**Fig. 9.18** Weakly-singular integral equation: condition numbers for the *hp*-version with geometric meshes ($\sigma = 0.5$). [Theory: Chapter 6, Section 6.2.]

# Part III
# Three-Dimensional Problems

# Chapter 10
# Two-Level Methods: the $hp$-Version on Rectangular Elements

In this chapter, we will define different two-level methods for the $hp$-version of the Galerkin method for both the hypersingular and weakly-singular integral equations on a screen in $\mathbb{R}^3$. As discussed in Chapter 2, it suffices to define a subspace decomposition for each preconditioner. The contents in this chapter which are related to the hypersingular integral equation are based on the papers [2, 3, 116, 118, 119, 223]. However, this chapter corrects a common error in these articles in the proof of coercivity of the decomposition; see Assumption 2.4. We use the newly-derived theorem, Theorem A.11, instead of Theorem A.10 which are used in the afore-cited papers. The results for the weakly-singular integral equation in Section 10.3 are new; see also Remark 10.1.

## 10.1 Preliminaries

### 10.1.1 Two-level meshes

**The coarse mesh.** We assume that the screen $\Gamma$ is quasi-uniformly partitioned into quadrilateral elements $\Gamma_i$, $i = 1,\dots,J$, and that the mesh is shape regular. The set of all edges of $\Gamma_i$ which are not on the boundary of $\Gamma$ is denoted by $\mathscr{E}_H$. We also denote by $H_i$ the maximum side length of $\Gamma_i$, and by $H$ the maximum value of $H_i$, $i = 1,\dots,J$.
**The fine mesh.** We subdivide each $\Gamma_i$ into quadrilateral elements $\Gamma_{ij}$, $j = 1,\dots,N_i$. In the whole chapter, we will use the generic notation $K$ to indicate $\Gamma_{ij}$ when it is not necessary to clarify the indices $i$ and $j$. We also assume that this mesh is shape regular and locally quasi-uniform. Let $h_K$ be the maximum side length of element $K$. We define $h_i = \max_{K\subset\Gamma_i} h_K$ and assume that, for all $i, j = 1,\dots,J$, if $\Gamma_i$ and $\Gamma_j$ are two adjacent subdomains then $h_i/h_j \simeq 1$.

Let $h$ be the maximum value of $h_i$, $i = 1,\dots,J$, and let $\mathscr{I}_h$ denote a suitable indexing set. We define $\mathscr{N}_h = \{\mathbf{x}_k : k \in \mathscr{I}_h\}$ to be the set of all element vertices which

 
E. P. Stephan and T. Tran, *Schwarz Methods and Multilevel Preconditioners for Boundary Element Methods*, https://doi.org/10.1007/978-3-030-79283-1_10

are not on the boundary of $\Gamma$, and $\mathscr{E}_h$ to be the set of all open element edges which are not on the boundary of $\Gamma$. We also define $\mathscr{I}_{H,v}$, $\mathscr{I}_{H,e}$, and $\mathscr{I}_{H,0}$ to be the set of indices corresponding to the fine grid nodes located at the subdomain vertices, the open subdomain edges, and the open subdomain interiors, respectively; see Figure 10.1. The $h$-wirebasket $\mathscr{W}_h$ is the set of all $\mathbf{x}_k \in \mathscr{N}_h$ located at the subdomain vertices or on the open subdomain edges, i.e.,

$$\mathscr{W}_h := \left\{ \mathbf{x}_k \in \mathscr{N}_h : k \in \mathscr{I}_{H,v} \cup \mathscr{I}_{H,e} \right\}.$$

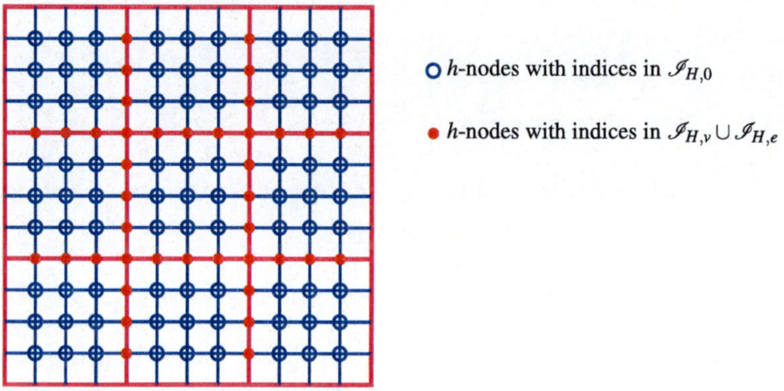


**Fig. 10.1** Fine-mesh nodal points ($h$-nodes) in a two-level mesh of $\Gamma$

The next subsection introduces shape functions that define the basis functions for the boundary element spaces to be used in the solutions of the hypersingular and weakly-singular integral equations.

## *10.1.2 Shape functions*

In order to define the shape functions on $\Gamma$, we first define the shape functions on the reference element $R_{\text{ref}} := (-1,1)^2$.

For each $p \in \mathbb{N}$, let $\tau_{p,j}$, $j = 0, \ldots, p$, be the solutions of

$$(1-t^2)L_p'(t) = 0,$$

ordered such that

$$-1 = \tau_{p,0} < \tau_{p,1} < \cdots < \tau_{p,p} = 1,$$

where $L_p$ is the Legendre polynomial normalised so that $L_p(1) = 1$. These points $\tau_{p,j}$ are called the Gauss–Lobatto–Legendre (GLL) points. The Lagrange polynomial associated with $\tau_{p,j}$ is the polynomial of degree $p$ satisfying

$$\ell_{p,i}(\tau_{p,j}) = \delta_{ij}, \qquad i, j = 0, \ldots, p.$$

Let $\hat{\gamma}_1, \ldots, \hat{\gamma}_4$ denote respectively the bottom, right, top, and left sides of $R_{\text{ref}}$, and let $\mathscr{P} = (p_1, \ldots, p_6)$ be the vector of polynomial degrees associated with $R_{\text{ref}}$, in which $p_1, \ldots, p_4$ are the degrees associated with $\hat{\gamma}_1, \ldots, \hat{\gamma}_4$, whereas $p_5$ and $p_6$ are degrees in the interior in the $\xi$- and $\eta$-directions, respectively. The GLL points induce the following nodes on $R_{\text{ref}}$ (see Figure 10.2)

1. Side nodes
   (a) on $\hat{\gamma}_1$: $(\tau_{p_1,j}, -1)$, $j = 0, \ldots, p_1$;
   (b) on $\hat{\gamma}_2$: $(1, \tau_{p_2,j})$, $j = 0, \ldots, p_2$;
   (c) on $\hat{\gamma}_3$: $(\tau_{p_3,j}, 1)$, $j = 0, \ldots, p_3$;
   (d) on $\hat{\gamma}_4$: $(-1, \tau_{p_4,j})$, $j = 0, \ldots, p_4$;
2. Interior nodes $(\tau_{p_5,i}, \tau_{p_6,j})$, $i = 1, \ldots, p_5 - 1$, $j = 1, \ldots, p_6 - 1$.

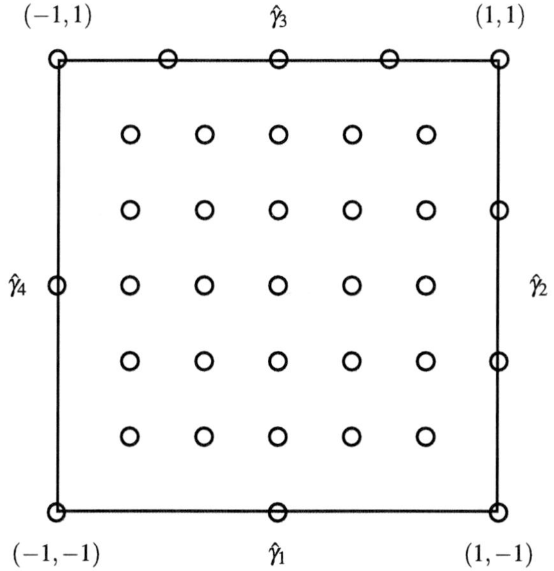


**Fig. 10.2** Side nodes and interior nodes ($p$-nodes) on $R_{\text{ref}}$ ($p_1 = p_4 = 2$, $p_2 = 3$, $p_3 = 4$, $p_5 = p_6 = 6$)

The nodal basis functions associated with these nodes are defined as follows.

1. Nodal modes, which are nodal basis functions at vertices of $R_{\text{ref}}$,
   (a) At $(-1,-1)$: $\ell_{p_1,0}(\xi)\ell_{p_4,0}(\eta)$;
   (b) At $(1,-1)$: $\ell_{p_1,p_1}(\xi)\ell_{p_2,0}(\eta)$;

(c) At $(1,1)$: $\ell_{p_3,p_3}(\xi)\ell_{p_2,p_2}(\eta)$;
(d) At $(-1,1)$: $\ell_{p_3,0}(\xi)\ell_{p_4,p_4}(\eta)$;

2. Side modes, which are nodal basis functions at nodes on sides of $R_{\text{ref}}$,

   (a) on $\hat{\gamma}_1$: $\ell_{p_1,j}(\xi)\ell_{p_6,0}(\eta)$, $\quad j = 1,\dots,p_1-1$;
   (b) on $\hat{\gamma}_2$: $\ell_{p_5,p_5}(\xi)\ell_{p_2,j}(\eta)$, $\quad j = 1,\dots,p_2-1$;
   (c) on $\hat{\gamma}_3$: $\ell_{p_3,j}(\xi)\ell_{p_6,p_6}(\eta)$, $\quad j = 1,\dots,p_3-1$;
   (d) on $\hat{\gamma}_4$: $\ell_{p_5,0}(\xi)\ell_{p_4,j}(\eta)$, $\quad j = 1,\dots,p_4-1$;

3. Internal modes, which are nodal basis functions in the interior of $R_{\text{ref}}$,

$$\ell_{p_5,i}(\xi)\ell_{p_6,j}(\eta), \quad i = 1,\dots,p_5-1, j = 1,\dots,p_6-1.$$

As a generic notation, we will denote in the sequel a node in $R_{\text{ref}}$ by $\hat{\mathbf{q}}$ and a nodal basis function at this node by $\hat{\ell}_{\hat{\mathbf{q}}}(\xi,\eta)$. In the following subsection, with the aid of bijections $\mathscr{F}_K$ that map $R_{\text{ref}}$ onto $K$, we shall transplant the shape functions defined on $R_{\text{ref}}$ onto the elements $K$; thereby define the boundary element space $\mathscr{V}$ (for the hypersingular integral equation) and $\widetilde{\mathscr{V}}$ (for the weakly-singular integral equation).

### 10.1.3 Boundary element spaces

Roughly speaking, the boundary element space $\mathscr{V}$ for the hypersingular integral equation consists of continuous functions defined on $\Gamma$ vanishing on the boundary of $\Gamma$, whose restrictions on each element $K$ are polynomials. Meanwhile the boundary element space $\widetilde{\mathscr{V}}$ for the weakly-singular integral equation consists of functions (not necessarily continuous) whose restriction on $K$ are polynomials.

We now define the basis functions in $\mathscr{V}$. For each element $K$ in the fine mesh, we associate a vector of polynomial degrees $\mathscr{P}_K = (p_1^K,\dots,p_6^K)$. Denoting

$$p_K = \max_{0\le i\le 6} p_i^K \quad \text{and} \quad \underline{p}_K = \min_{0\le i\le 6} p_i^K,$$

we assume that for any neighbouring elements $K$ and $K'$ the following property holds

$$\frac{p_K}{p_{K'}} \simeq 1.$$

The nodes on $K$ are the images (under the bijection $\mathscr{F}_K$) of nodes on $R_{\text{ref}}$ defined by the vector of polynomial degrees $\mathscr{P}_K$. This gives rise to a set of global nodes on $\Gamma$

$$\mathscr{N}_p = \{\mathbf{q}_k : k \in \mathscr{I}_p\},$$

where $\mathscr{I}_p$ denotes a suitable indexing set. This indexing set can be partitioned into disjoint subsets

$$\mathscr{I}_p = \mathscr{I}_{p,v} \cup \mathscr{I}_{p,e} \cup \mathscr{I}_{p,0}, \tag{10.1}$$

where $\mathscr{I}_{p,v}$, $\mathscr{I}_{p,e}$ and $\mathscr{I}_{p,0}$ consist of indices of nodes located at the element vertices, open element edges and open element interiors, respectively. The $p$-wire basket is defined as

$$\mathscr{W}_p = \{\mathbf{q}_k \in \mathscr{N}_p : k \in \mathscr{I}_{p,v} \cup \mathscr{I}_{p,e}\}.$$

A global nodal basis function $\phi_k^p$ associated with $\mathbf{q}_k$ is defined such that the restriction of $\phi_k^p$ to each element $K$ is of the form

$$\phi_k^p|_K = \begin{cases} \hat{\ell}_{\hat{\mathbf{q}}} \circ \mathscr{F}_K^{-1} & \text{if } \mathbf{q}_k \in \overline{K}, \\ 0 & \text{otherwise.} \end{cases} \tag{10.2}$$

Here $\hat{\mathbf{q}} = \mathscr{F}_K^{-1}(\mathbf{q}_k)$. The set of all these nodal basis functions is denoted by $\mathbf{\Phi}_p$ and the approximation space $\mathscr{V}$ is defined as

$$\mathscr{V} = \operatorname{span}(\mathbf{\Phi}_p) \cap \widetilde{H}^{1/2}(\Gamma).$$

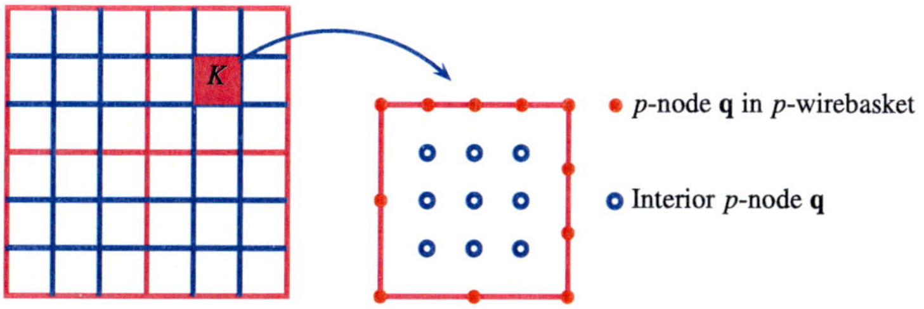


**Fig. 10.3** $p$-nodes $\mathbf{q}$ in one element $K$

## 10.2 The Hypersingular Integral Equation

We present in this section two preconditioners for the hypersingular integral equation: a non-overlapping additive Schwarz preconditioner and an overlapping additive Schwarz preconditioner.

### *10.2.1 A non-overlapping method*

#### 10.2.1.1 Subspace decomposition

As in [90] for the finite element method for partial differential equations, we consider a three-level decomposition of $\mathscr{V}$ consisting of a coarse level, an $h$ level, and a $p$ level.

The coarse mesh space is defined as

$$\mathcal{V}_H^* = \{v : (v|_{\Gamma_j}) \circ \mathcal{F}_j \in \mathbb{P}_1 \quad \forall j = 1, \dots, J\} \cap \widetilde{H}^{1/2}(\Gamma)$$

and the $h$-fine space as

$$\mathcal{V}_h = \{v : (v|_K) \circ \mathcal{F}_K \in \mathbb{P}_1 \quad \forall K\} \cap \widetilde{H}^{1/2}(\Gamma).$$

Here, $\mathbb{P}_1$ is the space of bilinear functions on $R_{\text{ref}}$.

Since $\mathcal{V}_H^*$ may not be a subspace of $\mathcal{V}_h$ (and of $\mathcal{V}$), we need to consider the interpolated subspace $\mathcal{V}_H = \Pi_h(\mathcal{V}_H^*)$, where $\Pi_h$ is the interpolation operator from the space of continuous functions into $\mathcal{V}_h$. The dimension of $\mathcal{V}_H$ equals that of $\mathcal{V}_H^*$. Indeed, if $\phi_k^H$ is the hat function in $\mathcal{V}_H^*$ which takes the value one at node $\mathbf{x}_k$ and zero at other nodes $\mathbf{x}_\ell$, $\ell \in \mathcal{I}_{H,v}$, then $\{\phi_k^H : k \in \mathcal{I}_{H,v}\}$ forms a basis for $\mathcal{V}_H^*$ while $\{\Pi_h \phi_k^H : k \in \mathcal{I}_{H,v}\}$ forms a basis for $\mathcal{V}_H$.

A nodal basis function $\phi_k^h \in \mathcal{V}_h$ associated with $\mathbf{x}_k \in \mathcal{N}_h$, the set of all element vertices not on the boundary of $\Gamma$, is a function in $\mathcal{V}_h$ which takes the value 1 at $\mathbf{x}_k$ and 0 at other nodes in $\mathcal{N}_h$. The set of all these functions is denoted by $\mathbf{\Phi}_h$.

Letting $\mathcal{V}_p = \mathcal{V}$, we decompose $\mathcal{V}$ as

$$\mathcal{V} = \mathcal{V}_H + \mathcal{V}_h + \mathcal{V}_p, \tag{10.3}$$

and further decompose $\mathcal{V}_h$ and $\mathcal{V}_p$ as

$$\mathcal{V}_h = \mathcal{V}_{\mathcal{W}_h} + \bigcup_{j=1}^{J} \mathcal{V}_{\Gamma_j} \quad \text{and} \quad \mathcal{V}_p = \mathcal{V}_{\mathcal{W}_p} + \bigcup_K \mathcal{V}_K, \tag{10.4}$$

where $\mathcal{V}_{\mathcal{W}_h}$ and $\mathcal{V}_{\mathcal{W}_p}$ are the $h$-wire basket and $p$-wire basket spaces defined by

$$\begin{aligned}
\mathcal{V}_{\mathcal{W}_h} &= \text{span}\{\phi_k^h \in \mathbf{\Phi}_h : \mathbf{x}_k \in \mathcal{W}_h\} = \text{span}\left\{\phi_k^h \in \mathbf{\Phi}_h : k \in \mathcal{I}_{H,v} \cup \mathcal{I}_{H,e}\right\},\\
\mathcal{V}_{\mathcal{W}_p} &= \text{span}\{\phi_k^p \in \mathbf{\Phi}_p : \mathbf{q}_k \in \mathcal{W}_p\} = \text{span}\left\{\phi_k^p \in \mathbf{\Phi}_p : k \in \mathcal{I}_{p,v} \cup \mathcal{I}_{p,e}\right\},
\end{aligned}$$

and $\mathcal{V}_{\Gamma_j}$ and $\mathcal{V}_K$ are the interior spaces defined by

$$\mathcal{V}_{\Gamma_j} = \text{span}\{\phi_k^h \in \mathbf{\Phi}_h : \mathbf{x}_k \in \Gamma_j\} \quad \text{and} \quad \mathcal{V}_K = \text{span}\{\phi_k^p \in \mathbf{\Phi}_p : \mathbf{q}_k \in K\}.$$

#### 10.2.1.2 Bilinear forms on subspaces

The choice of the bilinear forms in $\mathcal{V}_H$, $\mathcal{V}_{\Gamma_j}$ and $\mathcal{V}_K$ is straightforward

$$\begin{aligned}
b_H(v,w) &:= a_{\mathfrak{W}}(\Pi_{\mathcal{V}_H^*} v, \Pi_{\mathcal{V}_H^*} w) \quad && \forall v,w \in \mathcal{V}_H,\\
b_{\Gamma_j}(v,w) &:= a_{\mathfrak{W}}(v,w) && \forall v,w \in \mathcal{V}_{\Gamma_j},\\
b_K(v,w) &:= a_{\mathfrak{W}}(v,w) && \forall v,w \in \mathcal{V}_K,
\end{aligned} \tag{10.5}$$

where $\Pi_{\mathscr{V}_H^*}$ is the standard interpolation operator from the space of continuous functions into $\mathscr{V}_H^*$. We note that if $v^* \in \mathscr{V}_H^*$ and $v \in \mathscr{V}_H$ then

$$v = \Pi_{\mathscr{V}_h} v^* \iff v^* = \Pi_{\mathscr{V}_H^*} v. \tag{10.6}$$

Here, $\Pi_{\mathscr{V}_h} : C(\Gamma) \to \mathscr{V}_h$ is the standard interpolation operator. In the following, when there is no confusion, we shorten the notation to $\Pi_h$. We will also write $\Pi_H^*$ instead of $\Pi_{\mathscr{V}_H^*}$. Statement (10.6) implies

$$\Pi_H^* \Pi_h = Id \tag{10.7}$$

where $Id$ is the identity operator.

The bilinear form on the wire basket $\mathscr{V}_{\mathscr{W}_h}$ is chosen such that the problem involved is completely decoupled and only involves inverting a diagonal matrix:

$$b_{\mathscr{W}_h}(v,w) = \sum_{j=1}^{J} \sum_{x_k \in \partial\Gamma_j} h_j v(x_k) w(x_k), \qquad v,w \in \mathscr{V}_{\mathscr{W}_h}. \tag{10.8}$$

Similarly, we define a bilinear form on $\mathscr{V}_{\mathscr{W}_p}$ so that the problem involved is solved by inverting a diagonal matrix. This is done by using the quadrature rule defined in Section C.6 in Appendix C for the integral over a side $\gamma$ of an element $K$:

$$b_{\mathscr{W}_p}(v,w) = \sum_{\gamma \in \mathscr{E}_h} h_\gamma \sum_{\mathbf{q}_k \in \mathscr{W}_p \cap \overline{\gamma}} \omega_k v(\mathbf{q}_k) w(\mathbf{q}_k) \qquad \forall v,w \in \mathscr{V}_{\mathscr{W}_p}, \tag{10.9}$$

where we recall that $\mathscr{E}_h$ is the set of all element edges not on the boundary of $\Gamma$.

#### 10.2.1.3 Auxiliary lemmas

The first lemma establishes the equivalences of the bilinear forms on the wire baskets and the corresponding $L^2$-norms.

**Lemma 10.1.** *The bilinear form* $b_{\mathscr{W}_h}(\cdot,\cdot)$ *defined by* (10.8) *satisfies*

$$b_{\mathscr{W}_h}(v,v) \simeq \sum_{\gamma \in \mathscr{E}_H} \|v\|^2_{L_2(\gamma)}, \quad v \in \mathscr{V}_{\mathscr{W}_h}.$$

*Similarly, the bilinear form* $b_{\mathscr{W}_p}(\cdot,\cdot)$ *defined by* (10.9) *satisfies*

$$b_{\mathscr{W}_p}(v,v) \simeq \sum_{\gamma \in \mathscr{E}_h} \|v\|^2_{L_2(\gamma)}, \quad v \in \mathscr{V}_{\mathscr{W}_p}.$$

*The constants involved are independent of $v$, $H$, $h$ and $p$.*

*Proof.* This is a direct consequence of (C.89) in Appendix C. □

The next lemma gives a uniform estimate of an $h$-wire basket function extended by zero.

**Lemma 10.2.** *Let $\gamma \in \mathscr{E}_H$ be an open subdomain edge. We define $\Gamma_\gamma$ to be the union of $\gamma$ and the pair of subdomains sharing this edge. We also define $\mathscr{I}_{h,\gamma}$ to be the set of indices $k \in \mathscr{I}_h$ such that $\mathbf{x}_k \in \gamma$. If $v \in \mathscr{V}_{\mathscr{W}_h}$ is of the form*

$$v = \sum_{k \in \mathscr{I}_{h,\gamma}} v(\mathbf{x}_k)\phi_k^h,$$

*then*

$$\|v\|_{\widetilde{H}_{\mathrm{I}}^{1/2}(\Gamma_\gamma)} \lesssim \|v\|_{L_2(\gamma)}$$

*where the constant involved is independent of $H$, $h$ and $v$.*

*Proof.* This lemma is proved in [3, Lemma 15]. The proof is included here for completeness. By using transformations, we may assume that $\gamma = \{0\} \times (0,H)$ and $\Gamma_\gamma = (-H,H) \times (0,H)$. It suffices to show that

$$\|v\|_{\widetilde{H}_{\mathrm{I}}^{1/2}(\Gamma_\gamma)} \lesssim \|w\|_{L^2(\gamma)}$$

where $v$ has the form

$$v(x,y) = \begin{cases} (1+x/h)\, w(y), & (x,y) \in (-h,0) \times (0,H), \\ (1-x/h)\, w(y), & (x,y) \in (0,h) \times (0,H), \\ 0, & \text{otherwise}, \end{cases}$$

with $w$ being a continuous, piecewise linear function on $\gamma$ which satisfies $w(0) = w(H) = 0$. Simple calculations reveal

$$\|v\|^2_{\widetilde{H}_{\mathrm{I}}^0(\Gamma_\gamma)} = \|v\|^2_{L^2(\Gamma_\gamma)} \simeq h\|w\|^2_{L^2(\gamma)}$$

and

$$\|v\|^2_{\widetilde{H}_{\mathrm{I}}^1(\Gamma_\gamma)} = \|\frac{\partial v}{\partial x}\|^2_{\widetilde{H}_{\mathrm{I}}^1(\Gamma_\gamma)} + \|\frac{\partial v}{\partial y}\|^2_{\widetilde{H}_{\mathrm{I}}^1(\Gamma_\gamma)} \simeq h^{-1}\|w\|^2_{L^2(\gamma)} + h\|w'\|^2_{L^2(\gamma)}.$$

With the aid of the inverse inequality we deduce

$$\|v\|^2_{\widetilde{H}_{\mathrm{I}}^1(\Gamma_\gamma)} \lesssim h^{-1}\|w\|^2_{L^2(\gamma)}.$$

By interpolation,

$$\|v\|^2_{\widetilde{H}_{\mathrm{I}}^{1/2}(\Gamma_\gamma)} \lesssim \|w\|^2_{L^2(\gamma)}.$$ □

The following two lemmas are analogous to [2, Lemma 8] and [2, Lemma 7], respectively, where the $p$-version is considered with a uniform distribution of $p$. The current results assure us that the constants involved are independent of not only $p$ but also $h$ and $H$. We note a slight difference in the bilinear form $b_{\mathscr{W}_p}$ defined in [2].

**Lemma 10.3.** *Let $\gamma \in \mathscr{E}_h$ be an open element edge and let $K_\gamma$ be the pair of elements sharing $\gamma$. If $v \in \mathscr{V}_{\mathscr{W}_p}$ is of the form*

$$v = \sum_{\mathbf{q}_k \in \mathscr{W}_p \cap \gamma} v(\mathbf{q}_k)\phi_k^p,$$

*then*

$$\|v\|_{\widetilde{H}_{\mathrm{I}}^{1/2}(K_\gamma)} \lesssim \frac{p^*}{p_*}\|v\|_{L_2(\gamma)},$$

*where $p^*$ and $p_*$ are the maximum and minimum polynomial orders in $K_\gamma$.*

*Proof.* Let $K_\gamma^-$ and $K_\gamma^+$ be two elements such that $K_\gamma^- \cup \gamma \cup K_\gamma^+ = K_\gamma$. Also let $\hat{K}_\gamma^- = (-2,0)\times(-1,1)$, $\hat{K}_\gamma^+ = (0,2)\times(-1,1)$ and $\hat{\gamma} = \{0\}\times(-1,1)$ be the images of $K_\gamma^-$, $K_\gamma^+$ and $\gamma$ (respectively) under appropriate transformations. If $\hat{K}_\gamma = \hat{K}_\gamma^- \cup \hat{\gamma} \cup \hat{K}_\gamma^+$ and $\hat{v}$ is defined on $\hat{K}_\gamma$ from $v$ by using the above-mentioned transformations, then it follows from Lemma A.6 in Appendix A and a simple calculation that

$$\|v\|_{\widetilde{H}_{\mathrm{I}}^{1/2}(K_\gamma)} \simeq h_j^{1/2}\|\hat{v}\|_{\widetilde{H}_{\mathrm{I}}^{1/2}(\hat{K}_\gamma)} \quad \text{and} \quad \|v\|_{L_2(\gamma)} \simeq h_j^{1/2}\|\hat{v}\|_{L_2(\hat{\gamma})},$$

where $h_j$ is the maximum element size in $\Gamma_j$, the subdomain which contains $K_\gamma^-$ and $K_\gamma^+$. Therefore, it suffices to prove

$$\|\hat{v}\|_{\widetilde{H}_{\mathrm{I}}^{1/2}(\hat{K}_\gamma)} \lesssim \frac{p^*}{p_*}\|\hat{v}\|_{L_2(\hat{\gamma})}. \tag{10.10}$$

Let $\mathscr{P}^- = (p_1^-,\dots,p_6^-)$ and $\mathscr{P}^+ = (p_1^+,\dots,p_6^+)$ be the polynomial orders assigned to $K_\gamma^-$ and $K_\gamma^+$, respectively. Then, see Section 10.1.2,

$$\hat{v}(\xi,\eta) = \begin{cases} l_{p_5^-,p_5^-}(1+\xi)w(\eta) & \text{on } \hat{K}_\gamma^-, \\ l_{p_5^+,0}(-1+\xi)w(\eta) & \text{on } \hat{K}_\gamma^+, \end{cases}$$

where $w$ is a polynomial of degree at most $p := p_4^+ = p_2^-$ satisfying $w(\pm 1) = 0$. Therefore we deduce by using (C.88) and (C.90) in Appendix C

$$\|\hat{v}\|_{L_2(\hat{K}_\gamma)}^2 \lesssim \left(\|l_{p_5^-,p_5^-}\|_{L_2(\hat{\gamma})}^2 + \|l_{p_5^+,0}\|_{L_2(\hat{\gamma})}^2\right)\|w\|_{L_2(\hat{\gamma})}^2 \lesssim (p_*)^{-2}\|\hat{v}\|_{L_2(\hat{\gamma})}^2.$$

By using Schmidt's inequality (C.35) we obtain

$$\begin{aligned}
|\hat{v}|_{H^1(\hat{K}_\gamma)}^2 &\lesssim \left(\|l'_{p_5^-,p_5^-}\|_{L_2(\hat{\gamma})}^2 + \|l'_{p_5^+,0}\|_{L_2(\hat{\gamma})}^2\right)\|w\|_{L_2(\hat{\gamma})}^2 \\
&\quad + \left(\|l_{p_5^-,p_5^-}\|_{L_2(\hat{\gamma})}^2 + \|l_{p_5^+,0}\|_{L_2(\hat{\gamma})}^2\right)\|w'\|_{L_2(\hat{\gamma})}^2 \\
&\lesssim (p^*)^4\left(\|l_{p_5^-,p_5^-}\|_{L_2(\hat{\gamma})}^2 + \|l_{p_5^+,0}\|_{L_2(\hat{\gamma})}^2\right)\|w\|_{L_2(\hat{\gamma})}^2 \\
&\lesssim \frac{(p^*)^4}{(p_*)^2}\|\hat{v}\|_{L_2(\hat{\gamma})}^2.
\end{aligned}$$

Inequality (10.10) now follows from Lemma A.12 in Appendix A. □

**Lemma 10.4.** *For any $k \in \mathscr{I}_{p,v}$, let $M_k$ be the set of all element edges sharing $\mathbf{q}_k$ as one endpoint, and for each $\gamma \in M_k$, let $\omega_\gamma$ be the weight associated with $\mathbf{q}_k$ when considered as a quadrature point on $\overline{\gamma}$; see* (C.88) *in Appendix C. Then the nodal basis function $\phi_k^p$ associated with $\mathbf{q}_k$ satisfies*

$$\|\phi_k^p\|^2_{\widetilde{H}_\mathrm{I}^{1/2}(S_k)} \lesssim \frac{p^*}{p_*} \sum_{\gamma \in M_k} h_\gamma \omega_\gamma$$

*where $S_k = \operatorname{supp}(\phi_k^p)$ while $p^*$ and $p_*$ are the maximum and minimum polynomial orders in $S_k$.*

*Proof.* Thanks to the assumption on the mesh, the cardinality $M$ of $M_k$ is independent of $h$ and $H$. By indexing the edges $\gamma$ in $M_k$ in some order (e.g. clockwise) as $\gamma_1$, $\gamma_2, \ldots, \gamma_M$, we can represent $S_k$ as $S_k = \cup_{i=1}^M K_{i,i+1}$, where $K_{i,i+1}$ is the element in $S_k$ having $\gamma_i$ and $\gamma_{i+1}$ as two of its edges, see Figure 10.4, and where we use a periodic indexing, i.e. $\gamma_{M+1} = \gamma_1$ and $K_{M,M+1} = K_{M,1}$.

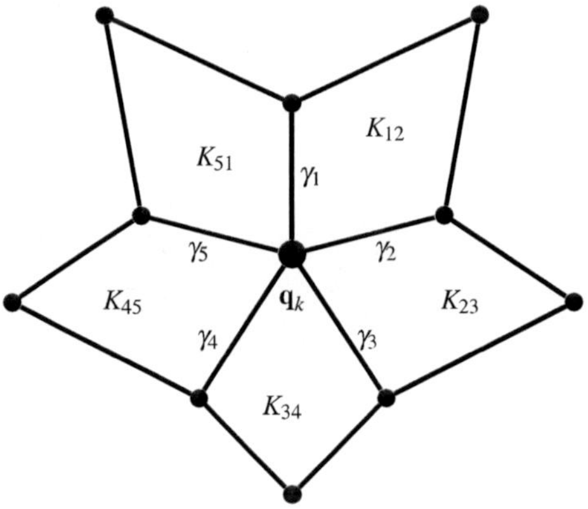


**Fig. 10.4** $S_k = \operatorname{supp} \phi_k^p$

Each element $K_{i,i+1}$ can be assumed to be the image of the reference element $\hat{K}$ by a bijection $\mathscr{F}_i$ such that $\mathscr{F}_i(\hat{\mathbf{q}}) = \mathbf{q}_k$, where $\hat{\mathbf{q}} = (-1,-1)$. Then, see (10.2), $\phi_k^p|_{K_{i,i+1}} = \hat{\ell}_{\hat{\mathbf{q}}} \circ \mathscr{F}_i^{-1}$, where

$$\hat{\ell}_{\hat{\mathbf{q}}}(\xi,\eta) = \ell_{p_i,0}(\xi)\ell_{p_{i+1},0}(\eta),$$

with $p_i$ being such that $p_i$ and $\omega_i$ are related by (C.88) in Appendix C. Here $\omega_i$ is the weight at $\mathbf{q}_k$ when considered as a node on $\overline{\gamma}_i$. A simple calculation reveals

$$\|\phi_k^p\|^2_{L_2(K_{i,i+1})} \simeq h_{\gamma_i} h_{\gamma_{i+1}} \|\hat{\ell}_{\hat{\mathbf{q}}}\|^2_{L_2(\hat{K})} \quad \text{and} \quad |\phi_k^p|^2_{H^1(K_{i,i+1})} \simeq |\hat{\ell}_{\hat{\mathbf{q}}}|^2_{H^1(\hat{K})}.$$

Reasoning as in the proof of Lemma 10.3 one can easily establish, noting (C.90),

$$\|\hat{\ell}_{\hat{\mathbf{q}}}\|^2_{L_2(\hat{K})} \lesssim \omega_i\omega_{i+1}$$

and

$$|\hat{\ell}_{\hat{\mathbf{q}}}|^2_{H^1(\hat{K})} \lesssim \frac{\omega_i}{\omega_{i+1}} + \frac{\omega_{i+1}}{\omega_i} \lesssim \left(\frac{p^*}{p_*}\right)^2,$$

resulting in

$$\|\phi_k^p\|^2_{L_2(K_{i,i+1})} \lesssim h_{\gamma_i}h_{\gamma_{i+1}}\,\omega_i\omega_{i+1} \quad \text{and} \quad |\phi_k^p|^2_{H^1(K_{i,i+1})} \lesssim \left(\frac{p^*}{p_*}\right)^2.$$

Recalling that by the periodic indexing $h_{\gamma_{M+1}} = h_{\gamma_1}$ and $\omega_{M+1} = \omega_1$, we obtain by summing over $i = 1,\dots,M$

$$\|\phi_k^p\|^2_{L_2(S_k)} = \sum_{i=1}^{M} \|\phi_k^p\|^2_{L_2(K_{i,i+1})} \lesssim \sum_{i=1}^{M} h_{\gamma_i}h_{\gamma_{i+1}}\,\omega_i\omega_{i+1}$$

and

$$|\phi_k^p|^2_{H^1(S_k)} = \sum_{i=1}^{M} |\phi_k^p|^2_{H^1(K_{i,i+1})} \lesssim \left(\frac{p^*}{p_*}\right)^2.$$

Applying Lemma A.12 in Appendix A gives

$$\|\phi_k^p\|^2_{\widetilde{H}_{\mathrm{I}}^{1/2}(S_k)} \lesssim \frac{p^*}{p_*}\left(\sum_{i=1}^{M} h_{\gamma_i}h_{\gamma_{i+1}}\,\omega_i\omega_{i+1}\right)^{1/2}.$$

An elementary calculation reveals that

$$\left(\sum_{i=1}^{M} h_{\gamma_i}h_{\gamma_{i+1}}\,\omega_i\omega_{i+1}\right)^{1/2} \le \sum_{i=1}^{M} h_{\gamma_i}\omega_i,$$

finishing the proof. □

Lemma 10.3 does not hold for the $H^{1/2}$-norm because this norm has different scaling properties. However, similar estimate holds on the reference element $\hat{K}$.

**Lemma 10.5.** *If $\hat{v}$ is of the form*

$$\hat{v} = \sum_{\hat{\mathbf{q}}\in\partial\hat{K}} \hat{v}(\hat{\mathbf{q}})\hat{\ell}_{\hat{\mathbf{q}}},$$

*then*

$$\|\hat{v}\|^2_{H_{\mathrm{I}}^{1/2}(\hat{K})} \lesssim \left(\frac{p^*}{p_*}\right)^2 \sum_{i=1}^{4} \|\hat{v}\|^2_{L_2(\hat{\gamma}_i)},$$

*where $p^*$ and $p_*$ are the maximum and minimum polynomial orders in $\hat{K}$, and $\hat{\gamma}_i$, $i = 1,\ldots,4$, denote the open edges of $\hat{K}$.*

*Proof.* Denoting $\hat{\mathbf{q}}_1 = (1,-1)$, $\hat{\mathbf{q}}_2 = (1,1)$, $\hat{\mathbf{q}}_3 = (-1,1)$ and $\hat{\mathbf{q}}_4 = (-1,-1)$, we can write $\hat{v}$ as $\hat{v} = \hat{v}_1 + \cdots + \hat{v}_4$, where

$$\hat{v}_i = \sum_{\hat{\mathbf{q}}\in\hat{\gamma}_i} \hat{v}(\hat{\mathbf{q}})\hat{\ell}_{\hat{\mathbf{q}}} + \hat{v}(\hat{\mathbf{q}}_i)\hat{\ell}_{\hat{\mathbf{q}}_i}.$$

Since

$$\|\hat{v}\|^2_{H_{\mathrm{I}}^{1/2}(\hat{K})} \lesssim \sum_{i=1}^{4} \|\hat{v}_i\|^2_{H_{\mathrm{I}}^{1/2}(\hat{K})},$$

it suffices to prove

$$\|\hat{v}_i\|^2_{H_{\mathrm{I}}^{1/2}(\hat{K})} \lesssim \left(\frac{p^*}{p_*}\right)^2 \|\hat{v}\|^2_{L_2(\hat{\gamma}_i)}.$$

We prove this inequality for $\hat{v}_1$ only. By definition, see Subsection 10.1.2,

$$\hat{v}_1(\xi,\eta) = \sum_{j=1}^{p_1-1} \hat{v}(\tau_{p_1,j},-1)\ell_{p_1,j}(\xi)\ell_{p_6,0}(\eta) + \hat{v}(\tau_{p_1,p_1},-1)\ell_{p_1,p_1}(\xi)\ell_{p_2,0}(\eta).$$

Therefore, with the aid of (C.88), (C.89) and (C.90), we infer

$$\|\hat{v}_1\|^2_{L_2(\hat{K})} \lesssim (p_*)^{-2} \sum_{j=1}^{p_1} \omega_{p_1,j}|\hat{v}(\tau_{p_1,j},-1)|^2 \le (p_*)^{-2}\|\hat{v}\|^2_{L_2(\hat{\gamma}_1)}.$$

Using Schmidt's inequality (C.35) we also obtain

$$\|\hat{v}_1\|^2_{H^1(\hat{K})} \lesssim \frac{(p^*)^4}{(p_*)^2} \sum_{j=1}^{p_1} \omega_{p_1,j}|\hat{v}(\tau_{p_1,j},-1)|^2 \le \frac{(p^*)^4}{(p_*)^2}\|\hat{v}\|^2_{L_2(\hat{\gamma}_1)}.$$

Applying Lemma A.12 gives the desired estimate, thus proves the lemma. □

As usual, we now study the coercivity and stability of the decomposition defined by (10.3) and (10.4).

#### 10.2.1.4 Coercivity and stability of the decomposition

**Lemma 10.6 (Coercivity of decomposition).** *For any*

$$v = v_H + v_{\mathscr{W}_h} + \sum_{j=1}^{J} v_{\Gamma_j} + v_{\mathscr{W}_p} + \sum_{K} v_K$$

*with $v_H \in \mathscr{V}_H$, $v_{\mathscr{W}_h} \in \mathscr{V}_{\mathscr{W}_h}$, $v_{\Gamma_j} \in \mathscr{V}_{\Gamma_j}$, $j = 1,\ldots,J$, $v \in \mathscr{V}_{\mathscr{W}_p}$, $v_K \in \mathscr{V}_K$, the following inequality holds*

$$a_{\mathfrak{W}}(v,v) \lesssim b_H(v_H,v_H) + b_{\mathscr{W}_h}(v_{\mathscr{W}_h},v_{\mathscr{W}_h}) + \sum_{j=1}^{J} b_{\Gamma_j}(v_{\Gamma_j},v_{\Gamma_j})$$
$$+ b_{\mathscr{W}_p}(v_{\mathscr{W}_p},v_{\mathscr{W}_p}) + \sum_K b_K(v_K,v_K).$$

*Proof.* Let

$$v_h := v_{\mathscr{W}_h} + \sum_{j=1}^{J} v_{\Gamma_j} \quad \text{and} \quad v_p := v_{\mathscr{W}_p} + \sum_K v_K.$$

Then $v = v_H + v_h + v_p$ so that by the Cauchy–Schwarz inequality

$$a_{\mathfrak{W}}(v,v) \simeq \|v\|^2_{\widetilde{H}_{\mathrm{I}}^{1/2}(\Gamma)} \lesssim \|v_H\|^2_{\widetilde{H}_{\mathrm{I}}^{1/2}(\Gamma)} + \|v_h\|^2_{\widetilde{H}_{\mathrm{I}}^{1/2}(\Gamma)} + \|v_p\|^2_{\widetilde{H}_{\mathrm{I}}^{1/2}(\Gamma)}$$
$$\simeq a_{\mathfrak{W}}(v_H,v_H) + a_{\mathfrak{W}}(v_h,v_h) + a_{\mathfrak{W}}(v_p,v_p).$$

Hence the lemma is proved if we prove the coercivity of the bilinear forms on $\mathscr{V}_H$, $\mathscr{V}_{\mathscr{W}_h}$, $\sum_j \mathscr{V}_{\Gamma_j}$, $\mathscr{V}_{\mathscr{W}_p}$, and $\sum_K \mathscr{V}_K$:

$$a_{\mathfrak{W}}(v_H,v_H) \lesssim b_H(v_H,v_H), \tag{10.11}$$

$$a_{\mathfrak{W}}(v_h,v_h) \lesssim b_{\mathscr{W}_h}(v_{\mathscr{W}_h},v_{\mathscr{W}_h}) + \sum_{j=1}^{J} b_{\Gamma_j}(v_{\Gamma_j},v_{\Gamma_j}), \tag{10.12}$$

$$a_{\mathfrak{W}}(v_p,v_p) \lesssim b_{\mathscr{W}_p}(v_{\mathscr{W}_p},v_{\mathscr{W}_p}) + \sum_K b_K(v_K,v_K). \tag{10.13}$$

This will be done in the following three lemmas (Lemmas 10.7–10.9), and thus on proving those lemmas we complete the proof of this lemma. □

Inequality (10.11) is proved in [3, Lemma 13]. We briefly sketch the proof here.

**Lemma 10.7.** *For any $v_H \in \mathscr{V}_H$ we have*

$$a_{\mathfrak{W}}(v_H,v_H) \lesssim b_H(v_H,v_H).$$

*Proof.* For any $v_H \in \mathscr{V}_H$, there exists $v_H^* \in \mathscr{V}_H^*$ such that $v_H = \Pi_h v_H^*$, so that

$$a_{\mathfrak{W}}(v_H,v_H) \simeq \|v_H\|^2_{\widetilde{H}_{\mathrm{I}}^{1/2}(\Gamma)} = \|\Pi_h v_H^*\|^2_{\widetilde{H}_{\mathrm{I}}^{1/2}(\Gamma)} \lesssim \|v_H^*\|^2_{\widetilde{H}_{\mathrm{I}}^{1/2}(\Gamma)},$$

where the last inequality is given by Lemma C.36. On the other hand, (10.6) gives $v_H^* = \Pi_H^* v_H$, and thus due to the definition of the bilinear form $b_H(\cdot,\cdot)$ in (10.5) we have

$$b_H(v_H,v_H) = a_{\mathfrak{W}}(\Pi_H^* v_H, \Pi_H^* v_H) = a_{\mathfrak{W}}(v_H^*,v_H^*) \simeq \|v_H^*\|^2_{\widetilde{H}_{\mathrm{I}}^{1/2}(\Gamma)}.$$

The required result then follows. □

Inequality (10.12) is proved by proving separately the coercivity of the bilinear form $b_{\mathscr{W}_h}(\cdot,\cdot)$ and the sum $\sum_{j=1}^{J} b_{\Gamma_j}(\cdot,\cdot)$. This is done in [3, Lemmas 14 and 16].

However, the proof of [3, Lemma 14] relies on (A.65) and the equivalence

$$b_{\Gamma_j}(v_{\Gamma_j}, v_{\Gamma_j}) = a_{\mathfrak{W}}(v_{\Gamma_j}, v_{\Gamma_j}) \simeq \|v_{\Gamma_j}\|^2_{\widetilde{H}_{\mathrm{I}}^{1/2}(\Gamma_j)}.$$

Lemma A.13 shows that the constants in this equivalence depend on $h$. We prove inequality (10.12) in the following lemma, where we use (A.86) instead of (A.65) and thus we are able to use the following equivalence to ensure that the equivalence constants are independent of $h$:

$$b_{\Gamma_j}(v_{\Gamma_j}, v_{\Gamma_j}) = a_{\mathfrak{W}}(v_{\Gamma_j}, v_{\Gamma_j}) \simeq \|v_{\Gamma_j}\|^2_{\widetilde{H}_{\mathrm{I}}^{1/2}(\Gamma)}.$$

**Lemma 10.8.** *For any $v_h = v_{\mathscr{W}_h} + \sum_{j=1}^J v_{\Gamma_j} \in \mathscr{V}_h$ with $v_{\mathscr{W}_h} \in \mathscr{V}_{\mathscr{W}_h}$ and $v_{\Gamma_j} \in \mathscr{V}_{\Gamma_j}$ we have*

$$a_{\mathfrak{W}}(v_h, v_h) \lesssim b_{\mathscr{W}_h}(v_{\mathscr{W}_h}, v_{\mathscr{W}_h}) + \sum_{j=1}^J b_{\Gamma_j}(v_{\Gamma_j}, v_{\Gamma_j}).$$

*Proof.* Let $v_\Gamma := \sum_{j=1}^J v_{\Gamma_j}$. Then $v_h = v_{\mathscr{W}_h} + v_\Gamma$ so that

$$a_{\mathfrak{W}}(v_h, v_h) \simeq \|v_h\|^2_{\widetilde{H}_{\mathrm{I}}^{1/2}(\Gamma)} \lesssim \|v_{\mathscr{W}_h}\|^2_{\widetilde{H}_{\mathrm{I}}^{1/2}(\Gamma)} + \|v_\Gamma\|^2_{\widetilde{H}_{\mathrm{I}}^{1/2}(\Gamma)}.$$

Hence it suffices to prove

$$\|v_{\mathscr{W}_h}\|^2_{\widetilde{H}_{\mathrm{I}}^{1/2}(\Gamma)} \lesssim b_{\mathscr{W}_h}(v_{\mathscr{W}_h}; v_{\mathscr{W}_h}) \tag{10.14}$$

and

$$\|v_\Gamma\|^2_{\widetilde{H}_{\mathrm{I}}^{1/2}(\Gamma)} \lesssim \sum_{j=1}^J b_{\Gamma_j}(v_{\Gamma_j}, v_{\Gamma_j}). \tag{10.15}$$

First we prove (10.14). This is done by decomposing $v_{\mathscr{W}_h}$ into edge and vertex components and prove the required inequality for each component as follows. Let

$$v_{\mathscr{W}_h,e} := \sum_{k \in \mathscr{I}_{H,e}} v(\mathbf{x}_k)\phi_k^h \quad \text{and} \quad v_{\mathscr{W}_h,v} := \sum_{k \in \mathscr{I}_{H,v}} v(\mathbf{x}_k)\phi_k^h.$$

Then $v_{\mathscr{W}_h} = v_{\mathscr{W}_h,e} + v_{\mathscr{W}_h,v}$ so that

$$\|v_{\mathscr{W}_h}\|^2_{\widetilde{H}_{\mathrm{I}}^{1/2}(\Gamma)} \lesssim \|v_{\mathscr{W}_h,e}\|^2_{\widetilde{H}_{\mathrm{I}}^{1/2}(\Gamma)} + \|v_{\mathscr{W}_h,v}\|^2_{\widetilde{H}_{\mathrm{I}}^{1/2}(\Gamma)}. \tag{10.16}$$

To deal with the first term on the right-hand side of (10.16), we partition the set $\mathscr{E}_H$ of all subdomain edges not on $\partial\Gamma$ into $\mathscr{E}_H = \mathscr{E}_{H,1} \cup \cdots \cup \mathscr{E}_{H,M}$ with the property that for all $m = 1, \ldots, M$

$$\gamma, \gamma' \in \mathscr{E}_{H,m} \iff \Gamma_\gamma \cap \Gamma_{\gamma'} = \varnothing \tag{10.17}$$

where $\Gamma_\gamma$ is the union of $\gamma$ and the pair of subdomains sharing this edge. The assumption on the mesh assures us that such a partition is possible with $M$ independent of $J$, $H$, and $h$. With this partition, we further decompose $v_{\mathscr{W}_h,e}$ as

$$v_{\mathscr{W}_h,e} = \sum_{m=1}^{M} v_{\mathscr{W}_h,e,m} \quad \text{where} \quad v_{\mathscr{W}_h,e,m} = \sum_{\gamma \in \mathscr{E}_{H,m}} v_\gamma \quad \text{with} \quad v_\gamma = \sum_{k \in \mathscr{I}_{h,\gamma}} v(\mathbf{x}_k)\phi_k^h.$$

Here $\mathscr{I}_{h,\gamma} = \{k \in \mathscr{I}_{H,e} : \mathbf{x}_k \in \gamma\}$. The triangle and Cauchy–Schwarz inequalities yield

$$\|v_{\mathscr{W}_h,e}\|^2_{\widetilde{H}_{\mathrm{I}}^{1/2}(\Gamma)} \lesssim M \sum_{m=1}^{M} \|v_{\mathscr{W}_h,e,m}\|^2_{\widetilde{H}_{\mathrm{I}}^{1/2}(\Gamma)}$$

Note that $\operatorname{supp}(v_\gamma) \subset \Gamma_\gamma$. Property (10.17) implies that for each $m = 1,\ldots,M$, the supports of $v_\gamma$ for $\gamma \in \mathscr{E}_{H,m}$ are disjoint. Hence we can invoke Theorem A.10 to obtain

$$\|v_{\mathscr{W}_h,e,m}\|^2_{\widetilde{H}_{\mathrm{I}}^{1/2}(\Gamma)} = \Big\| \sum_{\gamma \in \mathscr{E}_{H,m}} v_\gamma \Big\|^2_{\widetilde{H}_{\mathrm{I}}^{1/2}(\Gamma_\gamma)} \leq \sum_{\gamma \in \mathscr{E}_{H,m}} \|v_\gamma\|^2_{\widetilde{H}_{\mathrm{I}}^{1/2}(\Gamma_\gamma)}.$$

It then follows from Lemma 10.2 that

$$\|v_{\mathscr{W}_h,e,m}\|^2_{\widetilde{H}_{\mathrm{I}}^{1/2}(\Gamma)} \lesssim \sum_{\gamma \in \mathscr{E}_{H,m}} \|v_\gamma\|^2_{L^2(\gamma)}.$$

Consequently,

$$\|v_{\mathscr{W}_h,e}\|^2_{\widetilde{H}_{\mathrm{I}}^{1/2}(\Gamma)} \lesssim \sum_{m=1}^{M} \sum_{\gamma \in \mathscr{E}_{H,m}} \|v_\gamma\|^2_{L^2(\gamma)} = \sum_{\gamma \in \mathscr{E}_H} \|v_\gamma\|^2_{L^2(\gamma)}.$$

This and Lemma 10.1 imply

$$\|v_{\mathscr{W}_h,e}\|^2_{\widetilde{H}_{\mathrm{I}}^{1/2}(\Gamma)} \lesssim b_{\mathscr{W}_h}(v_{\mathscr{W}_h,e}, v_{\mathscr{W}_h,e}). \tag{10.18}$$

The second term on the right-hand side of (10.16) can be dealt with in the same manner. The index set $\mathscr{I}_{H,v}$ is decomposed into $\mathscr{I}_{H,v} = \mathscr{I}_{v,1} \cup \cdots \mathscr{I}_{v,L}$ such that for any $\ell = 1,\ldots,L$

$$k,k' \in \mathscr{I}_{v,\ell} \iff \operatorname{supp}(\phi_k^h) \cap \operatorname{supp}(\phi_{k'}^h) = \varnothing.$$

Denoting by $\gamma_k^h$ the support of $\phi_k^h$ and noting that $\operatorname{diam}(\gamma_k^h) \simeq h_k$, we deduce on invoking Lemma C.12 and recalling definition (10.8) of $b_{\mathscr{W}_h}(\cdot,\cdot)$

$$\|v_{\mathscr{W}_h,v}\|^2_{\widetilde{H}_{\mathrm{I}}^{1/2}(\Gamma)} \lesssim \sum_{k \in \mathscr{I}_{H,v}} \|v(\mathbf{x}_k)\phi_k^h\|^2_{\widetilde{H}_{\mathrm{I}}^{1/2}(\gamma_k^h)} \simeq \sum_{k \in \mathscr{I}_{H,v}} h_k |v(\mathbf{x}_k)|^2 = b_{\mathscr{W}_h}(v_{\mathscr{W}_h,v}, v_{\mathscr{W}_h,v}).$$

The above inequality, (10.18), and (10.16) yield (10.14).

Estimate (10.15) can be shown by invoking Theorem A.11 and then Lemma B.3 to obtain

$$\|v_\Gamma\|^2_{\widetilde{H}_{\mathrm{I}}^{1/2}(\Gamma)} \lesssim \sum_{j=1}^{J} \|v_{\Gamma_j}\|^2_{\widetilde{H}_{\mathrm{I}}^{1/2}(\Gamma)} \simeq \sum_{j=1}^{J} a_{\mathfrak{W}}(v_{\Gamma_j}, v_{\Gamma_j}) = \sum_{j=1}^{J} b_{\Gamma_j}(v_{\Gamma_j}, v_{\Gamma_j}).$$

This completes the proof of the lemma. □

The next lemma which establishes coercivity of the bilinear forms on the $p$-wire basket and $p$-interior spaces ($\mathscr{V}_{\mathscr{W}_p}$ and $\mathscr{V}_K$, respectively) are analogous to [2, Lemmas 9 and 12]. The current proofs assure us that the constants are independent of not only $p$ but also $H$ and $h$. We note here again that Theorem A.11 is used instead of Theorem A.10, as was done in [2].

**Lemma 10.9.** *For any $v_p = v_{\mathscr{W}_p} + \sum_K v_K$ with $v_{\mathscr{W}_p} \in \mathscr{V}_{\mathscr{W}_p}$ and $v_K \in \mathscr{V}_K$, the following inequality holds*

$$a_{\mathfrak{W}}(v_p, v_p) \lesssim b_{\mathscr{W}_p}(v_{\mathscr{W}_p}, v_{\mathscr{W}_p}) + \sum_K b_K(v_K, v_K).$$

*The constant involved is independent of $v_p$, $p$, $h$ and $H$.*

*Proof.* Let $v_{p,K} := \sum_K v_K$. Then $v_p = v_{\mathscr{W}_p} + v_{p,K}$ so that

$$a_{\mathfrak{W}}(v_p, v_p) \simeq \|v_p\|^2_{\widetilde{H}^{1/2}_{\mathrm{I}}(\Gamma)} \lesssim \|v_{\mathscr{W}_p}\|^2_{\widetilde{H}^{1/2}_{\mathrm{I}}(\Gamma)} + \|v_{p,K}\|^2_{\widetilde{H}^{1/2}_{\mathrm{I}}(\Gamma)}.$$

Since (due to Theorem A.11 and Lemma B.3)

$$\|v_{p,K}\|^2_{\widetilde{H}^{1/2}_{\mathrm{I}}(\Gamma)} \le \sum_K \|v_K\|^2_{\widetilde{H}^{1/2}_{\mathrm{I}}(\Gamma)} \simeq \sum_K b_K(v_K, v_K),$$

it suffices to prove

$$\|v_{\mathscr{W}_p}\|^2_{\widetilde{H}^{1/2}_{\mathrm{I}}(\Gamma)} \lesssim b_{\mathscr{W}_p}(v_{\mathscr{W}_p}, v_{\mathscr{W}_p}).$$

Similarly to $h$-wire basket functions, a function in the $p$-wire basket $v_{\mathscr{W}_p} \in \mathscr{V}_{\mathscr{W}_p}$ can be decomposed into two components, an edge component $v_{\mathscr{W}_{p,e}}$ and a vertex component $v_{\mathscr{W}_{p,v}}$, namely,

$$v_{\mathscr{W}_p} = v_{\mathscr{W}_{p,e}} + v_{\mathscr{W}_{p,v}}$$

where

$$v_{\mathscr{W}_{p,e}} = \sum_{k \in \mathscr{I}_{p,e}} v_p(\mathbf{q}_k)\phi_k^p \quad \text{and} \quad v_{\mathscr{W}_{p,v}} = \sum_{k \in \mathscr{I}_{p,v}} v_p(\mathbf{q}_k)\phi_k^p;$$

see (10.1) and Figure 10.3. The proof of

$$\|v_{\mathscr{W}_{p,e}}\|^2_{\widetilde{H}^{1/2}_{\mathrm{I}}(\Gamma)} \lesssim b_{\mathscr{W}_p}(v_{\mathscr{W}_{p,e}}, v_{\mathscr{W}_{p,e}})$$

follows exactly along the lines of the proof of the component $v_{\mathscr{W}_{h,e}}$ in (10.14), except that Lemma 10.3 is used in lieu of Lemma 10.2. Similarly, the proof of

$$\|v_{\mathscr{W}_{p,v}}\|^2_{\widetilde{H}^{1/2}_{\mathrm{I}}(\Gamma)} \lesssim b_{\mathscr{W}_p}(v_{\mathscr{W}_{p,v}}, v_{\mathscr{W}_{p,v}})$$

follows exactly along the lines of the proof of the component $v_{\mathscr{W}_{h,v}}$ in (10.14), except that Lemma 10.4 is used in lieu of Lemma C.12. The lemma is proved. □

**Stability of the decomposition**

We now prove the stability of the decomposition. First we prove that there exists a decomposition of a function in $\mathscr{V}$ which is stable under the defined bilinear forms. We note that because there is more than one way to define the component in the coarse space $\mathscr{V}_H$, the decomposition is not unique.

Recalling the definition and properties of Clément's interpolation operator in Subsection C.5.4, we introduce the notations

$$\Pi_h^0 : L_2(\Gamma) \longrightarrow \mathscr{V}_h \quad \text{and} \quad \Pi_{H^*}^0 : L_2(\Gamma) \longrightarrow \mathscr{V}_H^*;$$

We define for any $v \in \mathscr{V}$

$$v_h = \Pi_h^0 v \in \mathscr{V}_h, \quad v_H = \Pi_h \Pi_{H^*}^0 v_h \in \mathscr{V}_H, \quad v_p = v - v_h \in \mathscr{V}_p. \tag{10.19}$$

Letting

$$\widetilde{v}_h = v_h - v_H, \tag{10.20}$$

we can decompose $\widetilde{v}_h$ uniquely as

$$\widetilde{v}_h = v_{\mathscr{W}_h} + \sum_{j=1}^{J} v_{\Gamma_j}, \qquad v_{\mathscr{W}_h} \in \mathscr{V}_{\mathscr{W}_h}, v_{\Gamma_j} \in \mathscr{V}_{\Gamma_j}. \tag{10.21}$$

We can also decompose $v_p$ uniquely as

$$v_p = v_{\mathscr{W}_p} + \sum_K v_K, \qquad v_{\mathscr{W}_p} \in \mathscr{V}_{\mathscr{W}_p}, v_K \in \mathscr{V}_K. \tag{10.22}$$

Then it is clear that

$$v = v_H + v_{\mathscr{W}_h} + \sum_{j=1}^{J} v_{\Gamma_j} + v_{\mathscr{W}_p} + \sum_K v_K. \tag{10.23}$$

We now show stability for each component on the right-hand side of (10.23).

**Lemma 10.10.** *Let $v \in \mathscr{V}$ and $v_H$ be defined as in* (10.19). *Then*

$$b_H(v_H, v_H) \lesssim a_{\mathfrak{W}}(v, v).$$

*The constant involved is independent of $v$, $h$, $H$ and $p$.*

*Proof.* It follows from (10.7) that $\Pi_H^* v_H = \Pi_H^* \Pi_h \Pi_{H^*}^0 v_h = \Pi_{H^*}^0 v_h$. Therefore, the definition of the bilinear form $b_H(\cdot,\cdot)$ in (10.5) implies

$$b_H(v_H, v_H) = a_{\mathfrak{W}}(\Pi_{H^*}^0 v_h, \Pi_{H^*}^0 v_h) \simeq \|\Pi_{H^*}^0 v_h\|_{\widetilde{H}_{\mathrm{I}}^{1/2}(\Gamma)}^2 = \|\Pi_{H^*}^0 \Pi_h^0 v\|_{\widetilde{H}_{\mathrm{I}}^{1/2}(\Gamma)}^2.$$

It follows from Lemma C.46 that

$$b_H(v_H, v_H) \lesssim \|v\|_{\widetilde{H}_{\mathrm{I}}^{1/2}(\Gamma)}^2 \simeq a_{\mathfrak{W}}(v, v). \qquad \square$$

**Lemma 10.11.** *For any $v \in \mathscr{V}$, let $v_{\mathscr{W}_h}$ be defined as in* (10.21). *Then*

$$b_{\mathscr{W}_h}(v_{\mathscr{W}_h}, v_{\mathscr{W}_h}) \lesssim \max_{1 \le j \le J} (1 + \log(H_j/h_j))\, a_{\mathfrak{W}}(v, v).$$

*The constant involved is independent of v, h, H and p.*

*Proof.* It follows from Lemma 10.1, the definition of $v_{\mathscr{W}_h}$, and (10.21) that

$$\begin{aligned} b_{\mathscr{W}_h}(v_{\mathscr{W}_h}, v_{\mathscr{W}_h}) &\simeq \sum_{\gamma \in \mathscr{E}_H} \|v_{\mathscr{W}_h}\|^2_{L_2(\gamma)} \\ &\lesssim \sum_{\gamma \in \mathscr{E}_H} \|\widetilde{v}_h\|^2_{L_2(\gamma)} + \sum_{\gamma \in \mathscr{E}_H} \|\sum_{j=1}^J v_{\Gamma_j}\|^2_{L_2(\gamma)} \\ &= \sum_{\gamma \in \mathscr{E}_H} \|\widetilde{v}_h\|^2_{L_2(\gamma)}, \end{aligned}$$

where we note that $v_{\Gamma_j} \equiv 0$ on $\gamma$. Invoking Lemma C.24 we deduce

$$b_{\mathscr{W}_h}(v_{\mathscr{W}_h}, v_{\mathscr{W}_h}) \lesssim \max_{1 \le j \le J} \left(1 + \log \frac{H_j}{h_j}\right) \sum_{j=1}^J \|\widetilde{v}_h\|^2_{H^{1/2}_{\mathrm{w}}(\Gamma_j)}$$

where the weighted norm $\|\cdot\|_{H^{1/2}_{\mathrm{w}}(\Gamma_j)}$ is defined in Subsection A.2.6.1. Note that

$$\sum_{j=1}^J \|\widetilde{v}_h\|^2_{H^{1/2}_{\mathrm{w}}(\Gamma_j)} \le \sum_{j=1}^J H_j^{-1} \|\widetilde{v}_h\|^2_{L_2(\Gamma_j)} + |\widetilde{v}_h|^2_{H^{1/2}(\Gamma)}.$$

Since we can write $\widetilde{v}_h$ as $\widetilde{v}_h = \Pi_h(v_h - \Pi_H^0 v_h)$, by Lemma C.36 (more precisely, we use (C.64)) and Lemma C.46 we obtain

$$\begin{aligned} \sum_{j=1}^J \|\widetilde{v}_h\|^2_{H^{1/2}_{\mathrm{w}}(\Gamma_j)} &\lesssim \sum_{j=1}^J H_j^{-1} \|v_h - \Pi_H^0 v_h\|^2_{L_2(\Gamma_j)} + |v_h - \Pi_H^0 v_h|^2_{H^{1/2}(\Gamma)} \\ &\lesssim |v_h|^2_{H^{1/2}(\Gamma)} \le \|v_h\|^2_{\widetilde{H}^{1/2}_{\mathrm{S}}(\Gamma)} \simeq \|v_h\|^2_{\widetilde{H}^{1/2}_{\mathrm{I}}(\Gamma)}. \end{aligned}$$

Using again the boundedness of $\Pi_h^0$ given by Lemma C.46, we infer

$$\sum_{j=1}^J \|\widetilde{v}_h\|^2_{H^{1/2}_{\mathrm{w}}(\Gamma_j)} \lesssim \|v\|^2_{\widetilde{H}^{1/2}_{\mathrm{I}}(\Gamma)} \simeq a_{\mathfrak{W}}(v, v).$$

The desired result then follows. □

**Lemma 10.12.** *Let $v \in \mathscr{V}$ and $v_{\Gamma_j}$ be defined as in* (10.21). *Then*

$$\sum_{j=1}^J b_{\Gamma_j}(v_{\Gamma_j}, v_{\Gamma_j}) \lesssim \max_{1 \le j \le J} \left(1 + \log \frac{H_j}{h_j}\right)^2 a_{\mathfrak{W}}(v, v).$$

*The constant involved is independent of v, h, H and p.*

*Proof.* By the definition of the bilinear form $b_{\Gamma_j}(\cdot,\cdot)$, see (10.5), and by Lemma A.13 (i)

$$\sum_{j=1}^{J} b_{\Gamma_j}(v_{\Gamma_j}, v_{\Gamma_j}) = \sum_{j=1}^{J} a_{\mathfrak{W}}(v_{\Gamma_j}, v_{\Gamma_j}) \simeq \sum_{j=1}^{J} \|v_{\Gamma_j}\|^2_{\widetilde{H}_{\mathrm{I}}^{1/2}(\Gamma)} \lesssim \sum_{j=1}^{J} \|v_{\Gamma_j}\|^2_{\widetilde{H}_{\mathrm{w}}^{1/2}(\Gamma_j)}.$$

Due to the assumption on the coarse mesh, we can assume that $\Gamma_j = I \times I$ where $I = (0,H)$. Invoking inequality (A.99) in Lemma A.16 we deduce

$$\sum_{j=1}^{J} b_{\Gamma_j}(v_{\Gamma_j}, v_{\Gamma_j}) \lesssim \sum_{j=1}^{J} \Big( \int_I \|v_{\Gamma_j}(x,\cdot)\|^2_{\widetilde{H}_{\mathrm{w}}^{1/2}(I)}\, \mathrm{d}x + \int_I \|v_{\Gamma_j}(\cdot,y)\|^2_{\widetilde{H}_{\mathrm{w}}^{1/2}(I)}\, \mathrm{d}y \Big). \tag{10.24}$$

For any $x \in I$, Lemma C.21 gives

$$\int_I \|v_{\Gamma_j}(x,\cdot)\|^2_{\widetilde{H}_{\mathrm{w}}^{1/2}(I)}\, \mathrm{d}x \lesssim \max_{1\le j\le J} \Big(1 + \log\frac{H_j}{h_j}\Big) \int_I |v_{\Gamma_j}(x,\cdot)|^2_{H^{1/2}(I)}\, \mathrm{d}x.$$

Using (A.96) we infer

$$\int_I \|v_{\Gamma_j}(x,\cdot)\|^2_{\widetilde{H}_{\mathrm{w}}^{1/2}(I)}\, \mathrm{d}x \lesssim \max_{1\le j\le J} \Big(1 + \log\frac{H_j}{h_j}\Big) \|v_{\Gamma_j}\|^2_{H_{\mathrm{w}}^{1/2}(\Gamma_j)}.$$

The same estimate can be derived for the second integral on the right-hand side of (10.24), which then implies

$$\sum_{j=1}^{J} b_{\Gamma_j}(v_{\Gamma_j}, v_{\Gamma_j}) \lesssim \max_{1\le j\le J} \Big(1 + \log\frac{H_j}{h_j}\Big) \sum_{j=1}^{J} \|v_{\Gamma_j}\|^2_{H_{\mathrm{w}}^{1/2}(\Gamma_j)}.$$

Noting that $v_{\Gamma_j} = \widetilde{v}_h - v_{\mathscr{W}_h}$ on $\Gamma_j$, see (10.21), we deduce

$$\sum_{j=1}^{J} b_{\Gamma_j}(v_{\Gamma_j}, v_{\Gamma_j}) \lesssim \max_{1\le j\le J} \Big(1 + \log\frac{H_j}{h_j}\Big) \sum_{j=1}^{J} \Big( \|\widetilde{v}_h\|^2_{H_{\mathrm{w}}^{1/2}(\Gamma_j)} + \|v_{\mathscr{W}_h}\|^2_{H_{\mathrm{w}}^{1/2}(\Gamma_j)} \Big). \tag{10.25}$$

Let $H = \max_{1\le j\le J} H_j$. Since the coarse mesh is quasi-uniform, we deduce from the definition of the $\|\cdot\|_{H_{\mathrm{w}}^{1/2}(\Gamma_j)}$-norm and (10.20) that

$$\sum_{j=1}^{J} \|\widetilde{v}_h\|^2_{H_{\mathrm{w}}^{1/2}(\Gamma_j)} \lesssim \frac{1}{H}\|\widetilde{v}_h\|^2_{L^2(\Gamma)} + |\widetilde{v}_h|^2_{H^{1/2}(\Gamma)} = \frac{1}{H}\|v_h - v_H\|^2_{L^2(\Gamma)} + |v_h - v_H|^2_{H^{1/2}(\Gamma)}.$$

Noting (10.19) and the fact that the standard interpolation operator $\Pi_h$ is a projection, we can write $v_h - v_H = \Pi_h(v_h - \Pi_H^0 v_h)$, so that

$$\sum_{j=1}^{J} \|\widetilde{v}_h\|^2_{H^{1/2}_{\mathrm{w}}(\Gamma_j)} \lesssim \frac{1}{H}\|v_h - \Pi^0_H v_h\|^2_{L^2(\Gamma)} + |v_h - \Pi^0_H v_h|^2_{H^{1/2}(\Gamma)}.$$

Hence on invoking Lemma C.46 we deduce

$$\sum_{j=1}^{J} \|\widetilde{v}_h\|^2_{H^{1/2}_{\mathrm{w}}(\Gamma_j)} \lesssim |v_h|^2_{H^{1/2}(\Gamma)} \le \|v_h\|^2_{\widetilde{H}^{1/2}_{\mathrm{w}}(\Gamma)} \simeq \|v_h\|^2_{\widetilde{H}^{1/2}_{\mathrm{I}}(\Gamma)} \lesssim \|v\|^2_{\widetilde{H}^{1/2}_{\mathrm{I}}(\Gamma)}.$$

This inequality together with (10.25) and Lemma 10.11 yields the required result.
□

**Lemma 10.13.** *Let $v \in \mathscr{V}$ and $v_{\mathscr{W}_p}$ be defined as in* (10.22). *Then*

$$b_{\mathscr{W}_p}(v_{\mathscr{W}_p}, v_{\mathscr{W}_p}) \lesssim \max_K (1 + \log p_K)\, a_{\mathfrak{W}}(v, v),$$

*where the constant involved is independent of $v$, $H$, $h$ and $p$.*

*Proof.* Due to Lemma 10.1 it suffices to prove

$$\sum_K \sum_{\gamma \in \partial K} \|v_{\mathscr{W}_p}\|^2_{\widetilde{L}_2(\gamma)} \lesssim \max_K (1 + \log p_K)\|v\|^2_{\widetilde{H}^{1/2}_{\mathrm{I}}(\Gamma)}.$$

Noting that $v_{\mathscr{W}_p} = v_p$ on $\gamma$, by mapping to the reference element we obtain

$$\|v_{\mathscr{W}_p}\|^2_{\widetilde{L}_2(\gamma)} = \|v_p\|^2_{\widetilde{L}_2(\gamma)} \simeq h_\gamma \|\widehat{v}_p\|^2_{\widetilde{L}_2(\widehat{\gamma})}.$$

It follows from (C.57) (the proof of which is at the end of the proof of Lemma C.30) that

$$\|v_{\mathscr{W}_p}\|^2_{\widetilde{L}_2(\gamma)} \lesssim h_\gamma (1 + \log p_K) \|\widehat{v}_p\|^2_{H^{1/2}_{\mathrm{S}}(\widehat{K})} \simeq h_\gamma (1 + \log p_K) \|\widehat{v}_p\|^2_{H^{1/2}_{\mathrm{w}}(\widehat{K})}.$$

Invoking Lemma A.5 gives

$$\begin{aligned}\|v_{\mathscr{W}_p}\|^2_{\widetilde{L}_2(\gamma)} &\lesssim (1 + \log p_K)\|v_p\|^2_{H^{1/2}_{\mathrm{w}}(K)} \\ &= (1 + \log p_K)\left(|v_p|^2_{H^{1/2}(K)} + h_K^{-1}\|v_p\|^2_{\widetilde{L}_2(K)}\right)\end{aligned}$$

where the constant is independent of $p$, $H$, and $h$. Recall that $v_p = v - \Pi^0_{\mathscr{V}_h} v$; see (10.19). Summing over $\gamma$ and $K$ and using (C.80) and (C.81), we obtain the desired result. □

In the following lemma, to avoid a dependence of the constant on $h$, we employ an approach different from what is used in [2] where the pure $p$-version is studied.

**Lemma 10.14.** *Let $v \in \mathscr{V}$ and $v_K \in \mathscr{V}_K$ be defined as in* (10.22). *Then*

$$\sum_K b_K(v_K, v_K) \lesssim \max_K (1 + \log p_K)^2 a_{\mathfrak{W}}(v, v).$$

*The constant involved is independent of $v$, $H$, $h$ and $p$.*

*Proof.* By using successively Lemma B.3, Lemma A.13, and Lemma A.5, we obtain

$$b_K(v_K, v_K) = a_{\mathfrak{W}}(v_K, v_K) \simeq \|v_K\|^2_{\widetilde{H}_{\mathrm{I}}^{1/2}(\Gamma)} \lesssim \|v_K\|^2_{\widetilde{H}_{\mathrm{w}}^{1/2}(K)} \simeq h_K \|\widehat{v}_K\|^2_{\widetilde{H}_{\mathrm{w}}^{1/2}(\widehat{K})}.$$

Lemma 6 in [2] gives

$$b_K(v_K, v_K) \lesssim h_K \Big( \|\widehat{v}_K\|^2_{H_{\mathrm{w}}^{1/2}(\widehat{K})} + (1+\log p_K)^2 \|\widehat{v}_p\|^2_{H_{\mathrm{w}}^{1/2}(\widehat{K})} \Big).$$

Here we use the usual notation that if $u$ is a function defined in $K$, then $\widehat{u}$ is defined in the reference element $\widehat{K}$ by $\widehat{u}(\widehat{x}) = u(h_K \widehat{x})$, where $h_K$ is the scaling factor that defines $\widehat{K}$, namely, $\widehat{K} := \{x/h_K : x \in K\}$. Recalling that $v_K = v_p - v_{\mathscr{W}_p}$ on $K$, so that $\widehat{v}_K = \widehat{v}_p - \widehat{v}_{\mathscr{W}_p}$, we infer

$$b_K(v_K, v_K) \lesssim h_K \Big( \|\widehat{v}_{\mathscr{W}_p}\|^2_{H_{\mathrm{w}}^{1/2}(\widehat{K})} + (1+\log p_K)^2 \|\widehat{v}_p\|^2_{H_{\mathrm{w}}^{1/2}(\widehat{K})} \Big).$$

We note that a scaling argument at this stage will yield a constant growing as $h_K^{-1}$ for the first term in the sum on the right-hand side. To avoid this, we use Lemma 10.3 to obtain

$$\|\widehat{v}_{\mathscr{W}_p}\|^2_{H_{\mathrm{I}}^{1/2}(\widehat{K})} \lesssim \sum_{\widehat{\gamma} \in \partial \widehat{K}} \|\widehat{v}_{\mathscr{W}_p}\|^2_{L_2(\widehat{\gamma})},$$

which implies after a scaling argument and an application of Lemma A.5

$$b_K(v_K, v_K) \lesssim \sum_{\gamma \in \partial K} \|v_{\mathscr{W}_p}\|^2_{L_2(\gamma)} + (1+\log p_K)^2 \|v_p\|^2_{H_{\mathrm{w}}^{1/2}(K)}.$$

Summing over $K$ and proceeding as in the proof of Lemma 10.13, we obtain the desired result. □

We are now able to state the stability of the decomposition (10.3)–(10.4).

**Lemma 10.15 (Stability of decomposition).** *There exists a constant $C_0$ independent of $H$, $h$, and $p$ such that any $v \in \mathscr{V}$ admits a decomposition*

$$v = v_H + v_{\mathscr{W}_h} + \sum_{j=1}^{J} v_{\Gamma_j} + v_{\mathscr{W}_p} + \sum_K v_K$$

*satisfying*

$$\begin{aligned} b_H(v_H, v_H) + b_{\mathscr{W}_h}(v_{\mathscr{W}_h}, v_{\mathscr{W}_h}) + \sum_{j=1}^{J} b_{\Gamma_j}(v_{\Gamma_j}, v_{\Gamma_j}) + b_{\mathscr{W}_p}(v_{\mathscr{W}_p}, v_{\mathscr{W}_p}) + \sum_K b_K(v_K, v_K) \\ \lesssim \max_{1 \le j \le J} (1+\log(H_j p_j / h_j))^2 a_{\mathfrak{W}}(v, v) \end{aligned}$$

*where $p_j = \max_{K \subset \Gamma_j} p_K$.*

*Proof.* The lemma is a direct consequence of Lemmas 10.10, 10.11, 10.12, 10.13, and 10.14. □

As a consequence of Lemma 10.6, Lemma 10.15, and the general theory Theorem 2.1 we have

**Theorem 10.1.** *The condition number of the additive Schwarz operator $P_{\rm ad}$ is bounded as*

$$\kappa(P_{\rm ad}) \lesssim \max_{1\le j\le J}\left(1+\log\frac{H_j p_j}{h_j}\right)^2,$$

*where* $p_j = \max_{K\subset\Gamma_j} p_K$.

## 10.2.2 An overlapping method

### 10.2.2.1 The $h$-version

In this subsection, we first report on the result for the $h$-version in [230]. Recall the two-level mesh and the boundary element spaces $\mathscr{V}_0 := \mathscr{V}_H^*$ and $\mathscr{V} := \mathscr{V}_h$ defined in Subsection 10.1.1. We also recall the subspaces $\mathscr{V}_j := \mathscr{V}_{\Gamma_j}$, $j = 1,\dots,J$ defined in that subsection.

**Overlapping subdomains:**
We extend each subdomain $\Gamma_j$ in the following way. First we define, for some positive constant $\delta$, called the overlap size,

$$\widetilde{\mathscr{V}_j} = \operatorname{span}\{\phi_k : \mathbf{x}_k \notin \overline{\Gamma}_j, \operatorname{dist}(\mathbf{x}_k, \partial\Gamma_j) \le \delta\},$$

and denote

$$\widetilde{\Gamma_j} = \operatorname{supp}\{\phi_k : \phi_k \in \widetilde{\mathscr{V}_j}\},$$

which is the shaded area in Figure 10.5. Here the basis functions $\phi_k = \phi_k^h$ are defined in Subsection 10.2.1.1 and the distance is defined with the max norm $\|\mathbf{x}\| = \max\{|x_1|,|x_2|\}$ where $\mathbf{x} = (x_1,x_2)$. The extended subdomain $\Gamma_j'$ is then defined as $\Gamma_j' = \overline{\Gamma}_j \cup \widetilde{\Gamma_j}$. We note that $\Gamma_j'$ need not be a quadrilateral domain. Also, if $\delta$ is chosen such that $\delta \in (0,H]$, then

$$\operatorname{diam}(\Gamma_i') \simeq H. \tag{10.26}$$

**Subspace decomposition:**
The subspace decomposition

$$\mathscr{V} = \mathscr{V}_0' + \cdots + \mathscr{V}_J'$$

is performed with subspaces $\mathscr{V}_j'$, $j = 0,1,\dots,J$, defined by

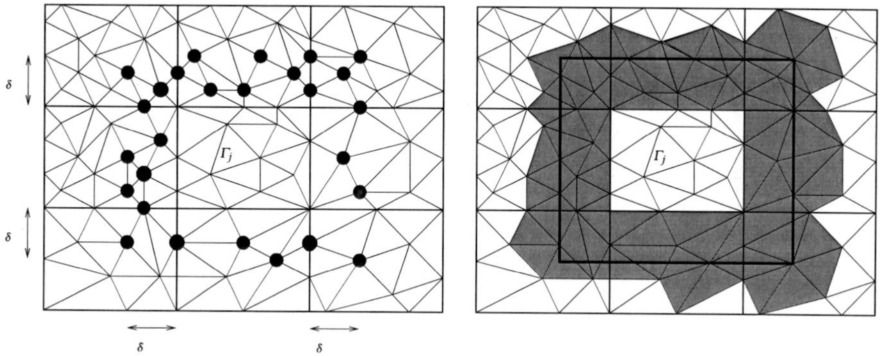


**Fig. 10.5** • vertex at a distance $\delta$ to $\overline{\Gamma}_j$, $\widetilde{\Gamma}_j$: shaded region, $\Gamma_j' = \Gamma_j \cup \widetilde{\Gamma}_j$: overlapping subdomain, $R_j$: (rectangle in shaded region) support of cutoff function $\theta_j$.

$$\mathscr{V}_0' := \mathscr{V}_H = \Pi_h \mathscr{V}_H^* = \Pi_h \mathscr{V}_0 \quad \text{and} \quad \mathscr{V}_j' := \mathscr{V}_j \cup \widetilde{\mathscr{V}_j} = \mathscr{V} \cap \widetilde{H}^{1/2}(\Gamma_j'),\ j = 1,\dots,J,$$

where $\Pi_h$ is the interpolation operator which interpolates continuous functions into functions in $\mathscr{V}$.

The bilinear forms $b_i(\cdot,\cdot)$ associated with the subspaces $\mathscr{V}_i'$ are defined by

$$b_0(v,w) = a_{\mathfrak{W}}(\Pi_H^* v, \Pi_H^* w) \quad \forall v,w \in \mathscr{V}_0',$$

and

$$b_j(v,w) = a_{\mathfrak{W}}(v,w) \quad \forall v,w \in \mathscr{V}_j',\ j = 1,\dots,J,$$

where $\Pi_H^*$ be the interpolation operator which interpolates continuous functions into functions in $\mathscr{V}_0 = \mathscr{V}_H^*$.

**Stability of the decomposition:**

We first prove the stability of the decomposition. The following results were proved in [72, Lemma 3.5]; however, for the convenience of the reader we present the detailed proof to justify that the results still hold with the weighted norm defined in Subsection A.2.6.1.

**Lemma 10.16.** *Let* $0 < \delta < \beta$ *and* $I_\beta := (0,\beta)$.

(i) *If* $v \in H^{1/2}(I_\beta)$, *then*

$$\int_\delta^\beta \frac{|v(x)|^2}{x}\,dx \lesssim \left(1 + \log^2 \frac{\beta}{\delta}\right) \|v\|^2_{H_w^{1/2}(I_\beta)}. \tag{10.27}$$

(ii) *If* $v \in H^{1/2}(I_\beta \times I_\beta)$, *then*

$$\int_0^\beta \Big( \int_\delta^\beta \frac{|v(x,y)|^2}{x}\,\mathrm{d}x \Big)\,\mathrm{d}y \lesssim \Big(1+\log^2\frac{\beta}{\delta}\Big)\|v\|^2_{H^{1/2}_{\mathrm{w}}(I_\beta\times I_\beta)}. \tag{10.28}$$

*Proof.* We first prove (10.27) for $v_\delta \in \mathscr{V}_\delta$, the space of continuous piecewise-linear functions on a uniform mesh of size $\delta$ on $(0,\beta)$. Simple calculation reveals

$$\int_\delta^\beta \frac{|v_\delta(x)|^2}{x}\,\mathrm{d}x \le \|v_\delta\|^2_{L_\infty(0,\beta)} \int_\delta^\beta \frac{\mathrm{d}x}{x} = \|v_\delta\|^2_{L_\infty(0,\beta)}\,\log\frac{\beta}{\delta}.$$

By invoking Lemma C.19 with $H=\beta$, $h=\delta$, and $I=(0,H)$, we deduce

$$\int_\delta^\beta \frac{|v_\delta(x)|^2}{x}\,\mathrm{d}x \lesssim \Big(1+\log^2\frac{\beta}{\delta}\Big)\|v_\delta\|^2_{H^{1/2}_{\mathrm{w}}(0,\beta)}. \tag{10.29}$$

Consider now $v\in H^{1/2}(0,\beta)$. Let $v_\delta\in\mathscr{V}_\delta$ be the $L^2$-projection of $v$ onto $\mathscr{V}_\delta$. It can be shown that, see (C.47) and (C.48) with $R=(0,\beta)$ and $h=\delta$,

$$\|v-v_\delta\|^2_{L^2(0,\beta)} \lesssim \delta|v|^2_{H^{1/2}(0,\beta)} \quad\text{and}\quad \|v_\delta\|_{H^{1/2}_{\mathrm{w}}(0,\beta)} \lesssim \|v\|_{H^{1/2}_{\mathrm{w}}(0,\beta)}. \tag{10.30}$$

Hence,

$$\begin{aligned}\int_\delta^\beta \frac{|v(x)|^2}{x}\,\mathrm{d}x &\lesssim \int_\delta^\beta \frac{|v(x)-v_\delta(x)|^2}{x}\,\mathrm{d}x + \int_\delta^\beta \frac{|v_\delta(x)|^2}{x}\,\mathrm{d}x\\ &\lesssim \frac{1}{\delta}\|v-v_\delta\|^2_{L^2(0,\beta)} + \int_\delta^\beta \frac{|v_\delta(x)|^2}{x}\,\mathrm{d}x.\end{aligned} \tag{10.31}$$

Inequality (10.27) follows immediately from (10.29), (10.30), and (10.31).

Inequality (10.28) is obtained by using successively (10.27), the definition of the weighted norm (see Subsection A.2.6.1), (A.97), Fubini's Theorem, and (A.98)

$$\begin{aligned}\int_0^\beta \Big( \int_\delta^\beta \frac{|v(x,y)|^2}{x}\,\mathrm{d}x \Big)\,\mathrm{d}y &\lesssim \Big(1+\log^2\frac{\beta}{\delta}\Big)\int_0^\beta \|v(\cdot,y)\|^2_{H^{1/2}_{\mathrm{w}}(0,\beta)}\,\mathrm{d}y\\ &= \Big(1+\log^2\frac{\beta}{\delta}\Big)\int_0^\beta \Big[\frac{1}{\beta}\|v(\cdot,y)\|^2_{L^2(0,\beta)}\\ &\quad + \int_0^\beta\int_0^\beta \frac{|v(x,y)-v(x',y)|^2}{|x-x'|^2}\,\mathrm{d}x\,\mathrm{d}x'\Big]\,\mathrm{d}y\end{aligned}$$

$$\lesssim \left(1+\log^2\frac{\beta}{\delta}\right)\left[\frac{1}{\beta}\|v\|^2_{L^2((0,\beta)^2)} + \int_0^\beta\int_0^\beta \frac{\|v(x,\cdot)-v(x',\cdot)\|^2_{L^2(0,\beta)}}{|x-x'|^2}\,dx\,dx'\right]$$

$$\lesssim \left(1+\log^2\frac{\beta}{\delta}\right)\|v\|^2_{H^{1/2}_{\mathrm{w}}((0,\beta)\times(0,\beta))},$$

finishing the proof of the lemma. □

The above results will be used to prove the following lemma, which is crucial to obtain the stability of the decomposition. Let $R$ be the union of overlapping rectangular subdomains $R_l$, $l=1,\dots,J$, of diameters $\beta_l$. Assume that the size of the overlap is $\delta$. Let $\{\theta_l : l=1,\dots,J\}$ be a partition of unity on $R$ defined by piecewise bilinear functions such that supp $\theta_l = \overline{R}_l$.

**Lemma 10.17.** *If $\delta \le \min_l \beta_l/2$, then for any $w \in \widetilde{H}^{1/2}(R)$ the following estimate holds*

$$\sum_{\ell=1}^{J}\|\theta_\ell w\|^2_{\widetilde{H}^{1/2}_{\mathrm{w}}(R_\ell)} \lesssim \sum_{\ell=1}^{J}\left(1+\log\frac{\beta_\ell}{\delta}\right)^2\|w\|^2_{H^{1/2}_{\mathrm{w}}(R_\ell)} + \iint_R \frac{|w(\mathbf{x})|^2}{\operatorname{dist}(\mathbf{x},\partial R)}\,d\mathbf{x}. \quad (10.32)$$

*Proof.* Without loss of generality, we can assume that $R_\ell = I\times I$, where $I=(0,\beta_\ell)$, so that $\theta_\ell$ can be defined as

$$\theta_\ell(x,y) = \eta_\ell(x)\eta_\ell(y) \quad\text{where}\quad \eta_\ell(t) = \begin{cases} t/\delta, & 0\le t\le\delta,\\ 1, & \delta<t<\beta_\ell-\delta,\\ (\beta_\ell-t)/\delta, & \beta_\ell-\delta\le t\le\beta_\ell.\end{cases}$$

In view of the definition of the norm $\|\cdot\|_{\widetilde{H}^{1/2}_{\mathrm{w}}(R_l)}$, see Subsection A.2.6.1) and Lemma A.16, in order to estimate $\|\theta_\ell w\|^2_{\widetilde{H}^{1/2}(R_\ell)}$ we will estimate

$$T_1 := \iint_{I\times I} \frac{\|(\theta_\ell w)(x,\cdot)-(\theta_\ell w)(x',\cdot)\|^2_{L^2(I)}}{|x-x'|^2}\,dx\,dx'$$

and

$$T_2 := \iint_{R_\ell} \frac{|(\theta_\ell w)(\mathbf{x})|^2}{\operatorname{dist}(\mathbf{x},\partial R_\ell)}\,d\mathbf{x}.$$

Splitting the integral over $I=(0,\beta_\ell)$ into three integrals over $I_1 := [0,\delta]$, $I_2 := [\delta,\beta_\ell-\delta]$, and $I_3 := [\beta_\ell-\delta,\beta_\ell]$, and denoting, for any $v$,

$$A_{ij}(v) := \iint_{I_i\times I_j} \frac{\|v(x,\cdot)-v(x',\cdot)\|^2_{L^2(I)}}{|x-x'|^2}\,dx\,dx',$$

we observe that, by symmetry, in order to estimate $T_1$ it suffices to consider $A_{11}(\theta_\ell w)$, $A_{12}(\theta_\ell w)$, $A_{13}(\theta_\ell w)$, $A_{23}(\theta_\ell w)$, and $A_{33}(\theta_\ell w)$. There is no need to consider $A_{22}(\theta_\ell w)$ because $A_{22}(\theta_\ell w) = A_{22}(w)$. The symmetric shape of $\theta_\ell$ implies the similarity of $A_{11}(\theta_\ell w)$ and $A_{33}(\theta_\ell w)$, and of $A_{12}(\theta_\ell w)$ and $A_{23}(\theta_\ell w)$. There remains three terms to be considered: $A_{11}(\theta_\ell w)$, $A_{12}(\theta_\ell w)$, and $A_{13}(\theta_\ell w)$.

Firstly, for $A_{11}$ we have

$$\begin{aligned}
A_{11}(\theta_\ell w) &\le 2 \iint_{I_1\times I_1} \frac{\|\theta_\ell(x,\cdot)-\theta_\ell(x',\cdot)\|^2_{L^2(I)}}{|x-x'|^2}\|w(x,\cdot)\|^2_{L^2(I)}\,\mathrm{d}x\,\mathrm{d}x' \\
&\quad + 2\iint_{I_1\times I_1} \frac{\|w(x,\cdot)-w(x',\cdot)\|^2_{L^2(I)}}{|x-x'|^2}\|\theta_\ell(x,\cdot)\|^2_{L^2(I)}\,\mathrm{d}x\,\mathrm{d}x' \\
&\le 2\iint_{I_1\times I_1} \frac{\sum_{i=1}^3 \|\frac{x}{\delta}\eta(\cdot)-\frac{x'}{\delta}\eta(\cdot)\|^2_{L^2(I_i)}}{|x-x'|^2}\|w(x,\cdot)\|^2_{L^2(I)}\,\mathrm{d}x\,\mathrm{d}x' + 2\beta_\ell A_{11}(w) \\
&\lesssim \frac{1}{\delta}\|w\|^2_{L^2(I_1\times I)} + A_{11}(w).
\end{aligned}$$

It is proved in Lemma C.25 that

$$\frac{1}{\delta}\|w\|^2_{L^2(I_1\times I)} \le c\Big(1+\log\frac{\beta_\ell}{\delta}\Big)\|w\|^2_{H^{1/2}_{\mathrm{w}}(R_\ell)}, \tag{10.33}$$

where $c$ is independent of $w$, $\beta_\ell$, and $\delta$. Thus $A_{11}(\theta_\ell w)$ is bounded by the right-hand side of (10.32).

For the term $A_{12}(\theta_\ell w)$ we have

$$\begin{aligned}
A_{12}(\theta_\ell w) &= \frac{1}{\delta^2}\int_0^\delta\int_\delta^{\beta_\ell-\delta} \frac{\|xw(x,\cdot)-\delta w(x',\cdot)\|^2_{L^2(I)}}{|x-x'|^2}\,\mathrm{d}x'\,\mathrm{d}x \\
&\le \frac{2}{\delta^2}\int_0^\delta\Big(\int_\delta^{\beta_\ell-\delta}\frac{dx'}{|x-x'|^2}\Big)|x-\delta|^2\|w(x,\cdot)\|^2_{L^2(I)}\,\mathrm{d}x + 2A_{12}(w) \\
&\le \frac{2}{\delta}\|w\|^2_{L^2(I_1\times I)} + 2A_{12}(w).
\end{aligned}$$

Using again (10.33) we deduce that $A_{12}(\theta_\ell w)$ is bounded by the right-hand side of (10.32).

Finally, for $A_{13}(\theta_\ell w)$ since the assumption $\delta \le \beta_\ell/2$ implies that $\frac{\beta_\ell}{2} - x \ge 0$ for $x \in I_1$ and $x' - \frac{\beta_\ell}{2} \ge 0$ for $x' \in I_3$, we find

$$A_{13}(\theta_\ell w) = \frac{1}{\delta^2}\int_0^\delta\int_{\beta_\ell-\delta}^{\beta_\ell} \frac{\|xw(x,\cdot)-(\beta_\ell-x')w(x',\cdot)\|^2_{L^2(I)}}{|x-x'|^2}\,\mathrm{d}x'\,\mathrm{d}x$$

$$\le \frac{2}{\delta^2}\int_0^{\delta}\int_{\beta_\ell-\delta}^{\beta_\ell} \frac{|x+x'-\beta_\ell|^2}{|x-x'|^2}\|w(x,\cdot)\|_{L^2(I)}^2\,dx'\,dx$$

$$+\frac{2}{\delta^2}\int_0^{\delta}\int_{\beta_\ell-\delta}^{\beta_\ell}|\beta_\ell-x'|^2\frac{\|w(x,\cdot)-w(x',\cdot)\|_{L^2(I)}^2}{|x-x'|^2}\,dx'\,dx$$

$$\le \frac{2}{\delta^2}\int_0^{\delta}\int_{\beta_\ell-\delta}^{\beta_\ell}\frac{|(x'-\frac{\beta_\ell}{2})-(\frac{\beta_\ell}{2}-x)|^2}{|(x'-\frac{\beta_\ell}{2})+(\frac{\beta_\ell}{2}-x)|^2}\|w(x,\cdot)\|_{L^2(I)}^2\,dx'\,dx+2A_{13}(w)$$

$$\le \frac{2}{\delta}\|w\|_{L^2(I_1\times I)}^2+2A_{13}(w).$$

Inequality (10.33) yields the estimate for $A_{13}(\theta_\ell w)$, which implies that $T_1$ is bounded by the right-hand side of (10.32).

To estimate $T_2$ we first assume, without loss of generality, that $\operatorname{dist}(\mathbf{x},\partial R_\ell)=x$, where $\mathbf{x}=(x,y)$. Then

$$T_2\le\int_0^{\beta_\ell}\Big[\int_0^{\delta}\frac{x^2}{\delta^2}\frac{|w(x,y)|^2}{x}\,dx+\int_{\delta}^{\beta_\ell}\frac{|w(x,y)|^2}{x}\,dx\Big]\,dy$$

$$\le\frac{1}{\delta}\|w\|_{L^2(I_1\times I)}^2+\int_0^{\beta_\ell}\int_{\delta}^{\beta_\ell}\frac{|w(x,y)|^2}{x}\,dx\,dy.$$

The desired estimate for $T_2$ now follows from (10.33) and (10.28). □

**Lemma 10.18 (Stability of decomposition).** *For any $u\in\mathcal{V}$ there exist $u_i\in\mathcal{V}_i'$ satisfying $u=\sum_{i=0}^J u_i$ such that*

$$\sum_{i=0}^{J} b_i(u_i,u_i)\lesssim\Big(1+\log^2\frac{H}{\delta}\Big)a_{\mathfrak{W}}(u,u), \qquad (10.34)$$

*where the constant is independent of u, H, h, and $\delta$.*

*Proof.* To define a decomposition for $u\in\mathcal{V}$ we need a projection and a partition of unity. Since the operator $-\Delta$ with domain of definition $\widetilde{H}^1(\Gamma)=H_0^1(\Gamma)$ is positive definite and self-adjoint, we can define $\Lambda=\sqrt{-\Delta}$ which in turn is self-adjoint as an operator from $\widetilde{H}_{\mathrm{I}}^{1/2}(\Gamma)$ to $H^{-1/2}(\Gamma)$. Moreover,

$$\langle\Lambda\xi,\xi\rangle\simeq\|\xi\|_{\widetilde{H}_{\mathrm{I}}^{1/2}(\Gamma)}^2 \quad \forall\xi\in\widetilde{H}_{\mathrm{I}}^{1/2}(\Gamma).$$

Let $P_H:\widetilde{H}_{\mathrm{I}}^{1/2}(\Gamma)\longrightarrow\mathcal{V}_0$ be the projection defined by the inner product $\langle\Lambda\cdot,\cdot\rangle$, *i.e.*

$$\langle\Lambda P_H v,w\rangle=\langle\Lambda v,w\rangle \quad \forall v\in\widetilde{H}_{\mathrm{I}}^{1/2}(\Gamma),\ w\in\mathcal{V}_0.$$

Using standard arguments one can prove that for any $v \in \widetilde{H}_{\mathrm{I}}^{1/2}(\Gamma)$ the following estimates hold

$$\|P_H v\|_{\widetilde{H}_{\mathrm{I}}^{1/2}(\Gamma)} \lesssim \|v\|_{\widetilde{H}_{\mathrm{I}}^{1/2}(\Gamma)} \quad \text{and} \quad \|P_H v - v\|_{L^2(\Gamma)} \lesssim H^{1/2} \|v\|_{\widetilde{H}_{\mathrm{I}}^{1/2}(\Gamma)}. \tag{10.35}$$

We next define a partition of unity having the properties of the partition in Lemma 10.17. For $j = 1, \ldots, J$, we define

$$R_j = \Gamma_j \cup \{\mathbf{x} \in \Gamma : \mathrm{dist}(\mathbf{x}, \partial\Gamma_j) \le \delta\},$$

which are quadrilateral domains. It is crucial to define $R_j$ as a quadrilateral domain so that Lemma 10.17 can be applied. The cutoff function $\theta_j$ is defined as in the proof of Lemma 10.17 so that $\operatorname{supp} \theta_j = \overline{R}_j$.

We are now ready to define a decomposition of $u \in \mathscr{V}$ such that (10.34) holds. For any $u \in \mathscr{V}$ let $u_0 = \Pi_h P_H u \in \mathscr{V}_0'$ and $u_j = \Pi_h(\theta_j w) \in \mathscr{V}_j'$, $j = 1, \ldots, J$, where $w = u - u_0$. The fact that $\operatorname{supp} u_j \subset \overline{\Gamma}_j'$ so that $u_j \in \mathscr{V}_j'$ for $j = 1, \ldots, J$ is clear from the definitions of $R_j$ and $\Gamma_j'$. It is also clear that $u = \sum_{i=0}^J u_i$.

We recall from Lemma 10.10 that

$$b_0(u_0, u_0) \lesssim a_{\mathfrak{W}}(u, u). \tag{10.36}$$

Moreover, by using Lemma C.36 and (10.35), we obtain the following estimates

$$\|u_i\|_{\widetilde{H}_{\mathrm{I}}^{1/2}(\Gamma)} \lesssim \|\theta_i w\|_{\widetilde{H}_{\mathrm{I}}^{1/2}(\Gamma)}, \tag{10.37}$$

$$\begin{aligned} \|w\|_{\widetilde{H}_{\mathrm{I}}^{1/2}(\Gamma)} &= \|\Pi_h(u - P_H u)\|_{\widetilde{H}_{\mathrm{I}}^{1/2}(\Gamma)} \lesssim \|u - P_H u\|_{\widetilde{H}_{\mathrm{I}}^{1/2}(\Gamma)} \\ &\lesssim \|u\|_{\widetilde{H}_{\mathrm{I}}^{1/2}(\Gamma)}, \end{aligned} \tag{10.38}$$

$$\begin{aligned} \|w\|_{L^2(\Gamma)} &= \|\Pi_h(u - P_H u)\|_{L^2(\Gamma)} \lesssim \|u - P_H u\|_{L^2(\Gamma)} \\ &\lesssim H^{1/2} \|u\|_{\widetilde{H}_{\mathrm{I}}^{1/2}(\Gamma)}. \end{aligned} \tag{10.39}$$

Moreover, by invoking Lemma A.13, we obtain

$$\|\theta_i w\|_{\widetilde{H}_{\mathrm{I}}^{1/2}(\Gamma)} \lesssim \|\theta_i w\|_{\widetilde{H}_{\mathrm{w}}^{1/2}(R_i)}. \tag{10.40}$$

Hence, by using successively Lemma B.3, (10.36), (10.37), (10.40), (10.32) we deduce

$$\begin{aligned} \sum_{i=0}^J b_i(u_i, u_i) &\simeq \|\Pi_H u_0\|^2_{\widetilde{H}_{\mathrm{I}}^{1/2}(\Gamma)} + \sum_{i=1}^J \|u_i\|^2_{\widetilde{H}_{\mathrm{I}}^{1/2}(\Gamma)} \lesssim \|u\|^2_{\widetilde{H}_{\mathrm{I}}^{1/2}(\Gamma)} + \sum_{i=1}^J \|\theta_i w\|^2_{\widetilde{H}_{\mathrm{I}}^{1/2}(\Gamma)} \\ &\lesssim \|u\|^2_{\widetilde{H}_{\mathrm{I}}^{1/2}(\Gamma)} + \sum_{i=1}^J \|\theta_i w\|^2_{\widetilde{H}_{\mathrm{w}}^{1/2}(R_i)} \end{aligned}$$

$$\lesssim \|u\|^2_{\widetilde{H}_{\mathrm{I}}^{1/2}(\Gamma)} + (1+\log^2\frac{H}{\delta})\sum_{i=1}^{J}\|w\|^2_{H_{\mathrm{w}}^{1/2}(R_i)} + \iint_{\Gamma} \frac{|w(\mathbf{x})|^2}{\operatorname{dist}(\mathbf{x},\partial\Gamma)}\,d\mathbf{x}. \tag{10.41}$$

Using (10.26) and the definition of the weighted norms in Subsection A.2.6.1, we can estimate the last two terms on the right-hand side by

$$\begin{aligned}
&(1+\log^2\frac{H}{\delta})\sum_{i=1}^{J}\|w\|^2_{H_{\mathrm{w}}^{1/2}(R_i)} + \iint_{\Gamma} \frac{|w(x)|^2}{\operatorname{dist}(\mathbf{x},\partial\Gamma)}\,d\mathbf{x} \\
&= (1+\log^2\frac{H}{\delta})\sum_{i=1}^{J}\Big(\frac{1}{H}\|w\|^2_{L^2(R_i)} + |w|^2_{H^{1/2}(R_i)}\Big) + \iint_{\Gamma} \frac{|w(x)|^2}{\operatorname{dist}(\mathbf{x},\partial\Gamma)}\,d\mathbf{x} \\
&\lesssim (1+\log^2\frac{H}{\delta})\Big(\frac{1}{H}\|w\|^2_{L^2(\Gamma)} + |w|^2_{H^{1/2}(\Gamma)}\Big) + \iint_{\Gamma} \frac{|w(\mathbf{x})|^2}{\operatorname{dist}(\mathbf{x},\partial\Gamma)}\,d\mathbf{x} \\
&\lesssim (1+\log^2\frac{H}{\delta})\Big(\frac{1}{H}\|w\|^2_{L^2(\Gamma)} + \|w\|^2_{\widetilde{H}_{\mathrm{w}}^{1/2}(\Gamma)}\Big) \\
&\simeq (1+\log^2\frac{H}{\delta})\Big(\frac{1}{H}\|w\|^2_{L^2(\Gamma)} + \|w\|^2_{\widetilde{H}_{\mathrm{I}}^{1/2}(\Gamma)}\Big).
\end{aligned}$$

Inequalities (10.38) and (10.39) then give

$$(1+\log^2\frac{H}{\delta})\sum_{i=1}^{J}\|w\|^2_{H_{\mathrm{w}}^{1/2}(R_i)} + \iint_{\Gamma} \frac{|w(x)|^2}{\operatorname{dist}(x,\partial\Gamma)}\,\mathrm{d}x \lesssim (1+\log^2\frac{H}{\delta})\|u\|^2_{\widetilde{H}_{\mathrm{I}}^{1/2}(\Gamma)}.$$

Hence we deduce from (10.41) the required estimate. □

**Coercivity of the decomposition:**

Finally we prove the coercivity of the decomposition in the next lemma.

**Lemma 10.19 (Coercivity of decomposition).** *For any $u \in \mathscr{V}$ if $u = \sum_{i=0}^{J} u_i$ for some $u_i \in \mathscr{V}_i'$, then*

$$a_{\mathfrak{W}}(u,u) \lesssim \sum_{i=0}^{J} b_i(u_i,u_i),$$

*where the constant is independent of $u$, $h$, $\delta$, and $H$.*

*Proof.* By construction there are at most $\nu$ subdomains $\overline{\Gamma}_i'$ to which any $x \in \Gamma$ can belong. (The utmost case happens when $\delta = H$.) A standard colouring argument yields, noting that $u_0 = \Pi_h\Pi_H u_0$,

$$\begin{aligned}
a_{\mathfrak{W}}(u,u) \simeq \|u\|^2_{\widetilde{H}_{\mathrm{I}}^{1/2}(\Gamma)} &\lesssim \|u_0\|^2_{\widetilde{H}_{\mathrm{I}}^{1/2}(\Gamma)} + \Big\|\sum_{i=1}^{J} u_i\Big\|^2_{\widetilde{H}_{\mathrm{I}}^{1/2}(\Gamma)} \\
&\lesssim \|\Pi_H u_0\|^2_{\widetilde{H}_{\mathrm{I}}^{1/2}(\Gamma)} + \nu\sum_{i=1}^{J}\|u_i\|^2_{\widetilde{H}_{\mathrm{I}}^{1/2}(\Gamma)} \simeq \sum_{i=0}^{J} b_i(u_i,u_i).
\end{aligned}$$

□

**Theorem 10.2.** *The condition number of the additive Schwarz operator* $P_{\text{ad}}$ *is bounded as*

$$\kappa(P_{\text{ad}}) \lesssim 1 + \log^2(H/\delta).$$

*Proof.* The above estimate is a result of Theorem 2.1, Lemmas 10.18 and 10.19. □

#### 10.2.2.2 The *hp*-version

**Subspace decomposition for the *hp*-version:**
To define an overlapping preconditioner, we will decompose $\mathscr{V}$ as in (10.3) with $\mathscr{V}_p$ decomposed by (10.4), while $\mathscr{V}_h$ is decomposed by subspaces defined on overlapping subdomains $\Gamma_j'$, as in Subsection 10.2.2.1.

The following estimate for the condition number can be shown in the same manner as for the *h*-version. We omit the proof.

**Theorem 10.3.** *The condition number of the additive Schwarz operator P is bounded as*

$$\kappa(P) \lesssim \max_{1\le j\le J} (1 + \log(H_j p_j/\delta))^2,$$

*where* $p_j := \max_{K\subset\Gamma_j} p_K$.

## 10.3 The Weakly-Singular Integral Equation

In this section we analyse a non-overlapping method for the weakly-singular integral equation. Recall the two-level mesh is defined in Subsection 10.1.1. On the fine mesh we define $\widetilde{\mathscr{V}}$ to be the space of piecewise polynomials of degree $p$ on each element $\Gamma_{ij}$, $i = 1,\dots,J$, $j = 1,\dots,N_i$, i.e.,

$$\widetilde{\mathscr{V}} := \left\{v : v|_{\Gamma_{ij}} \in \mathbb{P}_p \ \forall i = 1,\dots,J,\ j = 1,\dots,N_i\right\},$$

and on the coarse mesh, we define the space of piecewise constant functions , i.e.,

$$\widetilde{\mathscr{V}}_0 := \left\{v \in \widetilde{\mathscr{V}} : v|_{\Gamma_i} \in \mathbb{P}_0,\ i = 1,\dots,J\right\}.$$

We further define, for $i = 1,\dots,J$, $j = 1,\dots,N_i$,

$$\widetilde{\mathscr{V}}_i := \left\{v \in \widetilde{\mathscr{V}} : \operatorname{supp} v \subset \overline{\Gamma}_i,\ \langle v, 1\rangle_{L^2(\Gamma_i)} = 0,\ v|_{\Gamma_{ij}} \in \mathbb{P}_0,\ j = 1,\dots,N_i, \right\},$$
$$\widetilde{\mathscr{V}}_{ij} := \left\{v \in \widetilde{\mathscr{V}} : \operatorname{supp} v \subset \overline{\Gamma}_{ij},\ \langle v, 1\rangle_{L^2(\Gamma_{ij})} = 0\right\}.$$

A basis for $\widetilde{\mathscr{V}}_i$ is the set of tensor products of Haar basis functions, whereas a basis for $\widetilde{\mathscr{V}}_{ij}$ is the set of tensor products of Legendre polynomials of degree $q = 2,\dots,p$, scaled to the element $\Gamma_{ij}$.

Consider a function $v \in \widetilde{\mathscr{V}}$. Recalling the mean function defined by (3.13), we define, for $j = 1, \ldots, N_i$ and $i = 1, \ldots, J$,

$$\begin{aligned} v_{ij} &:= v|_{\Gamma_{ij}} - \mu_{\Gamma_{ij}}(v|_{\Gamma_{ij}}), \quad \widetilde{v}_i := \sum_{j=1}^{N_i} \mu_{\Gamma_{ij}}(v|_{\Gamma_{ij}}), \\ v_i &:= \widetilde{v}_i - \mu_{\Gamma_i}(\widetilde{v}_i), \quad v_0 := \sum_{i=1}^{J} \mu_{\Gamma_i}(\widetilde{v}_i). \end{aligned} \tag{10.42}$$

Then $v_0 \in \widetilde{\mathscr{V}}_0$, $v_i \in \widetilde{\mathscr{V}}_i$, $v_{ij} \in \widetilde{\mathscr{V}}_{ij}$, and

$$v = v_0 + \sum_{i=1}^{J} v_i + \sum_{i=1}^{J} \sum_{j=1}^{N_i} v_{ij}. \tag{10.43}$$

Therefore, the space $\widetilde{\mathscr{V}}$ can be decomposed by

$$\widetilde{\mathscr{V}} = \widetilde{\mathscr{V}}_0 \bigoplus_{i=1}^{J} \widetilde{\mathscr{V}}_i \bigoplus_{i=1}^{J} \bigoplus_{j=1}^{N_i} \widetilde{\mathscr{V}}_{ij}. \tag{10.44}$$

We use the same bilinear form $a_{\mathfrak{V}}(\cdot,\cdot)$ on each subspace.

**Lemma 10.20 (Coercivity of decomposition).** *For any $v \in \widetilde{\mathscr{V}}$ with decomposition* (10.43) *the following statement is true*

$$a_{\mathfrak{V}}(v,v) \lesssim a_{\mathfrak{V}}(v_0,v_0) + \sum_{i=1}^{J} a_{\mathfrak{V}}(v_i,v_i) + \sum_{i=1}^{J} \sum_{j=1}^{N_i} a_{\mathfrak{V}}(v_{ij},v_{ij})$$

*where the constant involved is independent of $v$, $H$, $h$, and $p$.*

*Proof.* First we note that $v_i$ and $v_{ij}$ have zero integral mean on $\Gamma_i$ and $\Gamma_{ij}$, respectively. Thanks to Lemma B.3, it suffices to prove

$$\|v\|^2_{\widetilde{H}_{\mathrm{I}}^{-1/2}(\Gamma)} \lesssim \|v_0\|^2_{\widetilde{H}_{\mathrm{I}}^{-1/2}(\Gamma)} + \sum_{i=1}^{J} \|v_i\|^2_{\widetilde{H}_{\mathrm{I}}^{-1/2}(\Gamma)} + \sum_{i=1}^{J} \sum_{j=1}^{N_i} \|v_{ij}\|^2_{\widetilde{H}_{\mathrm{I}}^{-1/2}(\Gamma)}. \tag{10.45}$$

This is obtained by first using the Cauchy–Schwarz inequality to have

$$\|v\|^2_{\widetilde{H}_{\mathrm{I}}^{-1/2}(\Gamma)} \lesssim \|v_0\|^2_{\widetilde{H}_{\mathrm{I}}^{-1/2}(\Gamma)} + \Big\|\sum_{i=1}^{J} v_i\Big\|^2_{\widetilde{H}_{\mathrm{I}}^{-1/2}(\Gamma)} + \Big\|\sum_{i=1}^{J} \sum_{j=1}^{N_i} v_{ij}\Big\|^2_{\widetilde{H}_{\mathrm{I}}^{-1/2}(\Gamma)}$$

and then invoking Theorem A.11 (A.88) to deduce (10.45). The constant depends only on the size of the screen $\Gamma$. □

We follow [108] to prove the stability of the decomposition. However, it is necessary to use the weighted norm ${}_*\|\cdot\|_{\widetilde{H}_{\mathrm{I}}^{-1/2}(\Omega)}$ which is scalable; see Lemma A.7 and compare it with Lemma A.6.

**Lemma 10.21 (Stability of decomposition).** *For any $v \in \widetilde{\mathscr{V}}$ the unique decomposition* (10.43) *satisfies*

$$a_{\mathfrak{V}}(v_0,v_0) + \sum_{i=1}^{J} a_{\mathfrak{V}}(v_i,v_i) + \sum_{i=1}^{J}\sum_{j=1}^{N_i} a_{\mathfrak{V}}(v_{ij},v_{ij}) \lesssim \left(1+\log^2\frac{H}{h}\right)\log^2(p+1)a_{\mathfrak{V}}(v,v)$$

*where the constant involved does not depend on $v$, $H$, $h$, and $p$.*

*Proof.* Again, due to Lemma B.3 it suffices to prove

$$_*\|v_0\|^2_{\widetilde{H}_{\mathrm{I}}^{-1/2}(\Gamma)} + \sum_{i=1}^{J} {}_*\|v_i\|^2_{\widetilde{H}_{\mathrm{I}}^{-1/2}(\Gamma)} + \sum_{i=1}^{J}\sum_{j=1}^{N_i} {}_*\|v_{ij}\|^2_{\widetilde{H}_{\mathrm{I}}^{-1/2}(\Gamma)} \lesssim \left(1+\log^2\frac{H}{h}\right)\log^2(p+1)\,{}_*\|v\|^2_{\widetilde{H}_{\mathrm{I}}^{-1/2}(\Gamma)}.$$

On the other hand, (A.82) and (A.83) imply

$$_*\|v_i\|_{\widetilde{H}_{\mathrm{I}}^{-1/2}(\Gamma)} \le {}_*\|v_i\|_{\widetilde{H}_{\mathrm{I}}^{-1/2}(\Gamma_i)} \quad\text{and}\quad {}_*\|v_{ij}\|_{\widetilde{H}_{\mathrm{I}}^{-1/2}(\Gamma)} \le {}_*\|v_{ij}\|_{\widetilde{H}_{\mathrm{I}}^{-1/2}(\Gamma_{ij})}.$$

Hence it suffices to prove

$$_*\|v_0\|^2_{\widetilde{H}_{\mathrm{I}}^{-1/2}(\Gamma)} + \sum_{i=1}^{J} {}_*\|v_i\|^2_{\widetilde{H}_{\mathrm{I}}^{-1/2}(\Gamma_i)} + \sum_{i=1}^{J}\sum_{j=1}^{N_i} {}_*\|v_{ij}\|^2_{\widetilde{H}_{\mathrm{I}}^{-1/2}(\Gamma_{ij})} \lesssim \left(1+\log^2\frac{H}{h}\right)\log^2(p+1)\,{}_*\|v\|^2_{\widetilde{H}_{\mathrm{I}}^{-1/2}(\Gamma)}. \tag{10.46}$$

Let

$$T_1 := \sum_{i=1}^{J} {}_*\|v_i\|^2_{\widetilde{H}_{\mathrm{I}}^{-1/2}(\Gamma_i)} \quad\text{and}\quad T_2 := \sum_{i=1}^{J}\sum_{j=1}^{N_i} {}_*\|v_{ij}\|^2_{\widetilde{H}_{\mathrm{I}}^{-1/2}(\Gamma_{ij})}.$$

We will first show that $T_1$ and $T_2$ are bounded by the right-hand side of (10.46) and then show that $_*\|v_0\|^2_{\widetilde{H}_{\mathrm{I}}^{-1/2}(\Gamma)}$ has the same property. Consider $T_1$. Lemma A.7 Part (ii) gives

$$_*\|v_i\|^2_{\widetilde{H}_{\mathrm{I}}^{-1/2}(\Gamma_i)} = H_i^3\,{}_*\|\widehat{v}_i\|^2_{\widetilde{H}_{\mathrm{I}}^{-1/2}(\widehat{\Gamma}_i)} \le H_i^3\,{}_*\|\widehat{v}_i\|^2_{\widetilde{H}_{\mathrm{I}}^{-1/2+\varepsilon}(\widehat{\Gamma}_i)} = H_i^{2\varepsilon}\,{}_*\|v_i\|^2_{\widetilde{H}_{\mathrm{I}}^{-1/2+\varepsilon}(\Gamma_i)}. \tag{10.47}$$

Now since $v_i = \widetilde{v}_i - \mu_{\Gamma_i}(\widetilde{v}_i)$, see (10.42), and $\langle \mu_{\Gamma_i}(\widetilde{v}_i), \varphi\rangle_{\Gamma_i} = \langle \widetilde{v}_i, \mu_{\Gamma_i}(\varphi)\rangle_{\Gamma_i}$ we have

$$_*\|v_i\|_{\widetilde{H}_{\mathrm{I}}^{-1/2+\varepsilon}(\Gamma_i)} = \sup_{\varphi\in H_{\mathrm{I}}^{1/2-\varepsilon}(\Gamma_i)} \frac{\langle \widetilde{v}_i - \mu_{\Gamma_i}(\widetilde{v}_i), \varphi\rangle_{\Gamma_i}}{_*\|\varphi\|_{H_{\mathrm{I}}^{1/2-\varepsilon}(\Gamma_i)}} = \sup_{\varphi\in H_{\mathrm{I}}^{1/2-\varepsilon}(\Gamma_i)} \frac{\langle \widetilde{v}_i, \varphi - \mu_{\Gamma_i}(\varphi)\rangle_{\Gamma_i}}{_*\|\varphi\|_{H_{\mathrm{I}}^{1/2-\varepsilon}(\Gamma_i)}}$$

$$\leq \|\widetilde{v}_i\|_{H_{\mathrm{I}}^{-1/2+\varepsilon}(\Gamma_i)} \sup_{\varphi\in H_{\mathrm{I}}^{1/2-\varepsilon}(\Gamma_i)} \frac{\|\varphi - \mu_{\Gamma_i}(\varphi)\|_{\widetilde{H}_{\mathrm{I}}^{1/2-\varepsilon}(\Gamma_i)}}{{}_*\|\varphi\|_{H_{\mathrm{I}}^{1/2-\varepsilon}(\Gamma_i)}}. \tag{10.48}$$

Lemma A.7 and Lemma A.10 yield

$$\begin{aligned}\|\varphi - \mu_{\Gamma_i}(\varphi)\|_{\widetilde{H}_{\mathrm{I}}^{1/2-\varepsilon}(\Gamma_i)} &\simeq H_i^{1/2+\varepsilon}\|\widehat{\varphi} - \widehat{\mu_{\Gamma_i}(\varphi)}\|_{\widetilde{H}_{\mathrm{I}}^{1/2-\varepsilon}(\widehat{\Gamma}_i)} \\ &\lesssim \frac{H_i^{1/2+\varepsilon}}{\varepsilon}\|\widehat{\varphi} - \widehat{\mu_{\Gamma_i}(\varphi)}\|_{H_{\mathrm{I}}^{1/2-\varepsilon}(\widehat{\Gamma}_i)} \\ &\simeq \frac{H_i^{1/2+\varepsilon}}{\varepsilon}\|\widehat{\varphi} - \mu_{\widehat{\Gamma}_i}(\widehat{\varphi})\|_{H_{\mathrm{S}}^{1/2-\varepsilon}(\widehat{\Gamma}_i)}\end{aligned} \tag{10.49}$$

where in the last step we used Theorem A.7 and Theorem A.9. The constants are independent of $H_i$ and $\varepsilon$ for $0 < \varepsilon \leq \varepsilon_0 < 1/2$. Since

$$\begin{aligned}\|\mu_{\widehat{\Gamma}_i}(\widehat{\varphi})\|_{H_{\mathrm{S}}^{1/2-\varepsilon}(\widehat{\Gamma}_i)} &= \left|\langle\widehat{\varphi}, 1\rangle_{\widehat{\Gamma}_i}\right| \, \|1\|_{H_{\mathrm{S}}^{1/2-\varepsilon}(\widehat{\Gamma}_i)} \\ &\leq \|\widehat{\varphi}\|_{H_{\mathrm{S}}^{1/2-\varepsilon}(\widehat{\Gamma}_i)} \, \|1\|_{\widetilde{H}_{\mathrm{S}}^{-1/2+\varepsilon}(\widehat{\Gamma}_i)} \, \|1\|_{H_{\mathrm{S}}^{1/2-\varepsilon}(\widehat{\Gamma}_i)} \\ &\simeq \|\widehat{\varphi}\|_{H_{\mathrm{S}}^{1/2-\varepsilon}(\widehat{\Gamma}_i)},\end{aligned}$$

we have $\|\widehat{\varphi} - \mu_{\widehat{\Gamma}_i}(\widehat{\varphi})\|_{H_{\mathrm{S}}^{1/2-\varepsilon}(\widehat{\Gamma}_i)} \lesssim \|\widehat{\varphi}\|_{H_{\mathrm{S}}^{1/2-\varepsilon}(\widehat{\Gamma}_i)} \simeq {}_*\|\widehat{\varphi}\|_{H_{\mathrm{I}}^{1/2-\varepsilon}(\widehat{\Gamma}_i)}$ which together with (10.49) implies

$$\|\varphi - \mu_{\Gamma_i}(\varphi)\|_{\widetilde{H}_{\mathrm{I}}^{1/2-\varepsilon}(\Gamma_i)} \lesssim \frac{H_i^{1/2+\varepsilon}}{\varepsilon} {}_*\|\widehat{\varphi}\|_{H_{\mathrm{I}}^{1/2-\varepsilon}(\widehat{\Gamma}_i)} = \frac{1}{\varepsilon} {}_*\|\varphi\|_{H_{\mathrm{I}}^{1/2-\varepsilon}(\Gamma_i)}$$

where in the last step we used the scaling property proved in Lemma A.7. Therefore, it follows from (10.47) and (10.48) that

$${}_*\|v_i\|^2_{\widetilde{H}_{\mathrm{I}}^{-1/2}(\Gamma_i)} \lesssim \frac{H_i^{2\varepsilon}}{\varepsilon^2}\|\widetilde{v}_i\|^2_{H_{\mathrm{I}}^{-1/2+\varepsilon}(\Gamma_i)}, \tag{10.50}$$

which implies, with the help of Theorem A.10 and Lemma C.18,

$$\begin{aligned}T_1 &\lesssim \frac{H^{2\varepsilon}}{\varepsilon^2}\Big\|\sum_{i=1}^{J}\widetilde{v}_i\Big\|^2_{H_{\mathrm{I}}^{-1/2+\varepsilon}(\Gamma)} \lesssim \left(\frac{H}{h}\right)^{2\varepsilon}\frac{1}{\varepsilon^2}\Big\|\sum_{i=1}^{J}\widetilde{v}_i\Big\|^2_{H_{\mathrm{I}}^{-1/2}(\Gamma)} \\ &\lesssim \left(\frac{H}{h}\right)^{2\varepsilon}\frac{1}{\varepsilon^2}\Big\|\sum_{i=1}^{J}\widetilde{v}_i\Big\|^2_{\widetilde{H}_{\mathrm{I}}^{-1/2}(\Gamma)}.\end{aligned}$$

By choosing

$$\varepsilon = \frac{1}{2\log(H/h)}$$

we deduce

$$T_1 \lesssim \log^2\frac{H}{h}\|\sum_{i=1}^{J}\widetilde{v}_i\|^2_{\widetilde{H}_{\mathrm{I}}^{-1/2}(\Gamma)} \le \log^2\frac{H}{h}\|v-\sum_{i=1}^{J}\sum_{j=1}^{N_i}v_{ij}\|^2_{\widetilde{H}_{\mathrm{I}}^{-1/2}(\Gamma)}$$

$$\lesssim \log^2\left(\frac{H}{h}\right)\left(\|v\|^2_{\widetilde{H}_{\mathrm{I}}^{-1/2}(\Gamma)}+\sum_{i=1}^{J}\sum_{j=1}^{N_i}\|v_{ij}\|^2_{\widetilde{H}_{\mathrm{I}}^{-1/2}(\Gamma_{ij})}\right).$$

Next we treat $T_2$ in the same manner. Instead of (10.50) we now have, noting (10.42),

$$_*\|v_{ij}\|^2_{\widetilde{H}_{\mathrm{I}}^{-1/2}(\Gamma_{ij})} \lesssim \frac{h_{ij}^{2\varepsilon}}{\varepsilon^2}\|v|_{\Gamma_{ij}}\|^2_{H_{\mathrm{I}}^{-1/2+\varepsilon}(\Gamma_{ij})}$$

so that, using again Theorem A.10 and Lemma C.18,

$$T_2 \lesssim \frac{h^{2\varepsilon}}{\varepsilon^2}\sum_{i=1}^{J}\sum_{j=1}^{N_i}\|v|_{\Gamma_{ij}}\|^2_{H_{\mathrm{I}}^{-1/2+\varepsilon}(\Gamma_{ij})} \le \frac{h^{2\varepsilon}}{\varepsilon^2}\|v\|^2_{H_{\mathrm{I}}^{-1/2+\varepsilon}(\Gamma)}$$

$$\le \frac{p^{4\varepsilon}}{\varepsilon^2}\|v\|^2_{H_{\mathrm{I}}^{-1/2}(\Gamma)} \lesssim \frac{p^{4\varepsilon}}{\varepsilon^2}\|v\|^2_{\widetilde{H}_{\mathrm{I}}^{-1/2}(\Gamma)}.$$

Choosing

$$\varepsilon = \frac{1}{2\log(1+p^2)}$$

we deduce

$$T_2 \lesssim \log^2(1+p)\|v\|^2_{\widetilde{H}_{\mathrm{I}}^{-1/2}(\Gamma)}.$$

Therefore,

$$T_1+T_2 \lesssim \log^2\left(\frac{H}{h}\right)\log^2(1+p)\|v\|^2_{\widetilde{H}_{\mathrm{I}}^{-1/2}(\Gamma)}.$$

By the triangle inequality, we have

$$\|v_0\|^2_{\widetilde{H}_{\mathrm{I}}^{-1/2}(\Gamma)} = \|v-\sum_{i=1}^{J}v_i-\sum_{i=1}^{J}\sum_{j=1}^{N_i}v_{ij}\|^2_{\widetilde{H}_{\mathrm{I}}^{-1/2}(\Gamma)}$$

$$\lesssim \|v\|^2_{\widetilde{H}_{\mathrm{I}}^{-1/2}(\Gamma)}+T_1+T_2$$

$$\lesssim \log^2\left(\frac{H}{h}\right)\log^2(1+p)\|v\|^2_{\widetilde{H}_{\mathrm{I}}^{-1/2}(\Gamma)}.$$

Finally, we need to prove $\|v\|^2_{\widetilde{H}_{\mathrm{I}}^{-1/2}(\Gamma)} \lesssim {}_*\|v\|^2_{\widetilde{H}_{\mathrm{I}}^{-1/2}(\Gamma)}$. This can be done by using duality arguments, noting that

$$\|v\|^2_{H_{\mathrm{I}}^{1/2}(\Gamma)} \ge |\Gamma|^2\|\widehat{v}\|^2_{H_{\mathrm{I}}^{1/2}(\widehat{\Gamma})} = |\Gamma|^2{}_*\|\widehat{v}\|^2_{H_{\mathrm{I}}^{1/2}(\widehat{\Gamma})} = |\Gamma|\,{}_*\|v\|^2_{H_{\mathrm{I}}^{1/2}(\Gamma)}.$$

Altogether we obtain (10.46). □

By using the general theory, we deduce from Lemma 10.20 and Lemma 10.21 the following theorem.

**Theorem 10.4.** *The condition number of the additive Schwarz operator $P_{\text{ad}}$ defined by the subspace decomposition* (10.44) *is bounded by*

$$\kappa(P_{\text{ad}}) \lesssim \left(1 + \log^2 \frac{H}{h}\right) \log^2(1+p).$$

*Remark 10.1.* It is noted that [108, Theorem 1] contains the same upper bound as that given in Theorem 10.4. However, the decomposition defined in this article, see [108, Equation (6)], results in an additive Schwarz operator for the $p$-version with a two-level mesh. (Compare that equation with (10.44).) For the $p$-version, it is a common practice to design additive Schwarz preconditioners on a one-level mesh. Therefore, the method proposed in this article can be considered as an additive Schwarz method for the $p$-version on a two-level mesh (if we keep $H$ and $h$ fixed), and it can also be considered as an additive Schwarz preconditioner for the $h$-version with a high polynomial order (if we keep $p$ fixed). The subproblems will be large if all $H$, $h$, and $p$ vary.

## 10.4 Numerical Results

The following numerical experiments were initially presented in [118] and [230]. The results are reprinted here with copyright permission. As in the case of two-dimensional problems (Chapter 9), the preconditioned conjugate gradient method is used to solve the preconditioned systems, and the Lanczos algorithm (the QR-method) is used in conjunction with the PCG to compute their eigenvalues.

### *10.4.1 The hypersingular integral equation*

In this section, we report on the numerical results when the non-overlapping additive Schwarz preconditioner is used to solve the hypersingular integral equation on the screen $(-1/2,1/2)^2 \times \{0\}$.

It is noted that the entries of the stiffness matrices are computed by analytical recurrence formulae; see [141]. For the CG method with the unpreconditioned system, we computed the stiffness matrices by using two different types of basis functions: the standard basis functions defined in Section 10.1 and the hierarchical basis functions with antiderivatives of Legendre polynomials; see Section C.3. For the PCG, we used two different bilinear forms on the wirebasket spaces $\mathscr{V}_{\mathscr{W}_h}$ and $\mathscr{V}_{\mathscr{W}_p}$, namely the bilinear form $a_{\mathfrak{W}}(\cdot,\cdot)$ and the bilinear forms (10.8) and (10.9). Both the CG and PCG stop when the residual is less than $10^{-6}$.

| $1/h$ | DoF | Number of iterations | | | | | |
|---|---|---|---|---|---|---|---|
| | | CG | | PCG | | | |
| | | Standard basis | Hierarchical basis | $H/h=1$ | | $H/h=2$ | $H/h=4$ |
| | | | | $a_{\mathfrak{W}}(\cdot,\cdot)$ | (10.8), (10.9) | | |
| 2 | 9 | 3 | 3 | 3 | 3 | 3 | |
| 4 | 49 | 10 | 10 | 10 | 9 | 10 | 9 |
| 6 | 121 | 25 | 16 | 12 | 13 | 10 | |
| 8 | 225 | 40 | 20 | 13 | 13 | 12 | 11 |
| 10 | 361 | 49 | 24 | 13 | 14 | 11 | |
| 12 | 529 | 56 | 27 | 13 | 14 | 12 | 12 |
| 14 | 729 | 58 | 29 | 13 | 14 | 12 | |
| 16 | 961 | 64 | 32 | 13 | 14 | 12 | 12 |

**Table 10.1** Number of iterations of CG and PCG, *hp*-version with $p=2$

| $p$ | DoF | Number of iterations | | | |
|---|---|---|---|---|---|
| | | CG | | PCG | |
| | | Standard basis | Hierarchical basis | $a_{\mathfrak{W}}(\cdot,\cdot)$ | $L^2$-inner product |
| 2 | 5 | 4 | 4 | 4 | 3 |
| 3 | 12 | 8 | 8 | 7 | 7 |
| 4 | 32 | 25 | 22 | 10 | 10 |
| 5 | 70 | 52 | 47 | 12 | 12 |
| 6 | 127 | 104 | 80 | 13 | 13 |
| 7 | 202 | 156 | 111 | 14 | 14 |
| 8 | 295 | 227 | 151 | 14 | 14 |

**Table 10.2** Number of iterations of CG and PCG, *hp*-version with $1/h=3$

### *10.4.2 The Lamé equation*

We finish this chapter with some numerical results for the crack problem (Lamé equation) to demonstrate that the overlapping preconditioner (Theorem 10.2) also work for systems of equations. The equation is of the form (1.7) with

$$\mathbf{g}(x,y)=(-y,x,0).$$

| DoF | $h$ | CG | PCG | | | |
|---|---|---|---|---|---|---|
| | | | $H=2h$ | | $H=4h$ | |
| | | | $\delta=h$ | $\delta=2h$ | $\delta=h$ | $\delta=2h$ |
| 147 | 1/8 | 4.14 | 2.38 | 4.11 | | |
| 675 | 1/16 | 8.24 | 2.78 | 4.27 | 5.15 | 9.01 |
| 2883 | 1/32 | 16.50 | 3.88 | 4.31 | 5.43 | 9.01 |
| 11907 | 1/64 | 33.07 | 5.36 | 5.88 | 5.56 | 9.01 |

**Table 10.3** Condition number for Lamé equation

| DoF | $h$ | CG | PCG | | | |
|---|---|---|---|---|---|---|
| | | | $H=2h$ | | $H=4h$ | |
| | | | $\delta=h$ | $\delta=2h$ | $\delta=h$ | $\delta=2h$ |
| 147 | 1/8 | 24 | 13 | 16 | | |
| 675 | 1/16 | 34 | 16 | 19 | 20 | 22 |
| 2883 | 1/32 | 48 | 19 | 21 | 22 | 25 |
| 11907 | 1/64 | 68 | 23 | 23 | 24 | 26 |

**Table 10.4** Number of iterations for Lamé equation

# Chapter 11
# Two-Level Methods: the $hp$-Version on Triangular Elements

In this chapter we report on the paper [111] by Heuer, Leydecker, and Stephan, which analyses a two-level additive Schwarz method for the $hp$-version on triangular elements applied to the hypersingular integral equation on a screen $\Gamma$.

## 11.1 Preliminaries

### *11.1.1 Sobolev spaces of functions vanishing on a part of the boundary of a domain*

In this chapter, besides the Sobolev spaces defined in Appendix A, we also use the following Sobolev spaces.

**Definition 11.1.** *Let $\Omega$ be a Lipschitz domain in $\mathbb{R}^2$ or a bounded interval in $\mathbb{R}$, and $\gamma$ be a subset of $\partial\Omega$, the boundary of $\Omega$. (In the case of a bounded interval $\Omega$, the subset $\gamma$ is one of the two endpoints of $\Omega$.) We define*

$$\widetilde{H}^{1/2}(\Omega,\gamma) := \left\{ v \in H^{1/2}(\Omega) : |||v|||_{\widetilde{H}^{1/2}(\Omega,\gamma)} < \infty \right\},$$

*where*

$$|||v|||^2_{\widetilde{H}^{1/2}(\Omega,\gamma)} := |v|^2_{H^{1/2}(\Omega)} + \iint\limits_{\Omega} \frac{|v(\mathbf{x})|^2}{\operatorname{dist}(\mathbf{x},\gamma)} d\mathbf{x}.$$

We note that if $v \in \widetilde{H}^{1/2}(\Omega,\gamma)$, then $v$ vanishes on $\gamma$ but not on the whole boundary $\partial\Omega$. So this norm is a special case of the weighted norm defined in Section A.2.6.1. A proof similar to that of Lemma A.5 gives

$$|||v|||^2_{\widetilde{H}^{1/2}(\Omega,\gamma)} = \tau |||\widehat{v}|||^2_{\widetilde{H}^{1/2}(\widehat{\Omega},\widehat{\gamma})} \tag{11.1}$$



E. P. Stephan and T. Tran, *Schwarz Methods and Multilevel Preconditioners for Boundary Element Methods*, https://doi.org/10.1007/978-3-030-79283-1_11

where $\widehat{\Omega} = \{\widehat{\mathbf{x}} \in \mathbb{R}^2 : \widehat{\mathbf{x}} = \mathbf{x}/\tau,\ \mathbf{x} \in \Omega\}$, $\widehat{\gamma} = \{\widehat{\mathbf{x}} \in \mathbb{R}^2 : \widehat{\mathbf{x}} = \mathbf{x}/\tau,\ \mathbf{x} \in \gamma\}$, and $\widehat{v}$ is defined by $\widehat{v}(\widehat{\mathbf{x}}) = v(\mathbf{x})$. Here $\tau = \mathrm{diam}(\Omega)$, and we assume that $\tau < 1$.

### 11.1.2 Extension operators

As in Chapter 10, the proposed preconditioner requires the use of special basis functions which are nodal basis functions, edge basis functions and interior or bubble functions (associated with vertices, edges and elements, respectively). The nodal and edge basis functions to be used in this chapter are constructed by extensions of polynomials from edges onto elements. The following specific extension operators will be used.

We first define these extensions on the reference triangle. Let

$$\begin{aligned}
\widehat{I} &:= (0,1), \\
\widehat{T} &:= \{(x,y) \in \mathbb{R}^2 : x \geq 0,\ y \geq 0,\ x+y \leq 1\}, \\
\mathbb{P}_p(\widehat{I}) &:= \mathrm{span}\{x^i : i = 0,1,\ldots,p\}, \\
\mathbb{P}_p^0(\widehat{I}) &:= \{v \in \mathbb{P}_p(\widehat{I} : v(0) = v(1) = 0\}, \\
\mathbb{P}_p(\widehat{T}) &:= \mathrm{span}\{x^i y^j : i,j = 0,1,\ldots,p,\ i+j \leq p\}.
\end{aligned}$$

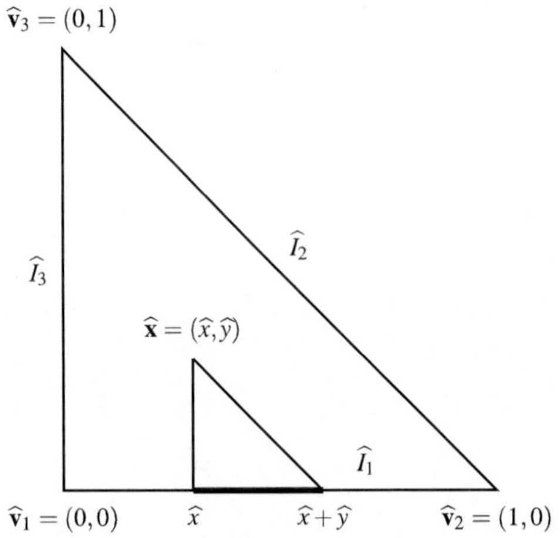


**Fig. 11.1** The reference triangle $\widehat{T}$

The vertices and edges of $\widehat{T}$ are denoted by $\widehat{\mathbf{v}}_1$, $\widehat{\mathbf{v}}_2$, $\widehat{\mathbf{v}}_3$, and $\widehat{I}_1$, $\widehat{I}_2$, $\widehat{I}_3$, respectively.

We first recall the following extension operators frequently used in the finite element analysis, see [18, 19, 139, 196],

$$\mathscr{F} : \mathbb{P}_p(\widehat{I}) \longrightarrow \mathbb{P}_p(\widehat{T}), \quad \mathscr{F}\varphi(\widehat{x},\widehat{y}) := \frac{1}{\widehat{y}} \int_{\widehat{x}}^{\widehat{x}+\widehat{y}} \varphi(t)\,\mathrm{d}t,$$

$$\mathscr{E} : \mathbb{P}_p(\widehat{I}) \longrightarrow \mathbb{P}_p(\widehat{T}), \quad \mathscr{E}\varphi(\widehat{x},\widehat{y}) := \frac{\widehat{x}}{\widehat{y}} \int_{\widehat{x}}^{\widehat{x}+\widehat{y}} \frac{\varphi(t)}{t}\,\mathrm{d}t \quad \text{if } \varphi(0)=0.$$

We now define the operator $\mathscr{E}^j : \mathbb{P}_p^0(\widehat{I}_j) \to \mathbb{P}_p(\widehat{T})$, $j=1,2,3$, which extends a polynomial $f \in \mathbb{P}_p(\widehat{I}_j)$ vanishing at the endpoints of $\widehat{I}_j$ into a complete polynomial in $\mathbb{P}_p(\widehat{T})$ vanishing on the other two edges $I_i$, $i \neq j$. More precisely,

$$\mathscr{E}^1 f(\widehat{x},\widehat{y}) := \frac{\widehat{x}(1-\widehat{x}-\widehat{y})}{\widehat{y}} \int_{\widehat{x}}^{\widehat{x}+\widehat{y}} \frac{f(t,0)}{t(1-t)}\,\mathrm{d}t, \quad \text{if } f(0,0)=f(1,0)=0,$$

$$\mathscr{E}^2 f(\widehat{x},\widehat{y}) := \frac{\widehat{xy}}{1-\widehat{x}-\widehat{y}} \int_{\widehat{x}}^{1-\widehat{y}} \frac{f(t,1-t)}{t(1-t)}\,\mathrm{d}t, \quad \text{if } f(0,1)=f(1,0)=0,$$

$$\mathscr{E}^3 f(\widehat{x},\widehat{y}) := \frac{\widehat{y}(1-\widehat{x}-\widehat{y})}{\widehat{x}} \int_{\widehat{x}}^{\widehat{x}+\widehat{y}} \frac{f(0,t)}{t(1-t)}\,\mathrm{d}t, \quad \text{if } f(0,0)=f(0,1)=0.$$

The above extension operators require that the given polynomial vanishes at both endpoints of the corresponding edge. We will need other extension operators which only require that the given polynomial vanishes at one endpoint of the edge. For a systematic description, we introduce the index addition $\dotplus$ which is a modular addition

$$i \dotplus k = i+k \,(\text{mod } 3), \quad i,k=1,2,3. \tag{11.2}$$

Under the convention (11.2), for $j=1,2,3$, an edge $\widehat{I}_j$ has as endpoints $\widehat{\mathbf{v}}_j$ and $\widehat{\mathbf{v}}_{j\dotplus 1}$, while a vertex $\widehat{\mathbf{v}}_j$ is shared by two edges $\widehat{I}_j$ and $\widehat{I}_{j\dotplus 1}$; see Figure 11.1.

For $j \in \{1,2,3\}$, let $i=j$ or $i=j\dotplus 1$ and

$$\mathbb{P}_p^i(\widehat{I}_j) = \{v \in \mathbb{P}_p(\widehat{I}_j) \,:\, v \text{ vanishes at the endpoint } \mathbf{v}_i \text{ of } \widehat{I}_j\}. \tag{11.3}$$

Two new types of extension operators are introduced: $\mathscr{E}_j^j : \mathbb{P}_p^j(\widehat{I}_j) \to \mathbb{R}$ and $\mathscr{E}_{j\dotplus 1}^j : \mathbb{P}_p^{j\dotplus 1}(\widehat{I}_j) \to \mathbb{R}$. These operators are explicitly defined by

$$\mathscr{E}_1^1 f(\widehat{x},\widehat{y}) := \frac{\widehat{x}}{\widehat{y}} \int_{\widehat{x}}^{\widehat{x}+\widehat{y}} \frac{f(t,0)}{t}\,\mathrm{d}t, \qquad \text{if } f(\widehat{\mathbf{v}}_1)=f(0,0)=0,$$

$$\mathscr{E}_2^1 f(\widehat{x},\widehat{y}) := \frac{1-\widehat{x}-\widehat{y}}{\widehat{y}} \int_{\widehat{x}}^{\widehat{x}+\widehat{y}} \frac{f(t,0)}{1-t}\,dt, \qquad \text{if } f(\widehat{\mathbf{v}}_2) = f(1,0) = 0,$$

$$\mathscr{E}_2^2 f(\widehat{x},\widehat{y}) := \frac{\widehat{y}}{1-\widehat{x}-\widehat{y}} \int_{\widehat{x}}^{1-\widehat{y}} \frac{f(t,1-t)}{1-t}\,dt, \quad \text{if } f(\widehat{\mathbf{v}}_2) = f(1,0) = 0,$$

$$\mathscr{E}_3^2 f(\widehat{x},\widehat{y}) := \frac{\widehat{x}}{1-\widehat{x}-\widehat{y}} \int_{\widehat{x}}^{1-\widehat{y}} \frac{f(t,1-t)}{t}\,dt, \quad \text{if } f(\widehat{\mathbf{v}}_3) = f(0,1) = 0,$$

$$\mathscr{E}_3^3 f(\widehat{x},\widehat{y}) := \frac{1-\widehat{x}-\widehat{y}}{\widehat{x}} \int_{\widehat{x}}^{\widehat{x}+\widehat{y}} \frac{f(0,t)}{1-t}\,dt, \qquad \text{if } f(\widehat{\mathbf{v}}_3) = f(0,1) = 0,$$

$$\mathscr{E}_1^3 f(\widehat{x},\widehat{y}) := \frac{\widehat{y}}{\widehat{x}} \int_{\widehat{x}}^{\widehat{x}+\widehat{y}} \frac{f(0,t)}{t}\,dt, \qquad \text{if } f(\widehat{\mathbf{v}}_1) = f(0,0) = 0.$$

We note that

$$\begin{aligned}
\mathscr{E}^1 f(\widehat{x},\widehat{y}) &= (1-\widehat{x}-\widehat{y})\mathscr{E}_1^1 f(\widehat{x},\widehat{y}) + \widehat{x}\mathscr{E}_2^1 f(\widehat{x},\widehat{y}),\\
\mathscr{E}^2 f(\widehat{x},\widehat{y}) &= \widehat{x}\mathscr{E}_2^2 f(\widehat{x},\widehat{y}) + \widehat{y}\mathscr{E}_3^2 f(\widehat{x},\widehat{y}),\\
\mathscr{E}^3 f(\widehat{x},\widehat{y}) &= \widehat{y}\mathscr{E}_3^3 f(\widehat{x},\widehat{y}) + (1-\widehat{x}-\widehat{y})\mathscr{E}_1^3 f(\widehat{x},\widehat{y}).
\end{aligned} \tag{11.4}$$

Moreover, for $j = 1,2,3$,

$$\begin{aligned}
\mathscr{E}^j f &= 0 \quad \text{on} \quad \widehat{I}_{j+1} \cup \widehat{I}_{j+2},\\
\mathscr{E}_j^j f &= 0 \quad \text{on} \quad \widehat{I}_{j+2},\\
\mathscr{E}_{j+1}^j f &= 0 \quad \text{on} \quad \widehat{I}_{j+1}.
\end{aligned} \tag{11.5}$$

It is easy to see that all the operators defined above are extension operators which extend polynomials of degree $p$ on an edge into polynomials of degree $p$ on $\widehat{T}$. Furthermore, all the operators which deal with polynomials that vanish at only one vertex, namely $\mathscr{E}_j^j$ and $\mathscr{E}_{j+1}^j$ for $j = 1,2,3$, are linear transformations of the operator $\mathscr{E} = \mathscr{E}_1^1$. To see that $\mathscr{E}_1^1 f$ is the extension of $f$ (defined on $\widehat{I}_1$) we use the mean value theorem to obtain

$$\begin{aligned}
\mathscr{E}_1^1 f(\widehat{x},\widehat{y})\Big|_{\widehat{I}_1} &= \lim_{\widehat{y}\to 0} \mathscr{E}_1^1 f(\widehat{x},\widehat{y})\\
&= \lim_{\widehat{y}\to 0} \frac{\widehat{x}}{\widehat{y}} \int_{\widehat{x}}^{\widehat{x}+\widehat{y}} \frac{f(t,0)}{t}\,dt = f(\widehat{x},0) \quad \forall \widehat{x} \in [0,1].
\end{aligned}$$

The main results concerning these operators are shown in the next theorems where bounds in different norms are obtained. The following theorem states the bounds in the $L^2$-norm.

**Theorem 11.1.** *For $j = 1, 2, 3$, the following estimates hold*

$$\|\mathscr{E}_j^j f\|_{L^2(\widehat{T})} \lesssim \|f\|_{L^2(\widehat{I}_j)} \quad \forall f \in \mathbb{P}_p^j(\widehat{I}_j), \tag{11.6}$$

$$\|\mathscr{E}_{j+1}^j f\|_{L^2(\widehat{T})} \lesssim \|f\|_{L^2(\widehat{I}_j)} \quad \forall f \in \mathbb{P}_p^{j+1}(\widehat{I}_j), \tag{11.7}$$

$$\|\mathscr{E}^j f\|_{L^2(\widehat{T})} \lesssim \|f\|_{L^2(\widehat{I}_j)} \quad \forall f \in \mathbb{P}_p^0(\widehat{I}_j). \tag{11.8}$$

*Moreover,*

$$\|\mathscr{E}_j^j f\|_{L^2(\widehat{I}_{j+1})} \lesssim \|f\|_{L^2(\widehat{I}_j)} \quad \forall f \in \mathbb{P}_p^j(\widehat{I}_j), \tag{11.9}$$

$$\|\mathscr{E}_{j+1}^j f\|_{L^2(\widehat{I}_{j+2})} \lesssim \|f\|_{L^2(\widehat{I}_j)} \quad \forall f \in \mathbb{P}_p^{j+1}(\widehat{I}_j). \tag{11.10}$$

*The constants are independent of $f$ and $p$.*

*Proof.* We prove the results for $j = 1$ only. The other cases can be proved similarly. For the reader's convenience, we first restate (11.6)–(11.8) for this value of $j$:

$$\|\mathscr{E}_1^1 f\|_{L^2(\widehat{T})} \lesssim \|f\|_{L^2(\widehat{I}_1)} \quad \forall f \in \mathbb{P}_p(\widehat{I}_1),\ f(\widehat{\mathbf{v}}_1) = 0,$$

$$\|\mathscr{E}_2^1 f\|_{L^2(\widehat{T})} \lesssim \|f\|_{L^2(\widehat{I}_1)} \quad \forall f \in \mathbb{P}_p(\widehat{I}_1),\ f(\widehat{\mathbf{v}}_2) = 0,$$

$$\|\mathscr{E}^1 f\|_{L^2(\widehat{T})} \lesssim \|f\|_{L^2(\widehat{I}_1)} \quad \forall f \in \mathbb{P}_p(\widehat{I}_1),\ f(\widehat{\mathbf{v}}_1) = f(\widehat{\mathbf{v}}_2) = 0,$$

By definition

$$\begin{aligned}\|\mathscr{E}_1^1 f\|_{L^2(\widehat{T})}^2 &= \int_0^1 \int_0^{1-\widehat{x}} \left(\frac{\widehat{x}}{\widehat{y}}\right)^2 \left(\int_{\widehat{x}}^{\widehat{x}+\widehat{y}} \frac{f(t,0)}{t}\,\mathrm{d}t\right)^2 d\widehat{y}\,d\widehat{x} \\ &\le \int_0^1 \int_0^{1-\widehat{x}} \left(\frac{1}{\widehat{y}}\right)^2 \left(\int_{\widehat{x}}^{\widehat{x}+\widehat{y}} f(t,0)\,\mathrm{d}t\right)^2 d\widehat{y}\,d\widehat{x}\end{aligned}$$

where we used $(\widehat{x}/t)^2 \le 1$ for $\widehat{x} \le t \le \widehat{x}+\widehat{y}$. Invoking Lemma C.52 we obtain (11.6). Analogous arguments yield (11.7), (11.8).

Next we prove (11.9)–(11.10) for $j = 1$, which now become

$$\|\mathscr{E}_1^1 f\|_{L^2(\widehat{I}_2)} \lesssim \|f\|_{L^2(\widehat{I}_1)} \quad \forall f \in \mathbb{P}_p(\widehat{I}_1),\ f(\widehat{\mathbf{v}}_1) = 0, \tag{11.11}$$

$$\|\mathscr{E}_2^1 f\|_{L^2(\widehat{I}_3)} \lesssim \|f\|_{L^2(\widehat{I}_1)} \quad \forall f \in \mathbb{P}_p(\widehat{I}_1),\ f(\widehat{\mathbf{v}}_2) = 0. \tag{11.12}$$

We first prove (11.11). We have

$$\|\mathscr{E}_1^1 f\|_{L^2(\widehat{I}_2)}^2 = \int_0^1 |\mathscr{E}_1^1 f(1-t,t)|^2\,\mathrm{d}t = \int_0^1 \left|\frac{1-t}{t}\int_{1-t}^1 \frac{f(s,0)}{s}\,\mathrm{d}s\right|^2 \mathrm{d}t$$

$$= \int_0^1 \left| \frac{1-t}{t} \int_0^t \frac{f(1-s,0)}{1-s}\, \mathrm{d}s \right|^2 \mathrm{d}t \le \int_0^1 \frac{1}{t^2} \left( \int_0^t |f(1-s,0)|\, \mathrm{d}s \right)^2 \mathrm{d}t.$$

Invoking Lemma C.52 we deduce

$$\|\mathscr{E}_1^1 f\|^2_{L^2(\widehat{I}_2)} \lesssim \int_0^1 |f(1-s,0)|^2\, \mathrm{d}s = \|f\|^2_{L^2(\widehat{I}_1)}.$$

Next we prove (11.12). We have, by using again Lemma C.52,

$$\|\mathscr{E}_2^1 f\|^2_{L^2(\widehat{I}_3)} = \int_0^1 |\mathscr{E}_2^1 f(0,\widehat{y})|^2\, d\widehat{y} = \int_0^1 \left| \frac{1-\widehat{y}}{\widehat{y}} \int_0^{\widehat{y}} \frac{f(t,0)}{1-t}\, \mathrm{d}t \right|^2 d\widehat{y}$$

$$\le \int_0^1 \frac{1}{\widehat{y}^2} \left| \int_0^{\widehat{y}} |f(t,0)|\, \mathrm{d}t \right|^2 d\widehat{y} \lesssim \|f\|^2_{L^2(\widehat{I}_1)}.$$

Results for the other values of $j$ can be obtained similarly. □

The next theorem states the bounds for the extension operators in the Sobolev norm defined in Definition 11.1.

**Theorem 11.2.** *For $j = 1,2,3$, the following estimates hold*

$$|||\mathscr{E}_j^j f|||_{\widetilde{H}^{1/2}(\widehat{T},\widehat{I}_{j\dot{+}2})} \lesssim (1+\log p)^{1/2} \|f\|_{L^2(\widehat{I}_j)} \quad \forall f \in \mathbb{P}_p^j(\widehat{I}_j), \tag{11.13}$$

$$|||\mathscr{E}_{j\dot{+}1}^j f|||_{\widetilde{H}^{1/2}(\widehat{T},\widehat{I}_{j\dot{+}1})} \lesssim (1+\log p)^{1/2} \|f\|_{L^2(\widehat{I}_j)} \quad \forall f \in \mathbb{P}_p^{j\dot{+}1}(\widehat{I}_j), \tag{11.14}$$

$$|||\mathscr{E}^j f|||_{\widetilde{H}^{1/2}(\widehat{T},\widehat{I}_{j\dot{+}1}\cup\widehat{I}_{j\dot{+}2})} \lesssim (1+\log p)^{1/2} \|f\|_{L^2(\widehat{I}_j)} \quad \forall f \in \mathbb{P}_p^0(\widehat{I}_j). \tag{11.15}$$

*The constants are independent of $f$ and $p$. (We recall* (11.2) *for the modular addition $\dot{+}$.)*

*Proof.* This theorem is proved in [110, Theorem 3.1]. We present the proof here for completeness, but only for the case $j = 1$. For convenience, we rewrite (11.13)–(11.15) for this case:

$$|||\mathscr{E}_1^1 f|||_{\widetilde{H}^{1/2}(\widehat{T},\widehat{I}_3)} \lesssim (1+\log p)^{1/2} \|f\|_{L^2(\widehat{I}_1)} \quad \forall f \in \mathbb{P}_p(\widehat{I}_1),\ f(\widehat{\mathbf{v}}_1) = 0, \tag{11.16}$$

$$|||\mathscr{E}_2^1 f|||_{\widetilde{H}^{1/2}(\widehat{T},\widehat{I}_2)} \lesssim (1+\log p)^{1/2} \|f\|_{L^2(\widehat{I}_1)} \quad \forall f \in \mathbb{P}_p(\widehat{I}_1),\ f(\widehat{\mathbf{v}}_2) = 0, \tag{11.17}$$

$$|||\mathscr{E}^1 f|||_{\widetilde{H}^{1/2}(\widehat{T},\widehat{I}_2\cup\widehat{I}_3)} \lesssim (1+\log p)^{1/2} \|f\|_{L^2(\widehat{I}_1)} \quad \forall f \in \mathbb{P}_p(\widehat{I}_1),\ f(\widehat{\mathbf{v}}_1) = f(\widehat{\mathbf{v}}_2) = 0. \tag{11.18}$$

We first prove (11.16). Recalling Definition 11.1 and noting that $\operatorname{dist}(\widehat{\mathbf{x}},\widehat{I}_3) = \widehat{x}$ for any $\widehat{\mathbf{x}} = (\widehat{x},\widehat{y}) \in \widehat{T}$, we have

$$|||\mathscr{E}_1^1 f|||^2_{\widetilde{H}^{1/2}(\widehat{T},\widehat{I}_3)} = |\mathscr{E}_1^1 f|^2_{H^{1/2}(\widehat{T})} + \int_{\widehat{T}} \frac{|\mathscr{E}_1^1 f(\widehat{\mathbf{x}})|^2}{\widehat{x}}\, d\widehat{\mathbf{x}}. \tag{11.19}$$

We will estimate the $H^{1/2}$-seminorm and the $L^2$-weighted norm on the right-hand side separately. It is proved in [110, Lemmas 3.3 and 3.5] that, for $f \in \mathbb{P}_p(\widehat{I}_1)$ satisfying $f(0,0) = 0$ and $\widehat{\gamma} = \{(0,0)\}$,

$$\|\mathscr{E}_1^1 f\|_{H^1(\widehat{T})} \lesssim \|f\|_{\widetilde{H}^{1/2}(\widehat{I}_1,\widehat{\gamma})},$$
$$\|\mathscr{E}_1^1 f\|_{L^2(\widehat{T})} \lesssim (1+\log p)\|f\|_{H^{-1/2}(\widehat{I}_1)}.$$

Interpolation then gives

$$\|\mathscr{E}_1^1 f\|_{H^{1/2}(\widehat{T})} \lesssim (1+\log p)^{1/2}\|f\|_{L^2(\widehat{I}_1)} \tag{11.20}$$

so that $|\mathscr{E}_1^1 f|_{H^{1/2}(\widehat{T})}$ is bounded by the right-hand side of (11.16). To estimate the weighted $L^2$-norm in (11.19) we note that

$$\int_{\widehat{T}} \frac{|\mathscr{E}_1^1 f(\widehat{\mathbf{x}})|^2}{\widehat{x}}\, d\widehat{\mathbf{x}} = \int_0^1 \int_0^{1-\widehat{x}} \frac{\widehat{x}}{\widehat{y}^2} \left( \int_{\widehat{x}}^{\widehat{x}+\widehat{y}} \frac{f(t,0)}{t}\, \mathrm{d}t \right)^2 d\widehat{y}\, d\widehat{x}. \tag{11.21}$$

Invoking Lemma C.52 we deduce

$$\int_{\widehat{T}} \frac{|\mathscr{E}_1^1 f(\widehat{\mathbf{x}})|^2}{\widehat{x}}\, d\widehat{\mathbf{x}} \le 4 \int_0^1 \int_{\widehat{x}}^1 |f(t,0)|^2\, \mathrm{d}t\, d\widehat{x} = 4 \int_0^1 t|f(t,0)|^2\, \mathrm{d}t \le 4\|f\|^2_{L^2(\widehat{I}_1)}, \tag{11.22}$$

proving (11.16). This result and a linear transformation yield (11.17).

To prove (11.18), the techniques proposed in [32] are used. Recalling (11.4) it suffices to prove the required estimate of $(1-\widehat{x}-\widehat{y})\mathscr{E}_1^1$ and $\widehat{x}\mathscr{E}_2^1$. We will prove only for the first term as the second can be dealt with analogously. Noting that

$$\operatorname{dist}(\widehat{\mathbf{x}}, \widehat{I}_2 \cup \widehat{I}_3) = \min\left\{ \operatorname{dist}(\widehat{\mathbf{x}}, \widehat{I}_2), \operatorname{dist}(\widehat{\mathbf{x}}, \widehat{I}_3) \right\} = \min\left\{ \frac{|\widehat{x}+\widehat{y}-1|}{\sqrt{2}}, \widehat{x} \right\},$$

we have, recalling Definition 11.1,

$$|||(1-\widehat{x}-\widehat{y})\mathscr{E}_1^1 f|||^2_{\widetilde{H}^{1/2}(\widehat{T},\widehat{I}_2\cup\widehat{I}_3)} \le |(1-\widehat{x}-\widehat{y})\mathscr{E}_1^1 f|^2_{H^{1/2}(\widehat{T})} + A_1 + A_2 \tag{11.23}$$

where

$$A_1 := \int_{\widehat{T}} \frac{\sqrt{2}\left|(1-\widehat{x}-\widehat{y})\mathscr{E}_1^1 f(\widehat{x},\widehat{y})\right|^2}{|\widehat{x}+\widehat{y}-1|}\, d\widehat{x}\, d\widehat{y},$$
$$A_2 := \int_{\widehat{T}} \frac{\left|(1-\widehat{x}-\widehat{y})\mathscr{E}_1^1 f(\widehat{x},\widehat{y})\right|^2}{\widehat{x}}\, d\widehat{x}\, d\widehat{y}. \tag{11.24}$$

We will prove that each term on the right-hand side of (11.23) is bounded by $(1+\log p)\|f\|^2_{L^2(\widehat{I}_1)}$. Consider the first term. Note that

$$\begin{aligned}
&\left|(1-\widehat{x}-\widehat{y})\mathscr{E}_1^1 f(\widehat{x},\widehat{y}) - (1-\widehat{x}'-\widehat{y}')\mathscr{E}_1^1 f(\widehat{x}',\widehat{y}')\right|^2 \\
&\quad = \Big|(1-\widehat{x}-\widehat{y})\big(\mathscr{E}_1^1 f(\widehat{x},\widehat{y}) - \mathscr{E}_1^1 f(\widehat{x}',\widehat{y}')\big) \\
&\qquad + \big((1-\widehat{x}-\widehat{y}) - (1-\widehat{x}'-\widehat{y}')\big)\mathscr{E}_1^1 f(\widehat{x}',\widehat{y}')\Big|^2 \\
&\quad \lesssim \left|\mathscr{E}_1^1 f(\widehat{x},\widehat{y}) - \mathscr{E}_1^1 f(\widehat{x}',\widehat{y}')\right|^2 + |\widehat{x}-\widehat{x}'+\widehat{y}-\widehat{y}'|^2|\mathscr{E}_1^1 f(\widehat{x}',\widehat{y}')|^2.
\end{aligned}$$

As a consequence, the definition of the $H^{1/2}$-seminorm and Fubini Theorem give

$$\begin{aligned}
|(1-\widehat{x}-\widehat{y})\mathscr{E}_1^1 f|^2_{H^{1/2}(\widehat{T})} &\lesssim |\mathscr{E}_1^1 f|^2_{H^{1/2}(\widehat{T})} \\
&\quad + \int_0^1\int_0^{1-\widehat{y}}\int_0^1\int_0^{1-\widehat{y}'} \frac{|\widehat{x}-\widehat{x}'+\widehat{y}-\widehat{y}'|^2|\mathscr{E}_1^1 f(\widehat{x}',\widehat{y}')|^2}{|(\widehat{x}-\widehat{x}')^2+(\widehat{y}-\widehat{y}')^2|^{3/2}}\, d\widehat{x}'\, d\widehat{y}'\, d\widehat{x} d\widehat{y} \\
&= |\mathscr{E}_1^1 f|^2_{H^{1/2}(\widehat{T})} + \int_0^1\int_0^{1-\widehat{y}'} A_3(\widehat{x}',\widehat{y}')|\mathscr{E}_1^1 f(\widehat{x}',\widehat{y}')|^2\, d\widehat{x}'\, d\widehat{y}'
\end{aligned}$$

where

$$A_3(\widehat{x}',\widehat{y}') := \int_0^1\int_0^{1-\widehat{y}} \frac{|\widehat{x}-\widehat{x}'+\widehat{y}-\widehat{y}'|^2}{|(\widehat{x}-\widehat{x}')^2+(\widehat{y}-\widehat{y}')^2|^{3/2}}\, d\widehat{x}\, d\widehat{y}.$$

The elementary inequality $(a+b)^2 \le 2(a^2+b^2)$ gives

$$\begin{aligned}
A_3(\widehat{x}',\widehat{y}') &\le \int_0^1\int_0^{1-\widehat{y}} \frac{2\, d\widehat{x}\, d\widehat{y}}{|(\widehat{x}-\widehat{x}')^2+(\widehat{y}-\widehat{y}')^2|^{1/2}} \\
&= \iint_{\widehat{T}} \frac{2\, d\widehat{x}\, d\widehat{y}}{|(\widehat{x}-\widehat{x}')^2+(\widehat{y}-\widehat{y}')^2|^{1/2}} \\
&\le \iint_{B_2(\widehat{x}',\widehat{y}')} \frac{2\, d\widehat{x}\, d\widehat{y}}{|(\widehat{x}-\widehat{x}')^2+(\widehat{y}-\widehat{y}')^2|^{1/2}} = 4\pi
\end{aligned}$$

for all $(\widehat{x}',\widehat{y}') \in \widehat{T}$, where $B_2(\widehat{x}',\widehat{y}')$ is the disk centered at $(\widehat{x}',\widehat{y}')$ and having radius 2, and where in the last step we changed to polar coordinates to calculate the integral. Consequently,

$$\begin{aligned}\left|(1-\widehat{x}-\widehat{y})\mathscr{E}_1^1 f\right|^2_{H^{1/2}(\widehat{T})} &\lesssim |\mathscr{E}_1^1 f|^2_{H^{1/2}(\widehat{T})} + \|\mathscr{E}_1^1 f\|^2_{L^2(\widehat{T})} \\ &\lesssim (1+\log p)\|f\|^2_{L^2(\widehat{I}_1)}\end{aligned}$$

where in the last step we used (11.20).

For the second term on the right-hand side of (11.23), we note that

$$A_1 \lesssim \|\mathscr{E}_1^1 f\|^2_{L^2(\widehat{T})} \le \|\mathscr{E}_1^1 f\|^2_{H^{1/2}(\widehat{T})}$$

so that $A_1$ satisfies the required estimate due to (11.20). It remains to prove the same estimate for $A_2$. We have, noting that $0 \le 1-\widehat{x}-\widehat{y} \le 1$,

$$\begin{aligned}A_2 &= \int_0^1\int_0^{1-\widehat{y}} \frac{\widehat{x}(1-\widehat{x}-\widehat{y})^2}{\widehat{y}^2}\left(\int_{\widehat{x}}^{\widehat{x}+\widehat{y}} \frac{f(t,0)}{t}\,\mathrm{d}t\right)^2 d\widehat{x}\,d\widehat{y} \\ &\le \int_0^1\int_0^{1-\widehat{y}} \frac{\widehat{x}}{\widehat{y}^2}\left(\int_{\widehat{x}}^{\widehat{x}+\widehat{y}} \frac{f(t,0)}{t}\,\mathrm{d}t\right)^2 d\widehat{x}\,d\widehat{y} \\ &\lesssim \|f\|^2_{L^2(\widehat{I}_1)}\end{aligned}$$

where in the last step we used (11.21) and (11.22). This proves (11.18) completing the proof for $j=1$. Similar arguments can be done for $j=2,3$. □

In the remainder of this section, for any $f \in C(\partial\widehat{T})$, the set of continuous function defined on $\partial\widehat{T}$, we denote by $f_i$ the restriction of $f$ on the edge $\widehat{I}_i$, $i=1,2,3$.

In order to construct the basis functions (Section 11.1.3) we need another type of extension operators which extend a continuous function defined on $\partial\widehat{T}$ whose restriction of one side vanishes and whose restrictions on the other two sides are polynomials of degree at most $p$. The next theorem defines these extension operators and provides the bounds in the $L^2$-norm.

**Theorem 11.3.** *For $j=1,2,3$, let*

$$\mathscr{P}_p(\partial\widehat{T},\widehat{I}_j) := \left\{f \in C(\partial\widehat{T}) : f_j = 0,\ f_i \in \mathbb{P}_p(\widehat{I}_i),\ i \in \{1,2,3\},\ i \ne j\right\}.$$

*We define $\underline{\mathscr{E}}^j : \mathscr{P}_p(\partial\widehat{T},\widehat{I}_j) \longrightarrow \mathbb{P}_p(\widehat{T})$ by*

$$\underline{\mathscr{E}}^j f := \mathscr{E}_j^{j+2} f_{j+2} + \mathscr{E}^{j+1}\left(f_{j+1} - (\mathscr{E}_j^{j+2} f_{j+2})|_{\widehat{I}_{j+1}}\right) \quad \forall f \in \mathscr{P}_p(\partial\widehat{T},\widehat{I}_j).$$

*Then $\underline{\mathscr{E}}^j f$ is the extension of $f$ to $\widehat{T}$, i.e., the restriction of $\underline{\mathscr{E}}^j f$ on $\partial\widehat{T}$ is $f$. Moreover,*

$$\|\underline{\mathscr{E}}^j f\|_{L^2(\widehat{T})} \lesssim \|f_{j+1}\|_{L^2(\widehat{I}_{j+1})} + \|f_{j+2}\|_{L^2(\widehat{I}_{j+2})}.$$

*and*

$$|\underline{\mathscr{E}}^j f|_{H^{1/2}(\widehat{T})} \lesssim (1+\log p)^{1/2} \left( \|f_{j+1}\|_{L^2(\widehat{I}_{j+1})} + \|f_{j+2}\|_{L^2(\widehat{I}_{j+2})} \right).$$

*The constant is independent of $f$ and $p$.*

*Proof.* To fix ideas, we explicitly write down the extension formula and proof for the case when $j = 2$, i.e., the extended function vanishes on the side $\widehat{I}_2$ of $\partial\widehat{T}$; see Figure 11.1. The operator $\underline{\mathscr{E}}^2 : \mathscr{P}_p(\partial\widehat{T}, \widehat{I}_2) \longrightarrow \mathbb{P}_p(\widehat{T})$ is defined by

$$\underline{\mathscr{E}}^2 f := \mathscr{E}_2^1 f_1 + \mathscr{E}^3 \left( f_3 - (\mathscr{E}_2^1 f_1)|_{\widehat{I}_3} \right) \quad \forall f \in \mathscr{P}_p(\partial\widehat{T}, \widehat{I}_2),$$

and the required estimates are

$$\|\underline{\mathscr{E}}^2 f\|_{L^2(\widehat{T})} \lesssim \|f_1\|_{L^2(\widehat{I}_1)} + \|f_3\|_{L^2(\widehat{I}_3)} \tag{11.25}$$

and

$$|\underline{\mathscr{E}}^2 f|_{H^{1/2}(\widehat{T})} \lesssim (1+\log p)^{1/2} \|f_1\|_{L^2(\widehat{I}_1)} + \|f_3\|_{L^2(\widehat{I}_3)}.$$

The idea behind the definition of $\underline{\mathscr{E}}^2$ is in order to extend $f$ to $\widehat{T}$, we extend $f_1$ and $f_3$ adjusted to avoid double counting on $\widehat{I}_3$.

Firstly, since $f_1 \in \mathbb{P}_p^2(\widehat{I}_1)$, see (11.3) and Figure 11.1, $\mathscr{E}_2^1 f_1$ is well defined. Next, due to the continuity of $f$, we have

$$f_3(\mathbf{v}_1) - \mathscr{E}_2^1 f_1(\mathbf{v}_1) = f(\mathbf{v}_1) - f(\mathbf{v}_1) = 0 \quad \text{and} \quad f_3(\mathbf{v}_3) - \mathscr{E}_2^1 f_1(\mathbf{v}_3) = 0 - 0 = 0.$$

Thus $\mathscr{E}^3 \left( f_3 - (\mathscr{E}_2^1 f_1)|_{\widehat{I}_3} \right)$ is also well defined. Clearly $\underline{\mathscr{E}}^2 f \in \mathbb{P}_p(\widehat{T})$. The restriction of $\underline{\mathscr{E}}^2 f$ on $\widehat{I}_i$ equals $f_i$, $i = 1, 2, 3$, due to (11.5).

In order to prove (11.25) we use the triangle inequality and invoke Theorem 11.1, using in particular (11.7), (11.8), and (11.10) to obtain

$$\begin{aligned}
\|\underline{\mathscr{E}}^2 f\|_{L^2(\widehat{T})} &\leq \|\mathscr{E}_2^1 f_1\|_{L^2(\widehat{T})} + \|\mathscr{E}^3 f_3\|_{L^2(\widehat{T})} + \left\| \mathscr{E}^3 \left( (\mathscr{E}_2^1 f_1)|_{\widehat{I}_3} \right) \right\|_{L^2(\widehat{T})} \\
&\lesssim \|f_1\|_{L^2(\widehat{I}_1)} + \|f_3\|_{L^2(\widehat{I}_3)} + \|(\mathscr{E}_2^1 f_1)|_{\widehat{I}_3}\|_{L^2(\widehat{I}_3)} \\
&\lesssim \|f_1\|_{L^2(\widehat{I}_1)} + \|f_3\|_{L^2(\widehat{I}_3)} + \|f_1\|_{L^2(\widehat{I}_1)}.
\end{aligned}$$

Finally, we invoke Theorem 11.2 to have

$$\begin{aligned}
|\underline{\mathscr{E}}^2 f|_{H^{1/2}(\widehat{T})} &\leq |\mathscr{E}_2^1 f_1|_{H^{1/2}(\widehat{T})} + |\mathscr{E}^3 f_3|_{H^{1/2}(\widehat{T})} + \left| \mathscr{E}^3 \left( (\mathscr{E}_2^1 f_1)|_{\widehat{I}_3} \right) \right|_{H^{1/2}(\widehat{T})} \\
&\lesssim (1+\log p)^{1/2} \left( \|f_1\|_{L^2(\widehat{I}_1)} + \|f_3\|_{L^2(\widehat{I}_3)} + \|(\mathscr{E}_2^1 f_1)|_{\widehat{I}_3}\|_{L^2(\widehat{I}_3)} \right) \\
&\lesssim (1+\log p)^{1/2} \left( \|f_1\|_{L^2(\widehat{I}_1)} + \|f_3\|_{L^2(\widehat{I}_3)} \right)
\end{aligned}$$

where in the last step we used (11.10). □

Using Theorem 11.2 and Theorem 11.3, we can prove the following result, which is proved in [110].

**Corollary 11.1.** *Let $T$ be a triangle and let $\gamma$ be one of its sides or the union of any two sides. If $f$ is a continuous function defined on $\partial T$ which vanishes on $\gamma$ and is a polynomial of degree up to $p$ on each remaining side in $\partial T \setminus \gamma$, then there exists an extension $U \in \mathbb{P}_p(T)$ such that $U = f$ on $\partial T$ and*

$$|||U|||_{\widetilde{H}^{1/2}(T,\gamma)} \lesssim (1+\log p)^{1/2} \|f\|_{L^2(\partial T)}.$$

*The constant is independent of $f$, $p$, and the size of $T$. Here the norm $|||\cdot|||_{\widetilde{H}^{1/2}(T,\gamma)}$ is defined in Definition 11.1.*

*Proof.* The result initially proved in [110, Theorem 2.1] is just a consequence of Theorem 11.2, Theorem 11.3 and a scaling argument. First we note the scaling properties, see (11.1),

$$|||U|||^2_{\widetilde{H}^{1/2}(T,\gamma)} = \tau |||\widehat{U}|||^2_{\widetilde{H}^{1/2}(\widehat{T},\widehat{\gamma})} \quad \text{and} \quad \|f\|^2_{L^2(\partial T)} = \tau \|\widehat{f}\|^2_{L^2(\partial \widehat{T})},$$

where $\tau$ is the diameter of $T$. (Here, for any entity, the hat version is its corresponding entity on the reference triangle $\widehat{T}$.) Hence, it suffices to prove

$$\left|\left|\left|\widehat{U}\right|\right|\right|_{\widetilde{H}^{1/2}(\widehat{T},\widehat{\gamma})} \lesssim (1+\log p)^{1/2} \|\widehat{f}\|_{L^2(\partial \widehat{T})}.$$

It also suffices to consider two cases: $\widehat{\gamma} = \widehat{I}_2$ and $\widehat{\gamma} = \widehat{I}_2 \cup \widehat{I}_3$. We denote by $\widehat{f}_i$ the restriction of $\widehat{f}$ on $\widehat{I}_i$, $i = 1,2,3$.

In the case $\widehat{\gamma} = \widehat{I}_2$ we define $\widehat{U} := \underline{\mathscr{E}}^2 f$ where $\underline{\mathscr{E}}^2$ is defined in Theorem 11.3 which also gives the required estimate.

Consider next the case when $\widehat{\gamma} = \widehat{I}_2 \cup \widehat{I}_3$. Then $f_2 \equiv 0$ and $f_3 \equiv 0$ while $f_1$ vanishes at the endpoints $\widehat{\mathbf{v}}_1$ and $\widehat{\mathbf{v}}_2$ of $\widehat{I}_1$. Thus we define $\widehat{U} := \mathscr{E}^1 f_1$ and obtain the required result from (11.15) in Theorem 11.2, noting that $\|\widehat{f}_1\|_{L^2(\widehat{I}_1)} \leq \|\widehat{f}\|_{L^2(\partial \widehat{T})}$. □

### *11.1.3 Construction of basis functions*

Vertex, edge, and interior basis functions are presented in detail in e.g., the book by Babuška and Szabó [215]. In this subsection we present how they are constructed as extensions of polynomials on edges.

The edge basis functions are defined as extensions of polynomials defined on an edge, vanishing at the endpoints of that edge. In Figure 11.2(a), a polynomial defined on $I$ vanishing at the endpoints of $I$ is extended to be a piecewise polynomial on $T_1 \cup T_2$ which vanishes at the non-common edges of these two triangles, so that it can be continuously extended by zero outside $T_1 \cup T_2$.

The vertex basis functions are defined on patches as in Figure 11.2(b). Their values at the nodal points on the edges of a patch (the skeleton) are given. These values include the value 1 at the center node and 0 at the nodes on the boundary of the patch.

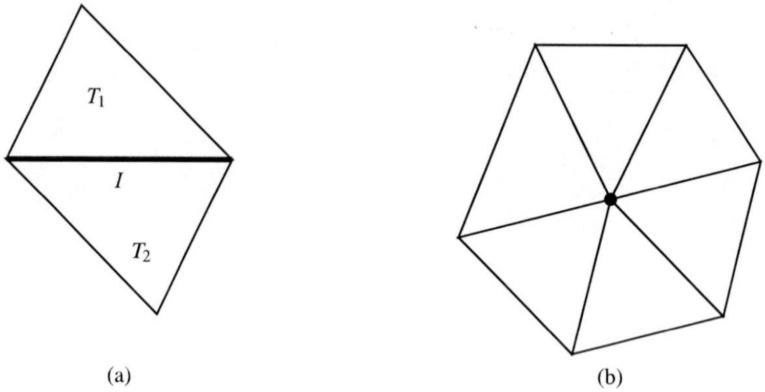


**Fig. 11.2** Patches of elements

For a given triangle $T$, affine transformations of the basis functions defined above on the reference triangle $\widehat{T}$ are used to span the polynomial space $\mathbb{P}_p(T)$. Therefore, it suffices to establish all required ingredients on the reference triangle $\widehat{T}$; see Figure 11.1.

#### 11.1.3.1 Construction of preliminary vertex basis functions

We first define three affine mappings $\alpha_j$ that maps $[0,1]$ onto the edges $\widehat{I}_j$ of $\widehat{T}$ (including the endpoints), $j = 1,2,3$,

$$\alpha_1(t) = (t,0), \quad \alpha_2(t) = (1-t,t), \quad \alpha_3(t) = (0,1-t), \quad t \in [0,1].$$

The construction of vertex basis functions uses the special low energy functions $\phi_0$ and $\phi_0^-$ defined in Section C.4.3 which were initially introduced in [169]. To define the vertex basis function $\widehat{\phi}_{\widehat{\mathbf{v}}_j}$ at the vertex $\widehat{\mathbf{v}}_j$ of $\widehat{T}$, $j = 1,2,3$, we first construct a function $f$ defined on the boundary $\partial\widehat{T}$ of $\widehat{T}$ as follows. Let $f_j$, $j = 1,2,3$, be the restriction of $f$ onto $\widehat{I}_j$. Then we define

$$f_j(\alpha_j(t)) = \phi_0(t), \quad f_{j+1}(\alpha_{j+1}(t)) = 0, \quad f_{j+2}(\alpha_{j+2}(t)) = \phi_0^-(t), \quad t \in [0,1].$$

The preliminary vertex basis function $\widetilde{\phi}_{\widehat{\mathbf{v}}_j}$ is defined by

$$\widetilde{\phi}_{\widehat{\mathbf{v}}_j} := \underline{\mathscr{E}}^{j+1} f$$

where the extension operator $\underline{\mathscr{E}}^{j+1}$ is defined in Theorem 11.3. As an example, the preliminary vertex basis function $\widetilde{\phi}_{\widehat{\mathbf{v}}_1}$ is defined by

$$\widetilde{\phi}_{\widehat{\mathbf{v}}_1} := \underline{\mathscr{E}}^2 f$$

where the restrictions of $f$ on $\widehat{I}_j$ are given by $f_1(t,0) = \phi_0(t)$, $f_2(1-t,t) = 0$, and $f_3(0,1-t) = \phi_0^-(t)$ or $f_3(0,t) = \phi_0(t)$, $t \in [0,1]$.

In Subsection 11.1.4 we will modify these preliminary basis functions to ensure the constant reproduction of the space.

#### 11.1.3.2 Construction of preliminary edge basis functions

Recalling the definition of the antiderivative $\mathscr{L}_q$ of the Legendre polynomial $L_{q-1}$ in (C.4), we consider the affine image $\mathscr{L}_{\widehat{I}_i,q}$ of $\mathscr{L}_q$ onto the edge $\widehat{I}_i$ of the reference triangle $\widehat{T}$, $i = 1,2,3$. These affine images vanish at the vertices of $\widehat{T}$. Therefore we can extend them by using the extension operators $\mathscr{E}^i$, $i = 1,2,3$, to define the edge basis functions. More precisely, we define

$$\widetilde{\phi}_{\widehat{I}_i,q} := \mathscr{E}^i \mathscr{L}_{\widehat{I}_i,q}, \quad q = 1,\dots,p-1.$$

There are $p-1$ basis functions on each edge, and altogether there are $3(p-1)$ edge basis functions on $\widehat{T}$.

#### 11.1.3.3 Construction of preliminary interior basis functions

The interior (or bubble) basis functions on the reference triangle $\widehat{T}$ are defined as tensor products of $\mathscr{L}_q$ by

$$\widehat{\phi}_{q,m}(\widehat{x},\widehat{y}) = \frac{\mathscr{L}_q(2\widehat{x}-1)}{1-\widehat{x}} \frac{\mathscr{L}_m(2\widehat{y}-1)}{1-\widehat{y}} (1-\widehat{x}-\widehat{y}), \quad q \geq 2,\ m \geq 2,\ q+m \leq p+1.$$

We note that $\widehat{\phi}_{q,m}$ vanishes on the boundary of the reference element $\widehat{T}$ and that it is a polynomial of degree $q+m-1$ with $q+m = 4,\dots,p+1$. There are $(p-1)(p-2)/2$ such interior functions.

Altogether, there are $(p+1)(p+2)/2$ basis functions for $\mathbb{P}_p(\widehat{T})$. A sample set of vertex, edge and interior basis functions is depicted in Figure 11.3.

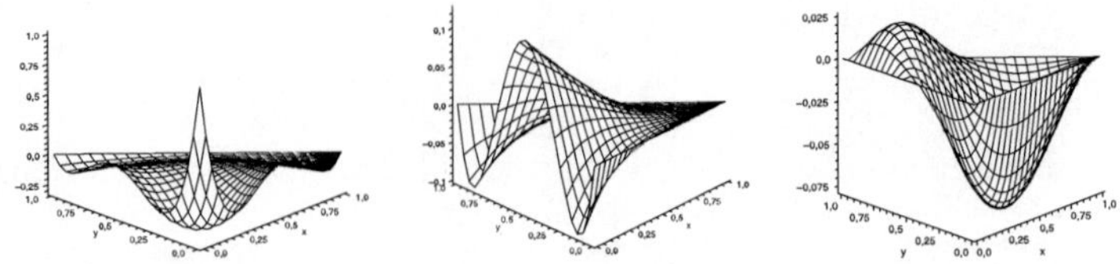

**Fig. 11.3** Sample basis functions (vertex, edge, interior) for $p = 4$.

#### 11.1.3.4 Representation of a polynomial by preliminary vertex, edge, and interior components

By using the preliminary basis functions $\widetilde{\phi}_{\widehat{\mathbf{v}}_i}$, $\widetilde{\phi}_{\widehat{I}_i,q}$, and $\widetilde{\phi}_{k,l}$, we can represent any function $u \in \mathbb{P}_p(\widehat{T})$ uniquely by

$$u = \sum_{i=1}^{3} \widetilde{u}_{\widehat{\mathbf{v}}_i} + \sum_{i=1}^{3} \widetilde{u}_{\widehat{I}_i} + \widetilde{u}_{\widehat{T}} \tag{11.26}$$

where $\widetilde{u}_{\widehat{\mathbf{v}}_i}$, $\widetilde{u}_{\widehat{I}_i}$, and $\widetilde{u}_{\widehat{T}}$ are the vertex, edge, and interior components, respectively, defined by

$$\begin{aligned}
&\widetilde{u}_{\widehat{\mathbf{v}}_i} \in \operatorname{span}\{\widetilde{\phi}_{\widehat{\mathbf{v}}_i}\}, \quad i = 1,2,3, \\
&\widetilde{u}_{\widehat{I}_i} \in \operatorname{span}\left\{\widetilde{\phi}_{\widehat{I}_i,q} : q = 2,\dots,p\right\}, \quad i = 1,2,3, \\
&\widetilde{u}_{\widehat{T}} \in \operatorname{span}\left\{\widetilde{\phi}_{k,l} : k \geq 2,\ l \geq 2,\ k+l \leq p+1\right\}.
\end{aligned}$$

The set of all vertices and edges is called the wire basket and is denoted by $\widehat{W}$. The space generated by all vertex and edge basis functions is called the wire basket space. Any function of the wire basket space is uniquely defined by its values on the wire basket $\widehat{W}$. An interpolation operator $\widetilde{\mathscr{I}}_{\widehat{W}}$ from $\mathbb{P}_p(\widehat{T})$ onto the space of wire basket functions is defined by

$$\widetilde{\mathscr{I}}_{\widehat{W}} u := \widetilde{u}_{\widehat{W}} := \sum_{i=1}^{3} \widetilde{u}_{\widehat{\mathbf{v}}_i} + \sum_{i=1}^{3} \widetilde{u}_{\widehat{I}_i} \quad \forall u \in \mathbb{P}_p(\widehat{T}). \tag{11.27}$$

Therefore we can represent any $u \in \mathbb{P}_p(\widehat{T})$ by

$$u = \widetilde{u}_{\widehat{W}} + \widetilde{u}_{\widehat{T}}. \tag{11.28}$$

### 11.1.4 Change of basis functions and change of the wire basket space

Since the wire basket space does not contain constants on $\widehat{T}$ we redefine the preliminary vertex and edge functions as follows. Writing the representation (11.28) for the constant function 1, we have

$$1 = \widetilde{\mathscr{I}}_{\widehat{W}} 1 + \widetilde{1}_{\widehat{T}} = \widetilde{1}_{\widehat{W}} + \widetilde{1}_{\widehat{T}}. \tag{11.29}$$

We define a new interpolation operator by

$$\mathscr{I}_{\widehat{W}} u := \widetilde{\mathscr{I}}_{\widehat{W}} u + \widetilde{1}_{\widehat{T}} \mu_{\partial \widehat{T}}(u) \quad \forall u \in \mathbb{P}_p(\widehat{T}) \tag{11.30}$$

where $\mu_{\partial \widehat{T}}(u) := \int_{\partial \widehat{T}} u / |\partial \widehat{T}|$ is the integral mean of $u$ over the boundary $\partial \widehat{T}$ of $\widehat{T}$. In particular,

$$\begin{aligned}
\mathscr{I}_{\widehat{W}} \widetilde{\phi}_{\widehat{\mathbf{v}}_i} &= \widetilde{\phi}_{\widehat{\mathbf{v}}_i} + \widetilde{1}_{\widehat{T}} \mu_{\partial \widehat{T}}(\widetilde{\phi}_{\widehat{\mathbf{v}}_i}), \\
\mathscr{I}_{\widehat{W}} \widetilde{\phi}_{\widehat{I}_i,q} &= \widetilde{\phi}_{\widehat{I}_i,q} + \widetilde{1}_{\widehat{T}} \mu_{\partial \widehat{T}}(\widetilde{\phi}_{\widehat{I}_i,q}), \\
\mathscr{I}_{\widehat{W}} \widetilde{\phi}_{k,l} &= 0,
\end{aligned}$$

where we recall that $\widetilde{\phi}_{k,l}$ vanishes on $\partial \widehat{T}$. We note that since $\widetilde{1}_{\widehat{T}}$ vanishes on $\partial \widehat{T}$, the interpolant $\mathscr{I}_{\widehat{W}} \widetilde{\phi}_{\widehat{\mathbf{v}}_i}$, as $\widetilde{\phi}_{\widehat{\mathbf{v}}_i}$, vanishes on the edge opposite to $\widehat{\mathbf{v}}_i$, while $\mathscr{I}_{\widehat{W}} \widetilde{\phi}_{\widehat{I}_i,q}$, as $\widetilde{\phi}_{\widehat{I}_i,q}$, vanishes on $\partial \widehat{T} \setminus \widehat{I}_i$. Hence we can define the new basis functions for $\mathbb{P}_p(\widehat{T})$ as

$$\phi_{\widehat{\mathbf{v}}_i} := \mathscr{I}_{\widehat{W}} \widetilde{\phi}_{\widehat{\mathbf{v}}_i}, \quad \phi_{\widehat{I}_i,q} := \mathscr{I}_{\widehat{W}} \widetilde{\phi}_{\widehat{I}_i,q}, \quad \phi_{k,l} := \widetilde{\phi}_{k,l}.$$

The new wire basket space is now the span of $\phi_{\widehat{\mathbf{v}}_i}$ and $\phi_{\widehat{I}_i,q}$, $i = 1,2,3$, $q = 2,\ldots,p$, which is also the range of the operator $\mathscr{I}_{\widehat{W}}$. The new vertex and edge components of $u$ with respect to the new basis functions are denoted by $u_{\widehat{\mathbf{v}}_i}$ and $u_{\widehat{I}_i}$, $i = 1,\ldots,3$. They are images under $\mathscr{I}_{\widehat{W}}$ of the preliminary components $\widetilde{u}_{\widehat{\mathbf{v}}_i}$ and $\widetilde{u}_{\widehat{I}_i}$; see (11.26). Meanwhile, the new interior component of $u$ is defined by

$$u_{\widehat{T}} := \widetilde{u}_{\widehat{T}} - \widetilde{1}_{\widehat{T}} \mu_{\partial \widehat{T}}(\widetilde{\mathscr{I}}_{\widehat{W}} u). \tag{11.31}$$

We note that both $\widetilde{u}_{\widehat{T}}$ and $\widetilde{1}_{\widehat{T}}$ are linear combinations of $\phi_{k,l}$ while $\mu_{\partial \widehat{T}}(\widetilde{\mathscr{I}}_{\widehat{W}}(u))$ is a constant, so $u_{\widehat{T}}$ is also a linear combination of $\phi_{k,l}$.

Consequently, if $u$ has the decomposition (11.26), then it also has a unique decomposition with respect to the new basis functions:

$$\begin{aligned}
u &= \left( \sum_{i=1}^{3} u_{\widehat{\mathbf{v}}_i} - \sum_{i=1}^{3} \widetilde{1}_{\widehat{T}} \mu_{\partial \widehat{T}}(\widetilde{u}_{\widehat{\mathbf{v}}_i}) \right) + \left( \sum_{i=1}^{3} u_{\widehat{I}_i} - \sum_{i=1}^{3} \widetilde{1}_{\widehat{T}} \mu_{\partial \widehat{T}}(\widetilde{u}_{\widehat{I}_i,q}) \right) + \widetilde{u}_{\widehat{T}} \\
&= \mathscr{I}_{\widehat{W}} u + u_{\widehat{T}}
\end{aligned} \tag{11.32}$$

noting that

$$\sum_{i=1}^{3} \mu_{\partial\widehat{T}}(\widetilde{u}_{\widehat{\mathbf{v}}_i}) + \sum_{i=1}^{3} \mu_{\partial\widehat{T}}(\widetilde{u}_{\widehat{I}_i,q}) = \mu_{\partial\widehat{T}}(\widetilde{\mathscr{I}}_{\widehat{W}}(u)).$$

In particular, writing the representations (11.31) and (11.32) for the constant function 1, we have

$$1_{\widehat{T}} = \widetilde{1}_{\widehat{T}} - \widetilde{1}_{\widehat{T}}\mu_{\partial\widehat{T}}(\widetilde{\mathscr{I}}_{\widehat{W}}1) = 0$$

and thus

$$\mathscr{I}_{\widehat{W}}1(\widehat{x}) = 1_{\widehat{W}}(\widehat{x}) = 1 \quad \forall \widehat{x} \in \operatorname{cl}(\widehat{T}) \tag{11.33}$$

where $\operatorname{cl}(\widehat{T})$ denotes the closure of $\widehat{T}$. In other words, the operator $\mathscr{I}_{\widehat{W}}$ reproduces constant functions.

### *11.1.5 The wire basket in three dimensions*

For technical reasons, we sometimes need to extend polynomials defined on two-dimensional surfaces onto three-dimensional domains. This subsection introduces the concepts of vertex functions and edge functions in three dimensions.

In the remainder of this chapter, we denote by $\widehat{\Omega}$ the reference tetrahedron

$$\widehat{\Omega} := \left\{(x,y,z) \in \mathbb{R}^3 \,:\, 0 \le x,y,z \le 1,\ x+y+z \le 1\right\}$$

and define for integer $p$

$$\mathbb{P}_p(\widehat{\Omega}) := \operatorname{span}\left\{x^i y^j z^k \,:\, 0 \le i+j+k \le p\right\}.$$

Not to be confused with the planar case, the vertices of $\widehat{\Omega}$ are denoted by $\widehat{\mathbf{w}}_i$, $i = 1,\dots,4$, edges by $\widehat{\mathbf{e}}_j$, $j = 1,\dots,6$, and faces by $\widehat{F}_k$, $k = 1,\dots,4$. To fix ideas, we introduce a numbering for these vertices and edges, as depicted in Figure 11.4, with face $\widehat{F}_k$ opposite to vertex $\widehat{\mathbf{w}}_k$, $k = 1,\dots,4$.

First we define the vertex basis functions $\widetilde{\Phi}_{\widehat{\mathbf{w}}_i}$, $i = 1,\dots,4$. We start by defining $\widetilde{\Phi}_{\widehat{\mathbf{w}}_i}$ on the faces $\widehat{F}_k$, $k = 1,\dots,4$ by using the method of Subsection 11.1.3.1 to construct three two-variable vertex basis functions on the three faces sharing this vertex, and set it to be zero on the remaining face. Consider for example the vertex $\widehat{\mathbf{w}}_1$. The vertex function $\widetilde{\Phi}_{\widehat{\mathbf{w}}_1}$ is defined on faces $\widehat{F}_2$, $\widehat{F}_3$, $\widehat{F}_4$, and is set to be zero on $\widehat{F}_1$. We note that $\widetilde{\Phi}_{\widehat{\mathbf{w}}_1}$ is continuous on $\partial\widehat{\Omega} = \widehat{F}_1 \cup \cdots \cup \widehat{F}_4$. The function $\widetilde{\Phi}_{\widehat{\mathbf{w}}_1}$ is then extended into a discrete harmonic function in $\widehat{\Omega}$; see Subsection C.4.4. The vertex functions at other vertices are defined similarly.

Similarly, we use the method in Subsection 11.1.3.2 to define the edge basis functions $\widetilde{\Phi}_{\widehat{\mathbf{e}}_j,q}$, $j = 1,\dots,6$ and $q = 1,\dots,p-1$. For example, consider the edge $\widehat{\mathbf{e}}_1$. We first identify the two faces sharing $\widehat{\mathbf{e}}_1$, namely $\widehat{F}_2$ and $\widehat{F}_3$, with $\widehat{T}$. We then define the edge basis functions $\widetilde{\Phi}_{\widehat{\mathbf{e}}_1,q}$, $q = 1,\dots,p-1$, on $\widehat{F}_2$ and $\widehat{F}_3$ and extend it by zero

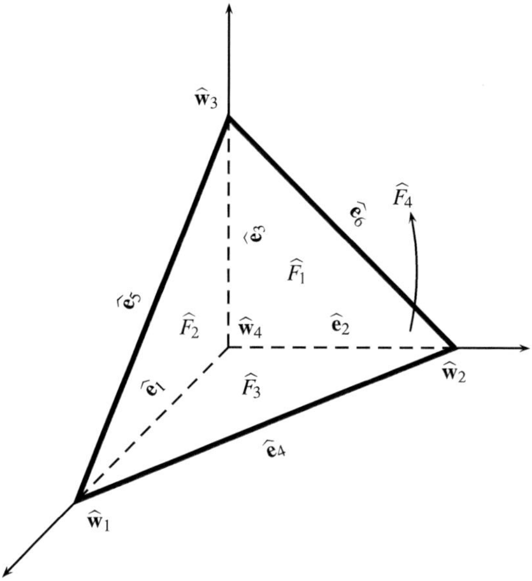


**Fig. 11.4** Vertices and edges of the reference tetrahedron $\widehat{\Omega}$

onto the remaining faces $\widehat{F}_1$ and $\widehat{F}_4$. Note again that $\widetilde{\Phi}_{\widehat{\mathbf{e}}_1,q}$ is continuous on $\partial\Omega$. We then extend these functions into $\widehat{\Omega}$ as discrete harmonic functions.

Finally, the face basis functions are defined using the method in Section 11.1.3.3 for each face $\widehat{F}_k$, $k = 1,\ldots,4$, which is now identified with the reference triangle $\widehat{T}$ by some affine transformation. The resulting basis function $\widetilde{\Phi}_{q,m}^{(k)}$ $(q,m \geq 2,\ q+m \leq p+1)$ on face $\widehat{F}_k$, $k = 1,\ldots,4$, is extended by zero onto the other faces. They are then extended to be discrete harmonic functions in $\widehat{\Omega}$. Note that corresponding to each face there are $(p-1)(p-2)/2$ face functions; see Section 11.1.3.3.

The vertex and edge components of $u \in \mathbb{P}_p(\widehat{\Omega})$ are denoted by $u_{\widehat{\mathbf{w}}_i}$, $i = 1,\ldots,4$, and $u_{\widehat{\mathbf{e}}_j}$, $j = 1,\ldots,6$. The wire basket $\widehat{W}(\widehat{\Omega})$ of $\widehat{\Omega}$ contains all four vertices and six edges. The corresponding wire basket interpolation operator $\widetilde{\mathscr{I}}_{\widehat{W}(\widehat{\Omega})}$ is defined by

$$\widetilde{\mathscr{I}}_{\widehat{W}(\widehat{\Omega})} u := \sum_{i=1}^{4} \widetilde{u}_{\widehat{\mathbf{w}}_i} + \sum_{j=1}^{6} \widetilde{u}_{\widehat{\mathbf{e}}_j} \quad \forall u \in \mathbb{P}_p(\widehat{\Omega}), \tag{11.34}$$

see (11.27), where

$$\widetilde{u}_{\widehat{\mathbf{w}}_i} = u(\widehat{\mathbf{w}}_i)\widetilde{\Phi}_{\widehat{\mathbf{w}}_i} \quad \text{and} \quad \widetilde{u}_{\widehat{\mathbf{e}}_j} = \sum_{q=1}^{p-1} \alpha_{j,q}\widetilde{\Phi}_{\widehat{\mathbf{e}}_j,q}, \quad \alpha_{j,q} \in \mathbb{R}.$$

It is known that the vertex, edge, and face basis functions

$$\begin{gathered} \widetilde{\Phi}_{\widehat{\mathbf{w}}_i}, \quad i = 1,\ldots,4, \\ \widetilde{\Phi}_{\widehat{\mathbf{e}}_j,q}, \quad j = 1,\ldots,6, \quad q = 1,\ldots,p-1, \\ \widetilde{\Phi}^{(k)}_{q,m}, \quad q,m \geq 2, \quad q+m \leq p+1, \end{gathered}$$

form a basis for the space of discrete harmonic functions in $\widehat{\Omega}$. In particular, since the constant function 1 in $\widehat{\Omega}$ is a discrete harmonic function we can write

$$1 = \widetilde{\mathscr{I}}_{\widehat{W}(\widehat{\Omega})} 1 + \sum_{k=1}^{4} \widetilde{1}_{\widehat{F}_k}$$

where the face components $\widetilde{1}_{\widehat{F}_k}$ of 1 are defined by

$$\widetilde{1}_{\widehat{F}_k} := \sum_{(q,m)\in\mathfrak{I}_p} \beta^{(k)}_{q,m}\widetilde{\Phi}^{(k)}_{q,m} \quad \text{for some } \beta^{(k)}_{q,m} \in \mathbb{R}. \tag{11.35}$$

Here $\mathfrak{I}_p := \{(q,m) : q,m \geq 2,\ q+m \leq p+1\}$.

As in the two-dimensional case, we define another interpolation operator which reproduces constants:

$$\mathscr{I}_{\widehat{W}(\widehat{\Omega})} u := \widetilde{\mathscr{I}}_{\widehat{W}(\widehat{\Omega})} u + \sum_{k=1}^{4} \widetilde{1}_{\widehat{F}_k} \mu_{\partial\widehat{F}_k}(u); \tag{11.36}$$

compare with (11.30).

### 11.1.6 Properties of the interpolation operators $\mathscr{I}_{\widehat{W}}$ and $\mathscr{I}_{\widehat{W}(\widehat{\Omega})}$

In this subsection, we prove some lemmas involving the properties of the wire basket interpolation operators defined above. First we prove an equivalence of norms which will be needed in the proof of Lemma 11.2 and Lemma 11.3.

**Lemma 11.1.** *Let $u \in \mathbb{P}_p(\widehat{T})$ and $\widetilde{u}_{\widehat{W}_j}$ be the preliminary wire basket component of $u$ associated with edge $\widehat{I}_j$, $j = 1,2,3$, i.e.,*

$$\widetilde{u}_{\widehat{W}_j} = \widetilde{u}_{\widehat{\mathbf{v}}_j} + \widetilde{u}_{\widehat{\mathbf{v}}_{j+1}} + \widetilde{u}_{\widehat{I}_j};$$

*see Figure 11.1. Then*

$$\frac{3}{4}\|\widetilde{u}_{\widehat{W}_j}\|_*^2 \le \|\widetilde{u}_{\widehat{W}_j}\|_{L^2(\widehat{I}_j)}^2 \le 2\|\widetilde{u}_{\widehat{W}_j}\|_*^2$$

*where*

$$\|\widetilde{u}_{\widehat{W}_j}\|_*^2 := \|\widetilde{u}_{\widehat{\mathbf{v}}_j}\|_{L^2(\widehat{I}_j)}^2 + \|\widetilde{u}_{\widehat{\mathbf{v}}_{j+1}}\|_{L^2(\widehat{I}_j)}^2 + \|\widetilde{u}_{\widehat{I}_j}\|_{L^2(\widehat{I}_j)}^2.$$

*Proof.* We prove the lemma for $j = 1$. In this case, $\widetilde{u}_{\widehat{W}_1} = \widetilde{u}_{\widehat{\mathbf{v}}_1} + \widetilde{u}_{\widehat{\mathbf{v}}_2} + \widetilde{u}_{\widehat{I}_1}$ and the required estimate is

$$\frac{3}{4}\|\widetilde{u}_{\widehat{W}_1}\|_*^2 \le \|\widetilde{u}_{\widehat{W}_1}\|_{L^2(\widehat{I}_1)}^2 \le 2\|\widetilde{u}_{\widehat{W}_1}\|_*^2$$

where

$$\|\widetilde{u}_{\widehat{W}_1}\|_*^2 = \|\widetilde{u}_{\widehat{\mathbf{v}}_1}\|_{L^2(\widehat{I}_1)}^2 + \|\widetilde{u}_{\widehat{\mathbf{v}}_2}\|_{L^2(\widehat{I}_1)}^2 + \|\widetilde{u}_{\widehat{I}_1}\|_{L^2(\widehat{I}_1)}^2.$$

Due to (C.59) we have

$$\begin{aligned}
\|\widetilde{u}_{\widehat{W}_1}\|_{L^2(\widehat{I}_1)}^2 &= \|\widetilde{u}_{\widehat{\mathbf{v}}_1}\|_{L^2(\widehat{I}_1)}^2 + \|\widetilde{u}_{\widehat{\mathbf{v}}_1}\|_{L^2(\widehat{I}_1)}^2 + \|\widetilde{u}_{\widehat{I}_1}\|_{L^2(\widehat{I}_1)}^2 \\
&\quad + 2\left\langle \widetilde{u}_{\widehat{\mathbf{v}}_1}, \widetilde{u}_{\widehat{\mathbf{v}}_2}\right\rangle_{L^2(\widehat{I}_1)} + 2\left\langle \widetilde{u}_{\widehat{\mathbf{v}}_1}, \widetilde{u}_{\widehat{\mathbf{e}}_1}\right\rangle_{L^2(\widehat{I}_1)} + 2\left\langle \widetilde{u}_{\widehat{\mathbf{v}}_2}, \widetilde{u}_{\widehat{\mathbf{e}}_1}\right\rangle_{L^2(\widehat{I}_1)} \\
&= \|\widetilde{u}_{\widehat{\mathbf{v}}_1}\|_{L^2(\widehat{I}_1)}^2 + \|\widetilde{u}_{\widehat{\mathbf{v}}_1}\|_{L^2(\widehat{I}_1)}^2 + \|\widetilde{u}_{\widehat{I}_1}\|_{L^2(\widehat{I}_1)}^2 + 2\left\langle \widetilde{u}_{\widehat{\mathbf{v}}_1}, \widetilde{u}_{\widehat{\mathbf{v}}_2}\right\rangle_{L^2(\widehat{I}_1)} \\
&\le 2\left(\|\widetilde{u}_{\widehat{\mathbf{v}}_1}\|_{L^2(\widehat{I}_1)}^2 + \|\widetilde{u}_{\widehat{\mathbf{v}}_1}\|_{L^2(\widehat{I}_1)}^2 + \|\widetilde{u}_{\widehat{I}_1}\|_{L^2(\widehat{I}_1)}^2\right).
\end{aligned}$$

On the other hand,

$$\|\widetilde{u}_{\widehat{W}_1}\|_{L^2(\widehat{I}_1)}^2 \ge \|\widetilde{u}_{\widehat{\mathbf{v}}_1}\|_{L^2(\widehat{I}_1)}^2 + \|\widetilde{u}_{\widehat{\mathbf{v}}_1}\|_{L^2(\widehat{I}_1)}^2 + \|\widetilde{u}_{\widehat{I}_1}\|_{L^2(\widehat{I}_1)}^2 - 2\left|\left\langle \widetilde{u}_{\widehat{\mathbf{v}}_1}, \widetilde{u}_{\widehat{\mathbf{v}}_2}\right\rangle_{L^2(\widehat{I}_1)}\right|. \quad (11.37)$$

We note that by the way that the vertex basis functions $\widetilde{\phi}_{\widehat{\mathbf{v}}_i}$, $i = 1, 2, 3$, are constructed (see Subsection 11.1.3.1) the restrictions of $\widetilde{u}_{\widehat{\mathbf{v}}_1}$ and $\widetilde{u}_{\widehat{\mathbf{v}}_2}$ on $\widehat{I}_1$ can be represented by

$$\widetilde{u}_{\widehat{\mathbf{v}}_1}\big|_{\widehat{I}_1}(t,0) = c\phi_0(t) \quad \text{and} \quad \widetilde{u}_{\widehat{\mathbf{v}}_2}\big|_{\widehat{I}_1}(t,0) = c^-\phi_0^-(t), \quad c,\, c^- \in \mathbb{R},\ t \in [0,1],$$

where $\phi_0$ and $\phi_0^-$ are defined in Definition C.1. Hence, by using (C.58) and noting that $\|\phi_0\|_{L^2(0,1)} = \|\phi_0^-\|_{L^2(0,1)}$ we deduce

$$\begin{aligned}
2\left|\left\langle \widetilde{u}_{\widehat{\mathbf{v}}_1}, \widetilde{u}_{\widehat{\mathbf{v}}_2}\right\rangle_{L^2(\widehat{I}_1)}\right| &= 2|c||c^-|\left|\left\langle \phi_0, \phi_0^-\right\rangle_{L^2(\widehat{I}_1)}\right| \\
&= \frac{1}{p+1}|c||c^-|\|\phi_0\|_{L^2(0,1)}\|\phi_0^-\|_{L^2(0,1)} \\
&= \frac{1}{p+1}\|\widetilde{u}_{\widehat{\mathbf{v}}_1}\|_{L^2(\widehat{I}_1)}\|\widetilde{u}_{\widehat{\mathbf{v}}_2}\|_{L^2(\widehat{I}_1)} \\
&\le \frac{1}{4}\left(\|\widetilde{u}_{\widehat{\mathbf{v}}_1}\|_{L^2(\widehat{I}_1)}^2 + \|\widetilde{u}_{\widehat{\mathbf{v}}_2}\|_{L^2(\widehat{I}_1)}^2\right) \quad \text{for } p \ge 1.
\end{aligned}$$

This inequality and (11.37) imply

$$\|\widetilde{u}_{\widehat{W}_1}\|^2_{L^2(\widehat{I}_1)} \geq \frac{3}{4}\left(\|\widetilde{u}_{\widehat{\mathbf{v}}_1}\|^2_{L^2(\widehat{I}_1)} + \|\widetilde{u}_{\widehat{\mathbf{v}}_1}\|^2_{L^2(\widehat{I}_1)} + \|\widetilde{u}_{\widehat{I}_1}\|^2_{L^2(\widehat{I}_1)}\right).$$

This completes the proof of the lemma. □

The next lemma gives some boundedness results for the operator $\mathscr{I}_{\widehat{W}}$.

**Lemma 11.2.** *The interpolation operator $\mathscr{I}_{\widehat{W}}$ defined by* (11.30) *satisfies, for all $u \in \mathbb{P}_p(\widehat{T})$,*

$$\|\mathscr{I}_{\widehat{W}} u\|_{L^2(\widehat{T})} \lesssim \|u\|_{L^2(\partial\widehat{T})}, \tag{11.38}$$

$$|\mathscr{I}_{\widehat{W}} u|_{H^{1/2}(\widehat{T})} \lesssim (1+\log p)^{1/2}\|u\|_{L^2(\partial\widehat{T})}, \tag{11.39}$$

$$\|\mathscr{I}_{\widehat{W}} u\|_{L^2(\widehat{T})} \lesssim (1+\log p)^{1/2}\|u\|_{H_S^{1/2}(\widehat{T})}, \tag{11.40}$$

$$|\mathscr{I}_{\widehat{W}} u|_{H^{1/2}(\widehat{T})} \lesssim (1+\log p)\|u\|_{H_S^{1/2}(\widehat{T})}. \tag{11.41}$$

*The constants are independent of u and p.*

*Proof.* First we note that (11.40) and (11.41) are direct consequences of (11.38), (11.39) and (C.55) in Lemma C.30. Hence we prove only (11.38) and (11.39). Recalling the definition of $\mathscr{I}_{\widehat{W}}$ in (11.30), it suffices to prove the required estimates for $\widetilde{\mathscr{I}}_{\widehat{W}} u$ and $\widetilde{1}_{\widehat{T}}\mu_{\partial\widehat{T}}(u)$, i.e., prove

$$\|\widetilde{\mathscr{I}}_{\widehat{W}} u\|_{L^2(\widehat{T})} \lesssim \|u\|_{L^2(\partial\widehat{T})}, \tag{11.42}$$

$$|\widetilde{\mathscr{I}}_{\widehat{W}} u|_{H^{1/2}(\widehat{T})} \lesssim (1+\log p)^{1/2}\|u\|_{L^2(\partial\widehat{T})}, \tag{11.43}$$

and

$$\|\widetilde{1}_{\widehat{T}}\mu_{\partial\widehat{T}}(u)\|_{L^2(\widehat{T})} \lesssim \|u\|_{L^2(\partial\widehat{T})}, \tag{11.44}$$

$$|\widetilde{1}_{\widehat{T}}\mu_{\partial\widehat{T}}(u)|_{H^{1/2}(\widehat{T})} \lesssim (1+\log p)^{1/2}\|u\|_{L^2(\partial\widehat{T})}. \tag{11.45}$$

First consider $\widetilde{\mathscr{I}}_{\widehat{W}} u$. Due to (11.27), we need to estimate each of $\widetilde{u}_{\widehat{\mathbf{v}}_i}$ and $\widetilde{u}_{\widehat{I}_i}$, the vertex and edge components of $u$. Since these components are linear combinations of, respectively, the basis functions $\widetilde{\phi}_{\widehat{\mathbf{v}}_i}$ and $\widetilde{\phi}_{\widehat{I}_i,q}$, $q = 1,\ldots,p-1$ (see Subsection 11.1.3.4) and since the extension operators are linear, we have

$$\widetilde{u}_{\widehat{\mathbf{v}}_i} = \underline{\mathscr{E}}^{i+1}\left(\widetilde{u}_{\widehat{\mathbf{v}}_i}|_{\partial\widehat{T}}\right) \quad \text{and} \quad \widetilde{u}_{\widehat{I}_i} = \mathscr{E}^i\left(\widetilde{u}_{\widehat{I}_i}|_{\widehat{I}_i}\right).$$

Consequently,

$$\widetilde{\mathscr{I}}_{\widehat{W}} u = \sum_{i=1}^{3} \underline{\mathscr{E}}^{i+1}\left(\widetilde{u}_{\widehat{\mathbf{v}}_i}|_{\partial\widehat{T}}\right) + \sum_{i=1}^{3} \mathscr{E}^i\left(\widetilde{u}_{\widehat{I}_i}|_{\widehat{I}_i}\right). \tag{11.46}$$

The triangle inequality, Theorem 11.1 and Theorem 11.3 yield

$$
\begin{aligned}
\|\widetilde{\mathscr{I}}_{\widehat{W}} u\|_{L^2(\widehat{T})} &= \sum_{i=1}^{3} \|\underline{\mathscr{E}}^{i+1}\left(\widetilde{u}_{\widehat{\mathbf{v}}_i}|_{\partial \widehat{T}}\right)\|_{L^2(\widehat{T})} + \sum_{i=1}^{3} \|\mathscr{E}^i\left(\widetilde{u}_{\widehat{I}_i}|_{\widehat{I}_i}\right)\|_{L^2(\widehat{T})} \\
&\lesssim \sum_{i=1}^{3} \left(\|\widetilde{u}_{\widehat{\mathbf{v}}_i}\|_{L^2(\widehat{I}_{i+2})} + \|\widetilde{u}_{\widehat{\mathbf{v}}_i}\|_{L^2(\widehat{I}_i)}\right) + \sum_{i=1}^{3} \|\widetilde{u}_{\widehat{I}_i}\|_{L^2(\widehat{I}_i)}.
\end{aligned}
$$

Note that the cyclic notation implies

$$
\sum_{i=1}^{3} \|\widetilde{u}_{\widehat{\mathbf{v}}_i}\|_{L^2(\widehat{I}_{i+2})} = \sum_{i=1}^{3} \|\widetilde{u}_{\widehat{\mathbf{v}}_{i+1}}\|_{L^2(\widehat{I}_i)}
$$

so that

$$
\|\widetilde{\mathscr{I}}_{\widehat{W}} u\|_{L^2(\widehat{T})} \lesssim \sum_{i=1}^{3} \left(\|\widetilde{u}_{\widehat{\mathbf{v}}_i}\|_{L^2(\widehat{I}_i)} + \|\widetilde{u}_{\widehat{\mathbf{v}}_{i+1}}\|_{L^2(\widehat{I}_i)} + \|\widetilde{u}_{\widehat{I}_i}\|_{L^2(\widehat{I}_i)}\right).
$$

Lemma 11.1 gives

$$
\|\widetilde{\mathscr{I}}_{\widehat{W}} u\|_{L^2(\widehat{T})} \lesssim \sum_{i=1}^{3} \|\widetilde{u}_{\widehat{W}_j}\|^2_{L^2(\widehat{I}_j)} = \|\widetilde{\mathscr{I}}_{\widehat{W}} u\|_{L^2(\partial \widehat{T})} = \|u\|_{L^2(\partial \widehat{T})}
$$

proving (11.42). To prove (11.43), we use (11.46) again, but this time we invoke Theorem 11.2 and Theorem 11.3 for the $H^{1/2}$-seminorm

$$
\begin{aligned}
|\mathscr{I}_{\widehat{W}} u|_{H^{1/2}(\widehat{T})} &\lesssim (1+\log p)^{1/2} \left(\|\widetilde{u}_{\widehat{\mathbf{v}}_i}\|_{L^2(\widehat{I}_{i+2})} + \|\widetilde{u}_{\widehat{\mathbf{v}}_i}\|_{L^2(\widehat{I}_i)} + \|\widetilde{u}_{\widehat{I}_i}\|_{L^2(\widehat{I}_i)}\right) \\
&\simeq (1+\log p)^{1/2} \|\widetilde{\mathscr{I}}_{\widehat{W}} u\|_{L^2(\partial \widehat{T})} = (1+\log p)^{1/2} \|u\|_{L^2(\partial \widehat{T})},
\end{aligned}
$$

proving (11.43).

Next we consider $\widetilde{1}_{\widehat{T}} \mu_{\partial \widehat{T}}(u)$. By the definition of $\widetilde{1}_{\widehat{T}}$ we have

$$
|\widetilde{1}_{\widehat{T}} \mu_{\partial \widehat{T}}(u)| = |1 - \widetilde{\mathscr{I}}_{\widehat{W}} 1| \, |\mu_{\partial \widehat{T}}(u)| \le \left(1 + |\widetilde{\mathscr{I}}_{\widehat{W}} 1|\right) |\mu_{\partial \widehat{T}}(u)|.
$$

Hence, by using Lemma C.31 and (11.42) (for the constant function 1) we deduce

$$
\begin{aligned}
\|\widetilde{1}_{\widehat{T}} \mu_{\partial \widehat{T}}(u)\|_{L^2(\widehat{T})} &\lesssim \left(\|1\|_{L^2(\widehat{T})} + \|\widetilde{\mathscr{I}}_{\widehat{W}} 1\|_{L^2(\widehat{T})}\right) \|u\|_{L^2(\partial \widehat{T})} \\
&\lesssim \|1\|_{L^2(\partial \widehat{T})} \|u\|_{L^2(\partial \widehat{T})} \simeq \|u\|_{L^2(\partial \widehat{T})},
\end{aligned}
$$

proving (11.44). Similarly, by using Lemma C.31 and (11.43) we deduce

$$
\begin{aligned}
|\widetilde{1}_{\widehat{T}} \mu_{\partial \widehat{T}}(u)|_{H^{1/2}(\widehat{T})} &\lesssim (1+\log p)^{1/2} \, |\widetilde{\mathscr{I}}_{\widehat{W}} 1|_{H^{1/2}(\widehat{T})} \|u\|_{L^2(\partial \widehat{T})} \\
&\lesssim (1+\log p) \, \|1\|_{L^2(\partial \widehat{T})} \, \|u\|_{L^2(\partial \widehat{T})} \\
&\simeq (1+\log p) \, \|u\|_{L^2(\partial \widehat{T})},
\end{aligned}
$$

proving (11.45). The lemma is proved. □

The following results concerning the interpolation operators on the three dimensional wire basket are initially proved in [32] (which deals with wire basket preconditioners for the $p$-version of the finite element method on tetrahedral meshes). However, several results in this reference contain an unknown constant $C(p)$ which depends on the polynomial degree $p$. This unknown constant stems from an unproved extension theorem. The results presented in the remainder this subsection, originally proved in [111], use Theorems 11.1–11.3 and provide an explicit dependency of the constant on $p$.

The following lemma can be compared with [32, Lemma 4.13 and Lemma 4.16].

**Lemma 11.3.** *The following estimate holds*

$$|\widetilde{\mathscr{I}}_{\widehat{W}(\widehat{\Omega})}u|^2_{H^1(\widehat{\Omega})} \lesssim (1+\log p)^3 \|u\|^2_{L^2(\widehat{W}(\widehat{\Omega}))} \quad \forall u \in \mathbb{P}_p(\widehat{\Omega}) \tag{11.47}$$

*and*

$$|\mathscr{I}_{\widehat{W}(\widehat{\Omega})}u|^2_{H^1(\widehat{\Omega})} \lesssim (1+\log p)^3 \|u\|^2_{L^2(\widehat{W}(\widehat{\Omega}))} \quad \forall u \in \mathbb{P}_p(\widehat{\Omega}) \tag{11.48}$$

*where $\widehat{W}(\widehat{\Omega})$ is the three-dimensional wire basket consisting of all vertices and edges of $\widehat{\Omega}$; see Subsection 11.1.5. The constants are independent of p and u.*

*Proof.* Recalling the definition of $\widetilde{\mathscr{I}}_{\widehat{W}(\widehat{\Omega})}u$ in (11.34) and the definitions of the components $\widetilde{u}_{\widehat{\mathbf{w}}_i}$ and $\widetilde{u}_{\widehat{\mathbf{e}}_i}$ in Subsection 11.1.5, we confirm that $\widetilde{\mathscr{I}}_{\widehat{W}(\widehat{\Omega})}u$ is the discrete harmonic extension of its boundary value. Lemma C.34 gives

$$\begin{aligned}
\|\widetilde{\mathscr{I}}_{\widehat{W}(\widehat{\Omega})}u\|^2_{H^1(\widehat{\Omega})} &\lesssim \|\widetilde{\mathscr{I}}_{\widehat{W}(\widehat{\Omega})}u\|^2_{H^{1/2}(\partial\widehat{\Omega})} \\
&\lesssim \sum_{i=1}^{4} \|\widetilde{u}_{\widehat{\mathbf{w}}_i}\|^2_{H^{1/2}(\partial\widehat{\Omega})} + \sum_{j=1}^{6} \|\widetilde{u}_{\widehat{\mathbf{e}}_j}\|^2_{H^{1/2}(\partial\widehat{\Omega})}.
\end{aligned}$$

Consider for example $\|\widetilde{u}_{\widehat{\mathbf{w}}_1}\|^2_{H^{1/2}(\partial\widehat{\Omega})}$. Noting the support of $\widetilde{u}_{\widehat{\mathbf{w}}_1}$ we have by using (A.106) and Lemma C.28

$$\begin{aligned}
\|\widetilde{u}_{\widehat{\mathbf{w}}_1}\|^2_{H^{1/2}(\partial\widehat{\Omega})} &\simeq \|\widetilde{u}_{\widehat{\mathbf{w}}_1}\|^2_{\widetilde{H}^{1/2}(\widehat{F}_2\cup\widehat{F}_3\cup\widehat{F}_4)} \\
&\lesssim (1+\log p)^2 \|\widetilde{u}_{\widehat{\mathbf{w}}_1}\|^2_{H^{1/2}(\widehat{F}_2\cup\widehat{F}_3\cup\widehat{F}_4)}.
\end{aligned} \tag{11.49}$$

Recalling the definition of the preliminary vertex basis functions in Section 11.1.3.1, the vertex component $\widetilde{u}_{\widehat{\mathbf{w}}_1}$ can be depicted as follows. First by using affine transformations we can map faces $\widehat{F}_2$, $\widehat{F}_3$, and $\widehat{F}_4$ onto triangles $\widehat{T}_2$, $\widehat{T}_3$, and $\widehat{T}_4$, as in Figure 11.5, such that the vertex $\widehat{\mathbf{w}}_1$ of the tetrahedron, which is also vertex $\widehat{\mathbf{v}}_1$ of the triangles, is located at the center.

On each triangle $\widehat{T}_k$, $k = 2,3,4$, the vertex function $\widetilde{u}_{\widehat{\mathbf{w}}_1}$ is the extension of the values of $u$ on the boundary of the triangles. The definition of the vertex basis function $\widetilde{\phi}_{\widehat{\mathbf{v}}_1}$ dictates that

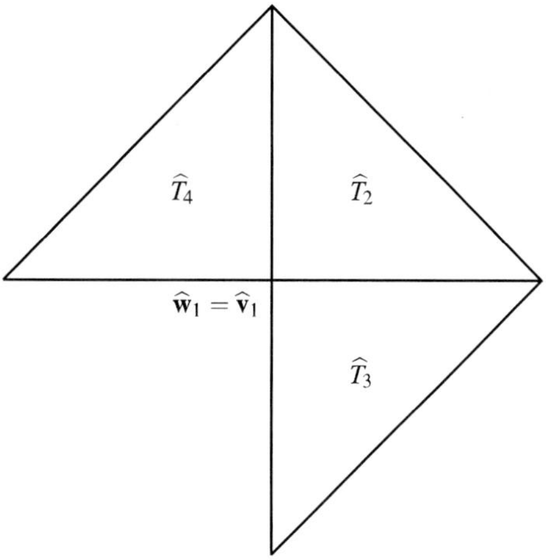


**Fig. 11.5** Affine images of $\widehat{F}_2$, $\widehat{F}_3$, and $\widehat{F}_4$

$$\widetilde{u}_{\widehat{\mathbf{w}}_1}(x,y) = \begin{cases} \underline{\mathscr{E}}^2 u(x,y), & (x,y) \in \widehat{T}_2, \\ \underline{\mathscr{E}}^2 u(x,-y), & (x,y) \in \widehat{T}_3, \\ \underline{\mathscr{E}}^2 u(-x,y), & (x,y) \in \widehat{T}_4. \end{cases}$$

Hence, by using Theorem 11.3 we deduce

$$\begin{aligned} \|\widetilde{u}_{\widehat{\mathbf{w}}_1}\|^2_{L^2(\widehat{F}_2 \cup \widehat{F}_3 \cup \widehat{F}_4)} &= \|\underline{\mathscr{E}}^2 u\|^2_{L^2(\widehat{T}_2)} + \|\underline{\mathscr{E}}^2 u(x,-y)\|^2_{L^2(\widehat{T}_3)} + \|\underline{\mathscr{E}}^2 u(-x,y)\|^2_{L^2(\widehat{T}_4)} \\ &\lesssim \|u\|^2_{L^2(\partial \widehat{T}_2)} + \|u\|^2_{L^2(\partial \widehat{T}_3)} + \|u\|^2_{L^2(\partial \widehat{T}_4)} \\ &\simeq \|u\|^2_{L^2(\widehat{W}(\widehat{\Omega}))}. \end{aligned} \tag{11.50}$$

On the other hand, for the $H^{1/2}$-seminorm we have

$$\begin{aligned} |\widetilde{u}_{\widehat{\mathbf{w}}_1}|^2_{H^{1/2}(\widehat{F}_2 \cup \widehat{F}_3 \cup \widehat{F}_4)} &\simeq |\underline{\mathscr{E}}^2 u|^2_{H^{1/2}(\widehat{T}_2)} + |\underline{\mathscr{E}}^2 u(x,-y)|^2_{H^{1/2}(\widehat{T}_3)} + |\underline{\mathscr{E}}^2 u(-x,y)|^2_{H^{1/2}(\widehat{T}_4)} \\ &\quad + \iint\limits_{\widehat{T}_2 \times \widehat{T}_3} \frac{|\underline{\mathscr{E}}^2 u(x,y) - \underline{\mathscr{E}}^2 u(x',-y')|^2}{|(x,y)-(x',-y')|^3} \, \mathrm{d}x\,\mathrm{d}y\,\mathrm{d}x'\,\mathrm{d}y' \\ &\quad + \iint\limits_{\widehat{T}_2 \times \widehat{T}_4} \frac{|\underline{\mathscr{E}}^2 u(x,y) - \underline{\mathscr{E}}^2 u(-x',y')|^2}{|(x,y)-(-x',y')|^3} \, \mathrm{d}x\,\mathrm{d}y\,\mathrm{d}x'\,\mathrm{d}y' \end{aligned}$$

$$+\iint_{\widehat{T}_3\times\widehat{T}_4} \frac{|\mathscr{E}^2u(x,-y)-\mathscr{E}^2u(-x',y')|^2}{|(-x,y)-(-x',y')|^3}\,\mathrm{d}x\,\mathrm{d}y\,\mathrm{d}x'\,\mathrm{d}y'$$
$$= |\mathscr{E}^2u|^2_{H^{1/2}(\widehat{T}_2)} + |\mathscr{E}^2u(x,-y)|^2_{H^{1/2}(\widehat{T}_3)} + |\mathscr{E}^2u(-x,y)|^2_{H^{1/2}(\widehat{T}_4)}$$
$$+3\iint_{\widehat{T}_2\times\widehat{T}_2} \frac{|\mathscr{E}^2u(x,y)-\mathscr{E}^2u(x',y')|^2}{|(x,y)-(x',y')|^3}\,\mathrm{d}x\,\mathrm{d}y\,\mathrm{d}x'\,\mathrm{d}y'$$
$$\simeq |\mathscr{E}^2u|^2_{H^{1/2}(\widehat{T}_2)} + |\mathscr{E}^2u(x,-y)|^2_{H^{1/2}(\widehat{T}_3)} + |\mathscr{E}^2u(-x,y)|^2_{H^{1/2}(\widehat{T}_4)}$$
$$\lesssim (1+\log p)\|u\|^2_{L^2(\widehat{W}(\widehat{\Omega}))} \tag{11.51}$$

where in the last step we used Theorem 11.3. It follows from (11.49), (11.50), and (11.51) that

$$\|\widehat{u}_{\widehat{\mathbf{w}}_1}\|^2_{H^{1/2}(\partial\widehat{\Omega})} \lesssim (1+\log p)^3\|u\|^2_{L^2(\widehat{W}(\widehat{\Omega}))}.$$

Similar results hold for $\widetilde{u}_{\widehat{\mathbf{w}}_i}$, $i=2,3,4$.

Next we consider the edge components $\widetilde{u}_{\widehat{\mathbf{e}}_j}$, $j=1,\ldots,6$. For example, consider $\widetilde{u}_{\widehat{\mathbf{e}}_1}$. We have by using again (A.106) and Lemma C.28

$$\|\widehat{u}_{\widehat{\mathbf{e}}_1}\|^2_{H^{1/2}(\widehat{\Omega})} \simeq \|\widehat{u}_{\widehat{\mathbf{e}}_1}\|^2_{\widetilde{H}^{1/2}(\widehat{F}_2\cup\widehat{F}_3)}$$
$$\lesssim (1+\log p)^2\|\widehat{u}_{\widehat{\mathbf{e}}_1}\|^2_{H^{1/2}(\widehat{F}_2\cup\widehat{F}_3)}.$$

Using again the affine mappings to change the geometries to $\widehat{T}_2$, $\widehat{T}_3$, and $\widehat{T}_4$ (see Figure 11.4 and Figure 11.5) we note that $\widehat{u}_{\widehat{\mathbf{e}}_1}|_{\widehat{T}_2}$ is the extension (by $\mathscr{E}^2$) of the values of $u$ on edge $\widehat{\mathbf{e}}_1$ and that

$$\widehat{u}_{\widehat{\mathbf{e}}_1}(x,y) = \mathscr{E}^2u(x,-y), \quad (x,y)\in\widehat{T}_3.$$

Using the same approach as for $\widetilde{u}_{\widehat{\mathbf{w}}_1}$ we can prove that

$$\|\widehat{u}_{\widehat{\mathbf{e}}_1}\|^2_{H^{1/2}(\partial\widehat{\Omega})} \lesssim (1+\log p)^3\|u\|^2_{L^2(\widehat{W}(\widehat{\Omega}))}.$$

The result also holds for $\widetilde{u}_{\widehat{\mathbf{e}}_j}$, $j=2,\ldots,6$. Altogether, we deduce (11.47).

In order to prove (11.48) it remains to prove that

$$\|\widetilde{1}_{\widehat{F}_k}\mu_{\partial\widehat{F}_k}(u)\|^2_{H^1(\widehat{\Omega})} \lesssim (1+\log p)^3\|u\|^2_{L^2(\partial\widehat{F}_k)}, \quad k=1,\ldots,4 \tag{11.52}$$

due to (11.35) and (11.36). It follows from Lemma C.31 and Lemma C.34 that

$$\|\widetilde{1}_{\widehat{F}_k}\mu_{\partial\widehat{F}_k}(u)\|^2_{H^1(\widehat{\Omega})} = \|\widetilde{1}_{\widehat{F}_k}\|^2_{H^1(\widehat{\Omega})}|\mu_{\partial\widehat{F}_k}(u)|^2$$
$$\lesssim \|\widetilde{1}_{\widehat{F}_k}\|^2_{H^{1/2}(\partial\widehat{\Omega})}\|u\|^2_{L^2(\partial\widehat{F}_k)}.$$

Noting the support of $\widetilde{1}_{\widehat{F}_k}$ and using (A.106) and Lemma C.28, we deduce

$$\begin{aligned}\|\widetilde{1}_{\widehat{F}_k}\|^2_{H^{1/2}(\partial\widehat{\Omega})} &\simeq \|\widetilde{1}_{\widehat{F}_k}\|^2_{\widetilde{H}^{1/2}(\widehat{F}_k)} \\ &\lesssim (1+\log p)^2\|\widetilde{1}_{\widehat{F}_k}\|^2_{H^{1/2}(\widehat{F}_k)}.\end{aligned}$$

Using (11.29), (11.42), and (11.43) (for the constant function 1) we infer

$$\begin{aligned}\|\widetilde{1}_{\widehat{F}_k}\|^2_{H^{1/2}(\partial\widehat{\Omega})} &\lesssim (1+\log p)^2\left(\|1\|^2_{H^{1/2}(\partial\widehat{\Omega})} + \|\widetilde{\mathscr{I}}_{\widehat{W}}1\|^2_{H^{1/2}(\partial\widehat{\Omega})}\right) \\ &\lesssim (1+\log p)^3\left(\|1\|^2_{H^{1/2}(\partial\widehat{\Omega})} + \|1\|^2_{H^{1/2}(\partial\widehat{F}_k)}\right) \\ &\simeq (1+\log p)^3.\end{aligned}$$

This proves (11.52), completing the proof of the lemma. □

We are now ready to present the preconditioners for the hypersingular integral equation.

## 11.2 Preconditioners for the Hypersingular Integral Equation

We recall that the boundary element space for solving the hypersingular integral equation (1.4) by the $hp$-version of the boundary element Galerkin method is

$$\mathscr{V} = \left\{v \in C(\Gamma) : v|_{\partial\Gamma} = 0 \quad \text{and} \quad v|_{\Gamma_i} \in \mathbb{P}_p(\Gamma_i),\ i = 1,\dots,n\right\}$$

where $\Gamma_1, \dots, \Gamma_n$ are triangles forming a conforming, regular, and quasi-uniform triangulation of $\Gamma$. In a standard way, we utilise the local basis functions defined in Section 11.1.3 to generate a basis for $\mathscr{V}$. In particular, we use the notation for components in (11.26), and the wire basket interpolation operators (11.27) and (11.30). Each triangular element $\Gamma_i$, $i = 1,\dots,n$, is mapped onto the reference triangle $\widehat{T}$. Thus we can define the global wire basket $W$ from the local wire baskets $W_j$, $j = 1,\dots,n$.

### 11.2.1 Subspace decomposition

As usual, the additive Schwarz preconditioners are defined from the subspace decomposition of $\mathscr{V}$

$$\mathscr{V} = \mathscr{V}_0 + \mathscr{V}_1 + \cdots + \mathscr{V}_n. \tag{11.53}$$

For our preconditioners, the subspace $\mathscr{V}_0$ is the space of wire basket functions with local basis functions $\phi_{\mathbf{v}_i}$ and $\phi_{I_i,q}$, $i = 1,2,3$ and $q = 2,\dots,p$ defined in Subsection 11.1.4, while $\mathscr{V}_j$ consists of functions in $\mathscr{V}$ supported in $\Gamma_j$, $j = 1,\dots,n$. Accordingly, any $v \in \mathscr{V}$ has a unique representation

$$v = v_W + \sum_{j=1}^{n} v_{\Gamma_j} \tag{11.54}$$

where $v_W \in \mathscr{V}_0$ and $v_{\Gamma_j} \in \mathscr{V}_j$, $j = 1, \ldots, n$. This is a non-overlapping decomposition, resulting in an iterative substructuring method. Preconditioners designed on this decomposition have a simple block-diagonal form when the basis functions are numbered appropriately. Depending on the choice of the local bilinear forms, see Subsection 2.3.2.2, these blocks are blocks of the stiffness matrix or specifically assembled blocks.

For $j = 1, \ldots, n$ the local bilinear form $a_j(\cdot,\cdot) : \mathscr{V}_j \times \mathscr{V}_j \longrightarrow \mathbb{R}$ inherits that of the bilinear form on $\mathscr{V}$, namely,

$$a_j(v,w) := a_{\mathfrak{W}}(v,w) \quad \forall v, w \in \mathscr{V}_j. \tag{11.55}$$

Two different choices of $a_0(\cdot,\cdot) : \mathscr{V}_0 \times \mathscr{V}_0 \longrightarrow \mathbb{R}$ will result in two different preconditioners.

## *11.2.2 Preconditioner I*

For this preconditioner, we follow [169]. The bilinear form $a_0(\cdot,\cdot)$ is defined by

$$a_0(v,w) := (1+\log p)^3 \sum_{j=1}^{n} \left\langle v - \mu_{\partial\Gamma_j}(v), w - \mu_{\partial\Gamma_j}(w) \right\rangle_{L^2(\partial\Gamma_j)} \qquad \forall v, w \in \mathscr{V}_0, \tag{11.56}$$

where we recall that $\mu_{\partial\Gamma_j}(v)$ is the integral mean of $v$ on $\partial\Gamma_j$. We note that, due to Lemma C.29,

$$\begin{aligned} a_0(v,v) &= (1+\log p)^3 \sum_{j=1}^{n} \|v_W - \mu_{\partial\Gamma_j}(v_W)\|^2_{L^2(\partial\Gamma_j)} \\ &= (1+\log p)^3 \sum_{j=1}^{n} \inf_{c_j \in \mathbb{R}} \|v - c_j\|^2_{L^2(\partial\Gamma_j)} \quad \forall v \in \mathscr{V}_0. \end{aligned} \tag{11.57}$$

We prove the following two lemmas.

**Lemma 11.4 (Coercivity of decomposition).** *For any $v \in \mathscr{V}$ satisfying* (11.54) *the following estimate holds*

$$a_{\mathfrak{W}}(v,v) \lesssim a_0(v_W, v_W) + \sum_{j=1}^{n} a_j(v_{\Gamma_j}, v_{\Gamma_j})$$

*where the constant is independent of $v$, $h$, and $p$.*

*Proof.* It follows from Theorem A.11 and Lemma B.3 that

$$a_{\mathfrak{W}}(v,v) \lesssim a_{\mathfrak{W}}(v_W,v_W) + \sum_{j=1}^{n} a_j(v_{\Gamma_j},v_{\Gamma_j}).$$

Hence it suffices to prove

$$a_{\mathfrak{W}}(v_W,v_W) \lesssim a_0(v_W,v_W)$$

or equivalently, see (11.57),

$$\|v_W\|^2_{\widetilde{H}_{\mathrm{I}}^{1/2}(\Gamma)} \lesssim (1+\log p)^3 \sum_{j=1}^{n} \inf_{c_j\in\mathbb{R}} \|v_W - c_j\|^2_{L^2(\partial\Gamma_j)}. \tag{11.58}$$

In order to prove this inequality we consider a three-dimensional domain $\Omega$ such that $\Gamma \subset \partial\Omega$. We decompose $\Omega$ into tetrahedra $\Omega_j$, $j=1,\ldots,N$, such that the trace of this mesh on $\Gamma$ is compatible with the mesh on $\Gamma$. Let $\Omega_i$ be a tetrahedron containing $\Gamma_i$ as a face. By an affine transformation, we map this tetrahedron onto the reference tetrahedron $\widehat{\Omega}$ such that $\Gamma_i$ is mapped onto the reference triangle $\widehat{T}$, a face of $\widehat{\Omega}$.

Let $\widehat{v}$ be the function defined on $\widehat{T}$ from $v_{W_i} := v_W\big|_{\Gamma_i}$ and the affine mapping, and let $\widehat{V}$ be the discrete harmonic extension of $\widehat{v}$ onto $\widehat{\Omega}$. It can be seen that $\widehat{v} = \mathscr{I}_{\widehat{T}}(\widehat{v}|_{\partial\widehat{T}})$ and $\widehat{V} = \mathscr{I}_{\widehat{W}(\widehat{\Omega})}(\widehat{v}|_{\partial\widehat{T}})$. Hence, Lemma 11.3 gives

$$|\widehat{V}|^2_{H^1(\widehat{\Omega})} \lesssim (1+\log p)^3 \|\widehat{v}\|^2_{L^2(\partial\widehat{T})}.$$

Since $\mathscr{I}_{\widehat{T}}$ and $\mathscr{I}_{\widehat{W}(\widehat{\Omega})}$ reproduce constants we deduce

$$|\widehat{V}|^2_{H^1(\widehat{\Omega})} = |\widehat{V}-c|^2_{H^1(\widehat{\Omega})} \lesssim (1+\log p)^3 \|\widehat{v}-c\|^2_{L^2(\partial\widehat{T})} \quad \forall c\in\mathbb{R}.$$

It follows from Lemma A.4 that

$$|V_{W_i}|^2_{H^1(\Omega_i)} \lesssim (1+\log p)^3 \|v_W - c_i\|^2_{L^2(\partial\Gamma_i)} \quad \forall c_i \in \mathbb{R}. \tag{11.59}$$

Here $V_{W_i}$ is the function defined on $\Omega_i$ corresponding to $\widehat{V}$ via the affine mapping. Patching these functions $V_{W_i}$, $i=1,\ldots,n$, together we obtain a function $V_W$ defined on the whole $\Omega$ whose trace on the boundary is $v_W$. The trace theorem gives

$$\|v_W\|^2_{\widetilde{H}_{\mathrm{I}}^{1/2}(\Gamma)} = \|v_W\|^2_{H^{1/2}(\partial\Omega)} \lesssim \|V_W\|^2_{H^1(\Omega)}.$$

Since $v_W$ vanishes on $\partial\Omega\setminus\Gamma$ we can invoke [181, Theorem 2.3.1] to obtain

$$\|V_W\|_{H^1(\Omega)} \simeq |V_W|_{H^1(\Omega)}$$

and thus

$$\|v_W\|^2_{\widetilde{H}^{1/2}_{\mathrm{I}}(\Gamma)} \lesssim |V_W|^2_{H^1(\Omega)} = \sum_{i=1}^{n} |V_{W_i}|^2_{H^1(\Omega_i)}.$$

This together with (11.59) implies

$$\|v_W\|^2_{\widetilde{H}^{1/2}_{\mathrm{I}}(\Gamma)} \lesssim (1+\log p)^3 \sum_{i=1}^{n} \|v_W - c_i\|^2_{L^2(\partial\Gamma_i)} \quad \forall c_i \in \mathbb{R},$$

resulting in (11.58). This completes the proof of the lemma. □

**Lemma 11.5 (Stability of decomposition).** *For any $v \in \mathscr{V}$ satisfying* (11.54) *the following estimate holds*

$$a_0(v_W, v_W) + \sum_{j=1}^{n} a_{\mathfrak{W}}(v_{\Gamma_j}, v_{\Gamma_j}) \lesssim (1+\log p)^4 a_{\mathfrak{W}}(v, v)$$

*where the constant is independent of v, h, and p.*

*Proof.* We first consider $a_0(v_W, v_W)$. Since

$$\begin{aligned} a_0(v_W, v_W) &= (1+\log p)^3 \sum_{i=1}^{n} \|v_W - \mu_{\partial\Gamma_i}(v_W)\|^2_{L^2(\partial\Gamma_i)} \\ &= (1+\log p)^3 \sum_{i=1}^{n} \|v - \mu_{\partial\Gamma_i}(v)\|^2_{L^2(\partial\Gamma_i)}, \end{aligned}$$

Lemma C.30 and a scaling argument imply

$$a_0(v_W, v_W) \lesssim (1+\log p)^4 |v|^2_{H^{1/2}(\Gamma)}.$$

Next we consider (see Lemma A.13)

$$a_{\mathfrak{W}}(v_{\Gamma_j}, v_{\Gamma_j}) \simeq \|v_{\Gamma_j}\|^2_{\widetilde{H}^{1/2}_{\mathrm{I}}(\Gamma)} \leq \|v_{\Gamma_j}\|^2_{\widetilde{H}^{1/2}_{\mathrm{I}}(\Gamma_j)}, \quad j = 1, \ldots, n.$$

By using an affine mapping which maps $\Gamma_j$ onto $\widehat{T}$, and which transformed $v$, $v_W$, and $v_{\Gamma_j}$ into $\widehat{v}$, $\widehat{v}_{\widehat{W}}$, and $\widehat{v}_{\widehat{T}}$, respectively, we can consider $\|\widehat{v}_{\widehat{T}}\|^2_{\widetilde{H}^{1/2}_{\mathrm{I}}(\widehat{T})}$ instead of $\|v_{\Gamma_j}\|^2_{\widetilde{H}^{1/2}_{\mathrm{I}}(\Gamma_i)}$. Since $\widehat{v}_{\widehat{T}} = (\widehat{v} - \widehat{v}_{\widehat{W}})|_{\widehat{T}} = (\widehat{v} - \mathscr{I}_{\widehat{W}}\widehat{v})|_{\widehat{T}}$ we have

$$\begin{aligned} \|\widehat{v}_{\widehat{T}}\|^2_{\widetilde{H}^{1/2}_{\mathrm{I}}(\widehat{T})} &= \|\widehat{v} - \mathscr{I}_{\widehat{W}}\widehat{v}\|^2_{\widetilde{H}^{1/2}_{\mathrm{I}}(\widehat{T})} \\ &\lesssim \|\widehat{v}\|^2_{\widetilde{H}^{1/2}_{\mathrm{I}}(\widehat{T})} + \|\mathscr{I}_{\widehat{W}}\widehat{v}\|^2_{\widetilde{H}^{1/2}_{\mathrm{I}}(\widehat{T})} \\ &\lesssim \|\widehat{v}\|^2_{\widetilde{H}^{1/2}_{\mathrm{I}}(\widehat{T})} + (1+\log p)^2 \|\mathscr{I}_{\widehat{W}}\widehat{v}\|^2_{H^{1/2}_{\mathrm{I}}(\widehat{T})} \end{aligned}$$

where in the last step we used Lemma C.28. It follows from (11.40) and (11.41) that

$$\|\widehat{v}_{\widehat{T}}\|^2_{\widetilde{H}^{1/2}_{\mathrm{I}}(\widehat{T})} = \|\widehat{v} - \mathscr{I}_{\widehat{W}}\widehat{v}\|^2_{\widetilde{H}^{1/2}_{\mathrm{I}}(\widehat{T})} \lesssim (1+\log p)^4 \|\widehat{v}\|^2_{H^{1/2}_{\mathrm{S}}(\widehat{T})}$$

where we used the fact that $\|\cdot\|^2_{H^{1/2}_{\mathrm{I}}(\widehat{T})} \simeq \|\cdot\|^2_{H^{1/2}_{\mathrm{S}}(\widehat{T})}$ with constants depending only on the diameter of $\widehat{T}$. Since $\mathscr{I}_{\widehat{W}}$ reproduces constants, we deduce

$$\|\widehat{v}_{\widehat{T}}\|^2_{\widetilde{H}^{1/2}_{\mathrm{I}}(\widehat{T})} = \|(\widehat{v}+c) - \mathscr{I}_{\widehat{W}}\widehat{(v+c)}\|^2_{\widetilde{H}^{1/2}_{\mathrm{I}}(\widehat{T})} \\ \lesssim (1+\log p)^4 \|\widehat{v}+c\|^2_{H^{1/2}_{\mathrm{S}}(\widehat{T})} \quad \forall c \in \mathbb{R}.$$

A quotient argument yields

$$\|\widehat{v}_{\widehat{T}}\|^2_{\widetilde{H}^{1/2}_{\mathrm{I}}(\widehat{T})} \lesssim (1+\log p)^4 |\widehat{v}|^2_{H^{1/2}(\widehat{T})}.$$

It follows from Lemma A.6 and (A.53) that

$$\|v_{\Gamma_j}\|^2_{\widetilde{H}^{1/2}_{\mathrm{I}}(\Gamma_j)} \lesssim (1+\log p)^4 |v|^2_{H^{1/2}(\Gamma_j)}$$

implying

$$\sum_{j=1}^{n} a_{\mathfrak{W}}(v_{\Gamma_j}, v_{\Gamma_j}) \lesssim \sum_{j=1}^{n} \|v_{\Gamma_j}\|^2_{\widetilde{H}^{1/2}_{\mathrm{I}}(\Gamma_j)} \lesssim (1+\log p)^4 \sum_{j=1}^{n} |v|^2_{H^{1/2}(\Gamma_j)} \\ \leq (1+\log p)^4 |v|^2_{H^{1/2}(\Gamma)} \lesssim (1+\log p)^4 a_{\mathfrak{W}}(v,v),$$

where in the penultimate step we used the definition of the $H^{1/2}$-seminorm to obtain $\sum_{j=1}^{n} |v|^2_{H^{1/2}(\Gamma_j)} \leq |v|^2_{H^{1/2}(\Gamma)}$. This completes the proof of the lemma. □

Lemma 11.4, Lemma 11.5, and Theorem 2.1 give the following result.

**Theorem 11.4.** *The condition number $\kappa(P_I)$ of the additive Schwarz operator $P_I$ associated with the subspace decomposition* (11.53) *and the choice of bilinear forms* (11.55) *and* (11.56) *is bounded by*

$$\kappa(P_I) \lesssim (1+\log p)^4$$

*where the constant is independent of $p$ and $h$.*

### *11.2.3 Preconditioner II*

This preconditioner uses the same bilinear form for the wire basket space $\mathscr{V}_0$, namely,

$$a_0(v,w) := a_{\mathfrak{W}}(v,w) \quad \forall v,w \in \mathscr{V}_0. \tag{11.60}$$

As usual, we prove the coercivity and stability of the decomposition (11.53).

**Lemma 11.6 (Coercivity of decomposition).** *For any $v \in \mathscr{V}$ satisfying* (11.54) *the following estimate holds*

$$a_{\mathfrak{W}}(v,v) \lesssim a_{\mathfrak{W}}(v_W,v_W) + \sum_{j=1}^{n} a_{\mathfrak{W}}(v_{\Gamma_j},v_{\Gamma_j})$$

*where the constant is independent of v, h, and p.*

*Proof.* The proof is standard as previously done by using Lemma B.3 and Theorem A.11. We omit the details. □

**Lemma 11.7 (Stability of decomposition).** *For any $v \in \mathscr{V}$ satisfying* (11.54) *the following estimate holds*

$$a_{\mathfrak{W}}(v_W,v_W) + \sum_{j=1}^{n} a_{\mathfrak{W}}(v_{\Gamma_j},v_{\Gamma_j}) \lesssim (1+\log p)^4 a_{\mathfrak{W}}(v,v)$$

*where the constant is independent of v, h, and p.*

*Proof.* This is a direct consequence of Lemma 11.5 and inequality (11.58). □

The above two lemmas and Theorem 2.1 prove the following theorem.

**Theorem 11.5.** *The condition number $\kappa(P_{II})$ of the additive Schwarz operator $P_{II}$ associated with the subspace decomposition* (11.53) *and the choice of bilinear forms* (11.55) *and* (11.60) *is bounded by*

$$\kappa(P_{II}) \lesssim (1+\log p)^4$$

*where the constant is independent of p and h.*

*Remark 11.1.* We finish this section by commenting on the costs of both preconditioners. For a given polynomial degree $p$, the dimension of the local subspace $\mathscr{V}_j$ associated with element $\Gamma_j$, $j = 1,\ldots,n$, equals the number of interior basis functions on $\Gamma_j$, which is $(p-1)(p-2)/2$; see Subsection 11.1.3. Thus the local problems associated with these local subspaces can be solved directly and in parallel. This is a procedure analogous to static condensation (albeit, it does not affect the wire basket components) and is standard in $p$-methods.

For refined meshes, the dimension of the wire basket space $\mathscr{V}_0$ can be much larger. It depends linearly in $p$ and $h^{-1}$. Therefore, the problem associated with this subspace using the bilinear form $a_{\mathfrak{W}}(\cdot,\cdot)$ (Preconditioner II) results in dense matrices and thus a direct solution is costly in general. One strategy is to solve this coarse problem approximately by, e.g., the conjugate gradient method. Preconditioner I on the other hand yields sparse matrices due to the use of the $L^2$-inner product in $a_0(\cdot,\cdot)$. Therefore, the problem can be solved efficiently. In this sense, Preconditioner II is rather of theoretical interest whereas Preconditioner I is the preferred method in practice.

## 11.3 Numerical Results

For our model problem we choose the domain $\Gamma = (-1/2, 1/2)^2 \times \{0\}$, use uniform triangular meshes of the type given in Figure 11.6, and select the right-hand side function $f = 1$ in the hypersingular integral equation. For this right-hand side function the *hp*-version with quasi-uniform meshes converges like $\mathcal{O}(h^{1/2}p^{-1})$ in the energy norm, see [29, 31].

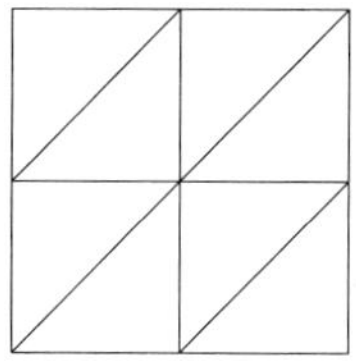


**Fig. 11.6** The mesh for the *p*-version.

Before presenting the numerical results, we comment on some computational issues. Firstly, we note that if the basis functions of the boundary element space $\mathcal{V}$ are appropriately order, then the matrix form of the preconditioners has a block diagonal form

$$\mathbf{C} := \begin{bmatrix} \mathbf{A}_W & \mathbf{0} & \cdots & \mathbf{0} & \mathbf{0} \\ \mathbf{0} & \mathbf{A}_{\Gamma_1} & \cdots & \mathbf{0} & \mathbf{0} \\ \vdots & \vdots & \ddots & \vdots & \vdots \\ \mathbf{0} & \mathbf{0} & \cdots & \mathbf{A}_{\Gamma_{n-1}} & \mathbf{0} \\ \mathbf{0} & \mathbf{0} & \cdots & \mathbf{0} & \mathbf{A}_{\Gamma_n} \end{bmatrix}$$

Here the submatrix $\mathbf{A}_W$ is the discretization of the bilinear form $a_0(\cdot,\cdot)$ and $\mathbf{A}_{\Gamma_j}$ is the discretization of the bilinear form $a_{\mathfrak{W}}(\cdot,\cdot)$ involving the interior basis functions defined on $\Gamma_j$, $j = 1,\ldots,n$. However, it is noted that in the implementation this matrix is not explicitly computed. One only needs to solve the local problem in $\mathcal{V}_j$ using the matrix $\mathbf{A}_{\Gamma_j}$, $j = 1,\ldots,n$, and solve the global problem in $\mathcal{V}_0$ using the matrix $\mathbf{A}_W$.

To construct the submatrix $\mathbf{A}_W$ it is necessary to compute

$$a_0(v,w) = (1+\log p)^3 \sum_{j=1}^{n} \left\langle v - \mu_{\partial\Gamma_j}(v), w - \mu_{\partial\Gamma_j}(w) \right\rangle_{L^2(\partial\Gamma_j)} \quad \forall v, w \in \mathcal{V}_0.$$

As usual, we carry out this computation on each element $\Gamma_j$ by computing on the reference element $\widehat{T}$ and using the affine transformation, namely, we compute

$$\langle \widehat{v} - \mu_{\partial\widehat{T}}(\widehat{v}), \widehat{w} - \mu_{\partial\widehat{T}}(\widehat{w}) \rangle_{L^2(\partial\widehat{T})} \quad \forall \widehat{v}, \widehat{w} \in \widehat{\mathcal{V}}_0.$$

First we recall that the wire basket space on $\widehat{T}$ has $P := 3 + 3(p-1)(p-2)/2$ basis functions, consisting of 3 vertex basis functions and $3(p-1)(p-2)/2$ edge basis functions; see Section 11.1.3. We denote these basis functions by $\widehat{\phi}_j$, $j = 1,\ldots,P$, and let $\mathbf{M}$ be the mass matrix defined by

$$M_{ij} := \int_{\partial\widehat{T}} \widehat{\phi}_i(x)\widehat{\phi}_j(x)\,\mathrm{d}x, \quad i,j = 1,\ldots,P.$$

Let $\widehat{\mathbf{v}}$, $\widehat{\mathbf{w}}$, and $\widehat{\mathbf{z}}$ (respectively) be the coefficient vectors in the representations (with respect to the basis function $\widehat{\phi}_j$) of the polynomials $\widehat{v}$, $\widehat{w}$, and the constant function 1 (respectively). Then, noting that $1_{\widehat{W}} = 1$ (since $\mathscr{I}_{\widehat{W}}$ reproduces constants; see (11.33)), we have

$$\begin{aligned}
\big\langle \widehat{v} - \mu_{\partial\widehat{T}}(\widehat{v}), \widehat{w} - \mu_{\partial\widehat{T}}(\widehat{w})\big\rangle_{L^2(\partial\widehat{T})} &= \langle \widehat{v}, \widehat{w}\rangle_{L^2(\partial\widehat{T})} \\
&\quad - \mu_{\partial\widehat{T}}(\widehat{w})\,\langle \widehat{v}, 1\rangle_{L^2(\partial\widehat{T})} - \mu_{\partial\widehat{T}}(\widehat{v})\,\langle \widehat{w}, 1\rangle_{L^2(\partial\widehat{T})} \\
&\quad + \mu_{\partial\widehat{T}}(\widehat{v})\mu_{\partial\widehat{T}}(\widehat{w})\,\langle 1, 1\rangle_{L^2(\partial\widehat{T})} \\
&= \widehat{\mathbf{w}}^\top \mathbf{M}\widehat{\mathbf{v}} - \mu_{\partial\widehat{T}}(\widehat{w})\widehat{\mathbf{z}}^\top \mathbf{M}\widehat{\mathbf{v}} - \mu_{\partial\widehat{T}}(\widehat{v})\widehat{\mathbf{z}}^\top \mathbf{M}\widehat{\mathbf{w}} \\
&\quad + \mu_{\partial\widehat{T}}(\widehat{v})\mu_{\partial\widehat{T}}(\widehat{w})\widehat{\mathbf{z}}^\top \mathbf{M}\widehat{\mathbf{z}}.
\end{aligned}$$

Simple calculations reveal

$$|\partial\widehat{T}| = \int_{\partial\widehat{T}} 1\,\mathrm{d}t = \int_{\partial\widehat{T}} 1_{\widehat{W}} \cdot 1_{\widehat{W}}\,\mathrm{d}t = \mathbf{z}^\top \mathbf{M}\mathbf{z}$$

so that

$$\mu_{\partial\widehat{T}}(\widehat{v}) = \frac{1}{|\partial\widehat{T}|}\int_{\partial\widehat{T}} 1 \cdot \widehat{v} = \frac{\widehat{\mathbf{z}}^\top \mathbf{M}\widehat{\mathbf{v}}}{\widehat{\mathbf{z}}^\top \mathbf{M}\widehat{\mathbf{z}}}.$$

Hence, since $\mathbf{M}$ is symmetric,

$$\begin{aligned}
\big\langle \widehat{v} - \mu_{\partial\widehat{T}}(\widehat{v}), \widehat{w} - \mu_{\partial\widehat{T}}(\widehat{w})\big\rangle_{L^2(\partial\widehat{T})} &= \widehat{\mathbf{w}}^\top \mathbf{M}\widehat{\mathbf{v}} - \frac{(\widehat{\mathbf{z}}^\top \mathbf{M}\widehat{\mathbf{v}})(\widehat{\mathbf{z}}^\top \mathbf{M}\widehat{\mathbf{w}})}{\widehat{\mathbf{z}}^\top \mathbf{M}\widehat{\mathbf{z}}} \\
&= \widehat{\mathbf{w}}^\top \mathbf{M}\widehat{\mathbf{v}} - \frac{\widehat{\mathbf{w}}^\top (\mathbf{M}\widehat{\mathbf{z}})(\mathbf{M}\widehat{\mathbf{z}})^\top \widehat{\mathbf{v}}}{\widehat{\mathbf{z}}^\top \mathbf{M}\widehat{\mathbf{z}}} \\
&= \widehat{\mathbf{w}}^\top \Big(\mathbf{M} - \frac{(\mathbf{M}\widehat{\mathbf{z}})(\mathbf{M}\widehat{\mathbf{z}})^\top}{\widehat{\mathbf{z}}^\top \mathbf{M}\widehat{\mathbf{z}}}\Big)\widehat{\mathbf{v}}.
\end{aligned}$$

Consequently, the $(i,j)$ entry of the submatrix $\mathbf{A}_{\widehat{W}}$ defined by $\widehat{a}_0(\cdot,\cdot)$ is computed as follows:

$$\big(\mathbf{A}_{\widehat{W}}\big)_{ij} = \widehat{a}_0(\widehat{\phi}_i, \widehat{\phi}_j) = (1 + \log p)^3 \Big\langle \widehat{\phi}_i - \mu_{\partial\widehat{T}}(\widehat{\phi}_i), \widehat{\phi}_j - \mu_{\partial\widehat{T}}(\widehat{\phi}_j)\Big\rangle_{L^2(\partial\widehat{T})}$$

$$= (1+\log p)^3 \mathbf{e}_j^\top \left( \mathbf{M} - \frac{(\mathbf{M}\widehat{\mathbf{z}})(\mathbf{M}\widehat{\mathbf{z}})^\top}{\widehat{\mathbf{z}}^\top \mathbf{M}\widehat{\mathbf{z}}} \right) \mathbf{e}_i,$$

where $\mathbf{e}_i = (0, \dots, 1, \dots, 0)$ denotes the $i^{\text{th}}$ canonical vector in $\mathbb{R}^P$. Thus

$$\mathbf{A}_{\widehat{W}} = (1+\log p)^3 \left( \mathbf{M} - \frac{(\mathbf{M}\widehat{\mathbf{z}})(\mathbf{M}\widehat{\mathbf{z}})^\top}{\widehat{\mathbf{z}}^\top \mathbf{M}\widehat{\mathbf{z}}} \right).$$

By using the affine transformations and assembly technique, the matrix $\mathbf{A}_W$ on the whole $\Gamma$ can be constructed.

*Remark 11.2.* To simplify the computation, the submatrix $\mathbf{A}_{\widehat{W}}$ can be defined by

$$\mathbf{A}_{\widehat{W}} = (1+\log p)^3 \left( \mathbf{D} - \frac{(\mathbf{D}\widehat{\mathbf{z}})(\mathbf{D}\widehat{\mathbf{z}})^\top}{\widehat{\mathbf{z}}^\top \mathbf{D}\widehat{\mathbf{z}}} \right)$$

where $\mathbf{D}$ is the diagonal matrix whose entries are the diagonal entries of $\mathbf{M}$. The matrix $\mathbf{A}_{\widehat{W}}$ in this case is a rank-one perturbation of the diagonal part $\mathbf{D}$ of the stiffness matrix $\mathbf{M}$. The two matrices $\mathbf{M}$ and $\mathbf{D}$ have equivalent spectra; see [169].

The following numerical results are reprinted from [111] with copyright permission. In Table 11.1, we present the condition numbers of the un-preconditioned stiffness matrix ($\kappa(\mathbf{A})$), and the condition numbers of the preconditioned matrices resulting from both preconditioners designed above ($\kappa(\mathbf{P}_I)$ and $\kappa(\mathbf{P}_{II})$). We also present the number of iterations (# iter) in each case. It is noted that it is simpler to implement Preconditioner I than Preconditioner II; see Remark 11.1.

| $p$ | DoF | CG | | PCG | | | |
|---|---|---|---|---|---|---|---|
| | | | | Preconditioner I | | Preconditioner II | |
| | | $\kappa(\mathbf{A})$ | # iter | $\kappa(\mathbf{P}_I)$ | # iter | $\kappa(\mathbf{P}_{II})$ | # iter |
| 1 | 1 | 0.100E+01 | 1 | 0.100E+01 | 1 | 0.100E+01 | 1 |
| 2 | 9 | 0.781E+01 | 6 | 0.561E+01 | 7 | 0.100E+01 | 1 |
| 3 | 25 | 0.734E+02 | 29 | 0.443E+02 | 18 | 0.547E+01 | 11 |
| 4 | 49 | 0.739E+03 | 100 | 0.745E+02 | 34 | 0.796E+01 | 17 |
| 5 | 81 | 0.102E+05 | 315 | 0.112E+03 | 44 | 0.103E+02 | 22 |
| 6 | 121 | 0.189E+06 | 993 | 0.150E+03 | 56 | 0.124E+02 | 27 |

**Table 11.1** Condition numbers and numbers of iterations for the $p$-version without and with preconditioning

In Figure 11.7 we present the condition numbers of the preconditioned matrix using Preconditioner I. The numerical results for Preconditioner II behave analogously. In this figure, we also plot the curve of $(1+\log p)^4$ and observe that the numerical results behave similarly in the given range of $p$.

In Figure 11.8 we present condition numbers of the preconditioned matrix $\mathbf{P}_I$ for the $h$-version and different polynomial degrees. The results confirm the asymptotic independence of the condition numbers on $h$.

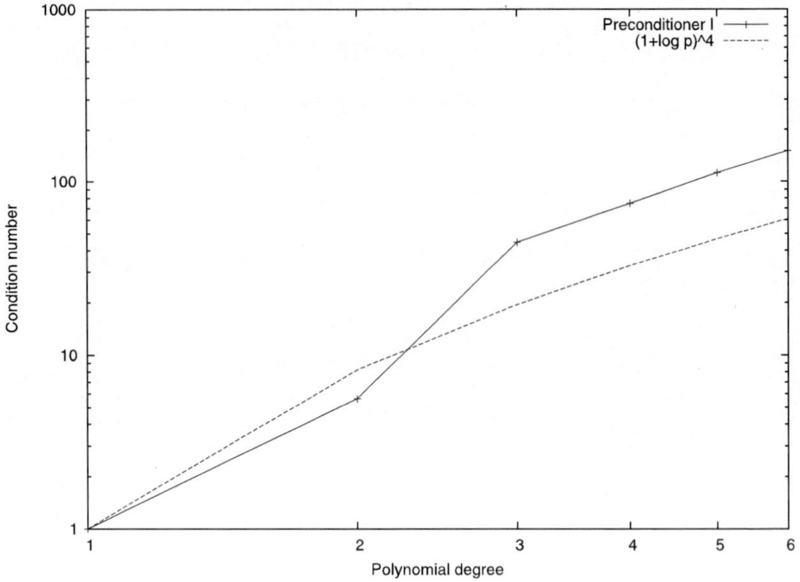


**Fig. 11.7** Condition number of the preconditioned Galerkin matrix, *p*-version.

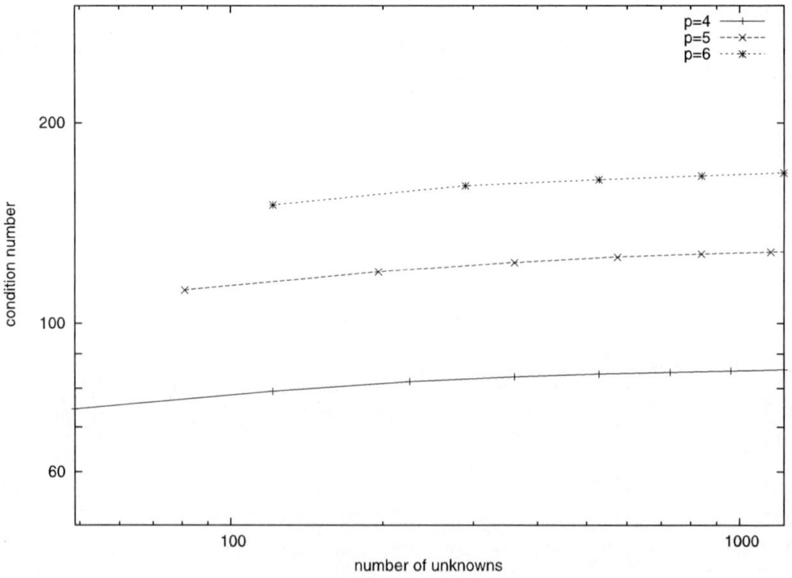


**Fig. 11.8** Condition number of the preconditioned Galerkin matrix, *h*-version

# Chapter 12
# Diagonal Scaling Preconditioner and Locally-Refined Meshes

In this chapter, we study the effect of mesh refinements on the condition numbers of the stiffness matrices arising from the $h$-version Galerkin boundary element discretisation of the weakly-singular and hypersingular integral equations, when quasi-uniform and locally refined meshes are used. The condition numbers of the diagonally scaled matrices are also studied.

So far in this monograph, our claim that the condition numbers for the $h$-version grow significantly with the degrees of freedom (the reason for the study of preconditioners) is not analytically supported. The first goal of this chapter is to address this issue. Moreover, when the solution to the boundary integral equation has singularities, it is more efficient to refine the mesh adaptively. Indeed, a uniform triangulation increases the degrees of freedom significantly, which is not suitable for dense matrices resulting from the boundary element discretisation. The adaptively refined partition may contain elements of very widely different sizes. As a consequence, the condition number may grow even more significantly, which means the advantages accrued by adaptivity are dissipated by the cost of solving a highly ill-conditioned linear system. This chapter also confirms the effect of local mesh refinement on the condition number. Furthermore, the chapter shows that a simple diagonal scaling preconditioner is a remedy. This preconditioner is in fact an additive Schwarz preconditioner with the subspaces in the decomposition being the spans of individual basis functions.

Section 12.3 presents results for shape-regular refinements. These results are first obtained in [5] where more general operators besides the hypersingular and weakly-singular integral operators are studied. We also present in Section 12.4 results obtained in [86] where anisotropic mesh refinements are studied. Numerical results are reported in Section 12.5.



E. P. Stephan and T. Tran, *Schwarz Methods and Multilevel Preconditioners for Boundary Element Methods*, https://doi.org/10.1007/978-3-030-79283-1_12

## 12.1 Problem Setting

In this chapter, $\Gamma$ is a manifold in $\mathbb{R}^d$, $d = 2,3$. When $d = 2$, the manifold $\Gamma$ is a bounded open curve in $\mathbb{R}^2$, and when $d = 3$ it is an open surface in $\mathbb{R}^3$. We consider the problem: Find $\varphi \in \widetilde{H}^m(\Gamma)$ satisfying

$$a(\varphi, \psi) = \langle \ell, \psi \rangle \quad \forall \psi \in \widetilde{H}^m(\Gamma) \tag{12.1}$$

where

$$a(\cdot,\cdot) = \begin{cases} a_{\mathfrak{V}}(\cdot,\cdot) & \text{if } m = -1/2, \\ a_{\mathfrak{W}}(\cdot,\cdot) & \text{if } m = 1/2. \end{cases} \tag{12.2}$$

Let $\mathscr{P}_h$ be a partition of $\Gamma$ into boundary elements $K$, as defined in Chapter 3 (for $d = 2$) and Chapters 10–11 (for $d = 3$). Let $h_K$ be the diameter of $K$. Let $\mathscr{N}_h$ be an indexing set of nodal points $x_k$ in the sense of Ciarlet [60]. This means, for piecewise constant elements ($m = -1/2$), $x_k$ is a midpoint of an element, while for continuous piecewise linear elements ($m = 1/2$), $x_k$ is a vertex of an element. Denoting $\{\phi_k\}_{k\in\mathscr{N}_h}$ to be the set of nodal basis functions defined on $\mathscr{P}_h$, i.e.,

$$\phi_k(x_\ell) = \delta_{k\ell}, \quad k,\ell \in \mathscr{N}_h. \tag{12.3}$$

we define $\Gamma_k$ to be the support of $\phi_k$ and $h_k$ to be the average of $h_K$, $K \subset \Gamma_k$. The boundary element space is defined by

$$\mathscr{V}_h := \operatorname{span}\{\phi_k : k \in \mathscr{N}_h\}.$$

Let $\mathbf{A}$ be the stiffness matrix of size $N \times N$ arising from the Galerkin approximation of (12.1). Here $N = \operatorname{card}(\mathscr{N})$. In the subsequent sections, we study the effect of local mesh refinements on the condition number of $\mathbf{A}$. Recall that

$$\kappa(\mathbf{A}) = \frac{\lambda_{\max}(\mathbf{A})}{\lambda_{\min}(\mathbf{A})}$$

where

$$\lambda_{\max}(\mathbf{A}) = \max_{\mathbf{v}\in\mathbb{R}^N} \frac{\langle \mathbf{A}\mathbf{v}, \mathbf{v}\rangle_{\mathbb{R}^N}}{\langle \mathbf{v}, \mathbf{v}\rangle_{\mathbb{R}^N}} \quad \text{and} \quad \lambda_{\min}(\mathbf{A}) = \min_{\mathbf{v}\in\mathbb{R}^N} \frac{\langle \mathbf{A}\mathbf{v}, \mathbf{v}\rangle_{\mathbb{R}^N}}{\langle \mathbf{v}, \mathbf{v}\rangle_{\mathbb{R}^N}}.$$

As in previous chapters, for the analysis it is convenient to evaluate the extremal eigenvalues of $\mathbf{A}$ in terms for functions in the boundary element spaces. The isomorphism $\mathbf{v} \mapsto v$ from $\mathbb{R}^N$ to $\mathscr{V}$ is defined as follows. For $k = 1,\dots,N$, let $c_k$ be the $k$-th component of the vector $\mathbf{v}$. We define

$$v_k := c_k\phi_k \quad \text{and} \quad v := \sum_{k=1}^N v_k = \sum_{k=1}^N c_k\phi_k. \tag{12.4}$$

Then, see (12.2),

$$\langle \mathbf{A}\mathbf{v}, \mathbf{v}\rangle_{\mathbb{R}^N} = a(v,v) \simeq \|v\|^2_{\widetilde{H}^m_{\mathrm{I}}(\Gamma)} \tag{12.5}$$

and, recalling (12.3),

$$\langle \mathbf{v}, \mathbf{v}\rangle_{\mathbb{R}^N} = \sum_{k=1}^{N} c_k^2 = \sum_{k=1}^{N} c_k^2 \phi_k(x_k)^2 = \sum_{k=1}^{N} v_k(x_k)^2 = \sum_{k=1}^{N} v(x_k)^2.$$

Therefore,

$$\lambda_{\max}(\mathbf{A}) = \max_{v\in\widetilde{H}^m(\Gamma)} \frac{\|v\|^2_{\widetilde{H}^m_{\mathrm{I}}(\Gamma)}}{\sum_{k=1}^{N} |v(x_k)|^2} \quad \text{and} \quad \lambda_{\min}(\mathbf{A}) = \min_{v\in\widetilde{H}^m(\Gamma)} \frac{\|v\|^2_{\widetilde{H}^m_{\mathrm{I}}(\Gamma)}}{\sum_{k=1}^{N} |v(x_k)|^2} \tag{12.6}$$

where we recall that $m = 1/2$ in the case of the hypersingular integral equation, and $m = -1/2$ in the case of the weakly-singular integral equation.

## 12.2 Preconditioning by Diagonal Scaling

We denote by $\mathbf{D}$ the diagonal matrix whose diagonal entries are those of $\mathbf{A}$, i.e.,

$$\mathbf{D}_{ij} = \mathbf{A}_{ij}\delta_{ij}, \quad i,j = 1,\ldots,N,$$

where $\delta_{ij}$ denotes the Kronecker delta. Taking $\mathbf{D}$ as the preconditioner, the preconditioned matrix becomes

$$\widetilde{\mathbf{A}} = \mathbf{D}^{-1/2}\mathbf{A}\mathbf{D}^{-1/2};$$

see Subsection 2.2.1. It is shown in Lemma 2.8 that

$$\kappa(\widetilde{\mathbf{A}}) = \frac{\lambda_{\max}(\widetilde{\mathbf{A}})}{\lambda_{\min}(\widetilde{\mathbf{A}})} = \frac{\lambda_{\max}(\widehat{\mathbf{A}})}{\lambda_{\min}(\widehat{\mathbf{A}})}$$

where

$$\widehat{\mathbf{A}} = \mathbf{D}^{-1}\mathbf{A}, \quad \lambda_{\max}(\widehat{\mathbf{A}}) = \max_{\mathbf{v}\in\mathbb{R}^N} \frac{\langle \mathbf{A}\mathbf{v}, \mathbf{v}\rangle_{\mathbb{R}^N}}{\langle \mathbf{D}\mathbf{v}, \mathbf{v}\rangle_{\mathbb{R}^N}}, \quad \lambda_{\min}(\widehat{\mathbf{A}}) = \min_{\mathbf{v}\in\mathbb{R}^N} \frac{\langle \mathbf{A}\mathbf{v}, \mathbf{v}\rangle_{\mathbb{R}^N}}{\langle \mathbf{D}\mathbf{v}, \mathbf{v}\rangle_{\mathbb{R}^N}}.$$

With $v$ and $v_k$ defined in (12.4) we have

$$\langle \mathbf{D}\mathbf{v}, \mathbf{v}\rangle_{\mathbb{R}^N} = \sum_{k=1}^{N} a(v_k, v_k) = \sum_{k=1}^{N} \|v_k\|^2_{\widetilde{H}^m_{\mathrm{I}}(\Gamma)}.$$

Therefore, due to (12.5),

$$
\begin{aligned}
\lambda_{\max}(\widehat{\mathbf{A}}) &= \max_{v \in \widetilde{H}^m(\Gamma)} \frac{\|v\|^2_{\widetilde{H}^m_{\rm I}(\Gamma)}}{\sum_{k=1}^N \|v_k\|^2_{\widetilde{H}^m_{\rm I}(\Gamma)}} \\
\lambda_{\min}(\widehat{\mathbf{A}}) &= \min_{v \in \widetilde{H}^m(\Gamma)} \frac{\|v\|^2_{\widetilde{H}^m_{\rm I}(\Gamma)}}{\sum_{k=1}^N \|v_k\|^2_{\widetilde{H}^m_{\rm I}(\Gamma)}}.
\end{aligned}
\tag{12.7}
$$

This preconditioner is in fact an additive Schwarz preconditioner. Indeed, let $\mathscr{V}_j$ be the space spanned by $\{\phi_j\}$, where we recall that $\{\phi_j : j = 1,\dots,N\}$ is the set of brick functions in the case $m = -1/2$, and hat functions in the case $m = 1/2$. Then the subspace decomposition

$$
\mathscr{V} = \mathscr{V}_1 + \cdots + \mathscr{V}_N \tag{12.8}
$$

defines an additive Schwarz preconditioner which yields exactly the diagonal scaling matrix $\mathbf{D}$.

In the remainder of this chapter, we find $c_{\max}$, $c_{\min}$, $C_{\max}$ and $C_{\min}$ such that for any $v \in \mathscr{V} \subset \widetilde{H}^m(\Gamma)$ of the form

$$
v = \sum_{k \in \mathscr{N}} v_k, \quad v_k := v(x_k)\phi_k, \tag{12.9}
$$

the following inequalities hold

$$
\begin{aligned}
\|v\|^2_{\widetilde{H}^m_{\rm I}(\Gamma)} &\le c_{\max} \sum_{k=1}^N |v(x_k)|^2, \\
c_{\min} \sum_{k=1}^N |v(x_k)|^2 &\le \|v\|^2_{\widetilde{H}^m_{\rm I}(\Gamma)},
\end{aligned}
\tag{12.10}
$$

and

$$
\begin{aligned}
\|v\|^2_{\widetilde{H}^m_{\rm I}(\Gamma)} &\le C_{\max} \sum_{k=1}^N \|v_k\|^2_{\widetilde{H}^m_{\rm I}(\Gamma)}, \\
C_{\min} \sum_{k=1}^N \|v_k\|^2_{\widetilde{H}^m_{\rm I}(\Gamma)} &\le \|v\|^2_{\widetilde{H}^m_{\rm I}(\Gamma)},
\end{aligned}
\tag{12.11}
$$

for $m = \pm 1/2$. Bounds for other values of $m$ can be found in [5]. We recall that (12.11) provides the coercivity and stability of the decomposition (12.8).

We will consider shape-regular mesh refinements in Section 12.3 and anisotropic mesh refinements in Section 12.4.

## 12.3 Shape-Regular Mesh Refinements

In this section, we consider mesh refinements which yield shape-regular elements. We present only results for the hypersingular and weakly-singular integral equations. More general results can be found in [5].

We make the following assumptions on $\mathscr{P}_h$:

**Assumption 12.1.** The family of partitions $\{\mathscr{P}_h : h > 0\}$ is nondegenerate or shape-regular, i.e., if $h_K$ is the diameter of an element $K$ and $\rho_K$ is the diameter of the largest inscribed ball in $K$, then

$$\frac{h_K}{\rho_K} \le c \quad \forall K \in \mathscr{P}_h,\ \forall h > 0.$$

Assumption 12.1 implies that the partition $\mathscr{P}_h$ is locally quasi-uniform: the ratio of the diameters of any pair of adjacent elements is uniformly bounded. We note that a locally quasi-uniform partition may still contain elements of greatly different size. In other words, if $h_{\max}$ (respectively, $h_{\min}$) denotes the diameter of the largest (respectively, smallest) element in $\mathscr{P}_h$, the ratio $h_{\max}/h_{\min}$ may be arbitrarily large.

**Assumption 12.2.** The number of elements $K$ intersecting $\Gamma_k$ is uniformly bounded, i.e., for any $K \in \mathscr{P}_h$, if

$$\begin{aligned} \mathscr{N}_h(K) &:= \{k \in \mathscr{N}_h : \Gamma_k \cap K \ne \varnothing\}, \\ M_h &:= \max\{\operatorname{card}(\mathscr{N}_h(K)) : K \in \mathscr{P}_h\}, \end{aligned} \tag{12.12}$$

then

$$M_h \le M \quad \forall h > 0. \tag{12.13}$$

Assumption 12.2 implies the following properties: The set $\mathscr{N}_h$ can be partitioned into disjoint subsets $\mathscr{N}_{h,1}, \ldots, \mathscr{N}_{h,L_h}$

$$\mathscr{N}_h = \mathscr{N}_{h,1} \cup \cdots \cup \mathscr{N}_{h,L_h}, \quad \mathscr{N}_{h,\ell} \cap \mathscr{N}_{h,\ell'} = \varnothing,\ \ell \ne \ell', \tag{12.14}$$

such that for all $\ell = 1, \ldots, L_h$

$$k, k' \in \mathscr{N}_{h,\ell} \implies \operatorname{interior}(\operatorname{supp}\phi_k) \cap \operatorname{interior}(\operatorname{supp}\phi_{k'}) = \varnothing. \tag{12.15}$$

The constants $L_h$ are uniformly bounded, i.e.,

$$L_h \le L \quad \forall h > 0.$$

For instance, if $\Gamma$ is an open curve and the Galerkin subspace consists of continuous piecewise polynomials of degree $p$, then $M = L = p + 1$. Due to the above uniform boundedness, in the sequel we omit the subscript $h$ for simplification of notations.

### 12.3.1 Coercivity of the decomposition

In this section, we find the constants $c_{\max}$ and $C_{\max}$ in (12.10) and (12.11). These estimates have the same effect as coercivity in the decomposition of $v$ by additive Schwarz operators.

First we need the following lemma.

**Lemma 12.1.** *Let $v \in \mathscr{V}$ be decomposed by* (12.9). *Then, for $q \in [1,\infty)$,*

$$\sum_{k\in\mathscr{N}} \|v\|_{L^q(\Gamma_k)}^q \lesssim \|v\|_{L^q(\Gamma)}^q \lesssim \sum_{k\in\mathscr{N}} \|v_k\|_{L^q(\Gamma)}^q.$$

*Proof.* The left inequality is a straightforward consequence of (12.14) and (12.15). Indeed

$$\sum_{k\in\mathscr{N}} \|v\|_{L^q(\Gamma_k)}^q = \sum_{\ell=1}^{L}\sum_{k\in\mathscr{N}_\ell} \|v\|_{L^q(\Gamma_k)}^q \le \sum_{\ell=1}^{L} \|v\|_{L^q(\Gamma)}^q = L\|v\|_{L^q(\Gamma)}^q.$$

To prove the right inequality, let $x \in K \in \mathscr{P}$. Then, noting (12.12),

$$|v(x)| = \Big|\sum_{k\in\mathscr{N}} v_k(x)\Big| = \Big|\sum_{k\in\mathscr{N}(K)} v_k(x)\Big| \le \sum_{k\in\mathscr{N}(K)} |v_k(x)|.$$

It follows from Hölder's inequality and (12.13) that

$$|v(x)|^q \lesssim \sum_{k\in\mathscr{N}(K)} |v_k(x)|^q$$

where the constant is independent of $q$. Integrating over $x \in K$ and summing over all $K \in \mathscr{P}$ give

$$\|v\|_{L^q(\Gamma)}^q \lesssim \sum_{k\in\mathscr{N}} \|v_k\|_{L^q(\Gamma)}^q,$$

completing the proof. □

The next lemma gives explicit forms of $c_{\max}$ and $C_{\max}$ in (12.10) and (12.11) for $m = 1/2$.

**Lemma 12.2.** *Let $v$ be decomposed by* (12.9). *Then for $d = 2, 3$*

$$\|v\|_{\widetilde{H}_{\mathrm{I}}^{1/2}(\Gamma)}^2 \lesssim \sum_{k\in\mathscr{N}} \|v_k\|_{\widetilde{H}_{\mathrm{I}}^{1/2}(\Gamma)}^2 \tag{12.16}$$

*and*

$$\|v\|_{\widetilde{H}_{\mathrm{I}}^{1/2}(\Gamma)}^2 \lesssim h_{\max}^{d-2} \sum_{k\in\mathscr{N}} |v(x_k)|^2. \tag{12.17}$$

*Inequality* (12.16) *is the coercivity of the decomposition* (12.8) *when $m = 1/2$.*

*Proof.* By (12.14) and the Cauchy–Schwarz inequality

$$\|v\|^2_{\widetilde{H}_{\rm I}^{1/2}(\Gamma)} = \|\sum_{\ell=1}^{L}\sum_{k\in\mathcal{N}_\ell} v_k\|^2_{\widetilde{H}_{\rm I}^{1/2}(\Gamma)} \le \left(\sum_{\ell=1}^{L}\|\sum_{k\in\mathcal{N}_\ell} v_k\|_{\widetilde{H}_{\rm I}^{1/2}(\Gamma)}\right)^2$$
$$\le L\sum_{\ell=1}^{L}\|\sum_{k\in\mathcal{N}_\ell} v_k\|^2_{\widetilde{H}_{\rm I}^{1/2}(\Gamma)}.$$

Noting (12.15), we can invoke Theorem A.11 to obtain

$$\|v\|^2_{\widetilde{H}_{\rm I}^{1/2}(\Gamma)} \lesssim \sum_{\ell=1}^{L}\sum_{k\in\mathcal{N}_\ell}\|v_k\|^2_{\widetilde{H}_{\rm I}^{1/2}(\Gamma)} = \sum_{k\in\mathcal{N}}\|v_k\|^2_{\widetilde{H}_{\rm I}^{1/2}(\Gamma)},$$

proving (12.16). The above inequality and Lemma C.13 also give

$$\|v\|^2_{\widetilde{H}_{\rm I}^{1/2}(\Gamma)} \lesssim \sum_{k\in\mathcal{N}}|v(x_k)|^2\|\phi_k\|^2_{\widetilde{H}_{\rm I}^{1/2}(\Gamma)} \simeq \sum_{k\in\mathcal{N}}|v(x_k)|^2\|\phi_k\|^2_{\widetilde{H}^{1/2}(\Gamma)}$$
$$\simeq \sum_{k\in\mathcal{N}}|v(x_k)|^2 h_k^{d-2} \le h_{\max}^{d-2}\sum_{k\in\mathcal{N}}|v(x_k)|^2,$$

proving (12.17). □

The following lemma gives explicit forms of $c_{\max}$ and $C_{\max}$ in (12.10) and (12.11) for $m=-1/2$.

**Lemma 12.3.** *Let $v$ be decomposed by* (12.9). *Then*

$$\|v\|^2_{\widetilde{H}_{\rm I}^{-1/2}(\Gamma)} \le c_{\max}\sum_{k\in\mathcal{N}}|v(x_k)|^2$$

*and*

$$\|v\|^2_{\widetilde{H}_{\rm I}^{-1/2}(\Gamma)} \le C_{\max}\sum_{k\in\mathcal{N}}\|v_k\|^2_{\widetilde{H}_{\rm I}^{-1/2}(\Gamma)}$$

*where*

$$c_{\max} \simeq \begin{cases} Nh_{\max}^2\Big(1+|\log(Nh_{\max})|\Big), & d=2,\\ N^{1/2}h_{\max}^3, & d=3,\end{cases} \tag{12.18}$$

*and*

$$C_{\max} \simeq \begin{cases} N\dfrac{1+|\log(Nh_{\min})|}{1+|\log(h_{\max})|}, & d=2,\\ N^{1/2}, & d=3.\end{cases} \tag{12.19}$$

*The inequality involving* (12.19) *is the coercivity of the decomposition* (12.8) *in the case $m=-1/2$.*

*Proof.* Consider first the case when $d=2$. Applying Lemma A.18 Part (iv) gives, for any $q\in(1,2)$,

$$\|v\|^2_{\widetilde{H}_{\mathrm{I}}^{-1/2}(\Gamma)} \simeq \|v\|^2_{\widetilde{H}^{-1/2}(\Gamma)} \lesssim \frac{q}{q-1}\|v\|^2_{L^q(\Gamma)} \lesssim \frac{q}{q-1}\left(\sum_{k\in\mathscr{N}} \|v_k\|^q_{L^q(\Gamma)}\right)^{2/q} \tag{12.20}$$

where in the last step we used Lemma 12.1. Here the constants are independent of $q$. Lemma C.12 gives

$$\sum_{k\in\mathscr{N}} \|v_k\|^q_{L^q(\Gamma)} = \sum_{k\in\mathscr{N}} |v(x_k)|^q \|\phi_k\|^q_{L^q(\Gamma)} \simeq \sum_{k\in\mathscr{N}} h_k |v(x_k)|^q \le h_{\max} \sum_{k\in\mathscr{N}} |v(x_k)|^q.$$

Using Hölder's inequality

$$\sum_{k=1}^N a_k b_k \le \left(\sum_{k=1}^N a_k^p\right)^{1/p} \left(\sum_{k=1}^N b_k^{p'}\right)^{1/p'} \tag{12.21}$$

with

$$a_k = 1, \quad b_k = |v(x_k)|^q, \quad p = \frac{2}{2-q}, \quad p' = \frac{2}{q}, \tag{12.22}$$

we deduce

$$\sum_{k\in\mathscr{N}} \|v_k\|^q_{L^q(\Gamma)} \le h_{\max} N^{1-q/2} \left(\sum_{k\in\mathscr{N}} |v(x_k)|^2\right)^{q/2}.$$

This and (12.20) yield

$$\|v\|^2_{\widetilde{H}_{\mathrm{I}}^{-1/2}(\Gamma)} \lesssim \frac{q}{q-1}\, h_{\max}^{2/q}\, N^{2/q-1} \sum_{k\in\mathscr{N}} |v(x_k)|^2.$$

Let $q'$ be the conjugate of $q$, i.e., $1/q + 1/q' = 1$. Then

$$\begin{aligned} \|v\|^2_{\widetilde{H}_{\mathrm{I}}^{-1/2}(\Gamma)} &\lesssim q'\, h_{\max}^2\, h_{\max}^{-2/q'}\, N^{1-2/q'} \sum_{k\in\mathscr{N}} |v(x_k)|^2 \\ &= q'\, \left(N h_{\max}\right)^{-2/q'} N h_{\max}^2 \sum_{k\in\mathscr{N}} |v(x_k)|^2. \end{aligned} \tag{12.23}$$

Let

$$A = q'\, \left(N h_{\max}\right)^{-2/q'} \quad \text{and} \quad q' = \max\{3, |\log(N h_{\max})|\}.$$

Note that this choice of $q'$ ensures that $q < 2$ as required in (12.22). We consider different cases depending on the value of $N h_{\max}$.

- If $N h_{\max} > e^3$, then $q' = \log(N h_{\max})$ so that $(N h_{\max})^{-2/q'} = e^{-2}$, implying

$$A \lesssim \log(N h_{\max}) = |\log(N h_{\max})|.$$

- If $1 < N h_{\max} \le e^3$, then $q' = 3$ so that $(N h_{\max})^{-2/q'} = (N h_{\max})^{-2/3} < 1$, implying

$$A < 3.$$

- If $e^{-3} < Nh_{\max} \le 1$, then $q' = 3$ so that $1 < (Nh_{\max})^{-2/q'} = (Nh_{\max})^{-2/3} < e^2$. Hence
$$A < 3e^2.$$
- If $0 < Nh_{\max} \le e^{-3}$, then $q' = |\log(Nh_{\max})|$ so that $(Nh_{\max})^{-2/q'} = e^2$, implying
$$A \lesssim |\log(Nh_{\max})|.$$

In all cases, by choosing $q' = \max\{3, |\log(Nh_{\max})|\}$ we obtain $A \lesssim 1 + |\log(Nh_{\max})|$, so that (12.18) (for $d = 2$) follows from (12.23).

To obtain (12.19) for $d = 2$, we use Lemma C.12 and Hölder'inequality again with $a_k = h_k^{1-q}$, $b_k = |v(x_k)|^q\, h_k^q$, $p = 2/(2-q)$, and $p' = 2/q$, see (12.21), to have

$$\begin{aligned}\sum_{k\in\mathcal{N}} \|v_k\|^q_{L^q(\Gamma)} &\simeq \sum_{k\in\mathcal{N}} |v(x_k)|^q\, h_k \\ &\le \left(\sum_{k\in\mathcal{N}} h_k^{-2(q-1)/(2-q)}\right)^{1-q/2} \left(\sum_{k\in\mathcal{N}} |v(x_k)|^2 h_k^2\right)^{q/2}. \qquad (12.24)\end{aligned}$$

Since $1 < q < 2$ we have $2(q-1)/(2-q) > 0$, so that the first factor on the right-hand side is bounded as

$$\left(\sum_{k\in\mathcal{N}} h_k^{-2(q-1)/(2-q)}\right)^{1-q/2} \lesssim N^{1-q/2}\, h_{\min}^{1-q} = N^{q/2}(Nh_{\min})^{1-q}$$

while the second factor is bounded by

$$\sum_{k\in\mathcal{N}} |v(x_k)|^2 h_k^2 \simeq \sum_{k\in\mathcal{N}} \frac{|v(x_k)|^2 \|\phi_k\|^2_{\widetilde{H}^{-1/2}(\Gamma)}}{1+|\log h_k|} \le \frac{1}{1+|\log(h_{\max})|} \sum_{k\in\mathcal{N}} \|v_k\|^2_{\widetilde{H}^{-1/2}(\Gamma)}$$

where we have used Lemma C.13. Therefore, (12.20) and (12.24) imply

$$\begin{aligned}\|v\|^2_{\widetilde{H}_{\mathrm{I}}^{-1/2}(\Gamma)} &\lesssim \frac{q}{q-1} N (Nh_{\min})^{2(1-q)/q} \frac{1}{1+|\log(h_{\max})|} \sum_{k\in\mathcal{N}} \|v_k\|^2_{\widetilde{H}^{-1/2}(\Gamma)} \\ &= q'(Nh_{\min})^{-2/q'} \frac{N}{1+|\log(h_{\max})|} \sum_{k\in\mathcal{N}} \|v_k\|^2_{H^{-1/2}(\Gamma)}, \qquad (12.25)\end{aligned}$$

where $q'$ is the conjugate of $q$, i.e., $1/q + 1/q' = 1$. Letting $q' = \max\{3, |\log(Nh_{\min})|\}$ and repeating the argument in the proof of (12.18) above, we deduce

$$q'(Nh_{\min})^{-2/q'} \lesssim 1 + |\log(Nh_{\min})|.$$

Thus (12.19) (for $d = 2$) follows from (12.25).

Now consider the case when $d = 3$. Lemma A.18 Part (iii) (with $q = 4/3$) and Lemma 12.1 give

$$\|v\|^2_{\widetilde{H}_I^{-1/2}(\Gamma)} \simeq \|v\|^2_{\widetilde{H}^{-1/2}(\Gamma)} \lesssim \|v\|^2_{L^{4/3}(\Gamma)} \lesssim \left( \sum_{k\in\mathcal{N}} \|v_k\|^{4/3}_{L^{4/3}(\Gamma)} \right)^{3/2}.$$

Using Hölder's inequality again yields

$$\sum_{k\in\mathcal{N}} \|v_k\|^{4/3}_{L^{4/3}(\Gamma)} \lesssim N^{1/3} \left( \sum_{k\in\mathcal{N}} \|v_k\|^2_{L^{4/3}(\Gamma)} \right)^{2/3}$$

so that

$$\|v\|^2_{\widetilde{H}_I^{-1/2}(\Gamma)} \lesssim N^{1/2} \sum_{k\in\mathcal{N}} \|v_k\|^2_{L^{4/3}(\Gamma)}.$$

On the one hand, Lemma C.12 gives

$$\|v_k\|^2_{L^{4/3}(\Gamma)} = |v(x_k)|^2 \|\phi_k\|^2_{L^{4/3}(\Gamma)} \simeq |v(x_k)|^2 h_k^3 \le |v(x_k)|^2 h_{\max}^3$$

and on the other hand Lemma C.13 gives

$$\|v_k\|^2_{L^{4/3}(\Gamma)} \simeq |v(x_k)|^2 h_k^3 \simeq |v(x_k)|^2 \|\phi_k\|^2_{\widetilde{H}^{-1/2}(\Gamma)} = \|v_k\|^2_{\widetilde{H}_I^{-1/2}(\Gamma)}.$$

Therefore,

$$\|v\|^2_{\widetilde{H}_I^{-1/2}(\Gamma)} \lesssim N^{1/2} h_{\max}^3 \sum_{k\in\mathcal{N}} |v(x_k)|^2$$

and

$$\|v\|^2_{\widetilde{H}_I^{-1/2}(\Gamma)} \lesssim N^{1/2} \sum_{k\in\mathcal{N}} \|v_k\|^2_{\widetilde{H}_I^{-1/2}(\Gamma)}. \qquad \square$$

### 12.3.2 Stability of the decomposition

We now find the constants $c_{\min}$ and $C_{\min}$ in (12.10) and (12.11). These estimates have the same effect as stability of decomposition in the decomposition of $v$ by additive Schwarz operators. First we prove the result for the $\widetilde{H}^{1/2}$-norm.

**Lemma 12.4.** *Let $v$ be decomposed by* (12.9). *Then*

$$c_{\min} \sum_{k\in\mathcal{N}} |v(x_k)|^2 \le \|v\|^2_{\widetilde{H}_I^{1/2}(\Gamma)} \quad \textit{and} \quad C_{\min} \sum_{k\in\mathcal{N}} \|v_k\|^2_{\widetilde{H}_I^{1/2}(\Gamma)} \le \|v\|^2_{\widetilde{H}_I^{1/2}(\Gamma)}$$

*where*

$$c_{\min}^{-1} \simeq \begin{cases} N\Big(1 + |\log(Nh_{\min})|\Big), & d = 2, \\ N^{1/2} h_{\min}^{-1}, & d = 3, \end{cases}$$

*and*

$$C_{\min}^{-1} \simeq \begin{cases} N\Big(1 + |\log(Nh_{\min})|\Big), & d = 2, \\ N^{1/2}, & d = 3. \end{cases}$$

*The inequality involving $C_{\min}$ gives the stability of the decomposition* (12.8).

*Proof.* Recalling that $h_k$ is the average of $h_K$ for $K \subset \Gamma_k$ and that the number of elements $K$ in $\Gamma_k$ is bounded, as assumed in (12.13), we have

$$\operatorname{meas}(\Gamma_k) \simeq h_k^{d-1}.$$

Consequently, if $\widehat{\Gamma}$ is a reference element and $\widehat{v}$ is defined by $\widehat{v}(\widehat{x}) = v(x)$ for $x \in \Gamma_k$ and $\widehat{x} \in \widehat{\Gamma}$, then by changing of variables in the integration we have

$$\|\widehat{v}\|_{L^q(\widehat{\Gamma})} \simeq h_k^{-(d-1)/q} \|v\|_{L^q(\Gamma_k)}, \quad 1 \le q < \infty,$$

with the constants independent of $q$. Therefore, using equivalence of norms in a finite-dimensional space, we obtain

$$|v(x_k)|^2 \le \|v\|_{L^\infty(\Gamma_k)}^2 = \|\widehat{v}\|_{L^\infty(\widehat{\Gamma})}^2 \simeq \|\widehat{v}\|_{L^q(\widehat{\Gamma})}^2 \simeq h_k^{-2(d-1)/q} \|v\|_{L^q(\Gamma_k)}^2, \tag{12.26}$$

where the constants are independent of $q$. This implies

$$\sum_{k \in \mathcal{N}} |v(x_k)|^2 \lesssim \sum_{k \in \mathcal{N}} h_k^{-2(d-1)/q} \|v\|_{L^q(\Gamma_k)}^2. \tag{12.27}$$

For $d = 2$, we deduce from (12.27) and Hölder's inequality (with $a_k = h_k^{-2/q}$, $b_k = \|v\|_{L^q(\Gamma)}^2$, $p = q/(q-2)$, $p' = q/2$; see (12.21))

$$\sum_{k \in \mathcal{N}} |v(x_k)|^2 \lesssim \left( \sum_{k \in \mathcal{N}} h_k^{-2/(q-2)} \right)^{1-2/q} \left( \sum_{k \in \mathcal{N}} \|v\|_{L^q(\Gamma_k)}^q \right)^{2/q}.$$

Lemma 12.1 and Lemma A.18 Part (ii) give

$$\left( \sum_{k \in \mathcal{N}} \|v\|_{L^q(\Gamma_k)}^q \right)^{2/q} \lesssim \|v\|_{L^q(\Gamma)}^2 \lesssim q \|v\|_{H^{1/2}(\Gamma)}^2 \le q \|v\|_{\widetilde{H}^{1/2}(\Gamma)}^2 \simeq q \|v\|_{\widetilde{H}_{\mathrm{I}}^{1/2}(\Gamma)}^2$$

so that

$$\begin{aligned}
\sum_{k \in \mathcal{N}} |v(x_k)|^2 &\lesssim q \left( \sum_{k \in \mathcal{N}} h_k^{-2/(q-2)} \right)^{1-2/q} \|v\|_{\widetilde{H}_{\mathrm{I}}^{1/2}(\Gamma)}^2 \\
&\le q N^{1-2/q} h_{\min}^{-2/q} \|v\|_{\widetilde{H}_{\mathrm{I}}^{1/2}(\Gamma)}^2 \\
&= q (N h_{\min})^{-2/q} N \|v\|_{\widetilde{H}_{\mathrm{I}}^{1/2}(\Gamma)}^2.
\end{aligned}$$

This holds for any $q > 2$. Repeating the same argument as in the proof of Lemma 12.3 and choosing $q = \max\{3, |\log(N h_{\min})|\}$ we deduce

$$q(Nh_{\min})^{-2/q} \lesssim 1 + |\log(Nh_{\min})|$$

and thus obtain the result for $c_{\min}$ when $d = 2$.

For $d = 3$ and $q = 4$, it follows from (12.27) that

$$\sum_{k\in\mathcal{N}} |v(x_k)|^2 \lesssim h_{\min}^{-1} \sum_{k\in\mathcal{N}} \|v\|_{L^4(\Gamma_k)}^2. \tag{12.28}$$

For the sum on the right-hand side, we use Hölder's inequality and Lemma 12.1 to obtain

$$\sum_{k\in\mathcal{N}} \|v\|_{L^4(\Gamma_k)}^2 \leq N^{1/2} \left( \sum_{k\in\mathcal{N}} \|v\|_{L^4(\Gamma_k)}^4 \right)^{1/2} \lesssim N^{1/2} \|v\|_{L^4(\Gamma)}^2,$$

which together with Lemma A.18 Part (ii) yields

$$\sum_{k\in\mathcal{N}} \|v\|_{L^4(\Gamma_k)}^2 \lesssim N^{1/2} \|v\|_{H^{1/2}(\Gamma)}^2 \leq N^{1/2} \|v\|_{\widetilde{H}^{1/2}(\Gamma)}^2 \simeq N^{1/2} \|v\|_{\widetilde{H}_{\mathrm{I}}^{1/2}(\Gamma)}^2. \tag{12.29}$$

Inequalities (12.28) and (12.29) imply $c_{\min}^{-1} \simeq N^{1/2} h_{\min}^{-1}$ for $d = 3$.

We now prove the result for $C_{\min}$. Due to Lemma C.12

$$\sum_{k\in\mathcal{N}} \|v_k\|_{\widetilde{H}^{1/2}(\Gamma)}^2 = \sum_{k\in\mathcal{N}} |v(x_k)|^2 \|\phi_k\|_{\widetilde{H}^{1/2}(\Gamma)}^2 \simeq \sum_{k\in\mathcal{N}} |v(x_k)|^2 h_k^{d-2}. \tag{12.30}$$

For $d = 2$, it follows from (12.30) that $C_{\min}^{-1} = c_{\min}^{-1}$. For $d = 3$, (12.30) and (12.26) yield

$$\sum_{k\in\mathcal{N}} \|v_k\|_{\widetilde{H}^{1/2}(\Gamma)}^2 \lesssim \sum_{k\in\mathcal{N}} h_k^{1-4/q} \|v\|_{L^q(\Gamma_k)}^2$$

for any $q \geq 1$. Choosing $q = 4$ gives

$$\sum_{k\in\mathcal{N}} \|v_k\|_{\widetilde{H}^{1/2}(\Gamma)}^2 \lesssim \sum_{k\in\mathcal{N}} \|v\|_{L^4(\Gamma_k)}^2,$$

which together with (12.29) implies $C_{\min}^{-1} = N^{1/2}$, completing the proof of the lemma. □

The next lemma provides similar results for the $\widetilde{H}^{-1/2}$-norm.

**Lemma 12.5.** *Let $v$ be decomposed by* (12.9). *Then*

$$c_{\min} \sum_{k\in\mathcal{N}} |v(x_k)|^2 \leq \|v\|_{\widetilde{H}_{\mathrm{I}}^{-1/2}(\Gamma)}^2 \quad \textit{and} \quad C_{\min} \sum_{k\in\mathcal{N}} \|v_k\|_{\widetilde{H}_{\mathrm{I}}^{-1/2}(\Gamma)}^2 \leq \|v\|_{\widetilde{H}_{\mathrm{I}}^{-1/2}(\Gamma)}^2$$

*where*

$$c_{\min} \simeq h_{\min}^d, \quad d = 2, 3,$$

*and*

$$C_{\min}^{-1} \simeq \begin{cases} 1 + |\log(h_{\min})|, & d = 2, \\ 1, & d = 3. \end{cases}$$

*Proof.* Let $x_k \in K$. Then by using Lemma A.6 Part (iii) and equivalence of norms in a finite-dimensional space, we obtain (noting that the dimension of $\Gamma$ is $d-1$)

$$|v(x_k)|^2 \leq \|v\|_{L^\infty(K)}^2 = \|\hat{v}\|_{L^\infty(\widehat{K})}^2 \simeq \|\hat{v}\|_{H_{\rm I}^{-1/2}(\widehat{K})}^2 \simeq h_k^{-d} \|v\|_{H_{\rm I}^{-1/2}(K)}^2. \tag{12.31}$$

Summing over all $k \in \mathcal{N}$ and using Theorem A.10 yield

$$\sum_{k\in\mathcal{N}} |v(x_k)|^2 \lesssim h_{\min}^{-d} \sum_{K\in\mathcal{P}} \|v\|_{H_{\rm I}^{-1/2}(K)}^2 \leq h_{\min}^{-d} \|v\|_{H_{\rm I}^{-1/2}(\Gamma)}^2 \lesssim h_{\min}^{-d} \|v\|_{\widetilde{H}_{\rm I}^{-1/2}(\Gamma)}^2,$$

giving the result for $c_{\min}$.

To prove the result for $C_{\min}$ we use (12.31) again to obtain

$$\|v_k\|_{\widetilde{H}_{\rm I}^{-1/2}(\Gamma)}^2 \simeq \|v_k\|_{\widetilde{H}^{-1/2}(\Gamma)}^2 = |v(x_k)|^2 \|\phi_k\|_{\widetilde{H}^{-1/2}(\Gamma)}^2 \lesssim h_k^{-d} \|\phi_k\|_{\widetilde{H}^{-1/2}(\Gamma)}^2 \|v\|_{H_{\rm I}^{-1/2}(K)}^2.$$

Lemma C.13 gives

$$\|v_k\|_{\widetilde{H}_{\rm I}^{-1/2}(\Gamma)}^2 \lesssim \begin{cases} \big(1 + |\log h_k|\big) \|v\|_{H_{\rm I}^{-1/2}(K)}^2, & d = 2, \\ \|v\|_{H_{\rm I}^{-1/2}(K)}^2, & d = 3. \end{cases}$$

Summing over $k \in \mathcal{N}$ and using Theorem A.10 give the desired result. □

### 12.3.3 Bounds for the condition numbers

We are now able to state the main theorems of this chapter. We denote by $\mathbf{A}_{\mathfrak{W}}$ (respectively, $\mathbf{A}_{\mathfrak{V}}$) the stiffness matrix arising from the Galerkin approximation of the hypersingular integral equation (respectively, weakly-singular integral equation). Their diagonally scaled matrices are denoted by $\widehat{\mathbf{A}}_{\mathfrak{W}}$ and $\widehat{\mathbf{A}}_{\mathfrak{V}}$, respectively.

**Theorem 12.1 (Hypersingular integral equation).** *The condition numbers of* $\mathbf{A}_{\mathfrak{W}}$ *and* $\widehat{\mathbf{A}}_{\mathfrak{W}}$ *are bounded by*

$$\kappa(\mathbf{A}_{\mathfrak{W}}) \lesssim \begin{cases} N\Big(1 + |\log(Nh_{\min})|\Big), & d = 2, \\ N^{1/2} \dfrac{h_{\max}}{h_{\min}}, & d = 3, \end{cases}$$

*and*

$$\kappa(\widehat{\mathbf{A}}_{\mathfrak{W}}) \lesssim \begin{cases} N\Big(1 + |\log(Nh_{\min})|\Big), & d = 2, \\ N^{1/2}, & d = 3. \end{cases}$$

*Proof.* The theorem is a consequence of Lemma 12.2 and Lemma 12.4, together with (12.6) and (12.7). □

**Theorem 12.2 (Weakly-singular integral equation).** *The condition numbers of the matrices $\mathbf{A}_{\mathfrak{V}}$ and $\widehat{\mathbf{A}}_{\mathfrak{V}}$ are bounded by*

$$\kappa(\mathbf{A}_{\mathfrak{V}}) \lesssim \begin{cases} N\Big(1+|\log(Nh_{\max})|\Big)\left(\dfrac{h_{\max}}{h_{\min}}\right)^2, & d=2,\\ N^{1/2}\left(\dfrac{h_{\max}}{h_{\min}}\right)^3, & d=3,\end{cases}$$

*and*

$$\kappa(\widehat{\mathbf{A}}_{\mathfrak{V}}) \lesssim \begin{cases} N\Big(1+|\log(Nh_{\min})|\Big)\dfrac{1+|\log h_{\min}|}{1+|\log h_{\max}|}, & d=2,\\ N^{1/2}, & d=3.\end{cases}$$

*Proof.* The theorem is a consequence of Lemma 12.3 and Lemma 12.5, together with (12.6) and (12.7). □

## 12.4 Anisotropic Mesh Refinements

In this section we report on the results in [86] on the extremal eigenvalues and condition numbers of the stiffness matrices arising from the Galerkin boundary element discretisation of the weakly-singular and hypersingular integral equations in the case of anisotropic mesh refinements of a screen $\Gamma$ in $\mathbb{R}^3$. Assumption 12.1 is not required, that means each partition $\mathscr{P}_h$ of $\Gamma$ may contain elements $K$ for which the aspect ratio $h_K/\rho_K$ approaches infinity as $h \to 0$.

Each partition $\mathscr{P}_h$ is conforming such that each element $K \in \mathscr{P}$ is the image of the reference element $\widehat{K}$ by a bijective map $\chi_K : \widehat{K} \to K$, which is an affine mapping if $K$ is a triangle, or a bilinear mapping if $K$ is a quadrilateral. Let $J_K$ denote the $2\times 2$ Jacobian matrix of $\chi_K$. Then, for any function $f$ defined on $K$,

$$\int_K f(x)\,\mathrm{d}x = \int_{\widehat{K}} f\big(\chi_K(\widehat{x})\big)\left[\det\Big(J_K^\top(x)J_K(x)\Big)\right]^{1/2} d\widehat{x}.$$

The following conditions are imposed on the mappings $\chi_K$. For any $K \in \mathscr{P}_h$, we denote by $|K|$ the two-dimensional measure of $K$.

**Assumption 12.3.** The following statements hold:

$$|K|^2 \lesssim \det\Big(J_K^\top(\widehat{x})J_K(\widehat{x})\Big) \lesssim |K|^2,$$

$$\rho_K^2 \lesssim \lambda_{\min}\Big(J_K^\top(\widehat{x})J_K(\widehat{x})\Big).$$

The constants are independent of $\widehat{x} \in \widehat{K}$, $K \in \mathscr{P}_h$, and $h > 0$.

It is noted that the above assumption holds when $\chi_K$ is an affine mapping. It is also satisfied when $\chi_K$ is a bilinear map if $K$ is not too far from a parallelogram. The next assumption describes how the size and shape of neighbouring elements may vary even though the meshes are neither required to be quasi-uniform nor shape-regular.

**Assumption 12.4.** The following inequalities hold

$$h_K \lesssim h_{K'}, \quad \rho_K \lesssim \rho_{K'}, \quad |K| \lesssim |K'|, \quad \forall K, K' \in \mathscr{P}_h, \ \overline{K} \cap \overline{K}' \neq \varnothing.$$

The constants are independent of $K$, $K' \in \mathscr{P}_h$ and of all $h > 0$.

We also assume the following assumption which is a variant of Assumption 12.2.

**Assumption 12.5.** There exists a positive constant $M$ such that

$$\max_{h>0} \max_{x \in \mathscr{N}_h} \operatorname{card}\left\{K \in \mathscr{P}_h : x \in \overline{K}\right\} \leq M.$$

Commonly used meshes in the $h$-version of the boundary element method which satisfy Assumption 12.4 and Assumption 12.5 are power-graded meshes. On the screen $\Gamma = (-1,1) \times (-1,1) \times \{0\}$ these meshes are defined as follows. Choosing a grading exponent $\beta \geq 1$, we first define points on $[-1,1]$ which are symmetric about 0:

$$t_j = \begin{cases} -1 + \left(\dfrac{2j}{n}\right)^\beta, & 0 \leq j \leq n/2, \\ -t_{n-j}, & n/2 < j \leq n. \end{cases} \tag{12.32}$$

When $\beta = 1$, these points are uniformly distributed on $[-1,1]$. As $\beta$ increases from 1, the points are more clustered at the endpoints $\pm 1$. The length $\Delta t_j = t_j - t_{j-1}$ of the $j$th interval satisfies

$$\Delta t_j = \Delta t_{n-j} \simeq \frac{1}{n}\left(\frac{j}{n}\right)^{\beta-1}, \quad j = 1, \ldots, n/2. \tag{12.33}$$

Using these points, we define vertices on $\Gamma$ by

$$(x_i, y_j) = (t_i, t_j), \quad i, j = 0, \ldots, n. \tag{12.34}$$

These vertices form $N$ rectangular elements $K$ on $\Gamma$, where $N \sim n^2$; see Figure 12.1. Near a corner, these elements are shape-regular with $h_K \simeq (1/n)^\beta \simeq \rho_K$. Away from the boundary, they are also shape-regular with $h_K \simeq 1/n \simeq \rho_K$. However, near the middle of an edge we have $h_K \simeq 1/n$ while $\rho_K \simeq (1/n)^\beta$. Hence, the maximum aspect ratio grows like $n^{\beta-1}$, so that if $\beta > 1$ then degeneracy occurs.

Similarly to Section 12.3, we denote by $\phi_k$ the nodal basis function centred at the nodal point $x_k$. In the case of the hypersingular integral equation, this is the hat

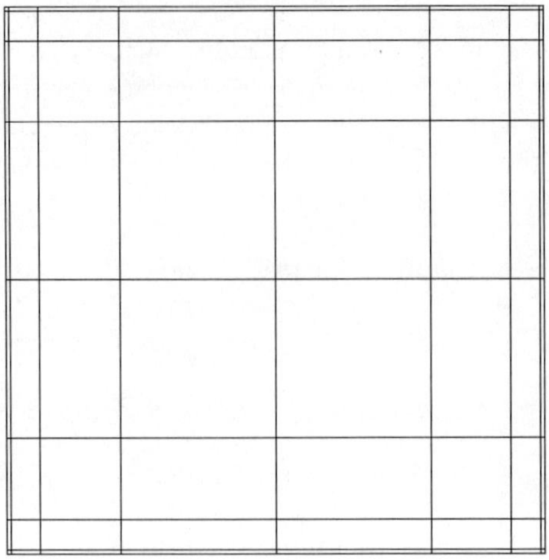

**Fig. 12.1** Power-graded tensor-product mesh with $\beta = 3$ and $n = 8$.

function while in the case of the weakly-singular integral equation it is the brick function.

We recall that $h_k$ is the average of $h_K$ for $K \subset \Gamma_k$, with $\Gamma_k$ being the support of $\phi_k$. We also define $\rho_k$ to be the average of $\rho_K$ with $K \subset \Gamma_k$. Assumption 12.4 implies that

$$\rho_k \simeq \rho_K, \quad K \subset \Gamma_k,\ k \in \mathcal{N}. \tag{12.35}$$

We also note that

$$N^{-\beta/2} \lesssim \rho_k \leq h_k \lesssim N^{-1/2}. \tag{12.36}$$

The reader is referred to [86, Example 5.5] for a generalisation of this mesh construction.

### 12.4.1 Technical results

We first show two inequalities involving $h_k$ and $\rho_k$.

**Lemma 12.6.** *For the mesh defined by* (12.32) *and* (12.34) *we have*

$$\sum_{k\in\mathscr{N}} h_k\rho_k^{-1} \lesssim \begin{cases} N, & 1\le\beta<2,\\ N(1+\log N), & \beta=2,\\ N^{\beta/2}, & \beta>2,\end{cases}$$
$$\sum_{k\in\mathscr{N}} \left(h_k\rho_k\right)^{-1} \lesssim \begin{cases} N^2, & 1\le\beta<2,\\ N^2(1+\log N)^2, & \beta=2,\\ N^{\beta}, & \beta>2.\end{cases} \tag{12.37}$$

*Proof.* Due to (12.35) we have

$$\sum_{k\in\mathscr{N}} h_k\,\rho_k^{-1} \simeq \sum_{K\in\mathscr{P}_h} h_K\,\rho_K^{-1}.$$

We number the elements $K$ from left to right, starting from the left bottom corner, so that $K_{ij}=(t_{i-1},t_i)\times(t_{j-1},t_j)$, $i,j=1,\dots,n$. Then, by symmetry (see Figure 12.1),

$$\sum_{k\in\mathscr{N}} h_k\,\rho_k^{-1} \simeq \sum_{i=1}^{n}\sum_{j=1}^{i} h_{K_{ij}}\,\rho_{K_{ij}}^{-1}.$$

Note that for $i=1,\dots,n$ and $j=1,\dots,i$ we have, recalling (12.33),

$$h_{K_{ij}}=\Delta t_i \simeq \frac{1}{n}\left(\frac{i}{n}\right)^{\beta-1} \quad\text{and}\quad \rho_{K_{ij}}=\Delta t_j \simeq \frac{1}{n}\left(\frac{j}{n}\right)^{\beta-1}.$$

Therefore,

$$\sum_{k\in\mathscr{N}} h_k\,\rho_k^{-1} \simeq \sum_{i=1}^{n}\sum_{j=1}^{i}\frac{(i/n)^{\beta-1}}{(j/n)^{\beta-1}} = n^2\sum_{i=1}^{n}\left(\frac{i}{n}\right)^{\beta-1}\frac{1}{n}\sum_{j=1}^{i}\left(\frac{j}{n}\right)^{1-\beta}\frac{1}{n}$$
$$\simeq n^2\int_{1/n}^{1} s^{\beta-1}\int_{1/n}^{s} t^{1-\beta}\,\mathrm{d}t\,\mathrm{d}s.$$

This yields the first inequality in (12.37), noting that $N=n^2$.

Similarly,

$$\sum_{k\in\mathscr{N}}\left(h_k\rho_k\right)^{-1} \simeq \sum_{i=1}^{n}\sum_{j=1}^{i}\frac{1}{\Delta t_i\,\Delta t_j} \simeq \sum_{i=1}^{n}\sum_{j=1}^{i} n^2\left(\frac{i}{n}\right)^{1-\beta}\left(\frac{j}{n}\right)^{1-\beta}$$
$$\simeq n^4\sum_{i=1}^{n}\left(\frac{i}{n}\right)^{1-\beta}\frac{1}{n}\sum_{j=1}^{i}\left(\frac{j}{n}\right)^{1-\beta}\frac{1}{n} \simeq n^4\int_{1/n}^{1} s^{1-\beta}\int_{1/n}^{s} t^{1-\beta}\,\mathrm{d}t\,\mathrm{d}s,$$

from which the second inequality in (12.37) follows. □

The following lemma is a special case of [86, Theorem 4.1].

**Lemma 12.7.** *The following statements hold*

(i) $\|\phi_k\|_{\widetilde{H}_{\mathrm{I}}^{1/2}(\Gamma)} = \|\phi_k\|_{\widetilde{H}_{\mathrm{I}}^{1/2}(\Gamma_k)} \lesssim \rho_k^{-1/2}\|\phi_k\|_{L^2(\Gamma_k)}$;

(ii) $\rho_k^{1/2}\|\phi_k\|_{L^2(\Gamma_k)} \lesssim \|\phi_k\|_{\widetilde{H}_{\mathrm{I}}^{-1/2}(\Gamma_k)}$.

*Proof.* The proof of more general results which follows the same lines as the proofs of [85, Theorems 3.2 and 3.6] is carried out in [86] and is omitted. □

Using the above lemma, one can prove the following lemma which is a partial generalisation of Lemma C.12 and Lemma C.13 for shape-regular meshes.

**Lemma 12.8.** *For any $k \in \mathscr{N}$ the following statements hold*

(i) $\|\phi_k\|_{L^p(\Gamma_k)} \simeq |\Gamma_k|^{1/p}$ *for* $1 \le p \le \infty$;

(ii) $\|\phi_k\|^2_{\widetilde{H}_{\mathrm{I}}^{1/2}(\Gamma)} \lesssim \|\phi_k\|^2_{\widetilde{H}_{\mathrm{I}}^{1/2}(\Gamma_k)} \lesssim |\Gamma_k|\rho_k^{-1}$;

(iii) $|\Gamma_k|\rho_k \lesssim \|\phi_k\|^2_{H_{\mathrm{I}}^{-1/2}(\Gamma_k)} \lesssim \|\phi_k\|^2_{H_{\mathrm{I}}^{-1/2}(\Gamma)}$;

(iv) $\|\phi_k\|^2_{\widetilde{H}_{\mathrm{I}}^{-1/2}(\Gamma)} \lesssim |\Gamma_k|^{3/2}$.

*Proof.* See [86, Theorem 4.2]. □

Finally, we need this result proved in [85].

**Lemma 12.9.** *For any $v \in \mathscr{V}$,*

$$\|v\|_{L^2(K)} \lesssim \rho_K^{-1/2}\|v\|_{\widetilde{H}_{\mathrm{I}}^{-1/2}(K)}, \quad K \in \mathscr{P}_h.$$

*Proof.* See [85, Theorem 3.6] and [85, Remark 3.8]. □

### *12.4.2 Coercivity of the decomposition*

In this section, we find the constants $c_{\max}$ and $C_{\max}$ in (12.10) and (12.11). These estimates have the same effect as the coercivity in the decomposition associated with additive Schwarz preconditioners.

The following lemma gives explicit forms of $c_{\max}$ and $C_{\max}$ in (12.10) and (12.11) when $m = 1/2$. It is noted that estimate (12.38) is an improvement over a similar result proved in [86, Lemma 5.2] in which the constant depends on the mesh type. Only in special cases (for example, when the mesh is a power-graded mesh) is the constant independent of all parameters; see [86, Theorem 7.5]. This improvement is due to the fact that in our proof we use the newly-derived estimate (A.86) instead of (A.65) as done in [86].

**Lemma 12.10.** *Let $v$ be decomposed by* (12.9). *Then*

$$\|v\|^2_{\widetilde{H}_{\mathrm{I}}^{1/2}(\Gamma)} \lesssim \sum_{k\in\mathcal{N}} \|v_k\|^2_{\widetilde{H}_{\mathrm{I}}^{1/2}(\Gamma)} \tag{12.38}$$

*and*

$$\|v\|^2_{\widetilde{H}_{\mathrm{I}}^{1/2}(\Gamma)} \lesssim \max_{k\in\mathcal{N}} \left(|\Gamma_k|\rho_k^{-1}\right) \sum_{k\in\mathcal{N}} |v(x_k)|^2. \tag{12.39}$$

*Proof.* The proof of (12.38) is exactly the same as that of (12.16) because the argument in that proof does not rely on the properties of the mesh and finite elements.

In order to prove (12.39), we further estimate $\|v_k\|^2_{\widetilde{H}_{\mathrm{I}}^{1/2}(\Gamma)}$ on the right-hand side of (12.38), using Lemma 12.8 (ii), as follows:

$$\|v_k\|^2_{\widetilde{H}_{\mathrm{I}}^{1/2}(\Gamma)} = |v(x_k)|^2 \|\phi_k\|^2_{\widetilde{H}_{\mathrm{I}}^{1/2}(\Gamma)} \lesssim |v(x_k)|^2 \|\phi_k\|^2_{\widetilde{H}_{\mathrm{I}}^{1/2}(\Gamma_k)} \lesssim |\Gamma_k|\rho_k^{-1}|v(x_k)|^2$$

Estimate (12.39) then follows. □

The next lemma shows similar results for the case when $m = -1/2$, which is related to the weakly singular integral equation.

**Lemma 12.11.** *Let $v$ be decomposed by* (12.9). *Then*

$$\|v\|^2_{\widetilde{H}_{\mathrm{I}}^{-1/2}(\Gamma)} \lesssim \Big(\sum_{k\in\mathcal{N}} |\Gamma_k|^3\Big)^{1/2} \sum_{k\in\mathcal{N}} |v(x_k)|^2 \tag{12.40}$$

*and*

$$\|v\|^2_{\widetilde{H}_{\mathrm{I}}^{-1/2}(\Gamma)} \lesssim \Big(\sum_{k\in\mathcal{N}} |\Gamma_k|\rho_k^{-2}\Big)^{1/2} \sum_{k\in\mathcal{N}} \|v_k\|^2_{\widetilde{H}_{\mathrm{I}}^{-1/2}(\Gamma)}. \tag{12.41}$$

*Proof.* A more general result is proved in [86, Lemma 5.1]. We only present the proof for the special case of our interest. By using Lemma A.18 part (iii) we obtain $\|v\|_{\widetilde{H}_{\mathrm{I}}^{-1/2}(\Gamma)} \lesssim \|v\|_{L^{4/3}(\Gamma)}$. This together with Hölder's inequality and (12.15) gives

$$\begin{aligned}\|v\|_{\widetilde{H}_{\mathrm{I}}^{-1/2}(\Gamma)} &\lesssim \|v\|_{L^{4/3}(\Gamma)} \lesssim \sum_{\ell=1}^{L_h} \Big\| \sum_{k\in\mathcal{N}_\ell} v_k \Big\|_{L^{4/3}(\Gamma)} \\ &\lesssim \Big( \sum_{\ell=1}^{L_h} \Big\| \sum_{k\in\mathcal{N}_\ell} v_k \Big\|^{4/3}_{L^{4/3}(\Gamma)} \Big)^{3/4} \lesssim \Big( \sum_{k\in\mathcal{N}} \|v_k\|^{4/3}_{L^{4/3}(\Gamma)} \Big)^{3/4}.\end{aligned}$$

Let $w_k$, $k \in \mathcal{N}$, be any positive weights. Then by using Hölder's inequality again with $1/p + 1/q = 1$ where $p = 3$ and $q = 3/2$, we have

$$\sum_{k\in\mathcal{N}} \|v_k\|^{4/3}_{L^{4/3}(\Gamma)} \le \Big(\sum_{k\in\mathcal{N}} w_k^2\Big)^{1/3} \Big(\sum_{k\in\mathcal{N}} w_k^{-1} \|v_k\|^2_{L^{4/3}(\Gamma)}\Big)^{2/3}.$$

This inequality and the one above yield

$$\|v\|^2_{\widetilde{H}_I^{-1/2}(\Gamma)} \lesssim \Big(\sum_{k\in\mathcal{N}} w_k^2\Big)^{1/2} \sum_{k\in\mathcal{N}} w_k^{-1} \|v_k\|^2_{L^{4/3}(\Gamma)}.$$

By choosing $w_k = |\Gamma_k|^{3/2}$ and applying Lemma 12.8 part (i), we obtain (12.40).

On the other hand, by choosing $w_k = |\Gamma_k|^{1/2}\, \rho_k^{-1}$ and applying Lemma 12.8 part (i) and part (iii), we obtain

$$\begin{aligned}
\|v\|^2_{\widetilde{H}_I^{-1/2}(\Gamma)} &\lesssim \Big(\sum_{k\in\mathcal{N}} |\Gamma_k|\, \rho_k^{-2}\Big)^{1/2} \sum_{k\in\mathcal{N}} |\Gamma_k|\, \rho_k |v(x_k)|^2 \\
&\lesssim \Big(\sum_{k\in\mathcal{N}} |\Gamma_k|\, \rho_k^{-2}\Big)^{1/2} \sum_{k\in\mathcal{N}} \|v_k\|^2_{H_I^{-1/2}(\Gamma)},
\end{aligned}$$

yielding (12.41).

□

### 12.4.3 *Stability of the decomposition*

In this section, we find the constants $c_{\min}$ and $C_{\min}$ in (12.10) and (12.11). These estimates have the same effect as the stability in the decomposition associated with additive Schwarz preconditioners.

The next lemma gives the results when $m = 1/2$. More general results can be found in [86, Lemma 5.2].

**Lemma 12.12.** *Let v be decomposed by* (12.9)*. Then*

$$\Big(\sum_{k\in\mathcal{N}} |\Gamma_k|^{-1}\Big)^{-1/2} \sum_{k\in\mathcal{N}} |v(x_k)|^2 \lesssim \|v\|^2_{\widetilde{H}_I^{1/2}(\Gamma)} \tag{12.42}$$

*and*

$$\Big(\sum_{k\in\mathcal{N}} |\Gamma_k|\, \rho_k^{-2}\Big)^{-1/2} \sum_{k\in\mathcal{N}} \|v_k\|^2_{\widetilde{H}_I^{1/2}(\Gamma)} \lesssim \|v\|^2_{\widetilde{H}_I^{1/2}(\Gamma)}. \tag{12.43}$$

*Proof.* Recalling the usual notation of the reference element $\widehat{K}$, for any $x_k \in K$ we have

$$|v(x_k)|^2 \le \|v\|^2_{L^\infty(K)} = \|\widehat{v}\|^2_{L^\infty(\widehat{K})}.$$

By equivalence of norms on finite-dimensional spaces, we have

$$\|\widehat{v}\|^2_{L^\infty(\widehat{K})} \simeq \|\widehat{v}\|^2_{L^4(\widehat{K})} \simeq |K|^{-1/2} \|v\|^2_{L^4(K)}.$$

Assumption 12.4 implies $|K|^{-1/2} \lesssim |\Gamma_k|^{-1/2}$ so that

$$|v(x_k)|^2 \lesssim |\Gamma_k|^{-1/2}\, \|v\|^2_{L^4(\Gamma_k)}. \tag{12.44}$$

Taking the sum over $k \in \mathcal{N}$ and using the Cauchy–Schwarz inequality and Assumption 12.5, we deduce

$$\sum_{k\in\mathcal{N}} |v(x_k)|^2 \lesssim \Big(\sum_{k\in\mathcal{N}} |\Gamma_k|^{-1}\Big)^{1/2} \Big(\sum_{k\in\mathcal{N}} \|v\|_{L^4(\Gamma_k)}^4\Big)^{1/2} \lesssim \Big(\sum_{k\in\mathcal{N}} |\Gamma_k|^{-1}\Big)^{1/2} \|v\|_{L^4(\Gamma)}^2.$$

Lemma A.18 part (i) then gives

$$\sum_{k\in\mathcal{N}} |v(x_k)|^2 \lesssim \Big(\sum_{k\in\mathcal{N}} |\Gamma_k|^{-1}\Big)^{1/2} \|v\|_{H_{\mathrm{I}}^{1/2}(\Gamma)}^2 \le \Big(\sum_{k\in\mathcal{N}} |\Gamma_k|^{-1}\Big)^{1/2} \|v\|_{\widetilde{H}_{\mathrm{I}}^{1/2}(\Gamma)}^2,$$

proving (12.42).

We can prove (12.43) in a similar manner. Indeed, Lemma 12.8 part (ii) and (12.44) imply

$$\|v_k\|_{\widetilde{H}_{\mathrm{I}}^{1/2}(\Gamma)}^2 = |v(x_k)|^2 \, \|\phi_k\|_{\widetilde{H}_{\mathrm{I}}^{1/2}(\Gamma)}^2 \lesssim |v(x_k)|^2 \, |\Gamma_k| \, \rho_k^{-1} \lesssim |\Gamma_k|^{1/2} \, \rho_k^{-1} \|v\|_{L^4(\Gamma_k)}^2.$$

Therefore, proceeding as above we obtain

$$\sum_{k\in\mathcal{N}} \|v_k\|_{\widetilde{H}_{\mathrm{I}}^{1/2}(\Gamma)}^2 \lesssim \Big(|\Gamma_k| \, \rho_k^{-2}\Big)^{1/2} \|v\|_{L^4(\Gamma)}^2 \lesssim \Big(|\Gamma_k| \, \rho_k^{-2}\Big)^{1/2} \|v\|_{\widetilde{H}_{\mathrm{I}}^{1/2}(\Gamma)}^2,$$

completing the proof of the lemma. □

The following lemma gives similar results when $m = -1/2$. More general results can be found in [86, Lemma 5.1].

**Lemma 12.13.** *Let $v$ be decomposed by* (12.9). *Then*

$$\min_{k\in\mathcal{N}} \big(|\Gamma_k| \, \rho_k\big) \sum_{k\in\mathcal{N}} |v(x_k)|^2 \lesssim \|v\|_{\widetilde{H}_{\mathrm{I}}^{-1/2}(\Gamma)}^2 \tag{12.45}$$

*and*

$$\min_{k\in\mathcal{N}} \big(|\Gamma_k|^{-1/2} \, \rho_k\big) \sum_{k\in\mathcal{N}} \|v_k\|_{\widetilde{H}_{\mathrm{I}}^{-1/2}(\Gamma)}^2 \lesssim \|v\|_{\widetilde{H}_{\mathrm{I}}^{-1/2}(\Gamma)}^2. \tag{12.46}$$

*Proof.* Choose $K \in \mathscr{P}_h$ such that $x_k \in \overline{K}$. Then, making the same arguments as in the proof of the previous lemma, we obtain

$$|v(x_k)|^2 \le \|v\|_{L^\infty(K)}^2 = \|\widehat{v}\|_{L^\infty(\widehat{K})}^2 \simeq \|\widehat{v}\|_{L^2(\widehat{K})}^2 \simeq |K|^{-1} \, \|v\|_{L^2(K)}^2.$$

Lemma 12.9 implies

$$|v(x_k)|^2 \lesssim |K|^{-1} \, \rho_K^{-1} \, \|v\|_{H_{\mathrm{I}}^{-1/2}(K)}^2 \lesssim |\Gamma_k|^{-1} \, \rho_k^{-1} \, \|v\|_{H_{\mathrm{I}}^{-1/2}(K)}^2 \tag{12.47}$$

so that, with the help of Theorem A.10,

$$\sum_{k\in\mathcal{N}} |v(x_k)|^2 \lesssim \max_{k\in\mathcal{N}} \big(|\Gamma_k|^{-1} \rho_k^{-1}\big) \|v\|_{H_{\mathrm{I}}^{-1/2}(\Gamma)}^2 \le \max_{k\in\mathcal{N}} \big(|\Gamma_k|^{-1} \rho_k^{-1}\big) \|v\|_{\widetilde{H}_{\mathrm{I}}^{-1/2}(\Gamma)}^2.$$

This proves (12.45).

To prove (12.46) it suffices to note that

$$\|v_k\|^2_{\widetilde{H}_I^{-1/2}(\Gamma)} = |v(x_k)|^2 \|\phi_k\|^2_{\widetilde{H}_I^{-1/2}(\Gamma)} \lesssim |\Gamma_k|^{1/2}\, \rho_k^{-1} \|v\|^2_{H_I^{-1/2}(K)}$$

where in the last step we used (12.47) and Lemma 12.8 part (iv). □

### 12.4.4 Bounds for the condition numbers

We are now able to state the main theorems of this section. Similarly to Theorem 12.1 and Theorem 12.2, we now have the following results for the anisotropic mesh refinement defined by (12.32) and (12.34). Other results for other anisotropic mesh refinements can be found in [86].

**Theorem 12.3 (Hypersingular integral equation).** *For the graded mesh defined by* (12.32) *and* (12.34)*,*

$$\lambda_{\max}(\mathbf{A}_{\mathfrak{W}}) \lesssim N^{-1/2}, \quad \lambda_{\min}(\mathbf{A}_{\mathfrak{W}}) \gtrsim \begin{cases} N^{-1}, & 1 \le \beta < 2, \\ N^{-1}(1+\log N)^{-1}, & \beta = 2, \\ N^{-\beta/2}, & \beta > 2, \end{cases}$$

$$\lambda_{\max}(\widehat{\mathbf{A}}_{\mathfrak{W}}) \lesssim 1, \qquad \lambda_{\min}(\widehat{\mathbf{A}}_{\mathfrak{W}}) \gtrsim \begin{cases} N^{-1/2}, & 1 \le \beta < 2, \\ N^{-1/2}(1+\log N)^{-1/2}, & \beta = 2, \\ N^{-\beta/4}, & \beta > 2. \end{cases}$$

*Consequently,*

$$\kappa(\mathbf{A}_{\mathfrak{W}}) \lesssim \begin{cases} N^{1/2}, & 1 \le \beta < 2, \\ N^{1/2}(1+\log N), & \beta = 2, \\ N^{(\beta-1)/2}, & \beta > 2, \end{cases}$$

*and*

$$\kappa(\widehat{\mathbf{A}}_{\mathfrak{W}}) \lesssim \begin{cases} N^{1/2}, & 1 \le \beta < 2, \\ N^{1/2}(1+\log N)^{1/2}, & \beta = 2, \\ N^{\beta/4}, & \beta > 2. \end{cases}$$

*Proof.* It follows from (12.6), (12.7), Lemma 12.10, and Lemma 12.12 that

$$\lambda_{\max}(\mathbf{A}_{\mathfrak{W}}) \lesssim \max_{k\in\mathcal{N}} \left(|\Gamma_k| \rho_k^{-1}\right), \qquad \lambda_{\max}(\widehat{\mathbf{A}}_{\mathfrak{W}}) \lesssim 1,$$

$$\lambda_{\min}(\mathbf{A}_{\mathfrak{W}}) \gtrsim \Big(\sum_{k\in\mathcal{N}} |\Gamma_k|^{-1}\Big)^{-1/2}, \quad \lambda_{\min}(\widehat{\mathbf{A}}_{\mathfrak{W}}) \gtrsim \Big(\sum_{k\in\mathcal{N}} |\Gamma_k|\, \rho_k^{-2}\Big)^{-1/2}.$$

Since $|\Gamma_k| \sim h_k \rho_k$ for $k \in \mathscr{N}$, we have

$$\lambda_{\max}(\mathbf{A}_{\mathfrak{W}}) \lesssim \max_{k \in \mathscr{N}} h_k \lesssim N^{-1/2}$$

where in the last step we used (12.36). Since $|\Gamma_k|^{-1} \sim (h_k \rho_k)^{-1}$ and $|\Gamma_k| \rho_k^{-2} \sim h_k \rho_k^{-1}$, the lower bounds for $\lambda_{\min}(\mathbf{A}_{\mathfrak{W}})$ and $\lambda_{\min}(\widehat{\mathbf{A}}_{\mathfrak{W}})$ follow from (12.37), completing the proof of the lemma. □

**Theorem 12.4 (Weakly-singular integral equation).** *For the graded mesh defined by* (12.32) *and* (12.34),

$$\lambda_{\min}(\mathbf{A}_{\mathfrak{V}}) \gtrsim N^{-3\beta/2}, \qquad \lambda_{\max}(\mathbf{A}_{\mathfrak{V}}) \lesssim N^{-1},$$

$$\lambda_{\min}(\widehat{\mathbf{A}}_{\mathfrak{V}}) \gtrsim N^{-(\beta-1)/2}, \quad \lambda_{\max}(\widehat{\mathbf{A}}_{\mathfrak{V}}) \lesssim \begin{cases} N^{1/2}, & 1 \le \beta < 2, \\ N^{1/2}(1+\log N)^{1/2}, & \beta = 2, \\ N^{\beta/4}, & \beta > 2. \end{cases}$$

*Consequently,*

$$\kappa(\mathbf{A}_{\mathfrak{V}}) \lesssim N^{(3\beta/2)-1} \quad \text{and} \quad \kappa(\widehat{\mathbf{A}}_{\mathfrak{V}}) \lesssim \begin{cases} N^{\beta/2}, & 1 \le \beta < 2, \\ N^{\beta/2}(1+\log N)^{1/2}, & \beta = 2, \\ N^{(3\beta-2)/4}, & \beta > 2. \end{cases}$$

*Proof.* It follows from (12.6), (12.7), Lemma 12.11, and Lemma 12.13 that

$$\lambda_{\min}(\mathbf{A}_{\mathfrak{V}}) \gtrsim \min_{k \in \mathscr{N}} \big(|\Gamma_k|\, \rho_k\big), \qquad \lambda_{\max}(\mathbf{A}_{\mathfrak{V}}) \lesssim \Big(\sum_{k \in \mathscr{N}} |\Gamma_k|^3\Big)^{1/2},$$

$$\lambda_{\min}(\widehat{\mathbf{A}}_{\mathfrak{V}}) \gtrsim \min_{k \in \mathscr{N}} \big(|\Gamma_k|^{-1/2}\, \rho_k\big), \qquad \lambda_{\max}(\widehat{\mathbf{A}}_{\mathfrak{V}}) \lesssim \Big(\sum_{k \in \mathscr{N}} |\Gamma_k| \rho_k^{-2}\Big)^{1/2}.$$

It also follows from (12.36) that

$$N^{-\beta} \lesssim |\Gamma_k| \lesssim N^{-1}.$$

The bounds for $\lambda_{\min}(\mathbf{A}_{\mathfrak{V}})$, $\lambda_{\max}(\mathbf{A}_{\mathfrak{V}})$, and $\lambda_{\min}(\widehat{\mathbf{A}}_{\mathfrak{V}})$ now follow from the above inequalities and (12.36). Noting that $|\Gamma_k| \rho_k^{-2} \sim h_k \rho_k^{-1}$, we also obtain the bound for $\lambda_{\max}(\widehat{\mathbf{A}}_{\mathfrak{V}})$ by using (12.37). The bounds for $\kappa(\mathbf{A}_{\mathfrak{V}})$ and $\kappa(\widehat{\mathbf{A}}_{\mathfrak{V}})$ then follow, completing the proof of the theorem. □

*Remark 12.1.* In the case of the weakly-singular integral equation, if one uses the Haar basis functions instead of the brick functions so that $v_k$ has integral mean zero, then by invoking Theorem A.11 one can show that the maximum eigenvalue of $\widehat{\mathbf{A}}_{\mathfrak{V}}$ is bounded by a constant as in the case of the hypersingular integral equation. See Chapter 3 and Chapter 10.

## 12.5 Numerical Results

We present in this section numerous numerical results that illustrate our theory. We considered the weakly-singular and hypersingular integral equations on various boundaries $\Gamma$. In each case, we computed the extremal eigenvalues $\lambda_{\min}(\mathbf{A})$, $\lambda_{\max}(\mathbf{A})$, $\lambda_{\min}(\widehat{\mathbf{A}})$, and $\lambda_{\max}(\widehat{\mathbf{A}})$, and the condition numbers $\kappa(\mathbf{A})$ and $\kappa(\widehat{\mathbf{A}})$ when the mesh is uniform and when it is refined to observe that diagonal scaling in fact restores the condition numbers of the matrix as is in the case of uniform meshes.

The following boundaries $\Gamma$ and equations are considered:

**Open curve:** $\Gamma$ is the open curve $(-1,1)\times\{0\}$ in $\mathbb{R}^2$. On this boundary a uniform mesh, a mesh graded (with grading parameter 3) at the endpoints $\pm 1$, and a geometric mesh are considered.
The weakly-singular and hypersingular integral equations on this boundary have the form

$$\mathfrak{V}u(x) = \frac{1}{\pi}\int_\Gamma u(y)\log\frac{1}{|x-y|}\,d\sigma_y = f(x), \quad x\in\Gamma,$$

$$\mathfrak{W}u(x) = -\frac{1}{\pi}\frac{\partial}{\partial \nu_x}\int_\Gamma u(x)\frac{\partial}{\partial \nu_y}\log\frac{1}{|x-y|}\,d\sigma_y = g(x), \quad x\in\Gamma.$$

**Open surface** $\Gamma$ is the open surface $(-1,1)\times(-1,1)\times\{0\}$ in $\mathbb{R}^3$. On this boundary a uniform mesh and a mesh that is refined at the corner $(-1,-1)$ are considered (see Figures 12.2, and 12.3).
The weakly-singular and hypersingular integral equations on this boundary have the form

$$\mathfrak{V}u(x) = \frac{1}{2\pi}\int_\Gamma \frac{u(y)}{|x-y|}\,d\sigma_y = f(x), \quad x\in\Gamma,$$

$$\mathfrak{W}u(x) = -\frac{1}{2\pi}\frac{\partial}{\partial \nu_x}\int_\Gamma u(x)\frac{\partial}{\partial \nu_y}\frac{1}{|x-y|}\,d\sigma_y = g(x), \quad x\in\Gamma.$$

The following tables and figures are reproduced from those in [5] (Copyright

| Geometric mesh ($d=2$) | | | | | | |
|---|---|---|---|---|---|---|
| $N$ | $\log \lambda_{\min}(\mathbf{A}_{\mathfrak{B}})$ | | $\lambda_{\max}(\mathbf{A}_{\mathfrak{B}})$ | | $\kappa(\mathbf{A}_{\mathfrak{B}})$ | |
| 8 | −0.5071e+1 | | 0.2525e+0 | | 0.4023e+02 | |
| 16 | −0.1061e+2 | 1.05 | 0.2523e+0 | 0.00 | 0.1021e+05 | 10.38 |
| 24 | −0.1615e+2 | 1.03 | 0.2523e+0 | 0.00 | 0.2611e+07 | 15.93 |
| 32 | −0.2170e+2 | 1.02 | 0.2523e+0 | 0.00 | 0.6685e+09 | 21.48 |
| 40 | −0.2724e+2 | 1.02 | 0.2523e+0 | 0.00 | 0.1712e+12 | 27.03 |
| 48 | −0.3279e+2 | 1.02 | 0.2523e+0 | 0.00 | 0.4398e+14 | 32.60 |
| theory | $\gtrsim -N$ | | $\lesssim N(1+\log N)$ | | $\lesssim 2^{2N} N(1+\log N)$ | |
| $N$ | $\lambda_{\min}(\widehat{\mathbf{A}}_{\mathfrak{B}})$ | | $\lambda_{\max}(\widehat{\mathbf{A}}_{\mathfrak{B}})$ | | $\kappa(\widehat{\mathbf{A}}_{\mathfrak{B}})$ | |
| 8 | 0.3133e+0 | | 0.2552e+1 | | 0.8146e+1 | |
| 16 | 0.1720e+0 | −0.94 | 0.4969e+1 | 0.98 | 0.2890e+2 | 1.92 |
| 24 | 0.1160e+0 | −0.99 | 0.7447e+1 | 1.01 | 0.6419e+2 | 2.00 |
| 32 | 0.8706e−1 | −1.00 | 0.9998e+1 | 1.03 | 0.1148e+3 | 2.03 |
| 40 | 0.6948e−1 | −1.02 | 0.1260e+2 | 1.04 | 0.1814e+3 | 2.06 |
| 48 | 0.5767e−1 | −1.01 | 0.1524e+2 | 1.04 | 0.2642e+3 | 2.06 |
| theory | $\gtrsim N^{-1}$ | | $\lesssim N^2$ | | $\lesssim N^3$ | |

**Table 12.1** Weakly-singular integral equation on open curve with geometric mesh: $h_{\min} = 2^{-(N-2)}$ and $h_{\max} = 1$. Theory: Lemma 12.3, Lemma 12.5, Theorem 12.2.

| Geometric mesh ($d=2$) | | | | | | |
|---|---|---|---|---|---|---|
| $N$ | $\log \lambda_{\min}(\mathbf{A}_{\mathfrak{W}})$ | | $\lambda_{\max}(\mathbf{A}_{\mathfrak{W}})$ | | $\kappa(\mathbf{A}_{\mathfrak{W}})$ | |
| 7 | 0.3784e+0 | | 0.1171e+1 | | 0.3094e+01 | |
| 15 | 0.2903e+0 | −0.30 | 0.1170e+1 | 0.00 | 0.4030e+01 | 0.29 |
| 23 | 0.2615e+0 | −0.21 | 0.1171e+1 | 0.00 | 0.4476e+01 | 0.21 |
| 31 | 0.2479e+0 | −0.16 | 0.1171e+1 | 0.00 | 0.4722e+01 | 0.16 |
| 39 | 0.2402e+0 | −0.12 | 0.1171e+1 | 0.00 | 0.4873e+01 | 0.13 |
| 47 | 0.2355e+0 | −0.10 | 0.1173e+1 | 0.00 | 0.4980e+01 | 0.09 |
| theory | $\gtrsim N^{-2}$ | | $\lesssim 1$ | | $\lesssim N^2$ | |
| $N$ | $\lambda_{\min}(\widehat{\mathbf{A}}_{\mathfrak{W}})$ | | $\lambda_{\max}(\widehat{\mathbf{A}}_{\mathfrak{W}})$ | | $\kappa(\widehat{\mathbf{A}}_{\mathfrak{W}})$ | |
| 7 | 0.4200e+00 | | 0.1296e+01 | | 0.3087e+01 | |
| 15 | 0.3197e+00 | −0.30 | 0.1296e+01 | 0.00 | 0.4055e+01 | 0.30 |
| 23 | 0.2874e+00 | −0.22 | 0.1297e+01 | 0.00 | 0.4511e+01 | 0.22 |
| 31 | 0.2722e+00 | −0.16 | 0.1297e+01 | 0.00 | 0.4763e+01 | 0.16 |
| 39 | 0.2637e+00 | −0.13 | 0.1297e+01 | 0.00 | 0.4918e+01 | 0.13 |
| 47 | 0.2584e+00 | −0.10 | 0.1298e+01 | −0.02 | 0.5024e+01 | 0.08 |
| theory | $\gtrsim N^{-2}$ | | $\lesssim 1$ | | $\lesssim N^2$ | |

**Table 12.2** Hypersingular integral equation on open curve with geometric mesh: $h_{\min} = 2^{-(N-1)}$ and $h_{\max} = 1$. Theory: Lemma 12.2, Lemma 12.4, Theorem 12.1.

| Uniform mesh ($d = 3$) | | | | | | |
|---|---|---|---|---|---|---|
| $N$ | $\lambda_{\min}(\mathbf{A}_{\mathfrak{B}})$ | | $\lambda_{\max}(\mathbf{A}_{\mathfrak{B}})$ | | $\kappa(\mathbf{A}_{\mathfrak{B}})$ | |
| 8 | 0.32e−1 | | 0.24e+0 | | 0.75e+1 | |
| 32 | 0.35e−2 | −1.59 | 0.60e−1 | −0.99 | 0.17e+2 | 0.60 |
| 128 | 0.42e−3 | −1.53 | 0.15e−1 | −1.00 | 0.36e+2 | 0.53 |
| 512 | 0.51e−4 | −1.51 | 0.38e−2 | −1.00 | 0.73e+2 | 0.51 |
| 2048 | 0.64e−5 | −1.50 | 0.94e−3 | −1.00 | 0.15e+3 | 0.50 |
| theory | $\gtrsim N^{-3/2}$ | | $\lesssim N^{-1}$ | | $\lesssim N^{1/2}$ | |
| $N$ | $\lambda_{\min}(\widehat{\mathbf{A}}_{\mathfrak{B}})$ | | $\lambda_{\max}(\widehat{\mathbf{A}}_{\mathfrak{B}})$ | | $\kappa(\widehat{\mathbf{A}}_{\mathfrak{B}})$ | |
| 8 | 0.40e+0 | | 0.30e+1 | | 0.75e+1 | |
| 32 | 0.35e+0 | −0.09 | 0.60e+1 | 0.51 | 0.17e+2 | 0.60 |
| 128 | 0.33e+0 | −0.03 | 0.12e+2 | 0.50 | 0.36e+2 | 0.53 |
| 512 | 0.33e+0 | −0.01 | 0.24e+2 | 0.50 | 0.73e+2 | 0.51 |
| 2048 | 0.33e+0 | 0.00 | 0.48e+2 | 0.50 | 0.15e+3 | 0.50 |
| theory | $\gtrsim 1$ | | $\lesssim N^{1/2}$ | | $\lesssim N^{1/2}$ | |

**Table 12.3** Weakly-singular integral equation on a screen with uniform mesh: $h_{\min} = h_{\max} = \mathcal{O}(N^{-1/2})$. Theory: Lemma 12.3, Lemma 12.5, Theorem 12.2.

| Non-uniform mesh ($d = 3$) | | | | | | |
|---|---|---|---|---|---|---|
| $N$ | $\lambda_{\min}(\mathbf{A}_{\mathfrak{B}})$ | | $\lambda_{\max}(\mathbf{A}_{\mathfrak{B}})$ | | $\kappa(\mathbf{A}_{\mathfrak{B}})$ | |
| 12 | 0.43e−2 | | 0.21e+0 | | 0.49e+2 | |
| 40 | 0.68e−4 | −3.45 | 0.59e−1 | −1.06 | 0.87e+3 | 2.39 |
| 140 | 0.11e−5 | −3.32 | 0.15e−1 | −1.09 | 0.14e+5 | 2.23 |
| 528 | 0.17e−7 | −3.13 | 0.38e−2 | −1.04 | 0.23e+6 | 2.09 |
| 2068 | 0.26e−9 | −3.05 | 0.94e−3 | −1.01 | 0.36e+7 | 2.03 |
| theory | $\gtrsim N^{-3}$ | | $\lesssim N^{-1}$ | | $\lesssim N^{2}$ | |
| $N$ | $\lambda_{\min}(\widehat{\mathbf{A}}_{\mathfrak{B}})$ | | $\lambda_{\max}(\widehat{\mathbf{A}}_{\mathfrak{B}})$ | | $\kappa(\widehat{\mathbf{A}}_{\mathfrak{B}})$ | |
| 12 | 0.37e+0 | | 0.36e+1 | | 0.98e+1 | |
| 40 | 0.35e+0 | −0.06 | 0.64e+1 | 0.47 | 0.18e+2 | 0.53 |
| 140 | 0.33e+0 | −0.03 | 0.12e+2 | 0.52 | 0.37e+2 | 0.55 |
| 528 | 0.33e+0 | −0.01 | 0.24e+2 | 0.52 | 0.73e+2 | 0.53 |
| 2068 | 0.33e+0 | 0.00 | 0.48e+2 | 0.51 | 0.15e+3 | 0.51 |
| theory | $\gtrsim 1$ | | $\lesssim N^{1/2}$ | | $\lesssim N^{1/2}$ | |

**Table 12.4** Weakly-singular integral equation on a screen with non-uniform mesh: $h_{\min} = \mathcal{O}(N^{-1})$ and $h_{\max} = \mathcal{O}(N^{-1/2})$. Theory: Lemma 12.3, Lemma 12.5, Theorem 12.2.

| Uniform mesh ($d = 3$) | | | | | | |
|---|---|---|---|---|---|---|
| $N$ | $\lambda_{\min}(\mathbf{A}_{\mathfrak{W}})$ | | $\lambda_{\max}(\mathbf{A}_{\mathfrak{W}})$ | | $\kappa(\mathbf{A}_{\mathfrak{W}})$ | |
| 9 | 0.20e+0 | | 0.36e+0 | | 0.18e+1 | |
| 49 | 0.56e−1 | −0.75 | 0.19e+0 | −0.37 | 0.35e+1 | 0.38 |
| 225 | 0.14e−1 | −0.89 | 0.99e−1 | −0.44 | 0.69e+1 | 0.45 |
| 961 | 0.36e−2 | −0.95 | 0.50e−1 | −0.47 | 0.14e+2 | 0.48 |
| theory | $\gtrsim N^{-1}$ | | $\lesssim N^{-1/2}$ | | $\lesssim N^{1/2}$ | |
| $N$ | $\lambda_{\min}(\widehat{\mathbf{A}}_{\mathfrak{W}})$ | | $\lambda_{\max}(\widehat{\mathbf{A}}_{\mathfrak{W}})$ | | $\kappa(\widehat{\mathbf{A}}_{\mathfrak{W}})$ | |
| 9 | 0.68e+0 | | 0.12e+1 | | 0.18e+1 | |
| 49 | 0.38e+0 | −0.34 | 0.13e+1 | 0.04 | 0.35e+1 | 0.38 |
| 225 | 0.20e+0 | −0.44 | 0.14e+1 | 0.02 | 0.69e+1 | 0.45 |
| 961 | 0.99e−1 | −0.47 | 0.14e+1 | 0.00 | 0.14e+2 | 0.48 |
| theory | $\gtrsim N^{-1/2}$ | | $\lesssim 1$ | | $\lesssim N^{1/2}$ | |

**Table 12.5** Hypersingular integral equation on a screen with uniform mesh: $h_{\min} = h_{\max} = \mathscr{O}(N^{-1/2})$. Theory: Lemma 12.2, Lemma 12.4, Theorem 12.1.

| Non-uniform mesh ($d = 3$) | | | | | | |
|---|---|---|---|---|---|---|
| $N$ | $\lambda_{\min}(\mathbf{A}_{\mathfrak{W}})$ | | $\lambda_{\max}(\mathbf{A}_{\mathfrak{W}})$ | | $\kappa(\mathbf{A}_{\mathfrak{W}})$ | |
| 2 | 0.33e+0 | | 0.53e+0 | | 0.16e+1 | |
| 11 | 0.81e−1 | −0.82 | 0.36e+0 | −0.23 | 0.44e+1 | 0.59 |
| 52 | 0.20e−1 | −0.89 | 0.19e+0 | −0.39 | 0.96e+1 | 0.50 |
| 229 | 0.51e−2 | −0.94 | 0.99e−1 | −0.45 | 0.20e+2 | 0.48 |
| 966 | 0.13e−2 | −0.96 | 0.50e−1 | −0.48 | 0.40e+2 | 0.49 |
| theory | $\gtrsim N^{-3/2}$ | | $\lesssim N^{-1/2}$ | | $\lesssim N$ | |
| $N$ | $\lambda_{\min}(\widehat{\mathbf{A}}_{\mathfrak{W}})$ | | $\lambda_{\max}(\widehat{\mathbf{A}}_{\mathfrak{W}})$ | | $\kappa(\widehat{\mathbf{A}}_{\mathfrak{W}})$ | |
| 2 | 0.94e+0 | | 0.10e+1 | | 0.11e+1 | |
| 11 | 0.68e+0 | −0.19 | 0.12e+1 | 0.09 | 0.18e+1 | 0.28 |
| 52 | 0.38e+0 | −0.37 | 0.13e+1 | 0.05 | 0.35e+1 | 0.41 |
| 229 | 0.20e+0 | −0.45 | 0.14e+1 | 0.02 | 0.69e+1 | 0.46 |
| 966 | 0.99e−1 | −0.48 | 0.14e+1 | 0.00 | 0.14e+2 | 0.48 |
| theory | $\gtrsim N^{-1/2}$ | | $\lesssim 1$ | | $\lesssim N^{1/2}$ | |

**Table 12.6** Hypersingular integral equation on a screen with non-uniform mesh of Figure 12.2: $h_{\min} = \mathscr{O}(N^{-1})$ and $h_{\max} = \mathscr{O}(N^{-1/2})$. Theory: Lemma 12.2, Lemma 12.4, Theorem 12.1.

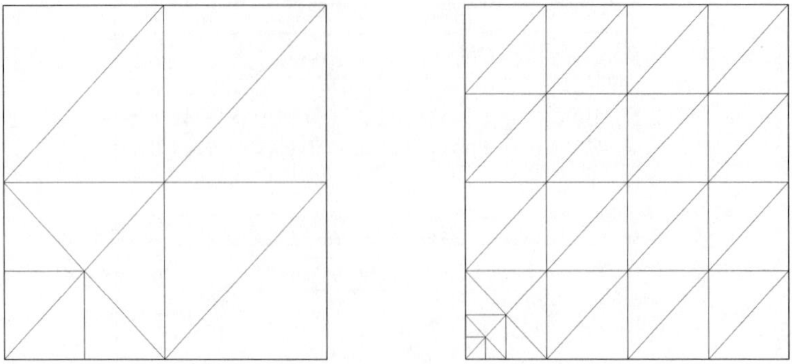

**Fig. 12.2** Successive meshes in a sequence of non-uniform meshes on the screen, with local refinement at one corner: $h_{\min} = \mathcal{O}(N^{-1})$ and $h_{\max} = \mathcal{O}(N^{-1/2})$.

| Non-uniform mesh ($d = 3$) | | | | | | |
|---|---|---|---|---|---|---|
| $N$ | $\lambda_{\min}(\mathbf{A}_{\mathfrak{W}})$ | | $\lambda_{\max}(\mathbf{A}_{\mathfrak{W}})$ | | $\kappa(\mathbf{A}_{\mathfrak{W}})$ | |
| 8 | 0.1900e+0 | | 0.3826e+0 | | 0.2014e+1 | |
| 35 | 0.6028e−1 | −0.78 | 0.3012e+0 | −0.16 | 0.4996e+1 | 0.62 |
| 142 | 0.1680e−1 | −0.91 | 0.2369e+0 | −0.17 | 0.1410e+2 | 0.74 |
| 548 | 0.4436e−2 | −0.99 | 0.1803e+0 | −0.20 | 0.4064e+2 | 0.78 |
| 2139 | 0.1141e−2 | −1.00 | 0.1308e+0 | −0.24 | 0.1146e+3 | 0.76 |
| theory | $\gtrsim N^{-1}$ | | $\lesssim N^{-1/4}$ | | $\lesssim N^{3/4}$ | |
| $N$ | $\lambda_{\min}(\widehat{\mathbf{A}}_{\mathfrak{W}})$ | | $\lambda_{\max}(\widehat{\mathbf{A}}_{\mathfrak{W}})$ | | $\kappa(\widehat{\mathbf{A}}_{\mathfrak{W}})$ | |
| 8 | 0.7113e+0 | | 0.1119e+1 | | 0.1573e+1 | |
| 35 | 0.4265e+0 | −0.35 | 0.1174e+1 | 0.03 | 0.2752e+1 | 0.38 |
| 142 | 0.2327e+0 | −0.43 | 0.1215e+1 | 0.02 | 0.5221e+1 | 0.46 |
| 548 | 0.1226e+0 | −0.47 | 0.1209e+1 | 0.00 | 0.9859e+1 | 0.47 |
| 2139 | 0.6329e−1 | −0.49 | 0.1225e+1 | 0.01 | 0.1936e+2 | 0.50 |
| theory | $\gtrsim N^{-1/2}$ | | $\lesssim 1$ | | $\lesssim N^{1/2}$ | |

**Table 12.7** Hypersingular integral equation on a screen with non-uniform mesh of Figure 12.3: $h_{\min} = \mathcal{O}(N^{-1/2})$ and $h_{\max} = \mathcal{O}(N^{-1/4})$. Theory: Lemma 12.2, Lemma 12.4, Theorem 12.1.

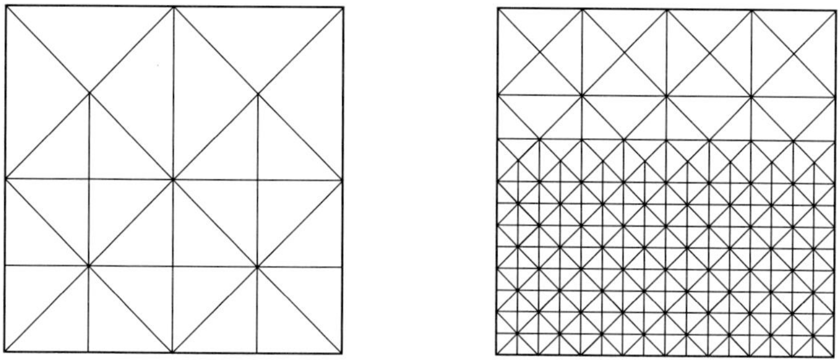

**Fig. 12.3** Successive meshes in a different sequence of non-uniformly refined meshes on the screen: $h_{\min} = \mathcal{O}(N^{-1/2})$ and $h_{\max} = \mathcal{O}(N^{-1/4})$.

# Chapter 13
# Multilevel Preconditioners with Adaptive Mesh Refinements

In Chapter 12 we showed that local mesh refinements have an adverse effect on the condition number of the boundary element matrix, namely, the condition number grows significantly if $h_{\max}/h_{\min}$ grows. We also proved that diagonal scaling at best reduces the condition number to that of a uniform triangulation, namely it still grows in the order of $O(N^{1/2})$ (in three dimensions), where $N$ is the degrees of freedom; see also [5] and [86].

In this chapter we study the use of multilevel methods to obtain optimal preconditioners in the case that the mesh is locally refined. Local mesh refinements, particularly important for boundary element discretisations, are carried out by using some a posteriori error estimators to mark elements to be refined; see e.g., [13, 14, 15, 50, 52, 54, 55, 92]. Different refinement strategies to partition these marked elements are available and the advantage of the so-called newest vertex bisection algorithm is explained in e.g., [130, 153]. In this chapter, we are interested in the newest vertex bisection algorithm.

It is noted that the article [4] also studies an optimal multilevel preconditioner for local mesh refinements. However, the analysis therein requires some restrictive conditions on the meshes which are not satisfied by the newest vertex bisection algorithm; see [4, page 391]. The analysis to be presented in this chapter extends that in [239] where the same bisection algorithm is studied for the V-cycle multigrid method for second order elliptic problems. As in previous chapters, the analysis in fractional order Sobolev spaces for boundary element methods requires different strategies than those used in finite element methods.

We report on the results in [75] where multilevel additive Schwarz algorithms are designed and analysed for the hypersingular integral equation. Even though we will focus on problems on surfaces in $\mathbb{R}^3$, the same analysis can be carried out for problems on curves in $\mathbb{R}^2$; see [75] for detail.

 
E. P. Stephan and T. Tran, *Schwarz Methods and Multilevel Preconditioners for Boundary Element Methods*, https://doi.org/10.1007/978-3-030-79283-1_13

## 13.1 Preliminaries

### 13.1.1 Mesh refinements and hierarchical structures

#### 13.1.1.1 Shape-regular triangulations

Let $\mathscr{T}_\ell$ denote a regular triangulation of $\Gamma$ into planar triangles and let $\mathscr{N}_\ell$ denote the set of nodes of the mesh $\mathscr{T}_\ell$. Each triangle $T \in \mathscr{T}_\ell$ is understood to be a closed set, i.e., $T$ contains its boundary.

For any subset $\tau \subset \Gamma$, we define

$$\mathscr{T}_\ell(\tau) := \{T \in \mathscr{T}_\ell : T \cap \tau \neq \varnothing\}. \tag{13.1}$$

We also define the patch $\omega_\ell^k(\tau) \subset \Gamma$ recursively by

$$\begin{aligned} \omega_\ell(\tau) := \omega_\ell^1(\tau) &:= \bigcup_{T \in \mathscr{T}_\ell(\tau)} T, \\ \omega_\ell^{k+1}(\tau) &:= \omega_\ell^k(\omega_\ell(\tau)) \quad \text{for } k \in \mathbb{N}. \end{aligned} \tag{13.2}$$

For any $z \in \Gamma$, we write $\mathscr{T}_\ell(z)$ and $\omega_\ell^k(z)$ as abbreviations for $\mathscr{T}_\ell(\{z\})$ and $\omega_\ell^k(\{z\})$, respectively.

Associated with the triangulation $\mathscr{T}_\ell$, we define the mesh-size function $h_\ell : \Gamma \to \mathbb{R}$ by

$$h_\ell(z) := \max_{T \in \mathscr{T}_\ell(z)} \operatorname{diam}(T) \quad \text{for all } z \in \Gamma.$$

We abuse the notation and write $h_\ell(T) = \operatorname{diam}(T)$ instead of $h_\ell(\mathring{T}) = \{\operatorname{diam}(T)\}$, where $\mathring{T}$ is the interior of $T$.

We suppose that $\mathscr{T}_\ell$ is uniformly shape-regular, i.e.,

$$\max\{h_\ell(T)^2/|T| : T \in \mathscr{T}_\ell\} \le \gamma. \tag{13.3}$$

where $|T|$ denotes the measure of the triangle $T$. It follows from this assumption that

$$h_\ell(T) \le h_\ell(z) \lesssim h_\ell(T) \quad \text{for all } z \in \mathscr{N}_\ell \text{ and } T \in \mathscr{T}_\ell(z), \tag{13.4}$$

where the hidden constant depends only on the $\gamma$-shape regularity of $\mathscr{T}_\ell$.

As usual, for $\ell \ge 0$, we consider the boundary element space

$$\mathscr{V}^\ell := \{v \in C(\Gamma) : v|_T \in \mathbb{P}_1(T)\ \forall T \in \mathscr{T}_\ell \quad \text{and} \quad v|_{\partial\Gamma} = 0\}. \tag{13.5}$$

The natural basis of $\mathscr{V}^\ell$ consists of hat functions $\eta_z^\ell$ which takes the value 1 at the node $z$ and zero at other nodal points:

$$\mathscr{V}^\ell = \operatorname{span}\{\eta_z : z \in \mathscr{N}_\ell\}.$$

These basis functions satisfy

$$\|\nabla \eta_z^\ell\|_{L^\infty(\Gamma)} \lesssim 1/h_\ell(z), \quad \mathrm{supp}(\eta_z^\ell) = \omega_\ell(z), \quad z \in \mathscr{N}_\ell \tag{13.6}$$

where the hidden constant depends only on the $\gamma$-shape regularity of $\mathscr{T}_\ell$.

#### 13.1.1.2 Newest vertex bisection algorithm (NVB algorithm)

The basic building block of an adaptive refinement algorithm is a method for dividing a triangle. In the newest vertex bisection algorithm, a reference edge $\mathbf{e}_T$ is chosen for each triangle $T$ which is called a parent triangle. This triangle $T$ is bisected by joining the midpoint of $\mathbf{e}_T$ with its opposite vertex, resulting in two child triangles $T_1$ and $T_2$. The midpoint of $\mathbf{e}_T$ becomes the newest vertex of both $T_1$ and $T_2$. The opposite side of the newest vertex of a new triangle becomes its reference edge; see Figure 13.1. We note that repeated applications of this bisection result in only two different triangles up to geometric similarity. This procedure of triangle bisection is described in Algorithm 13.1.

**Algorithm 13.1:** Bisection of a triangle in NVB

**Inputs:**
1: $T$: given triangle with a reference edge $\mathbf{e}_T$.

**Bisection:**
2: Bisect $T$ by joining the midpoint of $\mathbf{e}_T$ with its opposite vertex to obtain $T_1$ and $T_2$.
3: Mark the edge opposite to the new vertex to be the reference edge of $T_i$, $i = 1,2$.
4: In a local mesh refinement, bisect one or both child triangles $T_1$ and $T_2$ in the same manner only if it is required to avoid hanging nodes.
In a uniform refinement, bisect both $T_1$ are $T_2$.

**Outputs:**
5: Two, three or four child triangles. (Four child triangles in the case of uniform refinement.) Up to geometric similarity, there are only two different types of triangles. See Figure 13.1.

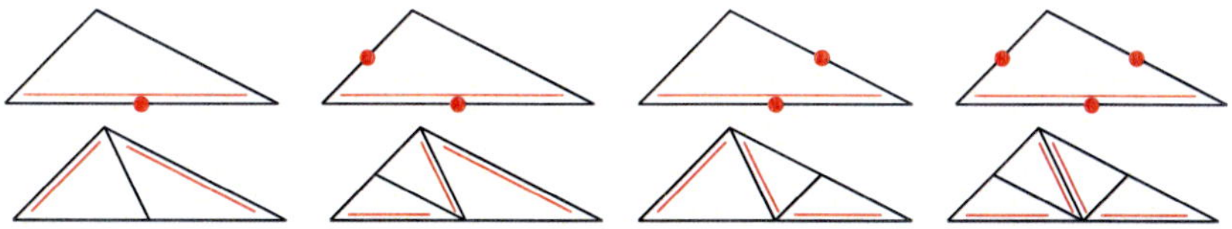

**Fig. 13.1** For each surface triangle $T \in \mathscr{T}_\ell$, there is one fixed *reference edge*, indicated by the double line (left, top). Refinement of $T$ is done by bisecting the reference edge, where its midpoint becomes a new node. The reference edges of the child triangles are opposite to this newest vertex (left, bottom). To avoid hanging nodes, iterated newest vertex bisection leads to 2, 3, or 4 child triangles if more than one edge, but at least the reference edge, is marked for refinement (right); see Algorithm 13.1.

We now explain the newest vertex bisection (NVB) algorithm. Let $\mathscr{T}_0$ be an initial mesh. For each element $T \in \mathscr{T}_0$ we choose one reference edge $\mathbf{e}_T$. There is no condition on the choice of reference edges in this initial mesh, but the longest edge is a good choice. Algorithm 13.2 explains how the mesh $\mathscr{T}_\ell$ is refined into $\mathscr{T}_{\ell+1}$.

It is proved in [20, 152, 153] that only a finite number of steps is required to remove hanging nodes. It is also proved in [20] that there exists a positive constant $\theta$ such that if $\theta_T$ is the smallest angle in $T$, then

$$\theta_T \ge \theta \quad \forall T \in \mathscr{T}_\ell,\ \ell = 0,1,2,\dots. \tag{13.7}$$

The presentation of the NVB algorithm in Algorithm 13.2 follows that of [130]. The while loop in this algorithm terminates after a finite number of iterates because $\widetilde{\mathscr{E}}_\ell^{(k-1)} \subset \widetilde{\mathscr{E}}_\ell^{(k)} \subset \mathscr{E}_\ell$. It is also clear that the for loop terminates since $\mathscr{M}_\ell \subset \mathscr{T}_\ell$. The right figures in Figure 13.2 demonstrates the steps in Algorithm 13.2.

**Algorithm 13.2:** Newest Vertex Bisection (NVB)

**Inputs:**

- $\mathscr{T}_\ell$ (a given refined mesh)
- $\mathscr{M}_\ell \subset \mathscr{T}_\ell$ (set of marked elements given by some error indicator)
- $\mathscr{E}_\ell := \{\mathbf{e}_T : T \in \mathscr{T}_\ell\}$ (set of all reference edges)
- $\widetilde{\mathscr{E}}_\ell^{(-1)} = \varnothing$;
- $\widetilde{\mathscr{E}}_\ell^{(0)} := \{\mathbf{e}_T \in \mathscr{E}_\ell : T \in \mathscr{M}_\ell\}$ (set of reference edges of marked elements)

$k \leftarrow 0$

**while** $\widetilde{\mathscr{E}}_\ell^{(k-1)} \subsetneq \widetilde{\mathscr{E}}_\ell^{(k)}$ **do**

  $k \leftarrow k+1$;

  $\bar{\mathscr{E}}_\ell^{(k)} := \{\mathbf{e}_T \in \mathscr{E}_\ell \setminus \widetilde{\mathscr{E}}_\ell^{(k-1)}$ : another edge of $T$ belongs to $\widetilde{\mathscr{E}}_\ell^{(k-1)}\}$ (set of reference edges of triangles marked with blue dashed lines in Figure 13.2)

  $\widetilde{\mathscr{E}}_\ell^{(k)} := \widetilde{\mathscr{E}}_\ell^{(k-1)} \cup \bar{\mathscr{E}}_\ell^{(k)}$

**end while**

$\widetilde{\mathscr{E}}_\ell := \widetilde{\mathscr{E}}_\ell^{(k)}$ (all reference (red) edges of triangles marked with a star or a blue edge)

$\widetilde{\mathscr{M}_\ell} := \{T \in \mathscr{T}_\ell : \mathbf{e}_T \in \widetilde{\mathscr{E}}_\ell\}$ (all $T$ marked with a star or a blue edge in Figure 13.2)

**for** $T \in \widetilde{\mathscr{M}_\ell}$ **do**

  Bisect $T$ by using Algorithm 13.1.

  If a new reference edge of a new triangle belongs to $\widetilde{\mathscr{E}}_\ell$ (blue edge in Figure 13.2), continue to bisect this new triangle by using this edge. (This step removes hanging nodes.)

**end for**

**Outputs:**

- Refined mesh $\mathscr{T}_{\ell+1}$
- A set of reference edges $\mathscr{E}_{\ell+1} := \{\mathbf{e}_T : T \in \mathscr{T}_{\ell+1}\}$

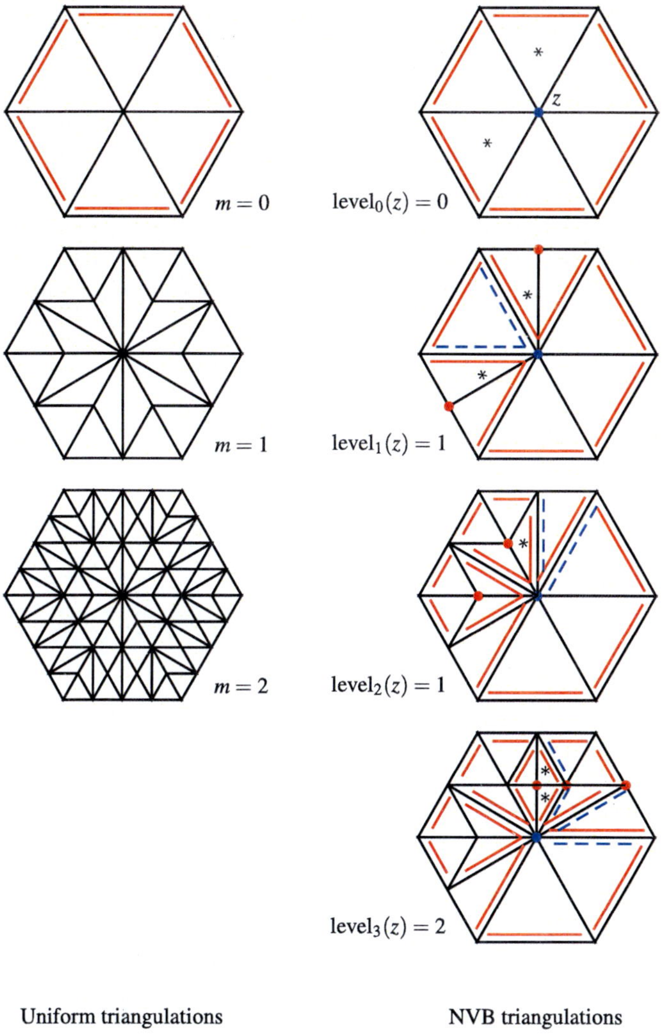


**Fig. 13.2** Left figures: Uniform triangulations. Right figures: NVB triangulations. Red dot: newest vertex. Asterisk: marked element. Red line: reference edge. Blue dashed line: marked edge. Triangles with dashed blue lines will also be bisected together with marked triangles. (The level function $\text{level}_\ell(z)$ is defined in Subsection 13.1.2 and the value of $m$ indicates the level of uniform refinements.)

#### 13.1.1.3 Hierarchical structures and the Scott–Zhang interpolation

It is clear from the above procedure that $\mathcal{N}_\ell \subset \mathcal{N}_{\ell+1}$. To design an optimal additive Schwarz scheme on locally refined meshes, we also define

$$
\begin{aligned}
\widetilde{\mathcal{N}_0} &:= \mathcal{N}_0, \\
\widetilde{\mathcal{N}_\ell} &:= \left(\mathcal{N}_\ell \setminus \mathcal{N}_{\ell-1}\right) \cup \left\{z \in \mathcal{N}_{\ell-1} : \omega_\ell(z) \subsetneq \omega_{\ell-1}(z)\right\}, \quad \ell \geq 1.
\end{aligned} \tag{13.8}
$$

Thus $\widetilde{\mathcal{N}_\ell}$ contains all new nodes and the neighbouring old nodes $z \in \mathcal{N}_{\ell-1}$ such that the support of the new basis function at $z$ is a strict subset of the support of the old basis function at that point, namely $\operatorname{supp}(\eta_z^\ell) \subsetneq \operatorname{supp}(\eta_z^{\ell-1})$; see Figure 13.3. It is noted that for $\ell = 0, 1, \ldots,$

$$
\begin{aligned}
\mathcal{N}_\ell &= \bigcup_{k=0}^{\ell} \widetilde{\mathcal{N}_k}, \\
\left(\mathcal{N}_\ell \setminus \widetilde{\mathcal{N}_\ell}\right) &\subset \mathcal{N}_{\ell-1} \\
\eta_z^{\ell-1} &= \eta_z^\ell \quad \forall z \in \mathcal{N}_\ell \setminus \widetilde{\mathcal{N}_\ell}.
\end{aligned} \tag{13.9}
$$

For $\ell \geq 0$ we define

$$
\begin{aligned}
\mathcal{V}_z^\ell &:= \operatorname{span}\{\eta_z^\ell\}, \quad z \in \mathcal{N}_\ell, \\
\widetilde{\mathcal{V}}^\ell &:= \operatorname{span}\left\{\eta_z^\ell : z \in \widetilde{\mathcal{N}_\ell}\right\} = \bigcup_{z \in \widetilde{\mathcal{N}_\ell}} \mathcal{V}_z^\ell.
\end{aligned} \tag{13.10}
$$

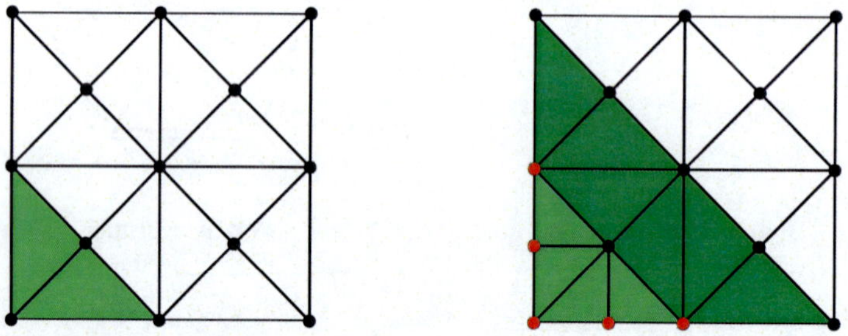

**Fig. 13.3** The left figure shows a mesh $\mathcal{T}_{\ell-1}$, where the two elements in the lower left corner are marked for refinement (green). Bisection of these two elements provides the mesh $\mathcal{T}_\ell$ (right), where two new nodes are created. The set $\widetilde{\mathcal{N}_\ell}$ consists of these new nodes plus those immediate neighbours, where $\operatorname{supp}(\eta_z^\ell) \subsetneq \operatorname{supp}(\eta_z^{\ell-1})$ (red). The union of the supports of basis functions in $\widetilde{\mathcal{V}}^\ell$ is given by the light- and dark-green areas in the right figure.

On each triangulation $\mathscr{T}_\ell$, we can define a Scott–Zhang interpolation operator $J_\ell : L^2(\Gamma) \longrightarrow \mathscr{V}^\ell$; see Subsection C.5.5. We note that in the definition of this interpolation, the choice of the triangle $T_z^\ell \in \mathscr{T}_\ell$ for each nodal point $z \in \mathscr{N}_\ell$ is arbitrary. Noting (13.9), we define $J_\ell$ by choosing $T_z^\ell$ such that

$$T_z^{\ell-1} = T_z^\ell \in \mathscr{T}_\ell \cap \mathscr{T}_{\ell-1} \quad \forall z \in \mathscr{N}_\ell \setminus \widetilde{\mathscr{N}_\ell}. \tag{13.11}$$

Note that we also have

$$\eta_z^\ell = \eta_z^{\ell-1} \quad \text{and} \quad \psi_z^\ell = \psi_z^{\ell-1} \quad \forall z \in \mathscr{N}_\ell \setminus \widetilde{\mathscr{N}_\ell} \tag{13.12}$$

where $\psi_z^\ell$ is the corresponding $L^2$-dual basis function to $\eta_z^\ell$; see Subsection C.5.5. For convenience, we rewrite the definition (C.84)

$$J_\ell v(y) = \sum_{z' \in \mathscr{N}_\ell} \eta_{z'}^\ell(y) \int_{T_{z'}^\ell} \psi_{z'}^\ell(x) v(x)\, \mathrm{d}s(x), \quad y \in \Gamma,\ \ell = 0, 1, 2, \ldots. \tag{13.13}$$

The sequence $\{J_\ell\}$ has the following properties.

**Lemma 13.1.** *For any $v \in L^2(\Gamma)$ and $\ell = 1, 2, \ldots$,*

$$(J_\ell - J_{\ell-1})v(z) = 0 \quad \forall z \in \mathscr{N}_\ell \setminus \widetilde{\mathscr{N}_\ell}, \tag{13.14}$$

$$(J_\ell - J_{\ell-1})v \in \widetilde{\mathscr{V}}^\ell, \tag{13.15}$$

$$|J_\ell v(z)| \lesssim \begin{cases} \dfrac{1}{h_\ell(z)} \|v\|_{L^2(\omega_\ell(z))} & \forall z \in \mathscr{N}_\ell, \\ \dfrac{1}{h_\ell(z)} \|v\|_{L^2(\omega_\ell^2(z))} & \forall z \in \mathscr{N}_{\ell+1} \setminus \mathscr{N}_\ell. \end{cases} \tag{13.16}$$

*In particular,*

$$|J_\ell v(z) - J_{\ell-1} v(z)| \lesssim \frac{1}{h_\ell(z)} \|v\|_{L^2(\omega_{\ell-1}^2(z))} \quad \forall z \in \widetilde{\mathscr{N}_\ell}. \tag{13.17}$$

*The constants depend only on the $\gamma$-shape regularity of $\mathscr{T}_\ell$.*

*Proof.* It follows from the property $\eta_{z'}^\ell(z) = \delta_{z,z'}$, (13.11), and (13.12) that, for any $z \in \mathscr{N}_\ell \setminus \widetilde{\mathscr{N}_\ell}$,

$$\begin{aligned}(J_\ell - J_{\ell-1})v(z) &= \sum_{z' \in \mathscr{N}_\ell} \eta_{z'}^\ell(z) \int_{T_{z'}^\ell} \psi_{z'}^\ell(x) v(x)\, \mathrm{d}s(x) \\ &\quad - \sum_{z'' \in \mathscr{N}_{\ell-1}} \eta_{z''}^{\ell-1}(z) \int_{T_{z''}^{\ell-1}} \psi_{z''}^{\ell-1}(x) v(x)\, \mathrm{d}s(x)\end{aligned}$$

$$= \int_{T_z^\ell} \psi_z^\ell(x)v(x)\,\mathrm{d}s(x) - \int_{T_z^{\ell-1}} \psi_z^{\ell-1}(x)v(x)\,\mathrm{d}s(x) = 0,$$

proving (13.14). It follows that

$$(J_\ell - J_{\ell-1})v = \sum_{z\in\mathcal{N}_\ell} (J_\ell - J_{\ell-1})v(z)\eta_z^\ell = \sum_{z\in\widetilde{\mathcal{N}_\ell}} (J_\ell - J_{\ell-1})v(z)\eta_z^\ell.$$

This means $(J_\ell - J_{\ell-1})v \in \operatorname{span}\{\eta_z : z \in \widetilde{\mathcal{N}_\ell}\}$, proving (13.15).

Next we prove (13.16). If $z \in \mathcal{N}_\ell$, we have from (C.83) that $\|\psi_z^\ell\|_{L^\infty(T_z^\ell)} \lesssim 1/|T_z^\ell|$. Moreover, $T_z^\ell \subset \omega_\ell(z)$, and thus

$$\begin{aligned}
|J_\ell v(z)| &\le \int_{T_z^\ell} |\psi_z^\ell(x)\, v(x)|\,\mathrm{d}s(x) \\
&\le \|\psi_z^\ell\|_{L^\infty(T_z^\ell)} |T_z^\ell|^{1/2} \|v\|_{L^2(T_z^\ell)} \\
&\lesssim |T_z^\ell|^{-1/2} \|v\|_{L^2(\omega_\ell(z))} \\
&\lesssim h_\ell(z)^{-1} \|v\|_{L^2(\omega_\ell(z))},
\end{aligned}$$

where in the last step we used (13.3) and (13.4). If $z \in \mathcal{N}_{\ell+1} \setminus \mathcal{N}_\ell$, then $z$ is the midpoint of an edge in $\mathcal{T}_\ell$. Therefore, there exist $z_1, z_2 \in \mathcal{N}_\ell$ such that

$$J_\ell v(z) = \eta_{z_1}^\ell(z) \int_{T_{z_1}^\ell} \psi_{z_1}^\ell(x)v(x)\,\mathrm{d}s(x) + \eta_{z_2}^\ell(z) \int_{T_{z_2}^\ell} \psi_{z_2}^\ell(x)v(x)\,\mathrm{d}s(x)$$

implying

$$|J_\ell v(z)| \le \int_{T_{z_1}^\ell} |\psi_{z_1}^\ell(x)v(x)|\,\mathrm{d}s(x) + \int_{T_{z_2}^\ell} |\psi_{z_2}^\ell(x)v(x)|\,\mathrm{d}s(x).$$

Repeating the above argument yields

$$\begin{aligned}
|J_\ell v(z)| &\lesssim h_\ell(z_1)^{-1} \|v\|_{L^2(\omega_\ell(z_1))} + h_\ell(z_2)^{-1} \|v\|_{L^2(\omega_\ell(z_2))} \\
&\le h_\ell(z)^{-1} \|v\|_{L^2(\omega_\ell^2(z))},
\end{aligned}$$

where in the last step we used $h_\ell(z) \le h_\ell(z_i)$ and $\omega_\ell(z_i) \subset \omega_\ell^2(z)$ for $i = 1, 2$, proving (13.16).

Finally, to prove (13.17) we invoke (13.16) to have, for $z \in \widetilde{\mathcal{N}_\ell}$,

$$\begin{aligned}
|J_\ell v(z) - J_{\ell-1}v(z)| &\le |J_\ell v(z)| + |J_{\ell-1}v(z)| \\
&\lesssim \frac{1}{h_\ell(z)} \|v\|_{L^2(\omega_\ell(z))} + \frac{1}{h_{\ell-1}(z)} \|v\|_{L^2(\omega_{\ell-1}^2(z))}
\end{aligned}$$

$$\lesssim \frac{1}{h_\ell(z)} \|v\|_{L^2(\omega_{\ell-1}^2(z))}$$

where in the last step we used $h_\ell(z) \le h_{\ell-1}(z)$ and $\omega_\ell(z) \subset \omega_{\ell-1}^2(z)$. □

### 13.1.2 Level functions and uniform mesh refinements

For each element $T \in \mathscr{T}_\ell$, let $T_0 \in \mathscr{T}_0$ be the unique ancestor element satisfying $T \subset T_0$. The generation of $T$ is defined by

$$\operatorname{gen}(T) := \log_2 \left(|T_0|/|T|\right) \in \mathbb{N}_0. \tag{13.18}$$

Recalling the definition (13.1) of $\mathscr{T}_\ell(z)$, we assign to each node $z \in \mathscr{N}_\ell$ the level

$$\operatorname{level}_\ell(z) := \left\lceil \max_{T \in \mathscr{T}_\ell(z)} \frac{\operatorname{gen}(T)}{2} \right\rceil, \tag{13.19}$$

where $\lceil \cdot \rceil$ denotes the Gaussian ceiling function, i.e., $\lceil x \rceil = \min\left\{n \in \mathbb{N}_0 : x \le n\right\}$ for $x \ge 0$. Note that (13.19) slightly differs from the level function in [239]. However, both definitions of the level function are equivalent.

Besides the sequence of locally refined triangulations $\mathscr{T}_\ell$, we consider a sequence of uniformly refined triangulations $\widehat{\mathscr{T}_m}$ defined as follows. We start with $\widehat{\mathscr{T}_0} := \mathscr{T}_0$ in which each triangle is assigned with a reference edge (usually the longest side). Having defined $\widehat{\mathscr{T}_m}$, we define $\widehat{\mathscr{T}_{m+1}}$ by bisecting each triangle $\widehat{T} \in \widehat{\mathscr{T}_m}$ three times to obtain four child triangles $\widehat{T}_j \in \widehat{\mathscr{T}_{m+1}}$, $j = 1, \dots, 4$; see the bottom right figure in Figure 13.1 and the left figures in Figure 13.2. As a consequence, if $\widehat{T} \in \widehat{\mathscr{T}_m}$ and $\widehat{T}_0 \in \widehat{\mathscr{T}_0}$ are such that $\widehat{T} \subset T_0$ ($\widehat{T}_0$ is the ancestor of $\widehat{T}$) then

$$\frac{|\widehat{T}_0|}{|\widehat{T}|} = 4^m, \quad m = 1, 2, \dots. \tag{13.20}$$

If we define

$$\widehat{h}_m := \max_{\widehat{T} \in \widehat{\mathscr{T}_m}} \operatorname{diam}(\widehat{T}), \quad m \ge 0,$$

then

$$\widehat{h}_m \simeq 2^{-m} \widehat{h}_0, \quad m \ge 0, \tag{13.21}$$

where the constant is independent of $m$. We note that $\widehat{h}_m \simeq \operatorname{diam}(\widehat{T})$ for all $\widehat{T} \in \widehat{\mathscr{T}_m}$, $m \ge 0$.

We define on each triangulation a boundary element space $\widehat{\mathscr{V}}^m$ of continuous piecewise linear functions on $\widehat{\mathscr{T}}^m$ vanishing on the boundary of $\Gamma$. We then have the following nested sequence

$$\widehat{\mathscr{V}}^0 \subset \widehat{\mathscr{V}}^1 \subset \cdots \subset \widehat{\mathscr{V}}^m \subset \cdots. \tag{13.22}$$

The following lemma gives the relation between this mesh size and the mesh-size function $h_\ell$ defined in Subsection 13.1.1.

**Lemma 13.2.** *For any* $z \in \mathscr{N}_\ell$, *let* $m = \text{level}_\ell(z)$. *Then*

(i) $z \in \widehat{\mathscr{N}_m}$,
(ii) $\eta_z^\ell \in \widehat{\mathscr{V}^m}$,
(iii) $\widehat{h}_m \lesssim h_\ell(z) \lesssim \widehat{h}_m$ *where the constants are independent of* $z$, $\ell$, *and* $L$.

*Proof.* First we prove (i). Note that for any $\widehat{T} \in \widehat{\mathscr{T}_m}$, (13.18) and (13.20) give

$$\text{gen}(\widehat{T}) = 2m.$$

Hence it follows from (13.19) that

$$\text{gen}(T) \le \text{gen}(\widehat{T}) \quad \forall T \in \mathscr{T}_\ell(z),\ \forall \widehat{T} \in \widehat{\mathscr{T}_m}. \tag{13.23}$$

Due to the way that both types of refinement (uniform and NVB) bisect triangles (Algorithm 13.1) and due to the initial choice $\widehat{\mathscr{T}_0} = \mathscr{T}_0$ with same reference edges, it follows from (13.23) that any $T \in \mathscr{T}_\ell(z)$ is a union of triangles $\widehat{T} \in \widehat{\mathscr{T}_m}$. In particular, with $T \in \mathscr{T}_\ell(z)$ there exists $\widehat{T} \in \widehat{\mathscr{T}_m}(z)$ such that $\widehat{T} \subset T$. This means $z$ is a vertex of $\widehat{T}$, i.e., $z \in \widehat{\mathscr{N}_m}$.

To prove Part (ii) we note that, for any $\widehat{T} \in \widehat{\mathscr{T}_m}$, either $\widehat{T} \subset \omega_\ell(z)$ or $\widehat{T} \subset \Gamma \setminus \omega_\ell(z)$ (i.e., no overlap of $\widehat{T}$ and $\omega_\ell(z)$ besides the triangle boundary). In the second case, clearly $\eta_z^\ell|_{\widehat{T}} \equiv 0$. In the first case

$$\eta_z^\ell|_{\widehat{T}} = \left(\eta_z^\ell|_T\right)\Big|_{\widehat{T}}$$

is a linear function. Therefore, $\eta_z^\ell \in \widehat{\mathscr{V}^m}$.

Finally, we prove (iii). For any $z \in \mathscr{N}_\ell$, there exists $T \in \mathscr{T}_\ell(z)$ such that

$$\text{gen}(T) \ge \text{gen}(T') \quad \forall T' \in \mathscr{T}_\ell(z).$$

It follows from (13.19) and the definition of $m$ that

$$\frac{\text{gen}(T)}{2} \le m < \frac{\text{gen}(T)}{2} + 1.$$

Noting that $|T| \simeq h_\ell(T)^2$ and $|T_0| \simeq h_0^2$, we deduce by using (13.18)

$$\log_2\left(\frac{h_0}{h_\ell(T)}\right) \lesssim m \lesssim \log_2\left(\frac{2h_0}{h_\ell(T)}\right)$$

implying

$$\frac{h_\ell(T)}{2} \lesssim 2^{-m} h_0 \lesssim h_\ell(T).$$

The required result follows from (13.4) and (13.21). $\square$

The analysis in this chapter requires some additional definitions and technical results. For a given node $z \in \widetilde{\mathcal{N}_\ell}$, it may happen that $z \in \widetilde{\mathcal{N}_{\ell+k}}$ with $\text{level}_\ell(z) = \text{level}_{\ell+k}(z)$ for some $k > 0$. We count how often a node $z \in \mathcal{N}_L$ with a fixed level $m \in \mathbb{N}_0$ shows up in the sets $\widetilde{\mathcal{N}_\ell}$. We therefore define, for $z \in \mathcal{N}_L$ and $m \in \mathbb{N}_0$,

$$\widetilde{\mathcal{K}_m}(z) := \big\{\ell \in \{0,1,\dots,L\} : z \in \widetilde{\mathcal{N}_\ell} \quad \text{and} \quad \text{level}_\ell(z) = m\big\}. \tag{13.24}$$

The following lemma from [239, Lemma 3.1] proves that the cardinality of the set $\widetilde{\mathcal{K}_m}(z)$ is uniformly bounded.

**Lemma 13.3.** *There exists a positive constant $d_0$ depending on $\theta$ in* (13.7) *such that for all positive integers $L$ and $m$, and for all $z \in \mathcal{N}_L$, the cardinality of $\widetilde{\mathcal{K}_m}(z)$ satisfies*

$$\text{card}(\widetilde{\mathcal{K}_m}(z)) \le d_0.$$

*Proof.* The result is obvious if $\widetilde{\mathcal{K}_m}(z)$ is empty. If it is not empty, we prove by contradiction. Assume that

$$\forall d_0 > 0,\ \exists L > 0,\ m > 0,\ z \in \mathcal{N}_L : \quad \text{card}(\widetilde{\mathcal{K}_m}(z)) > d_0.$$

Let $d := \text{card}(\widetilde{\mathcal{K}_m}(z))$. There exist $\ell_1,\dots,\ell_d \in \{0,1,\dots,L\}$ satisfying

$$\ell_1 < \ell_2 < \cdots < \ell_d, \quad z \in \widetilde{\mathcal{N}_{\ell_j}}, \quad \text{level}_{\ell_j}(z) = m, \quad j = 1,\dots,d.$$

It follows from Part (iii) of Lemma 13.2 that

$$h_{\ell_j}(z) \simeq \widehat{h}_m, \quad j = 1,\dots,d.$$

Recalling the definition of $h_\ell(z)$ in Subsection 13.1.1, the above statement implies that there exists a triangle $T \in \mathcal{T}_{\ell_1}(z)$ such that

$$\text{diam}(T) \simeq h_{\ell_1}(z) \simeq h_{\ell_d}(z).$$

In other words, this triangle $T$ also belongs to $\mathcal{T}_{\ell_d}(z)$, i.e., it is not bisected at all while the triangulation $\mathcal{T}_{\ell_1}$ is in the refinement process to yield $\mathcal{T}_{\ell_d}$. On the other hand, if the mesh $\mathcal{T}_{\ell_1}$ is refined to have $\mathcal{T}_{\ell_2}$, then at least one triangle in $\mathcal{T}_{\ell_1}(z)$ is bisected since $z \in \widetilde{\mathcal{N}_{\ell_1}} \cap \widetilde{\mathcal{N}_{\ell_2}}$. The process to go from $\mathcal{T}_{\ell_1}$ to $\mathcal{T}_{\ell_d}$ and the fact that $z \in \widetilde{\mathcal{N}_{\ell_j}}$ for all $j = 1,\dots,d$ imply that at least $d$ triangles in $\mathcal{T}_{\ell_1}(z)$ (the triangles can be repeated) are bisected. If $d_0$ is large enough, so is $d$, and consequently the triangle next to $T$ which shares the same edge with $T$ is bisected many times on the other two edges; see Figure 13.4. This contradicts the shape regular property (13.7). The above argument also implies that the constant $d_0$ depends on $\theta$. This completes the proof of the lemma. □

For each $z \in \mathcal{N}_\ell$, we further define the quantities

$$r_\ell(z) := \min_{T \in \mathcal{T}_{\ell-1}(\omega^2_{\ell-1}(z))} \text{gen}(T) \quad \text{and} \quad R_\ell(z) := \lfloor r_\ell(z)/2 \rfloor, \tag{13.25}$$

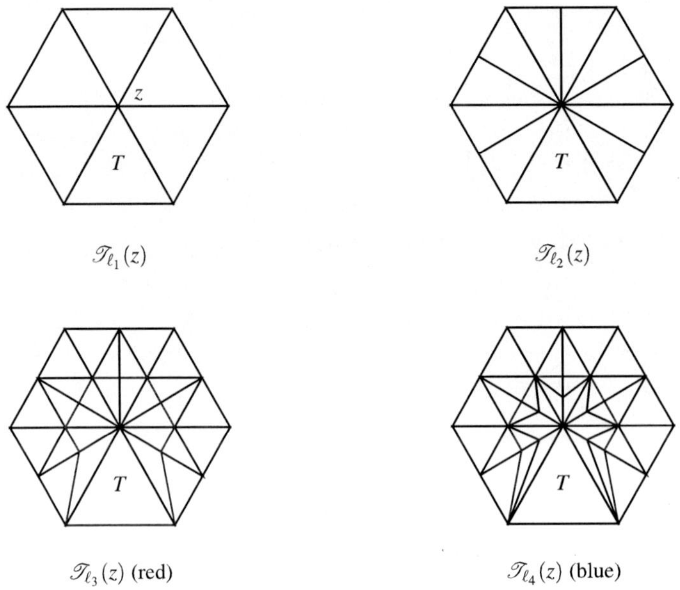


**Fig. 13.4** Non-shape regular triangulations. Note that $T \in \mathscr{T}_{\ell_j}(z)$ and that $h_{\ell_j}(z) \simeq \operatorname{diam}(T)$ for all $j = 1,\dots,d$.

where $\lfloor x \rfloor := \max\{n \in \mathbb{N}_0 : x \geq n\}$ is the Gaussian floor function for $x \geq 0$. Analogously to the patch $\omega_\ell^k(z)$ defined by (13.2), we define the patch $\widehat{\omega}_m^k(z)$ corresponding to the uniformly refined triangulation $\widehat{\mathscr{T}_m}$.

**Lemma 13.4.** (i) *There exists a constant* $c_0 \in \mathbb{N}$ *depending only on the initial triangulation* $\mathscr{T}_0$ *such that for all* $z \in \mathscr{N}_\ell$

$$\operatorname{level}_\ell(z) \leq R_\ell(z) + c_0.$$

(ii) *For all* $z \in \mathscr{N}_\ell$*, let* $m = R_\ell(z)$*. Then for any* $T \in \mathscr{T}_{\ell-1}(\omega_{\ell-1}^2(z))$*, there exists an element* $\widehat{T} \in \widehat{\mathscr{T}_m}$ *such that* $T \subset \widehat{T}$*.*

(iii) *There exists* $n \in \mathbb{N}_0$ *depending only on the initial triangulation* $\mathscr{T}_0$ *such that for all* $z \in \mathscr{N}_\ell$

$$\omega_\ell(z) \subseteq \omega_{\ell-1}^2(z) \subseteq \widehat{\omega}_{\operatorname{level}_\ell(z)}^n(z).$$

*Proof.* The proofs of (i) and (ii) follow the same lines with the proof of Lemma 3.3 in [239]. We reproduce them here for completeness.

Consider $z \in \mathcal{N}_\ell$. From the definition of $\text{level}_\ell(z)$ and $R_\ell(z)$ in (13.19) and (13.25), in order to prove Part (i) it suffices to prove

$$\text{gen}(T) \le r_\ell(z) + c_1 \quad \forall T \in \mathcal{T}_\ell(z) \tag{13.26}$$

where $c_1$ depends only on $\mathcal{T}_0$. For any $T \in \mathcal{T}_\ell(z)$, let $T' \in \mathcal{T}_{\ell-1}(z)$ be the father element of $T$. Then

$$|\text{gen}(T) - \text{gen}(T')| \le 2.$$

Furthermore, there exists a constant $k \in \mathbb{N}$ such that

$$|\text{gen}(T') - \text{gen}(T'')| \le k \quad \forall T'' \in \mathcal{T}_{\ell-1}(\omega_{\ell-1}^2(z)),$$

i.e., the difference of the generations of two elements of one patch is uniformly bounded due to [82, Theorem 3.3]. Therefore,

$$\begin{aligned}\text{gen}(T) &\le |\text{gen}(T) - \text{gen}(T')| + |\text{gen}(T') - \text{gen}(T'')| + \text{gen}(T'')\\ &\le c_1 + \text{gen}(T'').\end{aligned}$$

Taking the minimum over $T'' \in \mathcal{T}_{\ell-1}(\omega_{\ell-1}^2(z))$ and using the definition of $r_\ell(z)$ in (13.25), we deduce (13.26), and thus we prove (i).

To prove Part (ii) we note that for any $z \in \mathcal{N}_\ell$ and $T \in \mathcal{T}_{\ell-1}(\omega_{\ell-1}^2)$, definitions (13.18) and (13.25) give

$$\text{gen}(T) \ge r_\ell(z) \ge 2R_\ell(z) = 2m.$$

From the definition of the uniform triangulation, for all $\widehat{T} \in \widehat{\mathcal{T}_m}$ we have $\text{gen}(\widehat{T}) = 2m$. This implies

$$\text{gen}(\widehat{T}) \le \text{gen}(T) \quad \forall \widehat{T} \in \widehat{\mathcal{T}_m}$$

and thus

$$|T| \le |\widehat{T}| \quad \forall \widehat{T} \in \widehat{\mathcal{T}_m}.$$

Since both types of triangulations (the uniform triangulation and the newest vertex bisection algorithm) define new triangles by bisecting a triangle through a midpoint of an edge (resulting in similar triangles), the above inequality implies that there exists $\widehat{T} \in \widehat{\mathcal{T}_m}$ such that $T \subset \widehat{T}$.

Finally, to prove Part (iii) we first note that $\omega_\ell(z) \subset \omega_{\ell-1}(z) \subset \omega_{\ell-1}^2(z)$. It remains to prove that there exists an integer $n$ which depends only on $\mathcal{T}_0$ such that

$$\omega_{\ell-1}^2(z) \subset \widehat{\omega}_{\text{level}_\ell(z)}^n \tag{13.27}$$

Due to Part (ii) above, for any $T \in \mathcal{T}_{\ell-1}(\omega_{\ell-1}^2(z))$, there exists $\widehat{T} \in \widehat{\mathcal{T}_m}$ such that $T \subset \widehat{T}$. From the definition of the patch $\omega_{\ell-1}^2(z)$ and the patch $\widehat{\omega}_m^2(z)$, it follows that $\widehat{T} \subset \widehat{\omega}_m^2(z)$, i.e., $\widehat{T} \in \widehat{\mathcal{T}_m}(\widehat{\omega}_m^2(z))$. Hence $T \subset \widehat{\omega}_m^2(z)$, i.e.,

$$\omega_{\ell-1}^2(z) \subset \widehat{\omega}_m^2(z).$$

On the other hand, with $c_0$ being the integer given in Part (i), each triangle $\widehat{T} \in \widehat{\mathscr{T}_m}(\widehat{\omega}_m^2(z))$ is bisected into $4c_0$ triangles $\widehat{T}_j \in \widehat{\mathscr{T}_{m+c_0}}$, $j = 1,\dots,4c_0$ such that

$$\widehat{T} = \bigcup_{j=1}^{4c_0} \widehat{T}_j.$$

In particular, there exists $n \in \mathbb{N}$ with $n \leq 8c_0$ such that $\widehat{T} \in \mathscr{T}_{m+c_0}(\widehat{\omega}_{m+c_0}^n(z))$, so that

$$\omega_{\ell-1}^2(z) \subset \widehat{\omega}_{m+c_0}^n(z).$$

Part (i) states that $\text{level}_\ell(z) \leq m + c_0$. Hence by the definition of patches we have

$$\widehat{\omega}_{m+c_0}^n(z) \subset \widehat{\omega}_{\text{level}_\ell(z)}^n(z).$$

Both set inclusions above imply (13.27), completing the proof of the lemma. □

## 13.2 Multilevel Preconditioners for the Hypersingular Integral Equation

In this section we present two adaptive schemes for the hypersingular integral equation based on two multilevel additive Schwarz preconditioners. We recall the Galerkin boundary element approximation: For some fixed positive integer $L$, find $u_L \in \mathscr{V}^L$ satisfying

$$a_{\mathfrak{W}}(u_L, v) = \langle g, v \rangle \quad \forall v \in \mathscr{V}^L.$$

Here $\mathscr{V}^L$ is defined as in (13.5).

### *13.2.1 Local multilevel diagonal preconditioner (LMD preconditioner)*

A subspace decomposition of the boundary element space $\mathscr{V}^L$ in the form

$$\mathscr{V}^L = \sum_{\ell=0}^{L} \sum_{z \in \widetilde{\mathscr{N}_\ell}} \mathscr{V}_z^\ell \tag{13.28}$$

as usual defines an additive Schwarz preconditioner, which will be called the local multilevel diagonal (LMD) preconditioner. Here $\mathscr{V}_z^\ell$ and $\widetilde{\mathscr{N}_\ell}$ are defined by (13.10) and (13.8), respectively. We note that this subspace decomposition is valid due to (13.9).

Similarly to (5.5) in Chapter 5, the multilevel additive Schwarz operator is defined by

$$P_{LMD} = \sum_{\ell=0}^{L} \sum_{z\in\widetilde{\mathcal{N}_\ell}} P_{\mathscr{V}_z^\ell} \tag{13.29}$$

where $P_{\mathscr{V}_z^\ell} : \mathscr{V}^L \to \mathscr{V}_z^\ell$ is defined for all $v \in \mathscr{V}^L$ by

$$a_{\mathfrak{W}}(P_{\mathscr{V}_z^\ell} v, w) = a_{\mathfrak{W}}(v, w) \quad \forall w \in \mathscr{V}_z^\ell. \tag{13.30}$$

The matrix form of this preconditioner will be explained in Section 13.3.

The analysis also requires the $L^2$-projection $Q_j : L^2(\Gamma) \longrightarrow \widehat{\mathscr{V}}^j$ for $j = 0, 1, \ldots,$ defined by

$$\langle Q_j v, w\rangle_{L^2(\Gamma)} = \langle v, w\rangle_{L^2(\Gamma)} \quad \forall v \in L^2(\Gamma),\ w \in \widehat{\mathscr{V}}^j. \tag{13.31}$$

**Lemma 13.5 (Stability of decomposition).** *Any $v \in \mathscr{V}^\ell$ admits a decomposition*

$$v = \sum_{\ell=0}^{L} \sum_{z\in\widetilde{\mathcal{N}_\ell}} v_z^\ell, \quad v_z^\ell \in \mathscr{V}_z^\ell, \tag{13.32}$$

*that satisfies*

$$\sum_{\ell=0}^{L} \sum_{z\in\widetilde{\mathcal{N}_\ell}} a_{\mathfrak{W}}(v_z^\ell, v_z^\ell) \lesssim a_{\mathfrak{W}}(v, v). \tag{13.33}$$

*The constant is independent of $\ell = 0, \ldots, L$.*

*Proof.* Recalling the definition of $J_\ell$ by (13.13), for any $v \in \mathscr{V}^L$ with some fixed positive integer $L$, we can write

$$v = J_L v = \sum_{\ell=0}^{L} (J_\ell - J_{\ell-1}) v$$

where we set $J_{-1} = 0$. Here we used the fact that $J_\ell$ is a projection; see Lemma C.47. Let

$$v^\ell := (J_\ell - J_{\ell-1})v, \quad \ell = 0, \ldots, L.$$

Then it follows from (13.15) that $v^\ell \in \widetilde{\mathscr{V}}^\ell$. Hence we can write

$$v^\ell = \sum_{z\in\widetilde{\mathcal{N}_\ell}} v_z^\ell \quad \text{where} \quad v_z^\ell := v^\ell(z)\eta_z^\ell \in \mathscr{V}_z^\ell, \quad \ell = 0, \ldots, L,$$

so that $v$ admits the decomposition (13.32). Noting (13.6), in order to prove (13.33) we prove

$$\sum_{\ell=0}^{L} \sum_{z\in\widetilde{\mathcal{N}_\ell}} \|v_z^\ell\|^2_{\widetilde{H}_{\mathrm{I}}^{1/2}(\omega_\ell(z))} \lesssim \|v\|^2_{\widetilde{H}_{\mathrm{I}}^{1/2}(\Gamma)}. \tag{13.34}$$

We have by using Lemma A.13 and Lemma C.12

$$\|v_z^\ell\|^2_{\widetilde{H}_{\mathrm{I}}^{1/2}(\omega_\ell(z))} = |v_z^\ell|^2 \|\eta_z^\ell\|^2_{\widetilde{H}_{\mathrm{I}}^{1/2}(\omega_\ell(z))} \lesssim h_\ell(z)|v_z^\ell|^2. \tag{13.35}$$

Hence it suffices to estimate $|v_z^\ell|^2$ for $z \in \widetilde{\mathcal{N}_\ell}$ and $\ell = 0,\dots,L$. Recalling Lemma 13.4, we define $m_z = \mathrm{level}_\ell(z) - c_0$ which can be a negative integer. For negative integers $j$, we define $\widehat{\mathcal{V}}^j := \widehat{\mathcal{V}}^0$ and $Q_j := Q_0$; see (13.22) and (13.31). Denoting $r_z = R_\ell(z)$, we deduce from Lemma 13.4 Part (i) that $m_z \le r_z$ so that $Q_{m_z} v \in \widehat{\mathcal{V}}^{m_z} \subset \widehat{\mathcal{V}}^{r_z}$. It is also given by Part (ii) of this lemma that any $T \in \mathcal{T}_{\ell-1}(\omega^2_{\ell-1}(z))$ is a subset of some $\widehat{T} \in \widehat{\mathcal{T}_{r_z}}$. This implies that $(Q_{m_z} v)\big|_T$ is a linear function. Since $J_{\ell-1}$ is a projection, it follows that

$$J_{\ell-1} Q_{m_z} v = Q_{m_z} v \quad \text{on} \quad T$$

and in particular $J_{\ell-1} Q_{m_z} v(z) = Q_{m_z} v(z)$. On the other hand, since $\omega_\ell(z) \subset \omega_{\ell-1}(z) \subset \omega^2_{\ell-1}(z)$, the restriction of $Q_{m_z} v$ on any triangle $T \in \mathcal{T}_\ell(\omega_\ell(z))$ is also a linear function. Consequently, we also have $J_\ell Q_{m_z} v(z) = Q_{m_z} v(z)$ so that

$$\big(J_\ell - J_{\ell-1}\big) Q_{m_z} v(z) = 0.$$

Therefore, we can write

$$|v_z^\ell|^2 = \big|\big(J_\ell - J_{\ell-1}\big)\big(v - Q_{m_z} v\big)(z)\big|^2.$$

It follows from inequality (13.17) that

$$|v_z^\ell|^2 \lesssim h_\ell(z)^{-2} \|v - Q_{m_z} v\|^2_{L^2(\omega^2_{\ell-1}(z))}.$$

Altogether, Lemma 13.2 Part (iii), (13.35), and the above inequality yield

$$\|v\|^2_{\widetilde{H}_{\mathrm{I}}^{1/2}(\Gamma)} \lesssim \widehat{h}^{-1}_{\mathrm{level}_\ell(z)} \|v - Q_{m_z} v\|^2_{L^2(\omega^2_{\ell-1}(z))}.$$

We can now estimate the left-hand side of (13.33) as follows

$$\begin{aligned}\sum_{\ell=0}^{L} \sum_{z\in\widetilde{\mathcal{N}_\ell}} \|v_z^\ell\|^2_{\widetilde{H}_{\mathrm{I}}^{1/2}(\omega_\ell(z))} &\lesssim \sum_{\ell=0}^{L} \sum_{z\in\widetilde{\mathcal{N}_\ell}} \widehat{h}^{-1}_{\mathrm{level}_\ell(z)} \|v - Q_{m_z} v\|^2_{L^2(\omega^2_{\ell-1}(z))} \\ &= \sum_{m=0}^{\infty} \sum_{\ell=0}^{L} \sum_{\substack{z\in\mathcal{N}_\ell \\ \mathrm{level}_\ell(z)=m}} \widehat{h}_m^{-1} \|v - Q_{m_z} v\|^2_{L^2(\omega^2_{\ell-1}(z))}.\end{aligned}$$

Invoking Lemma 13.4 Part (iii) and then Lemma 13.3 we deduce

$$\sum_{\ell=0}^{L} \sum_{z\in\widetilde{\mathcal{N}_\ell}} \|v_z^\ell\|^2_{\widetilde{H}_{\mathrm{I}}^{1/2}(\omega_\ell(z))} \lesssim \sum_{m=0}^{\infty} \sum_{\ell=0}^{L} \sum_{\substack{z\in\mathcal{N}_\ell \\ \mathrm{level}_\ell(z)=m}} \widehat{h}_m^{-1} \|v - Q_{m_z} v\|^2_{L^2(\widehat{\omega}^n_m(z))}$$

$$= \sum_{m=0}^{\infty} \sum_{z \in \mathcal{N}_L} \sum_{\ell \in \widetilde{\mathcal{K}_m}(z)} \widehat{h}_m^{-1} \|v - Q_{m_z} v\|_{L^2(\widehat{\omega}_m^n(z))}^2$$

$$= \sum_{m=0}^{\infty} \sum_{z \in \mathcal{N}_L \cap \widehat{\mathcal{N}_m}} \sum_{\ell \in \widetilde{\mathcal{K}_m}(z)} \widehat{h}_m^{-1} \|v - Q_{m_z} v\|_{L^2(\widehat{\omega}_m^n(z))}^2$$

where in the last step we used the fact that $z \in \widehat{\mathcal{N}_m}$ for $z \in \mathcal{N}_\ell$ with $\mathrm{level}_\ell(z) = m$, which is given by Lemma 13.2 Part (i). Moreover, Lemma 13.3 gives $\mathrm{card}(\widetilde{\mathcal{K}_m}(z)) \le d_0$. Hence, the uniform shape regularity of $\widehat{T}_m$ and the definition $Q_j = Q_0$ for $j < 0$ yield

$$\begin{aligned}
\sum_{\ell=0}^{L} \sum_{z \in \widetilde{\mathcal{N}_\ell}} \|v_z^\ell\|_{\widetilde{H}_{\mathrm{I}}^{1/2}(\omega_\ell(z))}^2 &\lesssim \sum_{m=0}^{\infty} \sum_{z \in \mathcal{N}_L \cap \widehat{\mathcal{N}_m}} \widehat{h}_m^{-1} \|v - Q_{m_z} v\|_{L^2(\widehat{\omega}_m^n(z))}^2 \\
&\le \sum_{m=0}^{\infty} \sum_{z \in \widehat{\mathcal{N}_m}} \widehat{h}_m^{-1} \|v - Q_{m_z} v\|_{L^2(\widehat{\omega}_m^n(z))}^2 \\
&\le \sum_{m=0}^{\infty} \widehat{h}_m^{-1} \|v - Q_{m_z} v\|_{L^2(\Gamma)}^2 \\
&\lesssim \sum_{m=0}^{\infty} \widehat{h}_m^{-1} \|v - Q_m v\|_{L^2(\Gamma)}^2 .
\end{aligned}$$

Invoking Lemma C.35 we obtain (13.34) and complete the proof of the lemma. □

We now prove the Strengthened Cauchy–Schwarz inequality. Let

$$M := \max_{z \in \mathcal{N}_L} \mathrm{level}_L(z)$$

denote the maximal level of all nodes $z \in \mathcal{N}_L$. It follows from Lemma 13.2 that $\mathcal{N}_L \subset \widehat{\mathcal{N}_M}$ which then implies $\mathcal{V}^L \subset \widehat{\mathcal{V}}^M$. Defining

$$\widetilde{P}_m^L := \sum_{\ell=0}^{L} \sum_{\substack{z \in \widetilde{\mathcal{N}_\ell} \\ \mathrm{level}_\ell(z) = m}} P_{\mathcal{V}_z^\ell}, \tag{13.36}$$

we can write the additive Schwarz operator $P_{LMD}$ defined by (13.29) can be rewritten as

$$P_{LMD} = \sum_{m=0}^{M} \widetilde{P}_m^L. \tag{13.37}$$

**Lemma 13.6 (Strengthened Cauchy–Schwarz inequality).** *For $0 \le k \le m \le M$ and for any $v \in \widehat{\mathcal{V}}^k$*

$$a_{\mathfrak{W}}(\widetilde{P}_m^L v, v) \lesssim 2^{-(m-k)} a_{\mathfrak{W}}(v, v). \tag{13.38}$$

*The constant depends only on $\Gamma$ and the initial triangulation $\mathcal{T}_0$.*

*Proof.* The proof is somewhat a generalisation of Lemma 5.2. Since $\mathscr{V}_z^\ell = \operatorname{span}\{\eta_z^\ell\}$, the projection $P_{\mathscr{V}_z^\ell}$ can be explicitly represented as

$$P_{\mathscr{V}_z^\ell} v = \frac{a_{\mathfrak{W}}(v, \eta_z^\ell)}{a_{\mathfrak{W}}(\eta_z^\ell, \eta_z^\ell)} \eta_z^\ell$$

so that

$$a_{\mathfrak{W}}(\widetilde{P}_m^L v, v) = \sum_{\ell=0}^{L} \sum_{\substack{z \in \widetilde{\mathscr{N}_\ell} \\ \mathrm{level}_\ell(z)=m}} \frac{a_{\mathfrak{W}}(v, \eta_z^\ell)^2}{a_{\mathfrak{W}}(\eta_z^\ell, \eta_z^\ell)}.$$

Hölder's inequality, Lemma B.3 and Lemma C.12 give

$$\begin{aligned}
\frac{a_{\mathfrak{W}}(v, \eta_z^\ell)^2}{a_{\mathfrak{W}}(\eta_z^\ell, \eta_z^\ell)} &= \frac{\langle \mathfrak{W} v, \eta_z^\ell \rangle_{L^2(\Gamma)}^2}{a_{\mathfrak{W}}(\eta_z^\ell, \eta_z^\ell)} \lesssim \frac{\|\mathfrak{W} v\|_{L^2(\Gamma)}^2 \|\eta_z^\ell\|_{L^2(\Gamma)}^2}{\|\eta_z^\ell\|_{\widetilde{H}_{\mathrm{I}}^{1/2}(\omega_\ell(z))}^2} \\
&\lesssim h_\ell(z) \|\mathfrak{W} v\|_{L^2(\omega_\ell(z))}^2 \lesssim 2^{-(m-k)} \|\widehat{h}_k^{1/2} \mathfrak{W} v\|_{L^2(\omega_\ell(z))}^2
\end{aligned}$$

where in the last step we used Lemma 13.2 Part (iii) and (13.21). Lemma 13.4 Part (iii) implies

$$\frac{a_{\mathfrak{W}}(v, \eta_z^\ell)^2}{a_{\mathfrak{W}}(\eta_z^\ell, \eta_z^\ell)} \lesssim 2^{-(m-k)} \|\widehat{h}_k^{1/2} \mathfrak{W} v\|_{L^2(\widehat{\omega}_m^n(z))}^2.$$

Consequently, with the help of Lemma 13.3,

$$\begin{aligned}
a_{\mathfrak{W}}(\widetilde{P}_m^L v, v) &= \sum_{\ell=0}^{L} \sum_{\substack{z \in \widetilde{\mathscr{N}_\ell} \\ \mathrm{level}_\ell(z)=m}} 2^{-(m-k)} \|\widehat{h}_k^{1/2} \mathfrak{W} v\|_{L^2(\widehat{\omega}_m^n(z))}^2 \\
&= \sum_{z \in \widehat{\mathscr{N}_m} \cap \widetilde{\mathscr{N}_L}} \sum_{\ell \in \mathscr{K}_m(z)} 2^{-(m-k)} \|\widehat{h}_k^{1/2} \mathfrak{W} v\|_{L^2(\widehat{\omega}_m^n(z))}^2 \\
&\lesssim \sum_{z \in \widehat{\mathscr{N}_m}} 2^{-(m-k)} \|\widehat{h}_k^{1/2} \mathfrak{W} v\|_{L^2(\widehat{\omega}_m^n(z))}^2 \\
&\lesssim 2^{-(m-k)} \|\widehat{h}_k^{1/2} \mathfrak{W} v\|_{L^2(\Gamma)}^2.
\end{aligned} \tag{13.39}$$

It follows from [12, Corollary 3.2] (inverse estimate) that

$$a_{\mathfrak{W}}(\widetilde{P}_m^L v, v) \lesssim 2^{-(m-k)} \|v\|_{\widetilde{H}_{\mathrm{I}}^{1/2}(\Gamma)}^2 \simeq 2^{-(m-k)} a_{\mathfrak{W}}(v, v).$$

We note that the inverse property holds not only for uniform triangulations $\widehat{\mathscr{T}_k}$ but also for shape-regular triangulations and higher-order polynomials; see [12, Corollary 3.2]. □

Similarly to the case of two-dimensional problems, we deduce from the Strengthened Cauchy–Schwarz inequality the following result.

**Lemma 13.7.** *The multilevel additive Schwarz operator* $P_{LMD}$ *satisfies*

$$a_{\mathfrak{W}}(P_{LMD}v, v) \lesssim a_{\mathfrak{W}}(v, v) \quad \textit{for any } v \in \mathscr{V}^L$$

*where the constant is independent of the number of levels and the number of mesh points.*

*Proof.* The proof follows exactly along the same lines of Lemma 5.3 and is omitted. □

Lemma 13.5 and Proposition 2.1 yield $\lambda_{\min}(P_{LMD}) \gtrsim 1$. Moreover, Lemma 13.7 and Lemma 2.10 yield $\lambda_{\max}(P_{LMD}) \lesssim 1$. Therefore we have the following result.

**Theorem 13.1.** *The condition number of the multilevel additive Schwarz operator defined by* (13.29) *has a bounded condition number*

$$\kappa(P_{LMD}) \lesssim 1.$$

### *13.2.2 Global multilevel diagonal preconditioner (GMD preconditioner*

For this preconditioner, the boundary element space $\mathscr{V}^L$ is decomposed by

$$\mathscr{V}^L = \sum_{\ell=0}^{L} \sum_{z \in \mathscr{N}_\ell} \mathscr{V}_z^\ell. \tag{13.40}$$

See (5.2) for the corresponding two-dimensional version of this decomposition.

The global multilevel additive Schwarz operator $P_{GMD}$ is defined by

$$P_{GMD} = \sum_{\ell=0}^{L} \sum_{z \in \mathscr{N}_\ell} P_{\mathscr{V}_z^\ell} \tag{13.41}$$

where the projection $P_{\mathscr{V}_z^\ell}$ is defined by (13.30).

Since $\widetilde{\mathscr{N}_\ell} \subset \mathscr{N}_\ell$, see (13.9), the stability of the decomposition (13.40) can be derived from that of (13.28).

**Lemma 13.8 (Stability of decomposition).** *Any* $v \in \mathscr{V}^\ell$ *admits a decomposition*

$$v = \sum_{\ell=0}^{L} \sum_{z \in \mathscr{N}_\ell} v_z^\ell, \quad v_z^\ell \in \mathscr{V}_z^\ell, \tag{13.42}$$

*that satisfies*

$$\sum_{\ell=0}^{L} \sum_{z \in \mathcal{N}_\ell} a_{\mathfrak{W}}(v_z^\ell, v_z^\ell) \lesssim a_{\mathfrak{W}}(v, v). \tag{13.43}$$

*The constant is independent of* $\ell = 0, \ldots, L$.

*Proof.* The decomposition (13.42) can be chosen to be the same as (13.32) if one notes that $\widetilde{\mathcal{N}_\ell} \subset \mathcal{N}_\ell$. Thus (13.43) follows. □

A Strengthened Cauchy–Schwarz inequality can also be proved. However, this is not an optimal result. Similarly to (13.36) we define

$$P_m^L := \sum_{\ell=0}^{L} \sum_{\substack{z \in \mathcal{N}_\ell \\ \mathrm{level}_\ell(z) = m}} P_{\mathcal{V}_z^\ell}$$

and have

$$P_{GMD} = \sum_{m=0}^{M} P_m^L.$$

**Lemma 13.9 (Strengthened Cauchy–Schwarz inequality).** *For* $0 \le k \le m \le M$ *and for any* $v \in \widehat{\mathcal{V}}^k$

$$a_{\mathfrak{W}}(P_m^L v, v) \lesssim 2^{-(m-k)}(1+L) a_{\mathfrak{W}}(v, v). \tag{13.44}$$

*The constant depends only on* $\Gamma$ *and the initial triangulation* $\mathcal{T}_0$.

*Proof.* The proof follows along the same lines as that of Lemma 13.6 with some obvious modifications. The set $\widetilde{\mathcal{N}_\ell}$ is replaced by $\mathcal{N}_\ell$ and the set $\widetilde{\mathcal{K}_m}(z)$ is replaced by $\mathcal{K}_m(z)$ defined by, cf. (13.24),

$$\mathcal{K}_m(z) := \{\ell \in \{0, 1, \ldots, L\} : z \in \mathcal{N}_\ell \quad \text{and} \quad \mathrm{level}_\ell(z) = m\}.$$

Differently from Lemma 13.3 we now have

$$\mathrm{card}(\mathcal{K}_m) \le 1 + L.$$

This upper bound $1+L$ comes into play in the right-hand side of (13.39), resulting in the final result (13.44). □

As a result of this Strengthened Cauchy–Schwarz inequality, we have the following lemma which is similar to Lemma 13.7.

**Lemma 13.10.** *The multilevel additive Schwarz operator* $P_{GMD}$ *satisfies*

$$a_{\mathfrak{W}}(P_{GMD} v, v) \lesssim (1+L) a_{\mathfrak{W}}(v, v) \quad \textit{for any } v \in \mathcal{V}^L$$

*where the constant is independent of the number of levels and the number of mesh points.*

Thus we obtain the following bound for the condition number of $P_{GMD}$.

**Theorem 13.2.** *The condition number of the multilevel additive Schwarz operator defined by* (13.41) *has a bounded condition number*

$$\kappa(P_{GMD}) \lesssim (1+L).$$

*Remark 13.1.* The same preconditioner as the GMD preconditioner is studied in [4] where an optimal bound for the condition number is obtained. However, the local mesh refinement used in that study (which partitions a triangle into four subtriangles by connecting midpoints of its edges) requires some conditions which are not satisfied by the newest vertex bisection algorithm; see [4, pages 391 and 405]. For example, it is required in that study that if $T, T' \in \mathscr{T}_\ell$ share a common edge or node, then $|\mathrm{gen}(T) - \mathrm{gen}(T')| \le 1$. This condition is not satisfied by the NVB algorithm; see the last subfigure in the right column of Figure 13.2. (In the language of [4], $\mathrm{gen}(T)$ is the level of $T$.)

## 13.3 Numerical Experiments

We report here some numerical experiments carried out in [75]. They are reprinted with copyright permission.

**Experiment 1:** In this experiment, $\Gamma = (-1,1) \times \{0\}$ is a slit in $\mathbb{R}^2$. The right-hand side of the equation is $g(x) = 1$ and thus the exact solution is $u(x) = 2\sqrt{1-x^2}$. Figure 13.5 shows the condition numbers and numbers of iterations when different preconditioners are used: the local multilevel diagonal and global multilevel diagonal preconditioners, the hierarchical basis preconditioner; see e.g. [231]. These numbers are compared with the corresponding numbers when no preconditioner (denoted by **A**) is used.

**Experiment 2:** In this experiment, $\Gamma$ is the boundary of an L-shaped domain $\Omega$ in $\mathbb{R}^3$ defined by $\Omega = \big([-1,1] \times [-1,1] \times [0,1]\big) \setminus \big([0,1] \times [0,1] \times [0,1]\big)$. This is a case of closed boundary $\Gamma$. The analytical solution of the hypersingular integral equation is chosen to be the Neumann trace of the harmonic function given in the cylindrical coordinates by $u(r,\varphi,,z) = r^{2/3}\cos(2\varphi/3)$. Again, Figure 13.6 shows the condition numbers and numbers of iterations of the two multilevel preconditioners (LMD and GMD) together with the preconditioner by diagonal scaling. These numbers are compared with those of the unpreconditioned matrix **A**.

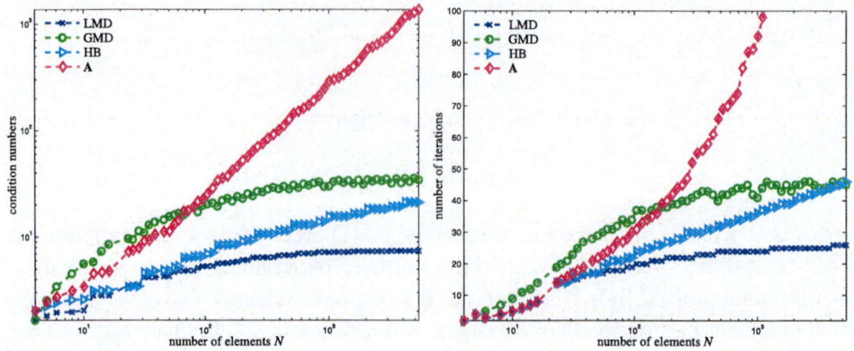


**Fig. 13.5** Experiment 1: Condition numbers (left) of the preconditioned matrices $P_{LMD}$, $P_{GMD}$, $P_{HB}$ and the unpreconditioned matrix **A** as well as the corresponding number of iterations (right) with $\Gamma = (-1,1) \times \{0\}$

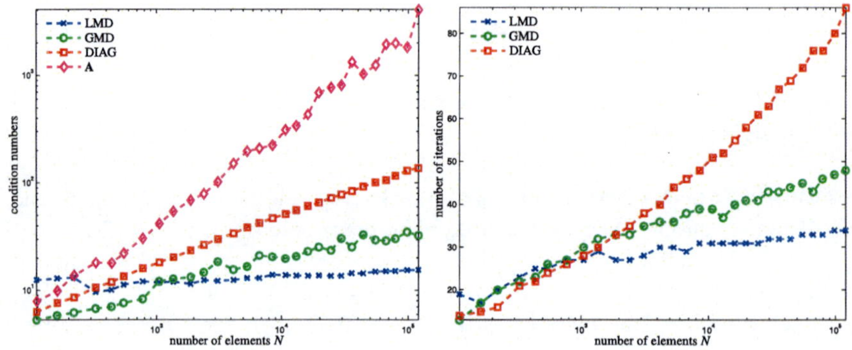


**Fig. 13.6** Experiment 2: Condition numbers (left) of the preconditioned matrices $P_{LMD}$, $P_{GMD}$, $P_{HB}$ and the unpreconditioned matrix **A** as well as the corresponding number of iterations (right) with $\Gamma$ being the boundary of an L-shaped domain in $\mathbb{R}^3$

## 13.4 A Remark on the Weakly-Singular Integral Equation

For the weakly-singular integral equation, a local multilevel additive Schwarz preconditioner for a sequence of locally-refined meshes is studied in [79]. This article also presents an adaptive algorithm which steers the local mesh refinement as well as the stopping of the PCG iteration. We refer interested readers to this paper for detail.

# Part IV
# FEM–BEM Coupling

# Chapter 14
# FEM-BEM Coupling

The coupling of boundary elements and finite elements is a powerful tool for the numerical treatment of boundary value problems on unbounded domains, and in particular, of interface problems. It allows us to benefit from the best of both worlds: boundary elements for treating unbounded domains and finite elements for treating non-homogeneity and non-linearity of the equations; see [245].

The first mathematical treatment of the FEM-BEM coupling is that of the so-called non-symmetric coupling which includes the Johnson–Nédélec coupling and the Bielak–MacCamy coupling; see [33, 44, 129]. Even though numerical experiments show that the method works well with general boundaries, the analysis in these papers is confined to smooth boundaries. In the search for coupling methods which are supported by mathematical analysis and suitable for Lipschitz boundaries, the symmetric coupling is originated independently in [62, 102]. As a consequence, plenty of studies on preconditioning of the symmetric coupling have been carried out [51, 81, 113, 117, 156]. Meanwhile, little has been known on preconditioning of the non-symmetric coupling. An optimal iterative process is suggested in [150], which requires smooth boundaries. Since the first analysis for the Johnson–Nédélec coupling on Lipschitz boundaries appeared in[187], researchers have paid more attention to preconditioning for this type of coupling.

In this chapter, we present preconditioners for the symmetric and non-symmetric FEM-BEM coupling methods. In Section 14.1 we first briefly introduce various variational formulations for the standard transmission problem which give rise to both coupling techniques. The symmetric coupling results from the use of all four operators in the Calderón projector, whereas in the non-symmetric coupling the hypersingular integral operator is not used. More information on FEM-BEM coupling can be found in [92, 205].

In the remaining sections, we discuss in detail the analysis for preconditioning each coupling technique. The symmetric coupling results in a symmetric matrix system with saddle point structure; see (14.14). We investigate how to precondition that matrix with the so-called three-block and two-block preconditioners. The HMCR, see Subsection 2.8.3, is used to solve this system. Interestingly, with a minor modification (by changing the sign in the second equation), the matrix turns out to be



E. P. Stephan and T. Tran, *Schwarz Methods and Multilevel Preconditioners for Boundary Element Methods*, https://doi.org/10.1007/978-3-030-79283-1_14

non-symmetric but positive definite and thus can be solved by the GMRES. Both approaches will be discussed in Section 14.2. Obviously, the non-symmetric coupling leads to a non-symmetric matrix system, which will be solved by the GMRES. This is discussed in Section 14.3.

Different results can be found in [51, 81, 113, 117, 156] and the references therein. In particular, a study of multilevel preconditioners is carried out in [76]. In this chapter, we present our results in the most general form so that any preconditioner (two-level or multilevel) designed for FEM and BEM separately can be combined to form a preconditioner for the coupled problem. In other words, various preconditioners developed in previous chapters can be applied. In particular, Subsection 14.2.2 presents a general approach in the matrix form, which can be used for any problem with any available preconditioners; see also Subsection 2.7.3.

## 14.1 The Interface Problem

Let $\Omega \subset \mathbb{R}^d$, $d=2,3$, be a bounded domain with Lipschitz boundary $\Gamma := \partial\Omega$, and let $\Omega^c := \mathbb{R}^d \setminus \bar{\Omega}$ with normal $\mathbf{n}$ on $\Gamma$ pointing into $\Omega^c$. Without loss of generality, when $d=2$ we assume that the logarithmic capacity $\mathrm{cap}(\Gamma)$ is less than one, which can always be obtained by scaling. This is to ensure positive definiteness of the weakly-singular integral operator; see e.g. [147].

Given $f \in L^2(\Omega_1)$, $u_0 \in H^{1/2}(\Gamma)$, $t_0 \in H^{-1/2}(\Gamma)$, and $k \in \mathbb{R}$ ($k \neq 0$), we consider the model interface problem: Find $u_1 \in H^1(\Omega)$ and $u_2 \in H^1_{\mathrm{loc}}(\Omega^c)$ such that

$$-\Delta u_1 + k^2 u_1 = f \qquad \text{in } \Omega, \tag{14.1a}$$

$$\Delta u_2 = 0 \qquad \text{in } \Omega^c, \tag{14.1b}$$

$$u_1 = u_2 + u_0 \qquad \text{on } \Gamma, \tag{14.1c}$$

$$\frac{\partial u_1}{\partial n} = \frac{\partial u_2}{\partial n} + t_0 \qquad \text{on } \Gamma, \tag{14.1d}$$

$$u_2(x) = \begin{cases} a + b\log|x| + o(1), & d=2, \\ O(1/|x|), & d=3, \end{cases} \qquad \text{as } |x| \to \infty, \tag{14.1e}$$

for some constants $a$ and $b$.

To ensure existence of solutions in the case $d=2$, we assume also the compatibility condition

$$\langle f, 1\rangle_\Omega + \langle t_0, 1\rangle_\Gamma = 0. \tag{14.2}$$

Since the late seventies of the last century, the coupling of the finite element method and the boundary element method has been widely used to solve (14.1) and other interface problems of similar types. The method was first proposed in 1979 [245] and first analysed in 1980 by Johnson and Nédélec [129]. It is therefore often referred to as the Johnson–Nédélec coupling. This coupling and the Bielak–MacCamy coupling [33] are non-symmetric coupling methods. Their analysis requires compactness of the double-layer potential operator $\mathfrak{K}$ to be defined in Section 14.1.1. This property fails to hold when the interface $\Gamma$ is a Lipschitz domain. Due to this lack of analysis, the symmetric coupling has been proposed independently by Costabel [62] and Han [102]. Not until 2009 was a mathematical analysis carried out by Sayas [187] for the Johnson–Nédélec coupling with $\Gamma$ being a polygonal or polyhedral boundary.

In the next two subsections, we briefly explain these two coupling methods.

## *14.1.1 The symmetric FEM-BEM coupling*

The mathematical boundary element community initially preferred the use of the symmetric coupling in the case of Lipschitz interfaces or for the elasticity system, thanks to the availability of a mathematical analysis. In this subsection, we present this coupling technique.

### 14.1.1.1 Existence and uniqueness of solutions

We introduce the following integral operators

$$\begin{aligned} \mathfrak{K}\psi(x) &:= 2\int_\Gamma \frac{\partial}{\partial n_y} G(x,y)\psi(y)\, d\sigma_y, \quad x\in\Gamma, \\ \mathfrak{K}'\psi(x) &:= 2\frac{\partial}{\partial n_x}\int_\Gamma G(x,y)\psi(y)\, d\sigma_y, \quad x\in\Gamma, \end{aligned} \tag{14.3}$$

and recall the definitions of the single-layer and double-layer potential operators

$$\begin{aligned} \mathfrak{S}v(x) &:= 2\int_\Gamma G(x,y)v(y)\, d\sigma_y, \quad x\in\mathbb{R}^d, \\ \mathfrak{D}v(x) &:= 2\int_\Gamma \frac{\partial G(x,y)}{\partial n_y} v(y)\, d\sigma_y, \quad x\in\mathbb{R}^d. \end{aligned}$$

The solution in the exterior domain can be represented by

$$u_2 = \frac{1}{2}\mathfrak{D}(\gamma u_2) - \frac{1}{2}\mathfrak{S}(\partial_n u_2) \quad \text{in } \Omega^c \tag{14.4}$$

where $\gamma$ and $\partial_n$ denote the trace and normal trace operators on $\Gamma$; see [147]. By taking the trace on $\Gamma$ and using the well-known jump relations, we deduce

$$\mathfrak{V}(\partial_n u_2) + (\mathfrak{I} - \mathfrak{K})u_2 = 0 \quad \text{on } \Gamma \tag{14.5}$$

where $\mathfrak{I}$ is the identity operator. On the other hand, by taking the normal trace on $\Gamma$, we obtain

$$2\partial_n u_2 = -\mathfrak{W}u_2 + (\mathfrak{I} - \mathfrak{K}')\partial_n u_2 \quad \text{on } \Gamma. \tag{14.6}$$

By multiplying equation (14.1a) by $2v$, integrating over $\Omega$, and formally using integration by parts, we deduce (noting (14.6), (14.1c)), and (14.1d))

$$\begin{aligned} &2a_\Delta(u_1, v) + \langle(\mathfrak{K}' - \mathfrak{I})(\partial_n u_2), v\rangle_\Gamma + a_{\mathfrak{W}}(u_1, v) \\ &= 2\langle f, v\rangle_\Omega + 2\langle t_0, v\rangle_\Gamma + a_{\mathfrak{W}}(u_0, v) \end{aligned}$$

where, for any $u$, $v$, and $w$,

$$a_\Delta(u,v) := \int_\Omega (\nabla u\cdot\nabla v + k^2 uv)\,\mathrm{d}x, \quad \langle v, w\rangle_\Omega := \int_\Omega vw\,\mathrm{d}x, \quad \langle v, w\rangle_\Gamma := \int_\Gamma vw\,\mathrm{d}\sigma.$$

Furthermore, equation (14.5) and conditions (14.1c), (14.1d) yield formally

$$a_{\mathfrak{V}}(\partial_n u_2, \psi) + \langle(\mathfrak{I} - \mathfrak{K})u_1, \psi\rangle_\Gamma = \langle(\mathfrak{I} - \mathfrak{K})u_0, \psi\rangle_\Gamma.$$

Therefore, the variational formulation to the interface problem (14.1) reads:

Find $(u, \phi) \in H^1(\Omega) \times H^{-1/2}(\Gamma)$ such that, for all $(v, \psi) \in H^1(\Omega) \times H^{-1/2}(\Gamma)$,

$$\begin{aligned} a_{\Delta\mathfrak{W}}(u, v) + \langle(\mathfrak{K}' - \mathfrak{I})\phi, v\rangle_\Gamma &= 2\langle f, v\rangle_\Omega + 2\langle t_0, v\rangle_\Gamma + a_{\mathfrak{W}}(u_0, v) \\ \langle(\mathfrak{K} - \mathfrak{I})u, \psi\rangle_\Gamma - a_{\mathfrak{V}}(\phi, \psi) &= \langle(\mathfrak{K} - \mathfrak{I})u_0, \psi\rangle_\Gamma. \end{aligned} \tag{14.7}$$

where $a_{\Delta\mathfrak{W}}(\cdot,\cdot) : H^1(\Omega) \times H^1(\Omega) \to \mathbb{R}$ is defined by

$$a_{\Delta\mathfrak{W}}(u, v) := 2a_\Delta(u, v) + a_{\mathfrak{W}}(u|_\Gamma, v|_\Gamma) \quad \forall u, v \in H^1(\Omega). \tag{14.8}$$

Unique solvability and equivalence of (14.1) and (14.7) are proved in [64, 66, 69] and is summarised in the following theorem.

**Theorem 14.1.** *Under the compatibility condition* (14.2) *when $d = 2$, and no condition for $d = 3$, the problem* (14.7) *is uniquely solvable. Moreover, if $u_1 \in H^1(\Omega)$ and $u_2 \in H^1_{\mathrm{loc}}(\Omega^c)$ satisfy* (14.1), *then*

$$u := u_1 \quad \textit{and} \quad \phi := \frac{\partial u_2}{\partial \mathbf{n}}$$

*satisfy* (14.7). *Conversely, if $u \in H^1(\Omega)$ and $\phi \in H^{-1/2}(\Gamma)$ satisfy* (14.7), *then*

$$u_1 := u \quad \textit{and} \quad u_2 := \frac{1}{2}\mathfrak{D}(u - u_0) - \frac{1}{2}\mathfrak{S}\phi$$

*solve the interface problem* (14.1).

#### 14.1.1.2 Galerkin solution

Let $\Omega_h$ be a mesh in $\Omega$ whose restriction on $\Gamma$ yields a mesh $\Gamma_h$. For the Galerkin scheme, we choose two finite dimensional subspaces (which appropriately define the $h$-, $p$-, or $hp$-version)

$$\mathscr{S}_{\overline{\Omega}} \subset H^1(\Omega) \quad \text{and} \quad \widetilde{\mathscr{S}_\Gamma} \subset H^{-1/2}(\Gamma),$$

and denote $\mathscr{X} := \mathscr{S}_{\overline{\Omega}} \times \widetilde{\mathscr{S}_\Gamma}$. For any function $v \in \mathscr{S}_{\overline{\Omega}}$ we denote by $v_\Omega$ (resp., $v_\Gamma$) the restriction of $v$ on the interior (resp., the boundary) of $\overline{\Omega}$. We define the Galerkin solution $(u,\phi) \in \mathscr{X}$ to (14.1) by

$$\begin{aligned} a_{\Delta\mathfrak{W}}(u,v) + \langle (\mathfrak{K}' - \mathfrak{I})\phi, v_\Gamma \rangle_\Gamma &= 2\langle f, v\rangle_\Omega + 2\langle t_0, v_\Gamma\rangle_\Gamma + a_{\mathfrak{W}}(u_0, v_\Gamma) \\ \langle (\mathfrak{K} - \mathfrak{I})u_\Gamma, \psi\rangle_\Gamma - a_{\mathfrak{V}}(\phi,\psi) &= \langle (\mathfrak{K} - \mathfrak{I})u_0, \psi\rangle_\Gamma \end{aligned} \tag{14.9}$$

for all $(v,\psi) \in \mathscr{X}$. Here, for any $v \in \mathscr{S}_{\overline{\Omega}}$, we denote by $v_\Gamma$ the trace of $v$ on $\Gamma$. We note that for notational simplicity we use the same notation for the Galerkin solution and the solution of (14.7).

#### 14.1.1.3 Matrix representation

Let $\{\eta_1,\dots,\eta_M\}$ be a basis for $\mathscr{S}_{\overline{\Omega}}$ consisting of continuous piecewise polynomial functions, and $\{\phi_1,\dots,\phi_N\}$ be a basis for $\widetilde{\mathscr{S}_\Gamma}$ consisting of piecewise polynomial functions (not necessarily continuous). The functions $\eta_j$ are supposed to be ordered such that $\{\eta_1,\dots,\eta_{M_\Omega}\}$ is the set of basis functions which vanish on the boundary $\Gamma$, and $\{\eta_{M_\Omega+1},\dots,\eta_{M_\Omega+M_\Gamma}\}$ (where $M_\Omega + M_\Gamma = M$) is the set of those basis functions $\eta_j$ which do not vanish on $\Gamma$. We define

$$\mathscr{I}_1 := \{1,\dots,M_\Omega\}, \quad \mathscr{I}_2 := \{M_\Omega+1,\dots,M_\Omega+M_\Gamma\}, \quad \mathscr{I}_3 := \{1,\dots,N\}. \tag{14.10}$$

Let $\underline{\mathbf{A}}_\Delta \in \mathbb{R}^{M\times M}$ be the stiffness matrix arising from the bilinear form $a_\Delta(\cdot,\cdot)$, i.e.,

$$\left(\underline{\mathbf{A}}_\Delta\right)_{ij} := 2a_\Delta(\eta_i,\eta_j), \quad i,j \in \mathscr{I}_1 \cup \mathscr{I}_2.$$

This matrix can be decomposed as

$$\underline{\mathbf{A}}_\Delta = \begin{pmatrix} \mathbf{A}_\Delta & \mathbf{B}^\top \\ \mathbf{B} & \mathbf{D} \end{pmatrix} \in \mathbb{R}^{M\times M} \tag{14.11}$$

where

$$\begin{aligned} &\mathbf{A}_\Delta \in \mathbb{R}^{M_\Omega\times M_\Omega}, && \left(\mathbf{A}_\Delta\right)_{ij} := 2a_\Delta(\eta_i,\eta_j), \quad i,j \in \mathscr{I}_1, \\ &\mathbf{B} \in \mathbb{R}^{M_\Gamma\times M_\Omega}, && \mathbf{B}_{ij} := 2a_\Delta(\eta_i,\eta_j), \quad i \in \mathscr{I}_2,\ j \in \mathscr{I}_1, \\ &\mathbf{D} \in \mathbb{R}^{M_\Gamma\times M_\Gamma}, && \mathbf{D}_{ij} := 2a_\Delta(\eta_i,\eta_j), \quad i,j \in \mathscr{I}_2. \end{aligned}$$

The matrix $\mathbf{A}_\Delta$ is in fact the discrete Laplacian for a Dirichlet problem, whereas $\underline{\mathbf{A}}_\Delta$ is the discrete Laplacian for a homogeneous Neumann problem.

Similarly, the stiffness matrices $\mathbf{W}$, $\mathbf{V}$, $\mathbf{K}$, and $\mathbf{M}$ arising from the bilinear forms defined by the operators $\mathfrak{W}$, $\mathfrak{V}$, $\mathfrak{K}$, and $\mathfrak{I}$, respectively, are given by

$$\begin{aligned}
&\mathbf{W} \in \mathbb{R}^{M_\Gamma \times M_\Gamma}, && \mathbf{W}_{ij} := a_{\mathfrak{W}}(\eta_{i,\Gamma}, \eta_{j,\Gamma}), && i,j \in \mathscr{I}_2,\\
&\mathbf{V} \in \mathbb{R}^{N \times N}, && \mathbf{V}_{ij} := a_{\mathfrak{V}}(\phi_i, \phi_j), && i,j \in \mathscr{I}_3,\\
&\mathbf{K} \in \mathbb{R}^{N \times M_\Gamma}, && \mathbf{K}_{ij} := \langle \mathfrak{K}\eta_{j,\Gamma}, \phi_i \rangle_\Gamma, && i \in \mathscr{I}_3, j \in \mathscr{I}_2,\\
&\mathbf{M} \in \mathbb{R}^{N \times M_\Gamma}, && \mathbf{M}_{ij} := \langle \eta_{j,\Gamma}, \phi_i \rangle_\Gamma, && i \in \mathscr{I}_3, j \in \mathscr{I}_2.
\end{aligned} \tag{14.12}$$

We note that $\mathbf{M}$ is the stiffness matrix defined from the bilinear form $\langle \mathfrak{I}\cdot, \cdot \rangle_\Gamma$ and we can scale the basis functions $\phi_i$ such that $\mathbf{M}_{ij} = \delta_{ij}$ (Kronecker's delta). However, even in this case $\mathbf{M}$ is a rectangular matrix and is not an identity matrix.

Defining

$$\begin{aligned}
&\mathbf{b}_\Omega \in \mathbb{R}^{M_\Omega}, && (\mathbf{b}_\Omega)_i := 2\langle f, \eta_i \rangle_\Omega, && i \in \mathscr{I}_1,\\
&\mathbf{b}_\Gamma \in \mathbb{R}^{M_\Gamma}, && (\mathbf{b}_\Gamma)_i := 2\langle f, \eta_i \rangle_\Gamma + 2\langle t_0, \eta_{i,\Gamma} \rangle_\Gamma + a_{\mathfrak{W}}(u_0, \eta_{i,\Gamma}), && i \in \mathscr{I}_2,\\
&\widetilde{\mathbf{b}}_\Gamma \in \mathbb{R}^{N}, && (\widetilde{\mathbf{b}}_\Gamma)_i := \langle (\mathfrak{K} - \mathfrak{I})u_0, \phi_i \rangle_\Gamma, && i \in \mathscr{I}_3,
\end{aligned} \tag{14.13}$$

we obtain from (14.9) the $(M+N) \times (M+N)$ linear system

$$\begin{pmatrix} \mathbf{A}_\Delta & \mathbf{B}^\top & \mathbf{0}_{M_\Omega N} \\ \mathbf{B} & \mathbf{D}+\mathbf{W} & \mathbf{K}^\top - \mathbf{M}^\top \\ \mathbf{0}_{N M_\Omega} & \mathbf{K}-\mathbf{M} & -\mathbf{V} \end{pmatrix} \begin{pmatrix} \mathbf{u}_\Omega \\ \mathbf{u}_\Gamma \\ \boldsymbol{\phi} \end{pmatrix} = \begin{pmatrix} \mathbf{b}_\Omega \\ \mathbf{b}_\Gamma \\ \widetilde{\mathbf{b}}_\Gamma \end{pmatrix}. \tag{14.14}$$

Here $\mathbf{0}_{KL}$ is the zero matrix of size $K \times L$. By denoting

$$\begin{aligned}
&\underline{\mathbf{W}} := \begin{pmatrix} \mathbf{0}_{M_\Omega M_\Omega} & \mathbf{0}_{M_\Omega M_\Gamma} \\ \mathbf{0}_{M_\Gamma M_\Gamma} & \mathbf{W} \end{pmatrix} \in \mathbb{R}^{M \times M}, && \mathbf{u}_{\Omega,\Gamma} := \begin{pmatrix} \mathbf{u}_\Omega \\ \mathbf{u}_\Gamma \end{pmatrix} \in \mathbb{R}^M,\\
&\underline{\mathbf{K}} := \begin{pmatrix} \mathbf{0}_{N M_\Omega} & \mathbf{K}-\mathbf{M} \end{pmatrix} \in \mathbb{R}^{N \times M}, && \mathbf{b}_{\Omega,\Gamma} := \begin{pmatrix} \mathbf{b}_\Omega \\ \mathbf{b}_\Gamma \end{pmatrix} \in \mathbb{R}^M,
\end{aligned} \tag{14.15}$$

and

$$\underline{\mathbf{A}}_{\Delta\mathfrak{W}} = \underline{\mathbf{A}}_\Delta + \underline{\mathbf{W}}, \tag{14.16}$$

we can rewrite the matrix equation (14.14) as

$$\begin{pmatrix} \underline{\mathbf{A}}_{\Delta\mathfrak{W}} & \underline{\mathbf{K}}^\top \\ \underline{\mathbf{K}} & -\mathbf{V} \end{pmatrix} \begin{pmatrix} \mathbf{u}_{\Omega,\Gamma} \\ \boldsymbol{\phi} \end{pmatrix} = \begin{pmatrix} \mathbf{b}_{\Omega,\Gamma} \\ \widetilde{\mathbf{b}}_\Gamma \end{pmatrix}. \tag{14.17}$$

We denote by $\mathbf{A}_\mathfrak{s}$ the symmetric $(M+N) \times (M+N)$-stiffness matrix in (14.14) (which is also the stiffness matrix in (14.17)), i.e.,

$$\mathbf{A}_{\mathfrak{s}} := \begin{pmatrix} \mathbf{A}_\Delta & \mathbf{B}^\top & \mathbf{0}_{M_\Omega N} \\ \mathbf{B} & \mathbf{D}+\mathbf{W} & \mathbf{K}^\top - \mathbf{M}^\top \\ \mathbf{0}_{NM_\Omega} & \mathbf{K}-\mathbf{M} & -\mathbf{V} \end{pmatrix} = \begin{pmatrix} \underline{\mathbf{A}}_{\Delta\mathfrak{W}} & \underline{\mathbf{K}}^\top \\ \underline{\mathbf{K}} & -\mathbf{V} \end{pmatrix}. \tag{14.18}$$

Then since $\underline{\mathbf{A}}_\Delta$ and $\underline{\mathbf{W}}$ are positive semi-definite and $\mathbf{V}$ is positive definite (provided that $\operatorname{cap}(\Gamma) < 1$ in the case $\Omega \subset \mathbb{R}^2$), the matrix $\mathbf{A}_{\mathfrak{s}}$ is indefinite. Indeed, Sylvester's Law of Inertia [83] and the congruence transform

$$\mathbf{A}_{\mathfrak{s}} = \begin{pmatrix} \mathbf{I}_M & -\underline{\mathbf{K}}^\top \mathbf{V}^{-1} \\ \mathbf{0}_{NM} & \mathbf{I}_N \end{pmatrix} \begin{pmatrix} \underline{\mathbf{A}}_{\Delta\mathfrak{W}} + \underline{\mathbf{K}}^\top \mathbf{V}^{-1}\underline{\mathbf{K}} & \mathbf{0}_{MN} \\ \mathbf{0}_{NM} & -\mathbf{V} \end{pmatrix} \begin{pmatrix} \mathbf{I}_M & \mathbf{0}_{MN} \\ -\mathbf{V}^{-1}\underline{\mathbf{K}} & \mathbf{I}_N \end{pmatrix}$$

imply that $\mathbf{A}_{\mathfrak{s}}$ has $m$ positive and $n$ negative eigenvalues. Here $\mathbf{I}_k$ denotes the identity matrix of size $k \times k$.

Since $\mathbf{A}_{\mathfrak{s}}$ is symmetric and indefinite, we will use a minimum residual method (MINRES) to solve (14.14) and (14.17). More precisely, we will use the HMCR method (Algorithm 2.15) defined in Subsection 2.8.3.

Another approach is to solve, instead of (14.17), the equivalent equation

$$\begin{pmatrix} \underline{\mathbf{A}}_{\Delta\mathfrak{W}} & \underline{\mathbf{K}}^\top \\ -\underline{\mathbf{K}} & \mathbf{V} \end{pmatrix} \begin{pmatrix} \mathbf{u}_{\Omega,\Gamma} \\ \boldsymbol{\phi} \end{pmatrix} = \begin{pmatrix} \mathbf{b}_{\Omega,\Gamma} \\ -\widetilde{\mathbf{b}}_\Gamma \end{pmatrix}. \tag{14.19}$$

This system is positive definite but not symmetric and will be solved by GMRES. We denote by $\mathbf{A}_{\mathfrak{u}}$ this non-symmetric but positive definite matrix, namely,

$$\mathbf{A}_{\mathfrak{u}} := \begin{pmatrix} \underline{\mathbf{A}}_{\Delta\mathfrak{W}} & \underline{\mathbf{K}}^\top \\ -\underline{\mathbf{K}} & \mathbf{V} \end{pmatrix}. \tag{14.20}$$

It is worth noting that the matrix $\underline{\mathbf{A}}_{\Delta\mathfrak{W}}$ defined by (14.16) is the stiffness matrix of the operator $\mathfrak{A}_{\Delta\mathfrak{W}} : H^1(\Omega) \to \widetilde{H}^{-1}(\Omega)$ defined by

$$\langle \mathfrak{A}_{\Delta\mathfrak{W}} v, w\rangle_\Gamma = a_{\Delta\mathfrak{W}}(v,w), \quad v,w \in H^1(\Omega), \tag{14.21}$$

where $a_{\Delta\mathfrak{W}}(\cdot,\cdot) : H^1(\Omega) \times H^1(\Omega) \to \mathbb{R}$ is defined in (14.8). The discretised operator $\mathfrak{A}_M : \mathscr{S}_{\overline{\Omega}} \to \mathscr{S}^*_{\overline{\Omega}}$ is given by, see Subsection B.2.2.2 in Appendix B,

$$\langle \mathfrak{A}_M v, w\rangle_{\mathscr{S}^*_{\overline{\Omega}}, \mathscr{S}_{\overline{\Omega}}} = \langle \mathfrak{A}_{\Delta\mathfrak{W}} v, w\rangle_\Gamma, \quad v,w \in \mathscr{S}_{\overline{\Omega}}.$$

Note that the coercivity condition (B.30) holds; thus, $\mathfrak{A}^{-1}_{\Delta\mathfrak{W}} : \widetilde{H}^{-1}(\Omega) \to H^1(\Omega)$ is well defined.

Similarly, the matrix $\underline{\mathbf{K}}$ is the stiffness matrix of the operator $\underline{\mathfrak{K}}$ defined by

$$\begin{aligned} &\underline{\mathfrak{K}} : H^1(\Omega) \to H^{1/2}(\Gamma), \\ &\underline{\mathfrak{K}}(v) := \mathfrak{K}(v|_\Gamma) - \mathfrak{I}(v|_\Gamma) = \mathfrak{K}(v|_\Gamma) - v|_\Gamma, \quad v \in H^1(\Omega). \end{aligned} \tag{14.22}$$

Indeed, by referring to Subsection B.2.2.1 with

$$\mathscr{X} = H^1(\Omega), \quad \mathscr{Y}^* = H^{1/2}(\Gamma), \quad \mathscr{X}_M = \mathscr{S}_{\overline{\Omega}}, \quad \mathscr{Y}_N = \widetilde{\mathscr{S}}_{\Gamma}, \quad \mathscr{Y}_N^* = \widetilde{\mathscr{S}}_{\Gamma}^*,$$

we define $\underline{\mathfrak{K}}_{NM} : \mathscr{X}_M \to \mathscr{Y}_N^*$ by

$$\underline{\mathfrak{K}}_{NM}(v) := i_{\mathscr{Y}_N}^* \underline{\mathfrak{K}} i_{\mathscr{X}_M}(v) = \big(\mathfrak{K}_{NM_\Gamma} - \mathfrak{I}_{NM_\Gamma}\big)(v|_\Gamma).$$

The stiffness matrix arising from $\underline{\mathfrak{K}}$ has entries

$$\underline{K}_{ij} = \langle \underline{\mathfrak{K}} \eta_j, \phi_i \rangle_{\mathscr{Y}_N^*, \mathscr{Y}_N}, \quad i = 1, \dots, N, \ j = 1, \dots, M.$$

Note that $\underline{\mathfrak{K}} \eta_j = 0$ for $j = 1, \dots, M_\Omega$. Hence, for any $i \in \mathscr{I}_3$,

$$\underline{K}_{ij} = \begin{cases} 0, & j \in \mathscr{I}_1, \\ \langle \mathfrak{K} \eta_{j,\Gamma} - \eta_{j,\Gamma}, \phi_i \rangle_{\mathscr{Y}_N^*, \mathscr{Y}_N}, & j \in \mathscr{I}_2, \end{cases}$$

Therefore, noting (14.12), we have

$$\underline{\mathbf{K}} = \big(\mathbf{0}_{NM_\Omega} \ \ \mathbf{K} - \mathbf{M}\big) \in \mathbb{R}^{N \times M}.$$

This means $\underline{\mathbf{K}}$ defined in (14.15) is the stiffness matrix of the operator $\underline{\mathfrak{K}}$ defined in (14.22).

Finally, for the weakly-singular integral operator, if the condition $\operatorname{cap}(\Gamma) < 1$ holds in the case that $\Omega$ is a two-dimensional domain, then (B.30) also holds for $a_{\mathfrak{V}}(\cdot,\cdot)$; see e.g. [92, 147]. Hence it is possible to define the inverse operator $\mathfrak{V}^{-1} : H^{1/2}(\Gamma) \to \widetilde{H}^{-1/2}(\Gamma)$ together with its discretised form and stiffness matrix.

## 14.1.2 Non-symmetric coupling methods

The symmetric coupling methods discussed in the previous subsection is not easy to implement since they involve all four boundary integral operators $\mathfrak{V}$, $\mathfrak{W}$, $\mathfrak{K}$, and $\mathfrak{K}'$ (whose stiffness matrices are, respectively, $\mathbf{V}$, $\mathbf{W}$, $\mathbf{K}$, and $\mathbf{K}^\top$). This is a costly computation; in particular the computation of the hypersingular integral operator is not favoured in the engineering community. The analysis in [187] supports the use of non-symmetric coupling methods, even when the interface $\Gamma$ is a polyhedron or a polygon. In this subsection we introduce two non-symmetric FEM-BEM coupling methods, namely, the Johnson–Nédélec coupling [129] and the Bielak–MacCamy coupling [33].

### 14.1.2.1 The Johnson–Nédélec coupling

Again we represent $u_2$ satisfying (14.1b) by (14.4). By multiplying (14.1a) by $v$ and formally using integration by parts, together with (14.1d), we obtain

$$a_\Delta(u_1,v) - \langle \partial_n u_2, v\rangle_\Gamma = \langle f,v\rangle_\Omega + \langle t_0,v\rangle_\Gamma .$$

Moreover, we deduce from (14.5)

$$a_{\mathfrak{V}}(\partial_n u_2,\psi) + \langle(\mathfrak{I}-\mathfrak{K})u_1,\psi\rangle_\Gamma = \langle(\mathfrak{I}-\mathfrak{K})u_0,\psi\rangle_\Gamma .$$

Therefore, the coupled variational formulation to the interface problem (14.1) reads:

Find $(u,\phi) \in H^1(\Omega)\times H^{-1/2}(\Gamma)$ such that, for all $(v,\psi)\in H^1(\Omega)\times H^{-1/2}(\Gamma)$,

$$\begin{aligned} a_\Delta(u,v) - \langle \phi, v\rangle_\Gamma &= \langle f,v\rangle_\Omega + \langle t_0,v\rangle_\Gamma ,\\ \langle(\mathfrak{I}-\mathfrak{K})u,\psi\rangle_\Gamma + a_{\mathfrak{V}}(\phi,\psi) &= \langle(\mathfrak{I}-\mathfrak{K})u_0,\psi\rangle_\Gamma . \end{aligned} \tag{14.23}$$

We note that even though the coupled system (14.23) is non-symmetric, it involves only two boundary integral operators, namely $\mathfrak{V}$ and $\mathfrak{K}$, while (14.7) involves four boundary integral operators $\mathfrak{V}$, $\mathfrak{W}$, $\mathfrak{K}$, and $\mathfrak{K}'$.

Unique solvability of (14.23) is proved in [187]; see also [188].

**Theorem 14.2.** *Under the compatibility condition* (14.2) *when* $d=2$, *and no condition for* $d=3$, *the problem* (14.23) *is uniquely solvable.*

*Moreover, if* $u_1\in H^1(\Omega)$ *and* $u_2 \in H^1_{\text{loc}}(\Omega^c)$ *satisfy* (14.1), *then*

$$u := u_1 \quad \textit{and} \quad \phi := \frac{\partial u_2}{\partial \mathbf{n}}$$

*satisfy* (14.23). *Conversely, if* $u\in H^1(\Omega)$ *and* $\phi\in H^{-1/2}(\Gamma)$ *satisfy* (14.23), *then*

$$u_1 := u \quad \textit{and} \quad u_2 := \frac{1}{2}\big(\mathfrak{D}(u-u_0) - \mathfrak{S}\phi\big)$$

*solve the interface problem* (14.1).

With the finite dimensional space $\mathscr{X}$ defined as in Subsection 14.1.1.2 we derive the following Galerkin equations for (14.23): $(u,\phi)\in\mathscr{X}$ satisfying, for all $(v,\psi)\in\mathscr{X}$,

$$\begin{aligned} a_\Delta(u,v) - \langle \phi, v\rangle_\Gamma &= \langle f,v\rangle_\Omega + \langle t_0,v\rangle_\Gamma ,\\ \langle(\mathfrak{I}-\mathfrak{K})u,\psi\rangle_\Gamma + a_{\mathfrak{V}}(\phi,\psi) &= \langle(\mathfrak{I}-\mathfrak{K})u_0,\psi\rangle_\Gamma . \end{aligned} \tag{14.24}$$

Here again, we use the same notation for the solution of (14.23) and (14.24).

Recalling the definitions of $\underline{\mathbf{A}}_\Delta$, $\mathbf{V}$, $\mathbf{M}$, and $\underline{\mathbf{K}}$, in (14.11), (14.12), and (14.15), we introduce further

$$\underline{\mathbf{M}} := \begin{pmatrix}\mathbf{0}_{NM_\Omega} & \mathbf{M}\end{pmatrix} \in \mathbb{R}^{N\times M}. \tag{14.25}$$

We note the factor 2 in the definition of (14.11). Hence, by multiplying the first equation in (14.24) by 2 we can write this system in matrix form as

$$\begin{pmatrix}\underline{\mathbf{A}}_\Delta & -2\underline{\mathbf{M}}^\top \\ -\underline{\mathbf{K}} & \mathbf{V}\end{pmatrix}\begin{pmatrix}\mathbf{u}_{\Omega,\Gamma}\\ \boldsymbol{\phi}\end{pmatrix} = \begin{pmatrix}\mathbf{b}_{\Omega,\Gamma}\\ \widetilde{\mathbf{b}}_\Gamma\end{pmatrix} \tag{14.26}$$

where, instead of (14.13), the right-hand side vector is defined by

$$\begin{aligned}
&\mathbf{b}_\Omega \in \mathbb{R}^{M_\Omega}, \quad (\mathbf{b}_\Omega)_i := 2\langle f, \eta_i\rangle_\Omega, && i \in \mathscr{I}_1,\\
&\mathbf{b}_\Gamma \in \mathbb{R}^{M_\Gamma}, \quad (\mathbf{b}_\Gamma)_i := 2\langle f, \eta_i\rangle_\Gamma + 2\langle t_0, \eta_{i,\Gamma}\rangle_\Gamma, && i \in \mathscr{I}_2,\\
&\widetilde{\mathbf{b}}_\Gamma \in \mathbb{R}^{N}, \quad (\widetilde{\mathbf{b}}_\Gamma)_i := \langle(\mathfrak{I} - \mathfrak{K})u_0, \phi_i\rangle_\Gamma, && i \in \mathscr{I}_3,
\end{aligned}$$

with $\mathbf{u}_{\Omega,\Gamma}$ and $\mathbf{b}_{\Omega,\Gamma}$ defined by (14.15). We denote

$$\mathbf{A}_{\mathfrak{JN}} := \begin{pmatrix} \underline{\mathbf{A}}_\Delta & -2\underline{\mathbf{M}}^\top \\ -\underline{\mathbf{K}} & \mathbf{V} \end{pmatrix}. \tag{14.27}$$

#### 14.1.2.2 The Bielak–MacCamy coupling

The Bielak–MacCamy coupling has the same property as the Johnson–Nédélec coupling in that it does not involve all four integral operators, in particular, not the hypersingular integral operator $\mathfrak{W}$. However, differently from the symmetric coupling and the Johnson–Nédélec coupling, its derivation is not initiated from (14.4). Instead, $u_2$ is represented indirectly by

$$u_2 = \frac{1}{2}\mathfrak{S}\phi \quad \text{in } \Omega^c \tag{14.28}$$

where $\phi \in H^{-1/2}(\Gamma)$ is some unknown function. This representation results in the indirect boundary integral equation method for solving the Laplace equation. Again, using the jump relation we deduce $2u_2 = \mathfrak{V}\phi$ on $\Gamma$, which together with (14.1c) yields

$$-2\langle u_1, \psi\rangle_\Gamma + a_{\mathfrak{V}}(\phi,\psi) = -2\langle u_0, \psi\rangle_\Gamma, \quad \psi \in H^{-1/2}(\Gamma).$$

Furthermore, taking the normal trace in (14.28) gives

$$2\partial_n u_2 = -\phi + \mathfrak{K}'\phi \quad \text{on } \Gamma.$$

Inserting this into the weak formulation of (14.1a) after using (14.1d) results in

$$2a_\Delta(u_1, v) + \langle(\mathfrak{I} - \mathfrak{K}')\phi, v\rangle_\Gamma = 2\langle f, v\rangle_\Omega + 2\langle t_0, v\rangle_\Gamma.$$

Therefore, the variational formulation of the Bielak–MacCamy coupling reads:

Find $(u,\phi) \in H^1(\Omega) \times H^{-1/2}(\Gamma)$ such that, for all $(v,\psi) \in H^1(\Omega) \times H^{-1/2}(\Gamma)$,

$$\begin{aligned}
2a_\Delta(u, v) + \langle(\mathfrak{I} - \mathfrak{K}')\phi, v\rangle_\Gamma &= 2\langle f, v\rangle_\Omega + 2\langle t_0, v\rangle_\Gamma,\\
-2\langle u, \psi\rangle_\Gamma + a_{\mathfrak{V}}(\phi,\psi) &= -2\langle u_0, \psi\rangle_\Gamma, \quad \psi \in H^{-1/2}(\Gamma).
\end{aligned} \tag{14.29}$$

**Theorem 14.3.** *Under the compatibility condition* (14.2) *when $d = 2$, and no condition for $d = 3$, the problem* (14.29) *is uniquely solvable.*

*Moreover, if* $u_1 \in H^1(\Omega)$ *and* $u_2 \in H^1_{\rm loc}(\Omega^{\rm c})$ *satisfy* (14.1), *then there exists* $\phi \in H^{-1/2}(\Gamma)$ *such that* $u_2 = \mathfrak{S}\phi/2$. *Moreover* $(u_1, \phi)$ *solves* (14.29). *Conversely, if* $u \in H^1(\Omega)$ *and* $\phi \in H^{-1/2}(\Gamma)$ *satisfy* (14.29), *then*

$$u_1 := u \quad \textit{and} \quad u_2 := \frac{1}{2}\mathfrak{S}\phi$$

*solve the interface problem* (14.1).

The Galerkin approximation to (14.29) can be proceeded as before on the space $\mathscr{X}$; see Subsection 14.1.1.2.

With the matrices $\underline{\mathbf{A}}_\Delta$, $\underline{\mathbf{M}}$, $\underline{\mathbf{K}}$, and $\mathbf{V}$ defined as in Subsection 14.1.2.1, we derive the matrix form of the Bielak–MacCamy coupling as

$$\begin{pmatrix} \underline{\mathbf{A}}_\Delta & -\underline{\mathbf{K}}^\top \\ -2\underline{\mathbf{M}} & \mathbf{V} \end{pmatrix} \begin{pmatrix} \mathbf{u}_{\Omega,\Gamma} \\ \boldsymbol{\phi} \end{pmatrix} = \begin{pmatrix} \mathbf{b}_{\Omega,\Gamma} \\ \widetilde{\mathbf{b}}_\Gamma \end{pmatrix} \tag{14.30}$$

where $\mathbf{b}_{\Omega,\Gamma} = (\mathbf{b}_\Omega \ \ \mathbf{b}_\Gamma)^\top$ and

$$\begin{aligned} &\mathbf{b}_\Omega \in \mathbb{R}^{M_\Omega}, && (\mathbf{b}_\Omega)_i := 2\langle f, \eta_i\rangle_\Omega, && i \in \mathscr{I}_1, \\ &\mathbf{b}_\Gamma \in \mathbb{R}^{M_\Gamma}, && (\mathbf{b}_\Gamma)_i := 2\langle f, \eta_i\rangle_\Gamma + 2\langle t_0, \eta_{i,\Gamma}\rangle_\Gamma, && i \in \mathscr{I}_2, \\ &\widetilde{\mathbf{b}}_\Gamma \in \mathbb{R}^{N}, && (\widetilde{\mathbf{b}}_\Gamma)_i := -2\langle u_0, \phi_i\rangle_\Gamma, && i \in \mathscr{I}_3. \end{aligned}$$

We denote

$$\mathbf{A}_{\mathfrak{BM}} := \begin{pmatrix} \underline{\mathbf{A}}_\Delta & -\underline{\mathbf{K}}^\top \\ -2\underline{\mathbf{M}} & \mathbf{V} \end{pmatrix}. \tag{14.31}$$

## 14.2 Preconditioning for the Symmetric Coupling

In this section, we design preconditioners for the matrix systems developed above. We adopt the following convention on notation of preconditioners and preconditioned matrices.

**Definition 14.1.**

1. *Let* $\mathbf{Q}$ *be a square matrix. A spectrally equivalent matrix to* $\mathbf{Q}$ *is denoted by* $\breve{\mathbf{Q}}$. *The symmetric preconditioned matrix is defined by*

$$\widetilde{\mathbf{Q}} := \breve{\mathbf{Q}}^{-1/2}\mathbf{Q}\breve{\mathbf{Q}}^{-1/2}.$$

2. *Let* $\mathbf{P}$, $\mathbf{Q}$, *and* $\mathbf{R}$ *be three matrices of appropriate sizes. The matrix* ${}_{\breve{\mathbf{P}}}\mathbf{Q}_{\breve{\mathbf{R}}}$ *is defined by*

$${}_{\breve{\mathbf{P}}}\mathbf{Q}_{\breve{\mathbf{R}}} := \breve{\mathbf{P}}^{-1/2}\mathbf{Q}\breve{\mathbf{R}}^{-1/2}.$$

### 14.2.1 Preconditioning with HMCR

In this subsection we design a three-block preconditioner for (14.14) and a two-block preconditioner for (14.17). Since these coefficient matrices are symmetric but indefinite, we solve the systems by the hybrid modified conjugate residual method (Algorithm 2.15). We will first present the general theory, and then apply the convergence result to the *h*-, *p*-, and *hp*-versions of FEM-BEM.

#### 14.2.1.1 Three-block preconditioner:

We now design a preconditioner for solving (14.14). The $3 \times 3$-block representation of $\mathbf{A}_{\mathfrak{S}}$ defined in (14.18), i.e.,

$$\mathbf{A}_{\mathfrak{S}} = \begin{pmatrix} \mathbf{A}_\Delta & \mathbf{B}^\top & \mathbf{0} \\ \mathbf{B} & \mathbf{D}+\mathbf{W} & \mathbf{K}^\top - \mathbf{I}^\top \\ \mathbf{0} & \mathbf{K}-\mathbf{I} & -\mathbf{V} \end{pmatrix},$$

is symmetrically preconditioned by

$$\breve{\mathbf{A}} := \begin{pmatrix} \breve{\mathbf{A}}_\Delta & 0 & 0 \\ 0 & \breve{\mathbf{D}} & 0 \\ 0 & 0 & \breve{\mathbf{V}} \end{pmatrix}. \tag{14.32}$$

The preconditioned matrix $\widetilde{\mathbf{A}} = \breve{\mathbf{A}}^{-1/2} \mathbf{A}_{\mathfrak{S}} \breve{\mathbf{A}}^{-1/2}$ then has the form

$$\widetilde{\mathbf{A}} = \begin{pmatrix} \widetilde{\mathbf{A}}_\Delta & {}_{\breve{\mathbf{A}}_\Delta}(\mathbf{B}^\top)_{\breve{\mathbf{D}}} & \mathbf{0} \\ {}_{\breve{\mathbf{D}}}\mathbf{B}_{\breve{\mathbf{A}}_\Delta} & \widetilde{\mathbf{D}} + {}_{\breve{\mathbf{D}}}\mathbf{W}_{\breve{\mathbf{D}}} & {}_{\breve{\mathbf{D}}}(\mathbf{K}^\top - \mathbf{I}^\top)_{\breve{\mathbf{V}}} \\ \mathbf{0} & {}_{\breve{\mathbf{V}}}(\mathbf{K}-\mathbf{I})_{\breve{\mathbf{D}}} & -\widetilde{\mathbf{V}} \end{pmatrix}$$

where we have used the notation convention in Definition 14.1. Defining

$$\widetilde{\underline{\mathbf{A}}} := \begin{pmatrix} \widetilde{\mathbf{A}}_\Delta & {}_{\breve{\mathbf{A}}_\Delta}(\mathbf{B}^\top)_{\breve{\mathbf{D}}} \\ {}_{\breve{\mathbf{D}}}\mathbf{B}_{\breve{\mathbf{A}}_\Delta} & \widetilde{\mathbf{D}} + {}_{\breve{\mathbf{D}}}\mathbf{W}_{\breve{\mathbf{D}}} \end{pmatrix} \quad \text{and} \quad \widetilde{\underline{\mathbf{K}}} := \begin{pmatrix} \mathbf{0} & {}_{\breve{\mathbf{V}}}(\mathbf{K}-\mathbf{I})_{\breve{\mathbf{D}}} \end{pmatrix} \tag{14.33}$$

and noting that $({}_{\breve{\mathbf{V}}}(\mathbf{K}-\mathbf{I})_{\breve{\mathbf{D}}})^\top = {}_{\breve{\mathbf{D}}}(\mathbf{K}^\top - \mathbf{I}^\top)_{\breve{\mathbf{V}}}$ we can rewrite $\widetilde{\underline{\mathbf{A}}}$ as

$$\widetilde{\mathbf{A}} = \begin{pmatrix} \widetilde{\underline{\mathbf{A}}} & \widetilde{\underline{\mathbf{K}}}^\top \\ \widetilde{\underline{\mathbf{K}}} & -\widetilde{\mathbf{V}} \end{pmatrix}. \tag{14.34}$$

In view of Lemma 2.22 we will find an inclusion set for the eigenvalues of $\widetilde{\mathbf{A}}$. Due to Lemma C.10 in the appendix, this set is well described if we can estimate the eigenvalues of $\widetilde{\underline{\mathbf{A}}}$ and $\widetilde{\mathbf{V}}$, and the singular values of $\widetilde{\underline{\mathbf{K}}}$. These eigenvalues and singular values are of course determined by the choice of the preconditioner $\breve{\mathbf{A}}$ in (14.32).

In the following, for any square matrix $\mathbf{Q}$, we denote by $\Lambda(\mathbf{Q})$ the set of all eigenvalues of $\mathbf{Q}$, and by $[\underline{\mu}(\mathbf{Q}),\overline{\mu}(\mathbf{Q})]$ a set that contains $\Lambda(\mathbf{Q})$. With these notations introduced, we assume

$$\begin{aligned} \Lambda(\mathbf{A}_\Delta) &\subset [\underline{\mu}(\mathbf{A}_\Delta),\overline{\mu}(\mathbf{A}_\Delta)], & \underline{\mu}(\mathbf{A}_\Delta) &> 0,\\ \Lambda(\mathbf{D}) &\subset [\underline{\mu}(\mathbf{D}),\overline{\mu}(\mathbf{D})], & \underline{\mu}(\mathbf{D}) &> 0,\\ \Lambda(\mathbf{V}) &\subset [\underline{\mu}(\mathbf{V}),\overline{\mu}(\mathbf{V})], & \underline{\mu}(\mathbf{V}) &> 0,\\ \Lambda(\mathbf{W}) &\subset [\underline{\mu}(\mathbf{W}),\overline{\mu}(\mathbf{W})], & \underline{\mu}(\mathbf{W}) &> 0. \end{aligned} \tag{14.35}$$

These matrices are symmetric and positive definite. If they are constructed from the finite element and boundary element approximations, these bounds are usually well known. We also assume that the preconditioners $\breve{\mathbf{A}}_\Delta$, $\breve{\mathbf{D}}$, and $\breve{\mathbf{V}}$ are chosen such that

$$\begin{aligned} \Lambda(\widetilde{\mathbf{A}}_\Delta) &\subset [\underline{\mu}(\widetilde{\mathbf{A}}_\Delta),\overline{\mu}(\widetilde{\mathbf{A}}_\Delta)], & \underline{\mu}(\widetilde{\mathbf{A}}_\Delta) &> 0,\\ \Lambda(\widetilde{\mathbf{D}}) &\subset [\underline{\mu}(\widetilde{\mathbf{D}}),\overline{\mu}(\widetilde{\mathbf{D}})], & \underline{\mu}(\widetilde{\mathbf{D}}) &> 0,\\ \Lambda(\widetilde{\mathbf{V}}) &\subset [\underline{\mu}(\widetilde{\mathbf{V}}),\overline{\mu}(\widetilde{\mathbf{V}})], & \underline{\mu}(\widetilde{\mathbf{V}}) &> 0. \end{aligned} \tag{14.36}$$

As a consequence of Lemma C.5 in Appendix C, we can choose

$$\begin{aligned} \underline{\mu}(\breve{\mathbf{A}}_\Delta) &= \frac{\underline{\mu}(\mathbf{A}_\Delta)}{\overline{\mu}(\widetilde{\mathbf{A}}_\Delta)}, & \overline{\mu}(\breve{\mathbf{A}}_\Delta) &= \frac{\overline{\mu}(\mathbf{A}_\Delta)}{\underline{\mu}(\widetilde{\mathbf{A}}_\Delta)},\\ \underline{\mu}(\breve{\mathbf{D}}) &= \frac{\underline{\mu}(\mathbf{D})}{\overline{\mu}(\widetilde{\mathbf{D}})}, & \overline{\mu}(\breve{\mathbf{D}}) &= \frac{\overline{\mu}(\mathbf{D})}{\underline{\mu}(\widetilde{\mathbf{D}})},\\ \underline{\mu}(\breve{\mathbf{V}}) &= \frac{\underline{\mu}(\mathbf{V})}{\overline{\mu}(\widetilde{\mathbf{V}})}, & \overline{\mu}(\breve{\mathbf{V}}) &= \frac{\overline{\mu}(\mathbf{V})}{\underline{\mu}(\widetilde{\mathbf{V}})}. \end{aligned} \tag{14.37}$$

With these bounds on eigenvalues, we are ready to estimate the extremal eigenvalues of $\widetilde{\underline{\mathbf{A}}}$ and singular values of $\widetilde{\underline{\mathbf{K}}}$.

**The extremal eigenvalues of $\widetilde{\underline{\mathbf{A}}}$:**

First we estimate the minimum eigenvalue of $\widetilde{\underline{\mathbf{A}}}$.

**Lemma 14.1.** *Assume that* (14.35) *and* (14.36) *hold. Assume further that* $\underline{\mu}(\mathbf{W})$ *is sufficiently small. Then the matrix* $\widetilde{\underline{\mathbf{A}}}$ *is positive semi-definite. Moreover, the following statement holds:*

$$\lambda_{\min}(\widetilde{\underline{\mathbf{A}}}) \ge \frac{1}{2}\left(\underline{\mu}(\widetilde{\mathbf{A}}_\Delta) + \underline{\mu}(\mathbf{W})\frac{\underline{\mu}(\widetilde{\mathbf{D}})}{\overline{\mu}(\mathbf{D})} + 1\right)$$

$$-\frac{1}{2}\sqrt{\left(\underline{\mu}(\widetilde{\mathbf{A}}_\Delta)+\underline{\mu}(\mathbf{W})\frac{\underline{\mu}(\widetilde{\mathbf{D}})}{\overline{\mu}(\mathbf{D})}+1\right)^2-4\underline{\mu}(\widetilde{\mathbf{A}}_\Delta)\underline{\mu}(\mathbf{W})\frac{\underline{\mu}(\widetilde{\mathbf{D}})}{\overline{\mu}(\mathbf{D})}}.$$

*Proof.* In this proof only we set

$$\mathbf{E}_1 := \widetilde{\mathbf{A}}_\Delta, \quad \mathbf{E}_2 := {}_{\breve{\mathbf{A}}_\Delta}(\mathbf{B}^\top)_{\breve{\mathbf{D}}}, \quad \mathbf{E}_3 := \widetilde{\mathbf{D}} + {}_{\breve{\mathbf{D}}}\mathbf{W}_{\breve{\mathbf{D}}} = {}_{\breve{\mathbf{D}}}(\mathbf{D}+\mathbf{W})_{\breve{\mathbf{D}}},$$

so that

$$\underline{\widetilde{\mathbf{A}}} = \begin{pmatrix} \mathbf{E}_1 & \mathbf{E}_2 \\ \mathbf{E}_2^\top & \mathbf{E}_3 \end{pmatrix}.$$

Let $\lambda$ be an eigenvalue of $\underline{\widetilde{\mathbf{A}}}$. Then there exist some vectors $\mathbf{x}$ and $\mathbf{y}$, $(\mathbf{x},\mathbf{y}) \neq (\mathbf{0},\mathbf{0})$, satisfying

$$\mathbf{E}_1\mathbf{x}+\mathbf{E}_2\mathbf{y} = \lambda\mathbf{x}, \tag{14.38a}$$

$$\mathbf{E}_2^\top\mathbf{x}+\mathbf{E}_3\mathbf{y} = \lambda\mathbf{y}. \tag{14.38b}$$

Since our aim in this lemma is to look for a lower bound for $\lambda_{\min}(\underline{\widetilde{\mathbf{A}}})$, it suffices to consider small values of $\lambda$, say

$$\lambda < \underline{\mu}(\widetilde{\mathbf{A}}_\Delta) = \underline{\mu}(\mathbf{E}_1). \tag{14.39}$$

For these values of $\lambda$, the matrix $\mathbf{E}_1 - \lambda\mathbf{I}$ is invertible, and it follows from (14.38a) that $\mathbf{x} = -(\mathbf{E}_1 - \lambda\mathbf{I})^{-1}\mathbf{E}_2\mathbf{y}$. Substituting this into (14.38b) and premultiply the resulting equation by $\mathbf{y}^\top$ give

$$-\mathbf{y}^\top\mathbf{E}_2^\top(\mathbf{E}_1-\lambda\mathbf{I})^{-1}\mathbf{E}_2\mathbf{y}+\mathbf{y}^\top\mathbf{E}_3\mathbf{y} = \lambda\mathbf{y}^\top\mathbf{y}. \tag{14.40}$$

It follows from

$$(\mathbf{E}_1-\lambda\mathbf{I})^{-1} = \left(\mathbf{E}_1^{1/2}(\mathbf{I}-\lambda\mathbf{E}_1^{-1})\mathbf{E}_1^{1/2}\right)^{-1} = \mathbf{E}_1^{-1/2}(\mathbf{I}-\lambda\mathbf{E}_1^{-1})^{-1}\mathbf{E}_1^{-1/2}$$

that

$$\begin{aligned}
-\mathbf{y}^\top\mathbf{E}_2^\top(\mathbf{E}_1-\lambda\mathbf{I})^{-1}\mathbf{E}_2\mathbf{y} &= -\underbrace{\mathbf{y}^\top\mathbf{E}_2^\top\mathbf{E}_1^{-1/2}}_{\mathbf{z}^\top=}(\mathbf{I}-\lambda\mathbf{E}_1^{-1})^{-1}\underbrace{\mathbf{E}_1^{-1/2}\mathbf{E}_2\mathbf{y}}_{\mathbf{z}:=} \\
&\geq -\lambda_{\max}\left((\mathbf{I}-\lambda\mathbf{E}_1^{-1})^{-1}\right)\mathbf{z}^\top\mathbf{z} \\
&= -\frac{1}{\lambda_{\min}\left(\mathbf{I}-\lambda\mathbf{E}_1^{-1}\right)}\mathbf{z}^\top\mathbf{z}
\end{aligned} \tag{14.41}$$

where we have used the symmetry of $\mathbf{I}-\lambda\mathbf{E}_1^{-1}$. This symmetry also implies

$$\frac{\mathbf{x}^\top(\mathbf{I}-\lambda\mathbf{E}_1^{-1})\mathbf{x}}{\mathbf{x}^\top\mathbf{x}} = 1-\lambda\frac{\mathbf{x}^\top\mathbf{E}_1^{-1}\mathbf{x}}{\mathbf{x}^\top\mathbf{x}} \geq 1-\lambda\lambda_{\max}\left(\mathbf{E}_1^{-1}\right) = 1-\frac{\lambda}{\lambda_{\min}(\mathbf{E}_1)} \geq 1-\frac{\lambda}{\underline{\mu}(\mathbf{E}_1)},$$

so that

$$\lambda_{\min}\left(\mathbf{I}-\lambda\mathbf{E}_1^{-1}\right)\geq 1-\frac{\lambda}{\underline{\mu}(\mathbf{E}_1)}=\frac{\underline{\mu}(\mathbf{E}_1)-\lambda}{\underline{\mu}(\mathbf{E}_1)}. \tag{14.42}$$

It follows from (14.40)–(14.42) that

$$-\left(\frac{\underline{\mu}(\mathbf{E}_1)}{\underline{\mu}(\mathbf{E}_1)-\lambda}\right)\mathbf{z}^\top\mathbf{z}+\mathbf{y}^\top\mathbf{E}_3\mathbf{y}\leq\lambda\mathbf{y}^\top\mathbf{y}. \tag{14.43}$$

By setting $\mathbf{w}:=\breve{\mathbf{D}}^{-1/2}\mathbf{y}$ and using the definitions of $\mathbf{z}$ in (14.41) and of $\mathbf{E}_i$, $i=1,2,3$, we obtain

$$\begin{aligned}
\mathbf{z}^\top\mathbf{z}&=\mathbf{y}^\top\mathbf{E}_2^\top\mathbf{E}_1^{-1}\mathbf{E}_2\mathbf{y}=\mathbf{y}^\top\breve{\mathbf{D}}^{-1/2}\left(\mathbf{B}\mathbf{A}_\Delta^{-1}\mathbf{B}^\top\right)\breve{\mathbf{D}}^{-1/2}\mathbf{y}=\mathbf{w}^\top\mathbf{B}\mathbf{A}_\Delta^{-1}\mathbf{B}^\top\mathbf{w},\\
\mathbf{y}^\top\mathbf{E}_3\mathbf{y}&=\mathbf{y}^\top\breve{\mathbf{D}}^{-1/2}(\mathbf{D}+\mathbf{W})\breve{\mathbf{D}}^{-1/2}\mathbf{y}=\mathbf{w}^\top(\mathbf{D}+\mathbf{W})\mathbf{w},\\
\mathbf{y}^\top\mathbf{y}&=\mathbf{w}^\top\breve{\mathbf{D}}\mathbf{w}.
\end{aligned}$$

Therefore, (14.43) yields

$$-\left(\frac{\underline{\mu}(\widetilde{\mathbf{A}}_\Delta)}{\underline{\mu}(\widetilde{\mathbf{A}}_\Delta)-\lambda}\right)\mathbf{w}^\top\mathbf{B}\mathbf{A}_\Delta^{-1}\mathbf{B}^\top\mathbf{w}+\mathbf{w}^\top(\mathbf{D}+\mathbf{W})\mathbf{w}\leq\lambda\mathbf{w}^\top\breve{\mathbf{D}}\mathbf{w}.$$

Then it follows from (14.39) that

$$-\underline{\mu}(\widetilde{\mathbf{A}}_\Delta)\mathbf{w}^\top\mathbf{B}\mathbf{A}_\Delta^{-1}\mathbf{B}^\top\mathbf{w}+\left(\underline{\mu}(\widetilde{\mathbf{A}}_\Delta)-\lambda\right)\mathbf{w}^\top(\mathbf{D}+\mathbf{W})\mathbf{w}\leq\lambda\left(\underline{\mu}(\widetilde{\mathbf{A}}_\Delta)-\lambda\right)\mathbf{w}^\top\breve{\mathbf{D}}\mathbf{w},$$

or

$$\begin{aligned}
\underline{\mu}(\widetilde{\mathbf{A}}_\Delta)\mathbf{w}^\top\left(\mathbf{D}-\mathbf{B}\mathbf{A}_\Delta^{-1}\mathbf{B}^\top\right)\mathbf{w}-\lambda\mathbf{w}^\top\mathbf{D}\mathbf{w}+\left(\underline{\mu}(\widetilde{\mathbf{A}}_\Delta)-\lambda\right)\mathbf{w}^\top\mathbf{W}\mathbf{w}&\\
\leq\lambda\left(\underline{\mu}(\widetilde{\mathbf{A}}_\Delta)-\lambda\right)\mathbf{w}^\top\breve{\mathbf{D}}\mathbf{w}&.
\end{aligned}$$

Moreover, (14.39) also implies $\mathbf{y}\neq\mathbf{0}$ because otherwise (14.38a) implies that $\lambda$ is an eigenvalue of $\widetilde{\mathbf{A}}_\Delta$, which is a contradiction. As a consequence, $\mathbf{w}\neq\mathbf{0}$. Thus by noting the symmetry of the matrices involved, we deduce

$$\begin{aligned}
\underline{\mu}(\widetilde{\mathbf{A}}_\Delta)\lambda_{\min}\left(\mathbf{D}-\mathbf{B}\mathbf{A}_\Delta^{-1}\mathbf{B}^\top\right)-\lambda\lambda_{\max}(\mathbf{D})+\left(\underline{\mu}(\widetilde{\mathbf{A}}_\Delta)-\lambda\right)\lambda_{\min}(\mathbf{W})&\\
\leq\lambda\left(\underline{\mu}(\widetilde{\mathbf{A}}_\Delta)-\lambda\right)\lambda_{\max}(\breve{\mathbf{D}})&.
\end{aligned} \tag{14.44}$$

Bounds for the extremal eigenvalues of $\mathbf{D}$, $\mathbf{W}$, and $\breve{\mathbf{D}}$ are given in (14.35)–(14.37). Let us consider $\lambda_{\min}\left(\mathbf{D}-\mathbf{B}\mathbf{A}_\Delta^{-1}\mathbf{B}^\top\right)$.

Recall the stiffness matrix $\underline{\mathbf{A}}_\Delta$ and its decomposition (14.11). This matrix is a discrete Laplacian for a homogeneous Neumann problem, and thus $\lambda_{\min}(\underline{\mathbf{A}}_\Delta)=0$. From the congruence transform

$$\underline{\mathbf{A}}_\Delta = \begin{pmatrix} \mathbf{A}_\Delta & \mathbf{B}^\top \\ \mathbf{B} & \mathbf{D} \end{pmatrix} = \begin{pmatrix} \mathbf{A}_\Delta^{1/2} & \mathbf{0} \\ \mathbf{B}\mathbf{A}_\Delta^{-1/2} & \mathbf{I} \end{pmatrix} \begin{pmatrix} \mathbf{I} & \mathbf{0} \\ \mathbf{0} & \mathbf{D} - \mathbf{B}\mathbf{A}_\Delta^{-1}\mathbf{B}^\top \end{pmatrix} \begin{pmatrix} \mathbf{A}_\Delta^{1/2} & \mathbf{A}_\Delta^{-1/2}\mathbf{B}^\top \\ \mathbf{0} & \mathbf{I} \end{pmatrix}$$

and Sylvester's Law of Inertia [83] it follows that

$$\mathbf{x}^\top(\mathbf{D} - \mathbf{B}\mathbf{A}_\Delta^{-1}\mathbf{B}^\top)\mathbf{x} \ge 0 \quad \forall \mathbf{x}. \tag{14.45}$$

In particular $\lambda_{\min}(\mathbf{D} - \mathbf{B}\mathbf{A}_\Delta^{-1}\mathbf{B}^\top) \ge 0$. Hence (14.44), (14.35), and (14.36) yield

$$-\lambda\overline{\mu}(\mathbf{D}) + \left(\underline{\mu}(\widetilde{\mathbf{A}}_\Delta) - \lambda\right)\underline{\mu}(\mathbf{W}) \le \lambda\left(\underline{\mu}(\widetilde{\mathbf{A}}_\Delta) - \lambda\right)\overline{\mu}(\check{\mathbf{D}}),$$

or, after rearranging the terms,

$$\lambda^2\overline{\mu}(\check{\mathbf{D}}) - \lambda\left(\underline{\mu}(\widetilde{\mathbf{A}}_\Delta)\overline{\mu}(\check{\mathbf{D}}) + \underline{\mu}(\mathbf{W}) + \overline{\mu}(\mathbf{D})\right) + \underline{\mu}(\widetilde{\mathbf{A}}_\Delta)\underline{\mu}(\mathbf{W}) \le 0.$$

Since $\overline{\mu}(\check{\mathbf{D}}) > 0$ (see (14.37)) this implies

$$\begin{aligned}\lambda &\ge \frac{1}{2\overline{\mu}(\check{\mathbf{D}})}\Bigg(\underline{\mu}(\widetilde{\mathbf{A}}_\Delta)\overline{\mu}(\check{\mathbf{D}}) + \underline{\mu}(\mathbf{W}) + \overline{\mu}(\check{\mathbf{D}}) \\ &\quad - \sqrt{\left(\underline{\mu}(\widetilde{\mathbf{A}}_\Delta)\overline{\mu}(\check{\mathbf{D}}) + \underline{\mu}(\mathbf{W}) + \overline{\mu}(\check{\mathbf{D}})\right)^2 - 4\underline{\mu}(\widetilde{\mathbf{A}}_\Delta)\overline{\mu}(\check{\mathbf{D}})\underline{\mu}(\mathbf{W})}\Bigg) \\ &= \frac{1}{2}\left(\underline{\mu}(\widetilde{\mathbf{A}}_\Delta) + \frac{\underline{\mu}(\mathbf{W})}{\overline{\mu}(\check{\mathbf{D}})} + 1\right) - \frac{1}{2}\sqrt{\left(\underline{\mu}(\widetilde{\mathbf{A}}_\Delta) + \frac{\underline{\mu}(\mathbf{W})}{\overline{\mu}(\check{\mathbf{D}})} + 1\right)^2 - 4\underline{\mu}(\widetilde{\mathbf{A}}_\Delta)\frac{\underline{\mu}(\mathbf{W})}{\overline{\mu}(\check{\mathbf{D}})}}.\end{aligned}$$

The right-hand side is non-negative when $\underline{\mu}(\mathbf{W})$ is sufficiently small. Hence $\widetilde{\underline{\mathbf{A}}}$ is positive semi-definite. The required lower bound for $\lambda_{\min}(\widetilde{\underline{\mathbf{A}}})$ is now a consequence of (14.37). □

*Remark 14.1.* In general, for the Galerkin boundary element method, $\underline{\mu}(\mathbf{W}) \to 0$ when $h$ goes to 0 and/or $p$ tends to $\infty$. Therefore, the assumption on $\underline{\mu}(\mathbf{W})$ in the above lemma holds.

Next we estimate the maximum eigenvalue of $\widetilde{\underline{\mathbf{A}}}$.

**Lemma 14.2.** *The following statement holds:*

$$\lambda_{\max}(\widetilde{\underline{\mathbf{A}}}) \le \max\left\{\overline{\mu}(\widetilde{\mathbf{A}}_\Delta), \overline{\mu}(\widetilde{\mathbf{D}}) + \frac{\overline{\mu}(\mathbf{W})\overline{\mu}(\widetilde{\mathbf{D}})}{\underline{\mu}(\mathbf{D})}\right\} + \sqrt{\frac{\overline{\mu}(\widetilde{\mathbf{A}}_\Delta)\overline{\mu}(\mathbf{D})\overline{\mu}(\widetilde{\mathbf{D}})}{\underline{\mu}(\mathbf{D})}}.$$

*Proof.* Writing $\widetilde{\underline{\mathbf{A}}}$ as $\widetilde{\underline{\mathbf{A}}} = \widetilde{\underline{\mathbf{A}}}_1 + \widetilde{\underline{\mathbf{A}}}_2$ where

$$\widetilde{\underline{\mathbf{A}}}_1 := \begin{pmatrix} \widetilde{\mathbf{A}}_\Delta & \mathbf{0} \\ \mathbf{0} & \widetilde{\mathbf{D}} + {}_{\check{\mathbf{D}}}\mathbf{W}_{\check{\mathbf{D}}} \end{pmatrix} \quad \text{and} \quad \widetilde{\underline{\mathbf{A}}}_2 := \begin{pmatrix} \mathbf{0} & {}_{\check{\mathbf{A}}_\Delta}(\mathbf{B}^\top)_{\check{\mathbf{D}}} \\ {}_{\check{\mathbf{D}}}\mathbf{B}_{\check{\mathbf{A}}_\Delta} & \mathbf{0} \end{pmatrix},$$

we have, due to the symmetry of $\widetilde{\underline{\mathbf{A}}}_1$ and $\widetilde{\underline{\mathbf{A}}}_2$, and Lemmas C.7, C.8, and C.9 in Appendix C,

$$\begin{aligned}\lambda_{\max}(\widetilde{\underline{\mathbf{A}}}) &\le \lambda_{\max}(\widetilde{\underline{\mathbf{A}}}_1) + \lambda_{\max}(\widetilde{\underline{\mathbf{A}}}_2)\\ &\le \max\left\{\lambda_{\max}(\widetilde{\mathbf{A}}_\Delta), \lambda_{\max}(\widetilde{\mathbf{D}}) + \lambda_{\max}({}_{\breve{\mathbf{D}}}\mathbf{W}_{\breve{\mathbf{D}}})\right\} + \sigma_{\max}({}_{\breve{\mathbf{A}}_\Delta}(\mathbf{B}^\top)_{\breve{\mathbf{D}}})\\ &\le \max\left\{\overline{\mu}(\widetilde{\mathbf{A}}_\Delta), \overline{\mu}(\widetilde{\mathbf{D}}) + \overline{\mu}({}_{\breve{\mathbf{D}}}\mathbf{W}_{\breve{\mathbf{D}}})\right\} + \sigma_{\max}({}_{\breve{\mathbf{A}}_\Delta}(\mathbf{B}^\top)_{\breve{\mathbf{D}}}). \qquad (14.46)\end{aligned}$$

The bounds $\overline{\mu}(\widetilde{\mathbf{A}}_\Delta)$ and $\overline{\mu}(\widetilde{\mathbf{D}})$ are given in (14.35) and (14.36). It follows from Lemma C.5 that

$$\overline{\mu}({}_{\breve{\mathbf{D}}}\mathbf{W}_{\breve{\mathbf{D}}}) \le \frac{\overline{\mu}(\mathbf{W})}{\underline{\mu}(\breve{\mathbf{D}})} \le \frac{\overline{\mu}(\mathbf{W})\overline{\mu}(\widetilde{\mathbf{D}})}{\underline{\mu}(\mathbf{D})}.$$

It remains to estimate $\sigma_{\max}({}_{\breve{\mathbf{A}}_\Delta}(\mathbf{B}^\top)_{\breve{\mathbf{D}}})$. The singular values of a matrix $\mathbf{E}$ are the square root of the eigenvalues of $\mathbf{E}^\top\mathbf{E}$; see [83]. Hence

$$\sigma_{\max}({}_{\breve{\mathbf{A}}_\Delta}(\mathbf{B}^\top)_{\breve{\mathbf{D}}}) = \sqrt{\lambda_{\max}\left({}_{\breve{\mathbf{D}}}\mathbf{B}_{\breve{\mathbf{A}}_\Delta}\,{}_{\breve{\mathbf{A}}_\Delta}(\mathbf{B}^\top)_{\breve{\mathbf{D}}}\right)} = \sqrt{\lambda_{\max}\left(\breve{\mathbf{D}}^{-1/2}\mathbf{B}\breve{\mathbf{A}}_\Delta^{-1}\mathbf{B}^\top\breve{\mathbf{D}}^{-1/2}\right)}.$$

For notational convenience, we set $\mathbf{E} = \breve{\mathbf{D}}^{-1/2}$, $\mathbf{F} = \breve{\mathbf{A}}_\Delta^{-1/2}$, and $\mathbf{G} = \mathbf{A}_\Delta^{-1}$. For any $\mathbf{x}$, if $\mathbf{B}^\top\mathbf{E}\mathbf{x} = \mathbf{0}$, then $\mathbf{E}\mathbf{B}\mathbf{F}^2\mathbf{B}^\top\mathbf{E}\mathbf{x} = \mathbf{0}$. If $\mathbf{x}$ is such that $\mathbf{B}^\top\mathbf{E}\mathbf{x} \ne \mathbf{0}$, then $\mathbf{x}^\top\mathbf{E}\mathbf{B}\mathbf{G}\mathbf{B}^\top\mathbf{E}\mathbf{x} \ne 0$ and $\mathbf{x}^\top\mathbf{E}\mathbf{E}\mathbf{x} \ne 0$. Therefore we can write

$$\frac{\mathbf{x}^\top\mathbf{E}\mathbf{B}\mathbf{F}^2\mathbf{B}^\top\mathbf{E}\mathbf{x}}{\mathbf{x}^\top\mathbf{x}} = \frac{\mathbf{x}^\top\mathbf{E}\mathbf{B}\mathbf{F}^2\mathbf{B}^\top\mathbf{E}\mathbf{x}}{\mathbf{x}^\top\mathbf{E}\mathbf{B}\mathbf{G}\mathbf{B}^\top\mathbf{E}\mathbf{x}} \cdot \frac{\mathbf{x}^\top\mathbf{E}\mathbf{B}\mathbf{G}\mathbf{B}^\top\mathbf{E}\mathbf{x}}{\mathbf{x}^\top\mathbf{E}\mathbf{E}\mathbf{x}} \cdot \frac{\mathbf{x}^\top\mathbf{E}^2\mathbf{x}}{\mathbf{x}^\top\mathbf{x}}.$$

With $\mathbf{y} := \mathbf{F}\mathbf{B}^\top\mathbf{E}\mathbf{x}$, $\mathbf{z} := \mathbf{E}\mathbf{x}$, and $\mathbf{w} := \widetilde{\mathbf{A}}_\Delta^{-1/2}\mathbf{y}$ we have

$$\begin{aligned}\frac{\mathbf{x}^\top\mathbf{E}\mathbf{B}\mathbf{F}^2\mathbf{B}^\top\mathbf{E}\mathbf{x}}{\mathbf{x}^\top\mathbf{x}} &= \frac{\mathbf{y}^\top\mathbf{y}}{\mathbf{y}^\top\mathbf{F}^{-1}\mathbf{G}\mathbf{F}^{-1}\mathbf{y}} \cdot \frac{\mathbf{z}^\top\mathbf{B}\mathbf{G}\mathbf{B}^\top\mathbf{z}}{\mathbf{z}^\top\mathbf{z}} \cdot \frac{\mathbf{x}^\top\mathbf{E}^2\mathbf{x}}{\mathbf{x}^\top\mathbf{x}}\\ &= \frac{\mathbf{y}^\top\mathbf{y}}{\mathbf{y}^\top\widetilde{\mathbf{A}}_\Delta^{-1}\mathbf{y}} \cdot \frac{\mathbf{z}^\top\mathbf{B}\mathbf{A}_\Delta^{-1}\mathbf{B}^\top\mathbf{z}}{\mathbf{z}^\top\mathbf{z}} \cdot \frac{\mathbf{x}^\top\breve{\mathbf{D}}^{-1}\mathbf{x}}{\mathbf{x}^\top\mathbf{x}}\\ &= \frac{\mathbf{w}^\top\widetilde{\mathbf{A}}_\Delta\mathbf{w}}{\mathbf{w}^\top\mathbf{w}} \cdot \frac{\mathbf{z}^\top\mathbf{B}\mathbf{A}_\Delta^{-1}\mathbf{B}^\top\mathbf{z}}{\mathbf{z}^\top\mathbf{z}} \cdot \frac{\mathbf{x}^\top\breve{\mathbf{D}}^{-1}\mathbf{x}}{\mathbf{x}^\top\mathbf{x}}.\end{aligned}$$

By using (14.45) we deduce

$$\frac{\mathbf{x}^\top\mathbf{E}\mathbf{B}\mathbf{F}^2\mathbf{B}^\top\mathbf{E}\mathbf{x}}{\mathbf{x}^\top\mathbf{x}} \le \frac{\mathbf{w}^\top\widetilde{\mathbf{A}}_\Delta\mathbf{w}}{\mathbf{w}^\top\mathbf{w}} \cdot \frac{\mathbf{z}^\top\mathbf{D}\mathbf{z}}{\mathbf{z}^\top\mathbf{z}} \cdot \frac{\mathbf{x}^\top\breve{\mathbf{D}}^{-1}\mathbf{x}}{\mathbf{x}^\top\mathbf{x}} \le \frac{\lambda_{\max}(\widetilde{\mathbf{A}}_\Delta)\lambda_{\max}(\mathbf{D})}{\lambda_{\min}(\breve{\mathbf{D}})} \le \frac{\overline{\mu}(\widetilde{\mathbf{A}}_\Delta)\overline{\mu}(\mathbf{D})}{\underline{\mu}(\breve{\mathbf{D}})}.$$

Thus

$$\sigma_{\max}(_{\breve{\mathbf{A}}_\Delta}(\mathbf{B}^\top)_{\breve{\mathbf{D}}}) = \sqrt{\lambda_{\max}\left(\breve{\mathbf{D}}^{-1/2}\mathbf{B}\breve{\mathbf{A}}_\Delta^{-1}\mathbf{B}^\top\breve{\mathbf{D}}^{-1/2}\right)} \le \sqrt{\frac{\overline{\mu}(\widetilde{\mathbf{A}}_\Delta)\overline{\mu}(\mathbf{D})}{\underline{\mu}(\breve{\mathbf{D}})}}.$$

The required result follows from the above inequality, (14.46), and (14.37). □

*Remark 14.2.* If we use the same approach to show a lower bound for $\lambda_{\min}(\underline{\widetilde{\mathbf{A}}})$, then analogously to (14.46) we have

$$\lambda_{\min}(\underline{\widetilde{\mathbf{A}}}) \ge \min\left\{\underline{\mu}(\widetilde{\mathbf{A}}_\Delta), \underline{\mu}(\widetilde{\mathbf{D}}) + \underline{\mu}(_{\breve{\mathbf{D}}}\mathbf{W}_{\breve{\mathbf{D}}})\right\} - \sigma_{\max}(_{\breve{\mathbf{A}}_\Delta}(\mathbf{B}^\top)_{\breve{\mathbf{D}}}),$$

so that

$$\lambda_{\min}(\underline{\widetilde{\mathbf{A}}}) \ge \min\left\{\underline{\mu}(\widetilde{\mathbf{A}}_\Delta), \underline{\mu}(\widetilde{\mathbf{D}}) + \underline{\mu}(_{\breve{\mathbf{D}}}\mathbf{W}_{\breve{\mathbf{D}}})\right\} - \sqrt{\frac{\overline{\mu}(\widetilde{\mathbf{A}}_\Delta)\overline{\mu}(\mathbf{D})\overline{\mu}(\widetilde{\mathbf{D}})}{\underline{\mu}(\mathbf{D})}}.$$

This lower bound could be negative.

**The extremal singular values of $\underline{\widetilde{\mathbf{K}}}$:**

We now turn to estimates for the extremal singular values of $\underline{\widetilde{\mathbf{K}}}$; see (14.33) and (14.34).

**Lemma 14.3.** *Assume that the preconditioner $\breve{\mathbf{D}}$ is chosen such that $\breve{\mathbf{D}}$ and $\mathbf{D}$ are spectrally equivalent. Then the maximum singular value of $\underline{\widetilde{\mathbf{K}}}$ defined in* (14.33) *is bounded independently of all parameters involved in the discretisation.*

*Proof.* Let $\mathscr{S} := \mathfrak{W} + (\mathfrak{K}^\top - \mathfrak{I}^\top)\mathfrak{V}^{-1}(\mathfrak{K} - \mathfrak{I})$ be the Steklov–Poincaré (or Dirichlet-to-Neumann) operator and $\mathbf{S} = \mathbf{W} + (\mathbf{K}^\top - \mathbf{I}^\top)\mathbf{V}^{-1}(\mathbf{K} - \mathbf{I})$ be its matrix representation. Then it is well known [53] that for any vector $\mathbf{v}$, if $v$ is the finite element function whose coefficient vector is $\mathbf{v}$ then $\mathbf{v}^\top\mathbf{S}\mathbf{v} \simeq \|v\|^2_{H^{1/2}(\Gamma)}$. It is also well known that

$$\mathbf{v}^\top\mathbf{D}\mathbf{v} \simeq \|v\|^2_{H^1(\Omega)} \gtrsim \|v|_\Gamma\|^2_{H^{1/2}(\Gamma)}.$$

Hence, if $\breve{\mathbf{D}}$ is chosen such that $\breve{\mathbf{D}}$ and $\mathbf{D}$ are spectrally equivalent, then $\mathbf{v}^\top\breve{\mathbf{D}}\mathbf{v} \gtrsim \|v|_\Gamma\|^2_{H^{1/2}(\Gamma)}$. As a consequence, we have

$$\begin{aligned}
\sigma_{\max}(\underline{\widetilde{\mathbf{K}}})^2 &\le \max_{\mathbf{v}\ne\mathbf{0}} \frac{\mathbf{v}^\top\breve{\mathbf{D}}^{-1/2}(\mathbf{K}^\top - \mathbf{I}^\top)\mathbf{V}^{-1}(\mathbf{K} - \mathbf{I})\breve{\mathbf{D}}^{-1/2}\mathbf{v}}{\mathbf{v}^\top\mathbf{v}} \\
&= \max_{\mathbf{w}\ne\mathbf{0}} \frac{\mathbf{w}^\top(\mathbf{K}^\top - \mathbf{I}^\top)\mathbf{V}^{-1}(\mathbf{K} - \mathbf{I})\mathbf{w}}{\mathbf{w}^\top\breve{\mathbf{D}}\mathbf{w}} \\
&\le \max_{\mathbf{w}\ne\mathbf{0}} \frac{\mathbf{w}^\top\mathbf{S}\mathbf{w}}{\mathbf{w}^\top\breve{\mathbf{D}}\mathbf{w}} \le C,
\end{aligned}$$

where in the penultimate step we have used the definition of $\mathbf{S}$ and the fact that $\mathbf{W}$ is positive semi-definite. □

**Convergence of the three-block preconditioner:**

Due to Lemma 2.22, the convergence rate of the three-block preconditioner is determined by $\lambda_{\min}(\widetilde{\mathbf{A}})$, $\lambda_{-}(\widetilde{\mathbf{A}})$, $\lambda_{+}(\widetilde{\mathbf{A}})$, and $\lambda_{\max}(\widetilde{\mathbf{A}})$, which in turn are determined by the extremal eigenvalues of $\widetilde{\underline{\mathbf{A}}}$ and extremal singular values of $\widetilde{\underline{\mathbf{K}}}$; see (14.34) and Lemma C.10.

**Theorem 14.4.** *Assume that the preconditioners* $\check{\mathbf{A}}_{\Delta}$, $\check{\mathbf{D}}$, *and* $\check{\mathbf{V}}$ *are chosen such that* (14.36) *hold. Then*

$$\Lambda(\widetilde{\mathbf{A}}) \subset [\underline{\mu}(\widetilde{\mathbf{A}}), \mu_{-}(\widetilde{\mathbf{A}})] \cup [\mu_{+}(\widetilde{\mathbf{A}}), \overline{\mu}(\widetilde{\mathbf{A}})]$$

*where*

$$\begin{aligned}
\underline{\mu}(\widetilde{\mathbf{A}}) &= \frac{1}{2}\left(\underline{\mu}(\widetilde{\underline{\mathbf{A}}}) - \overline{\mu}(\widetilde{\mathbf{V}}) - \sqrt{\left(\overline{\mu}(\widetilde{\underline{\mathbf{A}}}) + \overline{\mu}(\widetilde{\mathbf{V}})\right)^2 + 4\overline{\sigma}^2}\right),\\
\mu_{-}(\widetilde{\mathbf{A}}) &= -\underline{\mu}(\widetilde{\mathbf{V}}),\\
\mu_{+}(\widetilde{\mathbf{A}}) &= \underline{\mu}(\widetilde{\underline{\mathbf{A}}}),\\
\overline{\mu}(\widetilde{\mathbf{A}}) &= \frac{1}{2}\left(\overline{\mu}(\widetilde{\underline{\mathbf{A}}}) + \sqrt{\overline{\mu}(\widetilde{\underline{\mathbf{A}}})^2 + 4\overline{\sigma}^2}\right),
\end{aligned}$$

*with* $\underline{\mu}(\widetilde{\underline{\mathbf{A}}})$, $\overline{\mu}(\widetilde{\underline{\mathbf{A}}})$ *given by Lemmas 14.1, 14.2, and* $\underline{\mu}(\widetilde{\mathbf{V}})$, $\overline{\mu}(\widetilde{\mathbf{V}})$ *given by* (14.36). *Here* $\overline{\sigma}$ *is an upper bound of* $\sigma_{\max}(\widetilde{\underline{\mathbf{K}}})$ *given by Lemma 14.3.*

*Proof.* It follows from (14.34) and Lemma C.10 that

$$\begin{aligned}
\lambda_{\min}(\widetilde{\mathbf{A}}) &\geq \frac{1}{2}\left(\lambda_{\min}(\widetilde{\underline{\mathbf{A}}}) - \lambda_{\max}(\widetilde{\mathbf{V}}) - \sqrt{\left(\lambda_{\min}(\widetilde{\underline{\mathbf{A}}}) + \lambda_{\max}(\widetilde{\mathbf{V}})\right)^2 + 4\sigma_{\max}(\widetilde{\underline{\mathbf{K}}})^2}\right),\\
\lambda_{+}(\widetilde{\mathbf{A}}) &\geq \lambda_{\min}(\widetilde{\underline{\mathbf{A}}}),\\
\lambda_{\max}(\widetilde{\mathbf{A}}) &\leq \frac{1}{2}\left(\lambda_{\max}(\widetilde{\underline{\mathbf{A}}}) + \sqrt{\lambda_{\max}(\widetilde{\underline{\mathbf{A}}})^2 + 4\sigma_{\max}(\widetilde{\underline{\mathbf{K}}})^2}\right).
\end{aligned}$$

Bounds for the extremal eigenvalues of $\widetilde{\underline{\mathbf{A}}}$ are given in Lemma 14.1 and Lemma 14.2. Bounds for the extremal eigenvalues of $\widetilde{\mathbf{V}}$ are given in (14.36). A bound for $\sigma_{\max}(\widetilde{\underline{\mathbf{K}}})^2$ is given in Lemma 14.3.

In order to obtain an upper bound for $\lambda_{-}(\widetilde{\mathbf{A}})$, we use a different approach than that given in Lemma C.10, due to a lack of technique to estimate $\sigma_{\min}(\widetilde{\underline{\mathbf{K}}})$. Let $\lambda$ be a negative eigenvalue of $\widetilde{\mathbf{A}}$. Then there exist some vectors $\mathbf{x}$ and $\mathbf{y}$, $(\mathbf{x},\mathbf{y}) \neq (\mathbf{0},\mathbf{0})$, satisfying

$$\widetilde{\underline{\mathbf{A}}}\mathbf{x} + \widetilde{\underline{\mathbf{K}}}^{\top}\mathbf{y} = \lambda\mathbf{x} \tag{14.47a}$$

$$\widetilde{\underline{\mathbf{K}}}\mathbf{x} - \widetilde{\mathbf{V}}\mathbf{y} = \lambda\mathbf{y}. \tag{14.47b}$$

Since $\widetilde{\underline{\mathbf{A}}}$ is positive semi-definite and $\lambda < 0$, the matrix $\widetilde{\underline{\mathbf{A}}} - \lambda\mathbf{I}$ is invertible, so that (14.47a) implies

$$\mathbf{x} = -(\widetilde{\underline{\mathbf{A}}} - \lambda\mathbf{I})^{-1}\widetilde{\underline{\mathbf{K}}}^{\top}\mathbf{y}.$$

Substituting this into (14.47b) and premultiplying the resulting equation by $\mathbf{y}^\top$ yield

$$-\mathbf{y}^\top \underline{\widetilde{\mathbf{K}}}(\underline{\widetilde{\mathbf{A}}} - \lambda \mathbf{I})^{-1} \underline{\widetilde{\mathbf{K}}}^\top \mathbf{y} - \mathbf{y}^\top \widetilde{\mathbf{V}} \mathbf{y} = \lambda \mathbf{y}^\top \mathbf{y}.$$

Since $-\mathbf{y}^\top \underline{\widetilde{\mathbf{K}}}(\underline{\widetilde{\mathbf{A}}} - \lambda \mathbf{I})^{-1} \underline{\widetilde{\mathbf{K}}}^\top \mathbf{y} \le 0$, there follows

$$\lambda \mathbf{y}^\top \mathbf{y} \le -\mathbf{y}^\top \widetilde{\mathbf{V}} \mathbf{y}.$$

Note that $\mathbf{y} \neq 0$ because otherwise (14.47a) implies that $\lambda$ is an eigenvalue of $\underline{\widetilde{\mathbf{A}}}$, which contradicts the semi-definite property of $\underline{\widetilde{\mathbf{A}}}$. Hence it follows from the above inequality that $\lambda_-(\widetilde{\mathbf{A}}) \le -\lambda_{\min}(\widetilde{\mathbf{V}})$. This completes the proof of the theorem. □

*Remark 14.3.*

(i) Theorem 14.4 is purely at the matrix level, so it can be used for both two- and three-dimensional problems.

(ii) In [113] the *hp*-version of FEM-BEM coupling for the two-dimensional problem is considered. It is known that (cf. (14.35))

$$\Lambda(\mathbf{A}_\Delta) \subset [c_A h^2 p^{-4}, C_A], \qquad \Lambda(\mathbf{D}) \subset [c_D, C_D],$$
$$\Lambda(\mathbf{W}) \subset [c_W h p^{-2}, C_W], \qquad \Lambda(\mathbf{V}) \subset [c_V h^2 p^{-2}, C_V h],$$

where all the constants are positive and independent of $h$ and $p$. The results for the FEM part can be found in [146, 151, 166, 213] while those of the BEM part can be found in [113]. The authors of [113] employ spectrally equivalent preconditioners $\breve{\mathbf{A}}_\Delta$, $\breve{\mathbf{D}}$, and $\breve{\mathbf{V}}$. This means the sub-blocks $\mathbf{A}_\Delta$, $\mathbf{D}$, and $\mathbf{V}$ are exactly solved. This implies (14.36) holds with bounds independent of $h$ and $p$. As a consequence, Lemma 14.1 and Lemma 14.2 give

$$\underline{\mu}(\underline{\widetilde{\mathbf{A}}}) = c_{\underline{\widetilde{\mathbf{A}}}} h p^{-2} \quad \text{and} \quad \overline{\mu}(\underline{\widetilde{\mathbf{A}}}) = C_{\underline{\widetilde{\mathbf{A}}}}.$$

Theorem 14.4 yields the following inclusion set for the spectrum of $\widetilde{\mathbf{A}}$:

$$\Lambda(\widetilde{\mathbf{A}}) \subset [-a, -b] \cup [chp^{-2}, d]$$

where $a$, $b$, $c$, and $d$ are some positive constants independent of $h$ and $p$.

(iii) The case considered above shows that the three-block preconditioner is not a good choice for preconditioning of FEM-BEM coupling, because even though the diagonal blocks are inverted, the coupled preconditioned matrix still has eigenvalues going to zero quickly. This is due to the dependency of $\lambda_{\min}(\underline{\widetilde{\mathbf{A}}})$ on $\lambda_{\min}(\mathbf{W})$; see Lemma 14.1. Therefore, in the remainder of this chapter, we consider only two-block preconditioners.

#### 14.2.1.2 Two-block preconditioner:

In this subsection, we design a preconditioner for the $2 \times 2$-block representation of $\mathbf{A}_{\mathfrak{S}}$ defined in (14.18), i.e.,

$$\mathbf{A}_{\mathfrak{S}} = \begin{pmatrix} \underline{\mathbf{A}}_{\Delta} + \underline{\mathbf{W}} & \underline{\mathbf{K}}^{\top} \\ \underline{\mathbf{K}} & -\mathbf{V} \end{pmatrix}$$

where we recall that $\underline{\mathbf{A}}_{\Delta}$ and $\underline{\mathbf{W}}$ are symmetric positive semi-definite while $\mathbf{V}$ is symmetric positive definite. The matrix $\mathbf{A}_{\mathfrak{S}}$ is symmetrically preconditioned by

$$\breve{\mathbf{A}} := \begin{pmatrix} \breve{\mathbf{A}}_{\Delta} & \mathbf{0} \\ \mathbf{0} & \breve{\mathbf{V}} \end{pmatrix}$$

where $\breve{\mathbf{A}}_{\Delta}$ is spectrally equivalent to $\underline{\mathbf{A}}_{\Delta} + \underline{\mathbf{W}} + \beta\mathbf{M}$, $\beta > 0$, and $\breve{\mathbf{V}}$ is spectrally equivalent to $\mathbf{V}$. Here, $\mathbf{M}$ is an additional mass matrix which is added to make $\underline{\mathbf{A}}_{\Delta} + \underline{\mathbf{W}}$ positive definite. The preconditioned symmetric matrix then has the form

$$\widetilde{\mathbf{A}} := \breve{\mathbf{A}}^{-1/2}\mathbf{A}_{\mathfrak{S}}\breve{\mathbf{A}}^{-1/2} = \begin{pmatrix} \widetilde{\underline{\mathbf{A}}}_{\Delta} + {}_{\breve{\mathbf{A}}_{\Delta}}\underline{\mathbf{W}}_{\breve{\mathbf{A}}_{\Delta}} & {}_{\breve{\mathbf{A}}_{\Delta}}\underline{\mathbf{K}}^{\top}_{\breve{\mathbf{V}}} \\ {}_{\breve{\mathbf{V}}}\underline{\mathbf{K}}_{\breve{\mathbf{A}}_{\Delta}} & -\widetilde{\mathbf{V}} \end{pmatrix},$$

where we have used the notation convention in Definition 14.1. Similarly to Theorem 14.4 we have the following theorem.

**Theorem 14.5.** *Assume that there exist positive constants $\mu_0$, $\mu_1$, and $\mu_2$ satisfying*

$$\mu_0 \geq \frac{\mathbf{x}^{\top}(\underline{\mathbf{A}}_{\Delta} + \underline{\mathbf{W}})\mathbf{x}}{\mathbf{x}^{\top}(\underline{\mathbf{A}}_{\Delta} + \beta\mathbf{M})\mathbf{x}},$$
$$\mu_1 \leq \frac{\mathbf{x}^{\top}(\underline{\mathbf{A}}_{\Delta} + \underline{\mathbf{W}} + \underline{\mathbf{K}}^{\top}\mathbf{V}^{-1}\underline{\mathbf{K}})\mathbf{x}}{\mathbf{x}^{\top}(\underline{\mathbf{A}}_{\Delta} + \beta\mathbf{M})\mathbf{x}},$$
$$\mu_2 \geq \frac{\mathbf{x}^{\top}\underline{\mathbf{K}}^{\top}\mathbf{V}^{-1}\underline{\mathbf{K}}\mathbf{x}}{\mathbf{x}^{\top}(\underline{\mathbf{A}}_{\Delta} + \beta\mathbf{M})\mathbf{x}},$$

*for all nonzero vectors* $\mathbf{x}$. *Then*

$$\Lambda(\widetilde{\mathbf{A}}) \subset [\underline{\mu}(\widetilde{\mathbf{A}}), \mu_{-}(\widetilde{\mathbf{A}})] \cup [\mu_{+}(\widetilde{\mathbf{A}}), \overline{\mu}(\widetilde{\mathbf{A}})]$$

*where*

$$\underline{\mu}(\widetilde{\mathbf{A}}) = -\frac{1}{2}\left(\underline{\mu}(\widetilde{\mathbf{V}}) + \sqrt{\underline{\mu}(\widetilde{\mathbf{V}})^2 + 4\mu_2\overline{\mu}(\underline{\mathbf{A}})\overline{\mu}(\widetilde{\mathbf{V}})}\right),$$
$$\mu_{-}(\widetilde{\mathbf{A}}) = -\underline{\mu}(\widetilde{\mathbf{V}}),$$
$$\mu_{+}(\widetilde{\mathbf{A}}) = \frac{1}{2}\left(-\underline{\mu}(\widetilde{\mathbf{V}}) + \sqrt{\underline{\mu}(\widetilde{\mathbf{V}})^2 + 4\mu_1\underline{\mu}(\underline{\mathbf{A}})\underline{\mu}(\widetilde{\mathbf{V}})}\right),$$

$$\overline{\mu}(\widetilde{\mathbf{A}}) = \frac{1}{2}\left(-\underline{\mu}(\widetilde{\mathbf{V}}) + \mu_0\overline{\mu}(\underline{\underline{\mathbf{A}}}) + \sqrt{\left(\underline{\mu}(\widetilde{\mathbf{V}}) + \mu_0\overline{\mu}(\underline{\underline{\mathbf{A}}})\right)^2 + 4\mu_2\overline{\mu}(\underline{\underline{\mathbf{A}}})\overline{\mu}(\widetilde{\mathbf{V}})}\right),$$

*with* $\underline{\mu}(\widetilde{\mathbf{V}})$ *and* $\overline{\mu}(\widetilde{\mathbf{V}})$ *given by* (14.36). *Here* $\underline{\underline{\mathbf{A}}} = \breve{\mathbf{A}}_\Delta^{-1/2}(\underline{\mathbf{A}}_\Delta + \beta\mathbf{M})\breve{\mathbf{A}}_\Delta^{-1/2}$.

*Proof.* The theorem is a result of Lemma C.11 in Appendix C. □

In the following lemma, we show how the constants $\mu_0$, $\mu_1$, and $\mu_2$ behave.

**Lemma 14.4.** *The constants $\mu_0$, $\mu_1$, and $\mu_2$ in Theorem 14.5 can be chosen to be independent of all relevant parameters (namely, h and p).*

*Proof.* To determine $\mu_0$ we note that

$$\mathbf{x}^\top\underline{\mathbf{A}}_\Delta\mathbf{u} \le c\|\nabla u\|^2_{L^2(\Omega)} \quad \text{and} \quad \mathbf{x}^\top\mathbf{W}\mathbf{x} \simeq \|u\|^2_{H^{1/2}(\Gamma)/\mathbb{R}},$$

where $u$ is the piecewise-polynomial function the coefficient vector of which is $\mathbf{x}$. As a consequence, there exists a positive constant $c_1$ independent of all relevant parameters such that

$$\mathbf{x}^\top(\underline{\mathbf{A}}_\Delta + \mathbf{W})\mathbf{x} \le c_1\|u\|^2_{H^1(\Omega)}.$$

Furthermore, there exists a positive constant $c_2$ satisfying

$$\mathbf{x}^\top(\underline{\mathbf{A}}_\Delta + \beta\mathbf{M})\mathbf{x} \ge c_2\|u\|^2_{H^1(\Omega)}. \tag{14.48}$$

Therefore, $\mu_0 = c_1/c_2$.

On the other hand, under the assumption that the logarithmic capacity of the boundary $\Gamma$ is less than 1, the single-layer potential $\mathfrak{V}$ is positive definite on the space $H^{-1/2}(\Gamma)$; see [63, 65]. Moreover, $\mathfrak{V} : H^{-1/2}(\Gamma) \longrightarrow H^{1/2}(\Gamma)$ is bijective and $\mathscr{K} - \mathscr{I} : H^{1/2}(\Gamma) \longrightarrow H^{1/2}(\Gamma)$ is continuous, it follows that

$$\begin{aligned}\mathbf{x}^\top\underline{\mathbf{K}}^\top\mathbf{V}^{-1}\underline{\mathbf{K}}\mathbf{x} &\le \left\langle(\mathscr{K}^\top - \mathscr{I})\mathfrak{V}^{-1}(\mathscr{K} - \mathscr{I})u, u\right\rangle \\ &\le c\|u\|^2_{H^{1/2}(\Gamma)} \le c_3\|u\|^2_{H^1(\Omega)}.\end{aligned}$$

This together with (14.48) gives $\mu_2 = c_3/c_1$.

Finally, to determine $\mu_1$ we follow [63] to define an extension operator $\mathscr{E}$ with the following properties. Given $v \in H^{1/2}(\Gamma)$, then there exists $w := \mathscr{E}v$ satisfying $w = v$ on $\Gamma$ and

$$\|w\|_{H^1(\Omega)} \le c\|v\|_{H^{1/2}(\Gamma)}.$$

For all $u \in H^1(\Omega)$, by using the first Friedrich inequality and the above inequality we obtain

$$\begin{aligned}\|u\|^2_{H^1(\Omega)} &\le 2\|u - \mathscr{E}(u|_\Gamma)\|^2_{H^1(\Omega)} + 2\|\mathscr{E}(u|_\Gamma)\|^2_{H^1(\Omega)} \\ &\le c|u - \mathscr{E}(u|_\Gamma)|^2_{H^1(\Omega)} + 2\|\mathscr{E}(u|_\Gamma)\|^2_{H^1(\Omega)} \\ &\le c|u|^2_{H^1(\Omega)} + c\|\mathscr{E}(u|_\Gamma)\|^2_{H^1(\Omega)}\end{aligned}$$

$$\leq c\left(|u|^2_{H^1(\Omega)} + \|u\|^2_{H^{1/2}(\Gamma)}\right).$$

It is proved in [53] that the operator $\mathfrak{W} + (\mathscr{K}^\top - \mathscr{I})\mathfrak{V}^{-1}(\mathscr{K} - \mathscr{I}) : H^{1/2}(\Gamma) \longrightarrow H^{-1/2}(\Gamma)$ is elliptic. Hence the above inequality yields, for all finite element functions $u$,

$$\|u\|^2_{H^1(\Omega)} \leq c_4 \mathbf{x}^\top \left(\underline{\mathbf{A}}_\Delta + \mathbf{W} + \underline{\mathbf{K}}^\top \mathbf{V}^{-1} \underline{\mathbf{K}}\right)\mathbf{x}.$$

This together with (14.48) yields $\mu_1 = c_4/c_2$. □

*Remark 14.4.* Theorem 14.5 can be used for both two- and three-dimensional problems. For two-dimensional problems, different choices of $\breve{\mathbf{A}}_\Delta$ and $\breve{\mathbf{V}}$ have been studied with this two-block preconditioner.

(i) In [113] the authors study the *hp*-version of the coupled FEM-BEM with quasi-uniform meshes. They use preconditioners $\underline{\breve{\mathbf{A}}}_\Delta$ and $\breve{\mathbf{V}}$ which are spectrally equivalent to $\underline{\mathbf{A}}_\Delta + \underline{\mathbf{W}} + \beta\mathbf{M}$ and $\mathbf{V}$, respectively. Recalling that

$$\widetilde{\mathbf{V}} = \breve{\mathbf{V}}^{-1/2}\mathbf{V}\breve{\mathbf{V}}^{-1/2} \quad \text{and} \quad \underline{\underline{\mathbf{A}}} = \underline{\breve{\mathbf{A}}}_\Delta^{-1/2}\left(\underline{\mathbf{A}}_\Delta + \beta\mathbf{M}\right)\underline{\breve{\mathbf{A}}}_\Delta^{-1/2},$$

it follows that

$$\underline{\mu}(\widetilde{\mathbf{V}}) = c_V, \quad \overline{\mu}(\widetilde{\mathbf{V}}) = C_V, \quad \underline{\mu}(\underline{\underline{\mathbf{A}}}) = c_A, \quad \overline{\mu}(\underline{\underline{\mathbf{A}}}) = C_A,$$

where all the constants are positive and independent of $h$ and $p$. Theorem 14.5 and Lemma 14.4 yield

$$\Lambda(\widetilde{\mathbf{A}}) \subset [-a, -b] \cup [c, d]$$

where $a$, $b$, $c$, and $d$ are positive and independent of $h$ and $p$. This two-block preconditioner is clearly superior to the corresponding three-block preconditioner; see Remark 14.3.

(ii) The authors of [113] also use non-overlapping additive Schwarz methods for both preconditioners $\underline{\breve{\mathbf{A}}}_\Delta$ and $\breve{\mathbf{V}}$. As proved in Part II and Part III of the book,

$$\underline{\mu}(\widetilde{\mathbf{V}}) = c_V\left(1 + \log(Hp/h)\right)^{-2} \quad \text{and} \quad \overline{\mu}(\widetilde{\mathbf{V}}) = C_V,$$

where $c_V$ and $C_V$ are positive constants independent of $H$, $h$, and $p$. It is proved in [89, 90] that

$$\underline{\mu}(\underline{\underline{\mathbf{A}}}) = c_A\left(1 + \log(Hp/h)\right)^{-2} \quad \text{and} \quad \overline{\mu}(\underline{\underline{\mathbf{A}}}) = C_A,$$

where $c_A$ and $C_A$ are positive constants independent of $H$, $h$, and $p$. Theorem 14.5 and Lemma 14.4 yield

$$\Lambda(\widetilde{\mathbf{A}}) \subset [-a, -b\left(1 + \log(Hp/h)\right)^{-2}] \cup [c\left(1 + \log(Hp/h)\right)^{-2}, d],$$

where $a$, $b$, $c$, and $d$ are positive constants independent of the meshsize $h$ and polynomial degree $p$.

(iii) Theorem 14.5 can also be applied when multigrid methods are used instead of additive Schwarz preconditioners. Consider for simplicity the $h$-version only. In [81] the $V$-cycle multigrid method with Gauss–Seidel smoother for $\check{\underline{\mathbf{A}}}_\Delta$ and the $V$-cycle multigrid method for $\check{\mathbf{V}}$ are employed. It is proved in [38] that

$$\underline{\mu}(\underline{\mathbf{A}}) = c_A\left(1+\log h\right)^{-1} \quad \text{and} \quad \overline{\mu}(\underline{\mathbf{A}}) = C_A.$$

It is also proved in [172] that

$$\underline{\mu}(\widetilde{\mathbf{V}}) = c_V\left(1+\log h\right)^{-1} \quad \text{and} \quad \overline{\mu}(\widetilde{\mathbf{V}}) = C_V.$$

As a consequence of Theorem 14.5 and Lemma 14.4 we have

$$\Lambda(\widetilde{\mathbf{A}}) \subset \left[-a, -b\left(1+\log h\right)^{-1}\right] \cup \left[c\left(1+\log h\right)^{-1}, d\right].$$

### 14.2.2 Preconditioning with GMRES

In this subsection we use the GMRES method to solve the unsymmetric positive definite system (14.19), namely $\mathbf{A}_{\mathrm{u}}\mathbf{u} = \mathbf{b}$, which is rewritten here for convenience:

$$\begin{pmatrix} \underline{\mathbf{A}}_{\Delta\mathfrak{W}} & \underline{\mathbf{K}}^\top \\ -\underline{\mathbf{K}} & \mathbf{V} \end{pmatrix} \begin{pmatrix} \mathbf{u}_{\Omega,\Gamma} \\ \boldsymbol{\phi} \end{pmatrix} = \begin{pmatrix} \mathbf{b}_{\Omega,\Gamma} \\ -\widetilde{\mathbf{b}}_\Gamma \end{pmatrix}. \tag{14.49}$$

The analysis developed in Subsection 2.7.3 will be applied. Let $\mathbf{A}_{\mathrm{diag}}$ be the block-diagonal matrix of $\mathbf{A}_{\mathrm{u}}$ defined by (14.20), i.e.,

$$\mathbf{A}_{\mathrm{diag}} := \begin{pmatrix} \underline{\mathbf{A}}_{\Delta\mathfrak{W}} & \mathbf{0} \\ \mathbf{0} & \mathbf{V} \end{pmatrix}. \tag{14.50}$$

The preconditioner

$$\mathbf{C} := \begin{pmatrix} \mathbf{C}_{\Delta\mathfrak{W}} & \mathbf{0} \\ \mathbf{0} & \mathbf{C}_{\mathfrak{V}} \end{pmatrix} \tag{14.51}$$

is defined by applying various preconditioners $\mathbf{C}_{\Delta\mathfrak{W}}$ for $\underline{\mathbf{A}}_{\Delta\mathfrak{W}}$ and $\mathbf{C}_{\mathfrak{V}}$ for $\mathbf{V}$. As long as $\mathbf{C}_{\Delta\mathfrak{W}}$ and $\mathbf{C}_{\mathfrak{V}}$ satisfy (2.56), we can invoke Lemma 2.21 to obtain the convergence rate for the GMRES iteration.

The preconditioner $\mathbf{C}_{\mathfrak{V}}$ can be any preconditioner for the weakly-singular integral equation designed in Chapter 3, Chapter 4, Chapter 5, Chapter 6, Chapter 10, and Chapter 12. Let $\overline{\lambda}_{\mathfrak{V}}$ and $\underline{\lambda}_{\mathfrak{V}}$ be the upper bound and lower bound, respectively, of $\lambda_{\max}(P_{\mathfrak{V}})$ and $\lambda_{\min}(P_{\mathfrak{V}})$, respectively, where $P_{\mathfrak{V}}$ is the Schwarz operator associated with the weakly-singular integral operator $\mathfrak{V}$. This means

$$\underline{\lambda}_{\mathfrak{V}} a_{\mathfrak{V}}(\phi,\phi) \le a_{\mathfrak{V}}(P_{\mathfrak{V}}\phi,\phi) \le \overline{\lambda}_{\mathfrak{V}} a_{\mathfrak{V}}(\phi,\phi), \quad \phi \in \widetilde{\mathscr{S}_{\Gamma}}. \tag{14.52}$$

For example, if $P_{\mathfrak{V}}$ is the $h$-version two-level additive Schwarz operator, then it is proved in Subsection 3.1.1.2 that

$$\underline{\lambda}_{\mathfrak{V}} \simeq \frac{1}{\max\limits_{1\le i\le J}\left(1+H_i/h_i\right)} \quad \text{and} \quad \overline{\lambda}_{\mathfrak{V}} \simeq 1.$$

Recall that $\mathbf{P}_{\mathfrak{V}} = \mathbf{C}_{\mathfrak{V}}^{-1}\mathbf{V}$ with $\mathbf{P}_{\mathfrak{V}}$ being the matrix representation of $P_{\mathfrak{V}}$. Then (14.52) together with Lemma 2.8 and Lemma 2.9 implies

$$\underline{\lambda}_{\mathfrak{V}} \langle \mathbf{C}_{\mathfrak{V}}\mathbf{v},\mathbf{v}\rangle_{\mathbb{R}^N} \le \langle \mathbf{V}\mathbf{v},\mathbf{v}\rangle_{\mathbb{R}^N} \le \overline{\lambda}_{\mathfrak{V}} \langle \mathbf{C}_{\mathfrak{V}}\mathbf{v},\mathbf{v}\rangle_{\mathbb{R}^N} \quad \forall \mathbf{v} \in \mathbb{R}^N.$$

This is (2.56) for the $\mathbf{V}$ component in $\mathbf{A}_{\text{diag}}$.

It remains to show the same estimate for $\underline{\mathbf{A}}_{\Delta\mathfrak{W}}$ so that Lemma 2.21 can be invoked.

Assume that $P_{\Delta}$ is some Schwarz operator defined on some subspace decomposition of $\mathscr{S}_{\overline{\Omega}}$, namely $\mathscr{S}_{\overline{\Omega}} = \mathscr{S}_{\overline{\Omega},1} \oplus \cdots \oplus \mathscr{S}_{\overline{\Omega},J}$ for some positive integer $J$. Similarly, assume that $P_{\mathfrak{W}}$ is some Schwarz operator defined on some subspace decomposition of

$$\mathscr{S}_{\Gamma} := \operatorname{span}\left\{\eta_{j,\Gamma} : j \in \mathscr{I}_2\right\},$$

namely $\mathscr{S}_{\Gamma} = \mathscr{S}_{\Gamma,1} \oplus \cdots \oplus \mathscr{S}_{\Gamma,\widetilde{J}}$ for some positive integer $J_0$. Then

$$P_{\Delta\mathfrak{W}} := \begin{pmatrix} P_{\Delta} & 0 \\ 0 & P_{\mathfrak{W}} \end{pmatrix}$$

is the corresponding Schwarz operator defined on the subspace decomposition

$$\mathscr{S} = \bigoplus_{i=1}^{J}\bigoplus_{j=1}^{\widetilde{J}} \mathscr{S}_{\overline{\Omega},i} \times \mathscr{S}_{\Gamma,j}.$$

Their matrix representations satisfy

$$\mathbf{P}_{\Delta} = \mathbf{C}_{\Delta}^{-1}\underline{\mathbf{A}}_{\Delta}, \quad \mathbf{P}_{\mathfrak{W}} = \mathbf{C}_{\mathfrak{W}}^{-1}\mathbf{W}, \quad \mathbf{P}_{\Delta\mathfrak{W}} = \mathbf{C}_{\Delta\mathfrak{W}}^{-1}\underline{\mathbf{A}}_{\Delta\mathfrak{W}},$$

where

$$\mathbf{C}_{\Delta\mathfrak{W}} = \begin{pmatrix} \mathbf{C}_{\Delta} & \mathbf{0} \\ \mathbf{0} & \mathbf{C}_{\mathfrak{W}} \end{pmatrix}.$$

If the following estimates hold

$$\underline{\lambda}_{\Delta} a_{\Delta}(u,u) \le a_{\Delta}(P_{\Delta}u,u) \le \overline{\lambda}_{\Delta} a_{\Delta}(u,u), \quad u \in \mathscr{S}_{\overline{\Omega}}, \tag{14.53}$$

and

$$\underline{\lambda}_{\mathfrak{W}} a_{\mathfrak{W}}(u_{\Gamma},u_{\Gamma}) \le a_{\mathfrak{W}}(P_{\mathfrak{W}}u,u) \le \overline{\lambda}_{\mathfrak{W}} a_{\mathfrak{W}}(u_{\Gamma},u_{\Gamma}), \quad u_{\Gamma} \in \mathscr{S}_{\Gamma}, \tag{14.54}$$

then, with $\underline{\lambda}_{\Delta\mathfrak{W}} := \min\{\underline{\lambda}_{\Delta}, \underline{\lambda}_{\mathfrak{W}}\}$ and $\overline{\lambda}_{\Delta\mathfrak{W}} := \max\{\overline{\lambda}_{\Delta}, \overline{\lambda}_{\mathfrak{W}}\}$,

$$\underline{\lambda}_{\Delta\mathfrak{W}} a_{\Delta\mathfrak{W}}\big((u,v),(u,v)\big) \le a_{\Delta\mathfrak{W}}\big(P_{\Delta\mathfrak{W}}(u,v),(u,v)\big) \le \overline{\lambda}_{\Delta\mathfrak{W}} a_{\Delta\mathfrak{W}}\big((u,v),(u,v)\big)$$

for all $(u,v) \in \mathscr{S}$. Again, Lemma 2.8 and Lemma 2.9 imply

$$\underline{\lambda}_{\Delta\mathfrak{W}} \langle \mathbf{C}_{\Delta\mathfrak{W}} \mathbf{v}, \mathbf{v}\rangle_{\mathbb{R}^M} \le \langle \underline{\mathbf{A}}_{\Delta\mathfrak{W}} \mathbf{v}, \mathbf{v}\rangle_{\mathbb{R}^M} \le \overline{\lambda}_{\Delta\mathfrak{W}} \langle \mathbf{C}_{\Delta\mathfrak{W}} \mathbf{v}, \mathbf{v}\rangle_{\mathbb{R}^M} \quad \forall \mathbf{v} \in \mathbb{R}^M.$$

We need the following notations. For any $\mathbf{w} = (\mathrm{w}_1, \dots, \mathrm{w}_M) \in \mathbb{R}^M$, recalling the definition of $\mathscr{I}_1$ and $\mathscr{I}_2$ in (14.10), we define

$$\begin{aligned} 2\mathbf{w}_1 &:= (\mathrm{w}_1, \dots, , \mathrm{w}_{M_\Omega}) \in \mathbb{R}^{M_\Omega}, & \mathbf{w}_2 &:= (\mathrm{w}_{M_\Omega+1}, \dots, , \mathrm{w}_M) \in \mathbb{R}^{M_\Gamma}, \\ w &:= \sum_{i \in \mathscr{I}_1 \cup \mathscr{I}_2} \mathrm{w}_i \eta_i \in \mathscr{S}_{\overline{\Omega}} & w_\Gamma &:= w|_\Gamma = \sum_{i \in \mathscr{I}_2} \mathrm{w}_i \eta_{i,\Gamma} \in \mathscr{S}_\Gamma. \end{aligned} \tag{14.55}$$

We first prove the following lemma.

**Lemma 14.5.** *There exist positive constants $\widetilde{C}_1$ and $\widetilde{C}_2$ independent of $M$ and $N$ such that*

$$\Big\langle \underline{\mathbf{K}}^\top \mathbf{V}^{-1} \underline{\mathbf{K}} \mathbf{w}, \mathbf{w} \Big\rangle_{\mathbb{R}^M} \le \widetilde{C}_1 \langle \underline{\mathbf{A}}_{\Delta\mathfrak{W}} \mathbf{w}, \mathbf{w}\rangle_{\mathbb{R}^M} \quad \forall \mathbf{w} \in \mathbb{R}^M \tag{14.56}$$

*and*

$$\Big\langle \underline{\mathbf{K}}\, \underline{\mathbf{A}}_{\Delta\mathfrak{W}}^{-1} \underline{\mathbf{K}}^\top \mathbf{w}, \mathbf{w} \Big\rangle_{\mathbb{R}^N} \le \widetilde{C}_2 \langle \mathbf{V}\mathbf{w}, \mathbf{w}\rangle_{\mathbb{R}^N} \quad \forall \mathbf{w} \in \mathbb{R}^N. \tag{14.57}$$

*Proof.* We first prove (14.56). Let $\mathfrak{K}_{NM_\Gamma}$, $\mathfrak{K}'_{M_\Gamma N}$, $\mathfrak{I}_{NM_\Gamma}$, $\mathfrak{I}'_{M_\Gamma N}$, and $\mathfrak{V}_N^{-1}$, respectively, be discretised operators defined from $\mathfrak{K}$, $\mathfrak{K}'$, $\mathfrak{I}$, $\mathfrak{I}'$, and $\mathfrak{V}$; see Subsection B.2.2. Recall that $\mathbf{K}$, $\mathbf{K}^\top$, $\mathbf{M}$, $\mathbf{M}^\top$, and $\mathbf{V}^{-1}$, respectively, are the matrix representations of these operators. We define, see Subsection B.2.2,

$$\begin{aligned} \widetilde{\mathfrak{E}}_{M_\Gamma} &:= \big(\mathfrak{K}'_{M_\Gamma N} - \mathfrak{I}'_{M_\Gamma N}\big) \mathfrak{V}_N^{-1} \big(\mathfrak{K}_{NM_\Gamma} - \mathfrak{I}_{NM_\Gamma}\big) \in \mathscr{L}(\mathscr{S}_\Gamma, \mathscr{S}_\Gamma^*), \\ \mathfrak{E} &:= \big(\mathfrak{K}' - \mathfrak{I}'\big) \mathfrak{V}^{-1} \big(\mathfrak{K} - \mathfrak{I}\big) \in \mathscr{L}(H^{1/2}(\Gamma), H^{-1/2}(\Gamma)), \\ \mathfrak{E}_{M_\Gamma} &:= i^*_{\mathscr{S}_\Gamma} \mathfrak{E}\, i_{\mathscr{S}_\Gamma} \in \mathscr{L}(\mathscr{S}_\Gamma, \mathscr{S}_\Gamma^*). \end{aligned}$$

Note that in general $\mathfrak{E}_{M_\Gamma} \ne \widetilde{\mathfrak{E}}_{M_\Gamma}$. It follows from Lemma B.5 that the matrix representation of $\widetilde{\mathfrak{E}}_{M_\Gamma}$ is $\big(\mathbf{K}^\top - \mathbf{M}^\top\big)\mathbf{V}^{-1}\big(\mathbf{K} - \mathbf{M}\big)$. Hence, recalling (14.15) and (14.55), we deduce

$$\begin{aligned} \Big\langle \underline{\mathbf{K}}^\top \mathbf{V}^{-1} \underline{\mathbf{K}} \mathbf{w}, \mathbf{w} \Big\rangle_{\mathbb{R}^M} &= \Big\langle \big(\mathbf{K}^\top - \mathbf{M}^\top\big)\mathbf{V}^{-1}\big(\mathbf{K} - \mathbf{M}\big)\mathbf{w}_2, \mathbf{w}_2 \Big\rangle_{\mathbb{R}^{M_\Gamma}} \\ &= \Big\langle \widetilde{\mathfrak{E}}_{M_\Gamma} w_\Gamma, w_\Gamma \Big\rangle_{\mathscr{S}_\Gamma^*, \mathscr{S}_\Gamma} \\ &= \langle \mathfrak{E}_{M_\Gamma} w_\Gamma, w_\Gamma \rangle_{\mathscr{S}_\Gamma^*, \mathscr{S}_\Gamma} + \big\langle \big(\widetilde{\mathfrak{E}}_{M_\Gamma} - \mathfrak{E}_{M_\Gamma}\big) w_\Gamma, w_\Gamma \big\rangle_{\mathscr{S}_\Gamma^*, \mathscr{S}_\Gamma}. \end{aligned} \tag{14.58}$$

The first term on the right-hand side of (14.58) is easily estimated by

$$\langle \mathfrak{E}_{M_\Gamma} w_\Gamma, w_\Gamma \rangle_{\mathscr{S}_\Gamma^*, \mathscr{S}_\Gamma} = \langle \mathfrak{E} w_\Gamma, w_\Gamma \rangle_{H^{-1/2}(\Gamma), H^{1/2}(\Gamma)}$$

$$
\begin{aligned}
&= \left\langle \mathfrak{V}^{-1}\left(\mathfrak{K}-\mathfrak{I}\right)w_\Gamma, \left(\mathfrak{K}-\mathfrak{I}\right)w_\Gamma\right\rangle_{H^{-1/2}(\Gamma),H^{1/2}(\Gamma)} \\
&\leq c_{\mathfrak{V}^{-1}} \left\| \left(\mathfrak{K}-\mathfrak{I}\right)w_\Gamma\right\|^2_{H^{1/2}(\Gamma)} \\
&\lesssim c_{\mathfrak{V}^{-1}} c_{\mathfrak{K}} \|w_\Gamma\|^2_{H^{1/2}(\Gamma)} \leq c_{\mathfrak{V}}^{-1} c_{\mathfrak{K}} \|w\|^2_{H^1(\Omega)} \\
&\leq c_{\mathfrak{V}^{-1}} c_{\mathfrak{K}} c_{\Delta\mathfrak{W}}\, a_{\Delta\mathfrak{W}}(w,w) = \frac{\widetilde{C}_1}{2} \left\langle \underline{\mathbf{A}}_{\Delta\mathfrak{W}} \mathbf{w}, \mathbf{w}\right\rangle_{\mathbb{R}^M} \qquad (14.59)
\end{aligned}
$$

where $\widetilde{C}_1 := 2c_{\mathfrak{V}^{-1}} c_{\mathfrak{K}} c_{\Delta\mathfrak{W}}$. Here $c_{\mathfrak{V}^{-1}}$ and $c_{\mathfrak{K}}$ are the norm of $\mathfrak{V}^{-1}$ and $\mathfrak{K}-\mathfrak{I}$, respectively, and $c_{\Delta\mathfrak{W}}^{-1}$ is the coercivity constant of the bilinear form $a_{\Delta\mathfrak{W}}(\cdot,\cdot)$. For the second term on the right-hand side of (14.58), we need to invoke Lemma B.4 in Appendix B with

$$
\begin{gathered}
\mathfrak{A} = \mathfrak{V}, \quad \mathfrak{B} = \mathfrak{K}' - \mathfrak{I}', \quad \mathfrak{C} = \mathfrak{K} - \mathfrak{I}, \\
\mathscr{X} = H^{1/2}(\Gamma) = \mathscr{Y}^*, \quad \mathscr{Y} = H^{-1/2}(\Gamma) = \mathscr{X}^*, \\
\mathscr{X}_h = \mathscr{S}_\Gamma, \quad \mathscr{Y}_h = \widetilde{\mathscr{S}_\Gamma},
\end{gathered}
$$

to have

$$
\begin{aligned}
\left\langle \left(\widetilde{\mathfrak{E}}_{M_\Gamma} - \mathfrak{E}_{M_\Gamma}\right) w_\Gamma, w_\Gamma \right\rangle_{\mathscr{S}_\Gamma^*,\mathscr{S}_\Gamma} &\leq \left\| \left(\widetilde{\mathfrak{E}}_{M_\Gamma} - \mathfrak{E}_{M_\Gamma}\right) w_\Gamma \right\|_{\mathscr{S}_\Gamma^*} \|w_\Gamma\|_{\mathscr{S}_\Gamma} \\
&\leq \inf_{z\in\widetilde{\mathscr{S}_\Gamma}} \|\mathfrak{V}^{-1}(\mathfrak{K}-\mathfrak{I})w_\Gamma - z\|_{H^{-1/2}(\Gamma)} \|w_\Gamma\|_{\mathscr{S}_\Gamma} \\
&\leq \|\mathfrak{V}^{-1}(\mathfrak{K}-\mathfrak{I})w_\Gamma\|_{H^{-1/2}(\Gamma)} \|w_\Gamma\|_{\mathscr{S}_\Gamma} \\
&\leq c_{\mathfrak{V}}^{-1} \|(\mathfrak{K}-\mathfrak{I})w_\Gamma\|_{H^{1/2}(\Gamma)} \|w_\Gamma\|_{\mathscr{S}_\Gamma} \\
&\leq c_{\mathfrak{V}}^{-1} c_{\mathfrak{K}} \|w_\Gamma\|^2_{H^{1/2}(\Gamma)} \leq c_{\mathfrak{V}}^{-1} c_{\mathfrak{K}} c_{\Delta\mathfrak{W}} a_{\Delta\mathfrak{W}}(w,w) \\
&= \frac{\widetilde{C}_1}{2} \left\langle \underline{\mathbf{A}}_{\Delta\mathfrak{W}} \mathbf{w}, \mathbf{w}\right\rangle_{\mathbb{R}^M}. \qquad (14.60)
\end{aligned}
$$

Thus (14.56) follows from (14.58), (14.59), and (14.60).

We prove (14.57) in a similar manner. Recalling the definition of the linear bounded operator $\underline{\mathfrak{K}} \in \mathscr{L}(H^1(\Omega), H^{1/2}(\Gamma)$ by (14.22) and thus of its adjoint operator $\underline{\mathfrak{K}}' \in \mathscr{L}(H^{-1/2}(\Gamma), \widetilde{H}^{-1}(\Omega))$, we define

$$
\begin{aligned}
&\mathfrak{F} := \underline{\mathfrak{K}} \mathfrak{A}_{\Delta\mathfrak{W}}^{-1} \underline{\mathfrak{K}}' &&\in \mathscr{L}(H^{-1/2}(\Gamma), H^{1/2}(\Gamma)), \\
&\mathfrak{F}_N = i_{\mathscr{Y}_N}^* \mathfrak{F} i_{\mathscr{Y}_N} &&\in \mathscr{L}(\mathscr{Y}_N, \mathscr{Y}_N^*), \\
&\widetilde{\mathfrak{F}}_N := \underline{\mathfrak{K}}_{NM} \mathfrak{A}_M^{-1} \underline{\mathfrak{K}}'_{MN} &&\in \mathscr{L}(\mathscr{Y}_N, \mathscr{Y}_N^*),
\end{aligned}
$$

and note that the matrix representation of $\widetilde{\mathfrak{F}}_N$ is $\underline{\mathbf{K}}\mathbf{A}_{\Delta\mathfrak{W}}^{-1}\underline{\mathbf{K}}^\top$; see Lemma B.5. The proof follows along exactly the same lines as that of (14.56). □

The above results yield the following theorem.

**Theorem 14.6.** *Assume that $P_\Delta$, $P_{\mathfrak{W}}$, and $P_{\mathfrak{V}}$ are preconditioning operators defined with the corresponding bilinear form $a_\Delta(\cdot,\cdot)$, $a_{\mathfrak{W}}(\cdot,\cdot)$, and $a_{\mathfrak{V}}(\cdot,\cdot)$, respectively, such that* (14.52), (14.53), *and* (14.54) *hold. Then the preconditioned GMRES method applied to* (14.49) *with the preconditioner* $\mathbf{C}$ *defined by* (14.51) *has the following convergence rate*

$$\|\mathbf{r}_k\|_{\mathbf{C}} \le \left(1 - \frac{\widetilde{\alpha}^2}{\widetilde{\beta}^2}\right)^{k/2} \|\mathbf{r}_0\|_{\mathbf{C}}$$

*where $\mathbf{r}_0$ and $\mathbf{r}_k$ are, respectively, the initial and $k^{\text{th}}$ residuals of the iteration, and where*

$$\widetilde{\alpha} = \min\{\underline{\lambda}_\Delta, \underline{\lambda}_{\mathfrak{V}}, \underline{\lambda}_{\mathfrak{W}}\} \quad \text{and} \quad \widetilde{\beta} = \widetilde{C}_1^{1/2} \widetilde{C}_2^{1/2} \max\{\overline{\lambda}_\Delta, \overline{\lambda}_{\mathfrak{V}}, \overline{\lambda}_{\mathfrak{W}}\}$$

*with $\widetilde{C}_1$, $\widetilde{C}_2$ being the constants given in Lemma 14.5.*

*Proof.* The theorem is proved by invoking Lemma 2.21 with

$$\mathbf{A}_{11} = \underline{\mathbf{A}}_{\Delta\mathfrak{W}} \in \mathbb{R}^{M\times M}, \quad \mathbf{A}_{12} = \underline{\mathbf{K}}^\top = -\mathbf{A}_{21}^\top \in \mathbb{R}^{M\times N}, \quad \mathbf{A}_{22} = \mathbf{V} \in \mathbb{R}^{N\times N}.$$

The constants $\alpha$ and $\beta$ defined in (2.54) have the forms

$$\alpha = \min\{\underline{\lambda}_{\Delta\mathfrak{W}}, \underline{\lambda}_{\mathfrak{V}}\} = \min\{\underline{\lambda}_\Delta, \underline{\lambda}_{\mathfrak{V}}, \underline{\lambda}_{\mathfrak{W}}\}$$

and

$$\beta = C_1^{1/2} C_2^{1/2} \max\{\overline{\lambda}_{\Delta\mathfrak{W}}, \overline{\lambda}_{\mathfrak{V}}\} = C_1^{1/2} C_2^{1/2} \max\{\overline{\lambda}_\Delta, \overline{\lambda}_{\mathfrak{V}}, \overline{\lambda}_{\mathfrak{W}}\}$$

where

$$C_1 := \sup_{\substack{\mathbf{v}\in\mathbb{R}^M \\ \mathbf{v}\neq\mathbf{0}}} \frac{\langle \underline{\mathbf{K}}^\top \mathbf{V}^{-1} \underline{\mathbf{K}}\mathbf{v}, \mathbf{v}\rangle_{\mathbb{R}^M}}{\langle \underline{\mathbf{A}}_{\Delta\mathfrak{W}}\mathbf{v}, \mathbf{v}\rangle_{\mathbb{R}^M}} \quad \text{and} \quad C_2 := \sup_{\substack{\mathbf{v}\in\mathbb{R}^N \\ \mathbf{v}\neq\mathbf{0}}} \frac{\langle \underline{\mathbf{K}}\, \underline{\mathbf{A}}_{\Delta\mathfrak{W}}^{-1} \underline{\mathbf{K}}^\top \mathbf{v}, \mathbf{v}\rangle_{\mathbb{R}^N}}{\langle \mathbf{V}\mathbf{v}, \mathbf{v}\rangle_{\mathbb{R}^N}}.$$

It is clear that $\widetilde{\alpha}/\widetilde{\beta} \ge \alpha/\beta$, and thus the theorem follows from Lemma 2.21. □

*Remark 14.5.* Analogously to Remark 14.4, we collect here the different results obtained in various papers for various preconditioners $\mathbf{C}_{\Delta\mathfrak{W}}$ and $\mathbf{C}_{\mathfrak{V}}$.

(i) Consider the $p$-version for the solution of the interface problem (14.1). If the two-level additive Schwarz preconditioners for all three operators $\Delta$, $\mathfrak{W}$, and $\mathfrak{V}$ are used, then (14.52), (14.53), and (14.54) hold with

$$\overline{\lambda}_{\mathfrak{V}} \simeq 1, \qquad \underline{\lambda}_{\mathfrak{V}} \simeq \frac{1}{\log^2(p+1)},$$

$$\overline{\lambda}_{\mathfrak{W}} \simeq 1, \qquad \underline{\lambda}_{\mathfrak{W}} \simeq \frac{1}{\log^2 p},$$

$$\overline{\lambda}_\Delta \simeq 1, \qquad \underline{\lambda}_\Delta \simeq \frac{1}{\log^2 p},$$

see Lemma 4.3 and Lemma 4.4 for $\mathfrak{V}$, Lemma 4.1 and Lemma 4.2 for $\mathfrak{W}$, and [18, Lemmas 3.1–3.3] for $\Delta$. Moreover, the constants $\widetilde{C}_1$ and $\widetilde{C}_2$ given by Lemma 14.5 are independent of $p$. Therefore, Theorem 14.6 gives

$$\widetilde{\alpha} \simeq 1/(1+\log p)^2 \quad \text{and} \quad \widetilde{\beta} \simeq 1,$$

so that

$$\|\mathbf{r}_k\|_{\mathbf{C}} \le \left(1 - \frac{c}{(1+\log p)^4}\right)^{k/2} \|\mathbf{r}_0\|_{\mathbf{C}}.$$

This preconditioner is studied in [117].

(ii) Consider hierarchical basis block preconditioners for problem (14.1). The constants in (14.52), (14.53), and (14.54) are

$$\begin{aligned} \overline{\lambda}_{\mathfrak{V}} &\simeq 1, & \underline{\lambda}_{\mathfrak{V}} &\simeq (L+1)^{-2}, \\ \overline{\lambda}_{\mathfrak{W}} &\simeq 1, & \underline{\lambda}_{\mathfrak{W}} &\simeq (L+1)^{-2}, \\ \overline{\lambda}_{\Delta} &\simeq 1, & \underline{\lambda}_{\Delta} &\simeq (L+1)^{-2}, \end{aligned}$$

where $L$ is the number of levels. The results for $\Delta$ are proved in [243, Theorem 4.1] while those for $\mathfrak{W}$ and $\mathfrak{V}$ are proved in [231, Theorem 1 and Theorem 2]. Therefore, it follows from Theorem 14.6 that

$$\|\mathbf{r}_k\|_{\mathbf{C}} \le \left(1 - \frac{c}{(L+1)^4}\right)^{k/2} \|\mathbf{r}_0\|_{\mathbf{C}}.$$

This preconditioner is studied in [156].

(iii) In the same manner, other preconditioners for problem (14.1) can be derived as long as relevant preconditioners are studied for $\Delta$, $\mathfrak{W}$, and $\mathfrak{V}$.

*Remark 14.6.* As mentioned in the introduction of this chapter, our analysis holds not only for the particular problem considered in this chapter, but also for any problem whose FEM-BEM coupling matrix is of the form (14.18) or (14.20). Indeed, the same procedure can be applied to the time-harmonic eddy current problem, as studied in [145].

## 14.3 Preconditioning for the Non-Symmetric Coupling Methods

We recall that in the non-symmetric coupling methods, namely the Johnson–Nédélec coupling and the Bielak–MacCamy coupling, one has to solve the matrix system with the matrix being $\mathbf{A}_{\mathfrak{J}\mathfrak{N}}$ and $\mathbf{A}_{\mathfrak{B}\mathfrak{M}}$ defined by (14.27) and (14.31), respectively. Noting that

$$\mathbf{A}_{\mathfrak{B}\mathfrak{M}} = \mathbf{A}_{\mathfrak{J}\mathfrak{N}}^\top,$$

it suffices to consider $A_{\mathfrak{J}\mathfrak{N}}$. We define

$$A_{\mathrm{diag}} := \begin{pmatrix} \underline{\mathbf{A}}_\Delta & \mathbf{0} \\ \mathbf{0} & \mathbf{V} \end{pmatrix} \quad \text{and} \quad \mathbf{C} := \begin{pmatrix} \mathbf{C}_\Delta & \mathbf{0} \\ \mathbf{0} & \mathbf{C}_{\mathfrak{V}} \end{pmatrix} \tag{14.61}$$

where $\mathbf{C}_{\mathfrak{V}}$ and $\mathbf{C}_\Delta$ are designed such that (14.52) and (14.53) hold. As in Subsection 14.2.2, we will invoke Lemma 2.21 with

$$\mathbf{A} = \mathbf{A}_{\mathfrak{J}\mathfrak{N}}, \quad \mathbf{A}_{11} = \underline{\mathbf{A}}_\Delta, \quad \mathbf{A}_{12} = -2\underline{\mathbf{M}}^\top, \quad \mathbf{A}_{21} = -\underline{\mathbf{K}}, \quad \mathbf{A}_{22} = \mathbf{V},$$

and thus the task is to estimate the constants $c_{\mathbf{A}}$ and $C_{\beta_i}$, $i = 1,2,3$, defined in Lemma 2.21.

For any $\mathbf{v} = (v_1, \cdots, v_{M+N}) \in \mathbb{R}^{M+N}$ we define, recalling the definition of $\mathscr{I}_j$, $j = 1,2,3$, in (14.10),

$$\begin{aligned} \mathbf{v}_{11} &:= (\mathrm{v}_1, \ldots, \mathrm{v}_{M_\Omega}), & \mathbf{v}_{12} &:= (\mathrm{v}_{M_\Omega+1}, \ldots, \mathrm{v}_M), \\ \mathbf{v}_1 &:= (\mathbf{v}_{11}, \mathbf{v}_{12}), & \mathbf{v}_2 &:= (\mathrm{v}_{M+1}, \cdots, \mathrm{v}_{M+N}), \\ v_\Omega &:= \sum_{i\in\mathscr{I}_1} \mathrm{v}_i \eta_i, & v_\Gamma &:= \sum_{i\in\mathscr{I}_2} \mathrm{v}_i \eta_{i,\Gamma}, \\ \widetilde{v} &:= v_\Omega + v_\Gamma, & \overline{v} &:= \sum_{i\in\mathscr{I}_3} \mathrm{v}_{M+i} \phi_i, \end{aligned} \tag{14.62}$$

where we recall that $\eta_{i,\Gamma}$ is the restriction of $\eta_i$ on $\Gamma$. Then $\mathbf{v} = (\mathbf{v}_1, \mathbf{v}_2) \in \mathbb{R}^M \times \mathbb{R}^N$, $v_\Omega \in H_0^1(\Omega)$, $v_\Gamma \in H^{1/2}(\Gamma)$, $v \in H^1(\Omega)$, and $\psi_{\mathbf{v}} \in H^{-1/2}(\Gamma)$.

Since

$$\left\langle \underline{\mathbf{M}}^\top \mathbf{v}_2, \mathbf{v}_1 \right\rangle_{\mathbb{R}^M} = \begin{pmatrix} \mathbf{v}_{11}^\top & \mathbf{v}_{12}^\top \end{pmatrix} \begin{pmatrix} \mathbf{0}_{M_\Omega N} \\ \mathbf{M}^\top \end{pmatrix} \mathbf{v}_2 = \langle \mathbf{M}\mathbf{v}_{12}, \mathbf{v}_2 \rangle_{\mathbb{R}^N}$$

and

$$\begin{aligned} \langle \underline{\mathbf{K}}\mathbf{v}_1, \mathbf{v}_2 \rangle_{\mathbb{R}^N} &= \mathbf{v}_2^\top \begin{pmatrix} \mathbf{0}_{NM_\Omega} & \mathbf{K} - \mathbf{M} \end{pmatrix} \begin{pmatrix} \mathbf{v}_{11} \\ \mathbf{v}_{12} \end{pmatrix} = \mathbf{v}_2^\top (\mathbf{K} - \mathbf{M}) \mathbf{v}_{12} \\ &= \langle \mathbf{K}\mathbf{v}_{12}, \mathbf{v}_2 \rangle_{\mathbb{R}^N} - \langle \mathbf{M}\mathbf{v}_{12}, \mathbf{v}_2 \rangle_{\mathbb{R}^N}, \end{aligned}$$

we have

$$\begin{aligned} \langle \mathbf{A}_{12}\mathbf{v}_2, \mathbf{v}_1 \rangle_{\mathbb{R}^{n_1}} + \langle \mathbf{A}_{21}\mathbf{v}_1, \mathbf{v}_2 \rangle_{\mathbb{R}^{n_2}} &= -2\langle \underline{\mathbf{M}}^\top \mathbf{v}_2, \mathbf{v}_1 \rangle_{\mathbb{R}^M} - \langle \underline{\mathbf{K}}\mathbf{v}_1, \mathbf{v}_2 \rangle_{\mathbb{R}^N} \\ &= - \langle (\mathbf{M} + \mathbf{K})\mathbf{v}_{12}, \mathbf{v}_2 \rangle_{\mathbb{R}^N} \end{aligned}$$

Hence due to Lemma 2.21, it suffices to find constants $\widetilde{c}_{\mathbf{A}}$ and $\widetilde{C}_{\beta_i}$, $i = 1,2,3$, such that

$$\widetilde{c}_{\mathbf{A}} \left\langle \mathbf{A}_{\mathrm{diag}}\mathbf{v}, \mathbf{v} \right\rangle_{\mathbb{R}^{M+N}} \le \langle \mathbf{A}_{\mathfrak{J}\mathfrak{N}}\mathbf{v}, \mathbf{v} \rangle_{\mathbb{R}^{M+N}}, \qquad \forall \mathbf{v} \in \mathbb{R}^{M+N}, \tag{14.63}$$

$$\left\langle \underline{\mathbf{K}}^\top \mathbf{V}^{-1} \underline{\mathbf{K}}\mathbf{v}, \mathbf{v} \right\rangle_{\mathbb{R}^M} \le \widetilde{C}_{\beta_1} \langle \underline{\mathbf{A}}_\Delta \mathbf{v}, \mathbf{v} \rangle_{\mathbb{R}^M} \qquad \forall \mathbf{v} \in \mathbb{R}^M, \tag{14.64}$$

$$\left\langle \underline{\mathbf{M}}\,\underline{\mathbf{A}}_{\Delta}^{-1}\underline{\mathbf{M}}^{\top}\mathbf{v},\mathbf{v}\right\rangle_{\mathbb{R}^N} \le \widetilde{C}_{\beta_2}\left\langle \mathbf{V}\mathbf{v},\mathbf{v}\right\rangle_{\mathbb{R}^N} \qquad \forall \mathbf{v}\in\mathbb{R}^N, \tag{14.65}$$

$$\left|\left\langle (\mathbf{M}+\mathbf{K})\mathbf{v}_{12},\mathbf{v}_2\right\rangle_{\mathbb{R}^N}\right| \le \widetilde{C}_{\beta_3}\left\langle \mathbf{A}_{\mathrm{diag}}\mathbf{v},\mathbf{v}\right\rangle_{\mathbb{R}^{M+N}} \quad \forall \mathbf{v}\in\mathbb{R}^{M+N}. \tag{14.66}$$

We first deal with (14.66) because it will be useful to prove (14.63).

**Lemma 14.6.** *There exists a constant* $\widetilde{C}_{\beta_3}\in(0,1)$ *such that for any* $\mathbf{v}\in\mathbb{R}^{M+N}$ *we have*

$$\left|\left\langle (\mathbf{M}+\mathbf{K})\mathbf{v}_{12},\mathbf{v}_2\right\rangle_{\mathbb{R}^N}\right| \le \widetilde{C}_{\beta_3}\left\langle \mathbf{A}_{\mathrm{diag}}\mathbf{v},\mathbf{v}\right\rangle_{\mathbb{R}^{M+N}}.$$

*Here* $\mathbf{v}_{12}$ *and* $\mathbf{v}_2$ *are defined from* $\mathbf{v}$ *by* (14.62).

*Proof.* Recalling definition (14.62), we first note that

$$\left\langle (\mathbf{M}+\mathbf{K})\mathbf{v}_{12},\mathbf{v}_2\right\rangle_{\mathbb{R}^N} = \langle(\mathfrak{I}+\mathfrak{K})v_\Gamma,\overline{v}\rangle_\Gamma.$$

Let $\mathfrak{T}_{\mathrm{int}}: H^{1/2}(\Gamma)\to H^{-1/2}(\Gamma)$ be the interior Steklov–Poincaré operator defined by (B.11). Then it follows from Lemma B.2 that there exists a constant $c_{\mathfrak{K}}\in(1/2,1)$ such that

$$\|(\mathfrak{I}+\mathfrak{K})w\|_{\mathfrak{V}^{-1}}^2 \le 2c_{\mathfrak{K}}\langle\mathfrak{T}_{\mathrm{int}}w,w\rangle_\Gamma \quad \forall w\in H^{1/2}(\Gamma).$$

Here, the norm $\|\cdot\|_{\mathfrak{V}^{-1}}$ is defined by the inverse operator $\mathfrak{V}^{-1}$, i.e. $\|w\|_{\mathfrak{V}^{-1}} := \left\langle\mathfrak{V}^{-1}w,w\right\rangle_\Gamma^{1/2}$. It then follows that

$$\begin{aligned}
\left|\left\langle (\mathbf{M}+\mathbf{K})\mathbf{v}_{12},\mathbf{v}_2\right\rangle_{\mathbb{R}^N}\right| &\le \|(\mathfrak{I}+\mathfrak{K})v_\Gamma\|_{\mathfrak{V}^{-1}}\|\overline{v}\|_{\mathfrak{V}} \\
&\le \sqrt{2}\,c_{\mathfrak{K}}^{1/2}\,\langle\mathfrak{T}_{\mathrm{int}}v_\Gamma,v_\Gamma\rangle_\Gamma^{1/2}\,\|\overline{v}\|_{\mathfrak{V}} \\
&\le c_{\mathfrak{K}}\langle\mathfrak{T}_{\mathrm{int}}v_\Gamma,v_\Gamma\rangle_\Gamma + \frac{1}{2}\|\overline{v}\|_{\mathfrak{V}}^2 \\
&\le c_{\mathfrak{K}}\Big(\langle\mathfrak{T}_{\mathrm{int}}(\widetilde{v}|_\Gamma),\widetilde{v}|_\Gamma\rangle_\Gamma + \|\overline{v}\|_{\mathfrak{V}}^2\Big)
\end{aligned} \tag{14.67}$$

where $v_\Gamma$, $\overline{v}$ and $\widetilde{v}$ are defined in (14.62). Lemma B.1 in Appendix B states that there exists uniquely $v_D\in H^1(\Omega)$ such that

$$\begin{aligned}
-\Delta v_D &= 0 && \text{in } \Omega,\\
v_D &= \widetilde{v} && \text{on } \Gamma,\\
\frac{\partial v_D}{\partial n} &= \mathfrak{T}_{\mathrm{int}}(\widetilde{v}|_\Gamma) && \text{on } \Gamma,
\end{aligned}$$

and that $\|\nabla v_D\|_{L^2(\Omega)}^2 \le \|\nabla\widetilde{v}\|_{L^2(\Omega)}^2$. Therefore, by using Green's identity we can estimate the first term on the right-hand side of (14.67) by

$$\begin{aligned}
\langle\mathfrak{T}_{\mathrm{int}}(\widetilde{v}|_\Gamma),\widetilde{v}|_\Gamma\rangle_\Gamma &= \left\langle\frac{\partial v_D}{\partial n},v_D\right\rangle_\Gamma = \|\nabla v_D\|_{L^2(\Omega)}^2 \\
&\le \|\nabla\widetilde{v}\|_{L^2(\Omega)}^2 \le a_\Delta(\widetilde{v},\widetilde{v}) = \langle\underline{\mathbf{A}}_\Delta\mathbf{v}_1,\mathbf{v}_1\rangle_{\mathbb{R}^M}.
\end{aligned}$$

This and (14.67), together with the identity $\|\overline{v}\|_{\mathfrak{V}}^2 = \langle\mathbf{V}\mathbf{v}_2,\mathbf{v}_2\rangle_{\mathbb{R}^N}$, imply

$$\left|\left\langle (\mathbf{M}+\mathbf{K})\mathbf{v}_{12},\mathbf{v}_2\right\rangle_{\mathbb{R}^N}\right| \le c_{\mathfrak{K}}\Big( \langle \underline{\mathbf{A}}_\Delta \mathbf{v}_1,\mathbf{v}_1\rangle_{\mathbb{R}^M} + \langle \mathbf{V}\mathbf{v}_2,\mathbf{v}_2\rangle_{\mathbb{R}^N}\Big) = c_{\mathfrak{K}}\left\langle \mathbf{A}_{\mathrm{diag}}\mathbf{v},\mathbf{v}\right\rangle_{\mathbb{R}^{M+N}}.$$

This completes the proof with $\widetilde{C}_{\beta_3} := c_{\mathfrak{K}}$. □

We can now prove (14.63).

**Lemma 14.7.** *There exists a positive constant $\widetilde{c}_{\mathbf{A}}$ independent of $M$ and $N$ such that*

$$\langle \mathbf{A}_{\mathfrak{J}\mathfrak{N}}\mathbf{v},\mathbf{v}\rangle_{\mathbb{R}^{M+N}} \ge \widetilde{c}_{\mathbf{A}}\left\langle \mathbf{A}_{\mathrm{diag}}\mathbf{v},\mathbf{v}\right\rangle_{\mathbb{R}^{M+N}} \quad \forall \mathbf{v}\in\mathbb{R}^{M+N}.$$

*Proof.* Simple calculation, with definition (14.62), reveals that

$$\begin{aligned}\langle \mathbf{A}_{\mathfrak{J}\mathfrak{N}}\mathbf{v},\mathbf{v}\rangle_{\mathbb{R}^{M+N}} &= \langle \underline{\mathbf{A}}_\Delta \mathbf{v}_1,\mathbf{v}_1\rangle_{\mathbb{R}^M} - 2\left\langle \underline{\mathbf{M}}^\top \mathbf{v}_2,\mathbf{v}_1\right\rangle_{\mathbb{R}^M} - \langle \underline{\mathbf{K}}\mathbf{v}_1,\mathbf{v}_2\rangle_{\mathbb{R}^N} + \langle \mathbf{V}\mathbf{v}_2,\mathbf{v}_2\rangle_{\mathbb{R}^N}\\ &= \left\langle \mathbf{A}_{\mathrm{diag}}\mathbf{v},\mathbf{v}\right\rangle_{\mathbb{R}^{M+N}} - 2\left\langle \underline{\mathbf{M}}^\top \mathbf{v}_2,\mathbf{v}_1\right\rangle_{\mathbb{R}^M} - \langle \underline{\mathbf{K}}\mathbf{v}_1,\mathbf{v}_2\rangle_{\mathbb{R}^N}.\end{aligned}$$

Since

$$\left\langle \underline{\mathbf{M}}^\top \mathbf{v}_2,\mathbf{v}_1\right\rangle_{\mathbb{R}^M} = \begin{pmatrix}\mathbf{v}_{11}^\top & \mathbf{v}_{12}^\top\end{pmatrix}\begin{pmatrix}\mathbf{0}_{M_\Omega N}\\ \mathbf{M}^\top\end{pmatrix}\mathbf{v}_2 = \langle \mathbf{M}\mathbf{v}_{12},\mathbf{v}_2\rangle_{\mathbb{R}^N}$$

and

$$\begin{aligned}\langle \underline{\mathbf{K}}\mathbf{v}_1,\mathbf{v}_2\rangle_{\mathbb{R}^N} &= \mathbf{v}_2^\top\begin{pmatrix}\mathbf{0}_{NM_\Omega} & \mathbf{K}-\mathbf{M}\end{pmatrix}\begin{pmatrix}\mathbf{v}_{11}\\ \mathbf{v}_{12}\end{pmatrix} = \mathbf{v}_2^\top(\mathbf{K}-\mathbf{M})\mathbf{v}_{12}\\ &= \langle \mathbf{K}\mathbf{v}_{12},\mathbf{v}_2\rangle_{\mathbb{R}^N} - \langle \mathbf{M}\mathbf{v}_{12},\mathbf{v}_2\rangle_{\mathbb{R}^N},\end{aligned}$$

we have

$$\begin{aligned}\langle \mathbf{A}_{\mathfrak{J}\mathfrak{N}}\mathbf{v},\mathbf{v}\rangle_{\mathbb{R}^{M+N}} &= \left\langle \mathbf{A}_{\mathrm{diag}}\mathbf{v},\mathbf{v}\right\rangle_{\mathbb{R}^{M+N}} - \left\langle (\mathbf{M}+\mathbf{K})\mathbf{v}_{12},\mathbf{v}_2\right\rangle_{\mathbb{R}^N}\\ &\ge \left\langle \mathbf{A}_{\mathrm{diag}}\mathbf{v},\mathbf{v}\right\rangle_{\mathbb{R}^{M+N}} - \left|\left\langle (\mathbf{M}+\mathbf{K})\mathbf{v}_{12},\mathbf{v}_2\right\rangle_{\mathbb{R}^N}\right|.\end{aligned}$$

It follows from Lemma 14.6 that

$$\langle \mathbf{A}_{\mathfrak{J}\mathfrak{N}}\mathbf{v},\mathbf{v}\rangle_{\mathbb{R}^{M+N}} \ge (1-\widetilde{C}_{\beta_3})\left\langle \mathbf{A}_{\mathrm{diag}}\mathbf{v},\mathbf{v}\right\rangle_{\mathbb{R}^{M+N}},$$

completing the proof of the lemma with $\widetilde{c}_{\mathbf{A}} := 1-\widetilde{C}_{\beta_3} > 0$. □

Finally, the following lemma gives (14.64) and (14.65).

**Lemma 14.8.** *There exist positive constants $\widetilde{C}_1$ and $\widetilde{C}_2$ independent of $M$ and $N$ such that*

$$\left\langle \underline{\mathbf{K}}^\top \mathbf{V}^{-1}\underline{\mathbf{K}}\mathbf{w},\mathbf{w}\right\rangle_{\mathbb{R}^M} \le \widetilde{C}_1\,\langle \underline{\mathbf{A}}_\Delta \mathbf{w},\mathbf{w}\rangle_{\mathbb{R}^M} \quad \forall \mathbf{w}\in\mathbb{R}^M$$

*and*

$$\left\langle \underline{\mathbf{K}}\,\underline{\mathbf{A}}_\Delta^{-1}\underline{\mathbf{K}}^\top \mathbf{w},\mathbf{w}\right\rangle_{\mathbb{R}^N} \le \widetilde{C}_2\,\langle \mathbf{V}\mathbf{w},\mathbf{w}\rangle_{\mathbb{R}^N} \quad \forall \mathbf{w}\in\mathbb{R}^N.$$

*Proof.* Note that the stiffness matrix $\underline{\mathbf{A}}_\Delta$ differs from $\underline{\mathbf{A}}_{\Delta\mathfrak{W}}$ in that the former is defined with the bilinear from $a_\Delta(\cdot,\cdot)$ instead of $a_{\Delta\mathfrak{W}}(\cdot,\cdot)$. Thus, the proof follows exactly along the same lines of that of Lemma 14.5 and is omitted. □

We can now derive the convergence result for the Johnson–Nédélec and Bielak–MacCamy FEM-BEM coupling methods.

**Theorem 14.7.** *Assume that $P_\Delta$ and $P_{\mathfrak{V}}$ are preconditioning operators defined with the corresponding bilinear form $a_\Delta(\cdot,\cdot)$ and $a_{\mathfrak{V}}(\cdot,\cdot)$, respectively, such that* (14.52) *and* (14.53) *hold. Then the preconditioned GMRES method applied to* (14.26) *and to* (14.30) *with the preconditioner* $\mathbf{C}$ *defined by* (14.61) *has the following convergence rate*

$$\|\mathbf{r}_k\|_{\mathbf{C}} \leq \left(1 - \frac{\widetilde{\alpha}^2}{\widetilde{\beta}^2}\right)^{k/2} \|\mathbf{r}_0\|_{\mathbf{C}}$$

*where $\mathbf{r}_0$ and $\mathbf{r}_k$ are, respectively, the initial and $k^{\text{th}}$ residuals of the iteration, and where*

$$\widetilde{\alpha} = \widetilde{c}_{\mathbf{A}} d \quad \text{and} \quad \widetilde{\beta} = D^2\left(\max\{\widetilde{C}_{\beta_1}, \widetilde{C}_{\beta_2}\} + \widetilde{C}_{\beta_3} + 1\right)$$

*with $\widetilde{c}_{\mathbf{A}}$ and $\widetilde{C}_j$, $j = 1,2,3$, being the constants given in Lemma 14.6, Lemma 14.7, and Lemma 14.8.*

*Proof.* The theorem is a direct consequence of Lemma 2.21 in Chapter 2 and Lemma 14.6, Lemma 14.7, and Lemma 14.8. □

*Remark 14.7.* Analogously to Remark 14.4 and Remark 14.5, it is remarked here that Theorem 14.7 is a general result in that as soon as preconditioners are designed for $\Delta$ and $\mathfrak{V}$, then they can be applied for the non-symmetric coupling (Johnson–Nédélec or Bielak–MacCamy) and relevant convergence rates can be derived.

## 14.4 Numerical Results

In this section, we report on the numerical experiments carried out in [113], where HMCR is used as solver. The following tables are reprinted with copyright permission. Numerical results for the preconditioners with GMRES designed in Subsection 14.2.2 and preconditioners for non-symmetric coupling methods designed in Section 14.3 can be found in [76].

Problem (14.1) is solved with $\Omega$ being an L-shaped domain in $\mathbb{R}^2$ having vertices $(0,0)$, $(0,1/2)$, $(-1/2,1/2)$, $(-1/2,-1/2)$, $(1/2,-1/2)$, and $(1/2,0)$. The functions $f$, $u_0$, and $t_0$ are chosen such that

$$u_1(x,y) = \Im(z^{2/3}) \text{ for } z = x+iy \quad \text{and} \quad u_2(x,y) = \log|(x,y)+(0.3,0.3)|.$$

Detailed descriptions of mesh generation and boundary / finite element spaces are given in [113]. Tables 14.1–14.6 show bounds of the extreme eigenvalues of the unpreconditioned and preconditioned matrices with two-block and three-block preconditioners, preconditioner by additive Schwarz methods (see Remark 14.4 part (ii), and diagonal scaling.

**Table 14.1** The extreme eigenvalues of the unpreconditioned matrix $\mathbf{A}$

| $1/h$ | $p$ | $M+N$ | $\lambda_{\min}$ | $\alpha$ | $\lambda_{-}$ | $\alpha$ | $\lambda_{+}$ | $\alpha$ | $\lambda_{\max}$ |
|---|---|---|---|---|---|---|---|---|---|
| 2 | 2 | 37 | -0.9850 | | -0.1270E-1 | | 0.2472 | | 6.521 |
| 2 | 4 | 97 | -0.9857 | | -0.4967E-2 | -1.35 | 0.5687E-1 | -2.12 | 6.568 |
| 2 | 6 | 181 | -0.9857 | | -0.2624E-2 | -1.57 | 0.1834E-1 | -2.79 | 6.568 |
| 2 | 8 | 289 | -0.9857 | | -0.1615E-2 | -1.68 | 0.7487E-2 | -3.11 | 6.568 |
| 2 | 10 | 421 | -0.9857 | | -0.1092E-2 | -1.75 | 0.3592E-2 | -3.29 | 6.568 |
| 2 | 12 | 577 | -0.9857 | | -0.7867E-3 | -1.80 | 0.1932E-2 | -3.40 | 6.568 |
| 2 | 14 | 757 | -0.9857 | | -0.5933E-3 | -1.83 | 0.1130E-2 | -3.48 | 6.568 |
| 2 | 16 | 961 | -0.9857 | | -0.4646E-3 | -1.83 | 0.7045E-3 | -3.53 | 6.568 |
| 4 | 1 | 37 | -0.5643 | | -0.1531E-1 | | 0.3354 | | 6.724 |
| 8 | 1 | 97 | -0.2465 | 1.19 | -0.3996E-2 | 1.93 | 0.1229 | 1.44 | 7.526 |
| 16 | 1 | 289 | -0.1056 | 1.22 | -0.1021E-2 | 1.96 | 0.4092E-1 | 1.58 | 7.857 |
| 32 | 1 | 961 | -0.0455 | 1.21 | -0.2575E-3 | 1.98 | 0.1259E-1 | 1.70 | 7.961 |

**Table 14.2** The extreme eigenvalues of the three-block preconditioned matrix $\breve{\mathbf{A}}_3^{-1}\mathbf{A}$

| $1/h$ | $p$ | $M+N$ | $\lambda_{\min}$ | $\lambda_{-}$ | $\lambda_{+}$ | $\alpha$ | $\lambda_{\max}$ |
|---|---|---|---|---|---|---|---|
| 2 | 2 | 37 | -6.854 | -1.000 | 0.1909 | | 5.872 |
| 2 | 4 | 97 | -6.880 | -1.000 | 0.1040 | -0.87 | 5.899 |
| 2 | 6 | 181 | -6.883 | -1.000 | 0.6896E-1 | -1.01 | 5.902 |
| 2 | 8 | 289 | -6.884 | -1.000 | 0.5067E-1 | -1.07 | 5.903 |
| 2 | 10 | 421 | -6.884 | -1.000 | 0.3963E-1 | -1.10 | 5.904 |
| 2 | 12 | 577 | -6.884 | -1.000 | 0.3231E-1 | -1.12 | 5.904 |
| 2 | 14 | 757 | -6.884 | -1.000 | 0.2713E-1 | -1.13 | 5.904 |
| 2 | 16 | 961 | -6.885 | -0.999 | 0.2347E-1 | -1.08 | 5.904 |
| 4 | 1 | 37 | -1.655 | -1.000 | 0.4401 | | 5.350 |
| 8 | 1 | 97 | -1.433 | -1.000 | 0.2419 | 0.86 | 5.365 |
| 16 | 1 | 289 | -1.370 | -1.000 | 0.1308 | 0.88 | 5.396 |
| 32 | 1 | 961 | -1.356 | -1.000 | 0.6891E-1 | 0.92 | 5.410 |

**Table 14.3** The extreme eigenvalues of the two-block preconditioned matrix $\breve{\mathbf{A}}_2^{-1}\mathbf{A}$

| $1/h$ | $p$ | $M+N$ | $\lambda_{\min}$ | $\lambda_-$ | $\lambda_+$ | $\lambda_{\max}$ |
|---|---|---|---|---|---|---|
| 2 | 2 | 37 | -6.858 | -1.000 | 0.9853 | 5.877 |
| 2 | 4 | 97 | -6.884 | -1.000 | 0.9850 | 5.905 |
| 2 | 6 | 181 | -6.889 | -1.000 | 0.9849 | 5.910 |
| 2 | 8 | 289 | -6.891 | -1.000 | 0.9850 | 5.912 |
| 2 | 10 | 421 | -6.891 | -1.000 | 0.9850 | 5.913 |
| 2 | 12 | 577 | -6.892 | -1.000 | 0.9850 | 5.913 |
| 2 | 14 | 757 | -6.892 | -1.000 | 0.9850 | 5.914 |
| 2 | 16 | 961 | -6.893 | -0.999 | 0.9850 | 5.914 |
| 4 | 1 | 37 | -5.798 | -1.000 | 1.947 | 4.952 |
| 8 | 1 | 97 | -5.827 | -1.000 | 1.939 | 5.019 |
| 16 | 1 | 289 | -5.839 | -1.000 | 1.937 | 5.049 |
| 32 | 1 | 961 | -5.843 | -1.000 | 1.937 | 5.064 |

**Table 14.4** The extreme eigenvalues of the matrix $\breve{\mathbf{A}}_{ASM}^{-1}\mathbf{A}$ ($1/h = 2$, 3 elements)

| $p$ | $M+N$ | $\lambda_{\min}$ | $\lambda_-$ | $\alpha$ | $\lambda_+$ | $\alpha$ | $\lambda_{\max}$ |
|---|---|---|---|---|---|---|---|
| 2 | 37 | -6.815 | -0.4151 | | 0.2957 | | 5.836 |
| 4 | 97 | -6.818 | -0.2995 | -0.47 | 0.1969 | -0.58 | 5.839 |
| 6 | 181 | -6.819 | -0.2494 | -0.45 | 0.1519 | -0.64 | 5.839 |
| 8 | 289 | -6.820 | -0.2207 | -0.42 | 0.1290 | -0.56 | 5.840 |
| 10 | 421 | -6.820 | -0.2014 | -0.41 | 0.1151 | -0.51 | 5.840 |
| 12 | 577 | -6.821 | -0.1869 | -0.40 | 0.1057 | -0.46 | 5.840 |
| 14 | 757 | -6.821 | -0.1755 | -0.40 | 0.9879E-1 | -0.43 | 5.840 |
| 16 | 961 | -6.822 | -0.1663 | -0.40 | 0.9351E-1 | -0.41 | 5.841 |

**Table 14.5** The extreme eigenvalues of the matrix $\breve{\mathbf{A}}_{diag}^{-1}\mathbf{A}$ ($1/h = 2$, 3 elements)

| $p$ | $M+N$ | $\lambda_{\min}$ | $\lambda_-$ | $\alpha$ | $\lambda_+$ | $\alpha$ | $\lambda_{\max}$ |
|---|---|---|---|---|---|---|---|
| 2 | 37 | -6.815 | -0.4151 | | 0.2957 | | 5.836 |
| 4 | 97 | -6.817 | -0.2422 | -0.77 | 0.1781 | -0.73 | 5.838 |
| 6 | 181 | -6.817 | -0.1708 | -0.86 | 0.1271 | -0.83 | 5.838 |
| 8 | 289 | -6.817 | -0.1321 | -0.89 | 0.1002 | -0.82 | 5.838 |
| 10 | 421 | -6.817 | -0.1077 | -0.91 | 0.8354E-1 | -0.81 | 5.838 |
| 12 | 577 | -6.817 | -0.9089E-1 | -0.93 | 0.7214E-1 | -0.80 | 5.838 |
| 14 | 757 | -6.817 | -0.7857E-1 | -0.94 | 0.6377E-1 | -0.80 | 5.838 |
| 16 | 961 | -6.818 | -0.6935E-1 | -0.93 | 0.5732E-1 | -0.79 | 5.838 |

**Table 14.6** The numbers of minimum residual iterations which are required to reduce the initial residual by a factor of $10^{-3}$

| $1/h$ | $p$ | $M+N$ | $\mathbf{A}$ | $\breve{\mathbf{A}}_3^{-1}\mathbf{A}$ | $\breve{\mathbf{A}}_2^{-1}\mathbf{A}$ | $\breve{\mathbf{A}}_{ASM}^{-1}\mathbf{A}$ | $\breve{\mathbf{A}}_{diag}^{-1}\mathbf{A}$ |
|---|---|---|---|---|---|---|---|
| 2 | 2 | 37 | 36 | 21 | 11 | 30 | 30 |
| 2 | 4 | 97 | 138 | 34 | 12 | 47 | 52 |
| 2 | 6 | 181 | 285 | 40 | 13 | 54 | 67 |
| 2 | 8 | 289 | 536 | 45 | 13 | 62 | 77 |
| 2 | 10 | 421 | 892 | 50 | 13 | 66 | 94 |
| 2 | 12 | 577 | 1374 | 55 | 13 | 74 | 111 |
| 2 | 14 | 757 | 2328 | 60 | 13 | 82 | 135 |
| 2 | 16 | 961 | > 9999 | 79 | 17 | 102 | 197 |
| 4 | 1 | 37 | 17 | 15 | 10 | | |
| 8 | 1 | 97 | 38 | 23 | 13 | | |
| 16 | 1 | 289 | 67 | 32 | 13 | | |
| 32 | 1 | 961 | 130 | 45 | 14 | | |
| 64 | 1 | 3457 | 243 | 60 | 15 | | |
| 128 | 1 | 13057 | 476 | 75 | 15 | | |

# Part V
# Problems on the Sphere

# Chapter 15
# Pseudo-differential Equations with Spherical Splines

The aim of this chapter is to report on the design of additive Schwarz methods when spherical splines are used to solve the hypersingular integral equation on the sphere. On this geometry, this equation belongs to a wider class of the so-called *pseudo-differential equations* on the sphere; see Appendix B.1. These equations have long been used [126, 131] as a modern and powerful tool to tackle linear boundary-value problems. Svensson [214] introduces this approach to geodesists who study [77, 84] these problems on the sphere which is taken as a model of the earth. Efficient solutions to pseudodifferential equations on the sphere become more demanding when given data are collected by satellites.

For completeness, we will first introduce the general form of pseudodifferential equations, show how they can be solved approximately by spherical splines, and present additive Schwarz preconditioners for the hypersingular integral equation. Detailed results can be found in the articles [174, 175, 179]. More details on spherical splines can be found in the thesis [177].

Results for the Laplace–Beltrami equation can also be derived, but it is not in the scope of this book. The reader who is interested in this equation is referred to the article [175].

## 15.1 Pseudo-differential Operators and Sobolev Spaces

### *15.1.1 Sobolev spaces on the sphere*

Throughout this chapter and the next chapter, we denote by $\mathbb{S}$ the unit sphere in $\mathbb{R}^3$, i.e., $\mathbb{S} := \{\mathbf{x} \in \mathbb{R}^3 : |\mathbf{x}| = 1\}$ where $|\cdot|$ is the Euclidean norm in $\mathbb{R}^3$.

A spherical harmonic of order $\ell$ on $\mathbb{S}$ is the restriction to $\mathbb{S}$ of a homogeneous harmonic polynomial of degree $\ell$ in $\mathbb{R}^3$. The space of all spherical harmonics of order $\ell$ is the eigenspace of the Laplace–Beltrami operator $\Delta_{\mathbb{S}}$ (see Example 15.1)



E. P. Stephan and T. Tran, *Schwarz Methods and Multilevel Preconditioners for Boundary Element Methods*, https://doi.org/10.1007/978-3-030-79283-1_15

corresponding to the eigenvalue $\lambda_\ell = -\ell(\ell+1)$. The dimension of this space being $2\ell+1$, see e.g. [155], one may choose for it an orthonormal basis $\{Y_{\ell,m}\}_{m=-\ell}^{\ell}$.

The collection of all the spherical harmonics $Y_{\ell,m}$, $m=-\ell,\dots,\ell$ and $\ell=0,1,\dots,$ forms an orthonormal basis for $L_2(\mathbb{S})$. The basis $\{Y_{\ell,m}\}_{m=-\ell}^{\ell}$ allows us to define Sobolev spaces on spheres using Fourier coefficients of functions. Spaces defined this way are equivalent to those defined in Appendix A; see e.g., [138].

For $s \in \mathbb{R}$, the Sobolev space $H^s$ is defined by

$$H^s := \left\{ v \in \mathscr{D}'(\mathbb{S}) : \sum_{\ell=0}^{\infty} \sum_{m=-\ell}^{\ell} (\ell+1)^{2s} |\widehat{v}_{\ell,m}|^2 < \infty \right\}$$

where $\mathscr{D}'(\mathbb{S})$ is the space of distributions on $\mathbb{S}$ and $\widehat{v}_{\ell,m}$ are the Fourier coefficients of $v$ defined by

$$\widehat{v}_{\ell,m} = \int_{\mathbb{S}} v(\mathbf{x}) Y_{\ell,m}(\mathbf{x})\, d\sigma_{\mathbf{x}}, \quad m=-\ell,\dots,\ell, \quad \ell=0,1,\dots.$$

Here $d\sigma_{\mathbf{x}}$ is the element of surface area. The space $H^s$ is equipped with the following norm and inner product:

$$\begin{aligned} \|v\|_s &:= \left( \sum_{\ell=0}^{\infty} \sum_{m=-\ell}^{\ell} (\ell+1)^{2s} |\widehat{v}_{\ell,m}|^2 \right)^{1/2} \\ \langle v,w \rangle_s &:= \sum_{\ell=0}^{\infty} \sum_{m=-\ell}^{\ell} (\ell+1)^{2s} \widehat{v}_{\ell,m} \widehat{w}_{\ell,m}. \end{aligned} \tag{15.1}$$

When $s=0$ we write $\langle \cdot,\cdot \rangle$ instead of $\langle \cdot,\cdot \rangle_0$; this is in fact the $L_2$-inner product. We note that

$$|\langle v,w \rangle_s| \le \|v\|_s \|w\|_s \quad \forall v,w \in H^s,\ \forall s \in \mathbb{R}, \tag{15.2}$$

and

$$\|v\|_{s_1} = \sup_{\substack{w \in H^{s_2} \\ w \neq 0}} \frac{\langle v,w \rangle_{\frac{s_1+s_2}{2}}}{\|w\|_{s_2}} \quad \forall v \in H^{s_1},\ \forall s_1,s_2 \in \mathbb{R}. \tag{15.3}$$

When $s=1$, we can define an equivalent norm

$$\|v\|_{H^1(\mathbb{S})}^2 := \int_{\mathbb{S}} |\nabla^* v|^2 d\sigma + \int_{\mathbb{S}} |v|^2 d\sigma,$$

where

$$|v|_{H^1(\mathbb{S})}^2 := \int_{\mathbb{S}} |\nabla^* v|^2 d\sigma,$$

with $\nabla^*$, called the surface gradient, being the restriction of the gradient operator on $\mathbb{S}$, and $d\sigma$ being the Lebesgue measure on $\mathbb{S}$.

In the case $s = k \in \mathbb{Z}^+$, the set of non-negative integers, the Sobolev space $H^k$ can also be defined by using atlas for the unit sphere $\mathbb{S}$; see [159]. This construction uses a concept of homogeneous extensions.

**Definition 15.1.** *Given any spherical function $v$, its homogeneous extension of degree $n \in \mathbb{Z}^+$ to $\mathbb{R}^3 \backslash \{0\}$ is a function $v_n$ defined by*

$$v_n(\mathbf{x}) := |\mathbf{x}|^n v\left(\frac{\mathbf{x}}{|\mathbf{x}|}\right).$$

Let $\{(\Gamma_j, \phi_j)\}_{j=1}^J$ be an atlas for $\mathbb{S}$, i.e, a finite collection of charts $(\Gamma_j, \phi_j)$, where $\Gamma_j$ are open subsets of $\mathbb{S}$, covering $\mathbb{S}$, and where $\phi_j : \Gamma_j \to B_j$ are infinitely differentiable mappings whose inverses $\phi_j^{-1}$ are also infinitely differentiable. Here $B_j$, $j = 1, \ldots, J$, are open subsets in $\mathbb{R}^2$. Also, let $\{\alpha_j\}_{j=1}^J$ be a partition of unity subordinate to the atlas $\{(\Gamma_j, \phi_j)\}_{j=1}^J$, i.e., a set of infinitely differentiable functions $\alpha_j$ on $\mathbb{S}$ vanishing outside the sets $\Gamma_j$, such that $\sum_{j=1}^J \alpha_j = 1$ on $\mathbb{S}$. For any $k \in \mathbb{Z}^+$, the Sobolev space $H^k$ on the unit sphere is defined by

$$H^k := \{v : (\alpha_j v) \circ \phi_j^{-1} \in H^k(B_j),\ j = 1, \ldots, J\},$$

which is equipped with a norm defined by

$$\|v\|_k^* := \sum_{j=1}^J \|(\alpha_j v) \circ \phi_j^{-1}\|_{H^k(B_j)}. \tag{15.4}$$

This norm is equivalent to the norm defined in (15.1); see Ref. [138].

Let $v \in H^k$, $k \geq 1$. For each $\ell = 1, \ldots, k$, let $v_{\ell-1}$ denote the unique homogeneous extension of $v$ of degree $\ell - 1$. Then

$$|v|_\ell := \sum_{|\alpha|=\ell} \|D^\alpha v_{\ell-1}\|_{L_2(\mathbb{S})} \tag{15.5}$$

is a Sobolev-type seminorm of $v$ in $H^k$. Here $\|D^\alpha v_{\ell-1}\|_{L_2(\mathbb{S})}$ is understood as the $L_2$-norm of the restriction of the trivariate function $D^\alpha v_{\ell-1}$ to $\mathbb{S}$. When $\ell = 0$ we define

$$|v|_0 := \|v\|_{L_2(\mathbb{S})},$$

which can now be used together with (15.5) to define another norm in $H^k$:

$$\|v\|_k' := \sum_{\ell=0}^k |v|_\ell.$$

This norm turns out to be equivalent to the Sobolev norm defined in (15.4); see [159].

### 15.1.2 Pseudo-differential operators on the sphere

Let $\{\widehat{L}(\ell)\}_{\ell\geq 0}$ be a sequence of real numbers. A pseudo-differential operator $L$ is a linear operator that assigns to any distribution $v \in \mathscr{D}'(\mathbb{S})$ a distribution

$$\mathscr{L}v := \sum_{\ell=0}^{\infty} \sum_{m=-\ell}^{\ell} \widehat{L}(\ell)\widehat{v}_{\ell,m}Y_{\ell,m}. \tag{15.6}$$

The sequence $\{\widehat{L}(\ell)\}_{\ell\geq 0}$ is referred to as the *spherical symbol* of $L$. Let $\mathscr{K}(L) := \{\ell : \widehat{L}(\ell) = 0\}$. Then

$$\ker L = \operatorname{span}\{Y_{\ell,m} : \ell \in \mathscr{K}(L), m = -\ell, \ldots, \ell\}.$$

Denoting $M := \dim\ker L$, we assume that $0 \leq M < \infty$. We also assume that $L$ is a *strongly elliptic* pseudo-differential operator of order $2\alpha$, i.e.,

$$C_1(\ell+1)^{2\alpha} \leq \widehat{L}(\ell) \leq C_2(\ell+1)^{2\alpha} \quad \text{for all } \ell \notin \mathscr{K}(L), \tag{15.7}$$

for some positive constants $C_1$ and $C_2$.

More general pseudo-differential operators can be defined via Fourier transforms by using local charts; see e.g. Ref. [128, 173]. It can be easily seen by using (15.1), (15.2), and (15.3) that if $L$ is a pseudo-differential operator of order $2\alpha$ then $L$ maps $H^{s+\alpha}$ into $H^{s-\alpha}$ and is bounded for all $s \in \mathbb{R}$.

*Example 15.1.* The following pseudo-differential operators are strongly elliptic; see for example [214]. For the derivation of the symbols, see Appendix B.1.

(i) The Laplace–Beltrami operator is an operator of order 2 and has as symbol $\widehat{L}(\ell) = \ell(\ell+1)$. This operator is the restriction of the Laplacian on the sphere.
(ii) The hypersingular integral operator (without the minus sign) is an operator of order 1 and has as symbol $\widehat{L}(\ell) = \ell(\ell+1)/(2\ell+1)$. This operator arises from the boundary-integral reformulation of the Neumann problem with the Laplacian in the interior or exterior of the sphere.
(iii) The weakly-singular integral operator is an operator of order $-1$ and has as symbol $\widehat{L}(\ell) = 1/(2\ell+1)$. This operator arises from the boundary-integral reformulation of the Dirichlet problem with the Laplacian in the interior or exterior of the sphere.

The problem we consider in this chapter is posed as follows.

**Problem A:** *Let $L$ be a pseudo-differential operator of order $2\alpha$. Given*

$$g \in H^{-\alpha} \quad \textit{satisfying} \quad \widehat{g}_{\ell,m} = 0 \quad \textit{for all } \ell \in \mathscr{K}(L),\ m = -\ell, \ldots, \ell,$$

*find $u \in H^{\alpha}$ satisfying*

$$\begin{aligned} Lu &= g, \\ \langle \mu_i, u \rangle &= \gamma_i, \quad i = 1, \ldots, M, \end{aligned} \tag{15.8}$$

*where $\gamma_i \in \mathbb{R}$ and $\mu_i \in H^{-\alpha}$ are given. Here $\langle \cdot, \cdot \rangle$ denotes the duality inner product between $H^{-\alpha}$ and $H^{\alpha}$, which is the usual extension of the $H^0$-inner product.*

Assume that the functionals $\mu_1, \ldots, \mu_N$ are *unisolvent* with respect to $\ker L$, i.e.,

$$\text{for any } v \in \ker L, \text{ if } \langle \mu_i, v \rangle = 0 \text{ for all } i = 1, \ldots, M, \text{ then } v = 0.$$

Then Problem A has a unique solution; see [154]. Noting that

$$H^{\alpha} = \ker L \oplus (\ker L)^{\perp}_{H^{\alpha}},$$

the solution $u \in H^{\alpha}$ is sought in the form

$$u = u_0 + u_1 \quad \text{where} \quad u_0 \in \ker L \quad \text{and} \quad u_1 \in (\ker L)^{\perp}_{H^{\alpha}}, \tag{15.9}$$

where $u_0 \in \ker L$ satisfies

$$\langle \mu_i, u_0 \rangle = \gamma_i - \langle \mu_i, u_1 \rangle, \quad i = 1, \ldots, M,$$

and $u_1 \in (\ker L)^{\perp}_{H^{\alpha}}$ satisfies

$$\langle L u_1, v \rangle = \langle g, v \rangle \quad \text{for all } v \in H^{\alpha}.$$

## 15.2 Solving Pseudo-differential Equations by Spherical Splines

In this section, we will first recall the definition of spherical splines [6, 7, 8]. We will then present error estimates for the approximation of the solution of Problem A by spherical splines.

### *15.2.1 Spherical splines*

#### 15.2.1.1 Spherical triangulation

Let $\{\mathbf{x}_1, \mathbf{x}_2, \mathbf{x}_3\}$ be linearly independent vectors in $\mathbb{R}^3$. The *trihedron* $T$ generated by $\{\mathbf{x}_1, \mathbf{x}_2, \mathbf{x}_3\}$ is defined by

$$T := \{\mathbf{x} \in \mathbb{R}^3 : \mathbf{x} = b_1\mathbf{x}_1 + b_2\mathbf{x}_2 + b_3\mathbf{x}_3 \text{ with } b_i \geq 0,\ i = 1, 2, 3\}.$$

The intersection $\tau := T \cap \mathbb{S}$ is called a *spherical triangle*. Let $\Delta = \{\tau_i : i = 1, \ldots, \mathscr{T}\}$ be a set of spherical triangles. Then $\Delta$ is called a *spherical triangulation* of the sphere $\mathbb{S}$ if there hold

(i) $\bigcup_{i=1}^{\mathscr{T}} \tau_i = \mathbb{S}$;

(ii) Each pair of distinct triangles in $\Delta$ are either disjoint or share a common vertex or an edge.

A *spherical cap* centred at $\mathbf{x} \in \mathbb{S}$ and having radius $R$ is defined by

$$C(\mathbf{x},R) := \{\mathbf{y} \in \mathbb{S} : \cos^{-1}(\mathbf{x}\cdot\mathbf{y}) \le R\}.$$

For any spherical triangle $\tau$, let $|\tau|$ denote the diameter of the smallest spherical cap containing $\tau$, and $\rho_\tau$ denote the diameter of the largest spherical cap contained in $\tau$. We define

$$|\Delta| := \max\{|\tau| : \tau \in \Delta\}, \quad h := \tan(|\Delta|/2), \quad \rho_\Delta := \min\{\rho_\tau : \tau \in \Delta\}, \tag{15.10}$$

and refer to $|\Delta|$ as the mesh size. Throughout the chapter, we assume that $|\Delta| \le 1$ and that all triangulations are *shape-regular* and *quasi-uniform*, i.e., there exist $\beta > 1$ and $\gamma < 1$ such that

$$|\tau| \le \beta\rho_\tau \quad \text{and} \quad |\tau| \ge \gamma|\Delta| \quad \forall \tau \in \Delta.$$

Roughly speaking, in a shape-regular triangulation the smallest angle is not too small so that there are no long and thin triangles, while in a quasi-uniform triangulation the sizes of the triangles are not too much different.

#### 15.2.1.2 Spherical splines

Let $\mathscr{H}^d$ denote the space of trivariate homogeneous polynomials of degree $d$ restricted to $\mathbb{S}$. We define $S_d^r(\Delta)$ to be the space of homogeneous splines of degree $d$ and smoothness $r$ on a spherical triangulation $\Delta$, that is,

$$S_d^r(\Delta) := \{s \in C^r(\mathbb{S}) : s|_\tau \in \mathscr{H}^d, \tau \in \Delta\}.$$

In the whole chapter, for technical reasons we assume we assume that the integers $d$ and $r$ defining $S_d^r(\Delta)$ satisfy

$$\begin{cases} d \ge 3r+2 & \text{if } r \ge 1, \\ d \ge 1 & \text{if } r = 0; \end{cases} \tag{15.11}$$

see [6, 7, 8, 179].

Let $\tau$ be a spherical triangle with vertices $\mathbf{x}_1, \mathbf{x}_2$, and $\mathbf{x}_3$. Since $\mathbf{x}_1, \mathbf{x}_2$, and $\mathbf{x}_3$ are linearly independent, for any $\mathbf{x} \in \mathbb{S}$, there exist unique real numbers $b_{1,\tau}(\mathbf{x}), b_{2,\tau}(\mathbf{x})$, and $b_{3,\tau}(\mathbf{x})$ (called the *spherical barycentric coordinates* of $\mathbf{x}$ associated with $\tau$) which satisfy

$$\mathbf{x} = b_{1,\tau}(\mathbf{x})\mathbf{x}_1 + b_{2,\tau}(\mathbf{x})\mathbf{x}_2 + b_{3,\tau}(\mathbf{x})\mathbf{x}_3.$$

We define the homogeneous Bernstein basis polynomials of degree $d$ on $\tau$ to be the polynomials

$$B_{ijk}^{d,\tau}(\mathbf{x}) := \frac{d!}{i!j!k!} b_{1,\tau}(\mathbf{x})^i b_{2,\tau}(\mathbf{x})^j b_{3,\tau}(\mathbf{x})^k, \quad i+j+k=d.$$

As was shown in [6], we can use these polynomials as basis functions for $\mathscr{H}^d$.

The space $S_d^r(\Delta)$ has the following approximation property.

**Theorem 15.1.** *Assume that $\Delta$ is a $\beta$-quasiuniform spherical triangulation with $|\Delta| \le 1$, and that (15.11) holds. Then for any $v \in H^s$, there exists $\eta \in S_d^r(\Delta)$ satisfying*

$$\|v-\eta\|_t \le Ch^{s-t}\|v\|_s,$$

*where $t \le r+1$ and $t \le s \le d+1$. Here $C$ is a positive constant depending only on $d$ and the smallest angle in $\Delta$.*

*Proof.* See [179, Theorem 4.1]. □

We will need the following lemmas.

**Lemma 15.1.** *Let $p$ be a homogeneous polynomial of degree $d$ on a spherical triangle $\tau$, which is written in Bernstein–Bézier form as*

$$p(\mathbf{v}) = \sum_{i+j+k=d} c_{ijk} B_{ijk}^{d,\tau}(\mathbf{v}), \quad \mathbf{v} \in \tau.$$

*Then*

$$\|p\|_{L_2(\tau)} \simeq A_\tau^{1/2}\|\mathbf{c}_\tau\|,$$

*Here $A_\tau$ is the area of $\tau$ and $\mathbf{c}_\tau$ is the vector of components $c_{ijk}$, $i+j+k=d$.*

*Proof.* See [21, Lemma 5]. □

**Lemma 15.2.** *Let $p$ be a homogeneous polynomial of degree $d$ on a spherical triangle $\tau$ in $\Delta$. Then the following estimate holds*

$$|p|_{H^1(\tau)} \preceq \rho_\tau^{-1}\|p\|_{L_2(\tau)}.$$

*Proof.* See [21, Lemma 6]. □

#### 15.2.1.3 Quasi-interpolation

In this subsection we briefly discuss the construction of a quasi-interpolation operator $I_\Delta : L_2(\mathbb{S}) \to S_d^r(\Delta)$ which is defined in [159]. This operator will be used frequently in the remainder of this chapter. First we introduce the set of *domain points* of $\Delta$ to be

$$\mathscr{D} := \bigcup_{\tau=\langle \mathbf{x}_1,\mathbf{x}_2,\mathbf{x}_3\rangle \in \Delta} \left\{ \xi_{ijk}^\tau = \frac{i\mathbf{x}_1 + j\mathbf{x}_2 + k\mathbf{x}_3}{d} : i,j,k \in \mathbb{N} \text{ and } i+j+k=d \right\}.$$

Here, $\tau = \langle \mathbf{x}_1, \mathbf{x}_2, \mathbf{x}_3 \rangle$ denotes the spherical triangle whose vertices are $\mathbf{x}_1, \mathbf{x}_2, \mathbf{x}_3$. We denote the domain points by $\xi_1, \ldots, \xi_D$, where $D = \dim S_d^0(\Delta)$. Let $\{B_l : l = 1, \ldots, D\}$ be a basis for $S_d^0(\Delta)$ such that the restriction of $B_l$ on a triangle containing $\xi_l$ is the Bernstein-Bézier polynomial of degree $d$ associated with this point, and that $B_l$ vanishes on other triangles.

A set $\mathscr{M} := \{\zeta_l\}_{l=1}^N \subset \mathscr{D}$ is called a *minimal determining set* for $S_d^r(\Delta)$ if, for every $s \in S_d^r(\Delta)$, all the coefficients $v_l(s)$ in the expression $s = \sum_{l=1}^D v_l(s) B_l$ are uniquely determined by the coefficients corresponding to the basis functions which are associated with points in $\mathscr{M}$. Given a minimal determining set, we construct a basis $\{B_l^*\}_{l=1}^N$ for $S_d^r(\Delta)$ by requiring

$$v_{l'}(B_l^*) = \delta_{l,l'}, \quad 1 \le l, l' \le N.$$

The use of Hahn–Banach Theorem extends the linear functionals $v_l$, $l = 1, \ldots, N$, to be defined for all functions in $L_2(\mathbb{S})$. We continue to use the same symbol for these extensions.

The *quasi-interpolation operator* $I_\Delta : L_2(\mathbb{S}) \to S_d^r(\Delta)$ is now defined by

$$I_\Delta v := \sum_{l=1}^N v_l(v) B_l^*, \quad v \in L_2(\mathbb{S}).$$

Before introducing the properties of this operator, we need the following definition. For any $\tau \in \Delta$, we define

$$\omega_\tau := \bigcup_{i \in \mathscr{I}_\tau} \omega_i$$

where

$$\omega_i := \operatorname{supp}(B_i^*) \quad \text{and} \quad \mathscr{I}_\tau := \big\{ i \in \{1, \ldots, N : \} : \tau \subset \omega_i \big\}.$$

The following result can be found in [159, 174].

**Lemma 15.3.** *There exists a quasi-interpolation operator $I_\Delta : L^2(\mathbb{S}) \longrightarrow S_d^r(\Delta)$ satisfying, for any $v \in L^2(\mathbb{S})$,*

$$|I_\Delta v|_{H^k(\tau)} \lesssim h^{-k} \|v\|_{L^2(\omega_\tau)}, \quad k = 0, 1,$$

*and*

$$\|I_\Delta v\|_{H^k(\mathbb{S})} \lesssim h^{-k} \|v\|_{L^2(\mathbb{S})}, \quad k = 0, 1/2, 1.$$

*In particular, if the degree $d$ is even then for any $v \in H^{1/2}(\mathbb{S})$ we have*

$$\|I_\Delta v\|_{H^{1/2}(\mathbb{S})} \lesssim \|v\|_{H^{1/2}(\mathbb{S})}. \tag{15.12}$$

*The constants depend only on the smallest angle $\Theta_\Delta$ of the triangulation $\Delta$ and the polynomial degree $d$.*

*Proof.* See [174, Lemma 2.3]. We note here that the assumption $d$ even is to ensure that $S_d^r(\Delta)$ contains the constant functions and that $I_\Delta$ reproduces these functions. □

### 15.2.2 Approximate solutions and error estimates

We approximate the solution $u$ of Problem A by $\widetilde{u} \in H^\alpha$ having the form

$$\widetilde{u} = \widetilde{u}_0 + \widetilde{u}_1 \quad \text{where} \quad \widetilde{u}_0 \in \ker L \quad \text{and} \quad \widetilde{u}_1 \in S_d^r(\Delta). \tag{15.13}$$

The solution $\widetilde{u}_1$ will be found by solving the Galerkin equation

$$\langle L^* \widetilde{u}_1, v \rangle = \langle f, v \rangle \quad \forall v \in S_d^r(\Delta), \tag{15.14}$$

in which $L^*$ is a strongly elliptic pseudo-differential operator of order $2\alpha$ whose symbol is given by

$$\widehat{L^*}(\ell) = \begin{cases} \widehat{L}(\ell) & \text{if } \ell \notin \mathscr{K}(L) \\ (1+\ell)^{2\alpha} & \text{if } \ell \in \mathscr{K}(L). \end{cases}$$

Noting (15.7), we have

$$\widehat{L^*}(\ell) \simeq (1+\ell)^{2\alpha} \quad \forall \ell \in \mathbb{N}.$$

This assures us that the bilinear form $a^* : H^\alpha \times H^\alpha \to \mathbb{R}$ defined by

$$a^*(v,w) := \langle L^* v, w \rangle, \quad v, w \in H^\alpha,$$

is continuous and coercive in $H^\alpha$. The well known Lax–Milgram Theorem confirms the unique existence of $\widetilde{u}_1 \in S_d^r(\Delta)$.

Having found $\widetilde{u}_1$, we will find $\widetilde{u}_0 \in \ker L$ by solving the equations (cf. (15.8))

$$\langle \mu_i, \widetilde{u}_0 \rangle = \gamma_i - \langle \mu_i, \widetilde{u}_1 \rangle, \quad i = 1, \dots, M,$$

so that

$$\langle \mu_i, \widetilde{u} \rangle = \langle \mu_i, u \rangle, \quad i = 1, \dots, M.$$

The unique existence of $\widetilde{u}_0$ follows from Assumption B in exactly the same way as that of $u_0$ in (15.9); see [179, Theorem 2.1].

The exact solution $u$ to Problem A and the approximate solution $\widetilde{u}$ given by equation (15.13) satisfy the following estimate.

**Theorem 15.2.** *Assume that $\Delta$ is a $\beta$-quasiuniform spherical triangulation with $|\Delta| \le 1$ and that* (15.11) *hold. If the order $2\alpha$ of the pseudo-differential operator $L$ satisfies $\alpha \le r+1$, then $u$ and $\widetilde{u}$ satisfy*

$$\|u - \widetilde{u}\|_t \le C h_\Delta^{s-t} \|u\|_s,$$

*where $s \le d+1$ and $2\alpha - d - 1 \le t \le \min\{s, \alpha\}$. Here $C$ is a positive constant depending only on $d$ and the smallest angle in $\Delta$.*

*Proof.* See [179, Theorem 5.1]. □

In the next section, we design additive Schwarz preconditioners for solving (15.14) in the case of the hypersingular integral operator. Results for the Laplace–Beltrami operator can be found in [175].

## 15.3 Additive Schwarz Methods for Equations on the Sphere

### *15.3.1 Decomposition of the finite element space*

The construction of a preconditioner by additive Schwarz method, as is usual, requires a two-level mesh.

#### 15.3.1.1 Two-level meshes

Let $X := \{\mathbf{x}_1, \ldots, \mathbf{x}_K\}$ and $Y := \{\mathbf{y}_1, \ldots, \mathbf{y}_J\}$ be two sets of points on $\mathbb{S}$ with $J < K$. We denote by $\Delta_h$ (respectively, $\Delta_H$) the spherical triangulation generated by $X$ (respectively, $Y$), and recall the definitions

$$h = \tan(|\Delta_h|/2) \quad \text{and} \quad H = \tan(|\Delta_H|/2);$$

see (15.10). We assume that $|\Delta_H| > |\Delta_h|$. We also assume that both $\Delta_H$ and $\Delta_h$ are regular and quasi-uniform.

To distinguish triangles in the two different meshes, we denote a triangle in the fine mesh by $\tau$, and a triangle in the coarse mesh by $\tau_H$. For each $\tau \in \Delta_h$, we denote by $A_\tau$ its area.

To invoke results in [21, 159] we also define

$$\varsigma = \tan(\rho_{\Delta_h}/2), \quad h_\tau = \tan(|\tau|/2), \quad \text{and} \quad \varsigma_\tau = \tan(\rho_\tau/2) \quad \text{for } \tau \in \Delta_h$$

and, similarly for the coarse mesh,

$$\varsigma_H = \tan(\rho_{\Delta_H}/2), \quad H_{\tau_H} = \tan(|\tau_H|/2), \quad \text{and} \quad \varsigma_{\tau_H} = \tan(\rho_{\tau_H}/2) \quad \text{for } \tau_H \in \Delta_H.$$

It is straightforward to see that since $\Delta_h$ and $\Delta_H$ are regular and quasi-uniform, we have

$$\rho_\tau \simeq |\tau| \simeq |\Delta_h| \simeq h \simeq h_\tau \simeq \varsigma_\tau \simeq \varsigma$$
$$\rho_{\tau_H} \simeq |\tau_H| \simeq |\Delta_H| \simeq H \simeq H_{\tau_H} \simeq \varsigma_{\tau_H} \simeq \varsigma_H.$$

Denoting by $\operatorname{star}^1(\mathbf{x})$ the union of all triangles in $\Delta_h$ that share the vertex $\mathbf{x}$, we define

$$\operatorname{star}^k(\mathbf{x}) := \bigcup \{ \operatorname{star}^1(\mathbf{w}) \,:\, \mathbf{w} \text{ is a vertex of } \operatorname{star}^{k-1}(\mathbf{x}) \}, \quad k > 1,$$

and

$$\mathrm{star}^k(\tau) := \bigcup \{ \mathrm{star}^k(\mathbf{x}) \,:\, \mathbf{x} \text{ is a vertex of } \tau \}, \quad k \geq 1.$$

We make use of the radial projection defined in [159]. Let $\Omega$ be a subset of $\mathbb{S}$. We denote by $\mathbf{r}_\Omega$ the centre of a spherical cap of smallest possible radius containing $\Omega$, and by $\Pi_\Omega$ the tangential plane touching $\mathbb{S}$ at $\mathbf{r}_\Omega$. For each point $\mathbf{x} \in \Omega$, the intersection of $\Pi_\Omega$ and the ray passing through the origin and $\mathbf{x}$ is denoted by $\overline{\mathbf{x}}$. We define

$$R(\Omega) := \{ \overline{\mathbf{x}} \in \Pi_\Omega : \mathbf{x} \in \Omega \}. \tag{15.15}$$

The radial projection $\mathscr{R}_\Omega$ is defined by

$$\begin{aligned} \mathscr{R}_\Omega : R(\Omega) &\to \Omega \\ \overline{\mathbf{x}} &\mapsto \mathbf{x} := \overline{\mathbf{x}}/|\overline{\mathbf{x}}|. \end{aligned} \tag{15.16}$$

It is clear that $\mathscr{R}_\Omega$ is invertible.

The following result is proved in [21, 159].

**Lemma 15.4.** *Under the assumptions that $\Delta_h$ is a regular, quasi-uniform triangulation satisfying $|\Delta_h| \leq 1$ we have, for any $\tau \in \Delta_h$,*

(i) $A_\tau \simeq h^2$,
(ii) $\nu_k(\tau) \lesssim (2k+1)^2$

*where $A_\tau$ denotes the area of $\tau$, and $\nu_k(\tau)$ denotes the number of triangles in* $\mathrm{star}^k(\tau)$. *Here, the constants depend only on the smallest angle of the triangulation.*

#### 15.3.1.2 Domain decomposition

From the triangulations $\Delta_H$ and $\Delta_h$ we define the following subdomains. For each triangle $\tau_H^i \in \Delta_H$, $i = 1, \ldots, J$, let $\Omega_i$ be the union of all triangles in $\Delta_h$ that intersect $\tau_H^i$, i.e.,

$$\Omega_i = \bigcup_{\substack{\tau \in \Delta_h \\ \bar{\tau} \cap \bar{\tau}_H^i \neq \phi}} \tau.$$

Since a triangle in the fine mesh can intersect more than one triangle in the coarse mesh, the subdomains are overlapping. We denote the size of the overlapping by $\delta$, which is proportional to $h$ by our construction.

It will be assumed that we can partition the index set $\mathscr{I} = \{1, \ldots, J\}$ into $M$ sets $\mathscr{I}_k$ ($M$ independent of $J$) in which for any $1 \leq k \leq M$, if $i, j \in \mathscr{I}_k$ and $i \neq j$ then $\Omega_i \cap \Omega_j = \varnothing$.

*Remark 15.1.* The partition above is related to the graph colouring problem (see e.g. [34]). We define an undirected graph $G = (V, E)$ in which the set of vertices $V = \{V_1, \ldots, V_J\}$ is identified with the set of subdomains $\Omega_i$, and $E$ is the set of edges, where if $\Omega_i \cap \Omega_j \neq \varnothing$ then there is an edge between $V_i$ and $V_j$. A partition of the index set $\mathscr{I}$ satisfying the assumption above is equivalent to a colouring of the vertices of the graph $G$ in which adjacent vertices have different colours. The minimal number $\delta(G)$ of colours required is called the chromatic number of $G$. The number $M$ can be chosen to be $\delta(G)$.

When $G$ is neither a complete graph nor an odd cycle, an upper bound of $\delta(G)$ is the maximal degree $\mu(G)$ of $G$; see [34, Theorem 3, Chapter 5]. In terms of our subdomains, each subdomain intersects at most $\mu(G)$ other subdomains. It is clear that in our subdomain construction, the number $\mu(G)$ depends only on the smallest angle of the triangulation $\Delta_H$. Therefore, $M$ is independent of $J$.

We also assume, as in [219], that for $j = 1,...,J$, there exists $\delta_j > 0$ such that for any $\mathbf{x} \in \Omega_j$ there exists $i \in \{1,...,J\}$ satisfying

$$\mathbf{x} \in \Omega_i \quad \text{and} \quad \operatorname{dist}(\mathbf{x}, \partial\Omega_i) := \min_{\mathbf{y}\in\partial\Omega_i} \cos^{-1}(\mathbf{x}\cdot\mathbf{y}) \geq \delta_j. \tag{15.17}$$

Here $\delta_j$ measures the amount of overlap in $\Omega_j$ and this assumption ensures that the overlap of the subdomains is small. We denote

$$\Omega_{j,\delta_j} := \{\mathbf{x} \in \Omega_j : \operatorname{dist}(\mathbf{x}, \partial\Omega_j) \leq \delta_j\}.$$

#### 15.3.1.3 Subspace decomposition

Denoting by $\mathscr{V}$ the space $S_d^r(\Delta_h)$, we define

$$\mathscr{V}_i := \mathscr{V} \cap H_0^1(\Omega_i), \quad i = 1,\ldots,J.$$

We also denote by $\mathscr{V}_0'$ the space $S_d^r(\Delta_H)$. Since this space is not a subspace of $\mathscr{V}$ we use the quasi-interpolation $I_{\Delta_h}$ defined in Subsection 15.2.1.3 to define $\mathscr{V}_0 := I_{\Delta_h}\mathscr{V}_0'$. This results in the following subspace decomposition which will be used in the design of preconditioners for the hypersingular integral equation to be studied in the next subsection

$$\mathscr{V} = \mathscr{V}_0 + \mathscr{V}_1 + \cdots + \mathscr{V}_J. \tag{15.18}$$

As described in Section 2.3, this subspace decomposition will be used to define the additive Schwarz preconditioner for the hypersingular integral equation studied in the following subsection.

## 15.3.2 *The hypersingular integral equation*

#### 15.3.2.1 The equation and its Galerkin approximation

We recall that the hypersingular integral equation on the sphere is of the form

$$-\mathfrak{W}u + \omega^2 \int_{\mathbb{S}} u\,d\sigma = f \quad \text{on } \mathbb{S}, \tag{15.19}$$

where $\mathfrak{W}$ is the hypersingular integral operator given by

$$\mathfrak{W}v(\mathbf{x}) := -\frac{1}{2\pi}\frac{\partial}{\partial \nu_{\mathbf{x}}}\int_{\mathbb{S}} v(\mathbf{y})\frac{\partial}{\partial \nu_{\mathbf{y}}}\frac{1}{|\mathbf{x}-\mathbf{y}|}\,d\sigma_{\mathbf{y}}$$

and $\omega$ is some nonzero real constant. It can be shown that [161] the operator $-\mathfrak{W}$ can be represented in the form (15.6) with the symbol being $\widehat{L}(\ell) = \ell(\ell+1)/(2\ell+1)$; see Example 15.1. We note that the second term on the left-hand side of (15.19) is added to ensure strong ellipticity and thus unique existence of the solution.

Introducing the bilinear form

$$a(u,v) := -\langle \mathfrak{W}u, v\rangle + \omega^2 \langle u,1\rangle \langle v,1\rangle, \quad u,v \in H^{1/2}(\mathbb{S}),$$

we note that (see [161])

$$a(v,v) \simeq \|v\|^2_{H^{1/2}(\mathbb{S})} \quad \forall v \in H^{1/2}(\mathbb{S}). \tag{15.20}$$

A natural weak formulation of equation (15.19) is: Find $u \in H^{1/2}(\mathbb{S})$ satisfying

$$a(u,v) = \langle f, v\rangle \quad \forall v \in H^{1/2}(\mathbb{S}).$$

This bilinear form is clearly bounded and coercive (cf. [42]). This guarantees the unique solvability of the equation. The Ritz–Galerkin approximation problem is: Find $u_h \in S^r_d(\Delta_h)$ satisfying

$$a(u_h, v_h) = \langle f, v_h\rangle \quad \forall v_h \in S^r_d(\Delta_h).$$

The following proposition shows that the matrix system arising from this Galerkin approximation is ill-conditioned.

**Proposition 15.1.** *The condition number of the stiffness matrix* $\mathbf{A}$ *is bounded by*

$$\kappa(\mathbf{A}) \lesssim h^{-1}.$$

*Proof.* Recall that $\{\phi_i\}_{i=1}^N$ is a basis for $S^r_d(\Delta_h)$. Let $\mathbf{c} = (c_i)_{i=1}^N \in \mathbb{R}^N$. We define $v := \sum_{i=1}^N c_i\phi_i \in S^r_d(\Delta_h)$. Noting (15.20), we have

$$\mathbf{c}^T\mathbf{A}\mathbf{c} = a(v,v) \simeq \|v\|^2_{H^{1/2}(\mathbb{S})}. \tag{15.21}$$

We note that $I_{\Delta_h}v = v$ for all $v \in S^r_d(\Delta_h)$, where $I_{\Delta_h}$ is the quasi-interpolation operator defined in Subsection 15.2.1.3. It follows from (15.21) and Lemma 15.3 that

$$\mathbf{c}^T\mathbf{A}\mathbf{c} \simeq \|I_{\Delta_h}u\|^2_{H^{1/2}(\mathbb{S})} \lesssim h^{-1}\|u\|^2_{L_2(\mathbb{S})} \lesssim h^{-1}\sum_{\tau\in\Delta_h}\|u\|^2_{L_2(\tau)}.$$

By using Lemma 15.1 and noting $A_\tau \simeq |\tau|^2 \simeq h^2$ we obtain

$$\mathbf{c}^T \mathbf{A}\mathbf{c} \lesssim h^{-1} \sum_{\tau\in\Delta_h} A_\tau |\mathbf{c}_\tau|^2 \lesssim h \sum_{\tau\in\Delta_h} |\mathbf{c}_\tau|^2. \tag{15.22}$$

Here, $\mathbf{c}_\tau$ is a vector whose components are those of the vector $\mathbf{c}$ corresponding to the basis functions $\phi_i$ whose supports contain $\tau$. Proposition 5.1 in [12] ensures that the support of $\phi_i$ lies inside $\mathrm{star}^3(\tau)$ if the domain point corresponding to $\phi_i$ belongs to $\tau$. Therefore, if $\mu = \max\{\nu_3(\tau) : \tau \in \Delta_h\}$ then each component of $\mathbf{c}$ appears at most $\mu$ times in the sum on the right-hand side of (15.22). This together with Lemma 15.4 yields

$$\mathbf{c}^T \mathbf{A}\mathbf{c} \lesssim h\mu |\mathbf{c}|^2 \lesssim h|\mathbf{c}|^2 = h\mathbf{c}^T\mathbf{c},$$

which implies

$$\lambda_{\max}(\mathbf{A}) \lesssim h.$$

Using (15.21) and a similar argument as above, we have

$$\mathbf{c}^T \mathbf{A}\mathbf{c} \simeq \|u\|^2_{H^{1/2}(\mathbb{S})} \gtrsim \|u\|^2_{L_2(\mathbb{S})} = \sum_{\tau\in\Delta_h} \|u\|^2_{L_2(\tau)} \gtrsim \sum_{\tau\in\Delta_h} A_\tau |\mathbf{c}_\tau|^2 \simeq h^2 \mathbf{c}^T\mathbf{c},$$

implying

$$\lambda_{\min}(\mathbf{A}) \gtrsim h^2.$$

An upper bound for $\kappa(\mathbf{A})$ can now be derived. □

The additive Schwarz preconditioner based on the decomposition (15.18) will be used to precondition the above ill-conditioned system. In order to invoke Theorem 2.1 again we prove the stability and coercivity of the decomposition.

#### 15.3.2.2 Technical lemmas

To prove Assumption 2.3 (Stability of decomposition), standard analysis as used in previous chapters cannot be directly used. There are two levels of difficulties. First, we have to work with homogeneous polynomials on spherical triangles which do not have all properties of polynomials on planar triangles. Second, the space $\mathscr{V} = S_d^r(\Delta_h)$ does not contain constant functions when $d$ is odd.

Recalling the construction of the subdomains in Subsection 15.3.1, for each triangle $\tau_H^i$ in the coarse mesh $\Delta_H$, we defined a corresponding subdomain $\Omega_i$. We now define the triangle $\tau_{H,h}^i$ satisfying

$$\tau_H^i \subset \tau_{H,h}^i \subset \Omega_i, \quad i = 1,\dots,J. \tag{15.23}$$

Without loss of generality we can assume that the edges of $\tau_{H,h}^i$ are parallel and of distance $h$ from those of $\tau_H^i$; see Figure 15.1. (We can always choose $\Omega_i$ to be a bigger domain so that the assumption holds.)

For simplicity of presentation, we assume that the spherical triangles $\tau_H^i$, $i = 1,\dots,J$, are equilateral triangles. Then so are the spherical triangles $\tau_{H,h}^i$. The set $\{\tau_{H,h}^i : i = 1,\dots,J\}$ is a set of overlapping spherical triangles which covers the

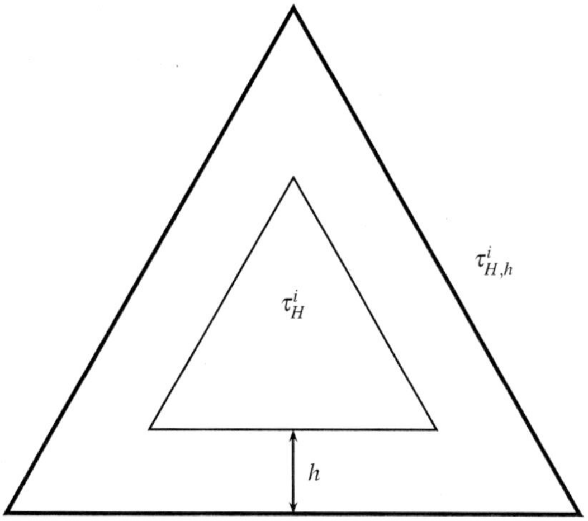


**Fig. 15.1** Extended spherical triangle $\tau^i_{H,h}$

sphere $\mathbb{S}$ (Figure 15.2). We will define a partition of unity $\{\theta_i\}_{i=1}^J$, whose definition is given in Appendix B, satisfying

$$\text{supp}(\theta_i) \subset \overline{\tau^i_{H,h}}, \quad i = 1, \ldots, J.$$

In the sequel, we denote

$$T_i := \text{supp}(\theta_i), \quad i = 1, \ldots, J,$$

and $W_i$ the smallest rectangle which contains $T_i$ and shares a common edge with $T_i$; see Figure 15.4.

We first define a partition of unity $\{\theta_i\}_{i=1}^J$. Note here that the partition of unity will be defined for the extended spherical triangles $\{\tau^i_{H,h} : i = 1, \ldots, J\}$ and it is also a partition of unity for the overlapping subdomains $\{\Omega_i : i = 1, \ldots, J\}$.

Let $\mathbf{x}$ be a point on the sphere $\mathbb{S}$. Then $\theta_i(\mathbf{x})$, for $i = 1, \ldots, J$, are defined as follows:

- Case 1: $\mathbf{x}$ belongs to only one triangle $\tau^{i_0}_{H,h}$ for some $i_0 \in \{1, \ldots, J\}$ (e.g. $\mathbf{x} \in A$ in Figure 15.3. Then

$$\theta_{i_0}(\mathbf{x}) := 1 \quad \text{and} \quad \theta_i(\mathbf{x}) := 0, \quad \text{for all } i \neq i_0.$$

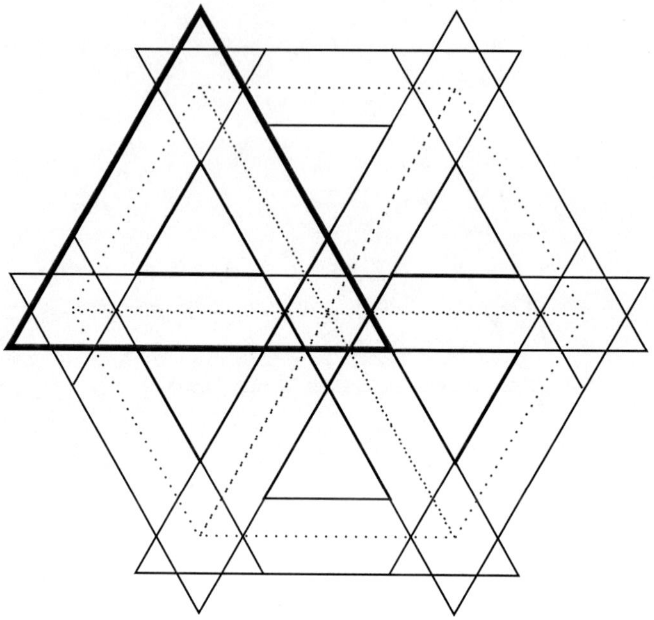

**Fig. 15.2** Six overlapping extended spherical triangles (Triangles with dotted edges: $\tau_H^i$)

- Case 2: **x** belongs to exactly two extended triangles $\tau_{H,h}^{i_0}$ and $\tau_{H,h}^{i_1}$ (e.g. $\mathbf{x} \in B$ in Figure 15.3. Then

$$\theta_{i_0}(\mathbf{x}) := \delta^{-1}\,\mathrm{dist}(\mathbf{x}, \partial\tau_{H,h}^{i_0})$$
$$\theta_{i_1}(\mathbf{x}) := \delta^{-1}\,\mathrm{dist}(\mathbf{x}, \partial\tau_{H,h}^{i_1})$$
$$\theta_i(\mathbf{x}) := 0, \quad i \notin \{i_0, i_1\},$$

  where $\partial\tau_{H,h}^i$ is the boundary of $\tau_{H,h}^i$.
- Case 3: **x** belongs to the intersection of more than two extended triangles (i.e. $\mathbf{x} \in U^\star$ in Figure 15.3. For simplicity of notation we denote the six triangles in Figure 15.3 by $\tau_1, \ldots, \tau_6$. Then

$$\theta_i(\mathbf{x}) := 0, \quad i \notin \{1,2,3,4,5,6\}$$

$$\theta_1(\mathbf{x}) = \begin{cases} \delta^{-1}\,\mathrm{dist}(\mathbf{x}, \partial\tau_1), & \text{if } \mathbf{x} \in U_3 \cup U_7 \\ \delta^{-2}\,\mathrm{dist}(\mathbf{x}, \partial\tau_1 \cap \mathrm{int}(\tau_5))\,\mathrm{dist}(\mathbf{x}, \partial\tau_1 \cap \mathrm{int}(\tau_3)), & \text{if } \mathbf{x} \in U_1 \cup U_2 \cup U_5 \\ 0, & \text{if } \mathbf{x} \in U_4 \cup U_6, \end{cases}$$

$$\theta_2(\mathbf{x}) = \begin{cases} \delta^{-2}\,\mathrm{dist}(\mathbf{x},\partial\tau_2\cap\mathrm{int}(\tau_6))\,\mathrm{dist}(\mathbf{x},\partial\tau_2\cap\mathrm{int}(\tau_4)), & \text{if } \mathbf{x}\in U_3 \\ \delta^{-2}\,\mathrm{dist}(\mathbf{x},\partial\tau_2)\,\mathrm{dist}(\mathbf{x},\partial\tau_1\cap\mathrm{int}(\tau_5)), & \text{if } \mathbf{x}\in U_2\cup U_4 \\ \delta^{-3}\,\mathrm{dist}(\mathbf{x},\partial\tau_2\cap\mathrm{int}(\tau_6))\times & \\ \quad \mathrm{dist}(\mathbf{x},\partial\tau_2\cap\mathrm{int}(\tau_4))\,\mathrm{dist}(\mathbf{x},\partial\tau_1\cap\mathrm{int}(\tau_5)), & \text{if } \mathbf{x}\in U_1 \\ 0, & \text{if } \mathbf{x}\in U_5\cup U_6\cup U_7, \end{cases}$$

Here, $\mathrm{int}(\tau_i)$ denotes the interior of $\tau_i$. The function $\theta_4(\mathbf{x})$ is defined similarly to $\theta_1(\mathbf{x})$, and $\theta_3(\mathbf{x})$, $\theta_5(\mathbf{x})$, $\theta_6(\mathbf{x})$ similarly to $\theta_2(\mathbf{x})$.

The supports $T_i$ of these functions $\theta_i$ can be one of the four shapes in Figure 15.4. The rectangles $W_i$ mentioned in Lemma 15.6 are also depicted in Figure 15.4.

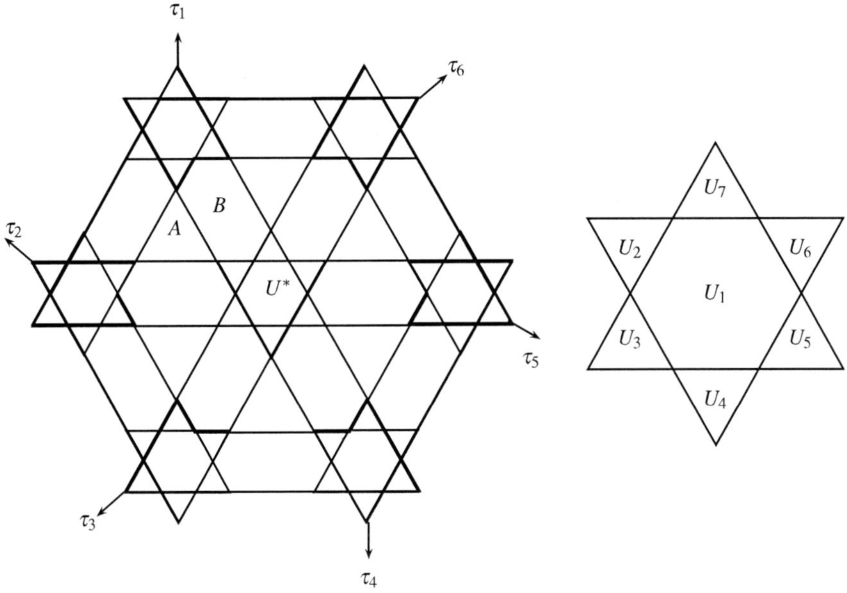


**Fig. 15.3** $\tau_i$: equilateral triangles, $i = 1,\ldots,6$

We will frequently use the following results.

**Lemma 15.5.** *Let* $0 < \alpha < \min\{\beta,\beta'\}$. *If* $v \in H^{1/2}([0,\beta']\times[0,\beta])$ *then*

$$\int_0^\beta \left( \int_\alpha^{\beta'} \frac{|v(x,y)|^2}{x}\,dx \right) dy \lesssim \left(1+\log^2\frac{\beta'}{\alpha}\right) \|v\|_{H_w^{1/2}([0,\beta']\times[0,\beta])}, \tag{15.24}$$

*and*

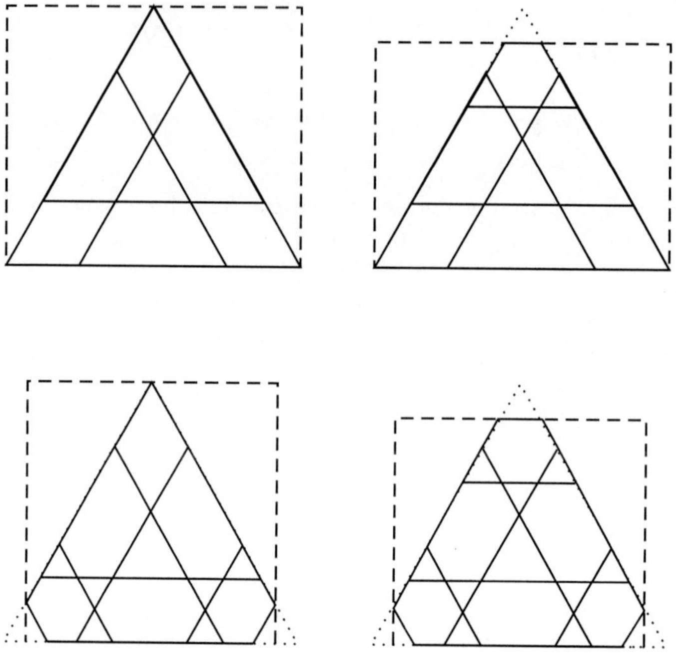

Fig. 15.4 Supports $T_i$ and rectangles $W_i$

$$\frac{1}{\alpha}\|v\|^2_{L_2([0,\alpha]\times[0,\beta])} \lesssim \left(1+\log\frac{\beta'}{\alpha}\right)^2 \|v\|_{H^{1/2}_{\mathrm{w}}([0,\beta']\times[0,\beta])}, \tag{15.25}$$

*Proof.* Estimates (15.24) and (15.25) are proved in Lemma 10.16 for $\beta=\beta'$; the results still hold for $\beta\neq\beta'$. □

Recall that for each subset $\Omega\subset\mathbb{S}$, the set $R(\Omega)$ is the image of $\Omega$ under the inverse of the radial projection $\mathscr{R}_\Omega$; see (15.15) and (15.16). It is shown in [159, Lemma 3.1] that

$$\|v\|_{L_2(\Omega)}\simeq\|\bar{v}_0\|_{L_2(R(\Omega))} \quad \text{and} \quad \|v\|_{H^1(\Omega)}\simeq\|\bar{v}_0\|_{H^1(R(\Omega))}$$

for any $v$ belongs to $L_2(\Omega)$ and $H^1(\Omega)$, respectively. Using interpolation (see [147, Theorem B2]), we obtain

$$\|v\|_{H^{1/2}(\Omega)}\simeq\|\bar{v}_0\|_{H^{1/2}(R(\Omega))}, \quad v\in H^{1/2}(\Omega). \tag{15.26}$$

From the definitions of $\|\cdot\|_{H^{1/2}_{\mathrm{w}}(\Omega)}$, $\|\cdot\|_{\tilde{H}^{1/2}_{\mathrm{w}}(\Omega)}$, and $|\cdot|_{H^{1/2}(\Omega)}$, we deduce by repeating the argument used in the proof of [159, Lemma 3.1] that for any $v\in H^{1/2}(\Omega)$

$$\begin{aligned} \|v\|_{H_w^{1/2}(\Omega)} &\simeq \|\bar{v}_0\|_{H_w^{1/2}(R(\Omega))}, \\ \|v\|_{\widetilde{H}_w^{1/2}(\Omega)} &\simeq \|\bar{v}_0\|_{\widetilde{H}_w^{1/2}(R(\Omega))}, \\ |v|_{H^{1/2}(\Omega)} &\simeq |\bar{v}_0|_{H^{1/2}(R(\Omega))}, \end{aligned} \tag{15.27}$$

where the constants are independent of the size of $\Omega$.

**Lemma 15.6.** *For any $v \in H^{1/2}(\mathbb{S})$ there holds*

$$\sum_{i=1}^{J} \|\theta_i v\|^2_{\widetilde{H}_w^{1/2}(T_i)} \lesssim \sum_{i=1}^{J} \left(1+\log^2\frac{H_j}{\delta}\right)^2 \|v\|^2_{H_w^{1/2}(W_i)} \tag{15.28}$$

*where $\delta$ is defined in* (15.17).

*Proof.* We prove (15.28) for $T_i$ being of the first shape in Figure 15.4. The cases when $T_i$ is of other shapes can be proved in the same manner.

Equivalences (15.27) and (15.26) allow us to prove instead of (15.28) the following inequality

$$\|\overline{\theta_i v_0}\|^2_{\widetilde{H}_w^{1/2}(R(T_i))} \lesssim \left(1+\log\frac{H_i}{\delta}\right)^2 \|\bar{v}_0\|^2_{\widetilde{H}_w^{1/2}(R(W_i))}.$$

For notational convenience, in this proof we write $T_i, W_i, \theta_i$, and $v$ instead of $R(T_i)$, $R(W_i)$, $(\bar{\theta}_i)_0$, and $\bar{v}_0$, namely we think of $T_i$ and $W_i$ as planar regions, and $\theta_i$ and $v$ as two variables functions. Here $H_i$ and $\delta$ are the size of $T_i$ and the size of the overlap.

It is noted that $W_i = [0,H_i]\times[0,H_i']$ where $H_i' = \sqrt{3}/2H_i$. Recall that

$$\|\theta_i v\|^2_{\widetilde{H}_w^{1/2}(T_i)} = |\theta_i v|^2_{H^{1/2}(T_i)} + \int_{T_i} \frac{[\theta_i v(\mathbf{x})]^2}{\operatorname{dist}(\mathbf{x},\partial T_i)}\,d\mathbf{x}, \tag{15.29}$$

where

$$|\theta_i v|^2_{H^{1/2}(T_i)} = \int_{T_i}\int_{T_i} \frac{|\theta_i v(\mathbf{x}) - \theta_i v(\mathbf{x}')|^2}{|\mathbf{x}-\mathbf{x}'|^3}\,d\mathbf{x}\,d\mathbf{x}'. \tag{15.30}$$

We first estimate the second term in the right hand side of (15.29), which is split into a sum of integrals over the triangles $T_i^\ell := A_\ell \cup B_\ell \cup C_\ell \cup D_\ell$ (see Figure 15.5), for $\ell = 1,2,3$, as follows:

$$\int_{T_i} \frac{[\theta_i v(\mathbf{x})]^2}{\operatorname{dist}(\mathbf{x},\partial T_i)}\,d\mathbf{x} = \sum_{\ell=1}^{3} \int_{T_i^\ell} \frac{[\theta_i v(\mathbf{x})]^2}{\operatorname{dist}(\mathbf{x},\partial T_i)}\,d\mathbf{x}.$$

We only need to estimate the integral over $T_i^1$, the other two can be bounded similarly. Recall that

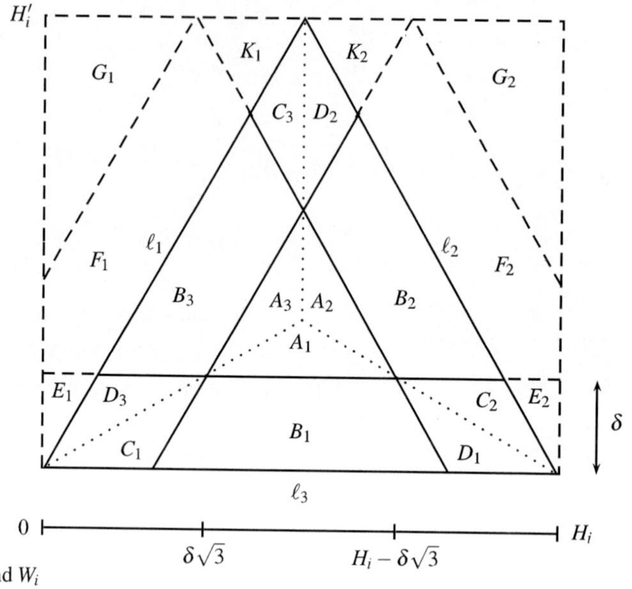


**Fig. 15.5** $T_i$ and $W_i$

$$\theta_i(\mathbf{x}) = \begin{cases} 1, & \text{if } \mathbf{x} \in A_1, \\ \delta^{-1} \operatorname{dist}(\mathbf{x}, \partial T_i), & \text{if } \mathbf{x} \in B_1, \\ \delta^{-2} \operatorname{dist}(\mathbf{x}, \partial T_i) \operatorname{dist}(\mathbf{x}, \ell_1), & \text{if } \mathbf{x} \in C_1, \\ \delta^{-2} \operatorname{dist}(\mathbf{x}, \partial T_i) \operatorname{dist}(\mathbf{x}, \ell_2), & \text{if } \mathbf{x} \in D_1. \end{cases}$$

Since $\delta^{-1} \operatorname{dist}(\mathbf{x}, \ell_k) \le 1$ for $k = 1, 2$ and $\mathbf{x}$ respectively belongs to $C_1$ and $D_1$, there holds

$$\theta_i(\mathbf{x}) \le \delta^{-1} \operatorname{dist}(\mathbf{x}, \partial T_i) \quad \forall \mathbf{x} \in B_1 \cup C_1 \cup D_1.$$

Hence

$$\begin{aligned} \int_{T_i^1} \frac{[\theta_i v(\mathbf{x})]^2}{\operatorname{dist}(\mathbf{x}, \partial T_i)} \, d\mathbf{x} &= \int_{A_1} \frac{[\theta_i v(\mathbf{x})]^2}{\operatorname{dist}(\mathbf{x}, \partial T_i)} \, d\mathbf{x} + \int_{B_1 \cup C_1 \cup D_1} \frac{[\theta_i v(\mathbf{x})]^2}{\operatorname{dist}(\mathbf{x}, \partial T_i)} \, d\mathbf{x} \\ &\lesssim \int_{A_1} \frac{|v(\mathbf{x})|^2}{y} \, d\mathbf{x} + \frac{1}{\delta} \int_{B_1 \cup C_1 \cup D_1} |v(\mathbf{x})|^2 \, d\mathbf{x} \\ &\le \int_0^{H_i} \left( \int_\delta^{H_i'} \frac{|v(x,y)|^2}{y} \, dy \right) dx + \frac{1}{\delta} \int_0^{H_i} \int_0^\delta |v(x,y)|^2 \, dy \, dx. \end{aligned}$$

It follows from Lemma 15.5 and $H_i' \simeq H_i$ that

$$\int_{T_i^1} \frac{[\theta_i v(\mathbf{x})]^2}{\operatorname{dist}(\mathbf{x}, \partial T_i)}\, d\mathbf{x} \lesssim \left(1 + \log \frac{H_i}{\delta}\right)^2 \|v\|^2_{H_w^{1/2}(W_i)}.$$

We now need to use Lemma A.16 to estimate the double integral in (15.30). This motivates us to transform the integral over the triangle $T_i$ into the integral over the rectangle $W_i$. To do so, we first introduce an extension $\widetilde{\theta}_i$ of $\theta_i$ over the rectangle $W_i$ as follows (see Figure 15.5):

$$\widetilde{\theta}_i(\mathbf{x}) := \begin{cases} \theta_i(\mathbf{x}), & \text{if } \mathbf{x} \in T_i, \\ \delta^{-1} \operatorname{dist}(\mathbf{x}, \ell_k), & \text{if } \mathbf{x} \in F_k, \quad k = 1, 2, \\ \delta^{-2} \operatorname{dist}(\mathbf{x}, \ell_k) \operatorname{dist}(\mathbf{x}, \ell_3), & \text{if } \mathbf{x} \in E_k, \quad k = 1, 2, \\ \delta^{-2} \operatorname{dist}(\mathbf{x}, \ell_1) \operatorname{dist}(\mathbf{x}, \ell_2), & \text{if } \mathbf{x} \in K_k, \quad k = 1, 2, \\ 1, & \text{if } \mathbf{x} \in G_k, \quad k = 1, 2. \end{cases}$$

The extension $\widetilde{\theta}_i$ is defined in order to preserve the continuity and symmetry of $\theta_i$ across the edges $\ell_1$ and $\ell_2$ of $T_i$. Since $T_i \subset W_i$ and $\widetilde{\theta}_i = \theta_i$ on $T_i$, there holds

$$\int_{T_i}\int_{T_i} \frac{|\theta_i v(\mathbf{x}) - \theta_i v(\mathbf{x}')|^2}{|\mathbf{x} - \mathbf{x}'|^3}\, d\mathbf{x}\, d\mathbf{x}' \le \int_{W_i}\int_{W_i} \frac{|\widetilde{\theta}_i v(\mathbf{x}) - \widetilde{\theta}_i v(\mathbf{x}')|^2}{|\mathbf{x} - \mathbf{x}'|^3}\, d\mathbf{x}\, d\mathbf{x}'.$$

Noting (A.97), we have

$$\begin{aligned} \int_{W_i}\int_{W_i} \frac{|\widetilde{\theta}_i v(\mathbf{x}) - \widetilde{\theta}_i v(\mathbf{x}')|^2}{|\mathbf{x} - \mathbf{x}'|^3}\, d\mathbf{x}\, d\mathbf{x}' &\lesssim \int_I\int_I \frac{\|\widetilde{\theta}_i v(x, \cdot) - \widetilde{\theta}_i v(x', \cdot)\|^2_{L_2(I')}}{|x - x'|^2}\, dx\, dx' \\ &\quad + \int_{I'}\int_{I'} \frac{\|\widetilde{\theta}_i v(\cdot, y) - \widetilde{\theta}_i v(\cdot, y')\|^2_{L_2(I)}}{|y - y'|^2}\, dy\, dy' \\ &=: \mathscr{I}_1 + \mathscr{I}_2, \end{aligned}$$

where $I = [0, H_i]$, $I' = [0, H_i']$, $\mathbf{x} = (x, y)$ and $\mathbf{x}' = (x', y')$. We will show that $\mathscr{I}_1$ is bounded by

$$\left(1 + \log \frac{H_i}{\delta}\right)^2 \|v\|^2_{H_w^{1/2}(W_i)}.$$

The term $\mathscr{I}_2$ can be estimated in the same manner. By using the triangle inequality, and noting that $\widetilde{\theta}_i \le 1$, we have

$$\begin{aligned} \mathscr{I}_1 &\le \int_I\int_I \frac{\|[\widetilde{\theta}_i(x, \cdot) - \widetilde{\theta}_i(x', \cdot)] v(x, \cdot)\|^2_{L_2(I')}}{|x - x'|^2}\, dx\, dx' \\ &\quad + \int_I\int_I \frac{\|v(x, \cdot) - v(x', \cdot)\|^2_{L_2(I')}}{|x - x'|^2}\, dx\, dx'. \end{aligned} \tag{15.31}$$

It follows from Lemma A.16 that the second integral on the right-hand side of inequality (15.31) is bounded by $\|v\|^2_{H_w^{1/2}(W_i)}$. We still need to estimate

$$A_{I,I}(\widetilde{\theta}_i v) := \int\limits_I \int\limits_I \frac{\|[\widetilde{\theta}_i(x,\cdot) - \widetilde{\theta}_i(x',\cdot)]v(x,\cdot)\|^2_{L_2(I')}}{|x-x'|^2}\, dx dx'.$$

We denote $I_1 := [0, \delta\sqrt{3}]$, $I_2 := [\delta\sqrt{3}, H_i - \delta\sqrt{3}]$, and $I_3 := [H_i - \delta\sqrt{3}, H_i]$. In order to show that $A_{I,I}(\widetilde{\theta}_i v)$ is bounded by $|v|^2_{H^{1/2}(W_i)}$, we will prove that $A_{I_k,I_\ell}(\widetilde{\theta}_i v)$ are bounded by $\|v\|^2_{H_w^{1/2}(W_i)}$, for $k,\ell = 1,2,3$. By the symmetry of $\widetilde{\theta}_i$, it is sufficient to prove the estimation for $A_{I_k,I_\ell}(\widetilde{\theta}_i v)$ for $(k,\ell) \in \{(1,1),(1,2),(1,3),(2,2)\}$.

Let $x,x' \in I_1$. Elementary calculation (though tedious) reveals

$$\max_{y\in I'} |\widetilde{\theta}_i(x,y) - \widetilde{\theta}_i(x',y)| \lesssim \delta^{-1}|x-x'|.$$

Thus

$$A_{I_1,I_1}(\widetilde{\theta}_i v) \lesssim \frac{1}{\delta} \int\limits_{I_1} \|v(x,\cdot)\|^2_{L_2(I')}\, dx = \frac{1}{\delta}\|v(\cdot,\cdot)\|^2_{L_2(I_1\times I')}.$$

Noting that the size of $I_1$ is defined to be proportional to $\delta$, and by using (15.25), we deduce

$$A_{I_1,I_1}(\widetilde{\theta}_i v) \lesssim \left(1+\log\frac{H_i}{\delta}\right)^2 \|v\|^2_{H_w^{1/2}(W_i)}.$$

We next estimate $A_{I_1,I_2\cup I_3}(\widetilde{\theta}_i v)$ by estimating the two integral $A_{I_1,[\delta\sqrt{3},2\delta\sqrt{3}]}(\widetilde{\theta}_i v)$ and $A_{I_1,[2\delta\sqrt{3},H_i]}(\widetilde{\theta}_i v)$. Repeating the argument used in estimating $A_{I_1,I_1}(\widetilde{\theta}_i v)$, we obtain

$$A_{I_1,[\delta\sqrt{3},\, 2\delta\sqrt{3}]}(\widetilde{\theta}_i v) \lesssim \left(1+\log\frac{H_i}{\delta}\right)^2 \|v\|^2_{H_w^{1/2}(W_i)}.$$

We note here that

$$\max_{y\in I'}[\widetilde{\theta}_i(x,y) - \widetilde{\theta}_i(x',y)]^2 \le 4, \quad \forall x,x' \in I, \tag{15.32}$$

and

$$|x - 2\delta\sqrt{3}| \ge \delta\sqrt{3} \quad \forall x \in I_1. \tag{15.33}$$

We then deduce from (15.32), (15.33) and (15.25) that

$$
\begin{aligned}
A_{I_1,[2\delta\sqrt{3},\,H_i]}(\widetilde{\theta}_i v) &\lesssim \int_{I_1} \|v(x,\cdot)\|^2_{L_2(I')} \int_{[2\delta\sqrt{3},\,H_i]} \frac{1}{(x'-x)^2}\,dx'\,dx \\
&\lesssim \int_{I_1} \frac{H_i - 2\delta\sqrt{3}}{(2\delta\sqrt{3}-x)(H_i - x)} \|v(x,\cdot)\|^2_{L_2(I')}\,dx \\
&\lesssim \frac{1}{\delta}\|v(\cdot,\cdot)\|^2_{L_2(I_1\times I')} \\
&\lesssim \left(1+\log\frac{H_i}{\delta}\right)^2 \|v\|^2_{H_w^{1/2}(W_i)}.
\end{aligned}
$$

We note here that the estimation for $A_{I_1,I_1}(\widetilde{\theta}_i v)$ is based on the fact that the size of $I_1$ is proportional to $\delta$ and

$$
|\widetilde{\theta}_i(x,y) - \widetilde{\theta}_i(x',y)| \lesssim \delta^{-1}|x-x'| \quad \forall x,x' \in I_1, \quad \forall y \in I'. \tag{15.34}
$$

The proof $A_{I_1,[\delta\sqrt{3},H_i]}(\widetilde{\theta}_i v)$ is then obtained by first splitting it into $A_{I_1,[\delta\sqrt{3},2\delta\sqrt{3}]}(\widetilde{\theta}_i v)$ and $A_{I_1,[2\delta\sqrt{3},H_i]}(\widetilde{\theta}_i v)$ in which the former is bounded by using similar argument as in the proof for $A_{I_1,I_1}(\widetilde{\theta}_i v)$, requiring the sizes of $I_1$ and $[\delta\sqrt{3},2\delta\sqrt{3}]$ to be proportional to $\delta$ and (15.34) to hold for $x \in I_1$, $x' \in [\delta\sqrt{3},2\delta\sqrt{3}]$. The latter is estimated by first using (15.32) to write it in the form containing the integral

$$
\int_{[2\delta\sqrt{3},H_i]} \frac{1}{|x-x'|^2}\,dx',
$$

which is then proved to be bounded by $c\delta^{-1}$ for some constant $c > 0$. This procedure can be used in estimating $A_{I_2,I_2}(\widetilde{\theta}_i v)$ by first writing

$$
A_{I_2,I_2}(\widetilde{\theta}_i v) = \int_{I_2}\int_{I_2}\int_{[0,H_i']} \frac{[\widetilde{\theta}_i(x,y) - \widetilde{\theta}_i(x',y)]^2 v^2(x,y)}{|x-x'|^2}\,dy\,dx\,dx'.
$$

The integral is then split into sum of integrals over subregions in which $(x,y)$ and $(x',y)$ can belong to one of the following sets $G_1$, $F_1 \cup H_1$, $B_3 \cup C_3 \cup D_2 \cup H_2$, $H_1 \cup C_3 \cup D_2 \cup B_2$, $H_2 \cup F_2$, $G_2$, and $A_1 \cup A_2 \cup A_3$ in Figure 15.5. The integral when $(x,y)$ and $(x',y)$ belong to $A_1 \cup A_2 \cup A_3$ is easily bounded by $\|v\|^2_{H_w^{1/2}(W_i)}$ by using Lemma A.16, noting that $\widetilde{\theta}_i(x,y) = \widetilde{\theta}_i(x',y) = 1$. The other integrals can be estimated by using similar argument as used in the proof for $A_{I_1,I}(\widetilde{\theta}_i v)$.

Combining all these we obtain the desired result. □

#### 15.3.2.3 Stability of decomposition: A general result for both odd and even degrees

We need to introduce an operator $P_H$ from $H^{1/2}(\mathbb{S})$ into $S^r_d(\Delta_H)$ defined by

$$a(P_H u, v) = a(u, v) \quad \forall v \in S^r_d(\Delta_H)$$

for any $u \in H^{1/2}(\mathbb{S})$. Standard finite element arguments yield

$$\begin{aligned} \|P_H u - u\|_{H_I^{1/2}(\mathbb{S})} &\lesssim \|u - v\|_{H_I^{1/2}(\mathbb{S})} \quad \forall v \in S^r_d(\Delta_H) \\ \|P_H u - u\|_{H_I^{1/2}(\mathbb{S})} &\lesssim \|u\|_{H_I^{1/2}(\mathbb{S})} \\ \|P_H u\|_{H_I^{1/2}(\mathbb{S})} &\lesssim \|u\|_{H_I^{1/2}(\mathbb{S})} \\ \|P_H u - u\|_{L_2(\mathbb{S})} &\lesssim H^{1/2} \|u\|_{H_I^{1/2}(\mathbb{S})}. \end{aligned} \tag{15.35}$$

**Lemma 15.7 (Stability of decomposition).** *There exists a positive constant $C$ depending on the smallest angle of $\Delta_h$ and the polynomial degree $d$ such that for any $u \in V$ there exist $u_i \in V_i, i = 0, \ldots, J$, satisfying $u = \sum_{i=0}^J u_i$ and*

$$\sum_{i=0}^J a(u_i, u_i) \lesssim \frac{H}{h} a(u, u).$$

*Proof.* Let $u_0' := P_H u$ and $u_0 = I_{\Delta_h} u_0'$. Let $\{\theta_i\}_{i=1}^J$ be a partition of unity defined on $\mathbb{S}$ satisfying $\mathrm{supp}(\theta_i) = \overline{\Omega}_i$, for $i = 1, \ldots, J$. We define $w := u - u_0$, and $u_i := I_{\Delta_h}(\theta_i w)$ for $i = 1, \ldots, J$. Then we can split $u \in V$ by

$$u = u_0 + I_{\Delta_h}(u - u_0) = u_0 + I_{\Delta_h}\left(\sum_{i=1}^J \theta_i w\right) = u_0 + u_1 + \cdots u_J.$$

It is clear that $u_i \in V_i$ for all $i = 0, \ldots, J$. By Lemma 15.3, we have

$$\|I_{\Delta_h}(P_H u - u)\|_{H_I^{1/2}(\mathbb{S})} \lesssim h^{-1/2} \|P_H u - u\|_{L_2(\mathbb{S})}. \tag{15.36}$$

By writing $u_0 = I_{\Delta_h}(P_H u - u) + u$ and using the triangle inequality together with inequalities (15.36) and (15.35), we have

$$\begin{aligned} a(u_0, u_0) \simeq \|u_0\|^2_{H_I^{1/2}(\mathbb{S})} &\le \left( \|I_{\Delta_h}(P_H u - u)\|_{H_I^{1/2}(\mathbb{S})} + \|u\|_{H_I^{1/2}(\mathbb{S})} \right)^2 \\ &\lesssim \left( h^{-1/2} \|P_H u - u\|_{L_2(\mathbb{S})} + \|u\|_{H_I^{1/2}(\mathbb{S})} \right)^2 \\ &\lesssim \frac{H}{h} \|u\|^2_{H_I^{1/2}(\mathbb{S})} \simeq \frac{H}{h} a(u, u). \end{aligned} \tag{15.37}$$

Applying Lemma 15.3 and noting that $\theta_i$ vanishes outside $\Omega_i$, we obtain

$$a(u_i,u_i) \simeq \|I_{\Delta_h}(\theta_i w)\|^2_{H_{\mathrm{I}}^{1/2}(\mathbb{S})} \lesssim h^{-1}\|\theta_i w\|^2_{L_2(\mathbb{S})} \simeq h^{-1}\|\theta_i w\|^2_{L_2(\Omega_i)}.$$

This together with the fact that $\|\theta_i w\|_{L_2(\Omega_i)} \le \|w\|_{L_2(\Omega_i)}$ implies

$$a(u_i,u_i) \lesssim h^{-1}\|w\|^2_{L_2(\Omega_i)}.$$

Summing up the above inequality over all subdomains, we obtain

$$\sum_{i=1}^{J} a(u_i,u_i) \lesssim h^{-1}\|w\|^2_{L_2(\mathbb{S})}. \tag{15.38}$$

Noting that $w \in S_d^r(\Delta_h)$ then $I_{\Delta_h} w = w$ and by applying Lemma 15.3 and the last inequality in (15.35), we infer

$$\begin{aligned}\|w\|_{L_2(\mathbb{S})} &= \|I_{\Delta_h}(u - I_{\Delta_h}P_H u)\|_{L_2(\mathbb{S})} = \|I_{\Delta_h}(u - P_H u)\|_{L_2(\mathbb{S})}\\ &\lesssim \|u - P_H u\|_{L_2(\mathbb{S})} \lesssim H^{1/2}\|u\|_{H_{\mathrm{I}}^{1/2}(\mathbb{S})}.\end{aligned}$$

This together with (15.38) gives

$$\sum_{i=1}^{J} a(u_i,u_i) \lesssim \frac{H}{h}\|u\|^2_{H_{\mathrm{I}}^{1/2}(\mathbb{S})} \simeq \frac{H}{h} a(u,u).$$

From this and (15.37) we infer

$$\sum_{i=0}^{J} a(u_i,u_i) \lesssim \frac{H}{h} a(u,u),$$

completing the proof. □

#### 15.3.2.4 A better result on stability of the decomposition for even degrees

In this subsection we assume that the degree $d$ of the spherical splines is even. We recall that this assumption on $d$ allows us to invoke (15.12) in Lemma 15.3.

**Lemma 15.8.** *Assume that the polynomial degree $d$ is even. For any $u \in V$, there exist $u_i \in V_i$ for $i = 0,\dots,J$ satisfying $u = \sum_{i=0}^{J} u_i$ and*

$$\sum_{i=0}^{J} a(u_i,u_i) \lesssim \left(1 + \log^2 \frac{H}{\delta}\right) a(u,u), \tag{15.39}$$

*where the constant depends only on the smallest angle of the triangulations.*

*Proof.* Recall that the support $T_i$ of the partition of unity function $\theta_i$ satisfies $T_i \subset \Omega_i$ (see (15.23)). We now define a decomposition of $u \in V$ such that (15.39) holds. For any $u \in V$ let $u_0 := I_{\Delta_h} P_H u \in V_0$ and $u_i = I_{\Delta_h}(\theta_i w) \in V_i$, $i = 1, \dots, J$, where $w = u - u_0$. It is clear that $u = \sum_{i=1}^J u_i$. By using Lemma 15.3 and (15.35) we obtain

$$\begin{aligned} a(u_0,u_0) &\simeq \|u_0\|^2_{H_I^{1/2}(\mathbb{S})} = \|I_{\Delta_h} P_H u\|^2_{H_I^{1/2}(\mathbb{S})} \lesssim \|u\|^2_{H_I^{1/2}(\mathbb{S})} \simeq a(u,u) \\ \|w\|_{H_I^{1/2}(\mathbb{S})} &= \|I_{\Delta_h}(u - P_H u)\|_{H_I^{1/2}(\mathbb{S})} \lesssim \|u - P_H u\|_{H_I^{1/2}(\mathbb{S})} \lesssim \|u\|_{H_I^{1/2}(\mathbb{S})} \end{aligned} \tag{15.40}$$

and

$$\|w\|_{L_2(\mathbb{S})} = \|I_{\Delta_h}(u - P_H u)\|_{L_2(\mathbb{S})} \lesssim \|u - P_H u\|_{L_2(\mathbb{S})} \lesssim H^{1/2} \|u\|_{H_I^{1/2}(\mathbb{S})}. \tag{15.41}$$

Lemma A.13 gives

$$\|\theta_i w\|_{H_I^{1/2}(\mathbb{S})} \lesssim \|\theta_i w\|_{\widetilde{H}_w^{1/2}(T_i)}. \tag{15.42}$$

By using successively Lemma 15.3, (15.42), (15.28) and the definition of the $\|\cdot\|_{\widetilde{H}_w^{1/2}(T_i)}$-norm we obtain

$$\begin{aligned} \sum_{i=1}^J a(u_i,u_i) &\simeq \sum_{i=1}^J \|u_i\|^2_{H_w^{1/2}(\mathbb{S})} \lesssim \sum_{i=1}^J \|\theta_i w\|^2_{H_w^{1/2}(\mathbb{S})} \lesssim \sum_{i=1}^J \|\theta_i w\|^2_{\widetilde{H}_w^{1/2}(T_i)} \\ &\lesssim \sum_{i=1}^J \left(1 + \log^2 \frac{H_j}{\delta}\right)^2 \|w\|^2_{H_w^{1/2}(W_i)} \\ &\simeq \left(1 + \log \frac{H}{\delta}\right)^2 \sum_{i=1}^J \left(\frac{1}{H} \|w\|^2_{L_2(W_i)} + |w|^2_{H^{1/2}(W_i)}\right). \end{aligned} \tag{15.43}$$

It is obvious that $\sum_{i=1}^J \|w\|^2_{L_2(W_i)} \simeq \|w\|_{L_2(\mathbb{S})}$ and by the definition of the seminorm $|\cdot|_{H^{1/2}(W_i)}$, it is clear that

$$\sum_{i=1}^J |w|^2_{H^{1/2}(W_i)} \lesssim |w|^2_{H^{1/2}(\mathbb{S})}.$$

This together with (15.43), (15.41) and (15.40) implies

$$\begin{aligned} \sum_{i=1}^J a(u_i,u_i) &\lesssim \left(1 + \log \frac{H}{\delta}\right)^2 \left(\frac{1}{H} \|w\|^2_{L_2(\mathbb{S})} + |w|^2_{H^{1/2}(\mathbb{S})}\right) \\ &\lesssim \left(1 + \log \frac{H}{\delta}\right)^2 \|u\|^2_{H^{1/2}(\mathbb{S})} \\ &\simeq \left(1 + \log \frac{H}{\delta}\right)^2 a(u,u), \end{aligned}$$

completing the proof of the lemma. □

#### 15.3.2.5 Bounds for the condition number of the preconditioned system

**Lemma 15.9 (Coercivity of the decomposition).** *There exists a positive constant C depending only on the smallest angle of the triangulation such that for any $u \in \mathscr{V}$ satisfying $u = \sum_{i=0}^{J} u_i$ with $u_i \in \mathscr{V}_i$ the following estimate holds*

$$a(v,v) \leq C \sum_{i=0}^{J} a(u_i,u_i).$$

*Proof.* By the assumption on the colouring of the subdomains there are at most $M$ subdomains to which any $\mathbf{x} \in \mathbb{S}$ can belong. By a standard colouring argument we have

$$\begin{aligned} a(v,v) \lesssim \|v\|^2_{H_{\rm I}^{1/2}(\mathbb{S})} &\leq \|v_0\|^2_{H_{\rm I}^{1/2}(\mathbb{S})} + \Big\|\sum_{i=1}^{J} v_i\Big\|^2_{H_{\rm I}^{1/2}(\mathbb{S})} \\ &\lesssim \|v_0\|^2_{H_{\rm I}^{1/2}(\mathbb{S})} + M\sum_{i=1}^{J} \|v_i\|^2_{H_{\rm I}^{1/2}(\mathbb{S})} \lesssim \sum_{i=0}^{J} a(v_i,v_i). \end{aligned}$$

The lemma is proved. □

Combining the results of Lemmas 15.7, 15.8, and 15.9 we obtain the following theorem.

**Theorem 15.3.** *The condition number of the additive Schwarz operator P is bounded by*

$$\kappa(P) \lesssim \frac{H}{h}.$$

*In particular, if the degree d is even then*

$$\kappa(P) \lesssim 1 + \log^2 \frac{H}{\delta}$$

*where the constants depend only of the smallest angle in $\Delta_h$ and the polynomial degree d.*

## 15.4 Numerical Results

The numerical results reported in this section are reprinted from [174] with copyright permission. In [174], we solved (15.19) with

$$f(\mathbf{x}) = f(x,y,z) = e^x(2x - xz).$$

by using spherical spline spaces $S_d^0(\Delta_h)$, in which $\Delta_h$ are spherical triangulations of the following types:

(i) Type 1: Uniform triangulations which are generated as follows. The initial (spherical) triangulation have vertices

$$(1,0,0),\ (0,1,0),\ (0,0,1),\ (-1,0,0),\ (0,-1,0),\ (0,0,-1).$$

The following finer meshes are obtained by dividing each triangle in the previous triangulation into four equal equilateral triangles (by connecting the midpoints of the three edges), resulting in triangulations with the number of vertices being 18, 66, 258, 1026, and 4098.

(ii) Type 2: Triangulations with vertices obtained from MAGSAT satellite data. The free STRIPACK package is used to generate the triangulations from these vertices. The number of vertices to be used are 204, 414, 836, and 1635.

Details of how to compute the stiffness matrix **A** and the right-hand side vector **b** of the matrix system arising from the Galerkin boundary element approximation can be found in [174, 179].

| | | $d=1$ $\omega^2=0.01$ | | $d=2$ $\omega^2=0.01$ | | $d=3$ | $\omega^2=0.1$ |
|---|---|---|---|---|---|---|---|
| $N$ | $h$ | $\kappa(\mathbf{A})$ | $\alpha$ | $\kappa(\mathbf{A})$ | $\alpha$ | $\kappa(\mathbf{A})$ | $\alpha$ |
| 18 | 0.7071 | 12.27 | | 469.36 | | 19109.90 | |
| 66 | 0.3536 | 20.90 | -0.77 | 736.53 | -0.65 | 25735.72 | -0.42 |
| 258 | 0.1768 | 39.22 | -0.91 | 1422.77 | -0.95 | 50877.82 | -0.98 |
| 1026 | 0.0883 | 75.93 | -0.95 | 2798.35 | -0.97 | 100876.63 | -0.99 |

**Table 15.1** Unpreconditioned systems with uniform triangulations; $\kappa(\mathbf{A}) = O(h^{\alpha})$

| DoF | $h$ | $\kappa(\mathbf{A})$ | $H$ | $\kappa(P)$ |
|---|---|---|---|---|
| 66 | 0.3536 | 20.90 | 0.7071 | 5.37 |
| 258 | 0.1768 | 39.22 | 0.3536 | 6.05 |
| | | | 0.7071 | 6.45 |
| 1026 | 0.0883 | 75.93 | 0.1768 | 6.46 |
| | | | 0.3536 | 6.68 |
| | | | 0.7071 | 7.45 |
| 4098 | 0.0442 | 149.40 | 0.0883 | 6.71 |
| | | | 0.1768 | 6.80 |
| | | | 0.3536 | 7.90 |
| | | | 0.7071 | 9.05 |

**Table 15.2** Condition numbers when $d=1$, $\omega^2=0.01$ with uniform triangulations

| DoF | $h$ | $\kappa(\mathbf{A})$ | $H$ | $\kappa(P)$ |
|---|---|---|---|---|
| 258 | 0.3536 | 736.53 | 0.7071 | 7.24 |
| 1026 | 0.1768 | 1422.77 | 0.3536 | 6.92 |
| | | | 0.7071 | 7.06 |
| 4098 | 0.0883 | 2798.35 | 0.1768 | 7.22 |
| | | | 0.3536 | 6.90 |
| | | | 0.7071 | 6.59 |

**Table 15.3** Condition numbers when $d = 2$, $\omega^2 = 0.01$ with uniform triangulations

| DoF | $h$ | $\kappa(\mathbf{A})$ | $H$ | $\kappa(P)$ |
|---|---|---|---|---|
| 578 | 0.3536 | 25735.72 | 0.7071 | 412.96 |
| 2306 | 0.1768 | 50877.82 | 0.3536 | 13.49 |
| | | | 0.7071 | 24.62 |
| 9218 | 0.0883 | 100876.63 | 0.1768 | 7.94 |
| | | | 0.3536 | 8.62 |
| | | | 0.7071 | 8.13 |

**Table 15.4** Condition numbers when $d = 3$, $\omega^2 = 0.1$ with uniform triangulations

| DoF | $h$ | $\kappa(\mathbf{A})$ | $H$ | $\kappa(P)$ |
|---|---|---|---|---|
| 204 | 0.511 | 390.4 | 1.307 | 19.5 |
| | | | 1.670 | 19.2 |
| | | | 2.292 | 26.7 |
| 414 | 0.370 | 509.9 | 1.307 | 17.2 |
| | | | 1.670 | 17.0 |
| | | | 2.292 | 27.1 |
| 836 | 0.280 | 785.2 | 1.307 | 13.8 |
| | | | 1.670 | 14.9 |
| | | | 2.292 | 28.3 |
| 1635 | 0.184 | 1048.5 | 1.307 | 11.1 |
| | | | 1.670 | 16.3 |
| | | | 2.292 | 31.5 |

**Table 15.5** Condition numbers when $d = 1$, $\omega^2 = 0.001$ with MAGSAT satellite data

| DoF | $h$ | $\kappa(\mathbf{A})$ | $H$ | $\kappa(P)$ |
|---|---|---|---|---|
| 810 | 0.511 | 1170.9 | 1.307 | 23.2 |
| | | | 1.670 | 11.7 |
| | | | 2.292 | 5.7 |
| 1650 | 0.370 | 1616.9 | 1.307 | 17.7 |
| | | | 1.670 | 9.0 |
| | | | 2.292 | 10.0 |
| 3338 | 0.280 | 2230.5 | 1.307 | 13.2 |
| | | | 1.670 | 11.1 |
| | | | 2.292 | 8.1 |
| 6534 | 0.184 | 3046.7 | 1.307 | 13.1 |
| | | | 1.670 | 9.0 |
| | | | 2.292 | 6.3 |

**Table 15.6** Condition numbers when $d = 2$, $\omega^2 = 0.01$ with MAGSAT satellite data

| DoF | $h$ | $\kappa(\mathbf{A})$ | $H$ | $\kappa(P)$ |
|---|---|---|---|---|
| 1820 | 0.511 | 3322.0 | 1.307 | 23.9 |
| | | | 1.670 | 10.9 |
| | | | 2.292 | 116.6 |
| 3710 | 0.370 | 4696.2 | 1.307 | 16.5 |
| | | | 1.670 | 7.7 |
| | | | 2.292 | 186.9 |
| 7508 | 0.280 | 6211.0 | 1.307 | 12.1 |
| | | | 1.670 | 12.0 |
| | | | 2.292 | 167.6 |

**Table 15.7** Condition numbers when $d = 3$, $\omega^2 = 1.2$ with MAGSAT satellite data

# Chapter 16
# Pseudo-differential Equations with Radial Basis Functions

In this chapter we consider again pseudo-differential equations on the sphere. However, the chapter is diverted from the main theme of other chapters in that boundary elements are not the tool used to solve the equations. Instead, we use radial basis functions (RBFs) which results in a meshless method. The benefit of RBFs lies in the fact that these functions allow us to deal with scattered data (obtained from satellites) which are usually unstructured. It is known that the RBF method converges fast but it is also known that it yields very ill-conditioned matrix systems. Therefore, preconditioners are particularly important. We present in this chapter the analysis of additive Schwarz preconditioners with RBFs, which is entirely different from the analysis usually seen in the case of boundary element methods.

We will consider the same problem and notations as introduced in Section 15.1.2, i.e., we will solve the following problem:

*Let $L$ be an elliptic pseudo-differential operator of order $2\alpha$. Given, for some $\sigma \geq 0$,*

$$g \in H^{\sigma-\alpha} \quad \textit{satisfying} \quad \widehat{g}_{\ell,m} = 0 \quad \textit{for all } \ell \in \mathscr{K}(L),\ m = 1, \ldots, N(n,\ell),$$

*find $u \in H^{\sigma+\alpha}$ satisfying*

$$\begin{aligned} Lu &= g, \\ \langle \mu_i, u \rangle &= \gamma_i, \quad i = 1, \ldots, M, \end{aligned} \tag{16.1}$$

*where $\gamma_i \in \mathbb{R}$ and $\mu_i \in H^{-\sigma-\alpha}$ are given.*

It should be noted that the inclusion of $\sigma \geq 0$ in the assumption on $g$ is to allow for the consideration of the collocation method. For the Galerkin method, it suffices to take $\sigma = 0$.

Existence and uniqueness of (16.1) has been discussed in Section 15.1.2. We define a bilinear form $a(\cdot,\cdot) : H^{\alpha+s} \times H^{\alpha-s} \to \mathbb{R}$, for any $s \in \mathbb{R}$, by

$$a(w,v) := \langle Lw, v \rangle \quad \text{for all } w \in H^{\alpha+s},\ v \in H^{\alpha-s}.$$

In particular, when $s = \sigma$ we have (see Section 15.1.2)



E. P. Stephan and T. Tran, *Schwarz Methods and Multilevel Preconditioners for Boundary Element Methods*, https://doi.org/10.1007/978-3-030-79283-1_16

$$a(u_1, v) = \langle g, v \rangle \quad \text{for all } v \in H^{\alpha-\sigma}. \tag{16.2}$$

## 16.1 Radial Basis Functions

In this section we introduce the definition of RBFs and present their basic properties. The finite-dimensional subspaces to be used in the approximation will be defined from spherical radial basis functions, which in turn are defined from kernels.

### *16.1.1 Positive-definite kernels*

A continuous function $\Theta : \mathbb{S}^{n-1} \times \mathbb{S}^{n-1} \to \mathbb{R}$ is called a *positive-definite kernel* on $\mathbb{S}^{n-1}$ if it satisfies

(i) $\Theta(\mathbf{x}, \mathbf{y}) = \Theta(\mathbf{y}, \mathbf{x})$ for all $\mathbf{x}, \mathbf{y} \in \mathbb{S}^{n-1}$,
(ii) for any positive integer $N$ and any set of distinct points $\{\mathbf{y}_1, \dots, \mathbf{y}_N\}$ on $\mathbb{S}^{n-1}$, the $N \times N$ matrix $\mathbf{B}$ with entries $\mathbf{B}_{i,j} = \Theta(\mathbf{y}_i, \mathbf{y}_j)$ is positive-semidefinite.

If the matrix $\mathbf{B}$ is positive-definite then $\Theta$ is called a *strictly positive-definite* kernel; see [189, 242].

We characterise the kernel $\Theta$ by a *shape function* $\theta$ as follows. Let $\theta : [-1, 1] \to \mathbb{R}$ be a univariate function having a series expansion in terms of Legendre polynomials,

$$\theta(t) = \sum_{\ell=0}^{\infty} \omega_{n-1}^{-1} N(n, \ell) \widehat{\theta}(\ell) P_\ell(n; t),$$

where $\omega_{n-1}$ is the surface area of the sphere $\mathbb{S}^{n-1}$, and $\widehat{\theta}(\ell)$ is the Fourier–Legendre coefficient,

$$\widehat{\theta}(\ell) = \omega_{n-1} \int_{-1}^{1} \theta(t) P_\ell(n; t) (1 - t^2)^{(n-3)/2} \, dt.$$

Here, $P_\ell(n; t)$ denotes the degree $\ell$ normalised Legendre polynomial in $n$ variables so that $P_\ell(n; 1) = 1$, as described in [155]. Using this shape function $\theta$, we define

$$\Theta(\mathbf{x}, \mathbf{y}) := \theta(\mathbf{x} \cdot \mathbf{y}) \quad \text{for all } \mathbf{x}, \mathbf{y} \in \mathbb{S}^{n-1}, \tag{16.3}$$

where $\mathbf{x} \cdot \mathbf{y}$ denotes the dot product between $\mathbf{x}$ and $\mathbf{y}$. By using the well-known addition formula for spherical harmonics [155],

$$\sum_{m=1}^{N(n,\ell)} Y_{\ell,m}(\mathbf{x}) Y_{\ell,m}(\mathbf{y}) = \omega_{n-1}^{-1} N(n, \ell) P_\ell(n; \mathbf{x} \cdot \mathbf{y}) \quad \text{for all } \mathbf{x}, \mathbf{y} \in \mathbb{S}^{n-1}, \tag{16.4}$$

we can write

$$\Theta(\mathbf{x},\mathbf{y}) = \sum_{\ell=0}^{\infty} \sum_{m=1}^{N(n,\ell)} \widehat{\theta}(\ell) Y_{\ell,m}(\mathbf{x}) Y_{\ell,m}(\mathbf{y}). \tag{16.5}$$

*Remark 16.1.* In [59], a complete characterisation of strictly positive-definite kernels is established: the kernel $\Theta$ is strictly positive-definite if and only if $\widehat{\theta}(\ell) \geq 0$ for all $\ell \geq 0$, and $\widehat{\theta}(\ell) > 0$ for infinitely many even values of $\ell$ and infinitely many odd values of $\ell$; see also [189] and [242].

In the remainder of this section, we shall define specific shape functions $\phi$ and $\psi$, and corresponding kernels $\Phi$ and $\Psi$, which will be used to define the approximation subspaces. The notations $\theta$ and $\Theta$ are reserved for future general reference.

### 16.1.2 Spherical radial basis functions

We choose a shape function $\phi$ satisfying for some $\tau \in \mathbb{R}$

$$\widehat{\phi}(\ell) \simeq (\ell+1)^{-2\tau} \quad \text{for all } \ell \geq 0. \tag{16.6}$$

A specific choice of $\phi$ will be clarified in Section 16.4. We also define from $\phi$ another shape function $\psi : [-1,1] \to \mathbb{R}$ by

$$\psi(t) = \sum_{\ell=0}^{\infty} \omega_{n-1}^{-1} N(n,\ell) \widehat{L}(\ell) \widehat{\phi}(\ell) P_\ell(n;t).$$

Since $\widehat{\psi}(\ell) = \widehat{L}(\ell)\widehat{\phi}(\ell)$, it follows from (15.7) and (16.6) that

$$|\widehat{\psi}(\ell)| \simeq (\ell+1)^{-2(\tau-\alpha)} \quad \text{for all } \ell \notin \mathscr{K}(L). \tag{16.7}$$

The corresponding kernels $\Phi$ and $\Psi$ are defined by (16.3), i.e.,

$$\Phi(\mathbf{x},\mathbf{y}) = \phi(\mathbf{x}\cdot\mathbf{y}) \quad \text{and} \quad \Psi(\mathbf{x},\mathbf{y}) = \psi(\mathbf{x}\cdot\mathbf{y}) \quad \text{for all} \quad \mathbf{x},\mathbf{y} \in \mathbb{S}^{n-1}.$$

We will define spherical radial basis functions by using the kernels $\Phi$ and $\Psi$. These functions are basis functions for the test and trial spaces in the Galerkin scheme for (16.2).

Let $X = \{\mathbf{x}_1, \ldots, \mathbf{x}_N\}$ be a set of data points on the sphere. Two important parameters characterising the set $X$ are the *mesh norm* $h_X$ and *separation radius* $q_X$, defined by

$$h_X := \sup_{\mathbf{y}\in\mathbb{S}^{n-1}} \min_{1\leq j\leq N} \cos^{-1}(\mathbf{x}_j \cdot \mathbf{y}) \quad \text{and} \quad q_X := \frac{1}{2} \min_{\substack{i\neq j \\ 1\leq i,j\leq N}} \cos^{-1}(\mathbf{x}_i \cdot \mathbf{x}_j). \tag{16.8}$$

The *spherical radial basis functions* $\Phi_j$ and $\Psi_j$, $j = 1, \ldots, N$, associated with $X$ and the kernels $\Phi$ and $\Psi$ (respectively) are defined by (see (16.5))

$$\Phi_j(\mathbf{x}) := \Phi(\mathbf{x},\mathbf{x}_j) = \sum_{\ell=0}^{\infty}\sum_{m=1}^{N(n,\ell)} \widehat{\phi}(\ell) Y_{\ell,m}(\mathbf{x}_j) Y_{\ell,m}(\mathbf{x}),$$

$$\Psi_j(\mathbf{x}) := \Psi(\mathbf{x},\mathbf{x}_j) = \sum_{\ell=0}^{\infty}\sum_{m=1}^{N(n,\ell)} \widehat{L}(\ell)\widehat{\phi}(\ell) Y_{\ell,m}(\mathbf{x}_j) Y_{\ell,m}(\mathbf{x}).$$

We note that for $j = 1,\dots,N$ there hold $\Psi_j = L\Phi_j$ and

$$\widehat{(\Phi_j)}_{\ell,m} = \widehat{\phi}(\ell) Y_{\ell,m}(\mathbf{x}_j) \quad \text{and} \quad \widehat{(\Psi_j)}_{\ell,m} = \widehat{L}(\ell)\widehat{\phi}(\ell) Y_{\ell,m}(\mathbf{x}_j),$$

for $m = 1,\dots,N(n,\ell)$, and $\ell = 0,1,2,\dots$. It follows from (16.6) that, for any $s \in \mathbb{R}$,

$$\sum_{\ell=0}^{\infty}\sum_{m=1}^{N(n,\ell)} (\ell+1)^{2s} \left|\widehat{(\Phi_j)}_{\ell,m}\right|^2 \simeq \sum_{\ell=0}^{\infty}\sum_{m=1}^{N(n,\ell)} (\ell+1)^{2(s-2\tau)} |Y_{\ell,m}(\mathbf{x}_j)|^2.$$

By using (16.4) and noting $P_\ell(n;\mathbf{x}_j\cdot\mathbf{x}_j) = P_\ell(n;1) = 1$ we obtain, recalling that $N(n,\ell) = O(\ell^{n-2})$,

$$\sum_{\ell=0}^{\infty}\sum_{m=1}^{N(n,\ell)} (\ell+1)^{2s} \left|\widehat{(\Phi_j)}_{\ell,m}\right|^2 \simeq \sum_{\ell=0}^{\infty} (\ell+1)^{2(s-2\tau)+n-2}.$$

The latter series converges if and only if $s < 2\tau + (1-n)/2$. Hence,

$$\Phi_j \in H^s \iff s < 2\tau + \frac{1-n}{2}. \tag{16.9}$$

Similarly, it follows from (16.7) that

$$\Psi_j \in H^s \iff s < 2(\tau-\alpha) + \frac{1-n}{2}.$$

Using $\Phi_i$ and $\Psi_i$, we are now able to define the finite dimensional subspaces for our approximation.

For our approximation schemes we consider a family of sets of data points $X_k = \{\mathbf{x}_1,\dots,\mathbf{x}_{N_k}\}$ for $k \in \mathbb{N}$ where $N_k < N_{k+1}$ so that $X_k \subset X_{k+1}$ and the corresponding mesh norm $h_k$, see (16.8), satisfies $h_k \to 0$ when $k \to \infty$.

The corresponding finite-dimensional spaces are defined by

$$\mathcal{V}_k^{\phi} := \operatorname{span}\{\Phi_1,\dots,\Phi_{N_k}\} \quad \text{and} \quad \mathcal{V}_k^{\psi} := \operatorname{span}\{\Psi_1,\dots,\Psi_{N_k}\}.$$

We note that $\mathcal{V}_k^{\phi} \subset \mathcal{V}_{k+1}^{\phi}$ and $\mathcal{V}_k^{\psi} \subset \mathcal{V}_{k+1}^{\psi}$ for all $k \in \mathbb{N}$.

The following approximation result is proved in [178, Proposition 3.3]; see also [176, Proposition 3.2 and Remark 3.3].

**Proposition 16.1.** *Assume that* (16.6) *holds for some* $\tau > (n-1)/2$. *Let* $s^*, t^* \in \mathbb{R}$ *satisfy* $t^* \le s^* \le 2\tau$ *and* $t^* \le \tau$. *There exists a positive constant* $C$ *such that, for any* $v \in H^{s^*}$ *and* $k > 0$, *there exists* $\eta_k \in \mathcal{V}_k^{\phi}$ *satisfying*

$$\|v - \eta_k\|_{t^*} \leq Ch_k^{s^*-t^*} \|v\|_{s^*}.$$

### 16.1.3 Native space and reproducing kernel property

In the analysis for the collocation method we will use the native space $\mathcal{N}_\phi$ associated with $\phi$ defined by

$$\mathcal{N}_\phi := \left\{ v \in \mathscr{D}'(\mathbb{S}^{n-1}) : \|v\|_\phi^2 = \sum_{\ell=0}^{\infty} \sum_{m=1}^{N(n,\ell)} \frac{|\widehat{v}_{\ell,m}|^2}{\widehat{\phi}(\ell)} < \infty \right\}.$$

We equip this space with an inner product and a norm

$$\langle v, w \rangle_\phi = \sum_{\ell=0}^{\infty} \sum_{m=1}^{N(n,\ell)} \frac{\widehat{v}_{\ell,m}\widehat{w}_{\ell,m}}{\widehat{\phi}(\ell)} \quad \text{and} \quad \|v\|_\phi = \left( \sum_{\ell=0}^{\infty} \sum_{m=1}^{N(n,\ell)} \frac{|\widehat{v}_{\ell,m}|^2}{\widehat{\phi}(\ell)} \right)^{1/2}.$$

Since $\widehat{\phi}(\ell)$ satisfies (16.6), the native space $\mathcal{N}_\phi$ can be identified with the Sobolev space $H^\tau$, and the corresponding norms are equivalent.

It is noted that for any function $v$ in the native space $\mathcal{N}_\phi$ there holds

$$v(\mathbf{x}_i) = \sum_{\ell=0}^{\infty} \sum_{m=1}^{N(n,\ell)} \frac{\widehat{v}_{\ell,m}\widehat{\phi}(\ell) Y_{\ell,m}(\mathbf{x}_i)}{\widehat{\phi}(\ell)} = \langle v, \Phi_i \rangle_\phi, \quad i = 1, \ldots, N;$$

see [234, page 134].

## 16.2 Solving Pseudo-differential Equations by Radial Basis Functions

In this section we briefly present how equation 16.2 is solved and state the convergence rate of the approximation by the Galerkin method. For the approximation by the collocation method and for the consideration of non strongly elliptic equations, the reader is referred to [178] and [176], respectively.

Recalling (16.9), we choose the shape functions $\phi$ in this subsection such that

$$\tau > \frac{1}{2}\left(\alpha + \frac{n-1}{2}\right), \tag{16.10}$$

so that $\mathcal{V}^\phi \subset H^\alpha$. We find $\widetilde{u}_1 \in \mathcal{V}^\phi$ by solving the Galerkin equation

$$a(\widetilde{u}_1, v) = \langle g, v \rangle \quad \text{for all } v \in \mathcal{V}^\phi. \tag{16.11}$$

Existence of a unique solution $\widetilde{u}_1$ follows from the positive definiteness of the kernel; see [178, Lemma 5.1]. The following error estimate is also proved in the same article.

**Theorem 16.1.** *Assume that the shape function $\phi$ is chosen to satisfy* (16.6), (16.10) *and $\tau \geq \alpha$, $\tau > (n-1)/2$. Assume further that $u \in H^s$ for some $s$ satisfying $\alpha \leq s \leq 2\tau$. If $\mu_i \in H^{-t}$ for $i = 1, \dots, M$ with $t \in \mathbb{R}$ satisfying $2(\alpha - \tau) \leq t \leq \alpha$, then for $h_X$ sufficiently small there holds*

$$\|u - \widetilde{u}\|_t \leq C h_X^{s-t} \|u\|_s.$$

*The constant $C$ is independent of $u$ and $h_X$.*

## 16.3 Additive Schwarz Methods

By writing $\widetilde{u}_1 = \sum_{i=1}^N c_i \Phi_i$ we derive from (16.11) the matrix equation $\mathbf{A}\mathbf{ç} = \mathbf{g}$, where

$$\mathbf{A}_{ij} = a(\Phi_i, \Phi_j) = \sum_{\ell=0}^{\infty} \sum_{m=1}^{N(n,\ell)} \widehat{L}(\ell) \, [\widehat{\phi}(\ell)]^2 \, Y_{\ell,m}(\mathbf{x}_i) Y_{\ell,m}(\mathbf{x}_j), \tag{16.12}$$

$\mathbf{c} = (c_1, \dots, c_N)$, and $\mathbf{g} = (\langle g, \Phi_1 \rangle, \dots, \langle g, \Phi_N \rangle)$. It is well known [157] that this matrix is ill-conditioned; see also the numerical results in Section 16.4. In this section we present how an overlapping Schwarz preconditioner is designed. We also carry out the analysis which, even though still follows the general framework of Theorem 2.1, requires a completely original approach due to the non-local structure of radial basis functions.

In the remainder of this chapter, we return to our standard notations in Chapter 2, namely we will write $\mathscr{V}$ for $\mathscr{V}^\phi$. Moreover, we will use a norm for Sobolev spaces on the sphere $\mathbb{S}$ which facilitates the analysis. This norm requires the introduction of the following concepts.

A spherical cap of radius $\alpha$ centered at $\mathbf{p} \in \mathbb{S}$ is defined by

$$C(\mathbf{p}, \alpha) := \{\mathbf{x} \in \mathbb{S} : \theta(\mathbf{p}, \mathbf{x}) < \alpha\},$$

where $\theta(\mathbf{p}, \mathbf{x}) = \cos^{-1}(\mathbf{p} \cdot \mathbf{x})$ is the geodesic distance between two points $\mathbf{x}, \mathbf{p} \in \mathbb{S}$. Let $\widehat{\mathbf{n}}$ and $\widehat{\mathbf{s}}$ denote the north and south poles of $\mathbb{S}$, respectively. Then a simple cover for the sphere is provided by

$$U_1 = C(\widehat{\mathbf{n}}, \theta_0) \quad \text{and} \quad U_2 = C(\widehat{\mathbf{s}}, \theta_0), \text{ where } \theta_0 \in (\pi/2, 2\pi/3).$$

The stereographic projection $\sigma_{\widehat{\mathbf{n}}}$ of the punctured sphere $\mathbb{S} \setminus \{\widehat{\mathbf{n}}\}$ onto $\mathbb{R}^n$ is defined as a mapping that maps $\mathbf{x} \in \mathbb{S} \setminus \{\widehat{\mathbf{n}}\}$ to the intersection of the equatorial hyperplane $\{z = 0\}$ and the extended line that passes through $\mathbf{x}$ and $\widehat{\mathbf{n}}$. The stereographic projection $\sigma_{\widehat{\mathbf{s}}}$ based on $\widehat{\mathbf{s}}$ can be defined analogously. We set

$$\psi_1 = \frac{1}{\tan(\theta_0/2)}\sigma_{\widehat{\mathbf{s}}}|_{U_1} \quad \text{and} \quad \psi_2 = \frac{1}{\tan(\theta_0/2)}\sigma_{\widehat{\mathbf{n}}}|_{U_2}, \tag{16.13}$$

so that $\psi_k$, $k = 1,2$, maps $U_k$ onto $B(0,1)$, the unit ball in $\mathbb{R}^n$. We conclude that $\mathscr{A} = \{U_k, \psi_k\}_{k=1}^2$ is a $C^\infty$ atlas of covering coordinate charts for the sphere. It is known [180] that the stereographic coordinate charts $\{\psi_k\}_{k=1}^2$ as defined in (16.13) map spherical caps to Euclidean balls, but in general concentric spherical caps are not mapped to concentric Euclidean balls. The projection $\psi_k$, for $k = 1,2$, does not distort too much the geodesic distance between two points $\mathbf{x},\mathbf{y} \in \mathbb{S}$, as shown in [136]. With the atlas so defined, we define the map $\pi_k$ which takes a real-valued function $g$ with compact support in $U_k$ into a real-valued function on $\mathbb{R}^n$ by

$$\pi_k(g)(\mathbf{x}) = \begin{cases} g \circ \psi_k^{-1}(\mathbf{x}), & \text{if } \mathbf{x} \in B(0,1), \\ 0, & \text{otherwise}. \end{cases}$$

Let $\{\chi_k : \mathbb{S} \to \mathbb{R}\}_{k=1}^2$ be a partition of unity subordinated to the atlas, i.e., a pair of non-negative infinitely differentiable functions $\chi_k$ on $\mathbb{S}$ with compact support in $U_k$, such that $\sum_k \chi_k = 1$. For any function $f : \mathbb{S} \to \mathbb{R}$, we can use the partition of unity to write

$$f = \sum_{k=1}^{2} (\chi_k f), \text{ where } (\chi_k f)(\mathbf{p}) = \chi_k(\mathbf{p}) f(\mathbf{p}), \quad \mathbf{p} \in \mathbb{S}.$$

The Sobolev space $H^s$ is defined to be the set

$$\{f \in L^2(\mathbb{S}) : \pi_k(\chi_k f) \in H^s(\mathbb{R}^n) \quad \text{for } k = 1,2\},$$

which is equipped with the norm

$$\|f\|_{H^s(\mathbb{S})} = \left( \sum_{k=1}^{2} \|\pi_k(\chi_k f)\|_{H^s(\mathbb{R}^n)}^2 \right)^{1/2}. \tag{16.14}$$

This $H^s(\mathbb{S})$-norm is equivalent to the $H^s$ norm given in Appendix A; see [138].

### *16.3.1 Subspace decomposition*

We assume that we are given a finite set $X = \{\mathbf{x}_1, \ldots, \mathbf{x}_N\}$ of points on $\mathbb{S}^n$. To define a subspace decomposition of $\mathscr{V}$, we first decompose the data set $X$ in the form $X = X_0 \cup \cdots \cup X_J$ as follows.

(1) Select $\alpha \in (0, \pi/3)$ and $\beta \in [\alpha, \pi]$.
(2) Choose first centre $\mathbf{p}_1 = \mathbf{x}_1 \in X$.
(3) Define $X_1 := \{\mathbf{x} \in X : \cos^{-1}(\mathbf{x} \cdot \mathbf{p}_1) \le \alpha\}$.
(4) Suppose $X_{j-1}$, $j > 1$, has been defined. Then $\mathbf{p}_j$ is chosen from $X \setminus \{\mathbf{p}_1, \cdots, \mathbf{p}_{j-1}\}$ such that $\cos^{-1}(\mathbf{p}_{j-1} \cdot \mathbf{p}_j) \ge \beta$ and $\cos^{-1}(\mathbf{p}_l \cdot \mathbf{p}_j) \ge \alpha$ for $l = 1, \ldots, j-2$.

(5) Define $X_j := \{\mathbf{x} \in X : \cos^{-1}(\mathbf{x}\cdot\mathbf{p}_j) \le \alpha\}$.
(6) Repeat (4) and (5) until every point in $X$ is in at least one $X_j$.
(7) Define $X_0 := \{\mathbf{p}_1,\dots,\mathbf{p}_J\}$.

The second condition in Step (4) ensures that the geodesic distance between centres is no less than $\alpha$, guaranteeing that the algorithm terminates. We note that the subsets $X_j$ overlap. The subspaces $\mathscr{V}_j$ can now be defined by

$$\mathscr{V}_j = \operatorname{span}\{\Phi_k : \mathbf{x}_k \in X_j\}, \quad j = 0,\dots,J,$$

and the following subspace decomposition is considered

$$\mathscr{V} = \mathscr{V}_0 + \cdots + \mathscr{V}_J.$$

Assume that the support of $\Phi(\mathbf{p},\cdot)$, which is a spherical cap centered at $\mathbf{p}$, has radius $\gamma$. (In case the spherical basis functions are unscaled Wendland's radial basis functions [234], we have $\gamma = \pi/3$). Then, functions in $\mathscr{V}_j$ have supports in $\Gamma_j$, where

$$\Gamma_j := C(\mathbf{p}_j, \alpha+\gamma), \quad j = 1,\dots,J.$$

As before, we assume that:

**Assumption 16.1.** We can partition the index set $\{1,\dots,J\}$ into $M$ (with $1 \le M \le J$) sets $J_m$, for $1 \le m \le M$, such that if $i, j \in J_m$ and $i \ne j$ then $\Gamma_i \cap \Gamma_j = \varnothing$.

### *16.3.2 Coercivity of the decomposition*

The proof for coercivity of decomposition is similar to previous proofs with different settings of problems and methods, except that we have to use the projected norm (16.14).

**Lemma 16.1 (Coercivity of decomposition).** *Under Assumption 16.1, there exists a positive constant $c$ independent of the set $X$ such that for every $v \in \mathscr{V}$ satisfying $v = \sum_{j=0}^J v_j$ with $v_j \in \mathscr{V}_j$ for $j = 0,\dots,J$ there holds*

$$a(v,v) \le cM \sum_{j=0}^{J} a(v_j,v_j).$$

*Proof.* Using the inequality $|a+b|^2 \le 2(|a|^2+|b|^2)$, we have

$$\|v\|_{H^\sigma(\mathbb{S})}^2 \le 2\left( \|v_0\|_{H^\sigma(\mathbb{S})}^2 + \left\| \sum_{j=1}^{J} v_j \right\|_{H^\sigma(\mathbb{S})}^2 \right).$$

Recalling the definition of the Sobolev norm (16.14), we have

$$\left\|\sum_{j=1}^{J} v_j\right\|_{H^\sigma(\mathbb{S})}^2 = \left\|\sum_{j=1}^{J} \pi_1(\chi_1 v_j)\right\|_{H^\sigma(\mathbb{R}^2)}^2 + \left\|\sum_{j=1}^{J} \pi_2(\chi_2 v_j)\right\|_{H^\sigma(\mathbb{R}^2)}^2.$$

Now, from the fact that $v_j \in V_j$ together with Assumption 16.1 we can partition the index set $\{1,\dots,J\}$ to $M$ sets of indices $J_m$ so that if $i,j \in J_m$ then supp $v_i \cap$ supp $v_j = \varnothing$. In this proof only, let $g_j = \pi_1(\chi_1 v_j)$ with $\Omega_j :=$ supp $g_j$. By using successively the triangle inequality and the Cauchy-Schwarz inequality, we have

$$\begin{aligned}\left\|\sum_{j=1}^{J} g_j\right\|_{H^\sigma(\mathbb{R}^2)}^2 &= \left\|\sum_{m=1}^{M}\sum_{j\in J_m} g_j\right\|_{H^\sigma(\mathbb{R}^2)}^2 \le \left(\sum_{m=1}^{M}\left\|\sum_{j\in J_m} g_j\right\|_{H^\sigma(\mathbb{R}^2)}\right)^2 \\ &\le M\sum_{m=1}^{M}\left\|\sum_{j\in J_m} g_j\right\|_{H^\sigma(\mathbb{R}^2)}^2.\end{aligned}$$

With the partition $\{\Omega_j : j \in J_m\}$ of $\Omega$ where $\Omega = \operatorname{supp}\left(\sum_{j\in J_m} g_j\right)$ we have, due to Theorem A.11 in Appendix A,

$$\|\sum_{j\in J_m} g_j\|_{H^\sigma(\mathbb{R}^2)}^2 = \|\sum_{j\in J_m} g_j\|_{\widetilde{H}^\sigma(\Omega)}^2 \le \sum_{j\in J_m} \|g_j\|_{\widetilde{H}^\sigma(\Omega)}^2 = \sum_{j\in J_m} \|g_j\|_{H^\sigma(\mathbb{R}^2)}^2.$$

Thus,

$$\left\|\sum_{j=1}^{J} g_j\right\|_{H^\sigma(\mathbb{R}^2)}^2 \le M\sum_{m=1}^{M}\sum_{j\in J_m} \|g_j\|_{H^\sigma(\mathbb{R}^2)}^2 = M\sum_{j=1}^{J} \|g_j\|_{H^\sigma(\mathbb{R}^2)}^2.$$

Hence, by using similar arguments for $\pi_2(\chi_2 v_j)$, we conclude

$$\begin{aligned}\left\|\sum_{j=1}^{J} v_j\right\|_{H^\sigma(\mathbb{S})}^2 &\le cM\left(\sum_{j=1}^{J} \|\pi_1(\chi_1 v_j)\|_{H^\sigma(\mathbb{R}^2)}^2 + \sum_{j=1}^{J} \|\pi_2(\chi_2 v_j)\|_{H^\sigma(\mathbb{R}^2)}^2\right) \\ &= cM\sum_{j=1}^{J} \|v_j\|_{H^\sigma(\mathbb{S})}^2.\end{aligned}$$

Therefore,

$$\|v\|_{H^\sigma(\mathbb{S})}^2 = \left\|\sum_{j=0}^{J} v_j\right\|_{H^\sigma(\mathbb{S})}^2 \le cM\sum_{j=0}^{J} \|v_j\|_{H^\sigma(\mathbb{S})}^2.$$

Using the fact that $a(v,v) \simeq \|v\|_{H^\sigma(\mathbb{S})}^2$ we obtain the result. □

### 16.3.3 Stability of the decomposition and bounds for the minimum eigenvalue of $P$

As we have mentioned, to prove stability in this case since radial basis functions are not local to the subdomains, we have to employ a different approach as with boundary element methods. We need the following concepts.;

Firstly, we recall the following definition of angle between closed subspaces of a general Hilbert space.

**Definition 16.1.** *Let $\mathfrak{V}$ be a Hilbert space with inner product and norm denoted by $\langle\cdot,\cdot\rangle$ and $\|\cdot\|$, respectively. Assume that $\mathfrak{U}_1$ and $\mathfrak{U}_2$ are two closed subspaces of $\mathfrak{V}$. The angle $\alpha$ between $\mathfrak{U}_1$ and $\mathfrak{U}_2$ is the angle in $[0,\pi/2]$ whose cosine is given by*

$$\cos\alpha = \sup\{\langle v,w\rangle : v\in\mathfrak{U}_1\cap\mathfrak{U}^\perp,\quad w\in\mathfrak{U}_2\cap\mathfrak{U}^\perp,\quad \|v\|\le 1,\ \|w\|\le 1\},$$

*where $\mathfrak{U}=\mathfrak{U}_1\cap\mathfrak{U}_2$, and $\mathfrak{U}^\perp$ is its orthogonal complement, namely,*

$$\mathfrak{U}^\perp := \{f\in\mathfrak{V} : \langle f,v\rangle = 0 \quad \forall v\in\mathfrak{U}\}.$$

It follows from the definition of orthogonal complement that

$$\begin{aligned}(\mathfrak{U}_1+\mathfrak{U}_2)^\perp &= \{f\in\mathscr{V} : \langle f,v\rangle = 0 \quad \forall v\in\mathfrak{U}_1+\mathfrak{U}_2\}\\ &= \{f\in\mathscr{V} : \langle f,v_1\rangle = 0 = \langle f,v_2\rangle \quad \forall v_1\in\mathfrak{U}_1 \text{ and } v_2\in\mathfrak{U}_2\}\\ &= \mathfrak{U}_1^\perp\cap\mathfrak{U}_2^\perp. \end{aligned} \tag{16.15}$$

The following theorem [195, Theorem 2.2] is crucial in our estimate of the minimum eigenvalue of $P$.

**Theorem 16.2.** *Let $\mathfrak{V}_1,\ \ldots,\ \mathfrak{V}_J$ be closed subspaces of a Hilbert space $\mathfrak{V}$, and $\mathfrak{W}_i := \cap_{j=i}^J\mathfrak{V}_j$, $i=1,\ldots,J$. If $Q_i:\mathfrak{V}\to\mathfrak{V}_i$ is the orthogonal projection onto $\mathfrak{V}_i$, $i=1,\ldots,J$, and $Q:\mathfrak{V}\to\mathfrak{W}_1$ is the orthogonal projection onto $\mathfrak{W}_1$, then*

$$\|\widetilde{Q}^l f - Qf\| \le c^l\|f-Qf\|,\quad \forall f\in\mathfrak{V},\quad l=1,2,\ldots,$$

*where $\widetilde{Q} := Q_J\cdots Q_1$ and*

$$c^2 = 1-\prod_{i=1}^{J-1}\sin^2\alpha_i,$$

*with $\alpha_i$ being the angle between $\mathfrak{V}_i$ and $\mathfrak{W}_{i+1}$.*

We shall apply Theorem 16.2 with $\mathfrak{V}$ being $\mathscr{V}$, which is equipped with the inner product $a(\cdot,\cdot)$ and induced norm $\|\cdot\|_a$, and $\mathfrak{V}_j$ being $\mathscr{V}_j^\perp$, $j=1,\ldots,J$. If $T$ is a linear operator on $\mathscr{V}$ we denote by $\|T\|_a$ the norm of $T$ defined by $\|\cdot\|_a$, i.e.,

$$\|T\|_a = \sup_{\substack{v\in\mathscr{V}\\ \|v\|_a\le 1}}\|Tv\|_a.$$

**Proposition 16.2.** *Let $\widetilde{Q} := Q_J \cdots Q_1$ where $Q_i$ is the orthogonal projection from $\mathscr{V}$ onto $\mathscr{V}_i^{\perp}$, and let $\mathscr{W}_i := \cap_{j=i}^{J} \mathscr{V}_j^{\perp}$, $i = 1, \ldots, J$. Then*

$$\|\widetilde{Q}\|_a \leq \left(1 - \prod_{i=1}^{J-1} \sin^2 \alpha_i\right)^{1/2} < 1,$$

*where $\alpha_i$ is the angle between $\mathscr{V}_i^{\perp}$ and $\mathscr{W}_{i+1}$.*

**Proof.** First we note that since $X_0 \subset X_1 \cup \cdots \cup X_J$, there holds $\mathscr{V} = \mathscr{V}_1 + \cdots + \mathscr{V}_J$. It follows from (16.15) that

$$\mathscr{W}_1^{\perp} = (\mathscr{V}_1 + \cdots + \mathscr{V}_J)^{\perp\perp} = \mathscr{V}^{\perp\perp} = \mathscr{V}.$$

Hence $\mathscr{W}_1 = \{0\}$ which implies that the orthogonal projection $Q$ from $\mathscr{V}$ onto $\mathscr{W}_1$ is identically zero. Theorem 16.2 then yields

$$\|\widetilde{Q}\|_a \leq \left(1 - \prod_{i=1}^{J-1} \sin^2 \alpha_i\right)^{1/2}.$$

It remains to show that $\alpha_i \neq 0$ for all $i = 1, \ldots, J-1$. Suppose that $\alpha_i = 0$ for some $i \in \{1, \ldots, J-1\}$. Then noting that

$$(\mathscr{V}_i^{\perp} \cap \mathscr{W}_{i+1})^{\perp} = W_i^{\perp},$$

we obtain from Definition 16.1

$$\sup\{a(v,w) : v \in \mathscr{V}_i^{\perp} \cap \mathscr{W}_i^{\perp},\ w \in \mathscr{W}_{i+1} \cap \mathscr{W}_i^{\perp},\ \|v\|_a \leq 1,\ \|w\|_a \leq 1\} = 1.$$

The spaces being finite dimensional, by compactness there exist $v \in \mathscr{V}_i^{\perp} \cap \mathscr{W}_i^{\perp}$ and $w \in \mathscr{W}_{i+1} \cap \mathscr{W}_i^{\perp}$ satisfying

$$\|v\|_a = \|w\|_a = 1 \quad \text{and} \quad a(v,w) = 1.$$

The condition for equality to occur in the Cauchy–Schwarz inequality implies $v = w$. Thus $v \in \mathscr{V}_i^{\perp} \cap \mathscr{W}_{i+1} = \mathscr{W}_i$. On the other hand $v \in \mathscr{W}_i^{\perp}$, which implies $v = 0$. This contradicts the fact that $\|v\|_a = 1$, proving the proposition. □

**Lemma 16.2.** *For any $v \in \mathscr{V}$ there exist $v_j \in \mathscr{V}_j$, $j = 0, \ldots, J$, satisfying $v = \sum_{j=0}^{J} v_j$ and*

$$\sum_{j=0}^{J} a(v_j, v_j) \leq \left(1 + \frac{J}{(1 - \|\widetilde{Q}\|_a)^2}\right) a(v,v),$$

*where $\widetilde{Q}$ is defined in Proposition 16.2.*

**Proof.** It follows from Proposition 16.2 that $I - \widetilde{Q}$ is invertible and satisfies

$$\|(I - \widetilde{Q})^{-1}\|_a \leq \frac{1}{1 - \|\widetilde{Q}\|_a},$$

where $I$ is the identity operator on $\mathscr{V}$. We define, for any $v \in \mathscr{V}$,

$$v_0 = P_0 v, \quad w = v - v_0,$$
$$v_1 = P_1 (I - \widetilde{Q})^{-1} w,$$
$$v_j = P_j Q_{j-1} \cdots Q_1 (I - \widetilde{Q})^{-1} w, \; j = 2, \dots, J,$$

where $P_j = I - Q_j$ is the orthogonal projection from $\mathscr{V}$ onto $\mathscr{V}_j$, $j = 0, \dots, J$, defined in Subsection 2.3.2.2. It is easy to check that $\sum_{j=1}^{J} v_j$ (being a telescoping sum) equals $w$, and therefore $\sum_{j=0}^{J} v_j = v$. A crude estimate yields, for $j = 1, \dots, J$,

$$a(v_j, v_j) \le \|(I - \widetilde{Q})^{-1}\|_a^2 a(v, v) \le \frac{1}{(1 - \|\widetilde{Q}\|_a)^2} a(v, v),$$

resulting in

$$\sum_{j=0}^{J} a(v_j, v_j) = a(v_0, v_0) + \sum_{j=1}^{J} a(v_j, v_j) \le \left(1 + \frac{J}{(1 - \|\widetilde{Q}\|_a)^2}\right) a(v, v),$$

proving the lemma. □

The above lemma with the help of the general theory in Chapter 2 yields the following estimate for the minimum eigenvalue of $P$

$$\lambda_{\min}(P) \ge \left(1 + \frac{J}{(1 - \|\widetilde{Q}\|_a)^2}\right)^{-1}. \tag{16.16}$$

This estimate is by no means sharp. We present in Table 16.1 three computed values:

$$\lambda_{\min}^{-1}(P), \quad 1 + \sum_{j=1}^{J} C_j^2, \quad \text{and} \quad 1 + \frac{J}{(1 - \|\widetilde{Q}\|_a)^2},$$

where for the middle term we explicitly compute $C_j$ as the norm of the operator defining $u_j$, namely,

$$C_1 = \|P_1 (I - \widetilde{Q})^{-1}\|_a \quad \text{and} \quad C_j = \|P_j Q_{j-1} \cdots Q_1 (I - \widetilde{Q})^{-1}\|_a, \; j = 2, \dots, J.$$

It is clear from Table 16.1 that $(1 + \sum_{j=1}^{J} C_j^2)^{-1}$ is a better approximant to $\lambda_{\min}(P)$. Our experiments show that the projection $P_j$ in the definition of $u_j$ plays a key role in reducing the norm of $(I - \widetilde{Q})^{-1}$ but we cannot account for this fact.

In Table 16.1 the norms of the operators were computed by using their matrix representations. For example, we computed $\|\widetilde{Q}\|_a$ as follows. Recalling the definition of the positive definite matrix $\mathbf{A}$, see (16.12), and using the Cholesky factorisation $\mathbf{A} = \mathbf{L}^T \mathbf{L}$, we obtain for any $u = \sum_{i=1}^{N} c_i \Phi_i \in V_X$

$$\|u\|_a^2 = \mathbf{c}^T \mathbf{A} \mathbf{c} = \|\mathbf{L}\mathbf{c}\|_{\ell_2}^2,$$

where $\mathbf{c} = (c_1 \cdots c_N)^T$. On the other hand, by writing

$$\widetilde{Q}\Phi_i = \sum_{k=1}^{N} d_{i,k}\Phi_k$$

one can easily see that

$$\|\widetilde{Q}u\|_a^2 = \mathbf{c}^T\widetilde{\mathbf{Q}}^T\mathbf{A}\widetilde{\mathbf{Q}}\mathbf{c} = \mathbf{c}^T\widetilde{\mathbf{Q}}^T\mathbf{L}^T\mathbf{L}\widetilde{\mathbf{Q}}\mathbf{c} = \|\mathbf{L}\widetilde{\mathbf{Q}}\mathbf{c}\|_{\ell_2}^2,$$

where $\widetilde{\mathbf{Q}}$ is the matrix representation of $\widetilde{Q}$ with column $i^{\text{th}}$ being $(d_{i,1} \cdots d_{i,N})^\top$. Therefore,

$$\|\widetilde{Q}\|_a = \sup_{u\in\mathscr{V}} \frac{\|\widetilde{Q}u\|_a}{\|u\|_a} = \sup_{\mathbf{c}\in\mathbb{R}^N} \frac{\|\mathbf{L}\widetilde{\mathbf{Q}}\mathbf{c}\|_{\ell_2}}{\|\mathbf{L}\mathbf{c}\|_{\ell_2}} = \sup_{\mathbf{c}\in\mathbb{R}^N} \frac{\|\mathbf{L}\widetilde{\mathbf{Q}}\mathbf{L}^{-1}\mathbf{c}\|_{\ell_2}}{\|\mathbf{c}\|_{\ell_2}} = \|\mathbf{L}\widetilde{\mathbf{Q}}\mathbf{L}^{-1}\|_{\ell_2}.$$

**Table 16.1** Upper bounds for $\lambda_{\min}^{-1}(P)$

| $N$ | $q_X$ | $\cos\alpha$ | $\cos\beta$ | $J$ | $\lambda_{\min}^{-1}(P)$ | $1+\sum_{j=1}^{J} C_j^2$ | $1+J/(1-\|\widetilde{Q}\|_a)^2$ |
|---|---|---|---|---|---|---|---|
| 1344 | $\pi/80$ | 0.90 | -0.90 | 44 | 120.41 | 1612.23 | 98914208.40 |
| | | 0.80 | -0.93 | 23 | 285.80 | 461.67 | 4003258.71 |
| | | 0.70 | -0.86 | 17 | 17.44 | 135.84 | 625605.77 |
| | | 0.60 | -0.87 | 13 | 13.03 | 33.11 | 7481.56 |
| | | 0.50 | -0.83 | 10 | 3.58 | 12.43 | 219.18 |
| 2133 | $\pi/100$ | 0.90 | -0.85 | 44 | 34.97 | 391.90 | 1684693.43 |
| | | 0.80 | -0.89 | 24 | 20.16 | 69.22 | 57266.95 |
| | | 0.70 | -0.89 | 17 | 6.56 | 24.09 | 712.68 |
| | | 0.60 | -0.85 | 13 | 8.18 | 26.44 | 4745.12 |
| | | 0.50 | -0.83 | 10 | 4.48 | 13.88 | 266.90 |
| 3458 | $\pi/140$ | 0.90 | -0.88 | 46 | 4.92 | 68.75 | 3662.59 |
| | | 0.80 | -0.80 | 22 | 5.31 | 41.58 | 2649.72 |
| | | 0.70 | -0.81 | 16 | 1.82 | 17.77 | 101.96 |
| | | 0.60 | -0.85 | 13 | 1.02 | 13.31 | 47.03 |
| | | 0.50 | -0.83 | 10 | 1.70 | 11.67 | 58.05 |
| 4108 | $\pi/160$ | 0.90 | -0.88 | 46 | 2.17 | 66.21 | 4023.85 |
| | | 0.80 | -0.81 | 24 | 6.41 | 36.54 | 1043.01 |
| | | 0.70 | -0.87 | 17 | 4.57 | 22.21 | 755.70 |
| | | 0.60 | -0.86 | 14 | 1.26 | 15.31 | 86.07 |
| | | 0.50 | -0.80 | 10 | 4.13 | 16.85 | 1388.24 |
| 7663 | $\pi/200$ | 0.80 | -0.89 | 23 | 1.37 | 27.47 | 154.84 |
| | | 0.70 | -0.88 | 17 | 1.68 | 19.32 | 130.77 |
| | | 0.60 | -0.81 | 11 | 5.54 | 17.35 | 609.86 |
| | | 0.50 | -0.84 | 10 | 3.27 | 12.78 | 170.82 |

### 16.3.4 Bounds for the condition number

Lemma 16.1 and Lemma 16.2 yield an upper bound for the condition number of the Schwarz operator.

**Theorem 16.3.** *The condition number of the additive Schwarz operator P is bounded by*

$$\kappa(P) \le cM\left(1 + \frac{J}{(1-\|\widetilde{Q}\|_a)^2}\right),$$

*where c is a constant independent of M, J and the set X, and $\widetilde{Q}$ is defined as in Proposition 16.2.*

## 16.4 Numerical Results

### 16.4.1 An algorithm

Suppose we number the scattered data following the satellite track as $\{1,\ldots,N\}$. Let $\alpha$ and $\beta$ be parameters satisfying $0 < \alpha < \pi/3$ and $\alpha \le \beta \le \pi$. The algorithm to partition $X$ can be described as follows.

(1) The first center is $p_1 = x_1$.
(2) Assume that the centers $p_1,\ldots,p_l$ have been chosen.
(3) Define $X_k$ as in Section 16.3.1 Item 5. If $X = \cup_{k=1}^{l} X_k$, then stop the algorithm and let $J = l$. Otherwise, choose the new center $p_{l+1}$ as a point in $X$ satisfying

$$\theta(p_{l+1}, p_l) \ge \beta \quad \text{and} \quad \theta(p_{l+1}, p_k) \ge \alpha, \quad k = 1,\ldots,l-1.$$

(4) Let $l = l+1$ and repeat Step (3).

Now $X_0$ is defined by $X_0 := \{p_1,\ldots,p_J\}$. The parameter $\beta$ is included so that the condition $X = \cup_{k=1}^{l} X_k$ in Step (3). value of $\alpha$ we chose an appropriate value of $\beta$ by starting with the value $\beta = \pi$, and decreased its value until the above-mentioned condition holds.

For $j = 0,\ldots,J$, let $I_j$ be a subset of the index set of $\{1,\ldots,N\}$ such that

$$m \in I_j \leftrightarrow x_m \in X_j.$$

The cardinality of the set $I_j$ is denoted by $s_j$ and the $m$-th element of the set $I_j$ is denoted by $I_j(m)$ . For a given vector $\mathbf{v} = (v_1,\ldots,v_N)^T$, the restriction of $\mathbf{v}$ on $X_j$ is the vector $\mathbf{u} = (u_1, u_2,\ldots,u_{s_j})^T$ defined as follows

$$u_m := v_{I_j(m)}, \qquad m = 1,\ldots,s_j,$$

and we write $\mathbf{u} = R_j(\mathbf{v})$, thus the restriction operation to $X_j$ is denoted by $R_j$. Conversely, for a vector $\mathbf{u} = (u_1, \ldots, u_{s_j})^T$, we extend $\mathbf{u}$ to $\mathbf{v} = (v_1, \ldots, v_N)^T$ as

$$v_k := \begin{cases} u_m & \text{if } k = I_j(m) \text{ for } 1 \le m \le s_j, \\ 0 & \text{otherwise.} \end{cases}$$

and write $\mathbf{v} = E_j(\mathbf{u})$, where $E_j$ denotes the extension operation.

**A pseudo-code.**

INPUT

Input the scattered set $X$ on the sphere, the right-hand side $f$, and the desired accuracy $\varepsilon$.

SETUP

(1) Partition the scattered set $X$ into $J$ overlapping subsets $X_j$ for $j = 1, \ldots, J$ and construct the global coarse set $X_0$.
(2) Set the initial residual vector $\mathbf{r} = \mathbf{f}$ where $\mathbf{f}$ is the vector defined from the right-hand side of (16.11).
(3) Set the initial pseudo-residual vector $\mathbf{p} = \mathbf{0}$.
(4) Set the initial solution vector $\mathbf{s} = \mathbf{0}$.
(5) Set the iteration counter iter $= 0$.

ITERATIVE SOLUTION

(1) **while** $\|\mathbf{r}\| > \varepsilon \|\mathbf{f}\|$
(2) Set $\mathbf{p} = \mathbf{0}$.
(3) **for** $j = 1$ **to** $J$
(4) Construct the local matrix $C$ with entries $C_{m,n} = A_{I_j(m), I_j(n)}$.
(5) Set the restriction residual vector $\mathbf{z} = R_j(\mathbf{r})$.
(6) Solve the linear system $C\mathbf{y} = \mathbf{z}$.
(7) Update the pseudo-residual vector $\mathbf{p} = \mathbf{p} + E_j(\mathbf{y})$.
(8) **end for**
(9) Construct the coarse global matrix $G$ with entries

$$G_{m,n} = A_{I_0(m), I_0(n)}.$$

(10) Set $\mathbf{z}_g = R_0(\mathbf{r})$.
(11) Solve the linear system $G\mathbf{y}_g = \mathbf{z}_g$.
(12) Update the pseudo-residual vector $\mathbf{p} = \mathbf{p} + E_0(\mathbf{y}_g)$.
(13) If iter $> 0$ then set $\zeta_0 = \zeta_1$.
(14) Set $\zeta_1 = \mathbf{p} \cdot \mathbf{r}$.
(15) Update the counter, iter $=$ iter $+1$.
(16) If iter $= 1$ then define $\overline{\mathbf{p}} = \mathbf{p}$ else $\overline{\mathbf{p}} = \mathbf{p} + (\zeta_1/\zeta_0)\overline{\mathbf{p}}$.
(17) Update the residual vector

$$\mathbf{r} = \mathbf{r} - \left( \frac{\mathbf{r} \cdot \mathbf{p}}{\overline{\mathbf{p}} \cdot \mathbf{A}\overline{\mathbf{p}}} \right) \mathbf{A}\overline{\mathbf{p}}.$$

(18) Update the solution vector

$$\mathbf{s} = \mathbf{s} + \left(\frac{\mathbf{r}\cdot\mathbf{p}}{\overline{\mathbf{p}}\cdot\mathbf{A}\overline{\mathbf{p}}}\right)\overline{\mathbf{p}}.$$

(19) **end while**

### *16.4.2 Numerical results*

In this section we reprint, with copyright permission, the numerical results obtained in [226]. The experiments are based on globally scattered data extracted from a very large data set collected by NASA satellite MAGSAT. The scattered data sets $X = \{\mathbf{x}_1, \mathbf{x}_2, \cdots, \mathbf{x}_N\}$ are extracted so that the separation radius $q_X$ defined by

$$q_X := \frac{1}{2}\min_{i\neq j}\cos^{-1}(\mathbf{x}_i\cdot\mathbf{x}_j)$$

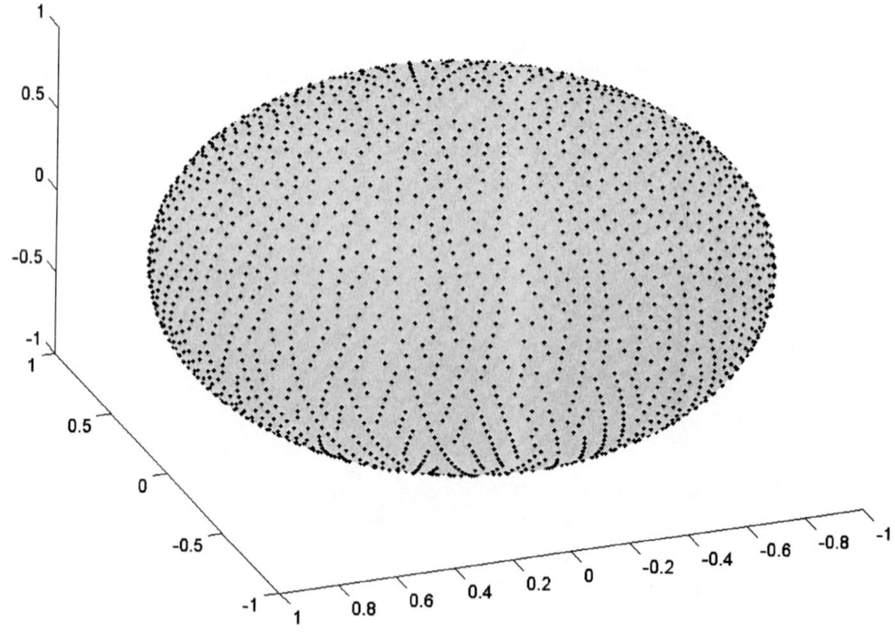


**Fig. 16.1** Global scattered MAGSAT satellite data

We solved the Dirichlet and Neumann problems in the unbounded domain $\mathbb{B} := \{\mathbf{x} \in \mathbb{R}^3 : \|\mathbf{x}\| > 1\}$, namely,

$$\begin{aligned} \Delta U &= 0 \qquad \text{in } \mathbb{B} \\ U &= U_D \qquad \text{on } \mathbb{S} \end{aligned}$$

and

$$\begin{aligned} \Delta U &= 0 \qquad \text{in } \mathbb{B} \\ \partial_\nu U &= U_N \qquad \text{on } \mathbb{S}, \end{aligned}$$

with a vanishing condition at infinity for both problems:

$$U(\mathbf{x}) = O(1/\|\mathbf{x}\|) \quad \text{as} \quad \|\mathbf{x}\| \to \infty.$$

Here $\nu$ is the outer normal vector to $\mathbb{S}$, and $U_D$ and $U_N$ are chosen such that the exact solutions to both problems are

$$U(\mathbf{x}) = \frac{1}{\|\mathbf{x} - \mathbf{p}\|} \quad \text{where } \mathbf{p} = (0,0,p) \text{ with } p \in (0,1). \tag{16.17}$$

Thus, for $\mathbf{x} = (x_1, x_2, x_3) \in \mathbb{S}$, we chose

$$U_D(\mathbf{x}) = \frac{1}{(1+p^2-2px_3)^{1/2}} \quad \text{and} \quad U_N(\mathbf{x}) = \frac{px_3 - 1}{(1+p^2-2px_3)^{3/2}}.$$

It is well known that the Dirichlet problem can be reformulated into a weakly-singular integral equation on $\mathbb{S}$, which can then be rewritten [225] as a pseudo-differential equation $Lu = f$ with

$$\widehat{L}(\ell) = \frac{1}{2\ell+1}$$

and

$$\begin{aligned} f(\mathbf{x}) &= -\frac{1}{2}U_D(\mathbf{x}) + \frac{1}{4\pi}\int_{\mathbb{S}} \frac{\partial}{\partial\nu(\mathbf{y})} \frac{U_D(\mathbf{y})}{\|\mathbf{x}-\mathbf{y}\|}\, d\sigma(\mathbf{y}) \\ &= \frac{1}{2}\sum_{\ell=1}^{\infty}\sum_{m=-\ell}^{\ell}\left(-1-\frac{1}{2\ell+1}\right)\widehat{(U_D)}_{\ell,m}Y_{\ell,m}(\mathbf{x}) \\ &= -\sum_{\ell=1}^{\infty}\sum_{m=-\ell}^{\ell}\frac{\ell+1}{2\ell+1}\widehat{(U_D)}_{\ell,m}Y_{\ell,m}(\mathbf{x}). \end{aligned}$$

In this case $L$ is a pseudo-differential operator of order $-1$ and hence $\sigma = -1/2$.

Similarly, the Neumann problem can be reformulated into a hypersingular integral equation on $\mathbb{S}$, which can then be rewritten [225] as a pseudo-differential equation $Lu = f$ with

$$\widehat{L}(\ell) = \frac{\ell(\ell+1)}{2\ell+1}$$

and

$$\begin{aligned} f(\mathbf{x}) &= \frac{1}{2}U_N(\mathbf{x}) + \frac{1}{4\pi}\frac{\partial}{\partial \nu(\mathbf{x})}\int_{\mathbb{S}} \frac{U_N(\mathbf{y})}{\|\mathbf{x}-\mathbf{y}\|}\, d\sigma(\mathbf{y}) \\ &= \frac{1}{2}\sum_{\ell=1}^{\infty}\sum_{m=-\ell}^{\ell}\left(1-\frac{1}{2\ell+1}\right)\widehat{(U_N)}_{\ell,m} Y_{\ell,m}(\mathbf{x}) \\ &= \sum_{\ell=1}^{\infty}\sum_{m=-\ell}^{\ell}\frac{\ell}{2\ell+1}\widehat{(U_N)}_{\ell,m} Y_{\ell,m}(\mathbf{x}). \end{aligned}$$

In this case $L$ is a pseudo-differential operator of order 1 and hence $\sigma = 1/2$.

The univariate function $\phi$ defining the kernel $\Phi$, see (16.5), is given by $\phi(t) = \rho_m(\sqrt{2-2t})$, where $\rho_m$ are Wendland's functions [234, page 128]. It is proved in [158, Proposition 4.6] that (16.6) holds with $\tau = m + 3/2$. Table 16.2 details the functions $\rho_m$ used in our experiments and the corresponding values for $\tau$. The spherical radial basis functions $\Phi_i$, $i = 1,\dots,N$, are computed by

| $m$ | $\rho_m(r)$ | $\tau$ |
|---|---|---|
| 0 | $(1-r)_+^2$ | 1.5 |
| 1 | $(1-r)_+^4(4r+1)$ | 2.5 |

**Table 16.2** Wendland's RBFs

$$\Phi_i(\mathbf{x}) = \rho_m(\sqrt{2-2\mathbf{x}\cdot\mathbf{x}_i}), \quad \mathbf{x} \in \mathbb{S}.$$

Tables 16.3–16.6 show the $\ell_2$-condition number $\kappa(\mathbf{A})$, the CPU time in seconds, and the number of iterations ITER for two different values of $p$, see (16.17), namely $p = 0.5$ and $p = 0.95$. The numbers suggest that when the point $\mathbf{p}$ is deeply buried in the sphere ($p = 0.5$), the unpreconditioned method can solve the problems with $m = 0$ but not with $m = 1$, though the solutions are not reliable due to large condition numbers of the stiffness matrix $\mathbf{A}$. The results are even worse when the point $\mathbf{p}$ is closer to the surface ($p = 0.95$); the unpreconditioned method can hardly solve the problems.

We then tested the overlapping additive Schwarz method as a preconditioner for the conjugate gradient method, with different values of $\alpha$ and $\beta$, see Section 16.3.1, and hence different values of $J$, the number of subproblems. The numbers in Tables 16.7–16.10 show the efficiency of the preconditioner. (In these tables $\kappa(P)$ is the condition number of the preconditioned systems.) The numbers suggest that the

| $m$ | $N$ | $\lambda_{\min}$ | $\lambda_{\max}$ | $\kappa(\mathbf{A})$ | CPU | ITER |
|---|---|---|---|---|---|---|
| 0 | 10443 | 2.729E-10 | 227.937 | 8.353E+11 | 168.7 | 259 |
| | 13897 | 1.182E-10 | 303.418 | 2.568E+12 | 98.2 | 85 |
| | 17262 | 5.359E-11 | 377.013 | 7.035E+12 | 227.7 | 128 |
| 1 | 10443 | -3.131E-14 | 167.478 | n/a | 133.2 | 209 |
| | 13897 | -1.866E-13 | 222.963 | n/a | 143.3 | 124 |

**Table 16.3** Unpreconditioned CG for the Dirichlet problem with $p = 0.5$

| $m$ | $N$ | $\lambda_{\min}$ | $\lambda_{\max}$ | $\kappa(\mathbf{A})$ | CPU | ITER |
|---|---|---|---|---|---|---|
| 0 | 10443 | 2.729E-10 | 227.937 | 8.353E+11 | >23000 | >41000 |
| | 13897 | 1.182E-10 | 303.418 | 2.568E+12 | >40000 | >33000 |
| | 17262 | 5.359E-11 | 377.013 | 7.035E+12 | >73000 | >48000 |
| 1 | 10443 | -3.131E-14 | 167.478 | n/a | >23000 | >41000 |
| | 13897 | -1.866E-13 | 222.963 | n/a | >65000 | >55000 |

**Table 16.4** Unpreconditioned systems for the Dirichlet problem with $p = 0.95$

| $m$ | $N$ | $\lambda_{\min}$ | $\lambda_{\max}$ | $\kappa(\mathbf{A})$ | CPU | ITER |
|---|---|---|---|---|---|---|
| 0 | 10443 | 5.449E-06 | 149.848 | 2.750E+07 | 775.0 | 1397 |
| | 13897 | 3.305E-06 | 204.544 | 6.189E+07 | 339.7 | 299 |
| | 17262 | 2.040E-06 | 259.912 | 1.274E+08 | 453.7 | 286 |
| 1 | 10443 | 1.207E-10 | 135.598 | 1.124E+12 | 447.3 | 611 |
| | 13897 | 3.694E-11 | 185.949 | 5.034E+12 | 195.6 | 195 |
| | 17262 | 1.217E-11 | 237.437 | 1.951E+13 | 400.6 | 237 |

**Table 16.5** Unpreconditioned systems for the Neumann problem with $p = 0.5$

| $m$ | $N$ | $\lambda_{\min}$ | $\lambda_{\max}$ | $\kappa(\mathbf{A})$ | CPU | ITER |
|---|---|---|---|---|---|---|
| 0 | 10443 | 5.449E-06 | 149.848 | 2.750E+07 | 4587.4 | 6398 |
| | 13897 | 3.305E-06 | 204.544 | 6.189E+07 | 6402.2 | 5772 |
| | 17262 | 2.040E-06 | 259.912 | 1.274E+08 | 9736.9 | 5743 |
| 1 | 10443 | 1.207E-10 | 135.598 | 1.124E+12 | >23000 | >41000 |
| | 13897 | 3.694E-11 | 185.949 | 5.034E+12 | >55000 | >55000 |
| | 17262 | 1.217E-11 | 237.437 | 1.951E+13 | >123000 | >69000 |

**Table 16.6** Unpreconditioned systems for the Neumann problem with $p = 0.95$

algorithm is not affected by the smoothness of the kernel. In all cases, for both the weakly-singular and hypersingular operators, the condition numbers and the CPU times are both much better than in the unpreconditioned case. The benefit is particularly great for the harder case $p = 0.95$.

It should be noted that when $\cos\alpha$ decreases (meaning that $\alpha$ increases) then $\lambda_{\max}$ decreases (whereas the behaviour of $\lambda_{\min}$ is influenced by the LANCZOS algorithm), but the CPU time first decreases then increases. We note that the larger value of $\alpha$ results in the larger size of the overlap and a smaller value of $J$ (the number of subproblems to be solved), which in turn implies larger sizes of the subproblems. As in the case of the differential operator considered in [137], this results in a smaller

| $m$ | $N$ | $\cos\alpha$ | $\cos\beta$ | $J$ | $\lambda_{\min}$ | $\lambda_{\max}$ | $\kappa(P)$ | CPU | ITER |
|---|---|---|---|---|---|---|---|---|---|
| 0 | 10443 | 0.95 | -0.07 | 86 | 2.110E-02 | 52.945 | 2.509E+03 | 86.0 | 62 |
| | 10443 | 0.90 | -0.63 | 45 | 3.048E-01 | 31.130 | 1.021E+02 | 101.0 | 29 |
| | 10443 | 0.80 | -0.87 | 20 | 1.199E-01 | 15.992 | 1.334E+02 | 316.4 | 31 |
| | 10443 | 0.60 | -0.87 | 11 | 5.191E-01 | 10.365 | 1.997E+01 | 716.0 | 18 |
| | 13897 | 0.95 | 0.03 | 90 | 6.043E-02 | 55.499 | 9.184E+02 | 168.1 | 59 |
| | 13897 | 0.90 | -0.76 | 46 | 1.961E-01 | 31.905 | 1.627E+02 | 271.4 | 35 |
| | 13897 | 0.80 | -0.67 | 26 | 1.081E+00 | 20.528 | 1.898E+01 | 555.5 | 19 |
| | 13897 | 0.60 | -0.69 | 13 | 1.088E+00 | 12.031 | 1.105E+01 | 1723.9 | 16 |
| | 17262 | 0.95 | -0.17 | 90 | 8.102E-02 | 55.624 | 6.866E+02 | 254.3 | 50 |
| | 17262 | 0.90 | -0.62 | 47 | 1.468E-02 | 32.666 | 2.226E+03 | 1118.4 | 76 |
| | 17262 | 0.80 | -0.76 | 24 | 9.821E-01 | 19.087 | 1.943E+01 | 1006.8 | 20 |
| | 17262 | 0.60 | -0.85 | 12 | 2.635E-01 | 11.240 | 4.266E+01 | 4507.6 | 23 |
| 1 | 10443 | 0.99 | 0.99 | 424 | 1.371E-02 | 223.980 | 1.634E+04 | 145.1 | 202 |
| | 10443 | 0.98 | 0.95 | 201 | 9.979E-03 | 113.033 | 1.133E+04 | 136.7 | 179 |
| | 10443 | 0.97 | 0.69 | 138 | 1.446E-02 | 81.024 | 5.605E+03 | 121.2 | 132 |
| | 10443 | 0.95 | -0.07 | 86 | 1.155E-02 | 53.828 | 4.658E+03 | 163.5 | 122 |
| | 10443 | 0.60 | -0.87 | 11 | 5.299E-02 | 10.402 | 1.963E+02 | 1202.7 | 31 |
| | 13897 | 0.99 | 0.98 | 395 | 8.076E-03 | 210.291 | 2.604E+04 | 327.7 | 259 |
| | 13897 | 0.98 | 0.74 | 217 | 1.169E-02 | 122.993 | 1.052E+04 | 248.8 | 168 |
| | 13897 | 0.97 | 0.68 | 138 | 3.374E-03 | 81.772 | 2.424E+04 | 414.9 | 234 |
| | 13897 | 0.95 | 0.03 | 90 | 9.140E-03 | 56.687 | 6.202E+03 | 389.4 | 134 |
| | 13897 | 0.60 | -0.69 | 13 | 1.054E+00 | 12.085 | 1.147E+01 | 1823.4 | 17 |

**Table 16.7** Preconditioned systems for the Dirichlet problem with $p = 0.5$

condition number $\kappa(P)$ because the preconditioner is closer to the inverse of the stiffness matrix. However, for an optimal value of $\alpha$ in terms of CPU time, one has to balance between the number of subproblems and their sizes. As our experiments show, any value of $\alpha$ so that $\cos\alpha \le 0.6$ is not recommended.

| $m$ | $N$ | $\cos\alpha$ | $\cos\beta$ | $J$ | $\lambda_{\min}$ | $\lambda_{\max}$ | $\kappa(P)$ | CPU | ITER |
|---|---|---|---|---|---|---|---|---|---|
| 0 | 10443 | 0.99 | 0.99 | 424 | 3.954E-03 | 223.130 | 5.644E+04 | 160.3 | 252 |
| | 10443 | 0.98 | 0.95 | 201 | 4.652E-03 | 111.679 | 2.400E+04 | 109.8 | 162 |
| | 10443 | 0.97 | 0.69 | 138 | 5.381E-02 | 79.763 | 1.482E+03 | 64.2 | 78 |
| | 10443 | 0.95 | -0.07 | 86 | 1.280E-02 | 52.945 | 4.135E+03 | 122.2 | 98 |
| | 10443 | 0.90 | -0.63 | 45 | 2.976E-01 | 31.130 | 1.046E+02 | 109.8 | 34 |
| | 13897 | 0.99 | 0.98 | 395 | 1.782E-02 | 208.653 | 1.171E+04 | 219.7 | 173 |
| | 13897 | 0.98 | 0.74 | 217 | 2.665E-02 | 120.889 | 4.536E+03 | 161.4 | 110 |
| | 13897 | 0.97 | 0.68 | 138 | 8.361E-03 | 80.043 | 9.574E+03 | 233.9 | 134 |
| | 13897 | 0.95 | 0.03 | 90 | 7.944E-02 | 55.499 | 6.987E+02 | 178.9 | 63 |
| | 13897 | 0.90 | -0.76 | 46 | 1.914E-01 | 31.905 | 1.667E+02 | 298.9 | 40 |
| | 17262 | 0.99 | 0.98 | 405 | 5.336E-02 | 214.422 | 4.019E+03 | 172.2 | 102 |
| | 17262 | 0.98 | 0.94 | 206 | 2.755E-02 | 115.125 | 4.179E+03 | 221.4 | 108 |
| | 17262 | 0.97 | 0.70 | 140 | 8.905E-03 | 81.374 | 9.139E+03 | 394.8 | 151 |
| | 17262 | 0.95 | -0.17 | 90 | 6.687E-02 | 55.624 | 8.318E+02 | 275.4 | 58 |
| | 17262 | 0.90 | -0.62 | 47 | 1.449E-02 | 32.666 | 2.255E+03 | 1209.6 | 84 |
| 1 | 10443 | 0.99 | 0.99 | 424 | 2.337E-03 | 224.019 | 9.585E+04 | 371.6 | 584 |
| | 10443 | 0.98 | 0.95 | 201 | 1.679E-03 | 113.144 | 6.740E+04 | 289.9 | 428 |
| | 10443 | 0.97 | 0.69 | 138 | 1.534E-03 | 81.042 | 5.284E+04 | 231.9 | 282 |
| | 10443 | 0.95 | -0.07 | 86 | 1.408E-03 | 53.828 | 3.822E+04 | 306.4 | 246 |
| | 10443 | 0.90 | -0.63 | 45 | 4.726E-02 | 31.444 | 6.653E+02 | 203.5 | 63 |
| | 13897 | 0.99 | 0.98 | 395 | 3.300E-04 | 210.991 | 6.395E+05 | 1228.8 | 1019 |
| | 13897 | 0.98 | 0.74 | 217 | 2.545E-03 | 122.884 | 4.829E+04 | 456.4 | 325 |
| | 13897 | 0.97 | 0.68 | 138 | 7.030E-04 | 82.394 | 1.172E+05 | 869.3 | 506 |
| | 13897 | 0.95 | 0.03 | 90 | 1.050E-02 | 56.681 | 5.401E+03 | 405.1 | 143 |
| | 13897 | 0.90 | -0.76 | 46 | 1.381E-02 | 32.438 | 2.349E+03 | 833.3 | 111 |

**Table 16.8** Preconditioned systems for the Dirichlet problem with $p = 0.95$

| $m$ | $N$ | $\cos\alpha$ | $\cos\beta$ | $J$ | $\lambda_{\min}$ | $\lambda_{\max}$ | $\kappa(P)$ | CPU | ITER |
|---|---|---|---|---|---|---|---|---|---|
| 0 | 10443 | 0.99 | 0.99 | 424 | 1.998E-03 | 56.451 | 2.825E+04 | 105.6 | 165 |
| | 10443 | 0.98 | 0.95 | 201 | 3.350E-04 | 31.681 | 9.446E+04 | 132.1 | 194 |
| | 10443 | 0.97 | 0.69 | 138 | 1.300E-04 | 24.397 | 1.870E+05 | 105.7 | 128 |
| | 10443 | 0.95 | -0.07 | 86 | 4.700E-05 | 18.374 | 3.917E+05 | 123.8 | 99 |
| | 10443 | 0.90 | -0.63 | 45 | 6.888E-01 | 12.711 | 1.845E+01 | 71.4 | 22 |
| | 13897 | 0.99 | 0.98 | 395 | 1.707E-02 | 52.756 | 3.091E+03 | 125.5 | 100 |
| | 13897 | 0.98 | 0.74 | 217 | 5.593E-02 | 34.114 | 6.100E+02 | 93.3 | 64 |
| | 13897 | 0.97 | 0.68 | 138 | 4.843E-02 | 24.713 | 5.102E+02 | 108.7 | 62 |
| | 13897 | 0.95 | 0.03 | 90 | 2.361E-02 | 19.316 | 8.182E+02 | 179.2 | 63 |
| | 13897 | 0.90 | -0.76 | 46 | 3.453E-01 | 12.978 | 3.758E+01 | 188.6 | 25 |
| | 17262 | 0.99 | 0.98 | 405 | 2.565E-02 | 54.441 | 2.122E+03 | 168.2 | 95 |
| | 17262 | 0.98 | 0.94 | 206 | 3.661E-02 | 32.891 | 8.984E+02 | 143.5 | 67 |
| | 17262 | 0.97 | 0.70 | 140 | 3.908E-02 | 25.302 | 6.475E+02 | 176.0 | 63 |
| | 17262 | 0.95 | -0.17 | 90 | 1.113E-01 | 19.347 | 1.738E+02 | 203.0 | 41 |
| | 17262 | 0.90 | -0.62 | 47 | 7.370E-02 | 13.570 | 1.841E+02 | 701.4 | 47 |
| 1 | 10443 | 0.99 | 0.99 | 424 | 1.204E-02 | 64.025 | 5.319E+03 | 128.3 | 156 |
| | 10443 | 0.98 | 0.95 | 201 | 5.453E-03 | 35.481 | 6.506E+03 | 147.1 | 170 |
| | 10443 | 0.97 | 0.69 | 138 | 5.919E-03 | 27.191 | 4.594E+03 | 154.6 | 152 |
| | 10443 | 0.95 | -0.07 | 86 | 8.853E-03 | 20.464 | 2.312E+03 | 146.4 | 101 |
| | 10443 | 0.90 | -0.63 | 45 | 1.181E-01 | 13.787 | 1.168E+02 | 121.8 | 35 |
| | 13897 | 0.99 | 0.98 | 395 | 1.110E-02 | 59.929 | 5.398E+03 | 184.9 | 165 |
| | 13897 | 0.98 | 0.74 | 217 | 1.145E-02 | 38.469 | 3.360E+03 | 165.2 | 125 |
| | 13897 | 0.97 | 0.68 | 138 | 5.694E-03 | 27.781 | 4.879E+03 | 255.5 | 158 |
| | 13897 | 0.95 | 0.03 | 90 | 1.569E-02 | 21.614 | 1.378E+03 | 253.0 | 93 |
| | 13897 | 0.90 | -0.76 | 46 | 3.217E-02 | 14.137 | 4.395E+02 | 376.3 | 51 |
| | 17262 | 0.99 | 0.98 | 405 | 1.402E-02 | 62.282 | 4.442E+03 | 272.2 | 145 |
| | 17262 | 0.98 | 0.94 | 206 | 8.439E-03 | 37.329 | 4.423E+03 | 355.7 | 159 |
| | 17262 | 0.97 | 0.70 | 140 | 1.098E-02 | 28.467 | 2.593E+03 | 341.5 | 120 |
| | 17262 | 0.95 | -0.17 | 90 | 2.380E-02 | 21.584 | 9.068E+02 | 350.9 | 72 |

**Table 16.9** Preconditioned systems for the Neumann problem with $p = 0.5$

| $m$ | $N$ | $\cos\alpha$ | $\cos\beta$ | $J$ | $\lambda_{\min}$ | $\lambda_{\max}$ | $\kappa(P)$ | CPU | ITER |
|---|---|---|---|---|---|---|---|---|---|
| 0 | 10443 | 0.99 | 0.99 | 424 | 1.982E-03 | 56.452 | 2.849E+04 | 181.3 | 224 |
| | 10443 | 0.98 | 0.95 | 201 | 3.340E-04 | 31.682 | 9.474E+04 | 162.9 | 222 |
| | 10443 | 0.97 | 0.69 | 138 | 1.300E-04 | 24.398 | 1.873E+05 | 115.8 | 126 |
| | 10443 | 0.95 | -0.07 | 86 | 4.700E-05 | 18.374 | 3.924E+05 | 148.7 | 105 |
| | 10443 | 0.90 | -0.63 | 45 | 1.100E-05 | 12.772 | 1.176E+06 | 173.4 | 50 |
| | 13897 | 0.99 | 0.98 | 395 | 9.992E-03 | 52.758 | 5.280E+03 | 144.9 | 121 |
| | 13897 | 0.98 | 0.74 | 217 | 9.638E-02 | 34.115 | 3.540E+02 | 88.5 | 62 |
| | 13897 | 0.97 | 0.68 | 138 | 4.898E-02 | 24.714 | 5.046E+02 | 119.5 | 66 |
| | 13897 | 0.95 | 0.03 | 90 | 2.462E-02 | 19.317 | 7.845E+02 | 170.0 | 59 |
| | 13897 | 0.90 | -0.76 | 46 | 3.435E-01 | 12.981 | 3.779E+01 | 223.8 | 29 |
| | 17262 | 0.99 | 0.98 | 405 | 3.767E-02 | 54.443 | 1.445E+03 | 160.8 | 85 |
| | 17262 | 0.98 | 0.94 | 206 | 2.676E-02 | 32.892 | 1.229E+03 | 209.8 | 93 |
| | 17262 | 0.97 | 0.70 | 140 | 8.573E-02 | 25.303 | 2.951E+02 | 175.4 | 61 |
| | 17262 | 0.95 | -0.17 | 90 | 8.531E-02 | 19.348 | 2.268E+02 | 230.5 | 46 |
| | 17262 | 0.90 | -0.62 | 47 | 8.307E-02 | 13.570 | 1.634E+02 | 636.0 | 43 |
| 1 | 10443 | 0.99 | 0.99 | 424 | 6.942E-05 | 64.025 | 9.223E+05 | 1467.0 | 2316 |
| | 10443 | 0.98 | 0.95 | 201 | 6.208E-06 | 35.481 | 5.715E+06 | 1087.1 | 1612 |
| | 10443 | 0.97 | 0.69 | 138 | 8.176E-07 | 27.191 | 3.326E+07 | 589.6 | 721 |
| | 10443 | 0.95 | -0.07 | 86 | 1.332E-07 | 20.464 | 1.537E+08 | 545.1 | 440 |
| | 10443 | 0.90 | -0.63 | 45 | 1.300E-08 | 13.787 | 1.061E+09 | 331.7 | 103 |
| | 13897 | 0.99 | 0.98 | 395 | 2.819E-03 | 59.929 | 2.126E+04 | 343.0 | 311 |
| | 13897 | 0.98 | 0.74 | 217 | 7.676E-03 | 38.469 | 5.011E+03 | 184.3 | 141 |
| | 13897 | 0.97 | 0.68 | 138 | 4.683E-03 | 27.781 | 5.932E+03 | 276.3 | 171 |
| | 13897 | 0.95 | 0.03 | 90 | 2.381E-02 | 21.614 | 9.078E+02 | 222.5 | 80 |
| | 13897 | 0.90 | -0.76 | 46 | 6.049E-02 | 14.137 | 2.337E+02 | 384.9 | 50 |
| | 17262 | 0.99 | 0.98 | 405 | 1.701E-02 | 62.282 | 3.662E+03 | 296.8 | 151 |
| | 17262 | 0.98 | 0.94 | 206 | 7.684E-03 | 37.329 | 4.858E+03 | 383.9 | 165 |
| | 17262 | 0.97 | 0.70 | 140 | 5.230E-03 | 28.467 | 5.443E+03 | 406.4 | 138 |
| | 17262 | 0.95 | -0.17 | 90 | 2.162E-02 | 21.584 | 9.981E+02 | 379.0 | 76 |

**Table 16.10** Preconditioned systems for the Neumann problem with $p = 0.95$

# Part VI
# Appendices

# Appendix A
# Interpolation Spaces and Sobolev Spaces

Throughout this book, Sobolev spaces of fractional orders are the major function space. Since these spaces have many properties which are not possessed by Sobolev spaces of integral orders, this appendix serves to provide the necessary background in this topic.

Since many properties of Sobolev spaces are derived from their interpolation property, we first recall in Section A.1 the concept of interpolation spaces. The main section is Section A.2 where different definitions of Sobolev spaces and norms (namely the Slobodetski norms, interpolation norms, and weighted norms) together with their properties are introduced. Each norm plays a different role in the analysis, which could have been observed in the previous chapters. The spatial domains in this consideration contain both the whole $\mathbb{R}^d$-space and Lipschitz domains in $\mathbb{R}^d$, $d \geq 1$. Equivalences between the different norms are established.

Another important property of fractional-ordered Sobolev spaces is the relation between the local norm and global norm. This property has to be derived via the interpolation norm; see Theorem A.10 and Theorem A.11. The results in the latter theorem are new; those for positive orders first appeared in [224] and those for negative orders are first presented in this chapter. This new development corrects a mistake in the derivation of the coercivity of the decomposition which is ubiquitously seen in the literature; see Subsection 2.6.4 for the definition of the coercivity of the decomposition.

The different uses of norms are possible due to equivalence of these norms. However, this equivalence may depend on the measure of the domain (on which the function is defined) and since this measure may depend on the problem parameters (mostly the meshsize $h$), care has to be taken. A scaling argument is required to only apply the norm equivalence on Sobolev spaces defined on a reference element (which means the constants are independent of $h$). The scaling property of norms when the domain is rescaled is therefore another important issue and will be discussed in this chapter.

The chapter also depicts other features of Sobolev spaces which are used throughout this monograph. The reader who is interested in the topic of Sobolev spaces can find more details in many other sources, e.g. [1, 88, 138, 147], to name a few.



E. P. Stephan and T. Tran, *Schwarz Methods and Multilevel Preconditioners for Boundary Element Methods*, https://doi.org/10.1007/978-3-030-79283-1

The chapter ends with a section on a generalised antiderivative operator in Sobolev spaces on curves; see Subsection A.2.12. This operator is a mechanism that allows us to go back and forth between the space $\widetilde{H}^{1/2}(\Gamma)$ and the space $\widetilde{H}^{-1/2}(\Gamma)$ so that results for the weakly-singular integral operator (defined on curves) can be derived from corresponding results for the hypersingular integral operator.

## A.1 Real Interpolation Spaces

### *A.1.1 Compatible couples and intermediate spaces*

Real interpolation spaces are defined from two normed spaces $A_0$ and $A_1$. These two spaces are said to be a *compatible couple* $\mathscr{A} = (A_0, A_1)$ if they are subspaces of a Hausdorff (or separated) topological vector space $\mathscr{U}$. Under this assumption the spaces $A_1 \cap A_2$ and $A_0 + A_1$ are well defined, in particular

$$A_0 + A_1 = \left\{ u \in \mathscr{U} \, : \, u = u_0 + u_1 \text{ for some } u_0 \in A_0 \text{ and } u_1 \in A_1 \right\}.$$

These spaces are equipped with the norms

$$\|u\|_{A_0 \cap A_1} = \left( \|u\|_{A_0}^2 + \|u\|_{A_1}^2 \right)^{1/2}$$

and

$$\|u\|_{A_0 + A_1} = \inf_{\substack{u_0 \in A_0, u_1 \in A_1 \\ u_0 + u_1 = u}} \left( \|u_0\|_{A_0}^2 + \|u_1\|_{A_1}^2 \right)^{1/2}.$$

If $A_0$ and $A_1$ are complete then $A_0 \cap A_1$ and $A_0 + A_1$ are also complete.

Consider another compatible couple for normed spaces $B_0$ and $B_1$, $\mathscr{B} = (B_0, B_1)$. The couple $\mathscr{T} = (T_0, T_1)$ of bounded operators

$$T_0 : A_0 \to B_0 \quad \text{and} \quad T_1 : A_1 \to B_1$$

is said to be *compatible* if $T_0 u = T_1 u$ for all $u \in A_0 \cap A_1$. We abbreviate this property by writing $\mathscr{T} : \mathscr{A} \to \mathscr{B}$.

Suppose that $\mathscr{A} = (A_0, A_1)$ is a compatible couple. A normed space $A$ is called an *intermediate space* between $A_0$ and $A_1$ (or with respect to $\mathscr{A}$) if

$$A_0 \cap A_1 \subset A \subset A_0 + A_1$$

with continuous inclusions. An intermediate space $A$ is called an *interpolation space* between $A_0$ and $A_1$ (or with respect to $\mathscr{A}$) if for any compatible couple $\mathscr{T} : \mathscr{A} \to \mathscr{A}$ there exists a bounded operator $T_A : A \to A$ such that $T_A u = T_0 u = T_1 u$ for all $u \in A_0 \cap A_1$.

If there are two compatible couples $\mathscr{A} = (A_0, A_1)$ and $\mathscr{B} = (B_0, B_1)$ then the intermediate spaces $A$ and $B$ are interpolation spaces with respect to $\mathscr{A}$ and $\mathscr{B}$, respectively, if for any compatible couple $\mathscr{T}:\mathscr{A}\to\mathscr{B}$ there exists uniquely a bounded operator $T_A : A \to B$ that coincides with $T_0$ and $T_1$ on $A_0 \cap A_1$.

We note that the existence of an operator $T_A : A \to B$ is easily seen. Indeed, for any $u \in A$ we can define $T_A u = T_0 u_0 + T_1 u_1$ where $u = u_0 + u_1$ is one representation of $u$ in $A_0 + A_1$. If there is another representation $u = u_0' + u_1'$, then $u_0 - u_0' = u_1' - u_1 \in A_0 \cap A_1$. Hence $T_0(u_0 - u_0') = T_1(u_1' - u_1)$ which implies $T_A(u_0 + u_1) = T_A(u_0' + u_1')$.

## A.1.2 The K-functional

For any compatible couple $\mathscr{A} = (A_0, A_1)$, the $K$-functional is defined for $t > 0$ and $u \in A_0 + A_1$ by

$$K(t,u,\mathscr{A}) := \inf_{\substack{u_0\in A_0,\ u_1\in A_1\\ u_0+u_1=u}} \left(\|u_0\|_{A_0}^2 + t^2\|u_1\|_{A_1}^2\right)^{1/2}.$$

The $K$-functional is used to define the interpolation space $[A_0,A_1]_{\theta,q}$, $\theta \in (0,1)$ and $q \in [1,\infty)$, as follows

$$[A_0,A_1]_{\theta,q} := \left\{u \in A_0 + A_1 \, : \, \|u\|_{[A_0,A_1]_{\theta,q}} < \infty\right\}$$

where

$$\|u\|_{[A_0,A_1]_{\theta,q}} := N_{\theta,q}\left(\int_0^\infty \left|t^{-\theta}K(t,u,\mathscr{A})\right|^q \frac{dt}{t}\right)^{1/q} \tag{A.1}$$

with the factor $N_{\theta,q}$ defined by

$$N_{\theta,q} := \left(q\theta(1-\theta)\right)^{1/q}; \tag{A.2}$$

see [147, page 319]. When $q = 2$ we write $[A_0,A_1]_\theta$ for $[A_0,A_1]_{\theta,2}$.

We state the following results for references.

**Theorem A.1.** *The following inclusions hold with continuous embedding:*

$$A_0 \cap A_1 \subset [A_0,A_1]_{\theta,q} \subset A_0 + A_1.$$

*Moreover, if the norm* $\|u\|_{[A_0,A_1]_{\theta,q}}$ *is defined by* (A.1) *with the factor* (A.2), *then*

$$\|u\|_{A_0+A_1} \le \|u\|_{[A_0,A_1]_{\theta,q}} \le \|u\|_{A_0}^{1-\theta}\,\|u\|_{A_1}^{\theta} \le \|u\|_{A_0\cap A_1}.$$

*Proof.* See [138, Proposition 2.3]. See also [147, Lemma B.1]. □

The following theorem is called the duality theorem.

**Theorem A.2.** *Assume that* $A_0 \cap A_1$ *is dense in* $A_0$ *and in* $A_1$. *If* $\theta \in (0,1)$ *and* $q \geq 1$, *then* $A_0 \cap A_1$ *is dense in* $[A_0, A_1]_{\theta,q}$ *and*

$$[A_0, A_1]^*_{\theta,q} = [A^*_0, A^*_1]_{\theta,q^*} \quad \text{where } \frac{1}{q} + \frac{1}{q^*} = 1$$

*with equal norms if proper factors* $N_{\theta,q}$ *and* $N_{\theta,q^*}$ *are used.*

*Proof.* See [25, Theorem 3.7.1] and [138, Theorem 6.2]. See also [57, Theorem 2.4] and [147, Theorem B.5] for the claim on equality of norms by choosing proper factors $N_{\theta,q}$ and $N_{\theta,q^*}$. □

The next theorem is known as the reiteration theorem.

**Theorem A.3.** *For any* $\theta, \theta_0, \theta_1 \in [0,1]$ *satisfying* $\theta_0 \leq \theta \leq \theta_1$, *if* $\eta \in [0,1]$ *is such that* $\theta = (1-\eta)\theta_0 + \eta\theta_1$, *then*

$$\big[[A_0, A_1]_{\theta_0}, [A_0, A_1]_{\theta_1}\big]_\eta = [A_0, A_1]_\theta$$

*with equality of norms, if a proper factor* $N_\theta$ *is used. The statement also holds if* $\theta_j \in \{0,1\}$ *for* $j = 0, 1$, *provided that in such case we define* $[A_0, A_1]_{\theta_j} = A_{\theta_j}$.

*Proof.* See [25, Theorem 3.5.3] and [138, Theorem 6.1]. See also [57, Theorem 3.7] and [147, Theorem B.6] for the claim on equality of norms. □

The next result establishes the interpolation property of $[A_0, A_1]_{\theta,q}$, regardless of the choice of $N_{\theta,q}$ (provided that it is the same for $\mathscr{A}$ and $\mathscr{B}$).

**Theorem A.4.** *Let* $\mathscr{A} = (A_0, A_1)$ *and* $\mathscr{B} = (B_0, B_1)$ *be two compatible couples of normed spaces. If* $\mathscr{T} = (T_0, T_1) : \mathscr{A} \to \mathscr{B}$ *is a compatible couple of linear operators, then there exists a unique bounded linear operator* $T_\theta : [A_0, A_1]_{\theta,q} \to [B_0, B_1]_{\theta,q}$ *satisfying*

$$T_\theta u = T_0 u = T_1 u, \quad u \in A_0 \cap A_1.$$

*Moreover, if* $M_0$ *and* $M_1$ *are positive constants such that*

$$\|T_j u\|_{B_j} \leq M_j \, \|u\|_{A_j}, \quad u \in A_j, \; j = 0, 1,$$

*then*

$$\|T_\theta u\|_{[B_0,B_1]_{\theta,q}} \leq M_0^{1-\theta} \, M_1^\theta \, \|u\|_{[A_0,A_1]_{\theta,q}}, \quad u \in [A_0, A_1]_{\theta,q}.$$

*In particular, if* $T : A_j \to B_j$ *is a linear operator that maps* $A_j$ *to* $B_j$, $j = 0, 1$, *then* $T$ *maps* $[A_0, A_1]_{\theta,q}$ *to* $[B_0, B_1]_{\theta,q}$ *and*

$$\|T\|_{[A_0,A_1]_{\theta,q} \to [B_0,B_1]_{\theta,q}} \leq \|T\|^{1-\theta}_{A_0 \to B_0} \, \|T\|^\theta_{A_1 \to B_1}.$$

Here $\|T\|_{X \to Y}$ denotes the norm of $T$ as an operator from $X$ to $Y$.

*Proof.* See [147, Theorem B.2]. See also [42, Proposition 14.1.5]. □

## A.2 Sobolev Spaces

In this section, we give detailed constructions of Sobolev spaces and their properties. References in this topic include [1, 43, 88, 138, 147].

We introduce in this section different types of Sobolev norms, each of them has its own use, even though they are equivalent. The Slobodetski norm with its integral form is easy for calculation, in particular to find bounds of some expressions. The interpolation norm is useful when properties of Sobolev spaces of integral orders (e.g., the $L^2$ and $H^1$ spaces) are derived for Sobolev spaces of fractional orders (e.g., the $H^{1/2}(\Gamma)$ and $\widetilde{H}^{1/2}(\Gamma)$ spaces). Weighted norms are used when a norm is not scalable under affine mappings.

### *A.2.1 Notations*

$\mathscr{D}(\mathbb{R}^d)$ : the space of all $C^\infty$ functions with compact supports in $\mathbb{R}^d$.

$\mathscr{S}^*(\mathbb{R}^d)$ : the space of temperate distributions in $\mathbb{R}^d$.

$\Omega$ : a Lipschitz open set or Lipschitz domain (open and connected) in $\mathbb{R}^d$.

$\mathscr{D}(\Omega)$ : the space of all $C^\infty$ functions with compact supports in $\Omega$.

$\mathscr{D}^*(\Omega)$ : the space of all distributions on $\Omega$.

$L^p(\Omega)$ : the Lebesgue spaces on $\Omega$, $1 \le p \le \infty$.

We also denote, for any closed subset $F$ of $\mathbb{R}^d$,

$$\mathscr{D}(F) = \{u : u = U|_F \text{ for some } U \in \mathscr{D}(\mathbb{R}^d)\}. \tag{A.3}$$

### *A.2.2 Sobolev spaces on $\mathbb{R}^d$*

#### A.2.2.1 The space $H^s(\mathbb{R}^d)$, $s \in \mathbb{R}$

For $s \in \mathbb{R}$, we define the Sobolev spaces $H^s(\mathbb{R}^d)$ by Fourier transforms

$$H^s(\mathbb{R}^d) := \{u \in \mathscr{S}^*(\mathbb{R}^d) : \int_{\mathbb{R}^d} (1+|\boldsymbol{\xi}|^2)^s |\widehat{u}(\boldsymbol{\xi})|^2 \, d\boldsymbol{\xi} < \infty\},$$

where $\widehat{u}$ is the Fourier transform of $u$ defined by

$$\widehat{u}(\boldsymbol{\xi}) := \int_{\mathbb{R}^d} e^{-i2\pi\boldsymbol{\xi}\cdot\mathbf{x}} u(\mathbf{x}) \, \mathrm{d}\mathbf{x} \quad \text{for} \quad \boldsymbol{\xi} \in \mathbb{R}^d.$$

The space $H^s(\mathbb{R}^d)$ is equipped with the norm

$$\|u\|_{H^s(\mathbb{R}^d)} = \|u\|_{H^s_{\mathrm{F}}(\mathbb{R}^d)} := \Big( \int_{\mathbb{R}^d} (1+|\boldsymbol{\xi}|^2)^s |\widehat{u}(\boldsymbol{\xi})|^2 \, d\boldsymbol{\xi} \Big)^{1/2}. \tag{A.4}$$

The following dual property is known; see e.g. [147, (3.22)],

$$H^{-s}(\mathbb{R}^d) = \Big( H^s(\mathbb{R}^d) \Big)^*, \quad s \in \mathbb{R}, \tag{A.5}$$

and

$$\|u\|_{H^{-s}(\mathbb{R}^d)} = \sup_{0 \neq v \in H^s(\mathbb{R}^d)} \frac{\langle u, v \rangle_{\mathbb{R}^d}}{\|v\|_{H^s(\mathbb{R}^d)}}, \quad u \in H^{-s}(\mathbb{R}^d).$$

Here $\langle \cdot, \cdot \rangle_{\mathbb{R}^d}$ is the dual pairing which extends the $L^2(\mathbb{R}^d)$-inner product.

It is known that the spaces $H^s(\mathbb{R}^d)$ have the exact interpolation property; see [138, Remark 7.5 page 32] and [147, Theorem B.7]. This means that if $s_0, s_1 \in \mathbb{R}$ and $\theta \in (0,1)$ are such that $s = (1-\theta)s_0 + \theta s_1$, then

$$H^s(\mathbb{R}^d) = [H^{s_0}(\mathbb{R}^d), H^{s_1}(\mathbb{R}^d)]_\theta \tag{A.6}$$

with equal norms

$$\|u\|_{H^s_{\mathrm{F}}(\mathbb{R}^d)} = \|u\|_{H^s_{\mathrm{I}}(\mathbb{R}^d)} \tag{A.7}$$

where $\|u\|_{H^s_{\mathrm{I}}(\mathbb{R}^d)}$ denotes $\|u\|_{[H^{s_0}(\mathbb{R}^d), H^{s_1}(\mathbb{R}^d)]_\theta}$ and where the interpolation space and interpolation norm are defined in Section A.1.2, provided that the normalisation factor $N_{\theta,q}$ in (A.2) is chosen to be, see [147, page 329],

$$N_{\theta,q} = \Big( \frac{2\sin(\pi\theta)}{\pi} \Big)^{1/2}. \tag{A.8}$$

#### A.2.2.2 The space $W^s(\mathbb{R}^d)$, $s \in \mathbb{R}$

Another Sobolev space on $\mathbb{R}^d$ can be defined with the Slobodetski norm as follows. For any non-negative number $s$

$$W^s(\mathbb{R}^d) = \{ u \in L^2(\mathbb{R}^d) : \|u\|_{H^s_{\mathrm{S}}(\mathbb{R}^d)} < \infty \} \tag{A.9}$$

where, for $r = 0, 1, 2, \ldots$ and $\mu \in (0,1)$, the Slobodetski norm $\|u\|_{H^s_{\mathrm{S}}(\mathbb{R}^d)}$ is defined by

$$\|u\|_{H^s_{\mathrm{S}}(\mathbb{R}^d)} := \begin{cases} \Big( \sum\limits_{|\alpha|=0}^{r} a_{\alpha,r} \, \|\partial^\alpha u\|^2_{L^2(\mathbb{R}^d)} \Big)^{1/2}, & s = r, \\ \Big( \|u\|^2_{H^r_{\mathrm{S}}(\mathbb{R}^d)} + \sum\limits_{|\alpha|=m} |\partial^\alpha u|^2_{H^\mu(\mathbb{R}^d)} \Big)^{1/2}, & s = r + \mu, \end{cases} \tag{A.10}$$

where $a_{\alpha,r} > 0$ is to be determined later. Here $\partial^\alpha$ is the differential operator defined by

$$\partial^\alpha u := \frac{\partial^{|\alpha|} u}{\partial x_1^{\alpha_1} \cdots \partial x_d^{\alpha_d}} \quad \text{with} \quad \mathbf{x} = (x_1, \cdots, x_d), \tag{A.11}$$
$$\alpha = (\alpha_1, \cdots, \alpha_d) \in \mathbb{N}^d \quad \text{and} \quad |\alpha| = \alpha_1 + \cdots + \alpha_d,$$

and $|\cdot|_{H^\mu(\mathbb{R}^d)}$ is the seminorm defined by

$$|u|_{H^\mu(\mathbb{R}^d)} := \Big( \iint\limits_{\mathbb{R}^d \times \mathbb{R}^d} \frac{|u(\mathbf{x}) - u(\mathbf{y})|^p}{|\mathbf{x} - \mathbf{y}|^{d+2\mu}} \, d\mathbf{x} \, d\mathbf{y} \Big)^{1/2}.$$

As will be seen in Theorem A.5, the two spaces $H^s(\mathbb{R}^d)$ and $W^s(\mathbb{R}^d)$ equal. This is the reason we denote the norm of $W^s(\mathbb{R}^d)$ by $\|\cdot\|_{H^s_{\mathrm{S}}(\mathbb{R}^d)}$ instead of $\|\cdot\|_{W^s(\mathbb{R}^d)}$.

The factor $a_{\alpha,r}$ in (A.10) is usually chosen to be one. However, as observed in [57]

$$\|u\|^2_{H^r_{\mathrm{F}}(\mathbb{R}^d)} = \sum_{|\alpha| \le r} \binom{r}{|\alpha|} \binom{|\alpha|}{\alpha} \|\partial^\alpha u\|^2_{L^2(\mathbb{R}^d)} \quad \text{where} \quad \binom{|\alpha|}{\alpha} := \frac{|\alpha|!}{\alpha_1! \cdots \alpha_d!}.$$

We choose, as in [57],

$$a_{\alpha,r} = \binom{r}{|\alpha|} \binom{|\alpha|}{\alpha} \tag{A.12}$$

to ensure that

$$\|u\|_{H^r_{\mathrm{F}}(\mathbb{R}^d)} = \|u\|_{H^r_{\mathrm{S}}(\mathbb{R}^d)}. \tag{A.13}$$

For $s < 0$, the space is defined by duality

$$W^s(\mathbb{R}^d) = \left(W^{-s}(\mathbb{R}^d)\right)^* \tag{A.14}$$

equipped with the dual norm

$$\|u\|_{W^s(\mathbb{R}^d)} = \|u\|_{H^s_{\mathrm{S}}(\mathbb{R}^d)} = \sup_{\substack{v \in W^{-s}(\mathbb{R}^d) \\ v \ne 0}} \frac{\langle u, v \rangle_{\mathbb{R}^d}}{\|v\|_{H^{-s}_{\mathrm{S}}(\mathbb{R}^d)}}.$$

The two spaces $H^s(\mathbb{R}^d)$ and $W^s(\mathbb{R}^d)$ coincide with equivalent norms, as shown in the following theorem.

**Theorem A.5.** *For any $s \in \mathbb{R}$, the spaces $H^s(\mathbb{R}^d)$ and $W^s(\mathbb{R}^d)$ coincide with equal norms if $s$ is an integer, and equivalent norms otherwise.*

*Proof.* Let $s = r + \mu$ where $r$ is a non-negative integer and $\mu \in [0, 1)$. If $\mu = 0$, then the result is given in (A.13). If $\mu > 0$, then it is shown in [147, Lemma 3.15] that

$$|u|^2_\mu = a_\mu \int\limits_{\mathbb{R}^d} |\xi|^{2\mu} |\widehat{u}(\xi)|^2 \, d\xi$$

where

$$a_\mu = \int_0^\infty t^{-2\mu-1}\Big(\int_{|\omega|=1} |e^{i2\pi\omega_1 t} - 1|^2\, d\omega\Big)\, \mathrm{d}t. \tag{A.15}$$

Note that

$$\int_{|\omega|=1} |e^{i2\pi\omega_1 t} - 1|^2\, d\omega = \begin{cases} O(t^2) & \text{as } t \to 0, \\ O(1) & \text{as } t \to \infty. \end{cases} \tag{A.16}$$

Hence, $a_\mu$ is bounded for $\mu \in (0,1)$ and therefore

$$\begin{aligned} \|u\|^2_{H^s_{\mathrm{S}}(\mathbb{R}^d)} &= \|u\|^2_{H^r_{\mathrm{F}}(\mathbb{R}^d)} + \sum_{|\alpha|=r} \int_{\mathbb{R}^d} a_\mu |\xi|^{2\mu} |(2\pi\xi)^\alpha \widehat{u}(\xi)|^2\, d\xi \\ &\simeq \int_{\mathbb{R}^d} (1+|\xi|^2)^{r+\mu} |\widehat{u}(\xi)|^2\, d\xi = \|u\|^2_{H^{r+\mu}_{\mathrm{F}}(\mathbb{R}^d)}. \end{aligned}$$

For $s < 0$, the required result is a consequence of duality arguments, noting (A.5) and (A.14). For $s = 0$ the spaces $H^0(\mathbb{R}^d)$ and $W^0(\mathbb{R}^d)$ are the Lebesgue space $L^2(\mathbb{R}^d)$ with norms equal the usual $L^2$-norm. □

It is noted that the equivalence constants in the above proof depend on $\mu$, and hence on $s$. The following result states the special case of $s$ independence.

**Corollary A.1.** *If $s = r + \mu$ for some non-negative integer $r$ and $\mu \in [\delta, 1/2)$ for some $\delta > 0$, then*

$$\|u\|^2_{H^s_{\mathrm{S}}(\mathbb{R}^d)} \simeq \|u\|^2_{H^s_{\mathrm{F}}(\mathbb{R}^d)} \quad \forall u \in H^s(\mathbb{R}^d) = W^s(\mathbb{R}^d)$$

*with the constants depending on $\delta$ but are independent of $s$.*

*Proof.* It suffices to estimate $a_\mu$ defined by (A.15). Due to (A.16),

$$a_\mu \simeq \int_0^1 t^{-2\mu+1}\, \mathrm{d}t + \int_1^\infty t^{-2\mu-1}\, \mathrm{d}t = \frac{1}{2\mu(1-\mu)}$$

(with the constants independent of $\mu$) so that $1/(1-\delta) \lesssim a_\mu \lesssim 1/\delta$. This proves the required result. □

### A.2.3 *Sobolev spaces on a Lipschitz domain*

Let $\Omega$ be an open subset of $\mathbb{R}^d$. The spaces $H^s(\Omega)$, $H^s_0(\Omega)$, and $\widetilde{H}^s(\Omega)$ can be defined either from $H^s(\mathbb{R}^d)$ or from $W^s(\mathbb{R}^d)$. Even though $H^s(\mathbb{R}^d) = W^s(\mathbb{R}^d)$, as proved in Theorem A.5, it is not necessary that both definitions give the same spaces on $\Omega$, unless $\Omega$ is a Lipschitz domain, the set of our main interest in this book. Section A.2.5 justifies the use of unique notations for these spaces. Nevertheless, since different definitions give different norms on the same space, we will differentiate these norms. Readers who are interested in more general sets $\Omega$ are referred to [88, 147].

### A.2.3.1 Lipschitz domains

For the reader's convenience, we first recall the definition of Lipschitz domains, following [147, Chapter 3].

**Definition A.1.** *Let $\Omega$ be an open subset of $\mathbb{R}^d$.*

(i) *$\Omega$ is called a Lipschitz hypograph if*

$$\Omega = \{\mathbf{x} = (\mathbf{x}', x_d) \in \mathbb{R}^d : x_d < \zeta(\mathbf{x}'),\ \mathbf{x}' = (x_1, \ldots, x_{d-1}) \in \mathbb{R}^{d-1}\}$$

*where $\zeta : \mathbb{R}^{d-1} \to \mathbb{R}$ is a Lipschitz function with a Lipschitz constant $L$.*

(ii) *$\Omega$ is a Lipschitz domain if its boundary $\partial\Omega$ is compact and if there exist finite families of open subsets $\{W_j\}_{j=1}^J$ and $\{\Omega_j\}_{j=1}^J$ having the following properties:*

(a) *The family $\{W_j\}_{j=1}^J$ is a finite cover of $\partial\Omega$, i.e., $\partial\Omega \subset \cup_{j=1}^J W_j$;*

(b) *Each $\Omega_j$ can be transformed, by a rotation and a translation, to be a Lipschitz hypograph such that $W_j \cap \Omega = W_j \cap \Omega_j$.*

*Remark A.1.* (i) A Lipschitz hypograph $\Omega$ is unbounded, its unbounded boundary being

$$\partial\Omega = \{\mathbf{x} = (\mathbf{x}', x_d) \in \mathbb{R}^d : x_d = \zeta(\mathbf{x}'),\ \mathbf{x}' = (x_1, \ldots, x_{d-1}) \in \mathbb{R}^{d-1}\}.$$

(ii) A Lipschitz domain $\Omega$ can be bounded or unbounded, but its boundary is bounded. If $\Omega$ is a bounded Lipschitz domain, then $\mathbb{R}^d \setminus \overline{\Omega}$ is an unbounded Lipschitz domain.

### A.2.3.2 The spaces $H^s(\Omega)$, $H_0^s(\Omega)$, and $\widetilde{H}^s(\Omega)$, $s \in \mathbb{R}$

We give the first definition of Sobolev spaces on a Lipschitz domain $\Omega$.

**Definition A.2.** *Let $\Omega$ be a Lipschitz domain in $\mathbb{R}^d$, $d \geq 1$. For $s \in \mathbb{R}$, we define $H^s(\Omega)$, $H_0^s(\Omega)$, and $\widetilde{H}^s(\Omega)$ by*

$$\begin{aligned} H^s(\Omega) &:= \{u \in \mathscr{D}^*(\Omega) : u = U|_\Omega \text{ for some } U \in H^s(\mathbb{R}^d)\}, \\ H_0^s(\Omega) &:= \overline{\mathscr{D}(\Omega)}^{H^s(\Omega)} = \text{closure of } \mathscr{D}(\Omega) \text{ in } H^s(\Omega), \\ \widetilde{H}^s(\Omega) &:= \{u \in H^s(\mathbb{R}^d) : \operatorname{supp} u \subset \overline{\Omega}\}, \end{aligned} \tag{A.17}$$

*with the corresponding norms*

$$\begin{aligned}
\|u\|_{H^s_{\mathrm{F}}(\Omega)} &:= \inf_{u=U|_\Omega} \|U\|_{H^s_{\mathrm{F}}(\mathbb{R}^d)}, && u \in H^s(\Omega),\\
\|u\|_{H^s_0(\Omega)} &:= \|u\|_{H^s_{\mathrm{F}}(\Omega)}, && u \in H^s_0(\Omega),\\
\|u\|_{\widetilde{H}^s_{\mathrm{F}}(\Omega)} &:= \|u\|_{H^s_{\mathrm{F}}(\mathbb{R}^d)}, && u \in \widetilde{H}^s(\Omega).
\end{aligned} \tag{A.18}$$

Notice that there exists $U^* \in \mathscr{X}(u) := \{U \in H^s(\mathbb{R}^d) : U|_\Omega = u\}$ satisfying

$$\|U^*\|_{H^s_{\mathrm{F}}(\mathbb{R}^d)} = \inf\{\|U\|_{H^s_{\mathrm{F}}(\mathbb{R}^d)} : U \in \mathscr{X}(u)\} = \|u\|_{H^s_{\mathrm{F}}(\Omega)}. \tag{A.19}$$

Indeed, let $m = \inf\{\|U\|_{H^s_{\mathrm{F}}(\mathbb{R}^d)} : U \in \mathscr{X}(u)\}$. Then there exists a sequence $\{U_k\}$ in $\mathscr{X}(u)$ satisfying $\lim_{k\to\infty}\|U_k\|_{H^s_{\mathrm{F}}(\mathbb{R}^d)} = m$. Since this sequence is bounded in the Hilbert space $H^s(\mathbb{R}^d)$, there exists a subsequence $\{U_{k_j}\}$ that converges weakly to $U^* \in H^s(\mathbb{R}^d)$. It can be seen that $U^* \in \mathscr{X}(u)$. Hence $m \le \|U^*\|_{H^s_{\mathrm{F}}(\mathbb{R}^d)}$. On the other hand, the weak convergence implies $\|U^*\|_{H^s_{\mathrm{F}}(\mathbb{R}^d)} \le \liminf_{j\to\infty}\|U_{k_j}\|_{H^s_{\mathrm{F}}(\mathbb{R}^d)} = m$. Hence, $U^*$ satisfies (A.19).

It is clear from (A.17) and (A.18) that

$$H^0(\Omega) = H^0_0(\Omega) = \widetilde{H}^0(\Omega)$$

and that

$$\|u|_\Omega\|_{H^s_{\mathrm{F}}(\Omega)} \le \|u\|_{\widetilde{H}^s_{\mathrm{F}}(\Omega)} \quad \text{for} \quad u \in \widetilde{H}^s(\Omega),\ s \in \mathbb{R}. \tag{A.20}$$

Moreover, we have the following relations among the three spaces (see [88, Theorem 1.4.2.4 and Theorem 1.4.5.2])

$$\begin{aligned}
&\widetilde{H}^s(\Omega) \subset H^s_0(\Omega) \subset H^s(\Omega), \quad s \in \mathbb{R}.\\
&\widetilde{H}^s(\Omega) = H^s_0(\Omega), \quad s \ge 0,\ s \ne \frac{1}{2}, \frac{3}{2}, \frac{5}{2}, \dots\\
&\widetilde{H}^s(\Omega) = H^s_0(\Omega) = H^s(\Omega), \quad 0 \le s < \frac{1}{2}.
\end{aligned} \tag{A.21}$$

*Remark A.2.* Assume that $\overline{\Omega} = \overline{\Omega}_1 \cup \dots \cup \overline{\Omega}_N$ with $\Omega_i \cap \Omega_j = \varnothing$ for $i \ne j$. The following properties are clear from definitions (A.17) and (A.18).

(i) $H^s(\Omega) \subset H^s(\Omega_i)$ and if $u \in H^s(\Omega)$, then $\|u|_{\Omega_i}\|_{H^s_{\mathrm{F}}(\Omega_i)} \le \|u\|_{H^s_{\mathrm{F}}(\Omega)}$ for $i = 1,\dots,N$. Consequently, the Cauchy–Schwarz inequality implies

$$\sum_{i=1}^N \|u|_{\Omega_i}\|^2_{H^s_{\mathrm{F}}(\Omega_i)} \le N\|u\|^2_{H^s_{\mathrm{F}}(\Omega)}. \tag{A.22}$$

(ii) $\widetilde{H}^s(\Omega_i) \subset \widetilde{H}^s(\Omega)$ and $\|u\|_{\widetilde{H}^s_{\mathrm{F}}(\Omega)} = \|u\|_{\widetilde{H}^s_{\mathrm{F}}(\Omega_i)}$ for all $u \in \widetilde{H}^s(\Omega_i)$, $i = 1,\dots,N$.

(iii) If $u \in \widetilde{H}^s(\Omega)$ is such that $u|_{\Omega_i} \in \widetilde{H}^s(\Omega_i)$, $i = 1,\dots,N$, then

$$\widehat{u}(\boldsymbol{\xi}) = \sum_{i=1}^{N} \widehat{(u|_{\Omega_i})}(\boldsymbol{\xi})$$

so that, see (A.4),

$$\|u\|^2_{\widetilde{H}^s_{\mathrm{F}}(\Omega)} \le N \sum_{i=1}^{N} \|u|_{\Omega_i}\|^2_{\widetilde{H}^s_{\mathrm{F}}(\Omega_i)}. \tag{A.23}$$

The constant $N$ on the right-hand sides of (A.22) and (A.23) is not satisfactory for the analysis of preconditioners. This is the main reason to introduce other spaces and norms in subsequent sections.

#### A.2.3.3 Density and duality properties

The following density properties hold

$$H^s(\mathbb{R}^d) = \overline{\mathscr{D}(\mathbb{R}^d)}^{H^s(\mathbb{R}^d)}, \qquad H^s(\Omega) = \overline{\mathscr{D}(\overline{\Omega})}^{H^s(\Omega)},$$
$$H^s_0(\Omega) = \overline{\mathscr{D}(\Omega)}^{H^s(\Omega)}, \qquad \widetilde{H}^s(\Omega) = \overline{\mathscr{D}(\Omega)}^{H^s(\mathbb{R}^d)},$$

where, for any subset $A$ of a normed vector space $X$, the set $\overline{A}^X$ denotes the closure of $A$ in $X$.

For any $s \in \mathbb{R}$, the spaces $H^s(\Omega)$ and $\widetilde{H}^s(\Omega)$ satisfy the duality relations

$$H^s(\Omega)^* = \widetilde{H}^{-s}(\Omega) \quad \text{and} \quad \widetilde{H}^s(\Omega)^* = H^{-s}(\Omega) \tag{A.24}$$

with respect to the dual pairing which is the extension of the $L^2(\Omega)$ inner product.

#### A.2.3.4 The spaces $W^s(\Omega)$ and $\widetilde{W}^s(\Omega)$, $s \in \mathbb{R}$

Sobolev spaces on $\Omega$ can also be constructed in the same manner as $W^s(\mathbb{R}^d)$ defined by (A.9):

$$W^s(\Omega) := \{u \in L^2(\Omega) : \|u\|_{H^s_{\mathrm{S}}(\Omega)} < \infty\},$$
$$\widetilde{W}^s(\Omega) := \{u \in L^2(\Omega) : \|u\|_{\widetilde{H}^s_{\mathrm{S}}(\Omega)} < \infty\},$$

where the Slobodetski norms $\|u\|_{H^s_{\mathrm{S}}(\Omega)}$ and $\|u\|_{\widetilde{H}^s_{\mathrm{S}}(\Omega)}$ are defined as follows are. For any $u \in H^s(\Omega)$, $s = r + \mu$ with $r = 0, 1, 2, \ldots$, and $\mu \in [0, 1)$ we define the seminorm

$$|u|^2_{H^s(\Omega)} := \begin{cases} \displaystyle\sum_{|\alpha|=r} a_{\alpha,r} \, \|\partial^\alpha u\|^2_{L^2(\Omega)}, & \mu = 0, \\ \displaystyle\sum_{|\alpha|=r} \iint_{\Omega\times\Omega} \frac{|\partial^\alpha u(\mathbf{x}) - \partial^\alpha u(\mathbf{y})|^2}{|\mathbf{x}-\mathbf{y}|^{d+2\mu}} \, d\mathbf{x}\, d\mathbf{y}, & 0 < \mu < 1, \end{cases} \tag{A.25}$$

where $\partial^\alpha u$ and $a_{\alpha,r}$ are defined in (A.11) and (A.12), respectively. The Slobodetski norm is defined for $u \in H^s(\Omega)$ by

$$\|u\|^2_{H^s_{\mathrm{S}}(\Omega)} := \begin{cases} \|u\|^2_{L^2(\Omega)} & s = 0, \\ \|u\|^2_{L^2(\Omega)} + |u|^2_{H^s(\Omega)} & s > 0. \end{cases} \tag{A.26}$$

For $u \in \widetilde{H}^s(\Omega)$ we define

$$\|u\|^2_{\widetilde{H}^s_{\mathrm{S}}(\Omega)} := \|u\|^2_{H^s_{\mathrm{S}}(\Omega)} + \sum_{|\alpha|=r} \int_\Omega \frac{|\partial^\alpha u(\mathbf{x})|^2}{\operatorname{dist}(\mathbf{x}, \partial\Omega)^{2\mu}} \, d\mathbf{x}. \tag{A.27}$$

In $H^{-s}(\Omega)$ and $\widetilde{H}^{-s}(\Omega)$ with $s > 0$, the norms $\|\cdot\|_{H^{-s}_{\mathrm{S}}(\Omega)}$ and $\|\cdot\|_{\widetilde{H}^{-s}_{\mathrm{S}}(\Omega)}$ can be defined by duality, see (A.24), as follows

$$\begin{aligned} \|u\|_{H^{-s}_{\mathrm{S}}(\Omega)} &:= \sup_{v \in \widetilde{H}^s(\Omega)} \frac{\langle u, v \rangle_\Omega}{\|v\|_{\widetilde{H}^s_{\mathrm{S}}(\Omega)}}, \\ \|u\|_{\widetilde{H}^{-s}_{\mathrm{S}}(\Omega)} &:= \sup_{v \in H^s(\Omega)} \frac{\langle u, v \rangle_\Omega}{\|v\|_{H^s_{\mathrm{S}}(\Omega)}}. \end{aligned} \tag{A.28}$$

Here $\langle \cdot, \cdot \rangle_\Omega$ is the dual pairing extension of the inner product in $L^2(\Omega)$.

The factor $a_{\alpha,r}$ in (A.25) ensures that

$$\|u\|_{H^r_{\mathrm{S}}(\Omega)} \le \|u\|_{H^r_{\mathrm{F}}(\Omega)}, \quad r = 0, 1, 2, \ldots. \tag{A.29}$$

#### A.2.3.5 The interpolation spaces $[H^{s_0}(\Omega), H^{s_1}(\Omega)]_\theta$ and $[\widetilde{H}^{s_0}(\Omega), \widetilde{H}^{s_1}(\Omega)]_\theta$

A third construction of Sobolev spaces on $\Omega$ is via interpolation introduced in Section A.1. In the literature, this interpolation is established with the $K$-functionals defined using the Sobolev norms $\|\cdot\|_{H^s_{\mathrm{F}}(\Omega)}$ and $\|\cdot\|_{\widetilde{H}^s_{\mathrm{F}}(\Omega)}$; see e.g. [57] and [147]. For completeness and also for better understanding, we first use the same definition. However, we will point out that this construction does not serve our purpose in the analysis of additive Schwarz preconditioners. Therefore, we will later introduce another construction.

For any integer $r$, we define the interpolation norm in $H^r(\Omega)$ and in $\widetilde{H}^r(\Omega)$ by

$$\|u\|_{H^r_{\mathrm{IF}}(\Omega)} := \|u\|_{H^r_{\mathrm{F}}(\Omega)} \quad \text{and} \quad \|u\|_{\widetilde{H}^r_{\mathrm{IF}}(\Omega)} := \|u\|_{\widetilde{H}^r_{\mathrm{F}}(\Omega)},$$

and the $K$-functionals

$$K_F(t,u,\mathscr{A}) := \inf_{\substack{u_0\in H^0,\ u_1\in H^r\\ u_0+u_1=u}} \left(\|u_0\|^2_{H^0_{\mathrm{IF}}(\Omega)} + t^2\|u_1\|^2_{H^r_{\mathrm{IF}}(\Omega)}\right)^{1/2},$$
$$K_F(t,u,\widetilde{\mathscr{A}}) := \inf_{\substack{u_0\in H^0,\ u_1\in \widetilde{H}^r\\ u_0+u_1=u}} \left(\|u_0\|^2_{\widetilde{H}^0_{\mathrm{IF}}(\Omega)} + t^2\|u_1\|^2_{\widetilde{H}^r_{\mathrm{IF}}(\Omega)}\right)^{1/2}, \tag{A.30}$$

with

$$\mathscr{A} = (H^0(\Omega), H^r(\Omega)) \quad \text{and} \quad \widetilde{\mathscr{A}} = (\widetilde{H}^0(\Omega), \widetilde{H}^r(\Omega)).$$

For $\theta \in (0,1)$ and $s = \theta r$, these $K$-functionals yield the following interpolation spaces

$$H^s_{\mathrm{IF}}(\Omega) := [H^0(\Omega), H^r(\Omega)]_\theta \quad \text{and} \quad \widetilde{H}^s_{\mathrm{IF}}(\Omega) := [\widetilde{H}^0(\Omega), \widetilde{H}^r(\Omega)]_\theta, \tag{A.31}$$

and their corresponding norms

$$\|u\|_{H^s_{\mathrm{IF}}(\Omega)} := \|u\|_{[H^0(\Omega),H^r(\Omega)]_\theta} \quad \text{and} \quad \|u\|_{\widetilde{H}^s_{\mathrm{IF}}(\Omega)} := \|u\|_{[\widetilde{H}^0(\Omega),\widetilde{H}^r(\Omega)]_\theta}. \tag{A.32}$$

We used the index $IF$ to make it clear that the interpolation spaces and norms are defined from the norms given in (A.18). We will later prove that if $\Omega$ is a Lipschitz domain, then $H^s(\Omega) = H^s_{\mathrm{IF}}(\Omega)$ and $\widetilde{H}^s(\Omega) = \widetilde{H}^s_{\mathrm{IF}}(\Omega)$ with equivalent norms.

As will be seen in Lemma A.1 and Theorem A.8, the interpolation property mentioned above hold. However, the norms $\|\cdot\|_{H^s_{\mathrm{F}}(\Omega)}$ and $\|\cdot\|_{\widetilde{H}^s_{\mathrm{F}}(\Omega)}$ used in the definition of the $K$-functionals in (A.30) are not suitable for the study of the scaling property and other properties of the interpolation norms; see Section A.2.7 and Section A.2.9. For this reason, we introduce another type of interpolation norms, starting with the Slobodetski norm.

For any non-negative integer $r$, we define

$$\|u\|_{H^r_{\mathrm{I}}(\Omega)} := \|u\|_{H^r_{\mathrm{S}}(\Omega)} \quad \text{and} \quad \|u\|_{\widetilde{H}^r_{\mathrm{I}}(\Omega)} := |u|_{H^r(\Omega)};$$

see (A.25). Note that, see (A.29),

$$\|u\|_{\widetilde{H}^r_{\mathrm{I}}(\Omega)} \le \|u\|_{H^r_{\mathrm{I}}(\Omega)} \le \|u\|_{H^r_{\mathrm{F}}(\Omega)}. \tag{A.33}$$

Instead of (A.30), the $K$-functionals are now defined by

$$K_S(t,u,\mathscr{A}) := \inf_{\substack{u_0\in H^0,\ u_1\in H^r\\ u_0+u_1=u}} \left(\|u_0\|^2_{H^0_{\mathrm{I}}(\Omega)} + t^2\|u_1\|^2_{H^r_{\mathrm{I}}(\Omega)}\right)^{1/2},$$
$$K_S(t,u,\widetilde{\mathscr{A}}) := \inf_{\substack{u_0\in H^0,\ u_1\in \widetilde{H}^r\\ u_0+u_1=u}} \left(\|u_0\|^2_{\widetilde{H}^0_{\mathrm{I}}(\Omega)} + t^2\|u_1\|^2_{\widetilde{H}^r_{\mathrm{I}}(\Omega)}\right)^{1/2}, \tag{A.34}$$

Hence, instead of (A.31) and (A.32) we now have

$$H^s_{\mathrm{I}}(\Omega) := [H^0(\Omega), H^r(\Omega)]_\theta \quad \text{and} \quad \widetilde{H}^s_{\mathrm{I}}(\Omega) := [\widetilde{H}^0(\Omega), \widetilde{H}^r(\Omega)]_\theta,$$

and their corresponding norms

$$\|u\|_{H_{\mathrm{I}}^s(\Omega)} := \|u\|_{[H^0(\Omega),H^r(\Omega)]_\theta} \quad \text{and} \quad \|u\|_{\widetilde{H}_{\mathrm{I}}^s(\Omega)} := \|u\|_{[\widetilde{H}^0(\Omega),\widetilde{H}^r(\Omega)]_\theta}.$$

The norm $\|\cdot\|_{H_{\mathrm{I}}^{-s}(\Omega)}$ and $\|\cdot\|_{\widetilde{H}_{\mathrm{I}}^{-s}(\Omega)}$ are defined by duality, see (A.24),

$$\begin{aligned}
\|u\|_{\widetilde{H}_{\mathrm{I}}^{-s}(\Omega)} &:= \sup_{0\neq v\in H^s(\Omega)} \frac{\langle u,v\rangle_\Omega}{\|v\|_{H_{\mathrm{I}}^s(\Omega)}} \\
\|u\|_{H_{\mathrm{I}}^{-s}(\Omega)} &:= \sup_{0\neq v\in \widetilde{H}^s(\Omega)} \frac{\langle u,v\rangle_\Omega}{\|v\|_{\widetilde{H}_{\mathrm{I}}^s(\Omega)}}.
\end{aligned} \tag{A.35}$$

In this book, we mostly use these interpolation norms and spaces, not the ones defined by (A.30). Therefore, for simplicity of notations, we do not include the index $S$ in these norms and spaces.

This consideration has not studied before in the literature, except in [42] for a special case when $s \in (0,1)$; see [42, Theorem 14.2.3]. It turns out that this construction is the correct construction for Section A.2.7 and Section A.2.9.

Given that the three different definitions of Sobolev spaces on $\mathbb{R}^d$ give the same space with equal norms or equivalent norms; see (A.6), (A.7), and Theorem A.5, what can we say about Sobolev spaces on $\Omega$? It turns out that thanks to the Lipschitz assumption of the domain $\Omega$, the same results hold. This is due to the existence of extension operators to be discussed in the next section.

### A.2.4 *Extension operators*

In this subsection, we collect some results on extension operators which extend functions defined on a Lipschitz domain $\Omega$ in $\mathbb{R}^d$ to functions defined on $\mathbb{R}^d$. First we recall that an operator $\mathscr{E} : W^s(\Omega) \to W^s(\mathbb{R}^d)$ is called an extension operator if $\mathscr{E}$ satisfies

$$\mathscr{E}u|_\Omega = u, \quad u \in W^s(\Omega).$$

(i) Calderón [47] constructed for each positive integer $r$ an extension operator $\mathscr{E}_r : W^r(\Omega) \to W^r(\mathbb{R}^d)$ that is bounded.
(ii) Stein [197, page 181] constructed an extension operator $\mathscr{E} : W^r(\Omega) \to W^r(\mathbb{R}^d)$ not depending on $r$ that is bounded for each positive integer $r$.
(iii) Other early developments of extension operators can be found, for example, in the book by Nečas [163].
(iv) Grisvard [88, Theorem 1.4.3.1] introduced for each positive real number $s$ a bounded extension operator $\mathscr{E}_S : W^s(\Omega) \to W^s(\mathbb{R}^d)$. Grisvard also mentioned that Aronszajn & Smith [10] and Seeley [192] showed that $\mathscr{E}_S$ can be chosen independently of $s$. This means the following estimate holds

$$\|\mathscr{E}_S u\|_{H^s_S(\mathbb{R}^d)} \lesssim \|u\|_{H^s_S(\Omega)}. \tag{A.36}$$

(v) McLean [147, Theorem A.4] modified Calderón's extension to show that for each non-negative integer $k$, there exists an extension operator $\mathscr{E}_{M_k} : W^s(\Omega) \to W^s(\mathbb{R}^d)$ that is bounded for $s \in [k, k+1)$. The proof of this theorem reveals that the norm of $\mathscr{E}_{M_k}$ depends on $k$ but not on $s$:

$$\|\mathscr{E}_{M_k} u\|_{H^s_S(\mathbb{R}^d)} \lesssim \|u\|_{H^s_S(\Omega)} \tag{A.37}$$

where the constant is independent of $u$ and $s \in [k, k+1)$.

(vi) Rychkov [182] further modified Calderón's extension and proved that there exists an extension operator $\mathscr{E}_R : H^s(\Omega) \to H^s(\mathbb{R}^d)$, not depending on $s \in \mathbb{R}$, that is continuous for all $s$:

$$\|\mathscr{E}_R u\|_{H^s_F(\mathbb{R}^d)} \lesssim \|u\|_{H^s_F(\Omega)}. \tag{A.38}$$

In particular, when $\Omega$ is a Lipschitz hypograph in $\mathbb{R}^d$, McLean [147, Theorem A.1] obtained a simple extension by reflection in the boundary of $\Omega$. Since we used this result in Section A.2.5, we state (without detailed proof) this result here.

**Theorem A.6.** *Let $\Omega$ be a Lipschitz hypograph defined by Definition A.1 and let $\mathscr{E}_M : W^s(\Omega) \to W^s(\mathbb{R}^d)$ be defined by*

$$\mathscr{E}_M u(\mathbf{x}) = \begin{cases} u(\mathbf{x}), & \mathbf{x} \in \Omega, \\ u(\mathbf{x}', 2\zeta(\mathbf{x}') - x_d), & \mathbf{x} \in \mathbb{R}^d \setminus \Omega, \end{cases} \tag{A.39}$$

*where $\mathbf{x}' \in \mathbb{R}^{d-1}$ and $\mathbf{x} = (\mathbf{x}', x_d)$. Then $\mathscr{E}_M$ is bounded for $s \in [0,1]$ and the norm of $\mathscr{E}_M$ is independent of $s$.*

*Proof.* This result is proved in [147, Theorem A.1]. The proof also reveals that

$$\|\mathscr{E}_M u\|_{H^s_S(\mathbb{R}^d)} \lesssim \|u\|_{H^s_S(\Omega)}, \quad 0 \le s \le 1,$$

with a constant independent of $u$ and $s$. □

### *A.2.5 Equivalence of norms*

Recall that in the case of Sobolev spaces of functions defined on the whole $\mathbb{R}^d$, the interpolation norm with a proper normalisation factor is equal to the norm defined by Fourier transforms, see (A.7) and (A.8), and both norms are equivalent to the Slobodetski norm; see Theorem A.5. In general, the equivalence constants depend on $s$, but when $s \in [\delta, 1/2)$ for some $\delta \in (0, 1/2)$, then the constants depend only on $\delta$; see Corollary A.1. We now use these results to study the equivalence of different Sobolev norms of functions defined on $\Omega$. The main tool is the extension operators discussed in Section A.2.4.

The following theorem shows that if $\Omega$ is a Lipschitz domain or a Lipschitz hypograph, then $H^s(\Omega) = W^s(\Omega)$ with equivalent norms.

**Theorem A.7.** (i) *Assume that $\Omega$ is a Lipschitz domain. For all $s \geq 0$, the two spaces $H^s(\Omega)$ and $W^s(\Omega)$ are equal with equivalent norms, i.e., for any $u \in H^s(\Omega)$,*

$$\|u\|_{H^s_{\mathrm{S}}(\Omega)} \simeq \|u\|_{H^s_{\mathrm{F}}(\Omega)}$$

*where the constants are independent of u but may depend on $\Omega$ and s.*

(ii) *Assume that $\Omega$ is a Lipschitz domain or Lipschitz hypograph. If $0 < \delta \leq s < 1/2$ for some $\delta \in (0, 1/2)$, then part* (i) *holds true with the equivalence constants independent of s and u, but possibly depending on $\delta$ and $\Omega$.*

*Proof.* Part (i) is proved in [138, 147]. The proof is presented here for completeness. Consider $u \in H^s(\mathbb{R}^d)$ and let $U \in H^s(\mathbb{R}^d)$ be such that $u = U|_\Omega$ and that $\|u\|_{H^s_{\mathrm{F}}(\Omega)} = \|U\|_{H^s(\mathbb{R}^d)}$; see (A.19). Then it follows successively from the definitions of the Slobodetski norms (A.10) and (A.26), and Theorem A.5 that

$$\|u\|_{H^s_{\mathrm{S}}(\Omega)} \leq \|U\|_{H^s_{\mathrm{S}}(\mathbb{R}^d)} \simeq \|U\|_{H^s_{\mathrm{F}}(\mathbb{R}^d)} = \|u\|_{H^s_{\mathrm{F}}(\Omega)}. \tag{A.40}$$

Therefore, $H^s(\Omega) \subset W^s(\Omega)$ with continuous inclusion.

Conversely, consider $u \in W^s(\Omega)$. Let $\mathscr{E}_S : W^s(\Omega) \to W^s(\mathbb{R}^d)$ be the extension operator discussed in Section A.2.4. Then (A.18), Theorem A.5, and the boundedness of $\mathscr{E}_S$ imply

$$\|u\|_{H^s_{\mathrm{F}}(\Omega)} \leq \|\mathscr{E}_S u\|_{H^s_{\mathrm{F}}(\mathbb{R}^d)} \simeq \|\mathscr{E}_S u\|_{H^s_{\mathrm{S}}(\mathbb{R}^d)} \lesssim \|u\|_{H^s_{\mathrm{S}}(\Omega)} \tag{A.41}$$

with constant depending on $s$ and the norm of $\mathscr{E}_S$. Therefore, $W^s(\Omega) \subset H^s(\Omega)$ with continuous inclusion.

To prove part (ii), we note that the equivalence constants in (A.40) and (A.41) are independent of $s$ if $s \in [\delta, 1/2)$; see Corollary A.1. If $\Omega$ is a Lipschitz domain, we choose the extension operator to be $\mathscr{E}_{M_0}$ in Section A.2.4 part (v), and if $\Omega$ is a Lipschitz hypograph, we choose $\mathscr{E}$ to be $\mathscr{E}_M$ defined by (A.39). Then by (A.37) and Theorem A.6, we obtain the last inequality in (A.41) with a constant independent of $u$ and $s \in [0, 1]$, but may depend on $\delta$. This proves the theorem. □

The next results show similar properties for $H^s_{\mathrm{IF}}(\Omega)$ and $\widetilde{H}^s_{\mathrm{IF}}(\Omega)$ which are defined by (A.31). First we prove a lemma.

**Lemma A.1.** *Assume that $\Omega$ is a Lipschitz domain in $\mathbb{R}^d$. For any integer r, if $s = \theta r$ for some $\theta \in [0, 1]$, then*

$$H^s(\Omega) = H^s_{\mathrm{IF}}(\Omega) \quad \textit{and} \quad \widetilde{H}^s(\Omega) = \widetilde{H}^s_{\mathrm{IF}}(\Omega)$$

*with equivalent norms:*

$$\|u\|_{H^s_{\mathrm{IF}}(\Omega)} \leq \|u\|_{H^s_{\mathrm{F}}(\Omega)} \leq C_r \|u\|_{H^s_{\mathrm{IF}}(\Omega)}, \quad u \in H^s(\Omega),$$
$$C_r^{-1} \|u\|_{\widetilde{H}^s_{\mathrm{IF}}(\Omega)} \leq \|u\|_{\widetilde{H}^s_{\mathrm{F}}(\Omega)} \leq \|u\|_{\widetilde{H}^s_{\mathrm{IF}}(\Omega)}, \quad u \in \widetilde{H}^s(\Omega).$$

*Proof.* Put

$$\begin{aligned} A_0 &= H^0(\Omega), \quad A_1 = H^r(\Omega), \quad A_s = [A_0, A_1]_\theta = H^s_{\rm IF}(\Omega), \quad \mathscr{A} = (A_0, A_1), \\ B_0 &= H^0(\mathbb{R}^d), \quad B_1 = H^r(\mathbb{R}^d), \quad B_s = [B_0, B_1]_\theta = H^s(\mathbb{R}^d), \quad \mathscr{B} = (B_0, B_1). \end{aligned}$$

For any $u \in H^s(\Omega)$, due to (A.19) and (A.7), there exists $U \in H^s(\mathbb{R}^d)$ such that

$$u = U|_\Omega \quad \text{and} \quad \|u\|_{H^s_{\rm F}(\Omega)} = \|U\|_{H^s_{\rm F}(\mathbb{R}^d)} = \|U\|_{H^s_{\rm I}(\mathbb{R}^d)}.$$

Since $U \in B_s$, there exist $U_j \in B_j$, $j = 0, 1$, satisfying $U = U_0 + U_1$. Define $u_j = U_j|_\Omega$. Then $u = u_0 + u_1$. Recalling the definition of the $K$-functionals in (A.30), we deduce

$$K_F(t, u, \mathscr{A})^2 \le \|u_0\|^2_{H^0_{\rm F}(\Omega)} + t^2 \|u_1\|^2_{H^r_{\rm F}(\Omega)} \le \|U_0\|^2_{H^0_{\rm F}(\mathbb{R}^d)} + t^2 \|U_1\|^2_{H^r_{\rm F}(\mathbb{R}^d)}$$

and thus $K_F(t, u, \mathscr{A}) \le K(t, U, \mathscr{B})$ with a constant depending on $s$ (and $\Omega$) only. (Note that it is not necessary to distinguish the different $K$-functionals for $U$ because they are all equal.) This together with (A.7) implies $H^s(\Omega) \subset A_s$ and

$$\|u\|_{H^s_{\rm IF}(\Omega)} \le \|U\|_{H^s_{\rm I}(\mathbb{R}^d)} = \|U\|_{H^s_{\rm F}(\mathbb{R}^d)} = \|u\|_{H^s_{\rm F}(\Omega)}. \tag{A.42}$$

Conversely, consider $u \in A_s$. Let $\mathscr{E}_S : A_j \to B_j$, $j = 0, 1$, be the extension operator defined in Section A.2.4, part (vi) satisfying, see (A.38),

$$\|\mathscr{E}_R u\|_{H^{r_j}_{\rm I}(\mathbb{R}^d)} = \|\mathscr{E}_R u\|_{B_j} \le \lambda_j \|u\|_{A_j} = \lambda_j \|u\|_{H^{r_j}_{\rm IF}(\Omega)}, \quad j = 0, 1,$$

where $r_0 = 0$ and $r_1 = r$. By Theorem A.4, the operator $\mathscr{E}_R$ maps $A_s$ into $B_s = H^s(\mathbb{R}^d)$ such that

$$\|\mathscr{E}_R u\|_{H^s_{\rm I}(\mathbb{R}^d)} \le \lambda_0^{1-\theta} \lambda_1^\theta \|u\|_{H^s_{\rm IF}(\Omega)} \le \max\{1, \lambda_0 \lambda_1\} \|u\|_{H^s_{\rm IF}(\Omega)}.$$

Consequently, with $C_r := \max\{1, \lambda_0 \lambda_1\}$,

$$\|u\|_{H^s_{\rm F}(\Omega)} \le \|\mathscr{E}_R u\|_{H^s_{\rm F}(\mathbb{R}^d)} = \|\mathscr{E}_R u\|_{H^s_{\rm I}(\mathbb{R}^d)} \le C_r \|u\|_{H^s_{\rm IF}(\Omega)}. \tag{A.43}$$

Thus, $A_s \subset H^s(\Omega)$. Therefore $A_s = H^s(\Omega)$ for all real number $s = \theta r$, $\theta \in [0, 1]$, and the two norms $\|\cdot\|_{H^s_{\rm F}(\Omega)}$ and $\|\cdot\|_{H^s_{\rm IF}(\Omega)}$ are equivalent. The constants depend on $r$ but not on $s$.

The results for $\widetilde{H}^s(\Omega)$ follows from duality by using Theorem A.2 and (A.24). Indeed, for any $s = \theta r$, it follows from above that $H^s(\Omega) = [H^0(\Omega), H^r(\Omega)]_\theta$, so that

$$\begin{aligned} \widetilde{H}^{-s}(\Omega) &= \Big([H^0(\Omega), H^r(\Omega)]_\theta\Big)^* = [(H^0(\Omega))^*, (H^r(\Omega))^*]_\theta \\ &= [\widetilde{H}^0(\Omega), \widetilde{H}^{-r}(\Omega)]_\theta. \end{aligned}$$

The norm equivalence can be seen from

$$\|u\|_{\widetilde{H}_{\mathrm{F}}^{-s}(\Omega)} = \sup_{0 \neq v \in H^s(\Omega)} \frac{\langle u, v \rangle_\Omega}{\|v\|_{H_{\mathrm{F}}^s(\Omega)}}$$

together with (A.42) and (A.43). □

We slightly extend the above lemma to have the following result.

**Lemma A.2.** *For any positive integer $r$ and $s \in [-r, r]$, if we define*

$$K_F(t, u, \mathscr{A}) := \inf_{\substack{u_0 \in H^{-r}(\Omega),\, u_1 \in H^r(\Omega) \\ u_0 + u_1 = u}} \left( \|u_0\|_{H_{\mathrm{F}}^{-r}(\Omega)}^2 + t^2 \|u_0\|_{H_{\mathrm{F}}^r(\Omega)}^2 \right)^{1/2}, \quad u \in H^{-r}(\Omega),$$

*where $\mathscr{A} := (H^{-r}(\Omega), H^r(\Omega))$, then the corresponding interpolation space satisfies*

$$[H^{-r}(\Omega), H^r(\Omega)]_\theta = H^s(\Omega)$$

*with equivalent norms, where $\theta = 1/2 + s/(2r)$.*

*Proof.* The proof follows exactly along the lines of that of Lemma A.1 if we replace $H^0(\Omega)$ in that lemma with $H^{-r}(\Omega)$. We omit the details. □

The above two lemmas can be extended as shown in the following theorem.

**Theorem A.8.** *Assume that $\Omega$ is a Lipschitz domain in $\mathbb{R}^d$. For any $s_0, s_1 \in \mathbb{R}$, $s_0 \leq s_1$, if $s = (1-\eta)s_0 + \eta s_1$ for some $\eta \in (0,1)$, then*

$$H^s(\Omega) = [H^{s_0}(\Omega), H^{s_1}(\Omega_1)]_{\eta,F} \quad \text{and} \quad \widetilde{H}^s(\Omega) = [\widetilde{H}^{s_0}(\Omega), \widetilde{H}^{s_1}(\Omega_1)]_{\eta,F}$$

*with equivalent norms where the constants are independent of $u$ and $s$, but may depend on $\Omega$, $s_0$, and $s_1$. Here, the index $F$ in the interpolation space indicates that the $K_F$-functional is used.*

*Proof.* The theorem is proved by using the reiteration theorem, Theorem A.3. We present the proof for $H^s(\Omega)$ only.

First consider $0 \leq s_0 \leq s_1$ and let $r$ be an integer satisfying $s_1 \leq r$. Invoking Lemma A.1 and Theorem A.3 with

$$A_0 = H^0(\Omega), \quad A_1 = H^r(\Omega), \quad \theta = s/r, \quad \theta_j = s_j/r, \ j = 0, 1,$$

we obtain the required result. Similar result holds for $r \leq s_0 \leq s_1 \leq 0$. Now consider the case when $s_0 < 0 \leq s_1$. Let $r > 0$ be an integer such that $-r \leq s_0 < 0 \leq s_1 \leq r$. It follows from Lemma A.2 that

$$H^{s_j}(\Omega) = [H^{-r}(\Omega), H^r(\Omega)]_{\theta_j,F}, \quad \theta_j = \frac{1}{2} + \frac{s_j}{2r}, \ j = 0, 1.$$

Invoking Theorem A.3 yields

$$[H^{s_0}(\Omega), H^{s_1}(\Omega)]_{\eta,F} = \Big[ [H^{-r}(\Omega), H^r(\Omega)]_{\theta_0,F}, [H^{-r}(\Omega), H^r(\Omega)]_{\theta_1,F} \Big]_{\eta,F}$$

$$= [H^{-r}(\Omega), H^r(\Omega)]_{(1-\eta)\theta_0+\eta\theta_1,F}$$
$$= [H^{-r}(\Omega), H^r(\Omega)]_{\frac{1}{2}+\frac{s}{2r},F} = H^s(\Omega)$$

where in the last step we used again Lemma A.2. This completes the proof of the theorem. □

We now prove similar results for the interpolation spaces $H^s_{\mathrm{I}}(\Omega)$ and $\widetilde{H}^s_{\mathrm{I}}(\Omega)$ constructed with the $K$-functional defined by (A.34).

**Lemma A.3.** *Assume that $\Omega$ is a Lipschitz domain in $\mathbb{R}^d$. For any integer $r$, if $s = \theta r$ for some $\theta \in (0,1)$, then*

$$H^s(\Omega) = H^s_{\mathrm{I}}(\Omega) \quad and \quad \widetilde{H}^s(\Omega) = \widetilde{H}^s_{\mathrm{I}}(\Omega)$$

*with equivalent norms:*

$$\|u\|_{H^s_{\mathrm{I}}(\Omega)} \le \|u\|_{H^s_{\mathrm{F}}(\Omega)} \le C_r \|u\|_{H^s_{\mathrm{I}}(\Omega)}, \quad u \in H^s(\Omega),$$
$$\|u\|_{\widetilde{H}^s_{\mathrm{I}}(\Omega)} \le \|u\|_{H^s_{\mathrm{F}}(\Omega)} \le C_r \|u\|_{\widetilde{H}^s_{\mathrm{I}}(\Omega)}, \quad u \in \widetilde{H}^s(\Omega).$$

*Proof.* The proof follows along the lines of that of Lemma A.1. A slight difference is the use of the Slobodetski norm in the definition of the $K$-functional. Therefore, we have to consider first the case when $r$ is a non-negative integer. Put

$$A_0 = H^0(\Omega), \quad A_1 = H^r(\Omega), \quad A_s = [A_0, A_1]_\theta, \quad \mathscr{A} = (A_0, A_1),$$
$$B_0 = H^0(\mathbb{R}^d), \quad B_1 = H^r(\mathbb{R}^d), \quad B_s = [B_0, B_1]_\theta, \quad \mathscr{B} = (B_0, B_1).$$

For any $u \in H^s(\Omega)$, due to (A.19) and (A.7), there exists $U \in H^s(\mathbb{R}^d)$ such that

$$u = U|_\Omega \quad \text{and} \quad \|u\|_{H^s_{\mathrm{F}}(\Omega)} = \|U\|_{H^s_{\mathrm{F}}(\mathbb{R}^d)} = \|U\|_{H^s_{\mathrm{I}}(\mathbb{R}^d)}.$$

Since $U \in B_s$, there exist $U_j \in B_j$, $j = 0,1$, satisfying $U = U_0 + U_1$. Define $u_j = U_j|_\Omega$. Then $u = u_0 + u_1$. Recalling the definition of the $K$-functionals in (A.34) and noting (A.33), we deduce

$$K_S(t,u,\mathscr{A})^2 \le \|u_0\|^2_{H^0_{\mathrm{F}}(\Omega)} + t^2\|u_1\|^2_{H^r_{\mathrm{F}}(\Omega)} \le \|U_0\|^2_{H^0_{\mathrm{F}}(\mathbb{R}^d)} + t^2\|U_1\|^2_{H^r_{\mathrm{F}}(\mathbb{R}^d)}$$

and thus $K_S(t,u,\mathscr{A}) \le K(t,U,\mathscr{B})$ with a constant depending on $s$ (and $\Omega$) only. (Here again, we do not distinguish the $K$-functionals for $U$ because they are all equal.) This together with (A.7) implies $H^s(\Omega) \subset A_s$ and

$$\|u\|_{H^s_{\mathrm{I}}(\Omega)} \le \|U\|_{H^s_{\mathrm{I}}(\mathbb{R}^d)} = \|U\|_{H^s_{\mathrm{F}}(\mathbb{R}^d)} = \|u\|_{H^s_{\mathrm{F}}(\Omega)}. \tag{A.44}$$

Conversely, consider $u \in A_s$. Let $\mathscr{E}_S : A_j \to B_j$, $j = 0,1$, be the extension operator defined in Section A.2.4, part (iv) satisfying

$$\|\mathscr{E}_S u\|_{H^{r_j}_{\mathrm{I}}(\mathbb{R}^d)} = \|\mathscr{E}_S u\|_{B_j} \le \lambda_j \|u\|_{A_j} = \lambda_j \|u\|_{H^{r_j}_{\mathrm{I}}(\Omega)}, \quad j = 0,1,$$

where $r_0 = 0$ and $r_1 = r$. By Theorem A.4, the operator $\mathscr{E}_S$ maps $A_s$ into $B_s = H^s(\mathbb{R}^d)$ such that

$$\|\mathscr{E}_S u\|_{H_{\mathrm{I}}^s(\mathbb{R}^d)} \le \lambda_0^{1-\theta} \lambda_1^{\theta} \|u\|_{H_{\mathrm{I}}^s(\Omega)} \le \max\{1, \lambda_0 \lambda_1\} \|u\|_{H_{\mathrm{I}}^s(\Omega)}.$$

Consequently,

$$\|u\|_{H_{\mathrm{F}}^s(\Omega)} \le \|\mathscr{E}_S u\|_{H_{\mathrm{F}}^s(\mathbb{R}^d)} = \|\mathscr{E}_S u\|_{H_{\mathrm{I}}^s(\mathbb{R}^d)} \le \max\{1, \lambda_0 \lambda_1\} \|u\|_{H_{\mathrm{I}}^s(\Omega)}. \tag{A.45}$$

Thus, $A_s \subset H^s(\Omega)$. Therefore $A_s = H^s(\Omega)$ for all real number $s \in [0, r]$ and the two norms $\|\cdot\|_{H_{\mathrm{F}}^s(\Omega)}$ and $\|\cdot\|_{H_{\mathrm{I}}^s(\Omega)}$ are equivalent. The constants depend on $r$ but not on $s$.

Differently from Lemma A.1, we cannot repeat the above argument for $r < 0$. We will proceed by proving the result for $\widetilde{H}_{\mathrm{I}}^s(\Omega)$ with $s \in [0, r]$. Put

$$\widetilde{A}_0 = \widetilde{H}^0(\Omega), \quad \widetilde{A}_1 = \widetilde{H}^r(\Omega), \quad \widetilde{A}_s = [\widetilde{A}_0, \widetilde{A}_1]_\theta, \quad \widetilde{\mathscr{A}} = (\widetilde{A}_0, \widetilde{A}_1),$$

where the $K$-functional is defined by (A.34). Proceeding as in the first part of this proof, with $U \in H^s(\mathbb{R}^d)$ being the zero extension of $u \in H^s(\Omega)$. Then we obtain, instead of (A.44),

$$\|u\|_{\widetilde{H}_{\mathrm{I}}^s(\Omega)} \le \|U\|_{H_{\mathrm{I}}^s(\mathbb{R}^d)} = \|U\|_{H_{\mathrm{F}}^s(\mathbb{R}^d)} = \|u\|_{\widetilde{H}_{\mathrm{F}}^s(\Omega)}.$$

Conversely, consider $u \in \widetilde{H}^s(\Omega)$. With $\mathscr{E}_S$ defined as above, we obtain

$$\|\mathscr{E}_S u\|_{H_{\mathrm{I}}^{r_j}(\mathbb{R}^d)} \le \lambda_j \|u\|_{H_{\mathrm{I}}^{r_j}(\Omega)} \lesssim \|u\|_{\widetilde{H}_{\mathrm{I}}^{r_j}(\Omega)}, \quad j = 0, 1,$$

where in the last step we used Poincaré's inequality, with the constant depending on $r$. Hence, by interpolation

$$\|\mathscr{E}_S u\|_{H_{\mathrm{I}}^s(\mathbb{R}^d)} \lesssim \|u\|_{\widetilde{H}_{\mathrm{I}}^s(\Omega)}.$$

The constant depends on $r$ but is independent of $s \in (0, r)$. Similarly to (A.45), we have, recalling that $U$ is the zero extension of $u$,

$$\|u\|_{\widetilde{H}_{\mathrm{F}}^s(\Omega)} = \|U\|_{H_{\mathrm{F}}^s(\mathbb{R}^d)} \simeq \|U\|_{H_{\mathrm{S}}^s(\mathbb{R}^d)} \le \|\mathscr{E}_S u\|_{H_{\mathrm{S}}^s(\mathbb{R}^d)} \simeq \|\mathscr{E}_S u\|_{H_{\mathrm{I}}^s(\mathbb{R}^d)} \lesssim \|u\|_{\widetilde{H}_{\mathrm{I}}^s(\Omega)}.$$

Thus, $\widetilde{A}_s \subset \widetilde{H}^s(\Omega)$. Therefore, $A_s = \widetilde{H}^s(\Omega)$ for all $s \in [0, r]$ with equivalence of norms.

The results for $r < 0$ for both spaces $H^s(\Omega)$ and $\widetilde{H}^s(\Omega)$ are derived by duality arguments. We omit the details. □

Similarly to Theorem A.8, we now state without proof the following theorem.

**Theorem A.9.** *Assume that $\Omega$ is a Lipschitz domain in $\mathbb{R}^d$. For any $s_0, s_1 \in \mathbb{R}$, $s_0 \le s_1$, if $s = (1-\eta)s_0 + \eta s_1$ for some $\eta \in (0, 1)$, then*

$$H^s(\Omega) = [H^{s_0}(\Omega), H^{s_1}(\Omega_1)]_\eta \quad \textit{and} \quad \widetilde{H}^s(\Omega) = [\widetilde{H}^{s_0}(\Omega), \widetilde{H}^{s_1}(\Omega_1)]_\eta$$

*with equivalent norms where the constants are independent of $u$ and $s$, but may depend on $\Omega$, $s_0$, and $s_1$. Here, we recall that the interpolation space is defined with the $K_S$-functional defined by* (A.34).

Thanks to TheoremsA.7, A.8, and A.9, in this book (apart from this subsection) we use only one notation $H^s(\Omega)$ to denote the space with different equivalent norms. Similarly for $\widetilde{H}^s(\Omega)$. We stress that the interpolation norms to be used are defined with the $K_S$-functional. This is the correct choice for the study of the scaling property in Section A.2.7 and the global-local property in Section A.2.9.

Finally, it is remarked that the recent paper [109] also shows some results on the equivalence of fractional-order Sobolev seminorms.

## A.2.6 The weighted norms

As will be seen in Section A.2.7, not all Sobolev norms are scalable. For this reason, we introduce weighted norms.

### A.2.6.1 The weighted norms in $H^s(\Omega)$ and $\widetilde{H}^s(\Omega)$ for $s > 0$

As will be seen in Section A.2.7, the $L^2$-norm and the seminorm defining the Slobodetski norm scale differently. In order to ensure scalability we introduce weighted norms. Different weights have different usage. Three different weighted norms are used in this book.

For $s = k + \mu > 0$ with $k = 0, 1, 2, \ldots$ and $\mu \in [0, 1)$ we define

$$
\begin{aligned}
\|v\|^2_{H^s_{\mathrm{w}}(\Omega)} &:= \frac{1}{\tau}\|v\|^2_{L^2(\Omega)} + |v|^2_{H^s(\Omega)},\\
{}_*\|v\|^2_{H^s_{\mathrm{w}}(\Omega)} &:= \frac{1}{\tau^d}\|v\|^2_{L^2(\Omega)} + |v|^2_{H^s(\Omega)},\\
\|v\|^2_{\widetilde{H}^s_{\mathrm{w}}(\Omega)} &:= \begin{cases} \|v\|^2_{H^s_{\mathrm{w}}(\Omega)} + \displaystyle\sum_{|\alpha|=k} \int_\Omega \frac{|\partial^\alpha v(\mathbf{x})|^2}{\operatorname{dist}(\mathbf{x}, \partial\Omega)^{2\mu}}\, d\mathbf{x}, & k \neq 0,\\[2ex] |v|^2_{H^s(\Omega)} + \displaystyle\int_\Omega \frac{|v(\mathbf{x})|^2}{\operatorname{dist}(\mathbf{x}, \partial\Omega)^{2s}}\, d\mathbf{x}, & k = 0, \end{cases}
\end{aligned}
\tag{A.46}
$$

where $\tau = \operatorname{diam}(\Omega)$ and $|v|_{H^s(\Omega)}$ is the Slobodetski seminorm defined in (A.25). Note that if $d = 1$ then $\|v\|_{H^s_{\mathrm{w}}(\Omega)} = {}_*\|v\|_{H^s_{\mathrm{w}}(\Omega)}$.

### A.2.6.2 The interpolation norms with weights in $H^s(\Omega)$ and $\widetilde{H}^{-s}(\Omega)$ for $s \in [0,1]$ and $\Omega \subset \mathbb{R}^2$

The non-scalability of the norms $\|\cdot\|_{H^s_{\mathrm{I}}(\Omega)}$ and $\|\cdot\|_{\widetilde{H}^{-s}_{\mathrm{I}}(\Omega)}$ for $s > 0$, as will be seen in Section A.2.7, prompts us to define another type of interpolation norms which uses the weighted norms defined in Subsection A.2.6.1, instead of the Slobodetski norm. We restrict ourselves to the case when $s \in [0,1]$ and $\Omega \subset \mathbb{R}^2$.

For any $u \in H^s(\Omega)$, $0 \le s \le 1$, recalling the definition of the weighted norm in (A.46), we define

$$
{}_*\|u\|_{H^s_{\mathrm{I}}(\Omega)} := \begin{cases} \|u\|_{L^2(\Omega)}, & s = 0, \\ {}_*\|u\|_{H^1_{\mathrm{w}}(\Omega)}, & s = 1, \\ \|u\|_{[H^0(\Omega),H^1(\Omega)]_{*,\theta}}, & s = \theta \in (0,1), \end{cases} \tag{A.47}
$$

where the interpolation norm $\|u\|_{[H^0(\Omega),H^1(\Omega)]_{*,\theta}}$ is defined as in (A.1) with the $K$-functional being

$$
K_*(t,u,\mathscr{A}) := \inf_{\substack{u_0 \in H^0,\ u_1 \in H^1 \\ u_0 + u_1 = u}} \left( {}_*\|u_0\|^2_{H^0_{\mathrm{I}}(\Omega)} + t^2\, {}_*\|u_1\|^2_{H^1_{\mathrm{I}}(\Omega)} \right)^{1/2}.
$$

The norm ${}_*\|u\|_{\widetilde{H}^{-s}_{\mathrm{I}}(\Omega)}$ is defined by duality:

$$
{}_*\|u\|_{\widetilde{H}^{-s}_{\mathrm{I}}(\Omega)} := \sup_{0 \ne v \in H^s(\Omega)} \frac{\langle u, v \rangle_\Omega}{{}_*\|v\|_{H^s_{\mathrm{I}}(\Omega)}}. \tag{A.48}
$$

We are particularly interested in the norm ${}_*\|\cdot\|_{\widetilde{H}^{-1/2}_{\mathrm{I}}(\Omega)}$ for the analysis of the weakly-singular integral equation.

## *A.2.7 Scaling properties*

The next lemma shows how the Slobodetski norms scale when the domain $\Omega$ is rescaled.

**Lemma A.4.** *Assume that $\Omega$ is an open and bounded domain in $\mathbb{R}^d$, $d \ge 1$, such that $\tau := \mathrm{diam}(\Omega) < 1$, and that $v$ is a function defined on $\Omega$. Let*

$$
\widehat{\Omega} := \left\{ \widehat{\mathbf{x}} \in \mathbb{R}^d : \widehat{\mathbf{x}} = \mathbf{x}/\tau,\ \mathbf{x} \in \Omega \right\},
$$
$$
\widehat{v}(\widehat{\mathbf{x}}) = v(\mathbf{x}) \quad \forall \widehat{\mathbf{x}} \in \widehat{\Omega},\ \mathbf{x} = \tau\widehat{\mathbf{x}} \in \Omega.
$$

*Then, for $s = k + \mu$, $k = 0,1,2,\dots$, and $\mu \in [0,1)$, the following relations hold*

$$
|v|^2_{H^s(\Omega)} = \tau^{d-2s} |\widehat{v}|^2_{H^s(\widehat{\Omega})},
$$

$$\tau^d \|\widehat{v}\|^2_{H^s_{\mathrm{S}}(\widehat{\Omega})} \le \|v\|^2_{H^s_{\mathrm{S}}(\Omega)} \le \tau^{d-2s} \|\widehat{v}\|^2_{H^s_{\mathrm{S}}(\widehat{\Omega})}.$$

*Proof.* We first note that

$$\|v\|^2_{L^2(\Omega)} = \tau^d \|\widehat{v}\|^2_{L^2(\widehat{\Omega})} \quad \text{and} \quad \partial^\alpha_{\widehat{x}} \widehat{v}(\widehat{x}) = \tau^{|\alpha|} \widehat{\partial^\alpha_x v}(\widehat{x}), \tag{A.49}$$

where $\partial^\alpha_x$ is defined in (A.11). Hence, if $s = k = 1, 2, \ldots$, then

$$\begin{aligned}|v|^2_{H^s(\Omega)} &= \sum_{|\alpha|=k} \|\partial^\alpha_x v\|^2_{L^2(\Omega)} = \tau^d \sum_{|\alpha|=k} \|\widehat{\partial^\alpha_x v}\|^2_{L^2(\widehat{\Omega})} = \tau^{d-2k} \sum_{|\alpha|=k} \|\partial^\alpha_{\widehat{x}} \widehat{v}\|^2_{L^2(\widehat{\Omega})} \\ &= \tau^{d-2s} |\widehat{v}|^2_{H^s(\widehat{\Omega})}.\end{aligned}$$

If $s = k + \mu$ with $k = 0, 1, 2, \ldots$ and $\mu \in (0,1)$ then

$$\begin{aligned}|v|^2_{H^s(\Omega)} &= \sum_{|\alpha|=k} \iint_{\Omega\times\Omega} \frac{|\partial^\alpha_x v(\mathbf{x}) - \partial^\alpha_x v(\mathbf{y})|^2}{|\mathbf{x}-\mathbf{y}|^{d+2\mu}} \, d\mathbf{x}\, d\mathbf{y} \\ &= \sum_{|\alpha|=k} \iint_{\widehat{\Omega}\times\widehat{\Omega}} \frac{|\widehat{\partial^\alpha_x v}(\widehat{\mathbf{x}}) - \widehat{\partial^\alpha_x v}(\widehat{\mathbf{y}})|^2}{\tau^{d+2\mu} |\widehat{\mathbf{x}}-\widehat{\mathbf{y}}|^{d+2\mu}} \, \tau^{2d} d\widehat{\mathbf{x}}\, d\widehat{\mathbf{y}} \\ &= \tau^{d-2k-2\mu} \sum_{|\alpha|=k} \iint_{\widehat{\Omega}\times\widehat{\Omega}} \frac{|\partial^\alpha_{\widehat{x}} \widehat{v}(\widehat{\mathbf{x}}) - \partial^\alpha_{\widehat{x}} \widehat{v}(\widehat{\mathbf{y}})|^2}{|\widehat{\mathbf{x}}-\widehat{\mathbf{y}}|^{d+2\mu}} \, d\widehat{\mathbf{x}}\, d\widehat{\mathbf{y}} \\ &= \tau^{d-2s} |\widehat{v}|^2_{H^s(\widehat{\Omega})}.\end{aligned}$$

This together with the first identity in (A.49) yields the result for the $\|\cdot\|_{H^s_{\mathrm{S}}(\Omega)}$-norm, completing the proof of the lemma. □

The non-scalability of the norm $\|\cdot\|_{H^s_{\mathrm{S}}(\Omega)}$ proved in Lemma A.4 dictates the introduction of another norm, the weighted norm defined in the next subsection.

The following lemma shows how these weighted norms scale when the domain $\Omega$ is rescaled. We will particularly be interested in the case when they are invariant under scaling, and that is the main reason for the different weights used in the definition above. Three most important cases are the invariance in (A.50), (A.51), and (A.52).

**Lemma A.5.** *Assume that $\Omega$ is an open and bounded domain in $\mathbb{R}^d$, $d \ge 1$, such that $\tau := \operatorname{diam}(\Omega) < 1$, and that $v$ is a function defined on $\Omega$. Let*

$$\begin{aligned}\widehat{\Omega} &:= \{\widehat{\mathbf{x}} \in \mathbb{R}^d : \widehat{\mathbf{x}} = \mathbf{x}/\tau, \ \mathbf{x} \in \Omega\}, \\ \widehat{v}(\widehat{\mathbf{x}}) &= v(\mathbf{x}) \quad \forall \widehat{\mathbf{x}} \in \widehat{\Omega}, \ \mathbf{x} = \tau\widehat{\mathbf{x}} \in \Omega.\end{aligned}$$

(i) *For $s = k + \mu$, $k = 0, 1, 2, \ldots$ and $\mu \in [0,1)$ the following statements hold*

$$\tau^{d-2s} \|\widehat{v}\|^2_{H^s_{\mathrm{w}}(\widehat{\Omega})} \le \|v\|^2_{H^s_{\mathrm{w}}(\Omega)} \le \tau^{d-1} \|\widehat{v}\|^2_{H^s_{\mathrm{w}}(\widehat{\Omega})} \qquad \textit{if } s \le \frac{1}{2},$$

$$\tau^{d-1}\|\widehat{v}\|^2_{H^s_w(\widehat{\Omega})} \le \|v\|^2_{H^s_w(\Omega)} \le \tau^{d-2s}\|\widehat{v}\|^2_{H^s_w(\widehat{\Omega})} \quad \text{if } s > \frac{1}{2},$$

$$\tau^{d-2s}{}_*\|\widehat{v}\|^2_{H^s_w(\widehat{\Omega})} \le {}_*\|v\|^2_{H^s_w(\Omega)} \le {}_*\|\widehat{v}\|^2_{H^s_w(\widehat{\Omega})} \quad \text{if } s \le \frac{d}{2},$$

$${}_*\|\widehat{v}\|^2_{H^s_w(\widehat{\Omega})} \le {}_*\|v\|^2_{H^s_w(\Omega)} \le \tau^{d-2s}{}_*\|\widehat{v}\|^2_{H^s_w(\widehat{\Omega})} \quad \text{if } s > \frac{d}{2}.$$

*In particular,*

$$\|v\|_{H^{1/2}_w(\Omega)} = \|\widehat{v}\|_{H^{1/2}_w(\widehat{\Omega})} \quad \text{if } d = 1 \text{ and } s = \frac{1}{2}, \tag{A.50}$$

*and*

$${}_*\|v\|_{H^1_w(\Omega)} = {}_*\|\widehat{v}\|_{H^1_w(\widehat{\Omega})} \quad \text{if } d = 2 \text{ and } s = 1. \tag{A.51}$$

(ii) *For $s = k + \mu$, $k = 1, 2, \ldots$, $\mu \in [0,1)$, we have*

$$\tau^{d-1}\|\widehat{v}\|^2_{\widetilde{H}^s_w(\widehat{\Omega})} \le \|v\|^2_{\widetilde{H}^s_w(\Omega)} \le \tau^{d-2s}\|\widehat{v}\|^2_{\widetilde{H}^s_w(\widehat{\Omega})}$$

*and for $s = \mu \in [0,1)$*

$$\|v\|^2_{\widetilde{H}^s_w(\Omega)} = \tau^{d-2s}\|\widehat{v}\|^2_{\widetilde{H}^s_w(\widehat{\Omega})}.$$

*In particular, if $d = 1$ and $s = 1/2$ then*

$$\|v\|_{\widetilde{H}^{1/2}_w(\Omega)} = \|\widehat{v}\|_{\widetilde{H}^{1/2}_w(\widehat{\Omega})}. \tag{A.52}$$

*Proof.* Lemma A.4 and (A.49) give

$$\|v\|^2_{L^2(\Omega)} = \tau^d\|\widehat{v}\|^2_{L^2(\widehat{\Omega})} \quad \text{and} \quad |v|^2_{H^s(\Omega)} = \tau^{d-2s}|\widehat{v}|^2_{H^s(\widehat{\Omega})}. \tag{A.53}$$

It then follows that

$$\|v\|^2_{H^s_w(\Omega)} = \tau^{d-1}\|\widehat{v}\|^2_{L^2(\widehat{\Omega})} + \tau^{d-2s}|\widehat{v}|^2_{H^s(\widehat{\Omega})}$$

and

$${}_*\|v\|^2_{H^s_w(\Omega)} = \|\widehat{v}\|^2_{L^2(\widehat{\Omega})} + \tau^{d-2s}|\widehat{v}|^2_{H^s(\widehat{\Omega})}.$$

Hence, noting that $\tau < 1$ and that

$$\|\widehat{v}\|^2_{H^s_w(\widehat{\Omega})} = {}_*\|\widehat{v}\|^2_{H^s_w(\widehat{\Omega})} = \|\widehat{v}\|^2_{L^2(\widehat{\Omega})} + |\widehat{v}|^2_{H^s(\widehat{\Omega})},$$

we can bound $\|v\|^2_{H^s_w(\Omega)}$ and ${}_*\|v\|^2_{H^s_w(\Omega)}$ below and above by $\|\widehat{v}\|^2_{H^s_w(\widehat{\Omega})}$ and ${}_*\|\widehat{v}\|^2_{H^s(\widehat{\Omega})}$, respectively, depending on the values of $s$. This proves Part (i).

Moreover, since $\partial^\alpha_{\widehat{x}}\widehat{v}(\widehat{\mathbf{x}}) = \tau^{|\alpha|}\widehat{\partial^\alpha_x v}(\widehat{\mathbf{x}})$ it is also easy to establish

$$\sum_{|\alpha|=k} \int_{\Omega} \frac{|\partial_x^\alpha v(\mathbf{x})|^2}{\operatorname{dist}(\mathbf{x},\partial\Omega)^{2\mu}}\, d\mathbf{x} = \tau^{d-2s} \sum_{|\alpha|=k} \int_{\widehat{\Omega}} \frac{|\partial_{\widehat{x}}^\alpha \widehat{v}(\widehat{\mathbf{x}})|^2}{\operatorname{dist}(\widehat{\mathbf{x}},\partial\widehat{\Omega})^{2\mu}}\, d\widehat{\mathbf{x}}.$$

Hence, for $k \neq 0$, we have

$$\|v\|^2_{\widetilde{H}^s_{\mathrm{w}}(\Omega)} = \tau^{d-1}\|\widehat{v}\|^2_{L^2(\widehat{\Omega})} + \tau^{d-2s}|\widehat{v}|^2_{H^s(\widehat{\Omega})} + \tau^{d-2s} \sum_{|\alpha|=k} \int_{\widehat{\Omega}} \frac{|\partial_{\widehat{x}}^\alpha \widehat{v}(\widehat{\mathbf{x}})|^2}{\operatorname{dist}(\widehat{\mathbf{x}},\partial\widehat{\Omega})^{2\mu}}\, d\widehat{\mathbf{x}}$$

and for $k = 0$ we have

$$\|v\|^2_{\widetilde{H}^s_{\mathrm{w}}(\Omega)} = \tau^{d-2s}|\widehat{v}|^2_{H^s(\widehat{\Omega})} + \tau^{d-2s} \int_{\widehat{\Omega}} \frac{|\widehat{v}(\widehat{\mathbf{x}})|^2}{\operatorname{dist}(\widehat{\mathbf{x}},\partial\widehat{\Omega})^{2s}}\, d\widehat{\mathbf{x}} = \tau^{d-2s}\|\widehat{v}\|^2_{\widetilde{H}^s_{\mathrm{w}}(\widehat{\Omega})}.$$

The required estimates in Part (ii) then follow. □

The next lemma shows how the interpolation norms scale when the bounded domain $\Omega$ is rescaled; see [5, Lemma 4.3].

**Lemma A.6.** *Assume that $\Omega$ is an open and bounded domain in $\mathbb{R}^d$, $d \geq 1$, such that $\tau := \operatorname{diam}(\Omega) < 1$, and that $v$ is a function defined on $\Omega$. Let $\widehat{\Omega} := \{\widehat{\mathbf{x}} \in \mathbb{R}^d : \widehat{\mathbf{x}} = \mathbf{x}/\tau,\ \mathbf{x} \in \Omega\}$ and*

$$\widehat{v}(\widehat{\mathbf{x}}) = v(\mathbf{x}) \quad \forall \widehat{\mathbf{x}} \in \widehat{\Omega},\ \mathbf{x} = \tau\widehat{\mathbf{x}} \in \Omega.$$

*Then, for $s \in [0, r]$,*

(i) *if $v \in \widetilde{H}^s(\Omega)$,*

$$\|v\|^2_{\widetilde{H}^s_{\mathrm{I}}(\Omega)} = \tau^{d-2s}\|\widehat{v}\|^2_{\widetilde{H}^s_{\mathrm{I}}(\widehat{\Omega})};$$

(ii) *if $v \in H^s(\Omega)$,*

$$\tau^d\|\widehat{v}\|^2_{H^s_{\mathrm{I}}(\widehat{\Omega})} \leq \|v\|^2_{H^s_{\mathrm{I}}(\Omega)} \leq \tau^{d-2s}\|\widehat{v}\|^2_{H^s_{\mathrm{I}}(\widehat{\Omega})};$$

(iii) *if $v \in H^{-s}(\Omega)$,*

$$\|v\|^2_{H^{-s}_{\mathrm{I}}(\Omega)} = \tau^{d+2s}\|\widehat{v}\|^2_{H^{-s}_{\mathrm{I}}(\widehat{\Omega})};$$

(iv) *if $v \in \widetilde{H}^{-s}(\Omega)$,*

$$\tau^{d+2s}\|\widehat{v}\|^2_{\widetilde{H}^{-s}_{\mathrm{I}}(\widehat{\Omega})} \leq \|v\|^2_{\widetilde{H}^{-s}_{\mathrm{I}}(\Omega)} \leq \tau^d\|\widehat{v}\|^2_{\widetilde{H}^{-s}_{\mathrm{I}}(\widehat{\Omega})}.$$

*Proof.* Since $\|v\|_{H^r_{\mathrm{I}}(\Omega)} = \|v\|_{H^r_{\mathrm{S}}(\Omega)}$ and $\|v\|_{\widetilde{H}^r_{\mathrm{I}}(\Omega)} = |v|_{H^r(\Omega)}$ for $r \in \mathbb{N}$, it follows from Lemma A.4 that

$$\|v\|^2_{\widetilde{H}^r_{\mathrm{I}}(\Omega)} = \tau^{d-2r}\|\widehat{v}\|^2_{\widetilde{H}^r_{\mathrm{I}}(\widehat{\Omega})}$$

and

$$\tau^d\|\widehat{v}\|^2_{H^r_{\mathrm{I}}(\widehat{\Omega})} \leq \|v\|^2_{H^r_{\mathrm{I}}(\Omega)} \leq \tau^{d-2r}\|\widehat{v}\|^2_{H^r_{\mathrm{I}}(\widehat{\Omega})}. \tag{A.54}$$

Since $\|v\|^2_{L^2(\Omega)} = \tau^d \|\widehat{v}\|^2_{L^2(\widehat{\Omega})}$, by interpolation between $L^2(\Omega)$ and $\widetilde{H}^r(\Omega)$) we prove Part (i). Similarly, by interpolation between $L^2(\Omega)$ and $H^r(\Omega)$) we prove Part (ii).

Part (iii) then follows from Part (i) by duality, thanks to $\langle v,w\rangle_\Omega = \tau^d \langle \widehat{v},\widehat{w}\rangle_{\widehat{\Omega}}$. Indeed, by duality we have

$$\|v\|^2_{H_{\mathrm{I}}^{-s}(\Omega)} = \sup_{w\in\widetilde{H}^s(\Omega)} \frac{\langle v,w\rangle^2_\Omega}{\|w\|^2_{\widetilde{H}_{\mathrm{I}}^s(\Omega)}} = \sup_{\widehat{w}\in\widetilde{H}^s(\widehat{\Omega})} \frac{\tau^{2d}\langle \widehat{v},\widehat{w}\rangle^2_{\widehat{\Omega}}}{\tau^{d-2s}\|\widehat{w}\|^2_{\widetilde{H}_{\mathrm{I}}^s(\widehat{\Omega})}} = \tau^{d+2s}\|\widehat{v}\|^2_{H_{\mathrm{I}}^{-s}(\widehat{\Omega})}.$$

Similarly, Part (iv) follows from Part (ii) by using the same argument. □

We now prove the scalability of ${}_*\|u\|_{H_{\mathrm{I}}^s(\Omega)}$ and ${}_*\|u\|_{\widetilde{H}_{\mathrm{I}}^{-s}(\Omega)}$ for $s \in [0,1]$.

**Lemma A.7.** *Under the assumptions of Lemma A.6, for $s \in [0,1]$ and $\Omega \subset \mathbb{R}^2$,*

(i) *if $v \in H^s(\Omega)$,*

$${}_*\|v\|^2_{H_{\mathrm{I}}^s(\Omega)} = \tau^{2-2s}\,{}_*\|\widehat{v}\|^2_{H_{\mathrm{I}}^s(\widehat{\Omega})};$$

(ii) *if $v \in \widetilde{H}^{-s}(\Omega)$,*

$${}_*\|v\|^2_{\widetilde{H}_{\mathrm{I}}^{-s}(\Omega)} = \tau^{2+2s}\,{}_*\|\widehat{v}\|^2_{\widetilde{H}_{\mathrm{I}}^{-s}(\widehat{\Omega})}.$$

*Proof.* By definition and (A.53)

$${}_*\|v\|^2_{H_{\mathrm{I}}^0(\Omega)} = \tau^2\,{}_*\|\widehat{v}\|^2_{H_{\mathrm{I}}^0(\widehat{\Omega})}.$$

This together with (A.51) and interpolation proves Part (i). Part (ii) follows from the definition of dual norms and Part (i), noting that $\langle v,w\rangle_\Omega = \tau^2 \langle \widehat{v},\widehat{w}\rangle_{\widehat{\Omega}}$,

$${}_*\|v\|^2_{\widetilde{H}_{\mathrm{I}}^{-s}(\Omega)} = \sup_{w\in H^s(\Omega)} \frac{\langle v,w\rangle^2_\Omega}{{}_*\|w\|^2_{H_{\mathrm{I}}^s(\Omega)}} = \sup_{\widehat{w}\in H^s(\widehat{\Omega})} \frac{\tau^4\langle \widehat{v},\widehat{w}\rangle^2_{\widehat{\Omega}}}{\tau^{2-2s}\,{}_*\|\widehat{w}\|^2_{H_{\mathrm{I}}^s(\widehat{\Omega})}} = \tau^{2+2s}\,{}_*\|\widehat{v}\|^2_{\widetilde{H}_{\mathrm{I}}^{-s}(\widehat{\Omega})}.$$

This proves the lemma. □

### A.2.8 Important results

It is sometimes useful to relate the interpolation norms and the weighted norms.

**Lemma A.8.** *Assume that $\Omega$ is an open bounded domain in $\mathbb{R}^d$ satisfying $\tau := \mathrm{diam}(\Omega) < 1$. Then, for $s = k+1/2$, $k = 0,1,2,\dots$,*

$$\tau^{2k}\|v\|^2_{\widetilde{H}_{\mathrm{I}}^s(\Omega)} \lesssim \|v\|^2_{\widetilde{H}_{\mathrm{w}}^s(\Omega)} \lesssim \|v\|^2_{\widetilde{H}_{\mathrm{I}}^s(\Omega)} \quad \forall v \in \widetilde{H}^s(\Omega)$$

*and*

$$\tau^{2k}\|v\|^2_{H^s_{\mathrm{I}}(\Omega)} \lesssim \|v\|^2_{H^s_{\mathrm{w}}(\Omega)} \lesssim \tau^{-2s}\|v\|^2_{H^s_{\mathrm{I}}(\Omega)} \quad \forall v \in H^s(\Omega).$$

*In particular, when $s = 1/2$ we have*

$$\|v\|_{\widetilde{H}^{1/2}_{\mathrm{I}}(\Omega)} \simeq \|v\|_{\widetilde{H}^{1/2}_{\mathrm{w}}(\Omega)} \quad \forall v \in \widetilde{H}^{1/2}(\Omega)$$

*and*

$$\|v\|^2_{H^{1/2}_{\mathrm{I}}(\Omega)} \lesssim \|v\|^2_{H^{1/2}_{\mathrm{w}}(\Omega)} \lesssim \tau^{-1}\|v\|^2_{H^{1/2}_{\mathrm{I}}(\Omega)} \quad \forall v \in H^{1/2}(\Omega)$$

*where the constants are independent of $v$ and $\tau$.*

*Proof.* The lemma is a direct consequence of Lemma A.5 and Lemma A.6, noting that

$$\|\widehat{v}\|^2_{\widetilde{H}^s_{\mathrm{w}}(\widehat{\Omega})} \simeq \|\widehat{v}\|^2_{\widetilde{H}^s_{\mathrm{I}}(\widehat{\Omega})} \quad \text{and} \quad \|\widehat{v}\|^2_{H^s_{\mathrm{w}}(\widehat{\Omega})} \simeq \|\widehat{v}\|^2_{H^s_{\mathrm{I}}(\widehat{\Omega})};$$

see [138, Theorem 11.7]. The constants in the above two equivalences are independent of $\tau$. □

Recall that if $\Omega$ is a Lipschitz domain in $\mathbb{R}^d$ and if $0 \le s < 1/2$, then $\widetilde{H}^s(\Omega) = H^s(\Omega)$ for $s \in (-1/2, 1/2)$; see (A.21). The next result, first studied in [108, Lemma 5], gives a bound for the norm of the injection $H^s(\Omega) \to \widetilde{H}^s(\Omega)$, $|s| < 1/2$.

**Lemma A.9.** *Let $\Omega$ be a Lipschitz domain in $\mathbb{R}^d$. Then for any $v \in \widetilde{H}^s(\Omega)$, $|s| < 1/2$,*

$$\|v\|_{\widetilde{H}^s_{\mathrm{S}}(\Omega)} \lesssim \frac{1}{1/2 - |s|}\|v\|_{H^s_{\mathrm{S}}(\Omega)}.$$

*The constant is independent of $v$. In particular, if $\delta > 0$ satisfies $\delta \le |s| < 1/2$ then the constant depends on $\delta$ but is independent of $s$.*

*Proof.* Consider first the case when $0 < s < 1/2$. The proof, carried out in [108, Lemma 5], follows along the lines of the proof of [88, Theorem 1.4.4.4]. See also the proofs of [147, Lemma 3.31 and Lemma 3.32]. None of these proofs gives detailed arguments. We provide them here.

Recall the definition of $\|v\|_{\widetilde{H}^s_{\mathrm{S}}(\Omega)}$ in (A.27):

$$\|v\|^2_{\widetilde{H}^s_{\mathrm{S}}(\Omega)} = \|v\|^2_{H^s_{\mathrm{S}}(\Omega)} + \int_{\Omega} \frac{|v(\mathbf{x})|^2}{\mathrm{dist}(\mathbf{x}, \partial\Omega)^{2s}}\, d\mathbf{x}. \tag{A.55}$$

It suffices to estimate the second term on the right-hand side.

First consider the case when $\Omega$ is a Lipschitz hypograph. Recall the definition of $\partial\Omega$ in Remark A.1. Then, for $\mathbf{x} = (\mathbf{x}', x_d) \in \Omega$ and $\mathbf{y} = (\mathbf{y}', y_d) \in \partial\Omega$,

$$|\zeta(\mathbf{x}') - x_d| \le |\zeta(\mathbf{x}') - \zeta(\mathbf{y}')| + |x_d - y_d| \le L|\mathbf{x}' - \mathbf{y}'| + |x_d - y_d| \le \sqrt{L^2+1}|\mathbf{x} - \mathbf{y}|$$

where we used the Lipschitz property of $\zeta$ in the penultimate step and the Cauchy–Schwarz inequality in the last step. Consequently,

$$|\zeta(\mathbf{x}') - x_d| \leq \sqrt{L^2+1} \inf_{\mathbf{y}\in\partial\Omega} |\mathbf{x}-\mathbf{y}| = \sqrt{L^2+1}\,\mathrm{dist}(\mathbf{x},\partial\Omega).$$

Therefore, we can estimate the second term on the right-hand side of (A.55) by using Fubini's Theorem and making a change of variables by setting $x = \zeta(\mathbf{x}') - x_d$

$$\begin{aligned} \int_\Omega \frac{|v(\mathbf{x})|^2}{\mathrm{dist}(\mathbf{x},\partial\Omega)^{2s}}\,d\mathbf{x} &\leq (L^2+1)^s \int_{\mathbb{R}^{d-1}} \int_{-\infty}^{\zeta(\mathbf{x}')} \frac{|v(\mathbf{x})|^2}{|\zeta(\mathbf{x}') - x_d|^{2s}}\,dx_d\,d\mathbf{x}' \\ &= (L^2+1)^s \int_{\mathbb{R}^{d-1}} \int_0^\infty \frac{|v(\mathbf{x}',\zeta(\mathbf{x}')-x)|^2}{x^{2s}}\,dx\,d\mathbf{x}' \\ &\lesssim \int_{\mathbb{R}^{d-1}} \int_0^\infty \frac{|v(\mathbf{x}',\zeta(\mathbf{x}')-x)|^2}{x^{2s}}\,dx\,d\mathbf{x}' \end{aligned} \quad \text{(A.56)}$$

where the constant is independent of $s$. Here, we note that $(L^2+1)^s \leq L^2+1$ because $0 < s < 1/2$. Fix $\mathbf{x}' \in \mathbb{R}^{d-1}$ and define $\widetilde{v}(x) := v(\mathbf{x}',\zeta(\mathbf{x}')-x)$. We want to estimate

$$\int_0^\infty \frac{|v(\mathbf{x}',\zeta(\mathbf{x}')-x)|^2}{x^{2s}}\,dx = \int_0^\infty \frac{|\widetilde{v}(x)|^2}{x^{2s}}\,dx. \quad \text{(A.57)}$$

Making use of the identity (see [88, equations (1,4,4,9)–(1,4,4,10)])

$$\widetilde{v}(x) = -w(x) + \int_x^\infty \frac{w(y)}{y}\,dy \quad \text{where} \quad w(x) = \frac{1}{x}\int_0^x \big(\widetilde{v}(t) - \widetilde{v}(x)\big)\,dt, \quad \text{(A.58)}$$

we deduce

$$\begin{aligned} \int_0^\infty \frac{|\widetilde{v}(x)|^2}{x^{2s}}\,dx &= \int_0^\infty \Big(-x^{-s}w(x) + x^{-s}\int_x^\infty \frac{w(y)}{y}\,dy\Big)^2\,dx \\ &\leq 2\int_0^\infty x^{-2s}w^2(x)\,dx + 2\int_0^\infty \Big(x^{-s}\int_x^\infty \frac{w(y)}{y}\,dy\Big)^2\,dx. \end{aligned}$$

Invoking Lemma C.52 part (i) (Hardy's inequality) with $r = 2s < 1$, $p = 2$, and with $f(y) = w(y)/y$, we obtain the following bound for the second term on the right-hand side above, with $\eta(s) := 1/(0.5-s)^2$,

$$\int_0^\infty \Big(x^{-s}\int_x^\infty \frac{w(y)}{y}\,dy\Big)^2\,dx \leq \eta(s)\int_0^\infty x^{-2s}w^2(x)\,dx.$$

This and (A.58) together with the Cauchy–Schwarz inequality imply

$$\int_0^\infty \frac{|\widetilde{v}(x)|^2}{x^{2s}}\,\mathrm{d}x \le 3\eta(s)\int_0^\infty x^{-2s}w^2(x)\,\mathrm{d}x = 3\eta(s)\int_0^\infty x^{-2(s+1)}\Big(\int_0^x \big(\widetilde{v}(t)-\widetilde{v}(x)\big)\,\mathrm{d}t\Big)^2\,\mathrm{d}x$$

$$\le 3\eta(s)\int_0^\infty x^{-2(s+1)}\Big(\int_0^x \mathrm{d}t\Big)\Big(\int_0^x |\widetilde{v}(x)-\widetilde{v}(t)|^2\,\mathrm{d}t\Big)\,\mathrm{d}x$$

$$= 3\eta(s)\int_0^\infty\int_0^x \frac{|\widetilde{v}(x)-\widetilde{v}(t)|^2}{x^{2s+1}}\,\mathrm{d}t\,\mathrm{d}x \le 3\eta(s)\int_0^\infty\int_0^x \frac{|\widetilde{v}(x)-\widetilde{v}(t)|^2}{|x-t|^{2s+1}}\,\mathrm{d}t\,\mathrm{d}x$$

$$\le 3\eta(s)\int_0^\infty\int_0^\infty \frac{|\widetilde{v}(x)-\widetilde{v}(t)|^2}{|x-t|^{2s+1}}\,\mathrm{d}t\,\mathrm{d}x$$

$$= 3\eta(s)\int_0^\infty\int_0^\infty \frac{|v(\mathbf{x}',\zeta(\mathbf{x}')-x)-v(\mathbf{x}',\zeta(\mathbf{x}')-t)|^2}{|x-t|^{2s+1}}\,\mathrm{d}t\,\mathrm{d}x.$$

Making another change of variables $x_d = \zeta(\mathbf{x}')-x$ and $y_d = \zeta(\mathbf{x}')-t$ we deduce

$$\int_0^\infty \frac{|\widetilde{v}(x)|^2}{x^{2s}}\,\mathrm{d}x \le 3\eta(s)\int_{-\infty}^{\zeta(\mathbf{x}')}\int_{-\infty}^{\zeta(\mathbf{x}')} \frac{|v(\mathbf{x}',x_d)-v(\mathbf{x}',y_d)|^2}{|x_d-y_d|^{2s+1}}\,\mathrm{d}x_d\,\mathrm{d}y_d.$$

This inequality, (A.56), (A.57), and Fubini's Theorem imply

$$\begin{aligned}
\int_\Omega \frac{|v(\mathbf{x})|^2}{\operatorname{dist}(\mathbf{x},\partial\Omega)^{2s}}\,d\mathbf{x} &\lesssim \eta(s)\int_{\mathbb{R}^{d-1}}\Big(\int_{-\infty}^{\zeta(\mathbf{x}')}\int_{-\infty}^{\zeta(\mathbf{x}')} \frac{|v(\mathbf{x}',x_d)-v(\mathbf{x}',y_d)|^2}{|x_d-y_d|^{2s+1}}\,\mathrm{d}x_d\,\mathrm{d}y_d\Big)\,d\mathbf{x}' \\
&\le \eta(s)\int_{\mathbb{R}^{d-1}}\Big(\int_{-\infty}^{\infty}\int_{-\infty}^{\infty} \frac{|V(\mathbf{x}',x_d)-V(\mathbf{x}',y_d)|^2}{|x_d-y_d|^{2s+1}}\,\mathrm{d}x_d\,\mathrm{d}y_d\Big)\,d\mathbf{x}' \\
&\le \eta(s)\int_{\mathbb{R}^{d-1}} \|V(\mathbf{x}',\cdot)\|^2_{H^s_S(\mathbb{R})}\,d\mathbf{x}' \qquad \text{(A.59)}
\end{aligned}$$

where $V:\mathbb{R}^d\to\mathbb{R}$ is such that $V|_\Omega = v$. It follows from (A.7) and Theorem A.5 that for almost all $\mathbf{x}'\in\mathbb{R}^{d-1}$

$$\|V(\mathbf{x}',\cdot)\|^2_{H^s_S(\mathbb{R})} \simeq \int_0^\infty |K(t,V(\mathbf{x}',\cdot))|^2\frac{\mathrm{d}t}{t^{2s+1}} \qquad \text{(A.60)}$$

where

$$K\big(t,V(\mathbf{x}',\cdot)\big) = \inf_{\big(V_1(\mathbf{x}'),V_2(\mathbf{x}')\big)\in\mathscr{X}(V,\mathbf{x}')} J^{1/2}\big(t,V_1(\mathbf{x}',\cdot),V_2(\mathbf{x}',\cdot)\big)$$

with

$$J\big(t,V_1(\mathbf{x}',\cdot),V_2(\mathbf{x}',\cdot)\big) := \Big(\|V_1(\mathbf{x}',\cdot)\|^2_{L^2(\mathbb{R})} + t^2\|V_2(\mathbf{x}',\cdot)\|^2_{H^1(\mathbb{R})}\Big)$$

and

$$\mathscr{X}(V,\mathbf{x}') := \big\{\big(V_1(\mathbf{x}',\cdot),V_2(\mathbf{x}',\cdot)\big) \in L^2(\mathbb{R})\times H^1(\mathbb{R}) : V(\mathbf{x}',\cdot) = V_1(\mathbf{x}',\cdot)+V_2(\mathbf{x}',\cdot)\big\}.$$

Notice that the constants in (A.60) are independent of $s$ if $0<\delta\le s<1/2$; see Corollary A.1. If we define

$$\mathscr{X}(V) := \big\{(V_1,V_2)\in L^2(\mathbb{R}^d)\times H^1(\mathbb{R}^d) : V = V_1+V_2\big\},$$

then $\mathscr{X}(V)\subset\mathscr{X}(V,\mathbf{x}')$ for almost all $\mathbf{x}'\in\mathbb{R}^{d-1}$. Hence (A.60) implies

$$\|V(\mathbf{x}',\cdot)\|^2_{H^s_S(\mathbb{R})}\, d\mathbf{x}' \lesssim \int_0^\infty \inf_{(V_1,V_2)\mathscr{X}(V)} J\big(t,V_1(\mathbf{x}',\cdot),V_2(\mathbf{x}',\cdot)\big)\frac{dt}{t^{2s+1}}$$

so that Fubini's Theorem gives

$$\begin{aligned}\int_{\mathbb{R}^{d-1}} \|V(\mathbf{x}',\cdot)\|^2_{H^s_S(\mathbb{R})}\, d\mathbf{x}' &\lesssim \int_0^\infty \inf_{(V_1,V_2)\mathscr{X}(V)} \Big(\int_{\mathbb{R}^{d-1}} J\big(t,V_1(\mathbf{x}',\cdot),V_2(\mathbf{x}',\cdot)\big)\, d\mathbf{x}'\Big)\frac{dt}{t^{2s+1}}\\ &\le \int_0^\infty \inf_{(V_1,V_2)\mathscr{X}(V)} \Big(\|V_1\|^2_{L^2(\mathbb{R}^d)} + t^2\|V_2\|^2_{H^1(\mathbb{R}^d)}\Big)\frac{dt}{t^{2s+1}}\\ &= \|V\|^2_{[L^2(\mathbb{R}^d),H^1(\mathbb{R}^d)]_s} = \|V\|^2_{H^s_F(\mathbb{R}^d)}\end{aligned}$$

where in the last step we used (A.7). Therefore, (A.59) implies

$$\int_\Omega \frac{|v(\mathbf{x})|^2}{\operatorname{dist}(\mathbf{x},\partial\Omega)^{2s}}\, d\mathbf{x} \lesssim \eta(s)\|V\|^2_{H^s_F(\mathbb{R}^d)}.$$

It follows from the definition of $\|v\|_{H^s_F(\Omega)}$, see (A.18), and Theorem A.7 part (ii) that

$$\int_\Omega \frac{|v(\mathbf{x})|^2}{\operatorname{dist}(\mathbf{x},\partial\Omega)^{2s}}\, d\mathbf{x} \lesssim \eta(s)\|v\|^2_{H^s_F(\Omega)} \simeq \eta(s)\|v\|^2_{H^s_S(\Omega)}$$

where the constants are independent of $v$ and $s\in[\delta,1/2)$. This proves the required result for a Lipschitz hypograph $\Omega$.

Next we consider the case when $\Omega$ is a general Lipschitz domain. Let $\{W_j : j = 1,\dots,J\}$ be the family of open sets given by Definition A.1, and let $W_0 := \{\mathbf{x}\in\Omega : \operatorname{dist}(\mathbf{x},\partial\Omega)>\beta\}$ with $\beta>0$ chosen sufficiently small such that $\overline{\Omega}\subset\cup_{j=0}^J W_j$.

Let $\{\varphi_j : j=0,\dots,J\}$ be a partition of unity for $\overline{\Omega}$ such that $\operatorname{supp}(\varphi_j)\subset W_j$, $j = 0,\dots,J$. The existence of such a partition of unity is given by [147, Corollary 3.22].

If $v_j := \varphi_j v$, then $\operatorname{supp}(v_j) \subset W_j$. Theorem 3.20 in [147] gives

$$\|v_j\|_{H^s_S(\Omega)} \lesssim \|v\|_{H^s_S(\Omega)}, \quad j = 0,\dots,J, \tag{A.61}$$

where the constant is independent of $s$ and $v$. Thus it follows from the Cauchy–Schwarz inequality that

$$\begin{aligned}
\|v\|^2_{\widetilde{H}^s_S(\Omega)} &= \|\sum_{j=0}^J v_j\|^2_{\widetilde{H}^s_S(\Omega)} \le J\sum_{j=0}^J \|v_j\|^2_{\widetilde{H}^s_S(\Omega)} \\
&\lesssim \sum_{j=0}^J \|v_j\|^2_{H^s_S(\Omega)} + \sum_{j=0}^J \int_\Omega \frac{|v_j(\mathbf{x})|^2}{\operatorname{dist}(\mathbf{x},\partial\Omega)^{2s}}\,d\mathbf{x} \\
&\lesssim \|v\|^2_{H^s_S(\Omega)} + \sum_{j=0}^J \int_\Omega \frac{|v_j(\mathbf{x})|^2}{\operatorname{dist}(\mathbf{x},\partial\Omega)^{2s}}\,d\mathbf{x}.
\end{aligned} \tag{A.62}$$

The constant is independent of $s$. Note that for $j = 1,\dots,J$

$$\int_\Omega \frac{|v_j(\mathbf{x})|^2}{\operatorname{dist}(\mathbf{x},\partial\Omega)^{2s}}\,d\mathbf{x} = \int_{\Omega\cap W_j} \frac{|v_j(\mathbf{x})|^2}{\operatorname{dist}(\mathbf{x},\partial\Omega)^{2s}}\,d\mathbf{x} = \int_{\Omega_j\cap W_j} \frac{|v_j(\mathbf{x})|^2}{\operatorname{dist}(\mathbf{x},\partial\Omega)^{2s}}\,d\mathbf{x}.$$

Since the boundary $\partial\Omega$ is compact by Definition A.1 part (ii), for each $\mathbf{x} \in \Omega \cap W_j$, there exists $\mathbf{x}^* \in \partial\Omega$ such that $|\mathbf{x} - \mathbf{x}^*| = \operatorname{dist}(\mathbf{x},\partial\Omega)$. By changing $W_j$ if necessary, we can assume that $\mathbf{x}^* \in W_j$. Since $\Omega \cap W_j = \Omega_j \cap W_j$, the part of $\partial\Omega$ contained in $W_j$ equals the part of $\partial\Omega_j$ contained in $W_j$. Hence $\mathbf{x}^* \in \partial\Omega_j$ and thus $\operatorname{dist}(\mathbf{x},\partial\Omega_j) \le \operatorname{dist}(\mathbf{x},\partial\Omega)$; see Figure A.1. This implies

$$\int_\Omega \frac{|v_j(\mathbf{x})|^2}{\operatorname{dist}(\mathbf{x},\partial\Omega)^{2s}}\,d\mathbf{x} \le \int_{\Omega_j\cap W_j} \frac{|v_j(\mathbf{x})|^2}{\operatorname{dist}(\mathbf{x},\partial\Omega_j)^{2s}}\,d\mathbf{x} \le \int_{\Omega_j} \frac{|v_j(\mathbf{x})|^2}{\operatorname{dist}(\mathbf{x},\partial\Omega_j)^{2s}}\,d\mathbf{x}.$$

Since $\Omega_j$ is a Lipschitz hypograph, the first part of the proof yields, with the help of (A.61),

$$\begin{aligned}
\int_{\Omega_j} \frac{|v_j(\mathbf{x})|^2}{\operatorname{dist}(\mathbf{x},\partial\Omega_j)^{2s}}\,d\mathbf{x} &\lesssim \eta(s)\|v_j\|^2_{H^s_S(\Omega_j)} = \eta(s)\|v_j\|^2_{H^s_S(\Omega_j\cap W_j)} = \eta(s)\|v_j\|^2_{H^s_S(\Omega\cap W_j)} \\
&\le \eta(s)\|v_j\|^2_{H^s_S(\Omega)} \lesssim \eta(s)\|v\|^2_{H^s_S(\Omega)},
\end{aligned} \tag{A.63}$$

for $j = 1,\dots,J$. The constant is independent of $s \in [\delta, 1/2)$. The case when $j = 0$ can be dealt with as follows:

$$\begin{aligned}
\int_\Omega \frac{|v_0(\mathbf{x})|^2}{\operatorname{dist}(\mathbf{x},\partial\Omega)^{2s}}\,d\mathbf{x} &= \int_{\Omega\cap W_0} \frac{|v_0(\mathbf{x})|^2}{\operatorname{dist}(\mathbf{x},\partial\Omega)^{2s}}\,d\mathbf{x} \le \frac{1}{\beta^{2s}} \int_{\Omega\cap W_0} |v_0(\mathbf{x})|^2\,d\mathbf{x} \\
&\lesssim \|v\|^2_{L^2(\Omega)} \le \|v\|^2_{H^s_S(\Omega)}
\end{aligned}$$

where the constant depends only on $\beta$. This inequality, (A.62), and (A.63) give the desired result for $s \in (0,1/2)$ and for any Lipschitz domain $\Omega$. The constant is independent of $s$ when $s \in [\delta, 1/2)$.

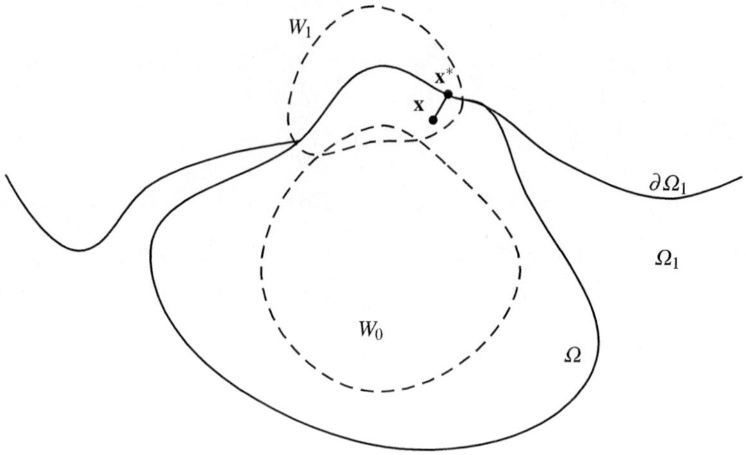


**Fig. A.1** Lipschitz domain $\Omega$, Lipschitz hypograph $\Omega_1$, open sets $W_0$ and $W_1$

The case when $s \in (-1/2, 0)$ is obtained by duality, see (A.28), as follows:

$$\|v\|_{\widetilde{H}^s_S(\Omega)} = \sup_{w \in H^{-s}(\Omega)} \frac{\langle v, w\rangle_\Omega}{\|w\|_{H_S^{-s}(\Omega)}} \lesssim \frac{1}{1+s} \sup_{w \in \widetilde{H}^{-s}(\Omega)} \frac{\langle v, w\rangle_\Omega}{\|w\|_{\widetilde{H}_S^{-s}(\Omega)}} = \frac{1}{1+s} \|v\|_{H^s_S(\Omega)}$$

where we used $H^s(\Omega) = \widetilde{H}^s(\Omega)$ for $s \in (0, 1/2)$. This completes the proof. □

In Lemma A.10 we will extend the above lemma to interpolation norms.
The following lemma is a consequence of Lemma A.9.

**Lemma A.10.** *Let $\Omega$ be a Lipschitz domain in $\mathbb{R}^d$. Then for any $v \in \widetilde{H}^s(\Omega)$, $|s| < 1/2$,*

$$\|v\|_{\widetilde{H}^s_I(\Omega)} \lesssim \frac{1}{1/2 - |s|} \|v\|_{H^s_I(\Omega)}.$$

*The constant is independent of $v$. In particular, if $\delta > 0$ satisfies $\delta \le |s| < 1/2$ then the constant depends on $\delta$ but is independent of $s$.*

*Proof.* The lemma is a consequence of Lemma A.9 and Theorem A.9. □

## A.2.9 Comparison of global and local norms

The following theorem concerning properties of the global and local $H_{\mathrm{I}}^s(\Omega)$ and $\widetilde{H}_{\mathrm{I}}^s(\Omega)$ norms is initially proved in [170, Lemma 3.2, p. 49]. An alternative proof is presented in [5, Theorem 4.1] and is included here for the reader's convenience.

**Theorem A.10.** *Let $\{\Omega_1,\dots,\Omega_N\}$ be a partition of a bounded Lipschitz domain $\Omega$ into nonoverlapping Lipschitz domains. For any $s\in\mathbb{R}$ the following inequalities hold (assuming that all the norms are well defined)*

$$\sum_{j=1}^N \|u\|^2_{H_{\mathrm{I}}^s(\Omega_j)} \le \|u\|^2_{H_{\mathrm{I}}^s(\Omega)} \tag{A.64}$$

*and*

$$\|u\|^2_{\widetilde{H}_{\mathrm{I}}^s(\Omega)} \le \sum_{j=1}^N \|u\|^2_{\widetilde{H}_{\mathrm{I}}^s(\Omega_j)}. \tag{A.65}$$

*Proof.* Let $r$ be an integer such that $s\in[-r,r]$. Introduce the product spaces

$$\Pi^s := H^s(\Omega_1)\times\cdots\times H^s(\Omega_N) \quad\text{and}\quad \widetilde{\Pi}^s := \widetilde{H}^s(\Omega_1)\times\cdots\times\widetilde{H}^s(\Omega_N)$$

with norms defined from the interpolation norms by

$$\|\mathbf{u}\|^2_{\Pi^s} := \sum_{j=1}^N \|u_j\|^2_{H_{\mathrm{I}}^s(\Omega_j)} \quad\text{and}\quad \|\mathbf{u}\|^2_{\widetilde{\Pi}^s} := \sum_{j=1}^N \|u_j\|^2_{\widetilde{H}_{\mathrm{I}}^s(\Omega_j)},$$

where $\mathbf{u}=(u_1,\dots,u_N)$. Let $\theta=s/r$. Then $\theta\in[0,1]$ and

$$\Pi^s = [\Pi^0,\Pi^r]_\theta \quad\text{and}\quad \widetilde{\Pi}^s = [\widetilde{\Pi}^0,\widetilde{\Pi}^r]_\theta.$$

Moreover,

$$\Pi^{-s} = \left(\widetilde{\Pi}^s\right)^* \quad\text{and}\quad \widetilde{\Pi}^{-s} = (\Pi^s)^*,$$

with the duality pairing

$$\langle \mathbf{u},\mathbf{v}\rangle := \sum_{j=1}^N \langle u_j, v_j\rangle_{\Omega_j},$$

where $\mathbf{v}=(v_1,\dots,v_N)$. Now consider the restriction and sum operators

$$\mathscr{R}: H^s(\Omega)\to\Pi^s \quad\text{and}\quad \mathscr{S}:\widetilde{\Pi}^s\to\widetilde{H}^s(\Omega)$$

defined by

$$\mathscr{R}u := \left(u|_{\Omega_1},\dots,u|_{\Omega_N}\right) \quad\text{and}\quad \mathscr{S}\mathbf{u} := \sum_{j=1}^N u_j.$$

Since

$$\|\mathscr{R}u\|_{\Pi^0} = \|u\|_{H^0_{\mathrm{I}}(\Omega)} \quad \text{and} \quad \|\mathscr{R}u\|_{\Pi^r} = \|u\|_{H^r_{\mathrm{I}}(\Omega)},$$

by interpolation

$$\|\mathscr{R}u\|_{\Pi^s} \le \|u\|_{H^s_{\mathrm{I}}(\Omega)} \quad \text{for } 0 \le s \le r. \tag{A.66}$$

Similarly, since

$$\|\mathscr{S}\mathbf{u}\|_{\widetilde{H}^0_{\mathrm{I}}(\Omega)} = \|\mathbf{u}\|_{\widetilde{\Pi}^0} \quad \text{and} \quad \|\mathscr{S}\mathbf{u}\|_{\widetilde{H}^r_{\mathrm{I}}(\Omega)} = \|\mathbf{u}\|_{\widetilde{\Pi}^r}$$

by interpolation

$$\|\mathscr{S}\mathbf{u}\|_{\widetilde{H}^s_{\mathrm{I}}(\Omega)} \le \|\mathbf{u}\|_{\widetilde{\Pi}^s} \quad \text{for } 0 \le s \le r. \tag{A.67}$$

The inequalities holds also for $-r \le s < 0$ by duality because

$$\langle \mathscr{R}u, \mathbf{v}\rangle_\Omega = \langle u, \mathscr{S}\mathbf{v}\rangle_\Omega .$$

Indeed,

$$\begin{aligned}\|\mathscr{R}u\|_{\Pi^s} &= \sup_{\mathbf{v}\in\widetilde{\Pi}^{-s}} \frac{\langle \mathscr{R}u, \mathbf{v}\rangle_\Omega}{\|\mathbf{v}\|_{\widetilde{\Pi}^{-s}}} = \sup_{\mathbf{v}\in\widetilde{\Pi}^{-s}} \frac{\langle u, \mathscr{S}\mathbf{v}\rangle_\Omega}{\|\mathbf{v}\|_{\widetilde{\Pi}^{-s}}} \le \sup_{\mathbf{v}\in\widetilde{\Pi}^{-s}} \frac{\|u\|_{H^s_{\mathrm{I}}(\Omega)}\|\mathscr{S}\mathbf{v}\|_{\widetilde{H}^{-s}_{\mathrm{I}}(\Omega)}}{\|\mathbf{v}\|_{\widetilde{\Pi}^{-s}}} \\ &\le \|\mathscr{S}\mathbf{v}\|_{\widetilde{H}^{-s}_{\mathrm{I}}(\Omega)} \quad \text{for } -r \le s < 0.\end{aligned} \tag{A.68}$$

Similarly,

$$\begin{aligned}\|\mathscr{S}\mathbf{u}\|_{\widetilde{H}^s_{\mathrm{I}}(\Omega)} &= \sup_{v\in H^{-s}_{\mathrm{I}}(\Omega)} \frac{\langle \mathscr{S}\mathbf{u}, v\rangle_\Omega}{\|v\|_{H^{-s}_{\mathrm{I}}(\Omega)}} = \sup_{v\in H^{-s}_{\mathrm{I}}(\Omega)} \frac{\langle \mathbf{u}, \mathscr{R}v\rangle_\Omega}{\|v\|_{H^{-s}_{\mathrm{I}}(\Omega)}} \\ &\le \sup_{v\in H^{-s}_{\mathrm{I}}(\Omega)} \frac{\|\mathbf{u}\|_{\widetilde{\Pi}^s}\|\mathscr{R}v\|_{\Pi^{-s}}}{\|v\|_{H^{-s}_{\mathrm{I}}(\Omega)}} \\ &\le \|\mathbf{u}\|_{\widetilde{\Pi}^s} \quad \text{for } -r \le s < 0.\end{aligned} \tag{A.69}$$

Now for any function $u$ we define $\mathbf{u} = (u|_{\Omega_1}, \dots, u|_{\Omega_N})$. Then $u = \mathscr{S}\mathbf{u}$ because $\{\Omega_1, \dots, \Omega_N\}$ is a partition of $\Omega$. Assuming that all the following norms are well defined, we deduce from (A.66) and (A.68)

$$\sum_{j=1}^N \|u\|^2_{H^s_{\mathrm{I}}(\Omega_j)} = \sum_{j=1}^N \|u|_{\Omega_j}\|^2_{H^s_{\mathrm{I}}(\Omega_j)} = \|\mathscr{R}u\|^2_{\Pi^s} \le \|u\|^2_{H^s_{\mathrm{I}}(\Omega)}$$

and from (A.67) and (A.69)

$$\|u\|^2_{\widetilde{H}^s_{\mathrm{I}}(\Omega)} = \|\mathscr{S}\mathbf{u}\|^2_{\widetilde{H}^s_{\mathrm{I}}(\Omega)} \le \|\mathbf{u}\|^2_{\widetilde{\Pi}^s} = \sum_{j=1}^N \|u|_{\Omega_j}\|^2_{\widetilde{H}^s_{\mathrm{I}}(\Omega_j)} = \sum_{j=1}^N \|u\|^2_{\widetilde{H}^s_{\mathrm{I}}(\Omega_j)},$$

proving (A.64) and (A.65). □

We now want to improve the result of Theorem A.10 to ensure that the norms on the right-hand side of (A.65) can be replaced with $\|u_j\|_{\widetilde{H}^s_{\mathrm{I}}(\Omega)}$ where $u_j$ is the zero extension of $u|_{\Omega_j}$ onto the exterior of $\Omega_j$. This is the purpose of Theorem A.11. We need to define some new interpolation spaces.

**The space $\widetilde{H}^s_*(\Omega_1)$ for $s \in (0,1)$:**
Let $\Omega_1 \subsetneq \Omega \subsetneq \mathbb{R}^n$ and $\Omega_2 = \Omega \setminus \Omega_1$. We define two spaces

$$A_0 := \{v \in L^2(\Omega) : \operatorname{supp}(v) \subseteq \overline{\Omega}_1\},$$
$$A_1 := \{v \in \widetilde{H}^1(\Omega) : \operatorname{supp}(v) \subseteq \overline{\Omega}_1\},$$

which form a compatible couple $\mathscr{A} = (A_0, A_1)$. For any $v \in A_0$ we define

$$\mathscr{X}(v) := \{(v_0, v_1) \in L^2(\Omega) \times \widetilde{H}^1(\Omega) : v_0 + v_1 = v\},$$
$$\mathscr{X}_1(v) := \{(v_0, v_1) \in A_0 \times A_1 : v_0 + v_1 = v\}.$$

We note that in the definition of $\mathscr{X}(v)$ with $v \in A_0$, the functions $v_0$ and $v_1$ do not have to be zero in $\Omega_2$. Instead, it suffices to have $v_0 = -v_1$ in $\overline{\Omega}_2$ so that $v = 0$ in $\Omega_2$. Hence, $\mathscr{X}_1(v) \subsetneq \mathscr{X}(v)$. Let us define two functionals

$$J : \mathbb{R}^+ \times L^2(\Omega) \times \widetilde{H}^1(\Omega) \to \mathbb{R} \quad \text{and} \quad J_1 : \mathbb{R}^+ \times A_0 \times A_1 \to \mathbb{R}$$

by

$$\begin{aligned} J(t, v_0, v_1) &:= \|v_0\|^2_{L^2(\Omega)} + t^2 |v_1|^2_{H^1_0(\Omega)}, \quad && v_0 \in L^2(\Omega),\ v_1 \in \widetilde{H}^1(\Omega), \\ J_1(t, v_0, v_1) &:= \|v_0\|^2_{L^2(\Omega_1)} + t^2 |v_1|^2_{H^1_0(\Omega_1)}, \quad && v_0 \in A_0,\ v_1 \in A_1. \end{aligned} \tag{A.70}$$

Then, for any $t > 0$,

$$\begin{aligned} \inf\{J(t, v_0, v_1) : (v_0, v_1) \in \mathscr{X}(v)\} &\le \inf\{J(t, v_0, v_1) : (v_0, v_1) \in \mathscr{X}_1(v)\} \\ &= \inf\{J_1(t, v_0, v_1) : (v_0, v_1) \in \mathscr{X}_1(v)\}. \end{aligned}$$

The infima are in fact minima. The following lemma shows that in general the above inequality is a strict inequality. It suffices to prove the following lemma with $r = 1$. First we recall the definition of the Hölder space $C^{0,\alpha}(\overline{\Omega})$ for $\alpha \in (0,1)$

$$C^{0,\alpha}(\overline{\Omega}) := \Big\{v \in C(\overline{\Omega}) : \sup_{\substack{x,y \in \Omega \\ x \ne y}} \frac{|v(x) - v(y)|}{|x - y|^\alpha} < \infty\Big\}.$$

**Lemma A.11.** *Assume that $\Omega$ has a smooth boundary and $\Omega_1 \subset \Omega$. Let $v \in A_0 \cap C^{0,\alpha}(\overline{\Omega})$ be such that $v \ge 0$ and $v \not\equiv 0$ in $\Omega_1$. Then, for any $t > 0$,*

$$\min\{J(t, v_0, v_1) : (v_0, v_1) \in \mathscr{X}(v)\} < \min\{J_1(t, v_0, v_1) : (v_0, v_1) \in \mathscr{X}_1(v)\}.$$

*Proof.* Fix $t > 0$. There exists $(v_0^*, v_1^*) \in \mathscr{X}(v)$ such that

$$(v_0^*, v_1^*) = \operatorname{argmin}\{J(t, v_0, v_1) : (v_0, v_1) \in \mathscr{X}(v)\}. \tag{A.71}$$

Indeed, let

$$m := \inf\{J(t,v_0,v_1) : (v_0,v_1) \in \mathscr{X}(v)\} = \inf\{J(t,v-v_1,v_1) : v_1 \in A_1\}$$

and let $\{v_1^{(k)}\}$ be a sequence in $A_1$ satisfying

$$\lim_{k\to\infty} J(t,v-v_1^{(k)},v_1^{(k)}) = m.$$

It follows from the definition of the functional $J$ that $\{v_1^{(k)}\}$ is bounded in $H_0^1(\Omega)$. Hence there exists a subsequence $\{v_1^{(k_j)}\}$ converging weakly in $H_0^1(\Omega)$ to $v_1^*$. It can be seen that $v_1^* \in A_1$. From the definition of $m$, it follows that

$$m \le J(t,v-v_1^*,v_1^*).$$

On the other hand, recall the property that if a sequence $\{v_j\}$ in a Banach space $E$ converges weakly to $v \in E$, then $\|v\|_E \le \liminf_{j\to\infty} \|v_j\|_E$. Hence, the weak convergence of $\{v_1^{(k_j)}\}$ to $v_1^*$ implies

$$J(t,v-v_1^*,v_1^*) \le \liminf_{j\to\infty} J(t,v-v_1^{(k_j)},v_1^{(k_j)}) = m.$$

Thus $J(t,v-v_1^*,v_1^*) = m$ and therefore $(v_0^*,v_1^*) := (v-v_1^*,v_1^*)$ is a minimiser.

To prove the required strict inequality, it suffices to show that $(v_0^*,v_1^*) \notin \mathscr{X}_1(v)$. The problem (A.71) is an optimisation problem with constraint, the constraint being $g(v_0,v_1) = 0$ where

$$g : L^2(\Omega) \times \widetilde{H}^1(\Omega) \to L^2(\Omega), \quad g(v_0,v_1) := v_0 + v_1 - v.$$

The Lagrangian functional $\mathscr{L} : L^2(\Omega) \times \widetilde{H}^1(\Omega) \times L^2(\Omega) \to \mathbb{R}$ is defined by

$$\mathscr{L}(v_0,v_1;p) := J(t,v_0,v_1) + G(v_0,v_1,p)$$

where $p$ is the Lagrangian multiplier and

$$G(v_0,v_1,p) := \langle p, g(v_0,v_1)\rangle_{L^2(\Omega)} = \langle p, v_0+v_1-v\rangle_{L^2(\Omega)}.$$

If $g(v_0,v_1) = 0$, then

$$\sup_{p\in L^2(\Omega)} G(v_0,v_1,p) = 0.$$

If $g(v_0,v_1) \ne 0$, then we can choose $p \in L^2(\Omega)$ such that $G(v_0,v_1,p) > 0$. Consequently

$$G(v_0,v_1,tp) = tG(v_0,v_1,p) \to \infty \quad \text{as } t \to \infty.$$

Hence

$$\sup_{p\in L^2(\Omega)} G(v_0,v_1,p) = \infty.$$

Therefore,

$$\sup_{p\in L^2(\Omega)} \mathscr{L}(v_0,v_1,p) = \begin{cases} J(t,v_0,v_1) & \text{if } g(v_0,v_1)=0,\\ \infty & \text{else.}\end{cases}$$

As a direct consequence, we conclude that

$$\min_{(v_0,v_1)\in\mathscr{X}(v)} J(t,v_0,v_1) = \inf_{v_0\in L^2(\Omega),\ v_1\in\widetilde{H}^1(\Omega)} \ \sup_{p\in L^2(\Omega)} \mathscr{L}(v_0,v_1,p). \tag{A.72}$$

For any functional $F:(v,p)\mapsto F(v,p)$, we denote by $\partial_v F(v,p)(\varphi)$ the $v$-Fréchet derivative of $F$ at $(v,p)$ acting on $\varphi$, and by $\partial_p F(v,p)(q)$ the $p$-Fréchet derivative of $F$ at $(v,p)$ acting on $q$. The minimiser $(v_0^*,v_1^*)$ of (A.71) or the solution $(v_0^*,v_1^*,p^*)$ to the minimax problem (A.72) solves the following equations

$$\partial_{v_0}\mathscr{L}(v_0,v_1,p)=0,\quad \partial_{v_1}\mathscr{L}(v_0,v_1,p)=0,\quad \partial_p\mathscr{L}(v_0,v_1,p)=0.$$

Since

$$\begin{aligned}
\partial_{v_0}J(t,v_0,v_1)(\varphi) &= 2\langle v_0,\varphi\rangle_{L^2(\Omega)} && \forall\varphi\in L^2(\Omega),\\
\partial_{v_1}J(t,v_0,v_1)(\psi) &= 2t^2\langle\nabla v_1,\nabla\psi\rangle_{L^2(\Omega)} && \forall\psi\in\widetilde{H}^1(\Omega),\\
\partial_{v_0}G(v_0,v_1,p)(\varphi) &= \langle p,\varphi\rangle_{L^2(\Omega)} && \forall\varphi\in L^2(\Omega),\\
\partial_{v_1}G(v_0,v_1,p)(\psi) &= \langle p,\psi\rangle_{L^2(\Omega)} && \forall\psi\in\widetilde{H}^1(\Omega),\\
\partial_p G(v_0,v_1,p)(q) &= \langle q,g(v_0,v_1)\rangle_{L^2(\Omega)} && \forall q\in L^2(\Omega),
\end{aligned}$$

we have

$$\begin{aligned}
\partial_{v_0}\mathscr{L}(v_0,v_1,p)(\varphi) &= 2\langle v_0,\varphi\rangle_{L^2(\Omega)}+\langle p,\varphi\rangle_{L^2(\Omega)} && \forall\varphi\in L^2(\Omega),\\
\partial_{v_1}\mathscr{L}(v_0,v_1,p)(\psi) &= 2t^2\langle\nabla v_1,\nabla\psi\rangle_{L^2(\Omega)}+\langle p,\psi\rangle_{L^2(\Omega)} && \forall\psi\in\widetilde{H}^1(\Omega),\\
\partial_p\mathscr{L}(v_0,v_1,p)(q) &= \langle q,v_0+v_1-v\rangle_{L^2(\Omega)} && \forall q\in L^2(\Omega).
\end{aligned}$$

Hence, $v_0^*$, $v_1^*$, and $p^*$ satisfy

$$2\langle v_0^*,\varphi\rangle_{L^2(\Omega)}+\langle p^*,\varphi\rangle_{L^2(\Omega)}=0\quad\forall\varphi\in L^2(\Omega), \tag{A.73}$$

$$2t^2\langle\nabla v_1^*,\nabla\psi\rangle_{L^2(\Omega)}+\langle p^*,\psi\rangle_{L^2(\Omega)}=0\quad\forall\psi\in\widetilde{H}^1(\Omega), \tag{A.74}$$

$$\langle q,v_0^*+v_1^*-v\rangle_{L^2(\Omega)}=0\quad\forall q\in L^2(\Omega). \tag{A.75}$$

It follows from (A.73) and (A.74) that

$$t^2\langle\nabla v_1^*,\nabla\psi\rangle_{L^2(\Omega)}-\langle v_0^*,\psi\rangle_{L^2(\Omega)}=0\quad\forall\psi\in\widetilde{H}^1(\Omega).$$

This equation and (A.75) give

$$t^2 \langle \nabla v_1^*, \nabla \psi \rangle_{L^2(\Omega)} + \langle v_1^*, \psi \rangle_{L^2(\Omega)} = \langle v, \psi \rangle_{L^2(\Omega)} \quad \forall \psi \in \widetilde{H}^1(\Omega).$$

This is a weak formulation of the following boundary value problem

$$\begin{aligned} -t^2 \Delta v_1^* + v_1^* &= v \quad \text{in} \quad \Omega, \\ v_1^* &= 0 \quad \text{on} \quad \partial\Omega. \end{aligned} \tag{A.76}$$

Since $\Omega$ has smooth boundary and $v \in C^{0,\alpha}(\overline{\Omega})$, we deduce that $v_1^* \in C(\overline{\Omega}) \cap C^2(\Omega)$. Moreover, since $v \geq 0$ in $\Omega$, due to the strong maximum principle, see e.g. [43, Corollary 9.37], either $v_1^* > 0$ in $\Omega$ or $v_1^* \equiv 0$ in $\Omega$. If $v_1^* \equiv 0$ then $v_0^* = v$. Moreover, (A.74) implies $p \equiv 0$ so that $v_0^* \equiv 0$ on $\Omega$ due to (A.73). This contradicts the assumption on $v$. Hence $v_1^* > 0$ on $\Omega$, which implies $(v_0^*, v_1^*) \notin \mathscr{X}_1(v)$. □

*Remark A.3.* The above result is consistent with the well-known fact that the solution $v_1^*$ of (A.76) in a subdomain $\Omega_2 \subsetneq \Omega$ depends on the values of $v$ in all of $\Omega$ and not only on the values of $v$ in $\Omega_2$; see e.g. [43, page 307].

On the compatible couple $\mathscr{A} = (A_0, A_1)$, see Section A.1.1, we define two $K$-functionals $K$ and $K_1$ by using the two functionals $J$ and $J_1$ defined in (A.70) as follows:

$$K(t, v, \mathscr{A}) := \inf_{(v_0, v_1) \in \mathscr{X}(v)} \left| J(v_0, v_1) \right|^{1/2} \quad \forall v \in A_0,$$

$$K_1(t, v, \mathscr{A}) := \inf_{(v_0, v_1) \in \mathscr{X}_1(v)} \left| J_1(v_0, v_1) \right|^{1/2} \quad \forall v \in A_0.$$

The corresponding interpolation norms defined from these $K$-functionals, see (A.1), are

$$\|v\|_{\widetilde{H}_1^s(\Omega)} := \Big( \int_0^\infty |K(t, u, \mathscr{A})|^2 \frac{dt}{t^{1+2s}} \Big)^{1/2},$$

$$\|v\|_{\widetilde{H}_1^s(\Omega_1)} := \Big( \int_0^\infty |K_1(t, u, \mathscr{A})|^2 \frac{dt}{t^{1+2s}} \Big)^{1/2}.$$

Lemma A.11 implies that in general $\|v\|_{\widetilde{H}_1^s(\Omega)} \leq \|v\|_{\widetilde{H}_1^s(\Omega_1)}$ for $v \in A_0$. Denote

$$\widetilde{H}_*^s(\Omega_1) := \left\{ v \in A_0 : \|v\|_{\widetilde{H}_1^s(\Omega)} < \infty \right\} \quad \text{and} \quad \|v\|_{\widetilde{H}_*^s(\Omega_1)} := \|v\|_{\widetilde{H}_1^s(\Omega)}. \tag{A.77}$$

It is clear that $\widetilde{H}^s(\Omega_1)$ is a proper subset of $\widetilde{H}_*^s(\Omega_1)$. (Here we extend functions in $\widetilde{H}^s(\Omega_1)$ by zero onto the exterior of $\Omega_1$.) Lemma A.11 shows that $\widetilde{H}_*^s(\Omega_1)$ is a proper subset of $\widetilde{H}^s(\Omega)$. This means

$$\widetilde{H}^s(\Omega_1) \subsetneq \widetilde{H}_*^s(\Omega_1) \subsetneq \widetilde{H}^s(\Omega). \tag{A.78}$$

The space $\widetilde{H}_*^s(\Omega_1)$ is first studied in [224].

**The space $\widetilde{H}_*^{-s}(\Omega_1)$ for $s \in [0,1]$:**

For $s \in (0,1)$, similarly to the definition of $\widetilde{H}_*^s(\Omega_1)$, we will in this section construct $\widetilde{H}_*^{-s}(\Omega_1)$ which contains $\widetilde{H}^{-s}(\Omega_1)$ as a subspace. First we recall from [138, Theorem 12.5, page 76] that

$$\widetilde{H}^{-s}(\Omega_1) = \left[L^2(\Omega_1), \widetilde{H}^{-1}(\Omega_1)\right]_s; \tag{A.79}$$

see also [57, Corollary 4.10] and [147, Theorem B.9]. Therefore we can equip the space $\widetilde{H}^{-s}(\Omega_1)$ with different norms, namely, the norm $\|\cdot\|_{\widetilde{H}_I^{-s}(\Omega_1)}$ defined by (A.35), the norm ${}_*\|\cdot\|_{\widetilde{H}_I^{-s}(\Omega_1)}$ defined by (A.48), and the interpolation norm intrinsic to the interpolation space on the right-hand side of (A.79). The non-scalability of $\|\cdot\|_{\widetilde{H}_I^{-s}(\Omega_1)}$, as shown in Lemma A.6 Part (iv), renders this norm least favourable. For this reason, we equip $\widetilde{H}^{-s}(\Omega_1)$ with the norm defined by (A.48). It is remarked that, by using interpolation, the interpolation norm scales in exactly the same way as ${}_*\|\cdot\|_{\widetilde{H}_I^{-s}(\Omega_1)}$; see Lemma A.7 Part (ii). Thus (up to some constant independent of the size of $\Omega_1$) these two norms are equal.

We define

$$\begin{aligned} A_0 &:= \left\{v \in L^2(\Omega) : \operatorname{supp}(v) \subseteq \overline{\Omega}_1,\ \langle v, 1\rangle_{\Omega_1} = 0\right\}, \\ A_1 &:= \left\{v \in \widetilde{H}^{-1}(\Omega) : \operatorname{supp}(v) \subseteq \overline{\Omega}_1,\ \langle v, 1\rangle_{\Omega_1} = 0\right\}, \end{aligned} \tag{A.80}$$

where the support is understood in the distributional sense. These two sets form a compatible couple $\mathscr{A} = (A_0, A_1)$. For any $v \in A_1$ we define

$$\begin{aligned} \mathscr{X}(v) &:= \left\{(v_0, v_1) \in L^2(\Omega) \times \widetilde{H}^{-1}(\Omega) : v_0 + v_1 = v\right\}, \\ K(t, v, \mathscr{A}) &:= \inf_{(v_0, v_1) \in \mathscr{X}(v)} \left|J(v_0, v_1)\right|^{1/2}, \end{aligned}$$

and

$$\begin{aligned} \mathscr{X}_1(v) &:= \left\{(v_0, v_1) \in A_0 \times A_1 : v_0 + v_1 = v\right\}, \\ K_1(t, v, \mathscr{A}) &:= \inf_{(v_0, v_1) \in \mathscr{X}_1(v)} \left|J_1(v_0, v_1)\right|^{1/2}, \end{aligned}$$

where

$$\begin{aligned} J(t, v_0, v_1) &:= \|v_0\|_{L^2(\Omega)}^2 + t^2\, {}_*\|v_1\|_{\widetilde{H}_I^{-1}(\Omega)}^2, \qquad v_0 \in L^2(\Omega),\ v_1 \in \widetilde{H}^{-1}(\Omega), \\ J_1(t, v_0, v_1) &:= \|v_0\|_{L^2(\Omega_1)}^2 + t^2\, {}_*\|v_1\|_{\widetilde{H}_I^{-1}(\Omega_1)}^2, \qquad v_0 \in A_0,\ v_1 \in A_1. \end{aligned}$$

We note that $\mathscr{X}_1(v) \subsetneq \mathscr{X}(v)$. If $v_1 \in A_1$ then recalling the definition of $A_1$ and using Hölder's inequality for dual norms we have

$${}_*\|v_1\|_{\widetilde{H}_I^{-1}(\Omega)} = \sup_{w \in H^1(\Omega)} \frac{\langle v_1, w\rangle_\Omega}{{}_*\|w\|_{H_I^1(\Omega)}} \le \sup_{w \in H^1(\Omega)} \inf_{c \in \mathbb{R}} \frac{\langle v_1, w + c\rangle_{\Omega_1}}{|w|_{H^1(\Omega)}}$$

$$\leq {}_*\|v_1\|_{\widetilde{H}_I^{-1}(\Omega_1)} \sup_{w\in H^1(\Omega)} \inf_{c\in\mathbb{R}} \frac{{}_*\|w+c\|_{H_I^1(\Omega_1)}}{|w|_{H^1(\Omega)}}.$$

Invoking Lemma A.15 Part (ii) we deduce

$${}_*\|v_1\|_{\widetilde{H}_I^{-1}(\Omega)} \lesssim {}_*\|v_1\|_{\widetilde{H}_I^{-1}(\Omega_1)} \sup_{w\in H^1(\Omega)} \frac{|w|_{H_I^1(\Omega_1)}}{|w|_{H_I^1(\Omega)}} \leq {}_*\|v_1\|_{\widetilde{H}_I^{-1}(\Omega_1)}.$$

This implies

$$J(t,v_0,v_1) \lesssim J_1(t,v_0,v_1), \quad (v_0,v_1)\in A_0\times A_1.$$

Thus, for $v\in A_1$,

$$|K(t,v,\mathscr{A})|^2 \leq \inf_{(v_0,v_1)\in\mathscr{X}_1(v)} J(t,v_0,v_1) \lesssim |K_1(t,v,\mathscr{A})|^2. \tag{A.81}$$

As a consequence, apart from the space $\widetilde{H}^{-s}(\Omega_1)$ with the norm ${}_*\|\cdot\|_{\widetilde{H}_I^{-s}(\Omega_1)}$ defined from the functional $K_1(t,v,\mathscr{A})$ as usual, we can also construct the space $\widetilde{H}_*^{-s}(\Omega_1)$ defined with the functional $K(t,v,\mathscr{A})$ by

$$\begin{aligned} \widetilde{H}_*^{-s}(\Omega_1) &:= \{v\in A_1 : \|v\|_{\widetilde{H}_*^{-s}(\Omega_1)} < \infty\}, \\ \|v\|_{\widetilde{H}_*^{-s}(\Omega_1)} &:= {}_*\|v\|_{\widetilde{H}_I^{-s}(\Omega)} = \left(\int_0^\infty |K(t,u,\mathscr{A})|^2 \frac{dt}{t^{1+2s}}\right)^{1/2}. \end{aligned} \tag{A.82}$$

It follows from (A.81) that

$$\|v\|_{\widetilde{H}_*^{-s}(\Omega_1)} \lesssim {}_*\|v\|_{\widetilde{H}_I^{-s}(\Omega_1)} \tag{A.83}$$

and hence

$$\widetilde{H}_0^{-s}(\Omega_1) \subset \widetilde{H}_*^{-s}(\Omega_1) \tag{A.84}$$

where, see (A.80),

$$\widetilde{H}_0^{\sigma}(\Omega_1) := \{v\in\widetilde{H}^{\sigma}(\Omega_1) : \langle v,1\rangle_{\Omega_1} = 0\}, \quad \sigma<0. \tag{A.85}$$

We can now generalise Theorem A.10 for the spaces $\widetilde{H}_*^s(\Omega_1)$, $s\in[-1,1]$, and obtain the following results. This generalisation is crucial for our proofs of coercivity of decompositions; see Assumption 2.4. For example, in the case of the hypersingular integral equation, since the bilinear form $a_j(u_j,u_j)$ on the right-hand side of this assumption is equivalent to the norm $\|u_j\|_{\widetilde{H}_I^{1/2}(\Gamma)}$, while inequality (A.65) corresponds to the norm $\|u_j\|_{\widetilde{H}_I^{1/2}(\Gamma_j)}$, which is a larger norm, Theorem A.10 cannot be used directly. Similar comment can also be made for the weakly-singular integral equation and the norm $\|u_j\|_{\widetilde{H}_I^{-1/2}(\Gamma)}$. The following theorem will be useful in these cases.

**Theorem A.11.** *Let $\{\Omega_1,\ldots,\Omega_N\}$ be a partition of a bounded Lipschitz domain $\Omega$ into nonoverlapping Lipschitz domains. For any function $u$ defined on $\Omega$, we denote by $u_j$ the zero extension of $u|_{\Omega_j}$ onto the exterior of $\Omega_j$.*

(i) *Assume that $u \in \widetilde{H}^s(\Omega)$ satisfies $u_j \in \widetilde{H}^s_*(\Omega_j)$ for $s \in [-1,1]$. Then*

$$\|u\|^2_{\widetilde{H}^s_{\rm I}(\Omega)} \le \sum_{j=1}^N \|u_j\|^2_{\widetilde{H}^s_{\rm I}(\Omega)}, \quad 0 \le s \le 1, \tag{A.86}$$

*and*

$$_*\|u\|^2_{\widetilde{H}^s_{\rm I}(\Omega)} \le \sum_{j=1}^N {}_*\|u_j\|^2_{\widetilde{H}^s_{\rm I}(\Omega)}, \quad -1 \le s \le 0. \tag{A.87}$$

(ii) *Assume that $u \in \widetilde{H}^s(\Omega)$ satisfies $u_j \in \widetilde{H}^s(\Omega_j)$ for $s \in [0,1]$ and $u_j \in \widetilde{H}^s_0(\Omega_j)$ for $s \in [-1,0)$; see* (A.85). *Then* (A.86) *and* (A.87) *also hold. Moreover, for $s \in [-1,0)$,*

$$\|u\|^2_{\widetilde{H}^s_{\rm I}(\Omega)} \lesssim \sum_{j=1}^N \|u_j\|^2_{\widetilde{H}^s_{\rm I}(\Omega)} \tag{A.88}$$

*where the constant depends on the size of $\Omega$, but not on the size of $\Omega_j$, $j = 1,\ldots,N$.*

*Proof.* Inequality (A.86) is proved in [224] following along the lines of the proof of Theorem A.10. We now prove both (A.86) and (A.87). Introduce the product space

$$\widetilde{\Pi}^s := \widetilde{H}^s_*(\Omega_1) \times \cdots \times \widetilde{H}^s_*(\Omega_N), \quad -1 \le s \le 1,$$

with norm defined from the interpolation norms by

$$\|\mathbf{u}\|^2_{\widetilde{\Pi}^s} := \sum_{j=1}^N \|u_j\|^2_{\widetilde{H}^s_*(\Omega_j)} = \begin{cases} \displaystyle\sum_{j=1}^N \|u_j\|^2_{\widetilde{H}^s_{\rm I}(\Omega)}, & 0 \le s \le 1, \\ \displaystyle\sum_{j=1}^N {}_*\|u_j\|^2_{\widetilde{H}^s_{\rm I}(\Omega)}, & -1 \le s < 0. \end{cases} \tag{A.89}$$

where $\mathbf{u} = (u_1,\ldots,u_N)$. Then

$$\widetilde{\Pi}^s = [\widetilde{\Pi}^0, \widetilde{\Pi}^1]_s.$$

Introduce also the space $X^s := \widetilde{H}^s(\Omega)$ equipped with the norm

$$\|u\|_{X^s} := \begin{cases} \|\cdot\|_{\widetilde{H}^s_{\rm I}(\Omega)}, & s \in [0,1], \\ {}_*\|\cdot\|_{\widetilde{H}^s_{\rm I}(\Omega)}, & s \in [-1,0), \end{cases} \tag{A.90}$$

and consider the sum operator $\mathscr{S} : \widetilde{\Pi}^s \to X^s$ defined by

$$\mathscr{S}\mathbf{u} := \sum_{j=1}^{N} u_j.$$

We recall that $X^s = [X^0, X^1]_s$ for $s \in (0,1)$ and $X^s = [X^0, X^{-1}]_{-s}$ for $s \in (-1,0)$; see (A.79). Since

$$\|\mathscr{S}\mathbf{u}\|_{X^0} = \|\mathscr{S}\mathbf{u}\|_{\widetilde{H}_\mathrm{I}^0(\Omega)} = \|\mathbf{u}\|_{\widetilde{\Pi}^0},$$
$$\|\mathscr{S}\mathbf{u}\|_{X^1} = \|\mathscr{S}\mathbf{u}\|_{\widetilde{H}_\mathrm{I}^1(\Omega)} = \|\mathbf{u}\|_{\widetilde{\Pi}^1},$$

by interpolation

$$\|\mathscr{S}\mathbf{u}\|_{X^s} \le \|\mathbf{u}\|_{\widetilde{\Pi}^s} \quad \text{for } 0 \le s \le 1.$$

On the other hand, if $v_j$ is the zero extension of $v|_{\Omega_j}$, then since

$$_*\|v\|^2_{H_\mathrm{I}^1(\Omega)} = \sum_{j=1}^{N} {_*\|v_j\|^2_{H_\mathrm{I}^1(\Omega)}}$$

we have from definition (A.48)

$$\|\mathscr{S}\mathbf{u}\|_{X^{-1}} = {_*\|\mathscr{S}\mathbf{u}\|_{\widetilde{H}_\mathrm{I}^{-1}(\Omega)}} = \sup_{v\in H^1(\Omega)} \frac{\langle \mathscr{S}\mathbf{u}, v\rangle_\Omega}{_*\|v\|_{H_\mathrm{I}^1(\Omega)}} = \sup_{v\in H^1(\Omega)} \frac{\sum_{j=1}^{N} \langle u_j, v_j\rangle_\Omega}{_*\|v\|_{H_\mathrm{I}^1(\Omega)}}$$
$$\le \sup_{v\in H^1(\Omega)} \frac{\|\mathbf{u}\|_{\widetilde{\Pi}^{-1}} \left(\sum_{j=1}^{N} {_*\|v_j\|^2_{H_\mathrm{I}^1(\Omega)}}\right)^{1/2}}{_*\|v\|_{H_\mathrm{I}^1(\Omega)}} = \|\mathbf{u}\|_{\widetilde{\Pi}^{-1}}.$$

By interpolation between $X^{-1}$ and $X^0$ we obtain

$$\|\mathscr{S}\mathbf{u}\|_{X^s} \le \|\mathbf{u}\|_{\widetilde{\Pi}^s} \quad \text{for } -1 \le s < 0.$$

Now for any function $u$ we define $\mathbf{u} = (u_1, \ldots, u_N)$. Then $u = \mathscr{S}\mathbf{u}$ and thus, for $s \in [-1,1]$,

$$\|u\|^2_{X^s} = \|\mathscr{S}\mathbf{u}\|^2_{X^s} \le \|\mathbf{u}\|^2_{\widetilde{\Pi}^s} = \sum_{j=1}^{N} \|u_j\|^2_{\widetilde{H}_*^s(\Omega_j)}.$$

This together with (A.89) and (A.90) proves Part (i).

For $j = 1, \ldots, N$, recall that $\widetilde{H}^s(\Omega_j) \subset \widetilde{H}_*^s(\Omega_j)$ for $s \in [0,1]$ and $\widetilde{H}_0^s(\Omega_j) \subset \widetilde{H}_*^s(\Omega_j)$ for $s \in [-1,0)$; see (A.78) and (A.84). Hence, the first statement in Part (ii) holds true. In particular when $s \in [-1,0)$, it follows from inequality (A.87) and a scaling argument that

$$\|u\|^2_{\widetilde{H}^s_{\mathrm{I}}(\Omega)} \simeq {}_*\|u\|^2_{\widetilde{H}^s_{\mathrm{I}}(\Omega)} \le \sum_{j=1}^{N} {}_*\|u_j\|^2_{\widetilde{H}^s_{\mathrm{I}}(\Omega)} \simeq \sum_{j=1}^{N} \|u_j\|^2_{\widetilde{H}^s_{\mathrm{I}}(\Omega)},$$

proving (A.88). The equivalence constants depend on the size of $\Omega$. □

### A.2.10 The special case $s = 1/2$

In this book, two frequently-used norms are the $H^{1/2}_{\mathrm{I}}(\Omega)$-norm and $\widetilde{H}^{1/2}_{\mathrm{I}}(\Omega)$-norm. We write their definitions here, see (A.1), for convenience

$$\begin{aligned} \|u\|_{H^{1/2}_{\mathrm{I}}(\Omega)} &= \left( \int_0^\infty \left| K(t,u,\mathscr{A}) \right|^2 \frac{\mathrm{d}t}{t^2} \right)^{1/2} \\ \|u\|_{\widetilde{H}^{1/2}_{\mathrm{I}}(\Omega)} &= \left( \int_0^\infty \left| K(t,u,\widetilde{\mathscr{A}}) \right|^2 \frac{\mathrm{d}t}{t^2} \right)^{1/2} \end{aligned} \tag{A.91}$$

where

$$K(t,u,\mathscr{A}) = \inf_{\substack{u_0 \in H^0,\ u_1 \in H^1 \\ u_0+u_1=u}} \left( \|u_0\|^2_{H^0_{\mathrm{I}}(\Omega)} + t^2 \|u_1\|^2_{H^1_{\mathrm{I}}(\Omega)} \right)^{1/2}$$

$$K(t,u,\widetilde{\mathscr{A}}) = \inf_{\substack{u_0 \in H^0,\ u_1 \in \widetilde{H}^1 \\ u_0+u_1=u}} \left( \|u_0\|^2_{\widetilde{H}^0_{\mathrm{I}}(\Omega)} + t^2 \|u_1\|^2_{\widetilde{H}^1_{\mathrm{I}}(\Omega)} \right)^{1/2}$$

with

$$\mathscr{A} = (H^0(\Omega), H^1(\Omega)) \quad \text{and} \quad \widetilde{\mathscr{A}} = (\widetilde{H}^0(\Omega), \widetilde{H}^1(\Omega)).$$

First we show how the interpolation norm $\|\cdot\|_{H^{1/2}_{\mathrm{I}}(\Omega)}$ (resp., $\|\cdot\|_{\widetilde{H}^{1/2}_{\mathrm{I}}(\Omega)}$) is related to the norms $\|\cdot\|_{L^2(\Omega)}$ and $\|\cdot\|_{H^1_{\mathrm{I}}(\Omega)}$ (resp., $\|\cdot\|_{L^2(\Omega)}$ and $\|\cdot\|_{\widetilde{H}^1_{\mathrm{I}}(\Omega)}$); cf. [138, Proposition 2.3].

**Lemma A.12.** *If $v \in H^1(\Omega)$ then*

$$\|v\|^2_{H^{1/2}_{\mathrm{I}}(\Omega)} \le 2\|v\|_{L_2(\Omega)} \|v\|_{H^1_{\mathrm{I}}(\Omega)}.$$

*Similarly, if $v \in \widetilde{H}^1_{\mathrm{I}}(\Omega) = H^1_0(\Omega)$ then*

$$\|v\|^2_{\widetilde{H}^{1/2}_{\mathrm{I}}(\Omega)} \le 2\|v\|_{L_2(\Omega)} \|v\|_{\widetilde{H}^1_{\mathrm{I}}(\Omega)}.$$

*Proof.* Suppose that $v \in H^1(\Omega)$. It is easy to see that the $K$-functional can be estimated as

$$K(t,v,\mathscr{A})^2 \le \min\left( \|v\|^2_{L_2(\Omega)}, t^2 \|v\|^2_{H^1_{\mathrm{I}}(\Omega)} \right).$$

Hence, with $\alpha = \|v\|_{L_2(\Omega)}/\|v\|_{H^1_{\mathrm{I}}(\Omega)}$, we deduce from (A.91)

$$\begin{aligned}\|v\|^2_{H^{1/2}_{\mathrm{I}}(\Omega)} &= \int_0^{\alpha} K(t,v,\mathscr{A})^2 \frac{dt}{t^2} + \int_{\alpha}^{\infty} K(t,v,\mathscr{A})^2 \frac{dt}{t^2} \\ &\le \int_0^{\alpha} t^2 \|v\|^2_{H^1_{\mathrm{I}}(\Omega)} \frac{dt}{t^2} + \int_{\alpha}^{\infty} \|v\|^2_{L_2(\Omega)} \frac{dt}{t^2} \\ &= 2\|v\|_{L_2(\Omega)}\|v\|_{H^1_{\mathrm{I}}(\Omega)}.\end{aligned}$$

The result for $\|v\|^2_{H^{1/2}_{\mathrm{I}}(\Omega)}$ can be proved similarly. □

The following lemma gives results on Sobolev norms of functions supported in a subset.

**Lemma A.13.** *Let $\Omega$ be a bounded Lipschitz domain in $\mathbb{R}^d$, $d = 1,2$, satisfying $\tau :=$ $\mathrm{diam}(\Omega) < 1$. Assume that $v \in \widetilde{H}^{1/2}(\Omega)$ satisfies $\mathrm{supp}(v) \subset \Omega' \subset \Omega$.*

(i) *The following relations between norms of v hold*

$$\|v\|_{\widetilde{H}^{1/2}_{\mathrm{I}}(\Omega)} \simeq \|v\|_{\widetilde{H}^{1/2}_{\mathrm{w}}(\Omega)} \lesssim \|v\|_{\widetilde{H}^{1/2}_{\mathrm{w}}(\Omega')} \simeq \|v\|_{\widetilde{H}^{1/2}_{\mathrm{I}}(\Omega')}$$

*where the constants are independent of v, $\tau$, and the size of $\Omega'$.*

(ii) *The opposite inequality is not true, i.e., there is no constant C independent of v, $\tau$, and the size of $\Omega'$ such that $\|v\|_{\widetilde{H}^{1/2}_{\mathrm{w}}(\Omega')} \le C\|v\|_{\widetilde{H}^{1/2}_{\mathrm{w}}(\Omega)}$.*

*Proof.* Lemma A.8 gives the equivalences

$$\|v\|_{\widetilde{H}^{1/2}_{\mathrm{I}}(\Omega)} \simeq \|v\|_{\widetilde{H}^{1/2}_{\mathrm{w}}(\Omega)} \quad \text{and} \quad \|v\|_{\widetilde{H}^{1/2}_{\mathrm{w}}(\Omega')} \simeq \|v\|_{\widetilde{H}^{1/2}_{\mathrm{I}}(\Omega')}.$$

It follows from Theorem A.10 that

$$\|v\|^2_{\widetilde{H}^{1/2}_{\mathrm{I}}(\Omega)} \le \|v\|^2_{\widetilde{H}^{1/2}_{\mathrm{I}}(\Omega')} + \|v\|^2_{\widetilde{H}^{1/2}_{\mathrm{I}}(\Omega\setminus\Omega')} = \|v\|^2_{\widetilde{H}^{1/2}_{\mathrm{I}}(\Omega')}.$$

Thus we also have

$$\|v\|_{\widetilde{H}^{1/2}_{\mathrm{w}}(\Omega)} \lesssim \|v\|_{\widetilde{H}^{1/2}_{\mathrm{w}}(\Omega')}.$$

We present here another proof of this inequality. Recall the definition of $\|v\|_{\widetilde{H}^{1/2}_{\mathrm{w}}(\Omega)}$:

$$\|v\|^2_{\widetilde{H}^{1/2}_{\mathrm{w}}(\Omega)} = |v|^2_{H^{1/2}(\Omega)} + \int_{\Omega} \frac{|v(\mathbf{x})|^2}{\mathrm{dist}(\mathbf{x},\partial\Omega)}\, d\mathbf{x}.$$

Clearly,

$$\int_{\Omega} \frac{|v(\mathbf{x})|^2}{\mathrm{dist}(\mathbf{x},\partial\Omega)}\, d\mathbf{x} = \int_{\Omega'} \frac{|v(\mathbf{x})|^2}{\mathrm{dist}(\mathbf{x},\partial\Omega)}\, d\mathbf{x} \le \int_{\Omega'} \frac{|v(\mathbf{x})|^2}{\mathrm{dist}(\mathbf{x},\partial\Omega')}\, d\mathbf{x}. \tag{A.92}$$

On the other hand, since $\mathrm{supp}(v) \subset \Omega'$

$$|v|^2_{H^{1/2}(\Omega)} = \iint\limits_{\Omega' \times \Omega'} \frac{|v(\mathbf{x}) - v(\mathbf{y})|^2}{|\mathbf{x}-\mathbf{y}|^{d+1}}\, d\mathbf{x}\, d\mathbf{y} + 2 \int\limits_{\Omega'} \Big( \int\limits_{\Omega \setminus \Omega'} \frac{d\mathbf{y}}{|\mathbf{x}-\mathbf{y}|^{d+1}} \Big) |v(\mathbf{x})|^2\, d\mathbf{x}. \quad \text{(A.93)}$$

Since

$$\int\limits_{\Omega \setminus \Omega'} \frac{d\mathbf{y}}{|\mathbf{x}-\mathbf{y}|^{d+1}} \lesssim \frac{1}{\mathrm{dist}(\mathbf{x}, \partial\Omega')}, \quad \text{(A.94)}$$

see Lemma A.14, the following holds

$$|v|^2_{H^{1/2}(\Omega)} \lesssim |v|^2_{H^{1/2}(\Omega')} + \int\limits_{\Omega'} \frac{|v(\mathbf{x})|^2}{\mathrm{dist}(\mathbf{x}, \partial\Omega')}\, d\mathbf{x} = \|v\|^2_{\widetilde{H}^{1/2}_{\mathrm{w}}(\Omega')}$$

so that, with the help of (A.92),

$$\|v\|^2_{\widetilde{H}^{1/2}_{\mathrm{w}}(\Omega)} \lesssim \|v\|^2_{\widetilde{H}^{1/2}_{\mathrm{w}}(\Omega')} + \int\limits_{\Omega'} \frac{|v(\mathbf{x})|^2}{\mathrm{dist}(\mathbf{x}, \Omega')}\, d\mathbf{x} \lesssim \|v\|^2_{\widetilde{H}^{1/2}_{\mathrm{w}}(\Omega')}.$$

This proves part (i).

Part (ii) is proved by the following counter-example, which is a modification of the counter-example in the Appendix of [2]. Consider $D$ to be the upper half of the unit disk in the $(x,y)$-plane, i.e.,

$$D = \big\{(r,\theta) : r \in [0,1],\ \theta \in [0,\pi]\big\}$$

where $(r,\theta)$ denote polar coordinates. Then define

$$\begin{aligned} \Omega &= \big\{(r,\theta) : r \in [0,1],\ \theta = 0 \text{ or } \theta = \pi\big\} = [-1,1] \times \{0\} \\ \Omega_\varepsilon &= \big\{(r,\theta) : r \in [\varepsilon, 3/4],\ \theta = 0\big\} = [\varepsilon, 3/4] \times \{0\}. \end{aligned}$$

For $\varepsilon \in (0,1/2)$, define $U_\varepsilon : D \longrightarrow \mathbb{R}$ and $U : D \longrightarrow \mathbb{R}$ by

$$U_\varepsilon(r,\theta) = \begin{cases} 0, & 0 \le r < \varepsilon, \\ (-\log r)^{-1/2} - (-\log\varepsilon)^{-1/2}, & \varepsilon \le r < 1/2, \\ (3-4r)\left((\log 2)^{-1/2} - (-\log\varepsilon)^{-1/2}\right), & 1/2 \le r < 3/4, \\ 0, & 3/4 \le r \le 1, \end{cases}$$

and

$$U(r,\theta) = \begin{cases} 0, & r = 0, \\ (-\log r)^{-1/2}, & 0 < r < 1/2, \\ (3-4r)(\log 2)^{-1/2}, & 1/2 \le r < 3/4, \\ 0, & 3/4 \le r \le 1, \end{cases}$$

These two functions are first studied in [2]. We now define $V_\varepsilon : D \longrightarrow \mathbb{R}$ by

$$V_\varepsilon(r,\theta) = \begin{cases} U_\varepsilon(r,\theta)\cos\theta, & 0 \le \theta < \pi/2, \\ 0, & \pi/2 \le \theta \le \pi. \end{cases}$$

Let $u_\varepsilon$ be the trace of $V_\varepsilon$ on the boundary of $D$. Then

$$\operatorname{supp}(u_\varepsilon) = [\varepsilon, 3/4] \times \{0\} = \Omega_\varepsilon.$$

For $(r,\theta) \in D$,

$$\begin{aligned} |V_\varepsilon(r,\theta)| &\le |U_\varepsilon(r,\theta)| &&\le |U(r,\theta)| \\ \left|\frac{\partial V_\varepsilon}{\partial r}(r,\theta)\right| &\le \left|\frac{\partial U_\varepsilon}{\partial r}(r,\theta)\right| &&\le \left|\frac{\partial U}{\partial r}(r,\theta)\right| \\ \left|\frac{\partial V_\varepsilon}{\partial \theta}(r,\theta)\right| &\le |U_\varepsilon(r,\theta)| &&\le |U(r,\theta)| \end{aligned} \tag{A.95}$$

We note that

$$\frac{\partial U}{\partial r}(r,\theta) = \begin{cases} \frac{1}{2}r^{-1}(-\log r)^{-3/2}, & 0 < r < 1/2, \\ -4(\log 2)^{-1/2}, & 1/2 < r < 3/4, \\ 0, & 3/4 < r < 1, \end{cases}$$

so that $U \in H^1(D)$. Indeed,

$$\iint_D |U(r,\theta)|^2 \,\mathrm{d}x\,\mathrm{d}y = \pi \int_0^{1/2} \frac{r}{-\log r}\,\mathrm{d}r + \pi \int_{1/2}^{3/4} \frac{1}{\log 2}(3-4r)^2 r\,\mathrm{d}r < \infty$$

and

$$\iint_D \left|\frac{\partial U}{\partial r}(r,\theta)\right|^2 \mathrm{d}x\,\mathrm{d}y = \frac{\pi}{4}\int_0^{1/2} \frac{dr}{r(-\log r)^3} + \pi \int_{1/2}^{3/4} \frac{16}{\log 2} r\,\mathrm{d}r < \infty.$$

It follows from (A.95) that $V_\varepsilon, U_\varepsilon \in H^1(D)$ and

$$\|V_\varepsilon\|_{H^1(D)} \lesssim \|U_\varepsilon\|_{H^1(D)} \le \|U\|_{H^1(D)}, \quad 0 < \varepsilon < 1.$$

Consequently, by the definition of the Slobodetski norm and the trace theorem

$$\|u_\varepsilon\|_{H_S^{1/2}(\Omega)} \le \|u_\varepsilon\|_{H_S^{1/2}(\partial D)} \simeq \|u_\varepsilon\|_{H^{1/2}(\partial D)} \lesssim \|V_\varepsilon\|_{H^1(D)} \lesssim 1.$$

Since $\operatorname{supp}(u_\varepsilon) = [\varepsilon, 3/4] \times \{0\}$, we have

$$\|u_\varepsilon\|^2_{\widetilde{H}^{1/2}_{\mathrm{w}}(\Omega)} = |u_\varepsilon|^2_{H^{1/2}(\Omega)} + \int_{-1}^{1} \frac{|u_\varepsilon(r,0)|^2}{\min\{1-r,1+r\}}\,\mathrm{d}r$$
$$= |u_\varepsilon|^2_{H^{1/2}(\Omega)} + \int_{\varepsilon}^{3/4} \frac{|u_\varepsilon(r,0)|^2}{\min\{1-r,1+r\}}\,\mathrm{d}r.$$

Due to

$$\|u_\varepsilon\|^2_{L^2(\Omega)} = \|u_\varepsilon\|^2_{L^2(\Omega_\varepsilon)} \simeq \int_{\varepsilon}^{3/4} \frac{|u_\varepsilon(r,0)|^2}{\min\{1-r,1+r\}}\,\mathrm{d}r,$$

we deduce

$$\|u_\varepsilon\|^2_{\widetilde{H}^{1/2}_{\mathrm{w}}(\Omega)} \simeq |u_\varepsilon|^2_{H^{1/2}(\Omega)} + \|u_\varepsilon\|^2_{L^2(\Omega)} = \|u_\varepsilon\|^2_{H^{1/2}_{\mathrm{S}}(\Omega)},$$

so that $\|u_\varepsilon\|^2_{\widetilde{H}^{1/2}_{\mathrm{w}}(\Omega)} \lesssim 1$. On the other hand, a simple calculation reveals that

$$\|u_\varepsilon\|^2_{\widetilde{H}^{1/2}_{\mathrm{w}}(\Omega_\varepsilon)} \ge \int_{\Omega_\varepsilon} \frac{|u_\varepsilon(r,0)|^2}{\mathrm{dist}(r,\partial\Omega_\varepsilon)}\,\mathrm{d}r > \int_{\varepsilon}^{1/2} \frac{|u_\varepsilon(r,0)|^2}{r}\,\mathrm{d}r$$
$$= \log|\log\varepsilon| + \frac{4(\log 2)^{1/2}}{(\log(1/\varepsilon))^{1/2}} - \frac{\log 2}{\log(1/\varepsilon)} - \log|\log 2| - 3.$$

Hence, $\|u_\varepsilon\|_{\widetilde{H}^{1/2}_{\mathrm{w}}(\Omega_\varepsilon)} \to \infty$ as $\varepsilon \to 0^+$, while $\|u_\varepsilon\|_{\widetilde{H}^{1/2}_{\mathrm{w}}(\Omega)}$ is bounded. This proves part (ii). □

We now prove the claim (A.94).

**Lemma A.14.** *Let $\Omega$ and $\Omega'$ be two open bounded domains in $\mathbb{R}^d$, $d = 1,2,3$, satisfying $\Omega' \subsetneq \Omega$, and let $\mathbf{x} \in \Omega'$.*

(i) *The following inequality holds*

$$\int_{\Omega\setminus\Omega'} \frac{d\mathbf{y}}{|\mathbf{x}-\mathbf{y}|^{d+1}} \lesssim \frac{1}{\mathrm{dist}(\mathbf{x},\partial\Omega')}.$$

(ii) *The opposite inequality is not true, i.e., there is no constant $C$ independent of $\mathbf{x}$ such that*

$$\frac{1}{\mathrm{dist}(\mathbf{x},\partial\Omega')} \le C \int_{\Omega\setminus\Omega'} \frac{d\mathbf{y}}{|\mathbf{x}-\mathbf{y}|^{d+1}}.$$

*Proof.* To prove part (i), we first observe that

$$\big(\Omega \setminus \Omega'\big) \subset \big(\mathbb{R}^n \setminus B_{\rho(\mathbf{x})}(\mathbf{x})\big), \quad \mathbf{x} \in \Omega',$$

where $\rho(\mathbf{x}) = \mathrm{dist}(\mathbf{x}, \partial\Omega')$ and $B_{\rho(\mathbf{x})}(\mathbf{x})$ is the ball centred at $\mathbf{x}$ having radius $\rho(\mathbf{x})$. By using spherical coordinates centred at $\mathbf{x}$ we obtain

$$\int_{\Omega\setminus\Omega'} \frac{d\mathbf{y}}{|\mathbf{x}-\mathbf{y}|^{n+1}} \le \int_{\mathbb{R}^n\setminus B_{\rho(\mathbf{x})}(\mathbf{x})} \frac{d\mathbf{y}}{|\mathbf{x}-\mathbf{y}|^{n+1}} = \omega_n \int_{\rho(\mathbf{x})}^{\infty} \frac{r^{n-1}\,dr}{r^{n+1}} = \frac{\omega_n}{\mathrm{dist}(\mathbf{x}, \partial\Omega')}.$$

To prove part (ii), we revisit the proof of part (i) of Lemma A.13. If the opposite of (A.94) holds, then

$$\int_{\Omega\setminus\Omega'} \frac{d\mathbf{y}}{|\mathbf{x}-\mathbf{y}|^{n+1}} \simeq \frac{1}{\mathrm{dist}(\mathbf{x}, \partial\Omega')}$$

with constants independent of $\mathbf{x}$ and $\Omega'$. Consider a function $u$ satisfying the assumptions in Lemma A.13. It follows from the definition of $\|u\|_{\widetilde{H}_{\mathrm{S}}^{1/2}(\Omega')}$ that

$$\|u\|^2_{\widetilde{H}_{\mathrm{w}}^{1/2}(\Omega')} \simeq |u|^2_{1/2,\Omega'} + \int_{\Omega'} \Big( \int_{\Omega\setminus\Omega'} \frac{d\mathbf{y}}{|\mathbf{x}-\mathbf{y}|^{n+1}} \Big) |u(\mathbf{x})|^2\, d\mathbf{x}.$$

This together with equation (A.93) yields

$$\|u\|^2_{\widetilde{H}_{\mathrm{w}}^{1/2}(\Omega')} \simeq |u|^2_{H^{1/2}(\Omega)} \le \|u\|^2_{\widetilde{H}_{\mathrm{w}}^{1/2}(\Omega)},$$

which contradicts part (ii) of Lemma A.13. □

The equivalence of the $H^s$-seminorm, $s = 1/2, 1$, and the quotient norm is useful for the analysis.

**Lemma A.15.**

(i) *Let $I$ be a bounded interval in $\mathbb{R}$. For any $v \in H^{1/2}(I)$, if we define*

$$\|v\|_{H_q^{1/2}(I)} := \inf_{\alpha\in\mathbb{R}} \|v+\alpha\|_{H_{\mathrm{w}}^{1/2}(I)},$$

*then*

$$|v|_{H^{1/2}(I)} \lesssim \|v\|_{H_q^{1/2}(I)} \lesssim |v|_{H^{1/2}(I)}.$$

(ii) *Let $R$ be a bounded domain in $\mathbb{R}^2$. For any $v \in H^1(R)$, if we define*

$$\|v\|_{H_q^1(R)} := \inf_{\alpha\in\mathbb{R}} {}_*\|v+\alpha\|_{H_{\mathrm{w}}^1(R)},$$

*then*

$$|v|_{H^1(R)} \lesssim \|v\|_{H_q^1(R)} \lesssim |v|_{H^1(R)}.$$

*The constants are independent of $v$ and the size of $I$ and $R$. Here the weighted norms $\|\cdot\|_{H_{\mathrm{w}}^{1/2}(I)}$ and ${}_*\|\cdot\|_{H_{\mathrm{w}}^1(R)}$ are defined in Subsection A.2.6.1.*

*Proof.* It suffices to prove Part (i). Part (ii) is proved in exactly the same manner. It follows from Lemma A.4 and Lemma A.5 that the seminorm $|\cdot|_{H^{1/2}(I)}$ and the quotient norm $\|\cdot\|_{H_q^{1/2}(I)}$ are invariant under scaling. Hence, to ensure that the constants are independent of the size of $I$, we prove the inequalities for $\widehat{I} = (0,1)$. The left inequality is obvious from the definition of the weighted norm $\|\cdot\|_{H_w^{1/2}(\widehat{I})}$. The right inequality can be obtained by using the same argument as in the proof of [60, Theorem 3.1.1]. □

We finish this subsection with a lemma on different forms of the $H^{1/2}$-norm.

**Lemma A.16.** *Let $I = (0,H)$ and $R = I \times I$ for some $H > 0$. Then for any $v \in H^{1/2}(R)$*

$$\int_I \|v(x,\cdot)\|^2_{H_w^{1/2}(I)}\,dx + \int_I \|v(\cdot,y)\|^2_{H_w^{1/2}(I)}\,dy \lesssim \|v\|^2_{H_w^{1/2}(R)}, \tag{A.96}$$

$$\begin{aligned} |v|^2_{H^{1/2}(R)} \le 4 \iint_R \frac{\|v(x_1,\cdot) - v(x_2,\cdot)\|^2_{L^2(I)}}{|x_1 - x_2|^2}\,dx_1\,dx_2 \\ + 4 \iint_R \frac{\|v(\cdot,y_1) - v(\cdot,y_2)\|^2_{L^2(I)}}{|y_1 - y_2|^2}\,dy_1\,dy_2, \end{aligned} \tag{A.97}$$

*and*

$$\begin{aligned} \|v\|^2_{H_w^{1/2}(R)} \simeq \|v\|^2_{L_w^2(R)} + \iint_R \frac{\|v(x_1,\cdot) - v(x_2,\cdot)\|^2_{L^2(I)}}{|x_1 - x_2|^2}\,dx_1\,dx_2 \\ + \iint_R \frac{\|v(\cdot,y_1) - v(\cdot,y_2)\|^2_{L^2(I)}}{|y_1 - y_2|^2}\,dy_1\,dy_2 \end{aligned} \tag{A.98}$$

*where*

$$\|v\|^2_{L_w^2(R)} := \frac{1}{H}\|v\|^2_{L^2(R)}.$$

*Moreover, if $v \in \widetilde{H}^{1/2}(R)$ then*

$$\|v\|^2_{\widetilde{H}_w^{1/2}(R)} \lesssim \int_I \|v(x,\cdot)\|^2_{\widetilde{H}_w^{1/2}(I)}\,dx + \int_I \|v(\cdot,y)\|^2_{\widetilde{H}_w^{1/2}(I)}\,dy \tag{A.99}$$

*and*

$$\int_I \|v(x,\cdot)\|^2_{\widetilde{H}_w^{1/2}(I)}\,dx + \int_I \|v(\cdot,y)\|^2_{\widetilde{H}_w^{1/2}(I)}\,dy \lesssim \|v\|^2_{L_w^2(R)} + \|v\|^2_{\widetilde{H}_w^{1/2}(R)}. \tag{A.100}$$

*The constants are independent of $H$ and $v$.*

*Proof.* Inequalities (A.98) is partially proved in [163, Lemma 5.3, page 95]; see also [163, Exercise 5.1, page 96]. We present the full proof here for completeness.

Let $\widehat{v}$ be defined by $\widehat{v}(\widehat{x},\widehat{y}) = v(H\widehat{x}, H\widehat{y})$ for all $(\widehat{x},\widehat{y}) \in \widehat{R} = (0,1)\times(0,1)$. It follows from Lemma A.5 (i) that

$$\|v(H\widehat{x},\cdot)\|^2_{H^{1/2}(I)} = \|\widehat{v}(\widehat{x},\cdot)\|^2_{H^{1/2}(\widehat{I})} \quad \text{and} \quad \|v\|^2_{H^{1/2}(R)} = H\|\widehat{v}\|^2_{H^{1/2}(\widehat{R})}.$$

Since

$$\int_I \|v(x,\cdot)\|^2_{H^{1/2}_{\mathrm{w}}(I)}\,\mathrm{d}x = H\int_{\widehat{I}} \|\widehat{v}(\widehat{x},\cdot)\|^2_{H^{1/2}_{\mathrm{w}}(\widehat{I})}\,d\widehat{x},$$

it suffices to prove (A.96) with $H = 1$. Similarly,

$$\|v\|^2_{L^2_{\mathrm{w}}(R)} = H\|\widehat{v}\|^2_{L^2_{\mathrm{w}}(\widehat{R})} = H\|\widehat{v}\|^2_{L^2(\widehat{R})},$$

$$\iint_R \frac{\|v(x_1,\cdot)-v(x_2,\cdot)\|^2_{L^2(I)}}{|x_1-x_2|^2}\,\mathrm{d}x_1\,\mathrm{d}x_2 = H\iint_{\widehat{R}} \frac{\|\widehat{v}(\widehat{x}_1,\cdot)-\widehat{v}(\widehat{x}_2,\cdot)\|^2_{L^2(\widehat{I})}}{|\widehat{x}_1-\widehat{x}_2|^2}\,d\widehat{x}_1\,d\widehat{x}_2,$$

$$\iint_R \frac{\|v(\cdot,y_1)-v(\cdot,y_2)\|^2_{L^2(I)}}{|y_1-y_2|^2}\,\mathrm{d}y_1\,\mathrm{d}y_2 = H\iint_{\widehat{R}} \frac{\|\widehat{v}(\cdot,\widehat{y}_1)-\widehat{v}(\cdot,\widehat{y}_2)\|^2_{L^2(\widehat{I})}}{|\widehat{y}_1-\widehat{y}_2|^2}\,d\widehat{y}_1\,d\widehat{y}_2.$$

Hence it suffices to prove (A.97) and (A.98) with $H = 1$.

Finally, Lemma A.5 (ii) gives $\|v\|^2_{\widetilde{H}^{1/2}_{\mathrm{w}}(R)} = H\|\widehat{v}\|^2_{\widetilde{H}^{1/2}_{\mathrm{w}}(\widehat{R})}$. On the other hand,

$$\int_I \|v(x,\cdot)\|^2_{\widetilde{H}^{1/2}_{\mathrm{w}}(I)}\,\mathrm{d}x = H\int_{\widehat{I}} \|\widehat{v}(\widehat{x},\cdot)\|^2_{\widetilde{H}^{1/2}_{\mathrm{w}}(\widehat{I})}\,d\widehat{x},$$

$$\int_I \|v(\cdot,y)\|^2_{\widetilde{H}^{1/2}_{\mathrm{w}}(I)}\,\mathrm{d}y = H\int_{\widehat{I}} \|\widehat{v}(\cdot,\widehat{y})\|^2_{\widetilde{H}^{1/2}_{\mathrm{w}}(\widehat{I})}\,d\widehat{y}.$$

Hence we will also prove (A.99) and (A.100) with $H = 1$.

In the remainder of this proof, $I = (0,1)$ and $R = I\times I$. We first prove (A.96). It follows from Lemma A.8 and the definition of interpolation norms that

$$\int_I \|v(x,\cdot)\|^2_{H^{1/2}_{\mathrm{w}}(I)}\,dx \simeq \int_I \|v(x,\cdot)\|^2_{H^{1/2}_{\mathrm{I}}(I)}\,dx$$

$$= \int_I \Big(\int_0^\infty |K(t,v(x,\cdot))|^2 \frac{\mathrm{d}t}{t^2}\Big)\,dx \qquad \text{(A.101)}$$

where, for almost all $x \in I$,

$$|K(t,v(x,\cdot))|^2 = \inf_{(w_1(x),w_2(x))\in\mathscr{X}(v,x)} \left(\|w_1(x,\cdot)\|^2_{L^2(I)} + t^2\|w_2(x,\cdot)\|^2_{H^1(I)}\right)$$

with

$$\mathscr{X}(v,x) := \left\{\left(w_1(x,\cdot), w_2(x,\cdot)\right) \in L^2(I)\times H^1(I) : v(x,\cdot) = w_1(x,\cdot) + w_2(x,\cdot)\right\}.$$

If we define

$$|K^*(t,v)|^2 = \inf_{(w_1,w_2)\in\mathscr{X}(v)} \left(\|w_1\|^2_{L^2(R)} + t^2\|w_2\|^2_{H^1(R)}\right)$$

with

$$\mathscr{X}(v) := \left\{(w_1,w_2) \in L^2(R)\times H^1(R) : v = w_1 + w_2\right\},$$

then $|K(t,v(x,\cdot))| \le |K^*(t,v)|$ for almost all $x \in I$. Hence (A.101) implies

$$\int_I \|v(x,\cdot)\|^2_{H^{1/2}_{\mathrm{w}}(I)}\,dx \le \int_0^1 \left(\int_0^\infty |K^*(t,v)|^2 \frac{\mathrm{d}t}{t^2}\right) dx = \|v\|^2_{H^{1/2}_{\mathrm{I}}(R)} \simeq \|v\|^2_{H^{1/2}_{\mathrm{w}}(R)}.$$

Similarly,

$$\int_I \|v(\cdot,y)\|^2_{H^{1/2}_{\mathrm{w}}(I)}\,dy \lesssim \|v\|^2_{H^{1/2}_{\mathrm{w}}(R)},$$

implying (A.96).

Next we prove (A.97). Recalling the definition of the $H^{1/2}$-seminorm and using the triangle and Cauchy–Schwarz inequalities we obtain, with $\mathbf{x}_1 = (x_1,y_1)$ and $\mathbf{x}_2 = (x_2,y_2)$,

$$|v|^2_{H^{1/2}(R)} = \iint_R \iint_R \frac{|v(\mathbf{x}_1) - v(\mathbf{x}_2)|^2}{|\mathbf{x}_1 - \mathbf{x}_2|^3}\,d\mathbf{x}_1\,d\mathbf{x}_2 \le 2(T_1 + T_2) \tag{A.102}$$

where

$$\begin{aligned} T_1 &= \iint_R \iint_R \frac{|v(x_1,y_1) - v(x_2,y_1)|^2}{|\mathbf{x}_1 - \mathbf{x}_2|^3}\,d\mathbf{x}_1\,d\mathbf{x}_2 \\ &= \iint_R \int_0^1 |v(x_1,y_1) - v(x_2,y_1)|^2 \left(\int_0^1 \frac{dy_2}{\left|(x_1-x_2)^2 + (y_1-y_2)^2\right|^{3/2}}\right) \mathrm{d}y_1\,\mathrm{d}x_1\,\mathrm{d}x_2 \end{aligned}$$

and

$$T_2 = \iint_R \iint_R \frac{|v(x_2,y_1) - v(x_2,y_2)|^2}{|\mathbf{x}_1 - \mathbf{x}_2|^3}\,d\mathbf{x}_1\,d\mathbf{x}_2$$

$$= \iint_R \int_0^1 |v(x_2,y_1) - v(x_2,y_2)|^2 \Big( \int_0^1 \frac{dx_1}{|(x_1-x_2)^2 + (y_1-y_2)^2|^{3/2}} \Big) \, dx_2 \, dy_1 \, dy_2.$$

Consider first $T_1$. Simple calculations reveal that

$$\int_0^1 \frac{dy_2}{|(x_1-x_2)^2 + (y_1-y_2)^2|^{3/2}} = \frac{1-y_1}{a^2\sqrt{a^2+(1-y_1)^2}} + \frac{y_1}{a^2\sqrt{a^2+y_1^2}}$$

where we shortened the notation by writing $a^2$ in lieu of $(x_1-x_2)^2$. Let

$$f(t) = \frac{t}{\sqrt{a^2+t^2}} \quad \text{and} \quad g(t) = f(t) + f(1-t), \quad t \in [0,1].$$

It can be easily shown that

$$\frac{1}{\sqrt{a^2+1}} = g(0) \leq g(t) \leq g(1/2) = \frac{1}{\sqrt{a^2+1/4}} \qquad \forall t \in [0,1].$$

Since $a^2 = (x_1-x_2)^2 \in [0,1]$ it follows that

$$1/\sqrt{2} \leq g(t) \leq 2 \quad \forall t \in [0,1], \; \forall a^2 \in [0,1].$$

Hence

$$\frac{1}{\sqrt{2}a^2} \leq \int_0^1 \frac{dy_2}{|(x_1-x_2)^2 + (y_1-y_2)^2|^{3/2}} = \frac{g(y_1)}{a^2} \leq \frac{2}{a^2}. \tag{A.103}$$

Consequently, recalling that $a^2 = (x_1-x_2)^2$,

$$T_1 \leq 2 \iint_R \frac{\|v(x_1,\cdot) - v(x_2,\cdot)\|^2_{L^2(I)}}{|x_1-x_2|^2} \, dx_1 \, dx_2.$$

Similarly,

$$\frac{1}{\sqrt{2}|y_1-y_2|^2} \leq \int_0^1 \frac{dx_1}{|(x_1-x_2)^2 + (y_1-y_2)^2|^{3/2}} \leq \frac{2}{|y_1-y_2|^2} \tag{A.104}$$

so that

$$T_2 \leq 2 \iint_R \frac{\|v(\cdot,y_1) - v(\cdot,y_2)\|^2_{L^2(I)}}{|y_1-y_2|^2} \, dy_1 \, dy_2.$$

It follows from (A.102) that

$$|v|^2_{H^{1/2}(R)} \le 4 \iint\limits_R \frac{\|v(x_1,\cdot) - v(x_2,\cdot)\|^2_{L^2(I)}}{|x_1 - x_2|^2}\, dx_1\, dx_2 + 4 \iint\limits_R \frac{\|v(\cdot,y_1) - v(\cdot,y_2)\|^2_{L^2(I)}}{|y_1 - y_2|^2}\, dy_1\, dy_2,$$

proving (A.97).

The $\lesssim$ inequality in (A.98) is a direct consequence of (A.97) and the definitions of the weighted norms. To prove the opposite inequality we use (A.96) to obtain

$$\begin{aligned}
&\iint\limits_R \frac{\|v(x_1,\cdot) - v(x_2,\cdot)\|^2_{L^2(I)}}{|x_1 - x_2|^2}\, dx_1\, dx_2 + \iint\limits_R \frac{\|v(\cdot,y_1) - v(\cdot,y_2)\|^2_{L^2(I)}}{|y_1 - y_2|^2}\, dy_1\, dy_2 \\
&= \int\limits_I |v(\cdot,y)|^2_{H^{1/2}(I)}\, dy + \int\limits_I |v(x,\cdot)|^2_{H^{1/2}(I)}\, dx \\
&\le \int\limits_I \|v(\cdot,y)\|^2_{H^{1/2}_{\mathrm{w}}(I)}\, dy + \int\limits_I \|v(x,\cdot)\|^2_{H^{1/2}_{\mathrm{w}}(I)}\, dx \lesssim \|v\|^2_{H^{1/2}_{\mathrm{w}}(R)}.
\end{aligned}$$

It is obvious that $\|v\|^2_{L^2_{\mathrm{w}}(R)} \le \|v\|^2_{H^{1/2}_{\mathrm{w}}(R)}$, yielding (A.98).

In order to prove (A.99) we note that the definition of the $\widetilde{H}^{1/2}$-norm and (A.97) yield

$$\begin{aligned}
\|v\|^2_{\widetilde{H}^{1/2}_{\mathrm{w}}(R)} \lesssim &\iint\limits_R \frac{\|v(x_1,\cdot) - v(x_2,\cdot)\|^2_{L^2(I)}}{|x_1 - x_2|^2}\, dx_1\, dx_2 \\
&+ \iint\limits_R \frac{\|v(\cdot,y_1) - v(\cdot,y_2)\|^2_{L^2(I)}}{|y_1 - y_2|^2}\, dy_1\, dy_2 \\
&+ \iint\limits_R \frac{|v(\mathbf{x})|^2}{\operatorname{dist}(\mathbf{x},\partial R)}\, d\mathbf{x}
\end{aligned}$$

where $\mathbf{x} = (x,y) \in R$. Since

$$\iint\limits_R \frac{\|v(x_1,\cdot) - v(x_2,\cdot)\|^2_{L^2(I)}}{|x_1 - x_2|^2}\, dx_1\, dx_2 = \int\limits_I |v(\cdot,y)|^2_{H^{1/2}(I)}\, dy,$$

$$\iint\limits_R \frac{\|v(\cdot,y_1) - v(\cdot,y_2)\|^2_{L^2(I)}}{|y_1 - y_2|^2}\, dy_1\, dy_2 = \int\limits_I |v(x,\cdot)|^2_{H^{1/2}(I)}\, dx,$$

it suffices to prove

$$\iint_R \frac{|v(\mathbf{x})|^2}{\text{dist}(\mathbf{x},\partial R)}\,d\mathbf{x} \simeq \int_I \Big(\int_I \frac{|v(x,y)|^2}{\text{dist}(x,\partial I)}\,\mathrm{d}x\Big)\,\mathrm{d}y + \int_I \Big(\int_I \frac{|v(x,y)|^2}{\text{dist}(y,\partial I)}\,\mathrm{d}y\Big)\,\mathrm{d}x. \tag{A.105}$$

We partition the square $R$ into four equal triangles by using the diagonals of $R$. Label them by $R_1$, $R_2$, $R_3$, and $R_4$, starting from the bottom triangle and moving counterclockwise; see Figure A.2.

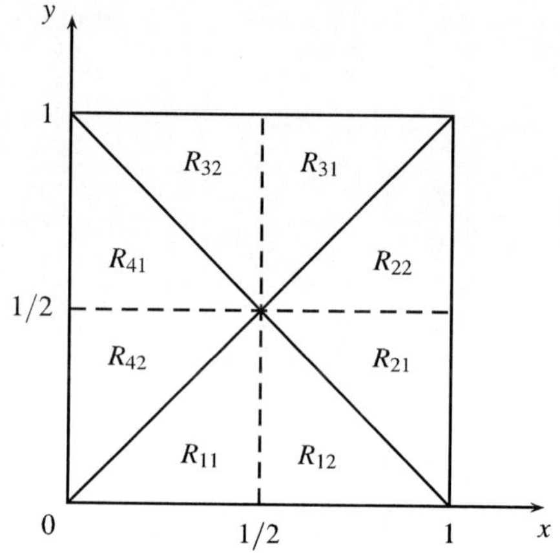


**Fig. A.2** $R = R_1 \cup \cdots \cup R_4$, $R_1 = R_{11} \cup R_{12}$, $R_2 = R_{21} \cup R_{22}$, $R_3 = R_{31} \cup R_{32}$, $R_4 = R_{41} \cup R_{42}$

Let

$$x_* := \min\{x, 1-x\}, \quad x^* := \max\{x, 1-x\},$$
$$y_* := \min\{y, 1-y\}, \quad y^* := \max\{y, 1-y\},$$

Then

$$\begin{aligned}\iint_R \frac{|v(\mathbf{x})|^2}{\text{dist}(\mathbf{x},\partial R)}\,d\mathbf{x} &= \iint_{R_1} \frac{|v(x,y)|^2}{y}\,\mathrm{d}x\,\mathrm{d}y + \iint_{R_2} \frac{|v(x,y)|^2}{1-x}\,\mathrm{d}x\,\mathrm{d}y \\ &\quad + \iint_{R_3} \frac{|v(x,y)|^2}{1-y}\,\mathrm{d}x\,\mathrm{d}y + \iint_{R_4} \frac{|v(x,y)|^2}{x}\,\mathrm{d}x\,\mathrm{d}y \\ &= \int_I \Big(\int_0^{x_*} \frac{|v(x,y)|^2}{y}\,\mathrm{d}y\Big)\,\mathrm{d}x + \int_I \Big(\int_{y^*}^1 \frac{|v(x,y)|^2}{1-x}\,\mathrm{d}x\Big)\,\mathrm{d}y\end{aligned}$$

$$+\int_I \Big(\int_{x^*}^1 \frac{|v(x,y)|^2}{1-y}\,\mathrm{d}y\Big)\,\mathrm{d}x + \int_I \Big(\int_0^{y_*} \frac{|v(x,y)|^2}{x}\,\mathrm{d}x\Big)\,\mathrm{d}y.$$

Noting that $0 \le x_*, y_* \le 1/2$ and $1/2 \le x^*, y^* \le 1$, we deduce

$$\iint_R \frac{|v(\mathbf{x})|^2}{\mathrm{dist}(\mathbf{x},\partial R)}\,d\mathbf{x} \le \int_I \Big(\int_0^{1/2} \frac{|v(x,y)|^2}{y}\,\mathrm{d}y\Big)\,\mathrm{d}x + \int_I \Big(\int_{1/2}^1 \frac{|v(x,y)|^2}{1-x}\,\mathrm{d}x\Big)\,\mathrm{d}y$$
$$+\int_I \Big(\int_{1/2}^1 \frac{|v(x,y)|^2}{1-y}\,\mathrm{d}y\Big)\,\mathrm{d}x + \int_I \Big(\int_0^{1/2} \frac{|v(x,y)|^2}{x}\,\mathrm{d}x\Big)\,\mathrm{d}y$$
$$= \int_I \Big(\int_I \frac{|v(x,y)|^2}{\mathrm{dist}(y,\partial I)}\,\mathrm{d}y\Big)\,\mathrm{d}x + \int_I \Big(\int_I \frac{|v(x,y)|^2}{\mathrm{dist}(x,\partial I)}\,\mathrm{d}x\Big)\,\mathrm{d}y.$$

It remains to prove the opposite inequality of (A.105). First consider the first term on the right-hand side of (A.105). We have by using Fubini's Theorem, recalling the definition of the triangles $R_1, \ldots, R_4$ in Figure A.2,

$$\int_I \Big(\int_I \frac{|v(x,y)|^2}{\mathrm{dist}(x,\partial I)}\,\mathrm{d}x\Big)\,\mathrm{d}y = \int_0^1 \Big(\int_0^{1/2} \frac{|v(x,y)|^2}{x}\,\mathrm{d}x + \int_{1/2}^1 \frac{|v(x,y)|^2}{1-x}\,\mathrm{d}x\Big)\,\mathrm{d}y$$
$$= \int_0^{1/2} \Big(\int_0^x \frac{|v(x,y)|^2}{x}\,\mathrm{d}y + \int_x^{1-x} \frac{|v(x,y)|^2}{x}\,\mathrm{d}y + \int_{1-x}^1 \frac{|v(x,y)|^2}{x}\,\mathrm{d}y\Big)\,\mathrm{d}x$$
$$+\int_{1/2}^1 \Big(\int_0^{1-x} \frac{|v(x,y)|^2}{1-x}\,\mathrm{d}y + \int_{1-x}^x \frac{|v(x,y)|^2}{1-x}\,\mathrm{d}y + \int_x^1 \frac{|v(x,y)|^2}{1-x}\,\mathrm{d}y\Big)\,\mathrm{d}x$$
$$\le \int_0^{1/2} \Big(\int_0^x \frac{|v(x,y)|^2}{y}\,\mathrm{d}y + \int_x^{1-x} \frac{|v(x,y)|^2}{x}\,\mathrm{d}y + \int_{1-x}^1 \frac{|v(x,y)|^2}{1-y}\,\mathrm{d}y\Big)\,\mathrm{d}x$$
$$+\int_{1/2}^1 \Big(\int_0^{1-x} \frac{|v(x,y)|^2}{y}\,\mathrm{d}y + \int_{1-x}^x \frac{|v(x,y)|^2}{1-x}\,\mathrm{d}y + \int_x^1 \frac{|v(x,y)|^2}{1-y}\,\mathrm{d}y\Big)\,\mathrm{d}x$$
$$= \iint_{R_{11}} \frac{|v(x,y)|^2}{y}\,\mathrm{d}x\,\mathrm{d}y + \iint_{R_4} \frac{|v(x,y)|^2}{x}\,\mathrm{d}x\,\mathrm{d}y + \iint_{R_{32}} \frac{|v(x,y)|^2}{1-y}\,\mathrm{d}x\,\mathrm{d}y$$
$$+\iint_{R_{12}} \frac{|v(x,y)|^2}{y}\,\mathrm{d}x\,\mathrm{d}y + \iint_{R_2} \frac{|v(x,y)|^2}{1-x}\,\mathrm{d}x\,\mathrm{d}y + \iint_{R_{31}} \frac{|v(x,y)|^2}{1-y}\,\mathrm{d}x\,\mathrm{d}y$$

$$= \sum_{i=1}^{4} \iint_{R_i} \frac{|v(\mathbf{x})|^2}{\operatorname{dist}(\mathbf{x}, \partial R)} \, d\mathbf{x} = \iint_{R} \frac{|v(\mathbf{x})|^2}{\operatorname{dist}(\mathbf{x}, \partial R)} \, d\mathbf{x}.$$

The same estimate can be derived for the second term on the right-hand side of (A.105), proving this equivalence. Inequality (A.100) is a direct consequence of (A.98) and (A.105). □

### *A.2.11 Sobolev spaces on curves and surfaces*

Since this book involves boundary integral equations, we mostly work with Sobolev spaces on curves and surfaces. It is worth stating the definition and properties of these spaces here.

Let $\Gamma$ be the boundary of a Lipschitz domain $\Omega$ in $\mathbb{R}^d$, $d = 2,3$. Then there is a surface measure $\sigma$ and the outward unit normal $\mathbf{n}$ exists $\sigma$-almost everywhere on $\Gamma$. For $s \in [0,1]$, the space $H^s(\Gamma)$ is defined via a partition of unity of subordinate to an atlas of local coordinate patches. Roughly speaking, as a point on $\Gamma$ can be represented (via local charts) by $(\mathbf{x}', \zeta(\mathbf{x}'))$ with $\mathbf{x}' \in \mathbb{R}^{d-1}$ where $\zeta : \mathbf{x}' \mapsto \zeta(\mathbf{x}') \in \mathbb{R}$ is a Lipschitz function, we can define

$$u_\zeta(\mathbf{x}') := u(\mathbf{x}', \zeta(\mathbf{x}')).$$

Then $u \in H^s(\Gamma)$ if $u \in L_2(\Gamma)$ and $u_\zeta \in H^s(\mathbb{R}^{d-1})$ (via local charts). A more precise definition can be found in [147].

The space $H^{-s}(\Gamma)$ is defined by duality:

$$\|u\|_{H^{-s}(\Gamma)} = \sup_{0 \neq v \in H^s(\Gamma)} \frac{|\langle u, v\rangle_\Gamma|}{\|v\|_{H^s(\Gamma)}}, \quad -1 \le s \le 1.$$

If $\Gamma$ is an open subset of a boundary $\widetilde{\Gamma}$ of a Lipschitz domain $\Omega$ (i.e., $\Omega$ is an open curve if $d = 2$ and open surface if $d = 3$) then we define

$$\begin{aligned} H^s(\Gamma) &:= \{U|_\Gamma : U \in H^s(\widetilde{\Gamma})\}, \\ \widetilde{H}^s(\Gamma) &:= \text{closure of } \mathscr{D}(\Gamma) \text{ in } H^s(\widetilde{\Gamma}), \\ H^s_0(\Gamma) &:= \text{closure of } \mathscr{D}(\Gamma) \text{ in } H^s(\Gamma). \end{aligned}$$

Here, $\mathscr{D}(\Gamma)$ is defined by

$$\mathscr{D}(\Gamma) := \{u \in \mathscr{D}(\widetilde{\Gamma}) : \operatorname{supp} u \subset \Gamma\}$$

while $\mathscr{D}(\widetilde{\Gamma})$ is defined by (A.3).

All the properties of Sobolev spaces on Lipschitz domains in $\mathbb{R}^d$ carry over to Sobolev spaces on $\Gamma$ when the order $s$ belongs to $[-1,1]$. In particular, in parallel to (A.18) we have

$$\|u\|_{\widetilde{H}^s(\Gamma)} \simeq \|u\|_{H^s(\widetilde{\Gamma})} \quad \forall u \in \widetilde{H}^s(\Gamma). \tag{A.106}$$

Finally, we state the following trace result proved by Costabel [63]

**Lemma A.17.** *Let $\Gamma$ be the boundary of a Lipschitz domain $\Omega$ in $\mathbb{R}^d$. Then the trace operator $\gamma : \mathscr{D}(\overline{\Omega}) \to \mathscr{D}(\Gamma)$ defined by $\gamma u := u|_\Gamma$ has a unique extension to a bounded operator $\gamma : H^s(\Omega) \to H^{s-1/2}(\Gamma)$ for $1/2 < s < 3/2$ which has a continuous right inverse.*

As a consequence of this lemma, we have in addition to (A.21)

$$H_0^s(\Omega) = \{u \in H^s(\Omega) : \gamma(u) = 0\}, \quad \frac{1}{2} < s \le 1.$$

We end this section by a result on Sobolev embedding.

**Lemma A.18.** *Let $\Gamma$ be a bounded, $(d-1)$-dimensional, open or closed manifold in $\mathbb{R}^d$ where $d \ge 2$.*

(i) *If $0 \le 2s < d-1$ and $q = 2(d-1)/(d-1-2s)$, then $q \in [2,\infty)$ and*

$$\|v\|_{L^q(\Gamma)} \lesssim \|v\|_{H^s(\Gamma)} \quad \forall v \in H^s(\Gamma).$$

(ii) *If $2s = d-1$, then for all $q \in [2,\infty)$,*

$$\|v\|_{L^q(\Gamma)} \lesssim \sqrt{q}\|v\|_{H^s(\Gamma)} \quad \forall v \in H^s(\Gamma),$$

*where the constant is independent of $q$.*

(iii) *If $0 \le 2s < d-1$ and $q = 2(d-1)/(d-1+2s)$, then $q \in (1,2]$ and*

$$\|v\|_{\widetilde{H}^{-s}(\Gamma)} \lesssim \|v\|_{L^q(\Gamma)} \quad \forall v \in L^q(\Gamma).$$

(iv) *If $2s = d-1$, then for all $q \in (1,2]$*

$$\|v\|_{\widetilde{H}^{-s}(\Gamma)} \lesssim \sqrt{\frac{q}{q-1}}\|v\|_{L^q(\Gamma)} \quad \forall v \in L^q(\Gamma),$$

*where the constant is independent of $q$.*

*Proof.* The result is proved in [5, Theorem 4.2]. We include the proof here for completeness. Part (i) follows immediately from the embedding $H^s(\Gamma) \hookrightarrow L^q(\Gamma)$ when $q = 2(d-1)/(d-1-2s)$; see e.g. [1].

To prove Part (ii), we first note that for any $q \ge 2$ the following estimate holds

$$\|v\|^2_{L^q(\mathbb{R}^{d-1})} \lesssim q\|v\|^2_{H^{(d-1)/2}(\mathbb{R}^{d-1})}$$

where the constant is independent of $q$. The analogous result holds for $\Gamma$ thanks to its construction in terms of local coordinate patches in $\mathbb{R}^{d-1}$.

Part (iii) (respectively, Part (iv)) is a result of Part (i) (respectively, Part (ii)) by using Hölder's inequality and duality properties (A.24). □

### A.2.12 A generalised antiderivative operator in Sobolev spaces

We need to create a mechanism to go back and forth between the $\widetilde{H}^{-1/2}$ norm and the $\widetilde{H}^{1/2}$ norm in order to carry over some results proved for the hypersingular integral equation to the weakly-singular integral equation. This can be done by introducing a generalised antiderivative operator, which was first introduced in [209] and further developed in [101]. We recall here the results on that operator.

**Lemma A.19.** *Let $\Gamma = (a,b)$ be an interval in $\mathbb{R}$. For all $0 \le s \le 1$ the following statements are valid*

(i) *The generalised differentiation operator $D_s : \widetilde{H}^s(\Gamma) \to \widetilde{H}^{s-1}(\Gamma)$ is bounded with norm independent of the measure of $\Gamma$.*
(ii) *There exists a bounded linear operator $I_s : \widetilde{H}_0^{s-1}(\Gamma) \to \widetilde{H}^s(\Gamma)$ satisfying*

$$D_s I_s \chi = \chi \text{ for any } \chi \in \widetilde{H}_0^{s-1}(\Gamma),$$

*where*

$$\widetilde{H}_0^{s-1}(\Gamma) := \{\chi \in \widetilde{H}^{s-1}(\Gamma) : \langle \chi, 1\rangle_{\widetilde{H}^{s-1}, H^{1-s}} = 0\}. \tag{A.107}$$

*Proof.* Part (i) is proved in [209, Lemma3.5] and Part (ii) in [101, Lemma 3]. The proof is included here for completeness. The idea is to define $D_0$, $D_1$, $I_0$, and $I_1$ by using the usual differentiation and integration operators, together with their adjoints, then use interpolation to define $D_s$ and $I_s$ for $s \in (0,1)$.

**Proof of** (i)**:** Let $D : H^1(\Gamma) \to H^0(\Gamma)$ be the weak derivative operator $D\chi = \chi'$. Then $D$ is bounded and $\|D\| = 1$. If $D^* : H^0(\Gamma) \to \widetilde{H}^{-1}(\Gamma)$ denotes the adjoint of $D$,

$$\langle D^*\chi, \varphi\rangle_{\widetilde{H}^{-1}, H^1} = \langle \chi, D\varphi\rangle_{H^0} \quad \forall \chi \in H^0(\Gamma),\ \varphi \in H^1(\Gamma),$$

then $D^*$ is bounded and $\|D^*\| = \|D\|$.

Similarly, let $I : H^0(\Gamma) \to H^1(\Gamma)$ be the integral operator defined for any $\chi \in H^0(\Gamma)$ by

$$I\chi(x) := \int_a^x \chi(y)\,\mathrm{d}y, \quad x \in \Gamma.$$

Since

$$\|I\chi\|_{H^0(\Gamma)}^2 = \int_a^b \left| \int_a^x \chi(y)\,\mathrm{d}y \right|^2 \mathrm{d}x \le \Big(\int_a^b (x-a)\,\mathrm{d}x\Big)\Big(\int_a^b |\chi(y)|^2\,\mathrm{d}y\Big)$$
$$= \frac{(b-a)^2}{2} \|\chi\|_{H^0(\Gamma)}^2,$$

we have

$$\|I\chi\|_{H^1(\Gamma)}^2 = \|I\chi\|_{H^0(\Gamma)}^2 + \|\chi\|_{H^0(\Gamma)}^2 \le \frac{(b-a)^2+2}{2} \|\chi\|_{H^0(\Gamma)}^2.$$

The adjoint $I^* : \widetilde{H}^{-1}(\Gamma) \to H^0(\Gamma)$ of $I$ defined by

$$\langle I^*\chi, \varphi\rangle_{H^0} = \langle \chi, I\varphi\rangle_{\widetilde{H}^{-1},H^1} \quad \forall \chi \in \widetilde{H}^{-1}(\Gamma),\ \varphi \in H^0(\Gamma)$$

is also bounded with $\|I^*\| = \|I\|$.

We now prove the first statement of the lemma. Consider the case when $s = 1$. The operator $D_1 : \widetilde{H}^1(\Gamma) \to \widetilde{H}^0(\Gamma)$ is defined to be $D_1 := D|_{\widetilde{H}^1(\Gamma)}$. Clearly, $D_1$ is bounded and $\|D_1\| = 1$.

$$\|D_1\chi\|_{\widetilde{H}^0_I(\Gamma)} = \|\chi\|_{\widetilde{H}^1_I(\Gamma)}, \tag{A.108}$$

implying $\|D_1\| = 1$. Now for $s = 0$ we define $D_0 : \widetilde{H}^0(\Gamma) \to \widetilde{H}^{-1}(\Gamma)$ by $D_0 := -D^*$. Then $\|D_0\| = \|D^*\| = 1$. Moreover, if $\chi, \varphi \in \widetilde{H}^1(\Gamma)$ then by using integration by parts we deduce

$$\begin{aligned}\langle \chi, D_0\varphi\rangle_{H^1,\widetilde{H}^{-1}} &= -\langle D_1\chi, \varphi\rangle_{\widetilde{H}^0} = -\int_\Gamma \chi'(x)\varphi(x)\,dx = \int_\Gamma \chi(x)\varphi'(x)\,dx \\ &= \langle \chi, D_1\varphi\rangle_{H^1,\widetilde{H}^{-1}}.\end{aligned}$$

This implies $D_0\varphi = D_1\varphi$ for all $\varphi \in \widetilde{H}^1(\Gamma)$, i.e., $D_1$ is the restriction of $D_0$ onto $\widetilde{H}^1(\Gamma)$. By interpolation, see Theorem A.4, for each $0 < s < 1$ there exists uniquely $D_s : \widetilde{H}^s(\Gamma) \to \widetilde{H}^{s-1}(\Gamma)$ with $\|D_s\chi\|_{\widetilde{H}^{s-1}_I(\Gamma)} \le \|\chi\|_{\widetilde{H}^s_I(\Gamma)}$, implying $\|D_s\| \le 1$. Moreover, $D_1\varphi = D_s\varphi$ for all $\varphi \in \widetilde{H}^1(\Gamma)$.

**Proof of** (ii)**:** To prove the second part of the lemma, we start again with $s = 1$ and define $I_1 : \widetilde{H}^0_0(\Gamma) \to \widetilde{H}^1(\Gamma)$ by $I_1 := I|_{\widetilde{H}^0_0(\Gamma)}$. Since

$$I_1\chi(a) = 0 \quad \text{and} \quad I_1\chi(b) = \langle \chi, 1\rangle_{H^0} = 0,$$

$I_1$ maps $\widetilde{H}^0_0(\Gamma)$ into $\widetilde{H}^1(\Gamma)$. By the definition of the $\widetilde{H}^s_I$-norm, we have

$$\|I_1\chi\|_{\widetilde{H}^1_I(\Gamma)} = |I_1\chi|_{H^1(\Gamma)} = \|\chi\|_{\widetilde{H}^0_I(\Gamma)}.$$

Hence, $I_1$ is bounded as an operator from $\widetilde{H}^0_0(\Gamma)$ into $\widetilde{H}^1(\Gamma)$ with $\|I_1\| = 1$.

For $s = 0$, we define $I_0 : \widetilde{H}^{-1}_0(\Gamma) \to \widetilde{H}^0(\Gamma)$ by $I_0 := I^*|_{\widetilde{H}^{-1}_0(\Gamma)}$. Then $\|I_0\| \le 1$. By considering the functional $\chi \mapsto \langle \chi, 1\rangle$ we can show that $\widetilde{H}^{s-1}_0(\Gamma)$ is the interpolation space between $\widetilde{H}^{-1}_0(\Gamma)$ and $\widetilde{H}^0_0(\Gamma)$. Hence, there exists $I_s : \widetilde{H}^{s-1}_0(\Gamma) \to \widetilde{H}^s(\Gamma)$ with $\|I_s\| \le 1$ such that $I_s\chi = I_1\chi$ for all $\chi \in \widetilde{H}^0_0(\Gamma)$.

Finally, we prove that $D_sI_s\chi = \chi$. First we note that $D_1I_1\chi = \chi$ for all $\chi \in \widetilde{H}^0_0(\Gamma)$ (because $DI\chi = \chi$ for all $\chi \in H^0(\Gamma)$). Next we show that $D_0I_0\chi = \chi$ for all $\chi \in \widetilde{H}^{-1}_0(\Gamma)$. By the definition of $D_0$ and $I_0$ we have, noting that $\widetilde{H}^{-1}_0 \subset \widetilde{H}^{-1} \subset H^{-1}$,

$$\langle D_0I_0\chi, \varphi\rangle_{H^{-1},\widetilde{H}^1} = \langle \chi, I_1D_1\varphi\rangle_{H^{-1},\widetilde{H}^1} \quad \forall \varphi \in \widetilde{H}^1(\Gamma).$$

Hence it suffices to prove $I_1 D_1 \varphi = \varphi$ for all $\varphi \in \widetilde{H}^1(\Gamma)$. Note that $I_1 D_1 \varphi$ is well defined for $\varphi \in \widetilde{H}^1(\Gamma)$ because

$$\langle D_1 \varphi, 1 \rangle_{H^0} = \varphi(b) - \varphi(a) = 0$$

so that $D_1 \varphi \in \widetilde{H}^0_0(\Gamma)$. Let $\psi = I_1 D_1 \varphi$. Then $D_1 \psi = D_1 \varphi$ and thus $\psi = \varphi$ because both functions vanish at the endpoints $a$ and $b$. By interpolation, $D_s I_s \chi = \chi$ for all $\chi \in \widetilde{H}^{s-1}_0(\Gamma)$. □

*Remark A.4.* 1. We note that for any $s \in [0,1)$, $I_s$ is the extension of $I_1$ onto $\widetilde{H}^{s-1}_0(\Gamma)$, i.e., $I_s v = I_1 v$ for any $v \in L^2_0(\Gamma)$.
2. Recall that $I_1 : \widetilde{H}^0_0(\Gamma) \to \widetilde{H}^1(\Gamma)$ is the usual integral operator. Therefore, if $v$ is a piecewise-constant function on $\Gamma$, then $I_1 v$ is piecewise linear on the same mesh.

# Appendix B
# Boundary Integral Operators

This chapter collects some important facts about boundary integral operators which have been used in the book. In the first section, Section B.1, we define the four operators that form the Calderón projector, namely, the weakly-singular integral operator, the hypersingular integral operator, the double-layer potential, and its adjoint. We show how they can be used to solve the Laplace equation on bounded and unbounded domains with the Dirichlet and Neumann boundary conditions. We also derive these operators as pseudo-differential operators when the surface is a sphere. In Lemma B.3 we present the coercivity and Gårding's inequalities of the bilinear forms defined from the weakly-singular integral operator and the hypersingular integral operator. This section will be useful for the reader who is not familiar with the integral equation method for the solution of boundary value problems.

In the second section of this appendix, Section B.2, we discuss general discretised operators and their matrix representations. The results will be applied to the boundary integral operators introduced in Section B.1.

## B.1 Boundary Integral Operators and Pseudo-differential Operators on the Sphere

In this section we first define the boundary potentials and boundary integral operators on general boundaries. We then relate these operators with pseudo-differential operators when the boundary is a sphere. These operators on the sphere can be represented in the general form

$$\mathscr{L}v := \sum_{\ell=0}^{\infty} \sum_{m=-\ell}^{\ell} \widehat{L}(\ell)\widehat{v}_{\ell,m}Y_{\ell,m};$$

see Section 15.1.2. We will determine the symbol $\widehat{L}(\ell)$ for each operator. More details can be found in the book [161].



E. P. Stephan and T. Tran, *Schwarz Methods and Multilevel Preconditioners for Boundary Element Methods*, https://doi.org/10.1007/978-3-030-79283-1

### B.1.1 Boundary potentials and boundary integral operators

Let $\Gamma$ be the boundary of a bounded domain $\Omega$ in $\mathbb{R}^d$, $d = 2,3$. We also denote $\Omega^- := \Omega$ and $\Omega^+ := \mathbb{R}^d \setminus \Omega$. Let $n$ denote the exterior normal to $\Gamma$.

The fundamental solution of the equation $(\Delta + \kappa^2)u = 0$ is given by

$$G(x,y) = \begin{cases} -\frac{1}{2\pi}\log|x-y|, & d=2,\ \kappa=0, \\ \frac{i}{4}H_0^{(1)}(\kappa|x-y|), & d=2,\ \kappa\neq 0, \\ \frac{1}{4\pi}\frac{e^{i\kappa|x-y|}}{|x-y|}, & d=3, \end{cases} \tag{B.1}$$

The single-layer potential $\mathfrak{S}$ and the double-layer potential $\mathfrak{D}$ are defined by

$$\mathfrak{S}v(x) := 2\int_\Gamma G(x,y)v(y)\,d\sigma_y, \qquad x\notin\Gamma,$$

$$\mathfrak{D}v(x) := 2\int_\Gamma v(y)\frac{\partial G(x,y)}{\partial n_y}\,d\sigma_y, \quad x\notin\Gamma,$$

where $n_y$ denotes the exterior normal to $\Gamma$ at $y\in\Gamma$, and $d\sigma_y$ denotes the arc-length element ($d=2$) or surface element ($d=3$).

Associated with these potentials, we define the following boundary integral operators

$$\begin{aligned} \mathfrak{V}v(x) &= 2\int_\Gamma G(x,y)v(y)\,d\sigma_y, && x\in\Gamma, \\ \mathfrak{K}v(x) &= 2\int_\Gamma v(y)\frac{\partial G(x,y)}{\partial n_y}\,d\sigma_y, && x\in\Gamma, \\ \mathfrak{K}'v(x) &= 2\frac{\partial}{\partial n_x}\int_\Gamma G(x,y)v(y)\,d\sigma_y, && x\in\Gamma, \\ \mathfrak{W}v(x) &= -2\frac{\partial}{\partial n_x}\int_\Gamma v(y)\frac{\partial G(x,y)}{\partial n_y}\,d\sigma_y, && x\in\Gamma. \end{aligned} \tag{B.2}$$

It is well-known that if $\Gamma$ is smooth then, for all $s\in\mathbb{R}$, the mappings

$$\begin{aligned} &\mathfrak{V}: H^{s-1/2}(\Gamma)\to H^{s+1/2}(\Gamma), && \mathfrak{W}: H^{s+1/2}(\Gamma)\to H^{s-1/2}(\Gamma), \\ &\mathfrak{K}: H^{s+1/2}(\Gamma)\to H^{s+1/2}(\Gamma), && \mathfrak{K}': H^{s-1/2}(\Gamma)\to H^{s-1/2}(\Gamma) \end{aligned}$$

are bounded operators. The operator $\mathfrak{V}$ is a weakly-singular integral operator and $\mathfrak{W}$ is a hypersingular integral operator. If $\Gamma$ is Lipschitz, then the above operators are bounded when $s\in(-1/2,1/2)$; see [63].

It is remarked that our definition (B.2) differs from relevant definitions in some other books and articles, e.g., [128], by a factor of 2. This is to avoid a factor of $1/2$ in the following well-known jump relations:

$$\begin{aligned} (\mathfrak{S}v)|_\Gamma &= \mathfrak{V}v, & \partial_n^{\pm}(\mathfrak{S}v) &= (\mp\mathfrak{I} + \mathfrak{K}')v, \\ (\mathfrak{D}v)|_\Gamma^{\pm} &= (\pm\mathfrak{I} + \mathfrak{K})v, & \partial_n^{\pm}(\mathfrak{D}v) &= -\mathfrak{W}v, \end{aligned} \tag{B.3}$$

where $+$ denotes the limit taken from the exterior domain, and $-$ the limit taken from the interior domain, and $\mathfrak{I}$ denotes the identity operator.

## B.1.2 Representation of harmonic functions by potentials

If $U \in H^1_{\text{loc}}(\mathbb{R}^3 \setminus \Gamma)$ satisfies $\Delta U = 0$ in $\mathbb{R}^3 \setminus \Gamma$ and $U(x) = O(1/\|x\|)$ as $\|x\| \to \infty$, then for $x \notin \Gamma$ the following representation formula holds

$$U(x) = \frac{1}{2}\mathfrak{S}([\partial_n U])(x) - \frac{1}{2}\mathfrak{D}([U])(x), \tag{B.4}$$

where $[v] := v^- - v^+$.

### B.1.2.1 Interior problems

It follows from (B.4) that

$$U(x) = \frac{1}{2}\mathfrak{S}(\partial_n^- U)(x) - \frac{1}{2}\mathfrak{D}(U)(x), \quad x \in \Omega. \tag{B.5}$$

**Dirichlet problem**

Let the Dirichlet data $U_D$ be given. By taking the limit on both sides of (B.5) as $x$ approaches a point on the boundary $\Gamma$ and using the jump relations (B.3) we obtain

$$U_D(x) = \frac{1}{2}\mathfrak{V}(\partial_n^- U)(x) + \frac{1}{2}U_D(x) - \frac{1}{2}\mathfrak{K}U_D(x), \quad x \in \Gamma,$$

implying

$$\mathfrak{V}(\partial_n^- U)(x) = U_D(x) + \mathfrak{K}U_D(x), \quad x \in \Gamma. \tag{B.6}$$

**Neumann problem**

Let the Neumann data $U_N$ be given. By taking the normal derivative $\partial_n^-$ on both sides of (B.5) and using the jump relations (B.3), we obtain

$$\partial_n^- U(x) = \frac{1}{2}\partial_n^- U(x) + \frac{1}{2}\mathfrak{K}'\partial_n^- U(x) + \frac{1}{2}\mathfrak{W}U(x), \quad x \in \Gamma.$$

implying

$$\mathfrak{W}U(x) = U_N(x) - \mathfrak{K}'U_N(x), \quad x \in \Gamma. \tag{B.7}$$

#### B.1.2.2 Exterior problems

It follows from (B.4) that

$$U(x) = \frac{1}{2}\mathfrak{D}(U)(x) - \frac{1}{2}\mathfrak{S}(\partial_n^+ U)(x), \quad x \in \mathbb{R}^3 \setminus \overline{\Omega}. \tag{B.8}$$

**Dirichlet problem**

By taking the limit on both sides of (B.8) as $x$ approaches a point on the boundary $\Gamma$ and using the jump relations (B.3) we obtain

$$U_D(x) = \frac{1}{2}U_D(x) + \frac{1}{2}\mathfrak{K}U_D(x) - \frac{1}{2}\mathfrak{V}(\partial_n^+ U)(x), \quad x \in \Gamma,$$

implying

$$\mathfrak{V}(\partial_n^+ U)(x) = -U_D(x) + \mathfrak{K}U_D(x), \quad x \in \Gamma. \tag{B.9}$$

**Neumann problem**

By taking the normal derivative $\partial_n^+$ on both sides of (B.8) as $x$ approaches a point on the boundary $\Gamma$ and using the jump relations (B.3) we obtain

$$\partial_n^+ U(x) = -\frac{1}{2}\mathfrak{W}U(x) + \frac{1}{2}\partial_n^+ U(x) - \frac{1}{2}\mathfrak{K}'(\partial_n^+ U), \quad x \in \Gamma,$$

implying

$$\mathfrak{W}U(x) = -U_N(x) - \mathfrak{K}'U_N(x), \quad x \in \Gamma, \tag{B.10}$$

### *B.1.3 Dirichlet-to-Neumann and Neumann-to-Dirichlet operators*

Given $u$ defined on $\Gamma$, we define the interior and exterior Dirichlet-to-Neumann (DtN) operators by

$$\mathfrak{T}_{\text{int}} : u \mapsto U_{\text{int}} \mapsto \partial_n^- U_{\text{int}}$$
$$\mathfrak{T}_{\text{ext}} : u \mapsto U_{\text{ext}} \mapsto \partial_n^+ U_{\text{ext}},$$

where $U_{\text{int}}$ (respectively, $U_{\text{ext}}$) is the solution to the interior (respectively, exterior) Dirichlet problem with $U_D = u$.

On the one hand, (B.6) gives

$$\partial_n^- U_{\text{int}} = \mathfrak{V}^{-1}(\mathfrak{I} + \mathfrak{K})u,$$

so that

$$\mathfrak{T}_{\text{int}} u = \mathfrak{V}^{-1}(\mathfrak{I} + \mathfrak{K})u.$$

On the other hand, it follows from (B.5) and (B.6) that

$$U_{\text{int}} = \frac{1}{2}\mathfrak{S}\mathfrak{V}^{-1}(\mathfrak{I}+\mathfrak{K})u - \frac{1}{2}\mathfrak{D}u,$$

so that

$$\mathfrak{T}_{\text{int}}u = \partial_n^- U_{\text{int}} = \frac{1}{2}(\mathfrak{I}+\mathfrak{K}')\mathfrak{V}^{-1}(\mathfrak{I}+\mathfrak{K})u + \frac{1}{2}\mathfrak{W}u.$$

Hence

$$\mathfrak{T}_{\text{int}} = \mathfrak{V}^{-1}(\mathfrak{I}+\mathfrak{K}) = \frac{1}{2}(\mathfrak{I}+\mathfrak{K}')\mathfrak{V}^{-1}(\mathfrak{I}+\mathfrak{K}) + \frac{1}{2}\mathfrak{W}. \tag{B.11}$$

Similarly, on the one hand, (B.9) gives

$$\partial_n^+ U_{\text{ext}} = \mathfrak{V}^{-1}(-\mathfrak{I}+\mathfrak{K})u,$$

so that

$$\mathfrak{T}_{\text{ext}}u = \mathfrak{V}^{-1}(-\mathfrak{I}+\mathfrak{K})u.$$

On the other hand, it follows from (B.8) and (B.9) that

$$U_{\text{ext}} = \frac{1}{2}\mathfrak{D}u + \frac{1}{2}\mathfrak{S}\mathfrak{V}^{-1}(\mathfrak{I}-\mathfrak{K})u,$$

so that

$$\mathfrak{T}_{\text{ext}}u = \partial_n^+ U_{\text{ext}} = -\frac{1}{2}\mathfrak{W}u - \frac{1}{2}(\mathfrak{I}-\mathfrak{K}')\mathfrak{V}^{-1}(\mathfrak{I}-\mathfrak{K})u.$$

Hence

$$\mathfrak{T}_{\text{ext}} = -\mathfrak{V}^{-1}(\mathfrak{I}-\mathfrak{K}) = -\frac{1}{2}\mathfrak{W} - \frac{1}{2}(\mathfrak{I}-\mathfrak{K}')\mathfrak{V}^{-1}(\mathfrak{I}-\mathfrak{K}). \tag{B.12}$$

It is noted that even though the boundary integral operators $\mathfrak{V}$, $\mathfrak{W}$, $\mathfrak{K}$, and $\mathfrak{K}'$ defined by (B.2) are twice the corresponding operators defined in some other books and articles, the DtN operators $\mathfrak{T}_{\text{int}}$ and $\mathfrak{T}_{\text{ext}}$ remain the same.

The inverses $\mathfrak{T}_{\text{int}}^{-1}$ and $\mathfrak{T}_{\text{ext}}^{-1}$ are the Neumann-to-Dirichlet (NtD) operators. The DtN operator is also called the Steklov–Poincaré operator. We summarise in the next two lemmas some of its important properties. We focus on $\mathfrak{T}_{\text{int}}$ only.

**Lemma B.1.** *Let $\Omega$ be a bounded domain with Lipschitz boundary $\Gamma$. For any $v_\Gamma \in H^{1/2}(\Gamma)$, there exists uniquely $v_D \in H^1(\Omega)$ satisfying*

$$\begin{aligned} \Delta v_D &= 0 && \text{in } \Omega, \\ v_D &= v_\Gamma && \text{on } \Gamma, \\ \frac{\partial v_D}{\partial n} &= \mathfrak{T}_{\text{int}}(v_\Gamma) && \text{on } \Gamma. \end{aligned} \tag{B.13}$$

*Moreover, if $v \in H^1(\Omega)$ is any extension of $v_\Gamma$ (i.e., $v|_\Gamma = v_\Gamma$), then*

$$\|\nabla v_D\|_{L^2(\Omega)}^2 \le \|\nabla v\|_{L^2(\Omega)}^2. \tag{B.14}$$

*Proof.* The unique existence of $v_D \in H^1(\Omega)$ satisfying (B.13) is clear, noting the definition of the DtN operator $\mathfrak{T}_{\text{int}}$. To prove (B.14), let $v \in H^1(\Omega)$ be such that its trace on $\Gamma$ equals $v_\Gamma$. Then $v_0 := v - v_D \in H_0^1(\Omega)$. It follows from Green's identity and the first equation in (B.13) that

$$\langle \nabla v_D, \nabla v_0 \rangle_\Omega = -\langle \Delta v_D, v_0 \rangle_\Omega = 0.$$

Therefore,

$$\|\nabla v\|_{L^2(\Omega)}^2 = \|\nabla v_D\|_{L^2(\Omega)}^2 + \|\nabla v_0\|_{L^2(\Omega)}^2 \geq \|\nabla v_D\|_{L^2(\Omega)}^2.$$

This completes the proof of the lemma. □

Before proving the next lemma, we recall the definition of the $\|\cdot\|_{\mathfrak{V}^{-1}}$-norm:

$$\|v\|_{\mathfrak{V}^{-1}} := \sqrt{\langle \mathfrak{V}^{-1} v, v \rangle_\Gamma} \quad \forall v \in H^{1/2}(\Gamma).$$

**Lemma B.2.** *There exist a constant $c_{\mathfrak{K}} \in (1/2, 1)$ such that for any $v \in H^{1/2}(\Gamma)$*

$$\frac{1}{2c_{\mathfrak{K}}} \|(\mathfrak{I} + \mathfrak{K})v\|_{\mathfrak{V}^{-1}}^2 \leq \langle \mathfrak{T}_{\text{int}} v, v \rangle_\Gamma \leq \frac{1}{2(1 - c_{\mathfrak{K}})} \|(\mathfrak{I} + \mathfrak{K})v\|_{\mathfrak{V}^{-1}}^2 \tag{B.15}$$

*Proof.* The result is proved in [164, Lemma 2.1]; see also [200, Proposition 5.2]. We include the proof here for the reader's convenience. It is noted that the factor 2 in (B.16) and (B.17) below is due to the factor 2 in our definition (B.2). One should also note that the $\|\cdot\|_{\mathfrak{V}^{-1}}$-norm differs by a factor of 1/2.

Theorem 5.6.11 in [128] and Theorem 5.1 in [200] give

$$\begin{aligned} \|(\mathfrak{I} + \mathfrak{K})v\|_{\mathfrak{V}^{-1}} &\leq 2c_{\mathfrak{K}} \|v\|_{\mathfrak{V}^{-1}} && \forall v \in H^{1/2}(\Gamma), \\ \|(\mathfrak{I} + \mathfrak{K})v\|_{\mathfrak{V}^{-1}} &\geq 2(1 - c_{\mathfrak{K}}) \|v\|_{\mathfrak{V}^{-1}} && \forall v \in H_0^{1/2}(\Gamma), \end{aligned} \tag{B.16}$$

where

$$c_{\mathfrak{K}} = \frac{1}{2} + \sqrt{\frac{1}{4} - c_{\mathfrak{V}} c_{\mathfrak{W}}} < 1.$$

Here, $c_{\mathfrak{V}} \in (0, 1/2]$ and $c_{\mathfrak{W}} \in (0, 1/2]$ are given in [128, Lemma 5.6.10] which satisfy

$$\begin{aligned} \langle \mathfrak{V} v, v \rangle_\Gamma &\geq 2c_{\mathfrak{V}} \|v\|_{H^{-1/2}(\Gamma)}^2 && \forall v \in H^{-1/2}(\Gamma), \\ \langle \mathfrak{W} w, w \rangle_\Gamma &\geq 2c_{\mathfrak{W}} \|w\|_{H^{1/2}(\Gamma)}^2 && \forall w \in H_0^{1/2}(\Gamma), \end{aligned} \tag{B.17}$$

with

$$H_0^{1/2}(\Gamma) := \{ w \in H^{1/2}(\Gamma) : \langle w, 1 \rangle_\Gamma = 0 \}.$$

For any $v \in H^{-1/2}(\Gamma)$, $v \neq 0$, we have

$$\|\mathfrak{V}v\|_{H^{1/2}(\Gamma)} = \sup_{\substack{u\in H^{-1/2}(\Gamma)\\ u\neq 0}} \frac{\langle \mathfrak{V}v, u\rangle_\Gamma}{\|u\|_{H^{-1/2}(\Gamma)}} \geq \frac{\langle \mathfrak{V}v, v\rangle_\Gamma}{\|v\|_{H^{-1/2}(\Gamma)}} \geq 2c_{\mathfrak{V}}\|v\|_{H^{-1/2}(\Gamma)}.$$

Hence, for any $w \in H^{1/2}(\Gamma)$ and $u \in H^{1/2}(\Gamma)$, by letting $v = \mathfrak{V}^{-1}u$ or $u = \mathfrak{V}v$ we deduce

$$\begin{aligned}
\|\mathfrak{V}^{-1}w\|_{H^{-1/2}(\Gamma)} &= \sup_{\substack{u\in H^{1/2}(\Gamma)\\ u\neq 0}} \frac{\left\langle \mathfrak{V}^{-1}w, u\right\rangle_\Gamma}{\|u\|_{H^{1/2}(\Gamma)}} = \sup_{\substack{\mathfrak{V}v\in H^{1/2}(\Gamma)\\ \mathfrak{V}v\neq 0}} \frac{\left\langle \mathfrak{V}^{-1}w, \mathfrak{V}v\right\rangle_\Gamma}{\|\mathfrak{V}v\|_{H^{1/2}(\Gamma)}}\\
&= \sup_{\substack{\mathfrak{V}v\in H^{1/2}(\Gamma)\\ \mathfrak{V}v\neq 0}} \frac{\langle w, v\rangle_\Gamma}{\|\mathfrak{V}v\|_{H^{1/2}(\Gamma)}} \leq \frac{1}{2c_{\mathfrak{V}}} \sup_{\substack{\mathfrak{V}v\in H^{1/2}(\Gamma)\\ \mathfrak{V}v\neq 0}} \frac{\langle w, v\rangle_\Gamma}{\|v\|_{H^{-1/2}(\Gamma)}}\\
&\leq \frac{1}{2c_{\mathfrak{V}}}\|w\|_{H^{1/2}(\Gamma)}.
\end{aligned}$$

Therefore,

$$\left\langle \mathfrak{V}^{-1}w, w\right\rangle_\Gamma \leq \|\mathfrak{V}^{-1}w\|_{H^{-1/2}(\Gamma)}\|w\|_{H^{1/2}(\Gamma)} \leq \frac{1}{2c_{\mathfrak{V}}}\|w\|^2_{H^{1/2}(\Gamma)}.$$

Noting that $c_{\mathfrak{V}}c_{\mathfrak{W}} = c_{\mathfrak{K}}(1-c_{\mathfrak{K}})$, we obtain by using successively (B.17), the above inequality, and (B.16) that

$$\begin{aligned}
\langle \mathfrak{W}w, w\rangle_\Gamma &\geq 4c_{\mathfrak{V}}c_{\mathfrak{W}}\left\langle \mathfrak{V}^{-1}w, w\right\rangle_\Gamma = 4c_{\mathfrak{K}}(1-c_{\mathfrak{K}})\|w\|^2_{\mathfrak{V}^{-1}}\\
&\geq \frac{1-c_{\mathfrak{K}}}{c_{\mathfrak{K}}}\|(\mathfrak{I}+\mathfrak{K})w\|^2_{\mathfrak{V}^{-1}}.
\end{aligned}$$

Hence, the left inequality in (B.15) is now easily seen by using the symmetric form in (B.11)

$$\langle \mathfrak{T}_{\text{int}}w, w\rangle_\Gamma = \frac{1}{2}\|(\mathfrak{I}+\mathfrak{K})w\|^2_{\mathfrak{V}^{-1}} + \frac{1}{2}\langle \mathfrak{W}w, w\rangle_\Gamma \geq \frac{1}{2c_{\mathfrak{K}}}\|(\mathfrak{I}+\mathfrak{K})w\|^2_{\mathfrak{V}^{-1}}.$$

The right inequality in (B.15) is derived from the non-symmetric representation of $\mathfrak{T}_{\text{int}}$ in (B.11) and (B.16)

$$\begin{aligned}
\langle \mathfrak{T}_{\text{int}}w, w\rangle_\Gamma &= \left\langle \mathfrak{V}^{-1}(\mathfrak{I}+\mathfrak{K})v, v\right\rangle_\Gamma \leq \|(\mathfrak{I}+\mathfrak{K})v\|_{\mathfrak{V}^{-1}}\|v\|_{\mathfrak{V}^{-1}}\\
&\leq \frac{1}{2(1-c_{\mathfrak{K}})}\|(\mathfrak{I}+\mathfrak{K})v\|^2_{\mathfrak{V}^{-1}}.
\end{aligned}$$

□

### *B.1.4 The weakly-singular and hypersingular bilinear forms*

The weakly-singular and hypersingular operators define the following bilinear forms

$$a_{\mathfrak{V}}(\cdot,\cdot) : \widetilde{H}^{-1/2}(\Gamma) \times \widetilde{H}^{-1/2}(\Gamma) \to \mathbb{R}, \quad a_{\mathfrak{V}}(v,w) := \langle \mathfrak{V}v, w\rangle \ \forall v,w \in \widetilde{H}^{-1/2}(\Gamma),$$
$$a_{\mathfrak{W}}(\cdot,\cdot) : \widetilde{H}^{1/2}(\Gamma) \times \widetilde{H}^{1/2}(\Gamma) \to \mathbb{R}, \quad a_{\mathfrak{W}}(v,w) := \langle \mathfrak{W}v, w\rangle \ \forall v,w \in \widetilde{H}^{1/2}(\Gamma).$$

These bilinear forms have the following properties.

**Lemma B.3.** *Recalling* (B.1), *when* $\kappa = 0$, *for any* $v \in \widetilde{H}^{1/2}(\Gamma)$ *and* $w \in \widetilde{H}_0^{-1/2}(\Gamma)$, *the following statements hold*

$$a_{\mathfrak{W}}(v,v) \simeq \|v\|^2_{\widetilde{H}_{\mathrm{I}}^{1/2}(\Gamma)},$$
$$a_{\mathfrak{V}}(w,w) \simeq \|w\|^2_{\widetilde{H}_{\mathrm{I}}^{-1/2}(\Gamma)} \simeq {}_*\|w\|^2_{\widetilde{H}_{\mathrm{I}}^{-1/2}(\Gamma)}.$$

*The constants are independent of* $v$ *and* $w$ *but may depend on the size of* $\Gamma$. *Here* $\widetilde{H}_0^{-1/2}(\Gamma)$ *defined in* (A.85) *is the subspace of* $\widetilde{H}^{-1/2}(\Gamma)$ *consisting of all functions of integral mean zero.*

*When* $\kappa \neq 0$, *we have*

$$a_{\mathfrak{V}}(\varphi,\varphi) \geq \gamma_{\mathfrak{V}}\|\varphi\|^2_{\widetilde{H}^{-1/2}(\Gamma)} - \eta_{\mathfrak{V}}\|\varphi\|^2_{\widetilde{H}^{-1}(\Gamma)} \quad \forall \varphi \in \widetilde{H}^{-1/2}(\Gamma),$$
$$a_{\mathfrak{W}}(\psi,\psi) \geq \gamma_{\mathfrak{W}}\|\psi\|^2_{\widetilde{H}^{1/2}(\Gamma)} - \eta_{\mathfrak{W}}\|\varphi\|^2_{L^2(\Gamma)} \quad \forall \psi \in \widetilde{H}^{1/2}(\Gamma),$$

*where* $\gamma_{\mathfrak{V}}$, $\gamma_{\mathfrak{W}}$, $\eta_{\mathfrak{V}}$, *and* $\eta_{\mathfrak{W}}$ *are constants independent of* $\varphi$ *and* $\psi$.

*Proof.* The equivalence between the bilinear forms and the non-weighted norms is given in [186] for $d = 3$. It is originally proved for $\mathfrak{V}$ and $d = 2$ by Hsiao and Wendland [127]; see also [147]. The equivalence between the non-weighted norms and weighted norms is obvious from their definitions, see Subsection A.2.5 and Subsection A.2.6, noting that the equivalence constants may depend on the size of $\Gamma$ which is fixed.

The result when $\kappa \neq 0$ can be found in [92, 202, 235]. □

## B.1.5 Representations of solutions to the Laplace equation by spherical harmonics

In the remainder of this section, $\Gamma$ is the unit sphere $\mathbb{S}$ in $\mathbb{R}^3$. Let $(r,\theta,\varphi)$ be the spherical coordinates of a point $\mathbf{x} = (x_1,x_2,x_3) \in \mathbb{R}^3$, where $r \geq 0$ is the radius and where $\theta \in [0,\pi]$, $\varphi \in [0,2\pi]$ are the two Euler angles

$$\begin{cases} x_1 = r\sin\theta\cos\varphi, \\ x_2 = r\sin\theta\sin\varphi, \\ x_3 = r\cos\theta. \end{cases}$$

The vectors

$$\mathbf{e}_r = \frac{\partial \mathbf{x}}{\partial r} = (\sin\theta\cos\varphi, \sin\theta\sin\varphi, \cos\theta),$$
$$\mathbf{e}_\theta = \frac{1}{r}\frac{\partial \mathbf{x}}{\partial \theta} = (\cos\theta\cos\varphi, \cos\theta\sin\varphi, -\sin\theta),$$
$$\mathbf{e}_\varphi = \frac{1}{r\sin\theta}\frac{\partial \mathbf{x}}{\partial \varphi} = (-\sin\varphi, \cos\varphi, 0)$$

are unitary, and they form an orthonormal basis for $\mathbb{R}^3$.

In these coordinates, the gradient operator and the Laplace operator have the form

$$\begin{aligned} \nabla &= \frac{\partial}{\partial r}\mathbf{e}_r + \frac{1}{r}\nabla_{\mathbb{S}}, \\ \Delta &= \frac{1}{r^2}\frac{\partial}{\partial r}\left(r^2\frac{\partial}{\partial r}\right) + \frac{1}{r^2}\Delta_{\mathbb{S}} \end{aligned} \tag{B.18}$$

where $\nabla_{\mathbb{S}}$ and $\Delta_{\mathbb{S}}$ are, respectively, the surface gradient and the Laplace–Beltrami operator on the unit sphere $\mathbb{S}$ defined by

$$\begin{aligned} \nabla_{\mathbb{S}} &:= \frac{1}{\sin\theta}\frac{\partial}{\partial\varphi}\mathbf{e}_\varphi + \frac{\partial}{\partial\theta}\mathbf{e}_\theta, \\ \Delta_{\mathbb{S}} &:= \frac{1}{\sin^2\theta}\frac{\partial^2}{\partial\varphi^2} + \frac{1}{\sin\theta}\frac{\partial}{\partial\theta}\left(\sin\theta\frac{\partial}{\partial\theta}\right). \end{aligned} \tag{B.19}$$

Define

$$H_{\ell,m}(r,\theta,\varphi) := r^\ell Y_{\ell,m}(\theta,\varphi) \quad \text{and} \quad K_{\ell,m}(r,\theta,\varphi) := \frac{1}{r^{\ell+1}}Y_{\ell,m}(\theta,\varphi)$$

where $Y_{\ell,m}$, $m = -\ell,\dots,\ell$ and $\ell = 0,1,\dots$, are spherical harmonics which form an orthonormal basis for $L_2(\mathbb{S})$; see e.g. [155]. Theorem 2.4.1 in [161] gives

$$\Delta_{\mathbb{S}}Y_{\ell,m} + \ell(\ell+1)Y_{\ell,m} = 0.$$

Hence by using (B.18) and (B.19) it is easy to check that $H_{\ell,m}$ is harmonic, smooth at the origin, and tends to $\infty$ as $r \to \infty$, whereas $K_{\ell,m}$ is harmonic, not smooth at the origin, and tends to 0 as $r \to \infty$; see e.g. [161]. Moreover, by using (B.19) one can show that, see [161, (2.5.36) & (2.5.37)],

$$\begin{aligned} \frac{\partial H_{\ell,m}}{\partial n} &= \ell Y_{\ell,m}, \\ \frac{\partial K_{\ell,m}}{\partial n} &= -(\ell+1)Y_{\ell,m}, \end{aligned} \tag{B.20}$$

For any $v \in \mathscr{D}'(\mathbb{S})$, the Fourier coefficients of $v$ are defined by

$$\widehat{v}_{\ell,m} = \int_{\mathbb{S}} v(\mathbf{x})Y_{\ell,m}(\mathbf{x})\,d\sigma_{\mathbf{x}}, \quad m = -\ell,\dots,\ell, \quad \ell = 0,1,\dots.$$

Here $d\sigma_{\mathbf{x}}$ is the element of surface area.

In the following we derive the representations of the solutions of the Dirichlet and Neumann problems in terms of the Fourier coefficients of the given data; see Subsection B.1.2.1 and Subsection B.1.2.2.

### B.1.5.1 Dirichlet problems

If the Dirichlet data $U_D$ has an expansion as a sum of spherical harmonics

$$U_D(\theta,\varphi) = \sum_{\ell=0}^{\infty} \sum_{m=-\ell}^{\ell} \widehat{(U_D)}_{\ell,m} Y_{\ell,m}(\theta,\varphi),$$

then the solution of the interior problem can be represented by

$$\begin{aligned} U(r,\theta,\varphi) &= \sum_{\ell=0}^{\infty} \sum_{m=-\ell}^{\ell} \widehat{(U_D)}_{\ell,m} H_{\ell,m}(r,\theta,\varphi) \\ &= \sum_{\ell=0}^{\infty} \sum_{m=-\ell}^{\ell} r^{\ell} \widehat{(U_D)}_{\ell,m} Y_{\ell,m}(\theta,\varphi), \end{aligned} \tag{B.21}$$

where the series is absolutely convergent for $r < 1$. The solution of the exterior problem can be represented by

$$\begin{aligned} U(r,\theta,\varphi) &= \sum_{\ell=0}^{\infty} \sum_{m=-\ell}^{\ell} \widehat{(U_D)}_{\ell,m} K_{\ell,m}(r,\theta,\varphi) \\ &= \sum_{\ell=0}^{\infty} \sum_{m=-\ell}^{\ell} \frac{1}{r^{\ell+1}} \widehat{(U_D)}_{\ell,m} Y_{\ell,m}(\theta,\varphi), \end{aligned} \tag{B.22}$$

where the series is absolutely convergent for $r > 1$.

### B.1.5.2 Neumann problems

If the Neumann data $U_N$ has an expansion as a sum of spherical harmonics

$$U_N(\theta,\varphi) = \sum_{\ell=0}^{\infty} \sum_{m=-\ell}^{\ell} \widehat{(U_N)}_{\ell,m} Y_{\ell,m}(\theta,\varphi),$$

then by using (B.20) one can show that for the interior problem

$$U(r,\theta,\varphi) = \sum_{\ell=0}^{\infty} \sum_{m=-\ell}^{\ell} \frac{1}{\ell} \widehat{(U_N)}_{\ell,m} H_{\ell,m}(r,\theta,\varphi)$$

$$= \sum_{\ell=0}^{\infty} \sum_{m=-\ell}^{\ell} \frac{r^{\ell}}{\ell} \widehat{(U_N)}_{\ell,m} Y_{\ell,m}(\theta,\varphi), \tag{B.23}$$

and for the exterior problem

$$\begin{aligned} U(r,\theta,\varphi) &= \sum_{\ell=0}^{\infty} \sum_{m=-\ell}^{\ell} -\frac{1}{\ell+1} \widehat{(U_N)}_{\ell,m} K_{\ell,m}(r,\theta,\varphi) \\ &= \sum_{\ell=0}^{\infty} \sum_{m=-\ell}^{\ell} -\frac{1}{(\ell+1)r^{\ell+1}} \widehat{(U_N)}_{\ell,m} Y_{\ell,m}(\theta,\varphi). \end{aligned} \tag{B.24}$$

## *B.1.6 Representations of boundary integral operators by spherical harmonics*

When $\Gamma = \mathbb{S}$, the unit sphere in $\mathbb{R}^3$, we can represent the operators $\mathfrak{T}_{\text{int}}$, $\mathfrak{T}_{\text{ext}}$, $\mathfrak{T}_{\text{int}}^{-1}$, $\mathfrak{T}_{\text{ext}}^{-1}$, $\mathfrak{V}$, $\mathfrak{K}$, $\mathfrak{K}'$, and $\mathfrak{W}$ in terms of spherical harmonics.

### B.1.6.1 DtN and NtD operators

By using (B.21) and (B.23) for the interior problems, (B.22) and (B.24) for the exterior problems, for any given function

$$u(\theta,\varphi) = \sum_{\ell=0}^{\infty} \sum_{m=-\ell}^{\ell} \hat{u}_{\ell,m} Y_{\ell,m}(\theta,\varphi),$$

we can easily derive that

$$\begin{aligned} \mathfrak{T}_{\text{int}} u &= \sum_{\ell=0}^{\infty} \sum_{m=-\ell}^{\ell} \ell \hat{u}_{\ell,m} Y_{\ell,m} \\ \mathfrak{T}_{\text{int}}^{-1} u &= \sum_{\ell=0}^{\infty} \sum_{m=-\ell}^{\ell} \frac{1}{\ell} \hat{u}_{\ell,m} Y_{\ell,m} \\ \mathfrak{T}_{\text{ext}} u &= -\sum_{\ell=0}^{\infty} \sum_{m=-\ell}^{\ell} (\ell+1) \hat{u}_{\ell,m} Y_{\ell,m} \\ \mathfrak{T}_{\text{ext}}^{-1} u &= -\sum_{\ell=0}^{\infty} \sum_{m=-\ell}^{\ell} \frac{1}{\ell+1} \hat{u}_{\ell,m} Y_{\ell,m}. \end{aligned} \tag{B.25}$$

### B.1.6.2 Weakly-singular integral operator

From the non-symmetric form in (B.11) and in (B.12) we deduce

$$\mathfrak{V}^{-1} = \frac{1}{2}\Big(\mathfrak{T}_{\text{int}} - \mathfrak{T}_{\text{ext}}\Big).$$

Therefore, by using (B.25) we have

$$\begin{aligned} \mathfrak{V}^{-1}u &= \sum_{\ell=0}^{\infty}\sum_{m=-\ell}^{\ell} (\ell+1/2)\,\hat{u}_{\ell,m}Y_{\ell,m}, \\ \mathfrak{V}u &= \sum_{\ell=0}^{\infty}\sum_{m=-\ell}^{\ell} \frac{1}{\ell+1/2}\hat{u}_{\ell,m}Y_{\ell,m}. \end{aligned} \tag{B.26}$$

### B.1.6.3 Hypersingular integral operator

From (B.7) and (B.10) we deduce

$$\begin{aligned} \mathfrak{T}_{\text{int}}^{-1}U_N &= U_{\text{int}} = \mathfrak{W}^{-1}(U_N) - \mathfrak{W}^{-1}\mathfrak{K}'(U_N) \\ \mathfrak{T}_{\text{ext}}^{-1}U_N &= U_{\text{ext}} = -\mathfrak{W}^{-1}(U_N) - \mathfrak{W}^{-1}\mathfrak{K}'(U_N), \end{aligned}$$

so that

$$\mathfrak{W}^{-1}(U_N) = \frac{1}{2}\Big(\mathfrak{T}_{\text{int}}^{-1}U_N - \mathfrak{T}_{\text{ext}}^{-1}U_N\Big).$$

Therefore,

$$\begin{aligned} \mathfrak{W}^{-1}u &= \sum_{\ell=0}^{\infty}\sum_{m=-\ell}^{\ell} \frac{\ell+1/2}{\ell(\ell+1)}\hat{u}_{\ell,m}Y_{\ell,m} \\ \mathfrak{W}u &= \sum_{\ell=0}^{\infty}\sum_{m=-\ell}^{\ell} \frac{\ell(\ell+1)}{\ell+1/2}\hat{u}_{\ell,m}Y_{\ell,m}. \end{aligned} \tag{B.27}$$

### B.1.6.4 The operator $\mathfrak{K}$ and its adjoint $\mathfrak{K}'$

It follows from (B.6) and (B.26)

$$\mathfrak{K}u = \mathfrak{V}\mathfrak{T}_{\text{int}}u - u = \sum_{\ell=0}^{\infty}\sum_{m=-\ell}^{\ell}\Big(\frac{\ell}{\ell+1/2} - 1\Big)\hat{u}_{\ell,m}Y_{\ell,m} = -\frac{1}{2}\mathfrak{V}u.$$

Note that we can use (B.9) and (B.26) to obtain the same formula.

Similarly, from (B.7) and (B.27) we deduce

$$\mathfrak{K}'u = -\mathfrak{W}\mathfrak{T}_{\text{int}}^{-1}u + u = \sum_{\ell=0}^{\infty}\sum_{m=-\ell}^{\ell}\Big(-\frac{\ell+1}{\ell+1/2} + 1\Big)\hat{u}_{\ell,m}Y_{\ell,m} = -\frac{1}{2}\mathfrak{V}u.$$

The same formula can be derived from (B.10) and (B.27).

## B.2 Discretised Operators

Let $\mathscr{X}$ and $\mathscr{Y}$ be two Banach spaces. Let $\mathscr{X}_M$ and $\mathscr{Y}_N$ be two finite-dimensional subspaces of $\mathscr{X}$ and $\mathscr{Y}$, respectively. Let $\dim(\mathscr{X}_M) = M$ and $\dim(\mathscr{Y}_N) = N$.

### *B.2.1 Natural embedding operators and biorthonormal bases*

The natural embedding operators are denoted by

$$i_{\mathscr{X}_M} : \mathscr{X}_M \to \mathscr{X}, \quad i_{\mathscr{X}_M} u = u, \quad \text{and} \quad i_{\mathscr{Y}_N} : \mathscr{Y}_N \to \mathscr{Y}, \quad i_{\mathscr{Y}_N} v = v.$$

Let $\mathscr{X}^*$, $\mathscr{Y}^*$, $\mathscr{X}_M^*$, and $\mathscr{Y}_N^*$ be, respectively, the dual spaces of $\mathscr{X}$, $\mathscr{Y}$, $\mathscr{X}_M$, and $\mathscr{Y}_N$. The adjoint operators of $i_{\mathscr{X}_M}$ and $\imath_{\mathscr{Y}_N}$ are defined by

$$\begin{aligned}
i^*_{\mathscr{X}_M} : \mathscr{X}^* \to \mathscr{X}^*_M, \qquad & \left\langle i^*_{\mathscr{X}_M} v, w \right\rangle_{\mathscr{X}^*_M, \mathscr{X}_M} = \left\langle v, i_{\mathscr{X}_M} w \right\rangle_{\mathscr{X}^*, \mathscr{X}}, \qquad v \in \mathscr{X}^*, \ w \in \mathscr{X}_M, \\
i^*_{\mathscr{Y}_N} : \mathscr{Y}^* \to \mathscr{Y}^*_N, \qquad & \left\langle i^*_{\mathscr{Y}_N} v, w \right\rangle_{\mathscr{Y}^*_N, \mathscr{Y}_N} = \left\langle v, i_{\mathscr{Y}_N} w \right\rangle_{\mathscr{Y}^*, \mathscr{Y}}, \qquad v \in \mathscr{Y}^*, \ w \in \mathscr{Y}_N,
\end{aligned}$$

where $\langle \cdot, \cdot \rangle_{\mathscr{X}^*, \mathscr{X}}$ and similar notations denote dual inner products.

The dual pair $(\mathscr{X}_M, \mathscr{X}^*_M)$ possesses the following biorthonormal bases

$$\mathscr{B}_{\mathscr{X}_M} := \{\eta_1, \ldots, \eta_M\} \quad \text{and} \quad \mathscr{B}_{\mathscr{X}^*_M} := \{\eta^*_1, \ldots, \eta^*_M\}$$

where

$$\left\langle \eta^*_i, \eta_j \right\rangle_{\mathscr{X}^*_M, \mathscr{X}_M} = \delta_{ij}, \quad i, j = 1, \ldots, M.$$

Similarly, for $(\mathscr{Y}_N, \mathscr{Y}^*_N)$ we have

$$\mathscr{B}_{\mathscr{Y}_N} := \{\phi_1, \ldots, \phi_N\} \quad \text{and} \quad \mathscr{B}_{\mathscr{Y}^*_N} := \{\phi^*_1, \ldots, \phi^*_N\}$$

with

$$\left\langle \phi^*_j, \phi_i \right\rangle_{\mathscr{Y}^*_N, \mathscr{Y}_N} = \delta_{ji}, \quad i, j = 1, \ldots, N.$$

It is known that any $x \in \mathscr{X}_M$, $x^* \in \mathscr{X}^*_M$, $y \in \mathscr{Y}_N$, and $y^* \in \mathscr{Y}^*_N$ can be represented as

$$\begin{aligned}
x &= \sum_{i=1}^{M} \langle \eta^*_i, x \rangle_{\mathscr{X}^*_M, \mathscr{X}_M} \eta_i, \qquad & x^* &= \sum_{i=1}^{M} \langle x^*, \eta_i \rangle_{\mathscr{X}^*_M, \mathscr{X}_M} \eta^*_i, \\
y &= \sum_{i=1}^{N} \langle \phi^*_i, y \rangle_{\mathscr{Y}^*_N, \mathscr{Y}_N} \phi_i, \qquad & y^* &= \sum_{i=1}^{N} \langle y^*, \phi_i \rangle_{\mathscr{Y}^*_N, \mathscr{Y}_N} \phi^*_i.
\end{aligned} \tag{B.28}$$

## B.2.2 Discretised operators and matrix representations

### B.2.2.1 Operators from $\mathscr{X}$ into $\mathscr{Y}^*$

Given $\mathfrak{A} \in \mathscr{L}(\mathscr{X}, \mathscr{Y}^*)$, we define $\mathfrak{A}_{NM} : \mathscr{X}_M \to \mathscr{Y}_N^*$ by

$$\langle \mathfrak{A}_{NM} v, w \rangle_{\mathscr{Y}_N^*, \mathscr{Y}_N} = \langle \mathfrak{A} i_{\mathscr{X}_M} v, i_{\mathscr{Y}_N} w \rangle_{\mathscr{Y}^*, \mathscr{Y}} \quad \forall v \in \mathscr{X}_M, \ \forall w \in \mathscr{Y}_N.$$

On the other hand,

$$\langle \mathfrak{A} i_{\mathscr{X}_M} v, i_{\mathscr{Y}_N} w \rangle_{\mathscr{Y}^*, \mathscr{Y}} = \langle i_{\mathscr{Y}_N}^* \mathfrak{A} i_{\mathscr{X}_M} v, w \rangle_{\mathscr{Y}_N^*, \mathscr{Y}_N} \quad \forall v \in \mathscr{X}_M, \ \forall w \in \mathscr{Y}_N.$$

Hence

$$\langle \mathfrak{A}_{NM} v, w \rangle_{\mathscr{Y}_N^*, \mathscr{Y}_N} = \langle i_{\mathscr{Y}_N}^* \mathfrak{A} i_{\mathscr{X}_M} v, w \rangle_{\mathscr{Y}_N^*, \mathscr{Y}_N} \quad \forall v \in \mathscr{X}_M, \ \forall w \in \mathscr{Y}_N.$$

This means

$$\mathfrak{A}_{NM} = i_{\mathscr{Y}_N}^* \mathfrak{A} i_{\mathscr{X}_M}. \tag{B.29}$$

The matrix representation of $\mathfrak{A}_{NM}$ has the form

$$\mathbf{A}_{NM}^{(j)} = [\mathfrak{A}_{NM} \eta_j]_{\mathscr{B}_{\mathscr{Y}_N^*}} = \begin{pmatrix} \langle \mathfrak{A}_{NM} \eta_j, \phi_1 \rangle_{\mathscr{Y}_N^*, \mathscr{Y}_N} \\ \vdots \\ \langle \mathfrak{A}_{NM} \eta_j, \phi_N \rangle_{\mathscr{Y}_N^*, \mathscr{Y}_N} \end{pmatrix} = \begin{pmatrix} \langle i_{\mathscr{Y}_N}^* \mathfrak{A} i_{\mathscr{X}_M} \eta_j, \phi_1 \rangle_{\mathscr{Y}_N^*, \mathscr{Y}_N} \\ \vdots \\ \langle i_{\mathscr{Y}_N}^* \mathfrak{A} i_{\mathscr{X}_M} \eta_j, \phi_N \rangle_{\mathscr{Y}_N^*, \mathscr{Y}_N} \end{pmatrix}$$
$$= \begin{pmatrix} \langle \mathfrak{A} i_{\mathscr{X}_M} \eta_j, i_{\mathscr{Y}_N} \phi_1 \rangle_{\mathscr{Y}_N^*, \mathscr{Y}_N} \\ \vdots \\ \langle \mathfrak{A} i_{\mathscr{X}_M} \eta_j, i_{\mathscr{Y}_N} \phi_N \rangle_{\mathscr{Y}^*, \mathscr{Y}} \end{pmatrix}$$

where $\mathbf{A}_{NM}^{(j)}$ denotes the $j^{\text{th}}$-column of the matrix $\mathbf{A}_{NM}$ and $[\mathfrak{A}_{NM} \eta_j]_{\mathscr{B}_{\mathscr{Y}_N^*}}$ denotes the vector coefficient of $\mathfrak{A}_{NM} \eta_j$ with respect to the basis $\mathscr{B}_{\mathscr{Y}_N^*}$; see (B.28).

### B.2.2.2 Operators from $\mathscr{X}$ into $\mathscr{X}^*$

In practice, $\mathfrak{A} = \mathfrak{W}$ (the hypersingular integral operator) or $\mathfrak{A} = \mathfrak{V}$ (the weakly-singular integral operator), we have $\mathscr{Y} = \mathscr{X}$ so that $\mathscr{Y}^* = \mathscr{X}^*$. In this case $M = N$, $\mathscr{Y}_N = \mathscr{X}_M$, and it follows from (B.29) that

$$\mathfrak{A}_M : \mathscr{X}_M \to \mathscr{X}_M^*, \quad \mathfrak{A}_M = i_{\mathscr{X}_M}^* \mathfrak{A} i_{\mathscr{X}_M}.$$

The Galerkin matrix, which is also the matrix representation of $\mathfrak{A}_M$, has entries

$$A_{ij} = \left\langle \mathfrak{A} i_{\mathscr{X}_M} \eta_j, i_{\mathscr{X}_M} \eta_i \right\rangle_{\mathscr{X}^*,\mathscr{X}} = \left\langle \mathfrak{A}_M \eta_j, \eta_i \right\rangle_{\mathscr{X}_M^*,\mathscr{X}_M}, \quad i,j = 1,\dots,M.$$

#### B.2.2.3 Operators from $\mathscr{X}$ into $\mathscr{X}$ where $\mathscr{X}$ is reflexive

Consider $\mathfrak{A} \in \mathscr{L}(\mathscr{X},\mathscr{X})$. Let $\mathscr{Y} = \mathscr{X}^*$ so that $\mathscr{Y}^* = \mathscr{X}$. Then $\mathfrak{A} \in \mathscr{L}(\mathscr{X},\mathscr{Y}^*)$. The discretised operator $\mathfrak{A}_{NM} : \mathscr{X}_M \to \mathscr{Y}_N^*$ defined by (B.29) satisfies

$$\langle \mathfrak{A}_{NM} v, w \rangle_{\mathscr{Y}_N^*,\mathscr{Y}_N} = \left\langle i_{\mathscr{Y}_N} w, \mathfrak{A} i_{\mathscr{X}_M} v \right\rangle_{\mathscr{X}^*,\mathscr{X}} \quad \forall v \in \mathscr{X}_M,\ \forall w \in \mathscr{Y}_N.$$

The $j^{\text{th}}$-column matrix representation of $\mathfrak{A}_{NM}$ with respect to the bases $\mathscr{B}_{\mathscr{X}_M}$ and $\mathscr{B}_{\mathscr{Y}_N^*}$ is

$$\mathbf{A}_{NM}^{(j)} = [\mathfrak{A}_{NM} \eta_j]_{\mathscr{B}_{\mathscr{Y}_N^*}} = \begin{pmatrix} \left\langle \mathfrak{A}_{NM} \eta_j, \phi_1 \right\rangle_{\mathscr{Y}_N^*,\mathscr{Y}_N} \\ \vdots \\ \left\langle \mathfrak{A}_{NM} \eta_j, \phi_N \right\rangle_{\mathscr{Y}_N^*,\mathscr{Y}_N} \end{pmatrix} = \begin{pmatrix} \left\langle i_{\mathscr{Y}_N} \phi_1, \mathfrak{A} i_{\mathscr{X}_M} \eta_j \right\rangle_{\mathscr{X}^*,\mathscr{X}} \\ \vdots \\ \left\langle i_{\mathscr{Y}_N} \phi_N, \mathfrak{A} i_{\mathscr{X}_M} \eta_j \right\rangle_{\mathscr{X}^*,\mathscr{X}} \end{pmatrix}.$$

### B.2.3 Compositions of operators

We consider the following linear operators and their discretisations

$$\begin{aligned} &\mathfrak{A} \in \mathscr{L}(\mathscr{Y},\mathscr{Y}^*), && \mathfrak{A}_N = i_{\mathscr{Y}_N}^* \mathfrak{A} i_{\mathscr{Y}_N} \in \mathscr{L}(\mathscr{Y}_N,\mathscr{Y}_N^*), \\ &\mathfrak{B} \in \mathscr{L}(\mathscr{Y},\mathscr{X}^*), && \mathfrak{B}_{MN} = i_{\mathscr{X}_M}^* \mathfrak{B} i_{\mathscr{Y}_N} \in \mathscr{L}(\mathscr{Y}_N,\mathscr{X}_M^*), \\ &\mathfrak{C} \in \mathscr{L}(\mathscr{X},\mathscr{Y}^*), && \mathfrak{C}_{NM} = i_{\mathscr{Y}_N}^* \mathfrak{C} i_{\mathscr{X}_M} \in \mathscr{L}(\mathscr{X}_M,\mathscr{Y}_N^*). \end{aligned}$$

Assume that

$$\langle \mathfrak{A} v, v \rangle_{\mathscr{Y}^*,\mathscr{Y}} \geq c_{\mathfrak{A}} \|v\|_{\mathscr{Y}}^2 \quad \forall v \in \mathscr{Y}. \tag{B.30}$$

According to the Lax–Milgram Theorem,

$$\mathfrak{A}^{-1} \in \mathscr{L}(\mathscr{Y}^*,\mathscr{Y}) \quad \text{and} \quad \mathfrak{A}_N^{-1} \in \mathscr{L}(\mathscr{Y}_N^*,\mathscr{Y}_N).$$

Hence we are able to define

$$\begin{aligned} \mathfrak{E} &:= \mathfrak{B}\mathfrak{A}^{-1}\mathfrak{C} && \in \mathscr{L}(\mathscr{X},\mathscr{X}^*), \\ \mathfrak{E}_M &:= i_{\mathscr{X}_M}^* \mathfrak{E} i_{\mathscr{X}_M} && \in \mathscr{L}(\mathscr{X}_M,\mathscr{X}_M^*), \\ \widetilde{\mathfrak{E}}_M &:= \mathfrak{B}_{MN}\mathfrak{A}_N^{-1}\mathfrak{C}_{NM} && \in \mathscr{L}(\mathscr{X}_M,\mathscr{X}_M^*). \end{aligned}$$

We note that in general $\mathfrak{E}_M \neq \widetilde{\mathfrak{E}}_M$. The following lemma gives an estimate for the difference.

**Lemma B.4.** *Under assumption* (B.30)*, there exists a positive constant* $c_{\mathfrak{E}}$ *independent of* $M$ *such that for any* $v \in \mathscr{X}_M$

$$\|(\mathfrak{E}_M - \widetilde{\mathfrak{E}}_M)v\|_{\mathscr{X}_M^*} \le c_{\mathfrak{E}} \inf_{w\in\mathscr{Y}_N} \|\mathfrak{A}^{-1}\mathfrak{C} i_{\mathscr{X}_M} v - i_{\mathscr{Y}_N} w\|_{\mathscr{Y}}.$$

*Proof.* This result is in fact [49, Lemma 2.4]. Recall that

$$\|(\mathfrak{E}_M - \widetilde{\mathfrak{E}}_M)v\|_{\mathscr{X}_M^*} = \sup_{\substack{w\in\mathscr{X}_M \\ w\neq 0}} \frac{\langle(\mathfrak{E}_M - \widetilde{\mathfrak{E}}_M)v, w\rangle_{\mathscr{X}_M^*,\mathscr{X}_M}}{\|w\|_{\mathscr{X}_M}}.$$

For any $v, w \in \mathscr{X}_M$, we have

$$\begin{aligned}
|\langle(\mathfrak{E}_M - \widetilde{\mathfrak{E}}_M)v, w\rangle_{\mathscr{X}_M^*,\mathscr{X}_M}| &= |\langle i_{\mathscr{X}_M}^* \mathfrak{B}(\mathfrak{A}^{-1} - i_{\mathscr{Y}_N}\mathfrak{A}_N^{-1} i_{\mathscr{Y}_N}^*)\mathfrak{C} i_{\mathscr{X}_M} v, w\rangle_{\mathscr{X}_M^*,\mathscr{X}_M}| \\
&= |\langle \mathfrak{B}(\mathfrak{A}^{-1} - i_{\mathscr{Y}_N}\mathfrak{A}_h^{-1} i_{\mathscr{Y}_N}^*)\mathfrak{C} i_{\mathscr{X}_M} v, i_{\mathscr{X}_M} w\rangle_{\mathscr{X}^*,\mathscr{X}}| \\
&= |\langle \mathfrak{B}^* i_{\mathscr{X}_M} w, (\mathfrak{A}^{-1} - i_{\mathscr{Y}_N}\mathfrak{A}_h^{-1} i_{\mathscr{Y}_N}^*)\mathfrak{C} i_{\mathscr{X}_M} v\rangle_{\mathscr{Y}^*,\mathscr{Y}}| \\
&\le \|\mathfrak{B}^* i_{\mathscr{X}_M} w\|_{\mathscr{Y}^*} \|(\mathfrak{A}^{-1} - i_{\mathscr{Y}_N}\mathfrak{A}_h^{-1} i_{\mathscr{Y}_N}^*)\mathfrak{C} i_{\mathscr{X}_M} v\|_{\mathscr{Y}} \\
&\le \|\mathfrak{B}^*\|_{\mathscr{X}\to\mathscr{Y}^*} \|i_{\mathscr{X}_M} w\|_{\mathscr{X}} \|(\mathfrak{A}^{-1} - i_{\mathscr{Y}_N}\mathfrak{A}_h^{-1} i_{\mathscr{Y}_N}^*)\mathfrak{C} i_{\mathscr{X}_M} v\|_{\mathscr{Y}} \\
&= \|\mathfrak{B}\|_{\mathscr{Y}\to\mathscr{X}^*} \|w\|_{\mathscr{X}_M} \|(\mathfrak{A}^{-1} - i_{\mathscr{Y}_N}\mathfrak{A}_h^{-1} i_{\mathscr{Y}_N}^*)\mathfrak{C} i_{\mathscr{X}_M} v\|_{\mathscr{Y}}.
\end{aligned}$$

Hence

$$\|(\mathfrak{E}_M - \widetilde{\mathfrak{E}}_M)v\|_{\mathscr{X}_M^*} \le \|\mathfrak{B}\|_{\mathscr{Y}\to\mathscr{X}^*} \|(\mathfrak{A}^{-1} - i_{\mathscr{Y}_N}\mathfrak{A}_h^{-1} i_{\mathscr{Y}_N}^*)\mathfrak{C} i_{\mathscr{X}_M} v\|_{\mathscr{Y}}.$$

It remains to estimate $\|(\mathfrak{A}^{-1} - i_{\mathscr{Y}_N}\mathfrak{A}_h^{-1} i_{\mathscr{Y}_N}^*)\mathfrak{C} i_{\mathscr{X}_M} v\|_{\mathscr{Y}}$. Let

$$f := \mathfrak{C} i_{\mathscr{X}_M} v \in \mathscr{Y}^* \quad \text{and} \quad u := \mathfrak{A}^{-1}\mathfrak{C} i_{\mathscr{X}_M} v \in \mathscr{Y},$$

so that $u$ is the solution to the equation $\mathfrak{A}u = f$ in $\mathscr{Y}^*$. The corresponding Galerkin equation is $\mathfrak{A}_h u_h = i_{\mathscr{Y}_N}^* f$ in $\mathscr{Y}_N^*$, and thus $u_h := \mathfrak{A}_h^{-1} i_{\mathscr{Y}_N}^* f = \mathfrak{A}_h^{-1} i_{\mathscr{Y}_N}^* \mathfrak{C} i_{\mathscr{X}_M} v \in \mathscr{Y}_N$ is the Galerkin approximation to $u$. By Céa's Lemma we have

$$\begin{aligned}
\|(\mathfrak{A}^{-1} - i_{\mathscr{Y}_N}\mathfrak{A}_h^{-1} i_{\mathscr{Y}_N}^*)\mathfrak{C} i_{\mathscr{X}_M} v\|_{\mathscr{Y}} &= \|u - i_{\mathscr{Y}_N} u_h\|_{\mathscr{Y}} \\
&\le \frac{\|\mathfrak{A}\|_{\mathscr{Y}\to\mathscr{Y}^*}}{c_{\mathfrak{A}}} \inf_{\substack{w\in\mathscr{Y}_N \\ w\neq 0}} \|u - i_{\mathscr{Y}_N} w\|_{\mathscr{Y}} \\
&= \frac{\|\mathfrak{A}\|_{\mathscr{Y}\to\mathscr{Y}^*}}{c_{\mathfrak{A}}} \inf_{\substack{w\in\mathscr{Y}_N \\ w\neq 0}} \|\mathfrak{A}^{-1}\mathfrak{C} i_{\mathscr{X}_M} v - i_{\mathscr{Y}_N} w\|_{\mathscr{Y}}.
\end{aligned}$$

This proves the lemma. □

We now consider the special case when $\mathfrak{B} = \mathfrak{C}^*$, i.e.,

$$\mathfrak{E} := \mathfrak{C}^*\mathfrak{A}^{-1}\mathfrak{C} \in \mathscr{L}(\mathscr{X}, \mathscr{X}^*),$$

$$\widetilde{\mathfrak{E}}_M := \mathfrak{C}_{MN}^* \mathfrak{A}_N^{-1} \mathfrak{C}_{NM} \in \mathscr{L}(\mathscr{X}_M, \mathscr{X}_M^*),$$

and denote by $\widetilde{\mathbf{E}}_M$ the matrix representation of $\widetilde{\mathfrak{E}}_M$. The next lemma shows how this matrix can be represented in terms of the matrix representations of $\mathfrak{A}_N^{-1}$, $\mathfrak{C}_{NM}$, and $\mathfrak{C}_{MN}^*$, namely $\mathbf{A}_N^{-1}$, $\mathbf{C}_{NM}$, and $\mathbf{C}_{MN}^\top$.

**Lemma B.5.** *The following identity holds*

$$\widetilde{\mathbf{E}}_M = \mathbf{C}_{MN}^\top \mathbf{A}_N^{-1} \mathbf{C}_{NM}.$$

*Proof.* This result is in fact [51, Theorem 5]. We reproduce the proof here for completeness. Since

$$\left(\widetilde{\mathbf{E}}_M\right)_{ij} = \left\langle \widetilde{\mathfrak{E}}_M \eta_j, \eta_i \right\rangle_{\mathscr{X}_M^*, \mathscr{X}_M}, \quad i, j = 1, \ldots, M,$$

it suffices to show that, for $i, j = 1, \ldots, M$,

$$\left\langle \widetilde{\mathfrak{E}}_M \eta_j, \eta_i \right\rangle_{\mathscr{X}_M^*, \mathscr{X}_M} = \left(\mathbf{C}_{MN}^\top \mathbf{A}_N^{-1} \mathbf{C}_{NM}\right)_{ij} = \left(\mathbf{C}^\top \mathbf{A}^{-1} \mathbf{C}\right)_{ij}, \tag{B.31}$$

In view of $\widetilde{\mathfrak{E}}_M \eta_j = \mathfrak{C}_{MN}^* \mathfrak{A}_N^{-1} \mathfrak{C}_{NM} \eta_j \in \mathscr{X}_M^*$, we define

$$\psi_j^* := \mathfrak{C}_{NM} \eta_j \in \mathscr{Y}_N^* \quad \text{and} \quad \xi_j := \mathfrak{A}_N^{-1} \psi_j^* \in \mathscr{Y}_N. \tag{B.32}$$

The definition of $\psi_j^*$ implies

$$\langle \psi_j^*, \phi_i \rangle_{\mathscr{Y}_N^*, \mathscr{Y}_N} = \langle \mathfrak{C}_{NM} \eta_j, \phi_i \rangle_{\mathscr{Y}_N^*, \mathscr{Y}_N} = C_{ij} \tag{B.33}$$

while that of $\xi_j$ implies

$$\psi_j^* = \mathfrak{A}_N \xi_j. \tag{B.34}$$

It follows from (B.28) that

$$\xi_j = \sum_{k=1}^N \langle \phi_k^*, \xi_j \rangle_{\mathscr{Y}_N^*, \mathscr{Y}_N} \phi_k$$

which together with (B.34) gives

$$\psi_j^* = \mathfrak{A}_N \Big( \sum_{k=1}^N \langle \phi_k^*, \xi_j \rangle_{\mathscr{Y}_N^*, \mathscr{Y}_N} \phi_k \Big) = \sum_{k=1}^N \langle \phi_k^*, \xi_j \rangle_{\mathscr{Y}_N^*, \mathscr{Y}_N} \mathfrak{A}_N \phi_k.$$

This implies

$$\langle \psi_j^*, \phi_i \rangle_{\mathscr{Y}_N^*, \mathscr{Y}_N} = \sum_{k=1}^N \langle \phi_k^*, \xi_j \rangle_{\mathscr{Y}_N^*, \mathscr{Y}_N} \langle \mathfrak{A}_N \phi_k, \phi_i \rangle_{\mathscr{Y}_N^*, \mathscr{Y}_N} = \sum_{k=1}^N A_{ik} \langle \phi_k^*, \xi_j \rangle_{\mathscr{Y}_N^*, \mathscr{Y}_N}$$

$$= \left( \mathbf{A} \begin{pmatrix} \langle \phi_1^*, \xi_j \rangle_{\mathscr{Y}_N^*, \mathscr{Y}_N} \\ \vdots \\ \langle \phi_N^*, \xi_j \rangle_{\mathscr{Y}_N^*, \mathscr{Y}_N} \end{pmatrix} \right)_i .$$

Comparing with (B.33) we deduce

$$\mathbf{C}^{(j)} = \mathbf{A} \begin{pmatrix} \langle \phi_1^*, \xi_j \rangle_{\mathscr{Y}_N^*, \mathscr{Y}_N} \\ \vdots \\ \langle \phi_N^*, \xi_j \rangle_{\mathscr{Y}_N^*, \mathscr{Y}_N} \end{pmatrix}$$

where $\mathbf{C}^{(j)}$ denotes the $j^{\text{th}}$ column of $\mathbf{C}$. Consequently

$$\mathbf{A}^{-1}\mathbf{C}^{(j)} = \begin{pmatrix} \langle \phi_1^*, \xi_j \rangle_{\mathscr{Y}_N^*, \mathscr{Y}_N} \\ \vdots \\ \langle \phi_N^*, \xi_j \rangle_{\mathscr{Y}_N^*, \mathscr{Y}_N} \end{pmatrix} .$$

Therefore

$$\left( \mathbf{C}^\top \mathbf{A}^{-1} \mathbf{C} \right)_{ij} = \sum_{k=1}^{N} C_{ki} \left( \mathbf{A}^{-1} \mathbf{C} \right)_{kj} = \sum_{k=1}^{N} C_{ki} \left\langle \phi_k^*, \xi_j \right\rangle_{\mathscr{Y}_N^*, \mathscr{Y}_N} . \tag{B.35}$$

On the other hand we have

$$\begin{aligned} \left\langle \widetilde{\mathfrak{E}}_M \eta_j, \eta_i \right\rangle_{\mathscr{X}_M^*, \mathscr{X}_M} &= \left\langle \mathfrak{C}_{MN}^* \mathfrak{A}_N^{-1} \mathfrak{C}_{NM} \eta_j, \eta_i \right\rangle_{\mathscr{X}_M^*, \mathscr{X}_M} \\ &= \left\langle \mathfrak{C}_{NM} \eta_i, \mathfrak{A}_N^{-1} \mathfrak{C}_{NM} \eta_j \right\rangle_{\mathscr{Y}_N^*, \mathscr{Y}_N} \\ &= \left\langle \psi_i^*, \xi_j \right\rangle_{\mathscr{Y}_N^*, \mathscr{Y}_N} . \end{aligned}$$

From (B.28) and (B.32) it follows that

$$\psi_i^* = \sum_{k=1}^{N} \langle \psi_i^*, \phi_k \rangle_{\mathscr{Y}_N^*, \mathscr{Y}_N} \phi_k^* = \sum_{k=1}^{N} \langle \mathfrak{C}_{NM} \eta_i, \phi_k \rangle_{\mathscr{Y}_N^*, \mathscr{Y}_N} \phi_k^* = \sum_{k=1}^{N} C_{ki} \phi_k^*,$$

implying

$$\left\langle \widetilde{\mathfrak{E}}_M \eta_j, \eta_i \right\rangle_{\mathscr{X}_M^*, \mathscr{X}_M} = \sum_{k=1}^{N} C_{ki} \left\langle \phi_k^*, \xi_j \right\rangle_{\mathscr{Y}_N^*, \mathscr{Y}_N} .$$

This identity and (B.35) yield (B.31), completing the proof of the lemma. □

# Appendix C
# Some Additional Results

In this chapter we collect all technical lemmas which are not important to understand the main arguments in the analysis of preconditioning. However, these lemmas are important for the rigorous analysis in the previous chapters.

It is noted that some results which have their own interests are first proved in this chapter. These include Lemma C.18 on inverse properties of polynomials.

## C.1 Conditioning of Matrices

In this section, many results on the extremal eigenvalues of the stiffness matrices arising from the Galerkin boundary element discretisation of boundary integral equations are derived.

First we study the condition numbers of the stiffness matrix $\mathbf{A}_{\mathfrak{V}}$ and $\mathbf{A}_{\mathfrak{W}}$ defined by the weakly-singular integral operator $\mathfrak{V}$ and hypersingular integral operator $\mathfrak{W}$. The results for the $h$-version have been studied in Chapter 12 and will not be repeated here. For the $p$-version, it is shown in [106, Lemma 2] that when the boundary $\Gamma$ is a curve in $\mathbb{R}^2$ then the condition numbers of $\mathbf{A}_{\mathfrak{V}}$ and $\mathbf{A}_{\mathfrak{W}}$ grow at most like $p^3$. It is then proved in [228, Lemma A.1] that they grow at most as $p^3$ and at least as $p^2$. Moreover, if the basis functions are properly scaled, these condition numbers grow as $p^2$ for both $\mathbf{A}_{\mathfrak{V}}$ and $\mathbf{A}_{\mathfrak{W}}$ in the case when $\Gamma = (-1,1) \times \{0\}$. We reproduce the proof of these results in Lemma C.4, where the scaled Legendre polynomials and their antiderivatives are used as basis functions. We also prove in this lemma that when $\Gamma = (-1,1) \times (-1,1) \times \{0\}$, the condition number of $\mathbf{A}_{\mathfrak{V}}$ still behaves like $p^2$ while that of $\mathbf{A}_{\mathfrak{W}}$ is bounded above by $p^6$. We first introduce these basis functions.



E. P. Stephan and T. Tran, *Schwarz Methods and Multilevel Preconditioners for Boundary Element Methods*, https://doi.org/10.1007/978-3-030-79283-1

### *C.1.1 The hierarchical basis functions for the p-version*

Let $I = (-1,1)$. For $q = 0,1,2,\ldots$, let

$$L_q^* = \sqrt{\frac{2q+1}{2}} L_q$$

where $L_q$ is the Legendre polynomial of degree $q$. Then

$$\langle L_q, L_r \rangle_I = \frac{2}{2q+1} \delta_{q,r} \quad \text{and} \quad \langle L_q^*, L_r^* \rangle_I = \delta_{q,r}, \tag{C.1}$$

where $\langle \cdot, \cdot \rangle_I$ is the $L^2$-inner product in $I$. Note also that (see [27, (3.8)]) for $1 \le k \le q$

$$L_q^{(k)}(1) = \frac{(q-k+1)(q-k+2)\cdots q(q+1)\cdots(q+k-1)(q+k)}{2^k k!}, \tag{C.2}$$

Moreover, the well-known property $L_q(-x) = (-1)^q L_q(x)$ for all $x \in \mathbb{R}$ implies

$$L_q^{(k)}(-x) = (-1)^{q+k} L_q^{(k)}(x), \quad x \in \mathbb{R}. \tag{C.3}$$

For $q = 2,3,\ldots$, we define

$$\begin{aligned} \mathscr{L}_q(x) &= \int_{-1}^{x} L_{q-1}(s)\,\mathrm{d}s, \\ \widehat{\mathscr{L}_q}(x) &= \frac{\mathscr{L}_q(x)}{\|\mathscr{L}_q\|_{L^2(I)}}, \\ \mathscr{L}_q^*(x) &= \int_{-1}^{x} L_{q-1}^*(s)\,\mathrm{d}s. \end{aligned} \tag{C.4}$$

Simple calculations give the following lemma.

**Lemma C.1.** *For $q, r = 2,3,\ldots$, we have*

$$\langle \mathscr{L}_q, \mathscr{L}_r \rangle_I = \begin{cases} \dfrac{4}{(2q-3)(2q-1)(2q+1)}, & q = r, \\ \dfrac{-2}{(2q-1)(2q+1)(2q+3)}, & r = q+2, \\ \dfrac{-2}{(2q-5)(2q-3)(2q-1)}, & r = q-2, \\ 0, & \textit{otherwise}, \end{cases} \tag{C.5}$$

$$\langle \mathscr{L}_q^*, \mathscr{L}_r^* \rangle_I = \begin{cases} \dfrac{2}{(2q-3)(2q+1)}, & q = r, \\ \dfrac{-1}{(2q+1)\sqrt{(2q-1)(2q+3)}}, & r = q+2, \\ \dfrac{-1}{(2q-3)\sqrt{(2q-5)(2q-1)}}, & r = q-2, \\ 0, & \textit{otherwise}, \end{cases} \tag{C.6}$$

$$\left\langle \widehat{\mathscr{L}_q}, \widehat{\mathscr{L}_r} \right\rangle_I = \begin{cases} 1, & q = r, \\ -\dfrac{1}{2}\sqrt{\dfrac{(2q-3)(2q+5)}{(2q-1)(2q+3)}}, & r = q+2, \\ -\dfrac{1}{2}\sqrt{\dfrac{(2q-7)(2q+1)}{(2q-5)(2q-1)}} & r = q-2, \\ 0, & \textit{otherwise}. \end{cases} \tag{C.7}$$

*Proof.* Due to the well-known recurrence formula, see e.g., [27],

$$(2q+1)L_q = L'_{q+1} - L'_{q-1}, \quad q = 1,2,3,\dots \tag{C.8}$$

and the fact that $L_q(-1) = (-1)^q$ we have

$$(2q+1)\mathscr{L}_{q+1} = L_{q+1} - L_{q-1}, \quad q = 1,2,3,\dots,$$

or

$$(2q-1)\mathscr{L}_q = L_q - L_{q-2}, \quad q = 2,3,\dots.$$

Hence for $q, r = 2,3,\dots$,

$$\begin{aligned} \langle \mathscr{L}_q, \mathscr{L}_r \rangle_I &= \frac{1}{(2q-1)(2r-1)} \langle L_q - L_{q-2}, L_r - L_{r-2} \rangle_I \\ &= \frac{1}{(2q-1)(2r-1)} \Big( \frac{2}{2q+1} (\delta_{qr} - \delta_{q,r-2}) \\ &\qquad - \frac{2}{2q-3} (\delta_{q-2,r} - \delta_{q-2,r-2}) \Big). \end{aligned}$$

This proves (C.5). This identity together with (C.4) implies (C.6) and (C.7). □

A similar result can be proved for $\{L'_k : k \geq 1\}$.

**Lemma C.2.** *For $k, \ell \geq 1$, we have*

$$\langle L'_k, L'_\ell \rangle_I = 2 \sum_{m=0}^{[(k-1)/2]} \sum_{n=0}^{[(\ell-1)/2]} (2(\ell-2n)-1)\delta_{k-1-2m,\ell-1-2n}.$$

*Proof.* From the recurrence formula (C.8) we have

$$L'_k = \sum_{m=0}^{[(k-1)/2]} (2(k-2m)-1) L_{k-1-2m}$$

so that

$$\begin{aligned}
\langle L'_k, L'_\ell \rangle_I &= \sum_{m=0}^{[(k-1)/2]} (2(k-2m)-1) \sum_{n=0}^{[(\ell-1)/2]} (2(\ell-2n)-1) \langle L_{k-1-2m}, L_{\ell-1-2n} \rangle_I \\
&= \sum_{m=0}^{[(k-1)/2]} (2(k-2m)-1) \sum_{n=0}^{[(\ell-1)/2]} (2(\ell-2n)-1) \frac{2\delta_{k-1-2m,\ell-1-2n}}{2(k-1-2m)+1} \\
&= 2 \sum_{m=0}^{[(k-1)/2]} \sum_{n=0}^{[(\ell-1)/2]} (2(\ell-2n)-1) \delta_{k-1-2m,\ell-1-2n}.
\end{aligned}$$

□

The following lemma concerning the extremal eigenvalues of the mass matrices defined by $\mathscr{L}_q$ and $\mathscr{L}_q^*$ is proved in [146]. The proof is reproduced here for convenience.

**Lemma C.3.** *Let $I := (-1,1)$ and let $\mathbf{M}, \mathbf{M}^* \in \mathbb{R}^{(p-1)\times(p-1)}$ be defined by*

$$M_{qr} = \langle \mathscr{L}_q, \mathscr{L}_r \rangle_I \quad \text{and} \quad M^*_{qr} = \langle \mathscr{L}^*_q, \mathscr{L}^*_r \rangle_I, \quad q, r = 2, \ldots, p.$$

*Then the extremal eigenvalues of the matrices $\mathbf{M}$ and $\mathbf{M}^*$ satisfy*

$$p^{-5} \lesssim \lambda_{\min}(\mathbf{M}) \le \lambda_{\max}(\mathbf{M}) \lesssim 1$$

*whereas*

$$\lambda_{\min}(\mathbf{M}^*) \simeq p^{-4} \quad \text{and} \quad \lambda_{\max}(\mathbf{M}^*) \simeq 1.$$

*Proof.* For any $\mathbf{x} = (c_2, \ldots, c_p) \in \mathbb{R}^{p-1}$, let $\phi = \sum_{q=2}^p c_q \mathscr{L}_q$ so that $\phi' = \sum_{q=2}^p c_q L_{q-1}$. Then

$$\|\phi\|^2_{\widetilde{H}^0(I)} = \mathbf{x}^\top \mathbf{M} \mathbf{x}$$

and, due to (C.1),

$$\|\phi\|^2_{\widetilde{H}^1(I)} = \|\phi'\|^2_{\widetilde{H}^0(I)} = \sum_{q,r=2}^p c_q c_r \langle L_{q-1}, L_{r-1} \rangle_{\widetilde{H}^0(I)} = \sum_{q=2}^p \frac{2}{2q-1} c_q^2.$$

The last identity implies

$$\frac{2}{2p-1} \mathbf{x}^\top \mathbf{x} \le \|\phi\|^2_{\widetilde{H}^1(I)} \le \frac{2}{3} \mathbf{x}^\top \mathbf{x}$$

or

$$\frac{3}{2} \|\phi\|^2_{\widetilde{H}^1(I)} \le \mathbf{x}^\top \mathbf{x} \le \frac{2p-1}{2} \|\phi\|^2_{\widetilde{H}^1(I)}. \tag{C.9}$$

Therefore,

$$p^{-5} \lesssim \frac{2}{2p-1} \frac{\|\phi\|^2_{\widetilde{H}^0(I)}}{\|\phi\|^2_{\widetilde{H}^1(I)}} \le \frac{\mathbf{x}^\top \mathbf{M}\mathbf{x}}{\mathbf{x}^\top \mathbf{x}} \le \frac{2}{3} \frac{\|\phi\|^2_{\widetilde{H}^0(I)}}{\|\phi\|^2_{\widetilde{H}^1(I)}} \lesssim 1$$

where in the first inequality we used the inverse inequality (C.35) in Subsection C.4.1. It follows that

$$p^{-5} \lesssim \lambda_{\min}(\mathbf{M}) \le \lambda_{\max}(\mathbf{M}) \lesssim 1.$$

The results for $\mathbf{M}^*$ can be obtained similarly, noting that $\langle L_q^*, L_r^* \rangle_I = \delta_{qr}$ so that instead of (C.9) we have $\mathbf{x}^\top \mathbf{x} = \|\phi\|^2_{\widetilde{H}^1(-1,1)}$. Hence

$$p^{-4} \lesssim \frac{\mathbf{x}^\top \mathbf{M}^* \mathbf{x}}{\mathbf{x}^\top \mathbf{x}} \simeq \frac{\|\phi\|^2_{\widetilde{H}^0(I)}}{\|\phi\|^2_{\widetilde{H}^1(I)}} \lesssim 1,$$

implying

$$p^{-4} \lesssim \lambda_{\min}(\mathbf{M}^*) \le \lambda_{\max}(\mathbf{M}^*) \lesssim 1.$$

To show an upper bound for $\lambda_{\min}(\mathbf{M}^*)$ we follow [26, 146] to choose

$$\phi = L''_{p+2} - \frac{(p+3)(p+4)}{4} L'_{p+1}. \tag{C.10}$$

First we note that $\phi$ is a polynomial of degree $p$ vanishing at $\pm 1$. Indeed, (C.2) gives $\phi(1) = 0$. Moreover, (C.3) gives

$$\begin{aligned}\phi(-1) &= (-1)^{p+4} L''_{p+2}(1) - \frac{(p+3)(p+4)}{4} (-1)^{p+2} L'_{p+1}(1) \\ &= (-1)^{p+2} \phi(1) = 0.\end{aligned}$$

Therefore, there exists $\mathbf{x} = (c_q)_{q=2}^p \in \mathbb{R}^{p-1}$ such that $\phi = \sum_{q=2}^p c_q \mathscr{L}_q^*$. It is shown in [146, Lemma 2] that $\|\phi\|^2_{\widetilde{H}^0(I)} \simeq p^{-4} \|\phi\|^2_{\widetilde{H}^1(I)}$, implying $\lambda_{\min}(\mathbf{M}^*) \lesssim p^{-4}$.

An upper bound for $\lambda_{\max}(\mathbf{M}^*)$ can be shown by choosing $\mathbf{x} = (1, 0, \ldots, 0)$ so that $\|\phi\|^2_{\widetilde{H}^0(I)} \simeq \|\phi\|^2_{\widetilde{H}^1(I)} \simeq 1$, implying $\lambda_{\max}(\mathbf{M}^*) \gtrsim 1$. This completes the proof of the lemma. □

## *C.1.2 The condition numbers of the weakly-singular and hypersingular stiffness matrices*

Since the number of elements in the $p$-version is fixed, in the following for simplicity we will assume that $\Gamma = (-1,1)^d$ has only one element, $d = 1, 2$. For $d = 1$, the boundary element spaces for the weakly-singular integral equation and hypersingular integral equation, respectively, are

$$\widetilde{\mathscr{V}} := \operatorname{span}\left\{L_q^* : q = 0,\dots,p\right\} \quad \text{and} \quad \mathscr{V} := \operatorname{span}\left\{\mathscr{L}_q^* : q = 2,\dots,p\right\}.$$

For $d = 2$,

$$\widetilde{\mathscr{V}} = \operatorname{span}\left\{L_{q_1}^* \otimes L_{q_2}^* : q_1, q_2 = 0,\dots,p\right\}$$

and

$$\mathscr{V} = \operatorname{span}\left\{\mathscr{L}_{q_1}^* \otimes \mathscr{L}_{q_2}^* : q_1, q_2 = 2,\dots,p\right\}$$

where the tensor product $\otimes$ is defined for any one-variable function $f$ and $g$ by

$$(f \otimes g)(x,y) = f(x)g(y), \quad (x,y) \in \Gamma.$$

When $d = 1$ we provide both upper and lower bounds for the condition numbers, while for $d = 2$ we only give an upper bound, as shown in the following lemma.

**Lemma C.4.** *Let $\mathbf{A}_{\mathfrak{V}}^d$ and $\mathbf{A}_{\mathfrak{W}}^d$ be the stiffness matrices defined by*

$$\mathbf{A}_{\mathfrak{V}}^d = \begin{cases} \left(a_{\mathfrak{V}}(L_q^*, L_r^*)\right)_{q,r=0}^p, & d = 1, \\ \left(a_{\mathfrak{V}}(L_{q_1}^* \otimes L_{q_2}^*, L_{r_1}^* \otimes L_{r_2}^*)\right)_{q_1,q_2,r_1,r_2=0}^p, & d = 2, \end{cases}$$

$$\mathbf{A}_{\mathfrak{W}}^d = \begin{cases} \left(a_{\mathfrak{W}}(\mathscr{L}_{q+1}, \mathscr{L}_{r+1})\right)_{q,r=2}^p, & d = 1, \\ \left(a_{\mathfrak{W}}(\mathscr{L}_{q_1}^* \otimes \mathscr{L}_{q_2}^*, \mathscr{L}_{r_1}^* \otimes \mathscr{L}_{r_2}^*)\right)_{q_1,q_2,r_1,r_2=2}^p, & d = 2. \end{cases}$$

*Then, for the weakly-singular integral equation,*

$$\lambda_{\min}(\mathbf{A}_{\mathfrak{V}}^d) \simeq p^{-2} \quad \text{and} \quad \lambda_{\max}(\mathbf{A}_{\mathfrak{V}}^d) \simeq 1, \quad d = 1,2.$$

*For the hypersingular integral equation,*

$$\lambda_{\min}(\mathbf{A}_{\mathfrak{W}}^1) \simeq p^{-2}, \quad p^{-6} \lesssim \lambda_{\min}(\mathbf{A}_{\mathfrak{W}}^2), \quad \text{and} \quad \lambda_{\max}(\mathbf{A}_{\mathfrak{W}}^d) \simeq 1, \quad d = 1,2.$$

*Consequently,*

$$\kappa(\mathbf{A}_{\mathfrak{V}}^d) \simeq \kappa(\mathbf{A}_{\mathfrak{W}}^1) \simeq p^2, \quad d = 1,2,$$

*and*

$$\kappa(\mathbf{A}_{\mathfrak{W}}^2) \lesssim p^6.$$

*Proof.* Consider first the matrix $\mathbf{A}_{\mathfrak{V}}^1$. For any $\mathbf{x} = (c_0,\dots,c_p) \in \mathbb{R}^{p+1}$ let $\phi = \sum_{q=0}^p c_q L_q^*$. Then since $\langle L_q^*, L_r^* \rangle_\Gamma = \delta_{qr}$, we have

$$\frac{\mathbf{x}^\top \mathbf{A}_{\mathfrak{V}}^1 \mathbf{x}}{\mathbf{x}^\top \mathbf{x}} \simeq \frac{\|\phi\|_{\widetilde{H}_{\mathrm{I}}^{-1/2}(\Gamma)}^2}{\|\phi\|_{\widetilde{H}_{\mathrm{I}}^0(\Gamma)}^2}.$$

The inverse inequality (see [27, Theorem 5.2] and Section C.4.1) and Lemma A.19 give

$$p^{-2}\|\phi\|^2_{\widetilde{H}^0_{\mathrm{I}}(\Gamma)} \lesssim \|\phi\|^2_{\widetilde{H}^{-1/2}_{\mathrm{I}}(\Gamma)} \le \|\phi\|^2_{\widetilde{H}^0_{\mathrm{I}}(\Gamma)}.$$

Hence

$$p^{-2} \lesssim \lambda_{\min}(\mathbf{A}^1_{\mathfrak{V}}) \le \lambda_{\max}(\mathbf{A}^1_{\mathfrak{V}}) \lesssim 1.$$

In particular, taking $\phi = L''_p$. Then (see [27, page 254])

$$\frac{\|\phi\|^2_{\widetilde{H}^{-1/2}_{\mathrm{I}}(\Gamma)}}{\|\phi\|^2_{\widetilde{H}^0_{\mathrm{I}}(\Gamma)}} \lesssim \frac{\|L'_p\|^2_{\widetilde{H}^{1/2}_{\mathrm{I}}(\Gamma)}}{\|L''_p\|^2_{\widetilde{H}^0_{\mathrm{I}}(\Gamma)}} \le \frac{\|L'_p\|_{\widetilde{H}^0_{\mathrm{I}}(\Gamma)}\|L'_p\|_{\widetilde{H}^1_{\mathrm{I}}(\Gamma)}}{\|L''_p\|^2_{\widetilde{H}^0_{\mathrm{I}}(\Gamma)}} \lesssim \frac{\|L'_p\|_{\widetilde{H}^0_{\mathrm{I}}(\Gamma)}}{\|L''_p\|_{\widetilde{H}^0_{\mathrm{I}}(\Gamma)}} \lesssim \frac{p}{p^3} = p^{-2}.$$

Hence $\lambda_{\min}(\mathbf{A}^1_{\mathfrak{V}}) \simeq p^{-2}$. Taking $\mathbf{x} = (1,0,\ldots,0) \in \mathbb{R}^{p+1}$ so that $\phi(x) \equiv 1$ we deduce $\lambda_{\max}(\mathbf{A}^1_{\mathfrak{V}}) \gtrsim 1$, implying $\lambda_{\max}(\mathbf{A}^1_{\mathfrak{V}}) \simeq 1$.

Similar arguments work for $\mathbf{A}^1_{\mathfrak{W}}$. For any $\mathbf{x} = (c_2,\ldots,c_p) \in \mathbb{R}^{p-1}$ let $\phi = \sum_{q=2}^{p} c_q \mathscr{L}^*_q$. Then $\phi' = \sum_{q=2}^{p} c_q L^*_{q-1}$, so that $\mathbf{x}^\top\mathbf{x} = \|\phi'\|^2_{\widetilde{H}^0_{\mathrm{I}}(\Gamma)} \simeq \|\phi\|^2_{\widetilde{H}^1_{\mathrm{I}}(\Gamma)}$. Therefore, by using the inverse inequality again we obtain

$$p^{-2} \lesssim \frac{\mathbf{x}^\top\mathbf{A}^1_{\mathfrak{W}}\mathbf{x}}{\mathbf{x}^\top\mathbf{x}} \simeq \frac{\|\phi\|^2_{\widetilde{H}^{1/2}_{\mathrm{I}}(\Gamma)}}{\|\phi\|^2_{\widetilde{H}^1_{\mathrm{I}}(\Gamma)}} \le 1.$$

This means

$$p^{-2} \lesssim \lambda_{\min}(\mathbf{A}^1_{\mathfrak{W}}) \le \lambda_{\max}(\mathbf{A}^1_{\mathfrak{W}}) \lesssim 1.$$

By choosing $\phi$ as in (C.10) we can show that

$$\frac{\mathbf{x}^\top\mathbf{A}^1_{\mathfrak{W}}\mathbf{x}}{\mathbf{x}^\top\mathbf{x}} \simeq \frac{\|\phi\|^2_{\widetilde{H}^{1/2}_{\mathrm{I}}(\Gamma)}}{\|\phi\|^2_{\widetilde{H}^1_{\mathrm{I}}(\Gamma)}} \lesssim \frac{\|\phi\|_{\widetilde{H}^0_{\mathrm{I}}(\Gamma)}}{\|\phi\|_{\widetilde{H}^1_{\mathrm{I}}(\Gamma)}} \lesssim p^{-2},$$

implying $\lambda_{\min}(\mathbf{A}^1_{\mathfrak{W}}) \lesssim p^{-2}$. Similarly, by choosing $\mathbf{x} = (1,0,\ldots,0) \in \mathbb{R}^{p-1}$ we can show that $\lambda_{\max}(\mathbf{A}^1_{\mathfrak{W}}) \gtrsim 1$. This proves the result for $\mathbf{A}^1_{\mathfrak{W}}$.

We now consider $\mathbf{A}^2_{\mathfrak{V}}$. For any $\mathbf{x} = (c_{q_1,q_2})_{q_1,q_2=0}^{p} \in \mathbb{R}^{(p+1)^2}$ let

$$\phi(x,y) = \sum_{q_1,q_2=0}^{p} c_{q_1,q_2}(L^*_{q_1} \otimes L^*_{q_2})(x,y).$$

Then

$$\|\phi\|^2_{\widetilde{H}^{-1/2}_{\mathrm{I}}(\Gamma\times\Gamma)} \simeq a_{\mathfrak{V}}(\phi,\phi) = \mathbf{x}^\top\mathbf{A}^2_{\mathfrak{V}}\mathbf{x}$$

and

$$\begin{aligned}\|\phi\|^2_{\widetilde{H}^0_{\mathrm{I}}(\Gamma\times\Gamma)} &= \sum_{q_1,q_2,r_1,r_2=0}^{p} c_{q_1,q_2}c_{r_1,r_2}\left\langle L^*_{q_1}\otimes L^*_{q_2}, L^*_{r_1}\otimes L^*_{r_2}\right\rangle_{\Gamma\times\Gamma}\\ &= \sum_{q_1,q_2,r_1,r_2=0}^{p} c_{q_1,q_2}c_{r_1,r_2}\left\langle L^*_{q_1}, L^*_{r_1}\right\rangle_{\Gamma}\left\langle L^*_{q_2}, L^*_{r_2}\right\rangle_{\Gamma}\end{aligned}$$

$$= \sum_{q_1,q_2,r_1,r_2=0}^{p} c_{q_1,q_2} c_{r_1,r_2} \delta_{q_1,r_1} \delta_{q_2,r_2} = \mathbf{x}^\top \mathbf{x}.$$

It follows that

$$p^{-2} \lesssim \frac{\mathbf{x}^\top \mathbf{A}^2_{\mathfrak{V}} \mathbf{x}}{\mathbf{x}^\top \mathbf{x}} \simeq \frac{\|\phi\|^2_{\widetilde{H}_{\mathrm{I}}^{-1/2}(\Gamma\times\Gamma)}}{\|\phi\|^2_{\widetilde{H}_{\mathrm{I}}^{0}(\Gamma\times\Gamma)}} \lesssim 1,$$

where we used again the inverse inequality. By choosing $\phi$ such that

$$\phi(x,y) = L_p''(x) \quad \forall x,y \in \Gamma,$$

we deduce, as in the case of $\mathbf{A}^1_{\mathfrak{V}}$, that $\lambda_{\min}(\mathbf{A}^2_{\mathfrak{V}}) \simeq p^{-2}$. Similarly, we can use the same argument as in the case $d=1$ to show $\lambda_{\max}(\mathbf{A}^2_{\mathfrak{V}}) \simeq 1$.

Finally, we prove the result for $\mathbf{A}^2_{\mathfrak{W}}$. For any $\mathbf{x} = (c_{q_1,q_2})_{q_1,q_2=2}^{p} \in \mathbb{R}^{(p-1)^2}$ let

$$\phi(x,y) = \sum_{q_1,q_2=2}^{p} c_{q_1,q_2} \mathscr{L}^*_{q_1} \otimes \mathscr{L}^*_{q_2}(x,y).$$

Then

$$\|\phi\|^2_{\widetilde{H}_{\mathrm{I}}^{1/2}(\Gamma\times\Gamma)} \simeq a_{\mathfrak{W}}(\phi,\phi) = \mathbf{x}^\top \mathbf{A}^2_{\mathfrak{W}} \mathbf{x}$$

and

$$\begin{aligned}
\|\phi\|^2_{\widetilde{H}_{\mathrm{I}}^{1}(\Gamma\times\Gamma)} &= \langle \nabla\phi, \nabla\phi \rangle_{\Gamma\times\Gamma} \\
&= \sum_{q_1,q_2,r_1,r_2=2}^{p} c_{q_1,q_2} c_{r_1,r_2} \left\langle \nabla(\mathscr{L}^*_{q_1} \otimes \mathscr{L}^*_{q_2}), \nabla(\mathscr{L}^*_{r_1} \otimes \mathscr{L}^*_{r_2}) \right\rangle_{\Gamma\times\Gamma} \\
&= \sum_{q_1,q_2,r_1,r_2=2}^{p} c_{q_1,q_2} c_{r_1,r_2} \Big( \langle L^*_{q_1-1}, L^*_{r_1-1} \rangle_\Gamma \langle \mathscr{L}^*_{q_2}, \mathscr{L}^*_{r_2} \rangle_\Gamma \\
&\qquad + \langle \mathscr{L}^*_{q_1}, \mathscr{L}^*_{r_1} \rangle_\Gamma \langle L^*_{q_2-1}, L^*_{r_2-1} \rangle_\Gamma \Big) \\
&= \underbrace{\sum_{q_1,q_2,r_2=2}^{p} c_{q_1,q_2} c_{q_1,r_2} \langle \mathscr{L}^*_{q_2}, \mathscr{L}^*_{r_2} \rangle_\Gamma}_{T_1:=} \\
&\quad + \underbrace{\sum_{q_1,q_2,r_1=2}^{p} c_{q_1,q_2} c_{r_1,q_2} \langle \mathscr{L}^*_{q_1}, \mathscr{L}^*_{r_1} \rangle_\Gamma}_{T_2:=}.
\end{aligned}$$

First we consider $T_1$. For any $q_1 \in \{2,\dots,p\}$, if we define $\mathbf{x}_{q_1} = (c_{q_1,2},\dots,c_{q_1,p})$ then, with $\mathbf{M}^*$ defined in Lemma C.3,

$$T_{1,q_1} := \sum_{q_2,r_2=2}^{p} c_{q_1,q_2} c_{q_1,r_2} \left\langle \mathscr{L}_{q_2}^*, \mathscr{L}_{r_2}^* \right\rangle_\Gamma = \mathbf{x}_{q_1}^\top \mathbf{M}^* \mathbf{x}_{q_1}$$

so that

$$p^{-4} \mathbf{x}_{q_1}^\top \mathbf{x}_{q_1} \simeq \lambda_{\min}(\mathbf{M}^*) \mathbf{x}_{q_1}^\top \mathbf{x}_{q_1} \le T_{1,q_1} \le \lambda_{\max}(\mathbf{M}^*) \mathbf{x}_{q_1}^\top \mathbf{x}_{q_1} \simeq \mathbf{x}_{q_1}^\top \mathbf{x}_{q_1}.$$

Since $\mathbf{x} = (\mathbf{x}_{q_1})_{q_1=2}^p$, it follows that

$$p^{-4} \mathbf{x}^\top \mathbf{x} = p^{-4} \sum_{q_1=2}^{p} \mathbf{x}_{q_1}^\top \mathbf{x}_{q_1} \lesssim T_1 = \sum_{q_1=2}^{p} T_{1,q_1} \lesssim \sum_{q_1=2}^{p} \mathbf{x}_{q_1}^\top \mathbf{x}_{q_1} = \mathbf{x}^\top \mathbf{x}.$$

Similar estimates hold for $T_2$. Consequently,

$$\|\phi\|_{\widetilde{H}_{\mathrm{I}}^1(\Gamma\times\Gamma)}^2 \lesssim \mathbf{x}^\top \mathbf{x} \lesssim p^4 \|\phi\|_{\widetilde{H}_{\mathrm{I}}^1(\Gamma\times\Gamma)}^2.$$

Therefore,

$$p^{-6} \lesssim p^{-4} \frac{\|\phi\|_{\widetilde{H}_{\mathrm{I}}^{1/2}(\Gamma\times\Gamma)}^2}{\|\phi\|_{\widetilde{H}_{\mathrm{I}}^1(\Gamma\times\Gamma)}^2} \lesssim \frac{\mathbf{x}^\top \mathbf{A}_{\mathfrak{W}}^2 \mathbf{x}}{\mathbf{x}^\top \mathbf{x}} \lesssim \frac{\|\phi\|_{\widetilde{H}_{\mathrm{I}}^{1/2}(\Gamma\times\Gamma)}^2}{\|\phi\|_{\widetilde{H}_{\mathrm{I}}^1(\Gamma\times\Gamma)}^2} \lesssim 1,$$

where we have used the inverse inequality

$$\|\phi\|_{\widetilde{H}_{\mathrm{I}}^{1/2}(\Gamma\times\Gamma)}^2 \gtrsim p^{-2} \|\phi\|_{\widetilde{H}_{\mathrm{I}}^1(\Gamma\times\Gamma)}^2.$$

A constant lower bound for $\lambda_{\max}(\mathbf{A}_{\mathfrak{W}}^2)$ can be shown as in the case $d = 1$ and is omitted. □

### *C.1.3 Extremal eigenvalues of equivalent matrices*

**Lemma C.5.** *Let* $\mathbf{A}$ *and* $\mathbf{C}$ *be two symmetric positive definite matrices of the same size and let* $\mathbf{B} := \mathbf{C}^{-1/2}\mathbf{A}\mathbf{C}^{-1/2}$. *Then*

$$\frac{\lambda_{\min}(\mathbf{A})}{\lambda_{\max}(\mathbf{B})} \le \lambda_{\min}(\mathbf{C}) \le \lambda_{\max}(\mathbf{C}) \le \frac{\lambda_{\max}(\mathbf{A})}{\lambda_{\min}(\mathbf{B})}$$

*and*

$$\frac{\lambda_{\min}(\mathbf{A})}{\lambda_{\max}(\mathbf{C})} \le \lambda_{\min}(\mathbf{B}) \le \lambda_{\max}(\mathbf{B}) \le \frac{\lambda_{\max}(\mathbf{A})}{\lambda_{\min}(\mathbf{C})}.$$

*Proof.* For any $\mathbf{x} \neq \mathbf{0}$ we have

$$\frac{\mathbf{x}^\top \mathbf{C}\mathbf{x}}{\mathbf{x}^\top \mathbf{x}} = \frac{\mathbf{x}^\top \mathbf{C}\mathbf{x}}{\mathbf{x}^\top \mathbf{A}\mathbf{x}} \cdot \frac{\mathbf{x}^\top \mathbf{A}\mathbf{x}}{\mathbf{x}^\top \mathbf{x}} = \frac{\mathbf{y}^\top \mathbf{y}}{\mathbf{y}^\top \mathbf{C}^{-1/2}\mathbf{A}\mathbf{C}^{-1/2}\mathbf{y}} \cdot \frac{\mathbf{x}^\top \mathbf{A}\mathbf{x}}{\mathbf{x}^\top \mathbf{x}} = \frac{\mathbf{y}^\top \mathbf{y}}{\mathbf{y}^\top \mathbf{B}\mathbf{y}} \cdot \frac{\mathbf{x}^\top \mathbf{A}\mathbf{x}}{\mathbf{x}^\top \mathbf{x}}$$

where $\mathbf{y} := \mathbf{C}^{1/2}\mathbf{x} \neq \mathbf{0}$. Since

$$\frac{1}{\lambda_{\max}(\mathbf{B})} \leq \frac{\mathbf{y}^\top\mathbf{y}}{\mathbf{y}^\top\mathbf{B}\mathbf{y}} \leq \frac{1}{\lambda_{\min}(\mathbf{B})} \quad \text{and} \quad \lambda_{\min}(\mathbf{A}) \leq \frac{\mathbf{x}^\top\mathbf{A}\mathbf{x}}{\mathbf{x}^\top\mathbf{x}} \leq \lambda_{\max}(\mathbf{A}),$$

the first result follows. The second result can be proved similarly. □

**Lemma C.6.** *If* $\mathbf{B}_1$ *and* $\mathbf{B}_2$ *are two symmetric, positive-definite matrices of the same size, then* $\mathbf{B}_2^{-1/2}\mathbf{B}_1\mathbf{B}_2^{-1/2}$ *and* $\mathbf{B}_1^{1/2}\mathbf{B}_2^{-1}\mathbf{B}_1^{1/2}$ *have the same eigenvalues.*

*Proof.* Assume that $\lambda$ is an eigenvalue of $\mathbf{B}_2^{-1/2}\mathbf{B}_1\mathbf{B}_2^{-1/2}$ with a corresponding eigenvector $\mathbf{v}$. Then with $\mathbf{w} = \mathbf{B}_1^{-1/2}\mathbf{B}_2^{1/2}\mathbf{v}$ we have

$$\begin{aligned}
\mathbf{B}_1^{1/2}\mathbf{B}_2^{-1}\mathbf{B}_1^{1/2}\mathbf{w} &= \mathbf{B}_1^{1/2}\mathbf{B}_2^{-1/2}\mathbf{v} \\
&= \mathbf{B}_1^{-1/2}\mathbf{B}_2^{1/2}\mathbf{B}_2^{-1/2}\mathbf{B}_1\mathbf{B}_2^{-1/2}\mathbf{v} \\
&= \mathbf{B}_1^{-1/2}\mathbf{B}_2^{1/2}(\lambda\mathbf{v}) \\
&= \lambda\mathbf{w},
\end{aligned}$$

i.e., $\lambda$ is an eigenvalue of $\mathbf{B}_1^{1/2}\mathbf{B}_2^{-1}\mathbf{B}_1^{1/2}$. Conversely, assume that $\lambda$ is an eigenvalue of $\mathbf{B}_1^{1/2}\mathbf{B}_2^{-1}\mathbf{B}_1^{1/2}$ with a corresponding eigenvector $\mathbf{v}$. Then with $\mathbf{w} = \mathbf{B}_2^{1/2}\mathbf{B}_1^{-1/2}\mathbf{v}$ we have

$$\begin{aligned}
\mathbf{B}_2^{-1/2}\mathbf{B}_1\mathbf{B}_2^{-1/2}\mathbf{w} &= \mathbf{B}_2^{-1/2}\mathbf{B}_1^{1/2}\mathbf{v} \\
&= \mathbf{B}_2^{1/2}\mathbf{B}_1^{-1/2}\mathbf{B}_1^{1/2}\mathbf{B}_2^{-1}\mathbf{B}_1^{1/2}\mathbf{v} \\
&= \mathbf{B}_2^{1/2}\mathbf{B}_1^{-1/2}(\lambda\mathbf{v}) \\
&= \lambda\mathbf{w},
\end{aligned}$$

i.e., $\lambda$ is an eigenvalue of $\mathbf{B}_2^{-1/2}\mathbf{B}_1\mathbf{B}_2^{-1/2}$. □

### C.1.4 Eigenvalues of block matrices

In the following lemma, we adopt the following notations. For any indefinite square matrix $\mathbf{Q}$, we denote the positive and negative eigenvalues of $\mathbf{Q}$ closest to zero by $\lambda_\pm(\mathbf{Q})$. Hence, if $\Lambda(\mathbf{Q})$ is the set of all eigenvalues of $\mathbf{Q}$, then

$$\Lambda(\mathbf{Q}) \subset [\lambda_{\min}(\mathbf{Q}), \lambda_-(\mathbf{Q})] \cup [\lambda_+(\mathbf{Q}), \lambda_{\max}(\mathbf{Q})].$$

If $\mathbf{P}$ is a rectangular matrix, then $\sigma_{\max}(\mathbf{P})$ and $\sigma_{\min}(\mathbf{P})$ denote the maximum and minimum singular values of $\mathbf{P}$.

The following lemma gives bounds for the eigenvalues of the sum of symmetric matrices.

**Lemma C.7.** *Assume that* $\mathbf{A} = \mathbf{A}_1 + \mathbf{A}_2$ *where* $\mathbf{A}_1$ *and* $\mathbf{A}_2$ *are two symmetric matrices. Let the eigenvalues of each matrix be arranged in non-increasing order. Then the* $s^{\text{th}}$*-eigenvalues of* $\mathbf{A}$ *and* $\mathbf{A}_1$*, denoted by* $\lambda_s(\mathbf{A})$ *and* $\lambda_s(\mathbf{A}_1)$*, respectively, satisfy*

$$\lambda_s(\mathbf{A}_1) + \lambda_{\min}(\mathbf{A}_2) \le \lambda_s(\mathbf{A}) \le \lambda_s(\mathbf{A}_1) + \lambda_{\max}(\mathbf{A}_2).$$

*Proof.* See [238, p.101]. The result can be interpreted as follows. When $\mathbf{A}_2$ is added to $\mathbf{A}_1$, the eigenvalues of $\mathbf{A}_1$ are changed by an amount which lies between the smallest and largest eigenvalues of $\mathbf{A}_2$. □

The next lemma represents the eigenvalues of a block-diagonal matrix in term of the eigenvalues of each block.

**Lemma C.8.** *Let* $\mathbf{A}$ *be a block diagonal matrix defined by*

$$\mathbf{A} = \begin{pmatrix} \mathbf{A}_1 & \mathbf{0} \\ \mathbf{0} & \mathbf{A}_2 \end{pmatrix}.$$

*Then*

$$\Lambda(\mathbf{A}) = \Lambda(\mathbf{A}_1) \cup \Lambda(\mathbf{A}_2).$$

*Proof.* This is a direct consequence of

$$\det(\mathbf{A} - \lambda \mathbf{I}) = \det(\mathbf{A}_1 - \lambda \mathbf{I}_1)\det(\mathbf{A}_2 - \lambda \mathbf{I}_2)$$

where $\mathbf{I}$, $\mathbf{I}_1$, and $\mathbf{I}_2$ are the identity matrices of the same size as $\mathbf{A}$, $\mathbf{A}_1$, and $\mathbf{A}_2$, respectively. □

**Lemma C.9.** *Let* $\mathbf{A}$ *be a block diagonal matrix defined by*

$$\mathbf{A} = \begin{pmatrix} \mathbf{0} & \mathbf{B}^\top \\ \mathbf{B} & \mathbf{0} \end{pmatrix}.$$

*Then* $\Lambda(\mathbf{A})$ *consists of plus and minus singular values of* $\mathbf{B}$ *and zeros. In particular,*

$$\lambda_{\min}(\mathbf{A}) = -\sigma_{\max}(\mathbf{B}) \quad \text{and} \quad \lambda_{\max}(\mathbf{A}) = \sigma_{\max}(\mathbf{B}).$$

*Proof.* See [83, p. 427]. □

**Lemma C.10.** *[193, Lemma 2.2] Let* $\mathbf{A}$ *be a symmetric and indefinite matrix of the form*

$$\mathbf{A} = \begin{pmatrix} \mathbf{A}_{11} & \mathbf{A}_{12} \\ \mathbf{A}_{21} & -\mathbf{A}_{22} \end{pmatrix}$$

*where* $\mathbf{A}_{ii}$*,* $i = 1, 2$*, are symmetric positive definite matrices and* $\mathbf{A}_{12} = \mathbf{A}_{21}^\top$ *are rectangular matrices of appropriate sizes. Then the spectrum of* $\mathbf{A}$ *satisfies*

$$\Lambda(\mathbf{A}) = [\lambda_{\min}(\mathbf{A}), \lambda_{-}(\mathbf{A})] \cup [\lambda_{+}(\mathbf{A}), \lambda_{\max}(\mathbf{A})]$$

*where*

$$\lambda_{\min}(\mathbf{A}) \geq \frac{1}{2}\Big(\lambda_{\min}(\mathbf{A}_{11}) - \lambda_{\max}(\mathbf{A}_{22}) - \sqrt{\left(\lambda_{\min}(\mathbf{A}_{11}) + \lambda_{\max}(\mathbf{A}_{22})\right)^2 + 4\sigma_{\max}(\mathbf{A}_{21})^2}\Big),$$

$$\lambda_{-}(\mathbf{A}) \leq \frac{1}{2}\Big(\lambda_{\max}(\mathbf{A}_{11}) - \sqrt{\lambda_{\max}(\mathbf{A}_{11})^2 - 4\sigma_{\min}(\mathbf{A}_{21})^2}\Big),$$

$$\lambda_{+}(\mathbf{A}) \geq \lambda_{\min}(\mathbf{A}_{11}),$$

$$\lambda_{\max}(\mathbf{A}) \leq \frac{1}{2}\Big(\lambda_{\max}(\mathbf{A}_{11}) + \sqrt{\lambda_{\max}(\mathbf{A}_{11})^2 + 4\sigma_{\max}(\mathbf{A}_{21})^2}\Big).$$

*Proof.* See [193]. □

**Lemma C.11.** *Let*

$$\mathbf{A} := \begin{pmatrix} \mathbf{A}_{11} & \mathbf{A}_{12} \\ \mathbf{A}_{21} & -\mathbf{A}_{22} \end{pmatrix} \quad \textit{and} \quad \mathbf{B} := \begin{pmatrix} \mathbf{B}_{11} & \mathbf{0} \\ \mathbf{0} & \mathbf{B}_{22} \end{pmatrix}$$

*where* $\mathbf{A}_{11}$ *is symmetric positive semi-definite,* $\mathbf{A}_{22}$, $\mathbf{B}_{11}$ *and* $\mathbf{B}_{22}$ *are symmetric positive definite, and* $\mathbf{A}_{12} = \mathbf{A}_{21}^{\top}$. *Assume that* $\mathbf{A}_{11}$ *can be written as* $\mathbf{A}_{11} = \mathbf{A}_{11}^{(1)} + \mathbf{A}_{11}^{(2)}$ *where* $\mathbf{A}_{11}^{(\ell)}$, $\ell = 1,2$, *are symmetric positive semi-definite. Assume further that there exist a positive definite matrix* $\mathbf{T}$ *and positive constants* $\theta_1$, $\theta_2$, $\theta_3$, $\Theta_0$, $\Theta_1$, $\Theta_2$, $\Theta_3$ *satisfying, for all vectors* $\mathbf{x}$ *of appropriate size,*

$$\Theta_0 \geq \frac{\mathbf{x}^{\top}\mathbf{A}_{11}\mathbf{x}}{\mathbf{x}^{\top}(\mathbf{A}_{11}^{(1)} + \mathbf{T})\mathbf{x}}$$

*and*

$$\theta_1 \leq \lambda_{\min}\left(\mathbf{B}_{11}^{-1/2}(\mathbf{A}_{11}^{(1)} + \mathbf{T})\mathbf{B}_{11}^{-1/2}\right), \qquad \Theta_1 \geq \lambda_{\max}\left(\mathbf{B}_{11}^{-1/2}(\mathbf{A}_{11}^{(1)} + \mathbf{T})\mathbf{B}_{11}^{-1/2}\right),$$

$$\theta_2 \leq \lambda_{\min}\left(\mathbf{B}_{22}^{-1/2}\mathbf{A}_{22}\mathbf{B}_{22}^{-1/2}\right), \qquad \Theta_2 \geq \lambda_{\max}\left(\mathbf{B}_{22}^{-1/2}\mathbf{A}_{22}\mathbf{B}_{22}^{-1/2}\right),$$

$$\theta_3 \leq \frac{\mathbf{x}^{\top}\left(\mathbf{A}_{11} + \mathbf{A}_{12}\mathbf{A}_{22}^{-1}\mathbf{A}_{21}\right)\mathbf{x}}{\mathbf{x}^{\top}(\mathbf{A}_{11}^{(1)} + \mathbf{T})\mathbf{x}}, \qquad \Theta_3 \geq \frac{\mathbf{x}^{\top}\mathbf{A}_{12}\mathbf{A}_{22}^{-1}\mathbf{A}_{21}\mathbf{x}}{\mathbf{x}^{\top}(\mathbf{A}_{11}^{(1)} + \mathbf{T})\mathbf{x}}.$$

*Then the spectrum of* $\widetilde{\mathbf{A}} := \mathbf{B}^{-1/2}\mathbf{A}\mathbf{B}^{-1/2}$ *satisfies*

$$\Lambda(\widetilde{\mathbf{A}}) = [\lambda_{\min}(\widetilde{\mathbf{A}}), \lambda_{-}(\widetilde{\mathbf{A}})] \cup [\lambda_{+}(\widetilde{\mathbf{A}}), \lambda_{\max}(\widetilde{\mathbf{A}})]$$

*where*

$$\lambda_{\min}(\widetilde{\mathbf{A}}) \geq -\frac{1}{2}\left(\theta_2 + \sqrt{\theta_2^2 + 4\Theta_1\Theta_2\Theta_3}\right),$$

$$\lambda_{-}(\widetilde{\mathbf{A}}) \leq -\theta_2,$$

$$\lambda_{+}(\widetilde{\mathbf{A}}) \geq \frac{1}{2}\left(-\theta_2 + \sqrt{\theta_2^2 + 4\theta_1\theta_2\theta_3}\right),$$

$$\lambda_{\max}(\widetilde{\mathbf{A}}) \leq \frac{1}{2}\left(-\theta_2 + \Theta_0\Theta_1 + \sqrt{(\theta_2 + \Theta_0\Theta_1)^2 + 4\Theta_1\Theta_2\Theta_3}\right).$$

*Proof.* Elementary calculations reveal

$$\widetilde{\mathbf{A}} = \begin{pmatrix} \mathbf{B}_{11}^{-1/2}\mathbf{A}_{11}\mathbf{B}_{11}^{-1/2} & \mathbf{B}_{11}^{-1/2}\mathbf{A}_{12}\mathbf{B}_{22}^{-1/2} \\ \mathbf{B}_{22}^{-1/2}\mathbf{A}_{21}\mathbf{B}_{11}^{-1/2} & -\mathbf{B}_{22}^{-1/2}\mathbf{A}_{22}\mathbf{B}_{22}^{-1/2} \end{pmatrix}.$$

Let

$$\widetilde{\mathbf{A}}_{11} := \mathbf{B}_{11}^{-1/2}\mathbf{A}_{11}\mathbf{B}_{11}^{-1/2}, \qquad \widetilde{\mathbf{A}}_{12} := \mathbf{B}_{11}^{-1/2}\mathbf{A}_{12}\mathbf{B}_{22}^{-1/2},$$
$$\widetilde{\mathbf{A}}_{21} := \mathbf{B}_{22}^{-1/2}\mathbf{A}_{21}\mathbf{B}_{11}^{-1/2}, \qquad \widetilde{\mathbf{A}}_{22} := \mathbf{B}_{22}^{-1/2}\mathbf{A}_{22}\mathbf{B}_{22}^{-1/2},$$

and let $\lambda$ be an eigenvalue of $\widetilde{\mathbf{A}}$. Then there exist some vectors $\mathbf{x}$ and $\mathbf{y}$ of appropriate size satisfying $(\mathbf{x},\mathbf{y}) \neq (\mathbf{0},\mathbf{0})$ and

$$\widetilde{\mathbf{A}}_{11}\mathbf{x} + \widetilde{\mathbf{A}}_{12}\mathbf{y} = \lambda\mathbf{x}, \tag{C.11a}$$
$$\widetilde{\mathbf{A}}_{21}\mathbf{x} - \widetilde{\mathbf{A}}_{22}\mathbf{y} = \lambda\mathbf{y}. \tag{C.11b}$$

Consider $\lambda < 0$. Then the matrix $\mathbf{I} - \lambda^{-1}\widetilde{\mathbf{A}}_{11}$ is invertible with eigenvalues greater than or equal to 1, which together with (C.11a) implies $\mathbf{x} = \lambda^{-1}(\mathbf{I} - \lambda^{-1}\widetilde{\mathbf{A}}_{11})^{-1}\widetilde{\mathbf{A}}_{12}\mathbf{y}$. It then follows from (C.11b) that

$$\mathbf{y}^\top\widetilde{\mathbf{A}}_{21}(\mathbf{I} - \lambda^{-1}\widetilde{\mathbf{A}}_{11})^{-1}\widetilde{\mathbf{A}}_{12}\mathbf{y} - \lambda\mathbf{y}^\top\widetilde{\mathbf{A}}_{22}\mathbf{y} = \lambda^2\mathbf{y}^\top\mathbf{y}.$$

Since the eigenvalues of $(\mathbf{I} - \lambda^{-1}\widetilde{\mathbf{A}}_{11})^{-1}$ are less than or equal to 1, the following inequality holds

$$\mathbf{y}^\top\widetilde{\mathbf{A}}_{21}(\mathbf{I} - \lambda^{-1}\widetilde{\mathbf{A}}_{11})^{-1}\widetilde{\mathbf{A}}_{12}\mathbf{y} \leq \mathbf{y}^\top\widetilde{\mathbf{A}}_{21}\widetilde{\mathbf{A}}_{12}\mathbf{y}.$$

We infer from the above two equations that

$$\mathbf{y}^\top\widetilde{\mathbf{A}}_{21}\widetilde{\mathbf{A}}_{12}\mathbf{y} - \lambda\mathbf{y}^\top\widetilde{\mathbf{A}}_{22}\mathbf{y} \geq \lambda^2\mathbf{y}^\top\mathbf{y}.$$

Note that $\mathbf{y} \neq \mathbf{0}$, because otherwise (C.11a) implies that $\lambda$ is an eigenvalue of $\widetilde{\mathbf{A}}_{11}$, which is a contradiction because $\widetilde{\mathbf{A}}_{11}$ is positive semi-definite while $\lambda$ is negative. Hence

$$\sigma_{\max}(\widetilde{\mathbf{A}}_{21})^2 - \lambda\lambda_{\min}(\widetilde{\mathbf{A}}_{22}) \geq \lambda^2,$$

implying

$$\lambda^2 + \theta_2\lambda - \sigma_{\max}(\widetilde{\mathbf{A}}_{21})^2 \leq 0.$$

This means

$$-\frac{1}{2}\left(\theta_2 + \sqrt{\theta_2 + 4\sigma_{\max}(\widetilde{\mathbf{A}}_{21})^2}\right) \leq \lambda < 0. \tag{C.12}$$

Let us now estimate $\sigma_{\max}(\widetilde{\mathbf{A}}_{21})$. For any nonzero vector $\mathbf{x}$ of appropriate size, we have, with $\mathbf{y} = \mathbf{B}_{22}^{-1/2}\mathbf{A}_{12}^\top\mathbf{B}_{11}^{-1/2}\mathbf{x}$ and $\mathbf{z} = \mathbf{B}_{11}^{-1/2}\mathbf{x}$,

$$\begin{aligned}
\frac{\mathbf{x}^\top\widetilde{\mathbf{A}}_{21}^\top\widetilde{\mathbf{A}}_{21}\mathbf{x}}{\mathbf{x}^\top\mathbf{x}} &= \frac{\mathbf{x}^\top\mathbf{B}_{11}^{-1/2}\mathbf{A}_{12}\mathbf{B}_{22}^{-1}\mathbf{A}_{12}^\top\mathbf{B}_{11}^{-1/2}\mathbf{x}}{\mathbf{x}^\top\mathbf{x}} \\
&= \frac{\mathbf{y}^\top\mathbf{y}}{\mathbf{y}^\top\widetilde{\mathbf{A}}_{22}^{-1}\mathbf{y}} \cdot \frac{\mathbf{z}^\top\mathbf{A}_{12}\mathbf{A}_{22}^{-1}\mathbf{A}_{12}^\top\mathbf{z}}{\mathbf{z}^\top\left(\mathbf{A}_{11}^{(1)}+\mathbf{T}\right)\mathbf{z}} \cdot \frac{\mathbf{x}^\top\mathbf{B}_{11}^{-1/2}\left(\mathbf{A}_{11}^{(1)}+\mathbf{T}\right)\mathbf{B}_{11}^{-1/2}\mathbf{x}}{\mathbf{x}^\top\mathbf{x}} \\
&\le \Theta_2\Theta_3\Theta_1.
\end{aligned} \tag{C.13}$$

This implies

$$\sigma_{\max}(\widetilde{\mathbf{A}}_{21})^2 \le \Theta_1\Theta_2\Theta_3.$$

This together with (C.12) yields the required estimate for $\lambda_{\min}(\widetilde{\mathbf{A}})$.

Next, by pre-multiplying (C.11a) by $\mathbf{x}^\top$, (C.11b) by $\mathbf{y}^\top$, subtracting the resulting equations, and noting that $\mathbf{x}^\top\widetilde{\mathbf{A}}_{12}\mathbf{y} = \mathbf{y}^\top\widetilde{\mathbf{A}}_{21}\mathbf{x}$, we obtain

$$\mathbf{x}^\top\widetilde{\mathbf{A}}_{11}\mathbf{x} + \mathbf{y}^\top\widetilde{\mathbf{A}}_{22}\mathbf{y} = \lambda\mathbf{x}^\top\mathbf{x} - \lambda\mathbf{y}^\top\mathbf{y}.$$

Since $\theta_2 \le \lambda_{\min}(\widetilde{\mathbf{A}}_{22})$ and $\mathbf{x}^\top\widetilde{\mathbf{A}}_{11}\mathbf{x} \ge 0$, it follows that $\theta_2\mathbf{y}^\top\mathbf{y} \le \lambda\mathbf{x}^\top\mathbf{x} - \lambda\mathbf{y}^\top\mathbf{y}$ or

$$(\theta_2 + \lambda)\mathbf{y}^\top\mathbf{y} \le \lambda\mathbf{x}^\top\mathbf{x} \le 0.$$

Since $\mathbf{y}^\top\mathbf{y} > 0$, this results in $\lambda \le -\theta_2$, yielding an upper bound for $\lambda_-(\widetilde{\mathbf{A}})$.

Consider now $\lambda > 0$ in (C.11). Then $\widetilde{\mathbf{A}}_{22} + \lambda\mathbf{I}$ is invertible. Thus by eliminating $\mathbf{y}$ in (C.11) and pre-multiplying the resulting equation by $\mathbf{x}^\top$ we obtain

$$\lambda\mathbf{x}^\top\mathbf{x} = \mathbf{x}^\top\widetilde{\mathbf{A}}_{11}\mathbf{x} + \mathbf{x}^\top\widetilde{\mathbf{A}}_{12}\left(\widetilde{\mathbf{A}}_{22} + \lambda\mathbf{I}\right)^{-1}\widetilde{\mathbf{A}}_{21}\mathbf{x}. \tag{C.14}$$

Therefore,

$$\begin{aligned}
\lambda\mathbf{x}^\top\mathbf{x} &= \mathbf{x}^\top\widetilde{\mathbf{A}}_{11}\mathbf{x} + \mathbf{x}^\top\widetilde{\mathbf{A}}_{12}\widetilde{\mathbf{A}}_{22}^{-1/2}\left(\mathbf{I} + \lambda\widetilde{\mathbf{A}}_{22}^{-1}\right)^{-1}\widetilde{\mathbf{A}}_{22}^{-1/2}\widetilde{\mathbf{A}}_{21}\mathbf{x} \\
&\ge \mathbf{x}^\top\widetilde{\mathbf{A}}_{11}\mathbf{x} + \lambda_{\min}\left(\mathbf{I} + \lambda\widetilde{\mathbf{A}}_{22}^{-1}\right)^{-1}\mathbf{x}^\top\widetilde{\mathbf{A}}_{12}\widetilde{\mathbf{A}}_{22}^{-1}\widetilde{\mathbf{A}}_{21}\mathbf{x}. \\
&\ge \mathbf{x}^\top\widetilde{\mathbf{A}}_{11}\mathbf{x} + \frac{1}{1 + \dfrac{\lambda}{\lambda_{\min}(\widetilde{\mathbf{A}}_{22})}}\mathbf{x}^\top\widetilde{\mathbf{A}}_{12}\widetilde{\mathbf{A}}_{22}^{-1}\widetilde{\mathbf{A}}_{21}\mathbf{x} \\
&\ge \mathbf{x}^\top\widetilde{\mathbf{A}}_{11}\mathbf{x} + \frac{\theta_2}{\theta_2 + \lambda}\mathbf{x}^\top\widetilde{\mathbf{A}}_{12}\widetilde{\mathbf{A}}_{22}^{-1}\widetilde{\mathbf{A}}_{21}\mathbf{x} \\
&\ge \frac{\theta_2}{\theta_2 + \lambda}\mathbf{x}^\top\left(\widetilde{\mathbf{A}}_{11} + \widetilde{\mathbf{A}}_{12}\widetilde{\mathbf{A}}_{22}^{-1}\widetilde{\mathbf{A}}_{21}\right)\mathbf{x} \\
&= \frac{\theta_2}{\theta_2 + \lambda}\mathbf{x}^\top\mathbf{B}_{11}^{-1/2}\left(\mathbf{A}_{11} + \mathbf{A}_{12}\mathbf{A}_{22}^{-1}\mathbf{A}_{21}\right)\mathbf{B}_{11}^{-1/2}\mathbf{x}
\end{aligned}$$

$$\geq \frac{\theta_2\theta_3}{\theta_2+\lambda}\,\mathbf{x}^\top\mathbf{B}_{11}^{-1/2}\left(\mathbf{A}_{11}^{(1)}+\mathbf{T}\right)\mathbf{B}_{11}^{-1/2}\mathbf{x}$$
$$\geq \frac{\theta_1\theta_2\theta_3}{\theta_2+\lambda}\,\mathbf{x}^\top\mathbf{x}.$$

If $\mathbf{x}=\mathbf{0}$ then (C.11b) implies that $-\lambda<0$ is an eigenvalue of $\widetilde{\mathbf{A}}_{22}$, which is a contradiction. Hence $\mathbf{x}^\top\mathbf{x}>0$, and the above inequality yields

$$\lambda^2+\theta_2\lambda-\theta_1\theta_2\theta_3\geq 0,$$

resulting in the required lower bound for $\lambda_+(\widetilde{\mathbf{A}})$.

Finally, it follows from (C.14) and (C.13) that

$$\lambda\mathbf{x}^\top\mathbf{x}\leq\Theta_0\mathbf{x}^\top\mathbf{B}_{11}^{-1/2}\left(\mathbf{A}_{11}^{(1)}+\mathbf{T}\right)\mathbf{B}_{11}^{-1/2}\mathbf{x}+\lambda_{\max}\left((\widetilde{\mathbf{A}}_{22}+\lambda\mathbf{I})^{-1}\right)\mathbf{x}^\top\widetilde{\mathbf{A}}_{12}\widetilde{\mathbf{A}}_{21}\mathbf{x}$$
$$\leq\left(\Theta_0\Theta_1+\frac{\Theta_1\Theta_2\Theta_3}{\theta_2+\lambda}\right)\mathbf{x}^\top\mathbf{x},$$

which implies

$$\lambda^2+(\theta_2-\Theta_0\Theta_1)\lambda-\theta_2\Theta_0\Theta_1-\Theta_1\Theta_2\Theta_3\leq 0.$$

This yields the required upper bound for $\lambda_{\max}(\widetilde{\mathbf{A}})$. □

## C.2 Norms of Nodal Basis Functions

In this section we present results on norms of the nodal basis functions of the boundary element spaces used to approximate the solution of the hypersingular integral equation. Let $\Gamma$ be the boundary of a Lipschitz domain in $\mathbb{R}^d$, $d=2,3$. Assume that $\Gamma$ is partitioned into triangular or rectangular elements. On this mesh one can define nodal basis functions $\phi_k$ which take the value 1 at one nodal point and zero at all other nodal points.

**Lemma C.12.** *Let $\Gamma_k$ be a $(d-1)$-dimensional manifold in $\mathbb{R}^d$, $d=2,3$, which is the support of a nodal basis function $\phi_k$, and let $h_k$ be the average size of the elements in $\Gamma_k$. Then for $s\geq 0$*

$$\|\phi_k\|^2_{H_{\mathrm{I}}^{-s}(\Gamma_k)}\simeq h_k^{d-1+2s},$$
$$\|\phi_k\|^2_{\widetilde{H}_{\mathrm{I}}^{s}(\Gamma_k)}\simeq h_k^{d-1-2s},$$

*and*

$$\|\phi_k\|_{L^q(\Gamma_k)}=\|\phi_k\|_{L^q(\Gamma)}\simeq h_k^{(d-1)/q},\quad 1\leq q\leq\infty.$$

*The constants are independent of $\Gamma_k$ and $q$.*

*Proof.* Let $\hat{\phi}$ be the scale of $\phi_k$ onto a reference element $\hat{\Gamma}$. Then, noting that the dimension of $\Gamma_k$ is $d-1$, it follows from Lemma A.6 that

$$\|\phi_k\|^2_{H_{\rm I}^{-s}(\Gamma_k)} = h_k^{d-1+2s}\|\hat{\phi}\|^2_{H_{\rm I}^{-s}(\hat{\Gamma})} \simeq h_k^{d-1+2s},$$
$$\|\phi_k\|^2_{\widetilde{H}_{\rm I}^{s}(\Gamma_k)} = h_k^{d-1-2s}\|\hat{\phi}\|^2_{\widetilde{H}_{\rm I}^{s}(\hat{\Gamma})} \simeq h_k^{d-1-2s}.$$

The final result can be shown by a simple calculation. □

As a consequence of the above lemma, we have the following results. More general results can be found in [5, Theorem 4.8].

**Lemma C.13.** *Let $h_k$ and $\phi_k$ be given as in Lemma C.12. Then*

$$\|\phi_k\|^2_{\widetilde{H}_{\rm F}^{1/2}(\Gamma)} \simeq \|\phi_k\|^2_{H_{\rm F}^{1/2}(\Gamma)} \simeq h_k^{d-2}, \quad d=2,3, \tag{C.15}$$

*and*

$$\|\phi_k\|^2_{H_{\rm F}^{-1/2}(\Gamma)} \simeq \|\phi_k\|^2_{\widetilde{H}_{\rm F}^{-1/2}(\Gamma)} \simeq \begin{cases} h_k^2(1+|\log h_k|), & d=2, \\ h_k^3, & d=3, \end{cases} \tag{C.16}$$

*where the norms are defined by* (A.18).

*Proof.* We recall that the dimension of $\Gamma$ is $d-1$. First we prove (C.15). It follows from (A.20) and Lemma A.3 that

$$\|\phi_k\|^2_{H_{\rm F}^{1/2}(\Gamma)} \le \|\phi_k\|^2_{\widetilde{H}_{\rm F}^{1/2}(\Gamma)} \simeq \|\phi_k\|^2_{\widetilde{H}_{\rm I}^{1/2}(\Gamma)}.$$

Since $\operatorname{supp}(\phi_k) = \overline{\Gamma}_k$, Theorem A.10 yields $\|\phi_k\|_{\widetilde{H}_{\rm I}^{1/2}(\Gamma)} \le \|\phi_k\|_{\widetilde{H}_{\rm I}^{1/2}(\Gamma_k)}$, so that the above inequality and Lemma C.12 imply

$$\|\phi_k\|^2_{H_{\rm F}^{1/2}(\Gamma)} \le \|\phi_k\|^2_{\widetilde{H}_{\rm F}^{1/2}(\Gamma)} \lesssim h_k^{d-2}. \tag{C.17}$$

On the other hand, the equivalence of norms given in Subsection A.2.5, the definition of the Slobodetski norm in (A.26), and Lemma A.4 on the scaling property of the $H^{1/2}$-seminorm give

$$\begin{aligned}\|\phi_k\|^2_{H_{\rm F}^{1/2}(\Gamma)} &\simeq \|\phi_k\|^2_{H_{\rm S}^{1/2}(\Gamma)} > |\phi_k|^2_{H^{1/2}(\Gamma)} \\ &\ge |\phi_k|^2_{H^{1/2}(\Gamma_k)} = h_k^{d-2}|\widehat{\phi}_k|^2_{H^{1/2}(\hat{\Gamma})} \simeq h_k^{d-2}.\end{aligned} \tag{C.18}$$

Inequalities (C.17) and (C.18) yield (C.15).

Next we prove (C.16). Noting that $\operatorname{supp}(\phi_k) = \overline{\Gamma}_k$, we deduce from Lemma C.12 and Theorem A.10

$$h_k^d \simeq \|\phi_k\|^2_{H_{\rm I}^{-1/2}(\Gamma_k)} \le \|\phi_k\|^2_{H_{\rm I}^{-1/2}(\Gamma)}.$$

The equivalence of norms given in Subsection A.2.5 and (A.20) yield

$$\|\phi_k\|^2_{H_{\rm I}^{-1/2}(\Gamma)} \simeq \|\phi_k\|^2_{H_{\rm F}^{-1/2}(\Gamma)} \le \|\phi_k\|^2_{\widetilde{H}_{\rm F}^{-1/2}(\Gamma)},$$

so that

$$h_k^d \lesssim \|\phi_k\|^2_{H_{\rm F}^{-1/2}(\Gamma)} \le \|\phi_k\|^2_{\widetilde{H}_{\rm F}^{-1/2}(\Gamma)}. \tag{C.19}$$

On the other hand, Lemma A.18 Parts (iii)–(iv), and Lemma C.12 give

$$\begin{aligned}
\|\phi_k\|^2_{\widetilde{H}_{\rm F}^{-1/2}(\Gamma)} &\lesssim \begin{cases} \|\phi_k\|^2_{L^{4/3}(\Gamma)}, & d=3,\\ \dfrac{q}{q-1}\|\phi_k\|^2_{L^q(\Gamma)}, \quad \text{for all } q\in(1,2], & d=2, \end{cases}\\
&\simeq \begin{cases} h_k^3, & d=3,\\ \dfrac{q}{q-1}h_k^{2/q} \quad \text{for all } q\in(1,2], & d=2. \end{cases}
\end{aligned}$$

Therefore, for $d=3$ the above inequality and (C.19) yield (C.16). For $d=2$, let $q'$ be the conjugate of $q$, i.e., $1/q+1/q'=1$. Then $q'\ge 2$ and

$$\|\phi_k\|^2_{\widetilde{H}_{\rm F}^{-1/2}(\Gamma)} \lesssim q' h_k^2 h_k^{-2/q'}.$$

If $e^{-2}\le h_k\le 1$ then $|\log h_k|\le 2$. We choose $q=2$ so that $q'=2$. Then $q'h_k^{-2/q'} = 2h_k^{-1}\le 2e^2$ so that

$$\|\phi_k\|^2_{\widetilde{H}_{\rm F}^{-1/2}(\Gamma)} \lesssim h_k^2.$$

If $0<h_k<e^{-2}$ then $|\log h_k|>2$. We choose $q$ such that $q'=|\log h_k|$. Then $q'h_k^{-2/q'} = |\log h_k| h_k^{-2/|\log h_k|} = e^2|\log h_k|$ so that

$$\|\phi_k\|^2_{\widetilde{H}_{\rm F}^{-1/2}(\Gamma)} \lesssim h_k^2|\log h_k|.$$

In any case, we choose $q$ such that $q'=\max\{2,|\log h_k|\}$ to obtain

$$\|\phi_k\|^2_{\widetilde{H}_{\rm F}^{-1/2}(\Gamma)} \lesssim h_k^2\big(1+|\log h_k|\big).$$

Combining this with (C.19) gives

$$h_k^2 \lesssim \|\phi_k\|^2_{\widetilde{H}_{\rm F}^{-1/2}(\Gamma)} \lesssim h_k^2\big(1+|\log h_k|\big).$$

By using [5, Lemma 4.5], a sharper left inequality can be obtained for sufficiently small $h_k$

$$h_k^2\big(1+|\log h_k|\big) \lesssim \|\phi_k\|^2_{\widetilde{H}_{\rm F}^{-1/2}(\Gamma)} \lesssim h_k^2\big(1+|\log h_k|\big),$$

see the proof of [5, Theorem 4.8]. This yields (C.16) for $d=2$, completing the proof of the lemma. □

## C.3 Further Properties of the Hierarchical Basis Functions for the $p$-Version

The following three lemmas show bounds for Sobolev norms of functions belonging to the spans of $L_k$, $\mathscr{L}_k$, and $\widehat{\mathscr{L}_k}$, respectively. These lemmas are used in Chapter 4 and Chapter 6.

**Lemma C.14.** *Let $I := (-1,1)$ and $u = \sum_{k=1}^{p} u_k L_k$ where $L_k$ is the Legendre polynomial of degree k. Then*

$$\|u\|^2_{H_I^{1/2}(I)} \lesssim \sum_{k=1}^{p} k u_k^2 \tag{C.20}$$

*and*

$$\|u\|^2_{\widetilde{H}_I^{-1/2}(I)} \gtrsim \sum_{k=1}^{p} \frac{u_k^2}{k^3}. \tag{C.21}$$

*The constant is independent of u and p.*

*Proof.* We first prove (C.20). From the definition of the interpolation norm by (A.91), noting that $u \in H^1(I)$, we have

$$\begin{aligned}\|u\|^2_{H_I^{1/2}(I)} &= \int_0^\infty \inf_{v \in H^1(I)} \left(\|v\|^2_{L^2(I)} + t^2 \|u - v\|^2_{H_I^1(I)}\right) \frac{dt}{t^2} \\ &\le \int_0^\infty \inf_{v \in \mathscr{P}} \left(\|v\|^2_{L^2(I)} + t^2 \|u - v\|^2_{H_I^1(I)}\right) \frac{dt}{t^2}\end{aligned} \tag{C.22}$$

where $\mathscr{P} := \operatorname{span}\{L_k : k = 1, \dots, p\}$. By the orthogonality property of the Legendre polynomials we have

$$\|v\|^2_{L^2(I)} = \sum_{k=1}^{p} \frac{2v_k^2}{2k+1}.$$

On the other hand

$$\begin{aligned}\|u' - v'\|^2_{L^2(I)} &= \sum_{k=1}^{p}\sum_{\ell=1}^{p} (u_k - v_k)(u_\ell - v_\ell)\left\langle L_k', L_\ell'\right\rangle_I \\ &= \sum_{k=1}^{p}\sum_{\ell=1}^{p} k^{3/2}(u_k - v_k)\, \ell^{3/2}(u_\ell - v_\ell)\, k^{-3/2}\ell^{-3/2}\left\langle L_k', L_\ell'\right\rangle_I \\ &\le \left(\sum_{k=1}^{p} k^3 (u_k - v_k)^2\right)^{1/2} \left\|\{k^{-3/2}\ell^{-3/2}\left\langle L_k', L_\ell'\right\rangle_I\}_{k,\ell=1,\dots,p}\right\|_2 \left(\sum_{\ell=1}^{p} \ell^3 (u_\ell - v_\ell)^2\right)^{1/2} \\ &= \left\|\{k^{-3/2}\ell^{-3/2}\left\langle L_k', L_\ell'\right\rangle_I\}_{k,\ell=1,\dots,p}\right\|_2 \sum_{k=1}^{p} k^3 (u_k - v_k)^2.\end{aligned} \tag{C.23}$$

It follows from the definition of matrix norms that

$$\left\|\{k^{-3/2}\ell^{-3/2}\left\langle L_k', L_\ell'\right\rangle_I\}_{k,\ell=1,\ldots,p}\right\|_2 \le \left\|\{k^{-3/2}\ell^{-3/2}\left\langle L_k', L_\ell'\right\rangle_I\}_{k,\ell=1,\ldots,p}\right\|_\infty$$
$$= \max_{1\le k\le p}\left(k^{-3/2}\sum_{\ell=1}^{p}\ell^{-3/2}\left\langle L_k', L_\ell'\right\rangle_I\right). \quad \text{(C.24)}$$

Lemma C.2 and a straightforward calculation give

$$\begin{aligned}
&k^{-3/2}\sum_{\ell=1}^{p}\ell^{-3/2}\left\langle L_k', L_\ell'\right\rangle_I \\
&= 2k^{-3/2}\sum_{\ell=1}^{p}\ell^{-3/2}\sum_{m=0}^{[(k-1)/2]}\sum_{n=0}^{[(\ell-1)/2]}\left(2(\ell-2n)-1\right)\delta_{k-1-2m,\ell-1-2n} \\
&\le 2k^{-3/2}\sum_{\ell=1}^{p}\ell^{-3/2}\sum_{m=0}^{k-1}\sum_{n=0}^{\ell-1}\left(2(\ell-n)-1\right)\delta_{k-1-m,\ell-1-n}.
\end{aligned}$$

Since, by change of variables,

$$\sum_{m=0}^{k-1}\sum_{n=0}^{\ell-1}\left(2(\ell-n)-1\right)\delta_{k-1-m,\ell-1-n} = \sum_{m=0}^{k-1}\sum_{n=0}^{\ell-1}(2n+1)\delta_{m,n} = \sum_{m=0}^{k-1}(2m+1)\sum_{n=0}^{\ell-1}\delta_{m,n}$$

we have

$$\begin{aligned}
k^{-3/2}\sum_{\ell=1}^{p}\ell^{-3/2}\left\langle L_k', L_\ell'\right\rangle_I &\le 2k^{-3/2}\sum_{m=0}^{k-1}(2m+1)\sum_{\ell=1}^{p}\ell^{-3/2}\sum_{n=0}^{\ell-1}\delta_{m,n} \\
&= 2k^{-3/2}\sum_{m=0}^{k-1}(2m+1)\sum_{\ell=m+1}^{p}\ell^{-3/2} \\
&\le 2k^{-3/2}\sum_{m=0}^{k-1}(2m+1)\int_{m+1}^{p}x^{-3/2}\,\mathrm{d}x \\
&\le 4k^{-3/2}\sum_{m=0}^{k-1}(2m+1)(m+1)^{-1/2} \\
&\le 8k^{-3/2}\sum_{m=0}^{k-1}(m+1)^{1/2} \le 8. \qquad \text{(C.25)}
\end{aligned}$$

Inequalities (C.23), (C.24), and (C.25) yield

$$\|u'-v'\|_{L^2(I)}^2 \lesssim \sum_{k=1}^{p}k^3(u_k-v_k)^2,$$

so that

$$\|u-v\|_{H_I^1(I)}^2 \lesssim \sum_{k=1}^{p}\left(\frac{1}{k}+k^3\right)v_k^2 \lesssim \sum_{k=1}^{p}k^3v_k^2.$$

This inequality and (C.22) imply

$$\|u\|^2_{H^{1/2}_{\mathrm{I}}(I)} \lesssim \int_0^\infty t^{-2} \inf_{(v_k)\in\mathbb{R}^p} \sum_{k=1}^p \left(\frac{v_k^2}{k} + t^2k^3(u_k - v_k)^2\right) \mathrm{d}t$$

$$= \int_0^\infty t^{-2} \sum_{k=1}^p \inf_{v_k\in\mathbb{R}} \left(\frac{v_k^2}{k} + t^2k^3(u_k - v_k)^2\right) \mathrm{d}t.$$

Note that if $f(x) = ax^2 + b(x-c)^2$ with $a,b > 0$, then $\min_{x\in\mathbb{R}} f(x) = f(bc/(a+b)) = abc^2/(a+b)$. Hence,

$$\|u\|^2_{H^{1/2}_{\mathrm{I}}(I)} \lesssim \sum_{k=1}^p \int_0^\infty \frac{k^3u_k^2}{1+k^4t^2}\,\mathrm{d}t = \sum_{k=1}^p \frac{\pi}{2} k u_k^2,$$

proving (C.20). Inequality (C.21) is proved by duality as follows:

$$\|u\|_{\widetilde{H}^{-1/2}_{\mathrm{I}}(I)} = \sup_{v\in H^{1/2}_{\mathrm{I}}(I)} \frac{\langle u,v\rangle_I}{\|v\|_{H^{1/2}_{\mathrm{I}}(I)}} \ge \frac{\langle u,v\rangle_I}{\|v\|_{H^{1/2}_{\mathrm{I}}(I)}} \quad \forall v\in\mathscr{P}.$$

In particular, for $v = \sum_{k=1}^p v_k L_k$ with $v_k = u_k/k^2$ we have, thanks to (C.1) and (C.20),

$$\|u\|_{\widetilde{H}^{-1/2}_{\mathrm{I}}(I)} \gtrsim \frac{\sum_{k=1}^p u_k^2/k^3}{\left(\sum_{k=1}^p u_k^2/k^3\right)^{1/2}} = \left(\sum_{k=1}^p u_k^2/k^3\right)^{1/2},$$

completing the proof of the lemma. □

**Lemma C.15.** *Let $I=(-1,1)$ and $u_k = c_k\widehat{\mathscr{L}_k}$, $k\ge 2$, where $\widehat{\mathscr{L}_k}$ is defined in (C.4). If $u = \sum_{k=2}^p u_k$ for some $p\ge 2$, then*

$$\sum_{k=2}^p c_k^2 \lesssim \|u\|^2_{\widetilde{H}^{1/2}_{\mathrm{I}}(I)} \lesssim \sum_{k=2}^p k\, c_k^2. \tag{C.26}$$

*Consequently,*

$$\|u\|^2_{\widetilde{H}^{1/2}_{\mathrm{I}}(I)} \lesssim \sum_{k=2}^p \|u_k\|^2_{\widetilde{H}^{1/2}_{\mathrm{I}}(I)} \lesssim p\,\|u\|^2_{\widetilde{H}^{1/2}_{\mathrm{I}}(I)}. \tag{C.27}$$

*The constants are independent of $u$ and $p$.*

*Proof.* Recalling the definition of $\widehat{\mathscr{L}_k}$ in Section C.1.1 and the orthogonality of the Legendre polynomials, we have

$$\|u_k\|^2_{\widetilde{H}^1_{\mathrm{I}}(I)} = c_k^2 \frac{\|L_{k-1}\|^2_{L^2(I)}}{\|\mathscr{L}_k\|^2_{L^2(I)}}$$

and

$$\|u\|^2_{\widetilde{H}^1_{\rm I}(I)} = \Big\|\sum_{k=2}^{p} \frac{c_k}{\|\mathscr{L}_k\|_{L^2(I)}} L_{k-1}\Big\|^2_{L^2(I)} = \sum_{k=2}^{p} \frac{c_k^2}{\|\mathscr{L}_k\|^2_{L^2(I)}} \|L_{k-1}\|^2_{L^2(I)}.$$

It follows from (C.1) and (C.5) that

$$\|u\|^2_{\widetilde{H}^1_{\rm I}(I)} \simeq \sum_{k=2}^{p} k^2 c_k^2. \tag{C.28}$$

Next we show that

$$\sum_{k=2}^{p} \frac{c_k^2}{k^2} \lesssim \|u\|^2_{L^2(I)} \lesssim \sum_{k=2}^{p} c_k^2. \tag{C.29}$$

Recalling the calculation of $\left\langle \widehat{\mathscr{L}_k}, \widehat{\mathscr{L}_l} \right\rangle_I$ in (C.7), we obtain

$$\begin{aligned}
\|u\|^2_{L^2(I)} &= \sum_{k=2}^{p}\sum_{l=2}^{p} c_k c_l \langle \widehat{\mathscr{L}_k}, \widehat{\mathscr{L}_l} \rangle_I \\
&= \sum_{k=2}^{p} c_k^2 + 2\sum_{k=2}^{p-2} c_k c_{k+2} \langle \widehat{\mathscr{L}_k}, \widehat{\mathscr{L}_{k+2}} \rangle_I \\
&= \sum_{k=2}^{p} c_k^2 - \sum_{k=2}^{p-2} c_k c_{k+2} \sqrt{\frac{(2k-3)(2k+5)}{(2k-1)(2k+3)}}.
\end{aligned} \tag{C.30}$$

Since $(2k-3)(2k+5) \le (2k-1)(2k+3)$ we deduce

$$\|u\|^2_{L^2(I)} \le \sum_{k=2}^{p} c_k^2 + \frac{1}{2}\sum_{k=2}^{p-2} c_k^2 + \frac{1}{2}\sum_{k=2}^{p-2} c_{k+2}^2 \le 2\sum_{k=2}^{p} c_k^2.$$

which proves the right inequality in (C.29). To prove the left inequality in (C.29) we observe that (C.30) and the Cauchy–Schwarz inequality give

$$\begin{aligned}
\|u\|^2_{L^2(I)} &\ge \sum_{k=2}^{p} c_k^2 \\
&\quad - \left(\sum_{k=2}^{p-2} c_k^2 \sqrt{\frac{(2k-3)(2k+5)}{(2k-1)(2k+3)}}\right)^{1/2} \left(\sum_{k=2}^{p-2} c_{k+2}^2 \sqrt{\frac{(2k-3)(2k+5)}{(2k-1)(2k+3)}}\right)^{1/2} \\
&= \sum_{k=2}^{p} c_k^2 \\
&\quad - \left(\sum_{k=2}^{p-2} c_k^2 \sqrt{\frac{(2k-3)(2k+5)}{(2k-1)(2k+3)}}\right)^{1/2} \left(\sum_{k=4}^{p} c_k^2 \sqrt{\frac{(2k-7)(2k+1)}{(2k-5)(2k-1)}}\right)^{1/2} \\
&\ge \sum_{k=2}^{p} c_k^2 - \frac{1}{2}\sum_{k=2}^{p-2} c_k^2 \sqrt{\frac{(2k-3)(2k+5)}{(2k-1)(2k+3)}} - \frac{1}{2}\sum_{k=4}^{p} c_k^2 \sqrt{\frac{(2k-7)(2k+1)}{(2k-5)(2k-1)}}.
\end{aligned}$$

It suffices to prove that

$$\sum_{k=2}^{p} c_k^2 - \frac{1}{2}\sum_{k=2}^{p-2} c_k^2 \sqrt{\frac{(2k-3)(2k+5)}{(2k-1)(2k+3)}} - \frac{1}{2}\sum_{k=4}^{p} c_k^2 \sqrt{\frac{(2k-7)(2k+1)}{(2k-5)(2k-1)}} \gtrsim \sum_{k=2}^{p} \frac{c_k^2}{k^2}.$$

This can be seen by noting that

$$\lim_{k\to\infty} k^2 \left(1 - \frac{1}{2}\sqrt{\frac{(2k-3)(2k+5)}{(2k-1)(2k+3)}} - \frac{1}{2}\sqrt{\frac{(2k-7)(2k+1)}{(2k-5)(2k-1)}}\right) = \frac{3}{2}.$$

Thus we showed the left inequality of (C.29). Inequality (C.26) is a consequence of (C.28) and (C.29) and interpolation, noting that the interpolation space between two spaces of polynomials is a space of polynomial; see [140, Theorem 1].

Inequality (C.27) follows from (C.26) if one notes that

$$\|u_k\|_{L^2(I)}^2 = c_k^2 \quad \text{and} \quad \|u_k\|_{\widetilde{H}_I^1(I)}^2 = c_k^2 \frac{\|L_{k-1}\|_{L^2(I)}^2}{\|\mathscr{L}_k\|_{L^2(I)}^2} \simeq k^2 c_k^2$$

so that $\|u_k\|_{\widetilde{H}_I^{1/2}(I)}^2 \simeq k c_k^2$. □

**Lemma C.16.** *Let $I = (-1,1)$ and $u = \sum_{k=2}^{p} c_k \mathscr{L}_k$, $p \ge 2$, where $\mathscr{L}_k$ is defined in (C.4). Then*

$$\sum_{k=2}^{p} \frac{c_k^2}{k^3} \lesssim \|u\|_{\widetilde{H}_I^{1/2}(I)}^2 \lesssim \sum_{k=2}^{p} \frac{c_k^2}{k^2}.$$

*The constants are independent of u and p.*

*Proof.* The lemma is a consequence of Lemma C.15, noting that

$$u = \sum_{k=2}^{p} \frac{c_k}{\|\mathscr{L}_k\|_{L^2(I)}} \widehat{\mathscr{L}_k}$$

and that $\|\mathscr{L}_k\|_{L^2(I)} \simeq k^{-3/2}$; see (C.5). □

The next lemma provides an upper bound for the bilinear form $a_{\mathfrak{V}}(\cdot,\cdot)$ acting on Legendre polynomials.

**Lemma C.17.** *Let I and J be two bounded and open intervals on the real line satisfying $\overline{I} \cap \overline{J} = \varnothing$. Let $L_{p_i,I}$ (respectively, $L_{p_j,J}$) be the transformation of the Legendre polynomial $L_{p_i}$ (respectively, $L_{p_j}$) onto I (respectively, J). The functions $L_{p_i,I}$ and $L_{p_j,J}$ are extended by zero outside I and J if necessary. Then*

$$a_{\mathfrak{V}}(L_{p_i,I}, L_{p_j,J}) \le \frac{1}{\pi} \frac{2|I|\,|J| \max(|I|/2, |J|/2)^{p_i+p_j}}{\operatorname{dist}(I,J)^{p_i+p_j}\sqrt{(2p_i+1)(2p_j+1)}}.$$

*Here $|I|$ and $|J|$ are the lengths of I and J.*

*Proof.* Let $x_0$ be the midpoint of $I$ and $y_0$ be the midpoint of $J$. Then there holds using the Taylor expansion of $\log|x-y|$

$$\log|x-y| = \sum_{\nu=0}^{p_i+p_j-1} \sum_{|\alpha|=\nu} \frac{\partial_x^{\alpha_1}\partial_y^{\alpha_2} \log|x_0-y_0|}{\nu!}(x-x_0)^{\alpha_1}(y-y_0)^{\alpha_2} + R_{p_i+p_j}(x,y)$$

and

$$R_{p_i+p_j}(x,y) = \sum_{|\alpha|=p_i+p_j} \frac{\partial_x^{\alpha_1}\partial_y^{\alpha_2} \log|x_0+\theta(x-x_0)-y_0-\theta(y-y_0)|}{(p_i+p_j)!}(x-x_0)^{\alpha_1}(y-y_0)^{\alpha_2},$$

with $\theta \in (0,1)$. Due to the orthogonal property of the Legendre polynomials the first sum vanishes and we have

$$\begin{aligned} a_{\mathfrak{V}}(L_{p_i,I}, L_{p_j,J}) &= -\frac{1}{\pi} \iint_{I\times J} L_{p_i,I}(x) L_{p_j,J}(y) \log|x-y|\, ds_y\, ds_x \\ &= -\frac{1}{\pi} \iint_{I\times J} L_{p_i,I}(x) L_{p_j,J}(y) R_{p_i+p_j}(x,y)\, ds_y\, ds_x. \end{aligned}$$

Applying the Cauchy–Schwarz inequality two times we obtain

$$\begin{aligned} a_{\mathfrak{V}}(L_{p_i,I}, L_{p_j,J}) &\le \frac{1}{\pi}\Big(\int_I L_{p_i,I}^2(x)\, \mathrm{d}s_x\Big)^{1/2}\Big(\int_J L_{p_j,J}^2(y)\, \mathrm{d}s_y\Big)^{1/2} \\ &\quad \times \Big(\iint_{I\times J} R_{p_i+p_j}^2(x,y)\, ds_y\, ds_x\Big)^{1/2} \\ &= \frac{|I|^{1/2}|J|^{1/2}}{\pi} \frac{2}{\sqrt{(2p_i+1)(2p_j+1)}} \Big(\iint_{I\times J} R_{p_i+p_j}^2(x,y)\, ds_y\, ds_x\Big)^{1/2} \\ &\le \frac{1}{\pi} \frac{2|I|\,|J|}{\sqrt{(2p_i+1)(2p_j+1)}} \sup_{(x,y)\in I\times J} |R_{p_i+p_j}(x,y)|. \end{aligned} \tag{C.31}$$

Since

$$|\partial_x^{\alpha_1}\partial_y^{\alpha_2} \log|x-y|| \le \frac{(\alpha_1+\alpha_2-1)!}{|x-y|^{\alpha_1+\alpha_2}}$$

we have

$$\begin{aligned} &\sup_{(x,y)\in I\times J} |R_{p_i+p_j}(x,y)| \\ &\le \sup_{(x,y)\in I\times J} \sum_{|\alpha|=p_i+p_j} \frac{|x-x_0|^{\alpha_1}|y-y_0|^{\alpha_2}}{|x_0+\theta(x-x_0)-y_0-\theta(y-y_0)|^{p_i+p_j}} \frac{(\alpha_1+\alpha_2-1)!}{(p_i+p_j)!} \end{aligned}$$

$$
\begin{aligned}
&\le \sup_{(x,y)\in I\times J} \sum_{|\alpha|=p_i+p_j} \frac{(|I|/2)^{\alpha_1}(|J|/2)^{\alpha_2}}{|x_0+\theta(x-x_0)-y_0-\theta(y-y_0)|^{p_i+p_j}} \frac{1}{p_i+p_j} \\
&\le \frac{1}{p_i+p_j} \frac{1}{\operatorname{dist}(I,J)^{p_i+p_j}} \sum_{|\alpha|=p_i+p_j} (|I|/2)^{\alpha_1}(|J|/2)^{\alpha_2} \\
&\le \frac{1}{\operatorname{dist}(I,J)^{p_i+p_j}} \max(|I|/2,|J|/2)^{p_i+p_j}. \qquad (C.32)
\end{aligned}
$$

Inequalities (C.31) and (C.32) imply

$$
a_{\mathfrak{W}}(L_{p_i,I},L_{p_j,J}) \le \frac{1}{\pi}\frac{2|I|\,|J|}{\sqrt{(2p_i+1)(2p_j+1)}}\frac{1}{\operatorname{dist}(I,J)^{p_i+p_j}}\max(|I|/2,|J|/2)^{p_i+p_j}.
$$

This completes the proof. □

## C.4 Some Properties of Polynomials

We collect in this section some useful results on polynomials.

### C.4.1 *Inverse properties*

In this subsection we present the inverse property of piecewise polynomial functions. First we state a classical theorem by Andrey Markov in the 1890s, the so-called Markov Theorem,

$$
\max_{-1\le x\le 1}|f'(x)| \le p^2 \max_{-1\le x\le 1}|f(x)| \qquad (C.33)
$$

where $f$ is a polynomial of degree $p$, $p\ge 1$. Later, in 1932 Erhard Schmidt stated (without proof) in the Sitzungsberichte der Preussicher Akademie the following inequality

$$
\Big(\int_{-1}^{1}|f'(x)|^2\,\mathrm{d}x\Big)^{1/2} \le k_p\, p^2\Big(\int_{-1}^{1}|f(x)|^2\,\mathrm{d}x\Big)^{1/2} \qquad (C.34)
$$

where

$$
\lim_{p\to\infty} k_p = 1.
$$

In their 1937 paper [122], Hille, Szegö, and Tamarkin gave various proofs for both inequalities (C.33) and (C.34). In particular, they proved that

$$
\lim_{p\to\infty} k_p = 1/\pi.
$$

R. Bellman, in his 1944 paper [22], provided an elegant proof for (C.34) with an explicit upper bound

$$\Big(\int_{-1}^{1} |f'(x)|^2 \, dx\Big)^{1/2} \le \frac{(p+1)^2}{\sqrt{2}} \Big(\int_{-1}^{1} |f(x)|^2 \, dx\Big)^{1/2}. \tag{C.35}$$

We now extend the above results to two-dimensional Lipschitz domains with fractional-order Sobolev norms. The generalisation also includes piecewise polynomial functions.

**Lemma C.18.** *Let $\Gamma$ be a rectangle in $\mathbb{R}^2$ which is partitioned by subrectangles $\Gamma_j$ whose side lengths are proportional to $H \in (0,1)$, $j = 1,\dots,J$. Let*

$$\mathscr{V} := \big\{v : v|_{\Gamma_j} \text{ is a polynomial of degree } p \ge 1,\ j = 1,\dots,J\big\}.$$

(i) *For any $0 \le s \le r \le 1$,*

$$\|v\|_{\widetilde{H}^r_{\mathrm{I}}(\Gamma)} \le H^{-(r-s)}\,(p+1)^{2(r-s)} \|v\|_{\widetilde{H}^s_{\mathrm{I}}(\Gamma)}, \quad v \in \mathscr{V} \cap \widetilde{H}^r(\Gamma),$$
$$_*\|v\|_{H^r_{\mathrm{I}}(\Gamma)} \le H^{-(r-s)}\,(p+2)^{2(r-s)}\, {}_*\|v\|_{H^s_{\mathrm{I}}(\Gamma)}, \quad v \in \mathscr{V} \cap H^r(\Gamma).$$

(ii) *For any $-1 \le s \le r \le 0$,*

$$_*\|v\|_{\widetilde{H}^r_{\mathrm{I}}(\Gamma)} \lesssim H^{-(r-s)}\,(p+2)^{2(r-s)}\, {}_*\|v\|_{\widetilde{H}^s_{\mathrm{I}}(\Gamma)}, \quad v \in \mathscr{V} \cap \widetilde{H}^r(\Gamma),$$
$$\|v\|_{H^r_{\mathrm{I}}(\Gamma)} \lesssim H^{-(r-s)}\,(p+1)^{2(r-s)} \|v\|_{H^s_{\mathrm{I}}(\Gamma)}, \quad v \in \mathscr{V} \cap H^r(\Gamma).$$

*The constants are independent of $v$, $H$, $p$, $s$, and $r$.*

*Proof.* This lemma was presented and proved in [108, Lemma 4] with slightly different norms. Our choice of norms is to ensure scalability. We first prove Part (i). By invoking Lemma A.6 Part (i), we obtain with $\widehat{\Gamma_j} = (-1,1)$ being the reference element that rescales $\Gamma_j$

$$\|v\|^2_{\widetilde{H}^1_{\mathrm{I}}(\Gamma)} = |v|^2_{H^1(\Gamma)} = \sum_{j=1}^{J} |v|_{\Gamma_j}|^2_{H^1(\Gamma_j)} = \sum_{j=1}^{J} |\widehat{v|_{\Gamma_j}}|^2_{H^1(\widehat{\Gamma}_j)}, \quad v \in \mathscr{V} \cap \widetilde{H}^1(\Gamma).$$

It follows from (C.35) and Lemma A.6 Part (i) that

$$|\widehat{v|_{\Gamma_j}}|^2_{H^1(\widehat{\Gamma}_j)} \le (p+1)^4 \|\widehat{v|_{\Gamma_j}}\|^2_{\widetilde{H}^0_{\mathrm{I}}(\widehat{\Gamma}_j)} = H^{-2}\,(p+1)^4 \|v|_{\Gamma_j}\|^2_{\widetilde{H}^0_{\mathrm{I}}(\Gamma_j)}.$$

Altogether, we deduce

$$\|v\|_{\widetilde{H}^1_{\mathrm{I}}(\Gamma)} \le H^{-1}\,(p+1)^2 \|v\|_{\widetilde{H}^0_{\mathrm{I}}(\Gamma)}, \quad v \in \mathscr{V} \cap \widetilde{H}^1(\Gamma).$$

Let

$$A_0 = \mathscr{V} \cap \widetilde{H}^0(\Gamma) \quad \text{and} \quad B_0 = A_1 = B_1 = \mathscr{V} \cap \widetilde{H}^1(\Gamma).$$

Note that $\mathscr{V} \cap \widetilde{H}^0(\Gamma)$ and $\mathscr{V} \cap \widetilde{H}^1(\Gamma)$ are algebraically equal, but they are equipped with different norms. Hence it is possible to define

$$T_j : A_j \to B_j, \quad T_j u = u,\ u \in A_j, \quad j = 0,1.$$

Making use of the fact that interpolating between spaces of polynomials yields spaces of polynomials (see [140] for the case of one-dimensional domains and [23] for higher dimensions) we deduce by invoking Theorem A.4, with $M_0 = H^{-1}p^2$, $M_1 = 1$, and $\theta = s$, that $A_s = [A_0, A_1]_\theta = \mathscr{V} \cap \widetilde{H}^s(\Gamma)$ and

$$\|v\|_{\widetilde{H}^1_\mathrm{I}(\Gamma)} \le H^{-(1-s)}\,(p+1)^{2(1-s)}\|v\|_{\widetilde{H}^s_\mathrm{I}(\Gamma)}, \quad v \in \mathscr{V} \cap \widetilde{H}^1(\Gamma),$$

noting that $T_\theta u = u$. In the same manner, let

$$A_0 = B_0 = A_1 = \mathscr{V} \cap \widetilde{H}^s(\Gamma) \quad \text{and} \quad B_1 = \mathscr{V} \cap \widetilde{H}^1(\Gamma).$$

Another application of Theorem A.4 with $\theta = (r-s)/(1-s)$, $M_0 = 1$, and $M_1 = H^{-(1-s)}p^{2(1-s)}$ yields

$$\|v\|_{\widetilde{H}^r_\mathrm{I}(\Gamma)} \le H^{-(r-s)}\,(p+1)^{2(r-s)}\|v\|_{\widetilde{H}^s_\mathrm{I}(\Gamma)}, \quad v \in \mathscr{V} \cap \widetilde{H}^r(\Gamma),$$

proving the first inequality in Part (i).

To prove the second inequality in Part (i), we use the same argument as above. Firstly, it follows from the definition of the ${}_*\|\cdot\|_{\widetilde{H}^1_\mathrm{I}(\Gamma)}$-norm in (A.46) and (A.47) and the scaling property proved in Lemma A.7 that

$${}_*\|v\|^2_{H^1_\mathrm{I}(\Gamma)} \le \sum_{j=1}^{J} {}_*\|v|_{\Gamma_j}\|^2_{H^1_\mathrm{I}(\Gamma_j)} = \sum_{j=1}^{J} {}_*\|\widehat{v|_{\Gamma_j}}\|^2_{H^1_\mathrm{I}(\widehat{\Gamma}_j)}.$$

It follows from (C.35) and Lemma A.7 Part (i) that

$$\begin{aligned}
{}_*\|\widehat{v|_{\Gamma_j}}\|^2_{H^1_\mathrm{I}(\widehat{\Gamma}_j)} &= \frac{1}{|\widehat{\Gamma}_j|}\|\widehat{v|_{\Gamma_j}}\|^2_{H^0_\mathrm{I}(\widehat{\Gamma}_j)} + |\widehat{v|_{\Gamma_j}}|^2_{H^1(\widehat{\Gamma}_j)} \\
&\le \big((p+1)^4+1\big)\|\widehat{v|_{\Gamma_j}}\|^2_{H^0_\mathrm{I}(\widehat{\Gamma}_j)} \le H^{-2}(p+2)^4\|v|_{\Gamma_j}\|^2_{H^0_\mathrm{I}(\Gamma_j)}.
\end{aligned}$$

Altogether we deduce

$$\begin{aligned}
{}_*\|v\|^2_{H^1_\mathrm{I}(\Gamma)} &\le H^{-2}(p+2)^4\|v\|^2_{H^0_\mathrm{I}(\Gamma)} \\
&= H^{-2}(p+2)^4\,{}_*\|v\|^2_{H^0_\mathrm{I}(\Gamma)}, \quad v \in \mathscr{V} \cap H^1(\Gamma).
\end{aligned}$$

Noting the definition of ${}_*\|v\|_{H^s_\mathrm{I}(\Gamma)}$ by (A.47) and invoking Theorem A.4, we obtain the required inequality.

Part (ii) is proved by duality arguments. Indeed, let $-1 \le s \le r \le 0$. Then

$${}_*\|w\|_{H^{-s}_\mathrm{I}(\Gamma)} \le H^{-(r-s)}(p+2)^{2(r-s)}\,{}_*\|w\|_{H^{-r}_\mathrm{I}(\Gamma)}$$

so that

$$
\begin{aligned}
{}_*\|v\|_{\widetilde{H}_{\mathrm{I}}^r(\Gamma)} &= \sup_{0\neq w\in H^{-r}(\Gamma)} \frac{\langle v,w\rangle_\Gamma}{{}_*\|w\|_{H_{\mathrm{I}}^{-r}(\Gamma)}} \\
&\le H^{-(r-s)}(p+2)^{2(r-s)} \sup_{0\neq w\in H^{-r}(\Gamma)} \frac{\langle v,w\rangle_\Gamma}{{}_*\|w\|_{H_{\mathrm{I}}^{-s}(\Gamma)}} \\
&\le H^{-(r-s)}(p+2)^{2(r-s)} \sup_{0\neq w\in H^{-s}(\Gamma)} \frac{\langle v,w\rangle_\Gamma}{{}_*\|w\|_{H_{\mathrm{I}}^{-s}(\Gamma)}} \\
&= H^{-(r-s)}(p+2)^{2(r-s)}\, {}_*\|v\|_{\widetilde{H}_{\mathrm{I}}^s(\Gamma)}.
\end{aligned}
$$

Similar arguments prove the second inequality in part (ii), completing the proof of the lemma. □

## C.4.2 *Some useful bounds for the $h$ and $p$ finite element functions*

### C.4.2.1 The $h$-version in one dimension

The following result gives a bound for the $L^\infty$-norm by the $H^{1/2}$-weighted norm. The corresponding $p$-version is Theorem C.1 (C.49) and the corresponding $h$-version in two dimensions is Lemma C.23 and Lemma C.25.

**Lemma C.19.** *Let $I$ be an interval of length $H$ which is partitioned into subintervals $J$ which form a quasi-uniform mesh of size $h$. If $v$ is a continuous function defined on $I$ whose restriction onto each subinterval is a polynomial of degree 1, then*

$$
\|v\|_{L^\infty(I)} \lesssim \left(1+\log\frac{H}{h}\right)^{1/2} \|v\|_{H_{\mathrm{w}}^{1/2}(I)} \tag{C.36}
$$

*where, see Subsection A.2.6.1,*

$$
\|v\|^2_{H_{\mathrm{w}}^{1/2}(I)} := \frac{1}{H}\|v\|^2_{L^2(I)} + |v|^2_{H^{1/2}(I)}.
$$

*Moreover, if $J$ is one of the subintervals with length $h$ which partition $I$, then*

$$
\|v\|_{L^2(J)} \lesssim h^{1/2}\left(1+\log\frac{H}{h}\right)^{1/2} \|v\|_{H_{\mathrm{w}}^{1/2}(I)}. \tag{C.37}
$$

*The constants are independent of $v$, $H$, and $h$.*

*Proof.* Inequality (C.36) is proved in [70, Lemma 1]. We include the proof here for completeness. As in the proof of Lemma C.23 we consider the reference element $\widehat{I} = (0,1)$ and $\widehat{v}(\widehat{\mathbf{x}}) = v(H\widehat{\mathbf{x}})$ for all $\widehat{\mathbf{x}} \in \widehat{I}$. The partition of $I$ is transplanted into a partition of $\widehat{I}$ with size $\widehat{h}$, which is then extended into a partition of $\widehat{\Omega}$, the triangle of vertices $(0,0)$, $(1,0)$, and $(0,1)$. We extend $\widehat{v}$ to be a continuous piecewise linear function $\widehat{V}$ on $\widehat{\Omega}$ such that

$$_*\|\widehat{V}\|_{H^1_{\mathrm{w}}(\widehat{\Omega})} \lesssim \|\widehat{v}\|_{H^{1/2}_{\mathrm{w}}(\widehat{I})}.$$

(Note that the weights in the definitions of $_*\|\cdot\|_{H^1_{\mathrm{w}}(\widehat{\Omega})}$ and $\|\cdot\|_{H^{1/2}_{\mathrm{w}}(\widehat{\Omega})}$ are 1.) It then follows from Lemma C.23, more precisely, from (C.42), that

$$\begin{aligned}\|v\|_{L^\infty(I)} = \|\widehat{v}\|_{L^\infty(\widehat{I})} &\le \|\widehat{V}\|_{L^\infty(\widehat{\Omega})} \lesssim \left(1+\log\frac{1}{\widehat{h}}\right)^{1/2} {}_*\|\widehat{V}\|_{H^1_{\mathrm{w}}(\widehat{\Omega})} \\ &\lesssim \left(1+\log\frac{1}{\widehat{h}}\right)^{1/2} \|\widehat{v}\|_{H^{1/2}_{\mathrm{w}}(\widehat{I})} = \left(1+\log\frac{H}{h}\right)^{1/2} \|v\|_{H^{1/2}_{\mathrm{w}}(I)}\end{aligned}$$

where in the last step we used (A.50) in Lemma A.5. This proves (C.36). Inequality (C.37) is a consequence of (C.36) and

$$\|v\|^2_{L^2(J)} \le h\|v\|^2_{L^\infty(J)} \le h\|v\|^2_{L^\infty(I)}. \qquad \square$$

The next lemma provides a useful bound for the weighted norm $\|\cdot\|_{\widetilde{H}^{1/2}_{\mathrm{w}}(I)}$. The analogous $p$-version is (C.50).

**Lemma C.20.** *Let $I=(0,H)$ be partitioned into subintervals which form a quasi-uniform mesh of size $h$. If $v$ is a continuous function defined on $I$ whose restriction onto each subinterval is a polynomial of degree 1, and if $v(0)=v(H)=0$, then*

$$\|v\|_{\widetilde{H}^{1/2}_{\mathrm{w}}(I)} \lesssim |v|_{H^{1/2}(I)} + \left(1+\log\frac{H}{h}\right)^{1/2} \|v\|_{L^\infty(I)} \qquad \text{(C.38)}$$

*where (see Subsection A.2.6.1)*

$$\|v\|^2_{\widetilde{H}^{1/2}_{\mathrm{w}}(I)} := |v|^2_{H^{1/2}(I)} + \int_I \frac{|v(x)|^2}{\mathrm{dist}(x,\partial I)}\,\mathrm{d}x.$$

*The constant is independent of $v$, $H$, and $h$.*

*Proof.* Inequality (C.38) is proved in [3, Lemma 10]; see also the proof of [236, Lemma 3.2]. We include the proof here for completeness. First we note that

$$\|v\|^2_{\widetilde{H}^{1/2}_{\mathrm{w}}(I)} = |v|^2_{H^{1/2}(I)} + \int_0^{H/2} \frac{|v(x)|^2}{x}\,\mathrm{d}x + \int_{H/2}^{H} \frac{|v(x)|^2}{H-x}\,\mathrm{d}x.$$

On the other hand

$$\int_0^{H/2} \frac{|v(x)|^2}{x}\,\mathrm{d}x = \int_0^{h} \frac{|v(x)|^2}{x}\,\mathrm{d}x + \int_h^{H/2} \frac{|v(x)|^2}{x}\,\mathrm{d}x$$

$$\leq \int_0^h \frac{|v(x)-v(0)|^2}{x}\,\mathrm{d}x + \|v\|^2_{L^\infty(I)} \int_h^H \frac{dx}{x}$$

$$\leq h^2 \|v'\|^2_{L^\infty(I)} + \|v\|^2_{L^\infty(I)} \log\frac{H}{h}.$$

By using the inverse inequality, see e.g. [28],

$$\|v'\|^2_{L^\infty(I)} \lesssim \frac{1}{h^2}\|v\|^2_{L^\infty(I)},$$

we deduce

$$\int_0^{H/2} \frac{|v(x)|^2}{x}\,\mathrm{d}x \lesssim \left(1+\log\frac{H}{h}\right)\|v\|^2_{L^\infty(I)}.$$

Similarly we can prove

$$\int_{H/2}^{H} \frac{|v(x)|^2}{H-x}\,\mathrm{d}x \lesssim \left(1+\log\frac{H}{h}\right)\|v\|^2_{L^\infty(I)},$$

completing the proof of (C.38). □

Combining Lemma C.19 and Lemma C.20 we have

**Lemma C.21.** *Let $I=(0,H)$ be partitioned into subintervals which form a quasi-uniform mesh of size $h$. If $v$ is a continuous function defined on $I$ whose restriction onto each subinterval is a polynomial of degree 1, and if $v(0)=v(H)=0$, then*

$$\|v\|_{\widetilde{H}^{1/2}_{\mathrm{w}}(I)} \lesssim \left(1+\log\frac{H}{h}\right)|v|_{H^{1/2}(I)} \tag{C.39}$$

*where we recall that, see Subsection A.2.6.1,*

$$|v|^2_{H^{1/2}(I)} := \iint_{I\times I} \frac{|v(x)-v(y)|^2}{|x-y|^2}\,\mathrm{d}x\,\mathrm{d}y$$

*and*

$$\|v\|^2_{\widetilde{H}^{1/2}_{\mathrm{w}}(I)} := |v|^2_{H^{1/2}(I)} + \int_I \frac{|v(x)|^2}{\mathrm{dist}(x,\partial I)}\,\mathrm{d}x.$$

*The constant is independent of $v$, $H$, and $h$.*

*Proof.* In view of Lemma C.20, in order to prove (C.39), it suffices to prove

$$\|v\|_{L^\infty(I)} \lesssim \left(1+\log\frac{H}{h}\right)^{1/2} |v|_{H^{1/2}(I)}. \tag{C.40}$$

Let $\widehat{v}$ be the scaled function defined from $v$ when $I$ is rescaled into $\widehat{I} = (0,1)$, i.e., $\widehat{v}(\widehat{x}) = v(H\widehat{x})$ for all $\widehat{x} \in \widehat{I}$. Then for any $\alpha \in \mathbb{R}$ we have, noting that $\widehat{v}(0) = v(0) = 0$,

$$\begin{aligned}\|v\|_{L^\infty(I)} &= \|\widehat{v}\|_{L^\infty(\widehat{I})} \le \|\widehat{v}+\alpha\|_{L^\infty(\widehat{I})} + |\alpha| \\ &= \|\widehat{v}+\alpha\|_{L^\infty(\widehat{I})} + |\widehat{v}(0)+\alpha| \le 2\|\widehat{v}+\alpha\|_{L^\infty(\widehat{I})}.\end{aligned}$$

The partition of $I$ induces a partition of $\widehat{I}$ with size $\widehat{h} = h/H$. Invoking Lemma C.19 for $\widehat{v}+\alpha$ defined on $\widehat{I}$ with mesh size $\widehat{h}$ we infer

$$\|v\|_{L^\infty(I)} \lesssim \left(1+\log\frac{1}{\widehat{h}}\right)^{1/2} \|\widehat{v}+\alpha\|_{H^{1/2}_{\mathrm{W}}(\widehat{I})} = \left(1+\log\frac{1}{\widehat{h}}\right)^{1/2} \|\widehat{v}+\alpha\|_{H^{1/2}_{\mathrm{S}}(\widehat{I})}. \quad \text{(C.41)}$$

Following along the lines of [60, Theorem 3.1.1] one can prove

$$\inf_{\alpha\in\mathbb{R}} \|\widehat{v}+\alpha\|_{H^{1/2}_{\mathrm{S}}(\widehat{I})} = |\widehat{v}|_{H^{1/2}(\widehat{I})}.$$

Hence, by taking the infimum of both sides of (C.41) over $\alpha \in \mathbb{R}$ and by using Lemma A.4, we infer

$$\|v\|_{L^\infty(I)} \lesssim \left(1+\log\frac{1}{\widehat{h}}\right)^{1/2} |\widehat{v}|_{H^{1/2}(\widehat{I})} = \left(1+\log\frac{H}{h}\right)^{1/2} |v|_{H^{1/2}(I)},$$

proving (C.40) and thus completing the proof of (C.39). □

### C.4.2.2 The $h$-version in two dimensions

We first present a Sobolev inequality, which is proved in [240] and will be useful for the proof of Lemma C.23. We recall that $W^{1,\infty}(D)$ is the Sobolev space of functions belonging to $L^\infty(D)$ whose weak first-order partial derivatives also belong to $L^\infty(D)$.

**Lemma C.22.** *Let $D$ be a bounded domain in $\mathbb{R}^2$ with Lipschitz boundary $\partial D$. Then*

$$\|v\|_{L^\infty(D)} \lesssim |\log\varepsilon|^{1/2}\|v\|_{H^1(D)} + \varepsilon\|v\|_{W^{1,\infty}(D)} \quad \forall v \in W^{1,\infty}(D),\ \varepsilon \in (0, e^{-2}).$$

*The constant depends of the size of $D$.*

*Proof.* The proof follows along the lines of the proofs of [240, Theorem 2.1 and Corollary 2.1]. Consider $x_0 \in D$ and let $D_\varepsilon = \{x \in D : |x-x_0| < \varepsilon\}$. We note that $|D_\varepsilon| \gtrsim \varepsilon^2$, where $|D_\varepsilon|$ is the measure of $D_\varepsilon$. The Sobolev inequality (see e.g. [43, Corollary 9.14])

$$|v(x) - v(x_0)| \lesssim \|v\|_{W^{1,\infty}(D)}|x-x_0| \quad \forall x \in D$$

implies

$$|v(x_0)| \lesssim |v(x)| + \|v\|_{W^{1,\infty}(D)}|x-x_0|.$$

In particular, for $x \in D_\varepsilon$

$$|v(x_0)| \lesssim |v(x)| + \varepsilon \|v\|_{W^{1,\infty}(D)}$$

so that, by taking the $L^2(D_\varepsilon)$-norm on both sides,

$$|D_\varepsilon|^{1/2}|v(x_0)| \lesssim \|v\|_{L^2(D_\varepsilon)} + \varepsilon \|v\|_{W^{1,\infty}(D)} |D_\varepsilon|^{1/2}.$$

This implies, noting that $|D_\varepsilon|^{-1/2} \lesssim \varepsilon^{-1}$,

$$|v(x_0)| \lesssim \varepsilon^{-1} \|v\|_{L^2(D_\varepsilon)} + \varepsilon \|v\|_{W^{1,\infty}(D)}.$$

Hölder's inequality yields, for $r \in [2, \infty)$,

$$\|v\|_{L^2(D_\varepsilon)} \le |D_\varepsilon|^{\frac{1}{2}-\frac{1}{r}} \|v\|_{L^r(D_\varepsilon)} \lesssim \varepsilon^{1-2/r} \|v\|_{L^r(D)}.$$

It is well known that $H^1(D)$ is continuously embedded into $L^r(D)$ for all $r \in [2, \infty)$. Moreover, by tracing the constant, see e.g. [43, Corollaries 9.11 and 9.13], we have

$$\|v\|_{L^r(D)} \lesssim r^{1/2} \|v\|_{H^1(D)}.$$

Therefore

$$\|v\|_{L^2(D_\varepsilon)} \lesssim \varepsilon^{1-2/r} r^{1/2} \|v\|_{H^1(D)}.$$

By choosing $r = |\log(1/\varepsilon)| \ge 2$ we have $\varepsilon^{-2/r} = 1$ so that

$$\|v\|_{L^2(D_\varepsilon)} \lesssim \varepsilon |\log \varepsilon|^{1/2} \|v\|_{H^1(D)}.$$

Altogether we deduce the required result. □

The following well-known result appears in various forms at various places; cf. [37, 36, 41, 218, 236, 243]. Its proof is included for completeness.

**Lemma C.23.** *Let $v$ be a finite element function defined on a substructure $\Omega \subset \mathbb{R}^2$ of diameter $H$, and let the minimum diameter of the elements in $\Omega$ be $h$. Then*

$$\|v\|_{L^\infty(\Omega)} \lesssim \left(1 + \log \frac{H}{h}\right)^{1/2} {}_*\|v\|_{H^1_{\mathrm{w}}(\Omega)}$$

*where (see Subsection A.2.6.1)*

$${}_*\|v\|^2_{H^1_{\mathrm{w}}(\Omega)} := \frac{1}{H^2} \|v\|^2_{L^2(\Omega)} + |v|^2_{H^1(\Omega)}.$$

*The constant is independent of $v$, $H$, and $h$, but may depend on the polynomial degree $p$ of $v$.*

*Proof.* Let $\widehat{\Omega}$ be the reference substructure defined by

$$\widehat{\Omega} := \{\widehat{\mathbf{x}} \in \mathbb{R}^2 : \widehat{\mathbf{x}} = \mathbf{x}/H,\ \mathbf{x} \in \Omega\}$$

and let $\widehat{v}(\widehat{\mathbf{x}}) = v(H\widehat{\mathbf{x}})$ for all $\widehat{\mathbf{x}} \in \widehat{\Omega}$. We partition $\widehat{\Omega}$ into elements of minimum diameter $\widehat{h} = h/H$. Then $\|v\|_{L^\infty(\Omega)} = \|\widehat{v}\|_{L^\infty(\widehat{\Omega})}$. Moreover, Lemma A.5 gives ${}_*\|v\|_{H^1_{\mathrm{w}}(\Omega)} = {}_*\|\widehat{v}\|_{H^1_{\mathrm{w}}(\widehat{\Omega})}$. Hence it suffices to prove

$$\|\widehat{v}\|_{L^\infty(\widehat{\Omega})} \lesssim \left(1 + \log\frac{1}{\widehat{h}}\right)^{1/2} {}_*\|\widehat{v}\|_{H^1_{\mathrm{w}}(\widehat{\Omega})}. \tag{C.42}$$

Invoking Lemma C.22 with $D = \widehat{\Omega}$, $\varepsilon = \widehat{h}$, and with $v$ being $\widehat{v}$, we deduce

$$\|\widehat{v}\|_{L^\infty(\widehat{\Omega})} \lesssim |\log \widehat{h}|^{1/2} \|\widehat{v}\|_{H^1(\widehat{\Omega})} + \widehat{h}\|\widehat{v}\|_{W^{1,\infty}(\widehat{\Omega})}.$$

The inverse property of the finite element functions (see e.g. [28, Proposition 5.1, page 157])

$$\|\varphi\|_{W^{1,\infty}(D)} \lesssim \widehat{h}^{-2/q} \|\varphi\|_{W^{1,q}(D)}, \quad 1 \le q \le \infty,$$

yields (C.42). □

A bound for the $L^2$-norm on an edge by the weighted $H^{1/2}$-norm on the domain is the subject of the next lemma.

**Lemma C.24.** *Let $G = (0,H) \times (0,H)$ and $\gamma$ be an edge of $G$. Assume that $G$ is partitioned into subsquares of size $h$. If $v$ is a continuous piecewise linear function on $G$ then*

$$\|v\|^2_{L^2(\gamma)} \lesssim \left(1 + \log\frac{H}{h}\right) \|v\|^2_{H^{1/2}_{\mathrm{w}}(G)}$$

*where $\|\cdot\|_{H^{1/2}_{\mathrm{w}}(G)}$ is defined in Section A.2.6.1. The constant is independent of $H$, $h$, and $v$.*

*Proof.* This result is proved in [3, Lemma 11], see also [2, Lemma 3], and is included here for completeness. Without loss of generality, we assume that $\gamma = (0,H) \times \{0\}$. Let $\gamma' = (0,H)$. Then

$$\|v\|^2_{L^2(\gamma)} = \int_0^H |v(x,0)|^2\,\mathrm{d}x \le \int_0^H \max_{y\in\gamma'} |v(x,y)|^2\,\mathrm{d}x.$$

It follows from (C.36) that

$$\|v\|^2_{L^2(\gamma)} \lesssim \left(1 + \log\frac{H}{h}\right) \int_0^H \|v(x,\cdot)\|^2_{H^{1/2}_{\mathrm{w}}(\gamma')}\,\mathrm{d}x.$$

Invoking Lemma A.16 (A.96) we obtain the required result. □

The following results are first proved in [72]. We state them here for convenience.

**Lemma C.25.** *Let* $R := (0,H) \times (0,H)$ *and* $R_h := (0,H) \times (0,h)$ *for* $0 < h < H$. *Then*

$$\|v\|^2_{L^2(R_h)} \lesssim \frac{h}{H} \|v\|^2_{L^2(R)} + h\,(h+H)\,|v|^2_{H^1(R)} \quad \forall v \in H^1(R) \tag{C.43}$$

*and*

$$\|v\|^2_{L^2(R_h)} \lesssim h\left(1+\log\frac{H}{h}\right) \|v\|^2_{H^{1/2}_{\mathrm{w}}(R)} \quad \forall v \in H^{1/2}(R). \tag{C.44}$$

*The constants are independent of v, H, and h.*

*Proof.* The lemma is proved in [72, Lemma 3.1 and Lemma 3.4]. We include the proof here for completeness. First we prove (C.43). For any $v \in H^1(R)$, since

$$v(x,y) = v(x,0) + \int_0^y \frac{\partial v(x,s)}{\partial s}\,\mathrm{d}s \tag{C.45}$$

we have, by using the Cauchy–Schwarz inequality and by integrating over $R_h$

$$\|v\|^2_{L^2(R_h)} \le 2h \int_0^H |v(x,0)|^2\,\mathrm{d}x + h^2\,|v|^2_{H^1(R)}. \tag{C.46}$$

On the other hand, it follows from (C.45) that

$$|v(x,0)|^2 \le 2|v(x,y)|^2 + 2y \int_0^H \left|\frac{\partial v(x,s)}{\partial s}\right|^2 \mathrm{d}s$$

so that, after integrating over $R$ and dividing by $H$,

$$\int_0^H |v(x,0)|^2\,\mathrm{d}x \le \frac{2}{H}\|v\|^2_{L^2(R)} + H\,|v|^2_{H^1(R)}.$$

This inequality and (C.46) imply (C.43).

Next we prove (C.44). Consider a mesh of size $h$ on $R$ and let $\mathcal{V}_h$ be the finite element space consisting of continuous piecewise linear functions defined on this mesh. Let $Q_h : L^2(R) \to \mathcal{V}_h$ be the $L^2$-projection onto $\mathcal{V}_h$, namely, for any $v \in L^2(R)$, we define $Q_h v$ by

$$\langle Q_h v, w\rangle_{L^2(R)} = \langle v, w\rangle_{L^2(R)} \quad \forall w \in \mathcal{V}_h.$$

Let $\widehat{R} := (0,1) \times (0,1)$. For any function $u$ defined on $R$, if $\widehat{u}(\widehat{x}) = u(H\widehat{x})$, then it can be easily checked that $\widehat{Q_h v}$ is the $L^2$-projection of $\widehat{v}$ onto $\widehat{\mathcal{V}}_h$, i.e, $\widehat{Q_h v} = \widehat{Q}_h\widehat{v}$. It is well known that $\widehat{Q}_h$ is bounded in $L^2(\widehat{R})$ and in $H^1(\widehat{R})$; see e.g. [41]. Hence, by interpolation, it is bounded in $H^{1/2}(\widehat{R})$, i.e., $\|\widehat{Q}_h\widehat{v}\|_{H^{1/2}_{\mathrm{I}}(\widehat{R})} \lesssim \|\widehat{v}\|_{H^{1/2}_{\mathrm{I}}(\widehat{R})}$. Hence, it follows from Lemma A.5 that

$$
\begin{aligned}
\|Q_h v\|^2_{H^{1/2}_{\mathrm{w}}(R)} &\le H\,\|\widehat{Q}_h\widehat{v}\|^2_{H^{1/2}_{\mathrm{w}}(\widehat{R})} \simeq H\,\|\widehat{Q}_h\widehat{v}\|^2_{H^{1/2}_{\mathrm{I}}(\widehat{R})} \\
&\lesssim H\,\|\widehat{v}\|^2_{H^{1/2}_{\mathrm{I}}(\widehat{R})} \simeq H\,\|\widehat{v}\|^2_{H^{1/2}_{\mathrm{w}}(\widehat{R})} \le \|v\|^2_{H^{1/2}_{\mathrm{w}}(R)}.
\end{aligned} \tag{C.47}
$$

On the other hand, a standard argument yields

$$
\|v - Q_h v\|^2_{L^2(R)} \lesssim h|v|^2_{H^{1/2}(R)}. \tag{C.48}
$$

Therefore,

$$
\|v\|^2_{L^2(R_h)} \lesssim \|Q_h v\|^2_{L^2(R_h)} + \|v - Q_h v\|^2_{L^2(R_h)} \lesssim \|Q_h v\|^2_{L^2(R_h)} + h|v|^2_{H^{1/2}(R)}.
$$

Since $Q_h v \in H^1(R)$, we can invoke (C.46) (with $v$ replaced by $Q_h v$) to have

$$
\begin{aligned}
\|v\|^2_{L^2(R_h)} &\lesssim h \int_0^H |Q_h v(x,0)|^2\,\mathrm{d}x + h^2\,|Q_h v|^2_{H^1(R)} + h\,|v|^2_{H^{1/2}(R)} \\
&\lesssim h \int_0^H |Q_h v(x,0)|^2\,\mathrm{d}x + h\,|Q_h v|^2_{H^{1/2}(R)} + h\,|v|^2_{H^{1/2}(R)}
\end{aligned}
$$

where in the last step we used the inverse inequality. This inequality and (C.47) yield

$$
\|v\|^2_{L^2(R_h)} \lesssim h \int_0^H |Q_h v(x,0)|^2\,\mathrm{d}x + h\,\|v\|^2_{H^{1/2}_{\mathrm{w}}(R)}.
$$

The first term on the right-hand side can be bounded by invoking Lemma C.24 and (C.47), completing the proof of (C.44).

The proof for the case when $R = (0,H)$ and $R_h = (0,h)$ is completely analogous and is omitted. □

### C.4.2.3 The $p$-version and $hp$-version in one dimension

We now present corresponding results for the $p$-version. The results in the following theorem are proved in [18]. We note that on a reference element $\widehat{I}$ the weight in the definition of $\|\cdot\|_{H^{1/2}_{\mathrm{w}}(\widehat{I})}$ can be chosen to be 1 and thus $\|\cdot\|_{H^{1/2}_{\mathrm{w}}(\widehat{I})} = \|\cdot\|_{H^{1/2}_{\mathrm{S}}(\widehat{I})}$. Note the analogy between (C.36), (C.38), (C.39) and (C.49), (C.50), (C.51).

**Theorem C.1.** *Let $\mathbb{P}_p(\widehat{I})$ be the space of polynomials of degree $p$ on the interval $\widehat{I} = (-1,1)$. Then for any $v \in \mathbb{P}_p(\widehat{I})$*

$$
\|v\|_{L^\infty(\widehat{I})} \lesssim (1 + \log p)^{1/2} \|v\|_{H^{1/2}_{\mathrm{w}}(\widehat{I})}. \tag{C.49}
$$

*Moreover, if $v(\pm 1) = 0$ then*

$$\|v\|_{\widetilde{H}^{1/2}_{\mathrm{w}}(\widehat{I})} \lesssim |v|_{H^{1/2}(\widehat{I})} + (1+\log p)^{1/2}\|v\|_{L^\infty(\widehat{I})} \tag{C.50}$$

*and*

$$\|v\|_{\widetilde{H}^{1/2}_{\mathrm{w}}(\widehat{I})} \lesssim (1+\log p)\|v\|_{H^{1/2}_{\mathrm{w}}(\widehat{I})}. \tag{C.51}$$

*In each case the constant is independent of v and p. Here, we recall that*

$$\|v\|^2_{H^{1/2}_{\mathrm{w}}(\widehat{I})} = \|v\|^2_{L^2(\widehat{I})} + \iint\limits_{\widehat{I}\times\widehat{I}} \frac{|v(x)-v(y)|^2}{|x-y|^2}\,\mathrm{d}x\,\mathrm{d}y,$$

$$\|v\|^2_{\widetilde{H}^{1/2}_{\mathrm{w}}(\widehat{I})} = \iint\limits_{\widehat{I}\times\widehat{I}} \frac{|v(x)-v(y)|^2}{|x-y|^2}\,\mathrm{d}x\,\mathrm{d}y + \int\limits_{\widehat{I}} \frac{|v(x)|^2}{\operatorname{dist}(x,\partial\widehat{I})}\,\mathrm{d}x;$$

*see Subsection A.2.6.1.*

*Proof.* These results are proved in Theorem 6.2, Theorem 6.5, and Theorem 6.6 in [18]. □

The following result which will be useful in our analysis for the $p$-version slightly extends (C.49) and (C.51) in the above theorem. (Compare this result with Lemma C.21.)

**Lemma C.26.** *Let $I$ be an interval in $\mathbb{R}$. If $v$ is a polynomial of degree at most $p$ in $I$, then*

$$\|v\|_{L^\infty(I)} \lesssim (1+\log p)^{1/2}\|v\|_{H^{1/2}_{\mathrm{w}}(I)}. \tag{C.52}$$

*Assume further that $v$ vanishes at the endpoints of $I$. Then*

$$\|v\|_{L^\infty(I)} \lesssim (1+\log p)^{1/2}|v|_{H^{1/2}(I)} \tag{C.53}$$

*and*

$$\|v\|_{\widetilde{H}^{1/2}_{\mathrm{w}}(I)} \lesssim (1+\log p)|v|_{H^{1/2}(I)}. \tag{C.54}$$

*The constants are independent of $v$, $p$ and of the length of $I$.*

*Proof.* Let $\widehat{v}$ be the scaled version of $v$ when $I$ is rescaled into $\widehat{I}=(-1,1)$, i.e., $\widehat{v}(\widehat{x}) = v(x)$ for all $x\in I$ where $\widehat{x}$ is the corresponding point on $\widehat{I}$ after the rescaling. Then $\|v\|_{L^\infty(I)} = \|\widehat{v}\|_{L^\infty(\widehat{I})}$ and $\|v\|_{H^{1/2}_{\mathrm{w}}(I)} = \|\widehat{v}\|_{H^{1/2}_{\mathrm{w}}(\widehat{I})}$; see Lemma A.5 (i). Hence (C.52) is a consequence of (C.49).

For any $\alpha\in\mathbb{R}$,

$$\begin{aligned}\|v\|_{L^\infty(I)} &= \|\widehat{v}\|_{L^\infty(\widehat{I})} \le \|\widehat{v}+\alpha\|_{L^\infty(\widehat{I})} + |\alpha| \\ &= \|\widehat{v}+\alpha\|_{L^\infty(\widehat{I})} + |\widehat{v}(1)+\alpha| \\ &\le 2\|\widehat{v}+\alpha\|_{L^\infty(\widehat{I})}.\end{aligned}$$

It follows from (C.49) that

$$\|v\|_{L^\infty(I)} \lesssim (1+\log p)^{1/2} \|\widehat{v}+\alpha\|_{H_S^{1/2}(\widehat{I})}.$$

Following along the lines of [60, Theorem 3.1.1] one can prove

$$\inf_{\alpha\in\mathbb{R}} \|\widehat{v}+\alpha\|_{H_S^{1/2}(\widehat{I})} = |\widehat{v}|_{H^{1/2}(\widehat{I})}.$$

The above results and Lemma A.4 give (C.53). To prove (C.54), we invoke successively Lemma A.5, Theorem C.1, and Lemma A.4 to have

$$\begin{aligned}\|v\|_{\widetilde{H}_w^{1/2}(I)} = \|\widehat{v}\|_{\widetilde{H}_w^{1/2}(\widehat{I})} &\lesssim |\widehat{v}|_{H^{1/2}(\widehat{I})} + (1+\log p)^{1/2} \|\widehat{v}\|_{L^\infty(\widehat{I})} \\ &= |v|_{H^{1/2}(I)} + (1+\log p)^{1/2} \|v\|_{L^\infty(I)}.\end{aligned}$$

Inequality (C.54) now follows from (C.53). □

Combining Lemma C.19 and Lemma C.26 we obtain the following result.

**Lemma C.27.** *Let $I$ be an interval of length $H$. Assume that $I$ is partitioned into subintervals of size $h$ such that the mesh is quasi-uniform. Let $v$ be a continuous function defined on $I$ such that its restriction on each subinterval is a polynomial of degree $p$. Then*

$$\|v\|^2_{L^\infty(I)} \lesssim \left(1+\log\frac{H}{h}\right)(1+\log p)\|v\|^2_{H_w^{1/2}(I)}.$$

*The constant is independent of $H$, $h$, $p$, and $v$.*

*Proof.* First we note that there exists a subinterval $J$ (of $I$) of length $h$ such that

$$\|v\|_{L^\infty(I)} = \|v\|_{L^\infty(J)} = \|\widehat{v}\|_{L^\infty(\widehat{J})}$$

where $\widehat{v}$ is the affine image of $v$ onto the reference interval $\widehat{J}$ of length 1. Since $v$ is a polynomial of degree $p$ on $J$, so is $\widehat{v}$, and thus Theorem C.1 and Lemma A.5 give

$$\|v\|^2_{L^\infty(I)} \lesssim (1+\log p)\|\widehat{v}\|^2_{H_w^{1/2}(\widehat{J})} = (1+\log p)\|v\|^2_{H_w^{1/2}(J)}.$$

It follows from the definition of the $\|\cdot\|_{H_w^{1/2}(J)}$ and (C.37) that

$$\|v\|^2_{H_w^{1/2}(J)} = \frac{1}{h}\|v\|^2_{L^2(J)} + |v|^2_{H^{1/2}(J)} \lesssim \left(1+\log\frac{H}{h}\right)\|v\|^2_{H_w^{1/2}(I)}.$$

Altogether, we have the desired result. □

#### C.4.2.4 The $p$-version and $hp$-version in two dimensions

A two-dimensional analogue of Lemma C.26 is the following lemma.

**Lemma C.28.** *Let $\widehat{T}$ be the reference triangle. Then*

$$\|v\|_{\widetilde{H}_{\mathrm{I}}^{1/2}(\widehat{T})} \lesssim (1+\log p)\|v\|_{H_{\mathrm{I}}^{1/2}(\widehat{T})} \quad \forall v \in \mathbb{P}_p(\widehat{T}).$$

*The constant is independent of $v$ and $p$.*

*Proof.* See [108, Lemma 6]. □

### C.4.2.5 Functions of zero mean

Functions of zero means are frequently considered in the study of the weakly-singular integral equation; see Subsection A.2.12. The following property is elementary.

**Lemma C.29.** *For any function $v \in L^2(D)$, $D \subset \mathbb{R}^d$, $d \geq 1$, with finite measure $|D|$, let $\mu_D(v)$ denote the integral mean of $v$ defined by*

$$\mu_D(v) := \int_D v(x)\,\mathrm{d}x.$$

*Then*

$$\|v - \mu_D(v)\|_{L^2(D)} = \min_{c\in\mathbb{R}} \|v - c\|_{L^2(D)}.$$

*Proof.* Defining $f(c) := \|v-c\|_{L^2(D)}^2$, we can write $f$ as a quadratic function of $c$ as

$$f(c) = |D|c^2 - 2|D|\mu_D(v)c + \|v\|_{L^2(D)}^2.$$

The result follows by finding the minimum value of this quadratic function. □

**Lemma C.30.** *Let $\widehat{I}$ be one side of the reference triangle $\widehat{T}$. Then for any polynomial $v$ of degree $p$ on $\widehat{T}$ the following inequality holds*

$$\|v\|_{L^2(\widehat{I})}^2 \lesssim (1+\log p)\|v\|_{H_S^{1/2}(\widehat{T})}^2. \tag{C.55}$$

*Moreover,*

$$\|v - \mu_{\partial\widehat{T}}(v)\|_{L^2(\partial\widehat{T})}^2 \lesssim (1+\log p)|v|_{H^{1/2}(\widehat{T})}^2 \tag{C.56}$$

*where $\mu_{\partial\widehat{T}}(v)$ is defined in Lemma C.29. The constants are independent of $v$ and $p$.*

*Proof.* Without loss of generality, we can assume that $\widehat{I} = (0,1)\times\{0\}$ and that the reference triangle $\widehat{T}$ has vertices $(0,0)$, $(1,0)$, and $(0,1)$. Let $\widehat{Q}$ denote the reference square $(0,1)\times(0,1)$. First, assume that for any polynomial $v$ of degree $p$ on $\widehat{Q}$, the following estimate holds

$$\|v\|_{L^2(\widehat{I})}^2 \lesssim (1+\log p)\|v\|_{H_S^{1/2}(\widehat{Q})}^2. \tag{C.57}$$

Now for any $v \in \mathbb{P}_p(\widehat{T})$ we extend $v$ from $\widehat{T}$ onto $\widehat{Q}$ by reflecting it about $\widehat{I}_2$; see Figure 11.1. The reflected function on the reflected triangle $\widetilde{T}$ is denoted by $\widetilde{v}$. By symmetry the coupling term between $v$ and $\widetilde{v}$ in the $H^{1/2}$-seminorm can be shown to satisfy

$$\int_{\widehat{T}}\int_{\widetilde{T}} \frac{|v(x)-\widetilde{v}(\widetilde{y})|^2}{|x-\widetilde{y}|^3}\,\mathrm{d}x\,d\widetilde{y} = \int_{\widehat{T}}\int_{\widehat{T}} \frac{|v(x)-v(y)|^2}{|x-y|^3}\,\mathrm{d}x\,\mathrm{d}y = |v|^2_{H^{1/2}(\widehat{T})}.$$

Therefore we deduce from (C.57) that

$$\begin{aligned}\|v\|^2_{L^2(\widehat{I})} &\lesssim (1+\log p)\left(\|v\|^2_{L^2(\widehat{T})} + \|\widetilde{v}\|^2_{L^2(\widetilde{T})} + |v|^2_{H^{1/2}(\widehat{T})} + |\widetilde{v}|^2_{H^{1/2}(\widetilde{T})}\right)\\ &\simeq (1+\log p)\|v\|^2_{H^{1/2}_{\mathrm{S}}(\widehat{T})}.\end{aligned}$$

To prove (C.56) we use the minimizing property of $\mu_{\partial\widehat{T}}(v)$ shown in Lemma C.29, (C.55), and a quotient space argument as follows:

$$\begin{aligned}\|v-\mu_{\partial\widehat{T}}(v)\|^2_{L^2(\partial\widehat{T})} &= \inf_{c\in\mathbb{R}} \|v-c\|^2_{L^2(\partial\widehat{T})} = \inf_{c\in\mathbb{R}} \sum_{i=1}^{3} \|v-c\|^2_{L^2(\partial\widehat{I}_i)}\\ &\lesssim (1+\log p) \inf_{c\in\mathbb{R}} \|v-c\|^2_{H^{1/2}_{\mathrm{S}}(\widehat{T})}\\ &= (1+\log p)|v|^2_{H^{1/2}(\widehat{T})}.\end{aligned}$$

It remains to prove (C.57). For $v \in \mathbb{P}_p(\widehat{Q})$ we have

$$\begin{aligned}\|v\|^2_{L^2(\widehat{I})} &= \int_0^1 |v(x,0)|^2\,\mathrm{d}x \le \int_0^1 \|v(x,\cdot)\|^2_{L^\infty(0,1)}\,\mathrm{d}x\\ &\lesssim (1+\log p)\int_0^1 \|v(x,\cdot)\|^2_{H^{1/2}_{\mathrm{S}}(0,1)}\,\mathrm{d}x\end{aligned}$$

where in the last step we used Theorem C.1. Invoking Lemma A.16 (A.96) we deduce

$$\begin{aligned}\|v\|^2_{L^2(\widehat{I})} &\lesssim (1+\log p)\int_0^1 \|v(x,\cdot)\|^2_{H^{1/2}_{\mathrm{w}}(0,1)}\,\mathrm{d}x \lesssim (1+\log p)\|v\|^2_{H^{1/2}_{\mathrm{w}}(\widehat{Q})}\,\mathrm{d}x\\ &= (1+\log p)\|v\|^2_{H^{1/2}_{\mathrm{S}}(\widehat{Q})}\,\mathrm{d}x.\end{aligned}$$

This proves (C.57) and completes the proof of the lemma. □

**Lemma C.31.** *Let $T_h$ be a triangle of diameter $h$ and $\widehat{T}$ be the reference triangle. Then*

$$\mu^2_{\partial T_h}(v) \lesssim h^{-1}\|v\|^2_{L^2(\partial T_h)} \quad \forall v \in H^{1/2}(T_h)$$

*and*

$$\mu^2_{\partial\widehat{T}}(v) \lesssim (1+\log p)\|v\|^2_{H_S^{1/2}(\widehat{T})} \quad \forall v \in \mathbb{P}_p(\widehat{T}).$$

*The constants are independent of v, h, and p. Here the integral mean* $\mu_{\partial T_h}(v)$ *and* $\mu_{\partial\widehat{T}}(v)$ *are defined in Lemma C.29.*

*Proof.* For any $v \in H^{1/2}(T_h)$, Hölder's inequality gives

$$\mu^2_{\partial T_h}(v) \le \frac{1}{|\partial T_h|^2} \int_{T_h} ds \int_{\partial T_h} |v(s)|^2\, ds \simeq h^{-1}\|v\|^2_{L^2(\partial T_h)}$$

which is the first assertion of the lemma. This result applied to the reference triangle $\widehat{T}$ and Lemma C.30 give the second assertion. □

### *C.4.3 Polynomials of low energies*

The following space introduced in [169] is useful in the definitions of basis functions in Chapter 11.

**Definition C.1.** *Let* $\mathbb{P}_p(\widehat{I})$ *be the space of polynomials of degree at most p in* $\widehat{I} := (0,1)$. *The low energy polynomials* $\phi_0, \phi_0^- \in \mathbb{P}_p(\widehat{I})$ *are defined by*

$$\begin{aligned}
\phi_0 &= \arg\min\left\{\|\phi\|_{L^2(\widehat{I})} : \phi \in \mathbb{P}_p(\widehat{I}),\ \phi(0)=1,\ \phi(1)=0\right\},\\
\phi_0^- &= \arg\min\left\{\|\phi\|_{L^2(\widehat{I})} : \phi \in \mathbb{P}_p(\widehat{I}),\ \phi(0)=0,\ \phi(1)=1\right\}.
\end{aligned}$$

We note that $\phi_0^-(x) = \phi_0(1-x)$. For illustration, see Figure C.1 for the plot of $\phi_0$ with $p=10$.

The following results are proved in [169, Lemmas 4.1–4.2].

**Lemma C.32.** *Let* $\phi_0$ *and* $\phi_0^-$ *be defined as in Definition C.1. Then*

$$\phi_0(x) = \sum_{k=0}^{p} \frac{r_k}{2} L_k(1-2x), \quad x \in [0,1],$$

*where* $L_k$ *is the Legendre polynomial of degree k and*

$$r_k = \begin{cases} (2k+1)r_0 & \text{if } k \text{ is even},\\ (2k+1)r_1/3 & \text{if } k \text{ is odd}, \end{cases}$$

*with*

$$r_0 = \frac{2}{(p+1)(p+2)} \quad \text{and} \quad \frac{r_1}{3} = \frac{2}{p(p+1)}.$$

*Moreover,*

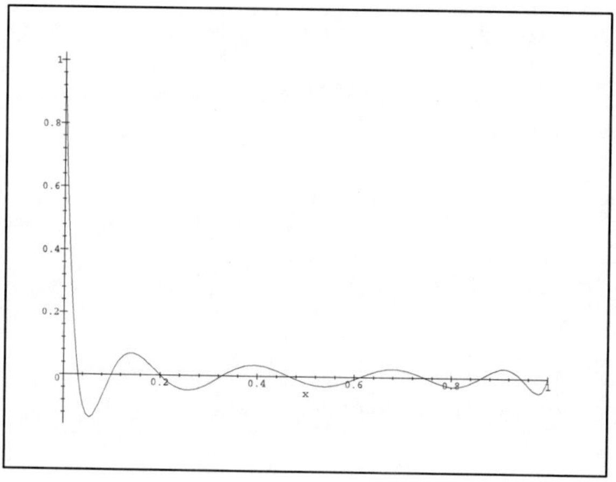


**Fig. C.1** Low energy function $\phi_0$ with $p = 10$.

$$\|\phi_0\|^2_{L^2(0,1)} = \frac{1}{p(p+1)}, \qquad \langle \phi_0, \phi_0^- \rangle_{L^2(0,1)} = \frac{(-1)^{p+1}}{2(p+1)} \|\phi_0\|^2_{L^2(0,1)}, \tag{C.58}$$

*and*

$$\langle \phi_0, v \rangle_{L^2(0,1)} = \langle \phi_0^-, v \rangle_{L^2(0,1)} = 0 \quad \forall v \in \mathbb{P}_p^0(\widehat{I}) \tag{C.59}$$

*where* $\mathbb{P}_p^0(\widehat{I})$ *is the space of polynomials in* $\mathbb{P}_p(\widehat{I})$ *which vanish at the endpoints* 0 *and* 1.

*Proof.* See [169, Lemma 4.1]. □

### C.4.4 Discrete harmonic functions and discrete harmonic extension

In this subsection we explain the concept of discrete harmonic functions and introduce the discrete harmonic extension; see e.g. [107]. Let $\widehat{\Omega}$ be the reference tetrahedron in $\mathbb{R}^3$, $\mathbb{P}_p(\widehat{\Omega})$ be the space of polynomials of degree at most $p$ in three variables, and $\mathbb{P}_p^0(\widehat{\Omega})$ be the space of polynomials in $\mathbb{P}_p(\widehat{\Omega})$ which vanish on the boundary $\partial\widehat{\Omega}$ of $\widehat{\Omega}$. The space of discrete harmonic functions $\widetilde{\mathscr{V}}_p$ is defined by

$$\widetilde{\mathscr{V}}_p := \{ v \in \mathbb{P}_p(\widehat{\Omega}) \, : \, a(v,w) = 0 \quad \forall w \in \mathbb{P}_p^0(\widehat{\Omega}) \}$$

where

$$a(v,w) := \int_{\widehat{\Omega}} \nabla v(x) \cdot \nabla w(x) \, dx.$$

If $f$ is a polynomial of degree $p$ defined on $\partial\widehat{\Omega}$, then the discrete harmonic extension of $f$ is a function $u \in \widetilde{\mathscr{V}}_p$ whose restriction on $\partial\widehat{\Omega}$ is $f$.

A discrete harmonic function $v \in \widetilde{\mathscr{V}}_p$ fulfils the minimizing property as stated in the following lemma.

**Lemma C.33.** *If $u \in \widetilde{\mathscr{V}}_p$ then for any $v \in \mathbb{P}_p(\widehat{\Omega})$ satisfying $v = u$ on $\partial\widehat{\Omega}$ the following inequality holds*

$$|u|_{H^1(\widehat{\Omega})} \le |v|_{H^1(\widehat{\Omega})}.$$

*Proof.* Let $w := u - v$. Then $w \in \mathbb{P}_p^0(\widehat{\Omega})$ and therefore $a(u,w) = 0$. Consequently,

$$\begin{aligned} |v|^2_{H^1(\widehat{\Omega})} &= a(v,v) = a(u-w,u-w) \\ &= a(u,u) + a(w,w) \\ &\ge a(u,u) = |u|^2_{H^1(\widehat{\Omega})}. \end{aligned}$$

□

The next lemma shows that for discrete harmonic functions, the $H^1$-norm can be bounded by the $H^{1/2}$-norm on the boundary.

**Lemma C.34.** *If $u$ is a discrete harmonic function in $\widehat{\Omega}$, then*

$$\|u\|_{H^1(\widehat{\Omega})} \lesssim \|u\|_{H^{1/2}(\partial\widehat{\Omega})}.$$

*The constant is independent of $u$ and $p$.*

*Proof.* Let $v$ be the Muñoz–Sola extension of $u|_{\partial\widehat{\Omega}}$f onto $\widehat{\Omega}$. By [196, Theorem 1],

$$\|v\|_{H^1(\widehat{\Omega})} \lesssim \|u\|_{H^{1/2}(\partial\widehat{\Omega})},$$

which implies, due to Lemma C.33,

$$|u|_{H^1(\widehat{\Omega})} \lesssim \|u\|_{H^{1/2}(\partial\widehat{\Omega})}. \tag{C.60}$$

On the other hand, the triangle inequality and Poincaré–Friedrich's inequality give

$$\begin{aligned} \|u\|_{L^2(\widehat{\Omega})} &\le \|u-v\|_{L^2(\widehat{\Omega})} + \|v\|_{L^2(\widehat{\Omega})} \\ &\lesssim |u-v|_{H^1(\widehat{\Omega})} + \|v\|_{L^2(\widehat{\Omega})} \\ &\le |u|_{H^1(\widehat{\Omega})} + \|v\|_{H^1(\widehat{\Omega})} \\ &\lesssim \|u\|_{H^{1/2}(\partial\widehat{\Omega})}. \end{aligned} \tag{C.61}$$

The required result follows from (C.60) and (C.61). □

## C.5 Some Useful Projections or Projection-like Operators

In this section, we present some results on the different operators which have been used in the book.

### C.5.1 The $L^2$-projection

The following result on the $L^2$-projection is used in Chapter 13. Let $\widehat{\mathscr{T}_0}$ be an initial triangulation on $\Gamma \subset \mathbb{R}^2$ with mesh size $\widehat{h}_0$ being the maximum diameter of triangles in $\mathscr{T}_0$. We consider a sequence of uniform triangulations $\{\widehat{\mathscr{T}_j}\}_{j\geq 0}$ with mesh size $\widehat{h}_j = 2^{-j}\widehat{h}_0$. On each mesh $\widehat{\mathscr{T}_j}$ let $\widehat{\mathscr{V}}^j$ be the boundary element space of continuous piecewise linear functions vanishing on the boundary of $\Gamma$. These spaces form a nested sequence $\widehat{\mathscr{V}}^0 \subset \widehat{\mathscr{V}}^1 \subset \cdots \subset \widehat{\mathscr{V}}^j \subset \cdots$.

**Lemma C.35.** *Let $Q_j : L^2(\Gamma) \longrightarrow \widehat{\mathscr{V}}^j$ be the $L^2$-orthogonal projection onto $\widehat{\mathscr{V}}^j$ defined by*

$$\langle Q_j v, w\rangle_{L^2(\Gamma)} = \langle v, w\rangle_{L^2(\Gamma)} \quad \forall v \in L^2(\Gamma),\ \forall w \in \widehat{\mathscr{V}}^j.$$

*Then for all $v \in \widetilde{H}^{1/2}(\Gamma)$ the following statement holds*

$$\sum_{j=0}^{\infty} \widehat{h}_j^{-1} \|v - Q_j v\|_{L^2(\Gamma)}^2 \lesssim \|v\|_{\widetilde{H}_{\mathrm{I}}^{1/2}(\Gamma)}^2.$$

*The constant depends only on $\Gamma$ and the initial triangulation $\widehat{\mathscr{T}_0}$.*

*Proof.* First we note that

$$\lim_{k\to\infty} \|v - Q_k v\|_{L^2(\Gamma)} = 0 \quad \forall v \in L^2(\Gamma).$$

Next we note that $Q_j - Q_{j-1}$ maps $L^2(\Gamma)$ into $\widehat{\mathscr{V}}^j$ for all $j = 1, 2, \ldots$. Consequently, if $i < j$ then $\widehat{\mathscr{V}}_i \subseteq \widehat{\mathscr{V}}_{j-1} \subset \widehat{\mathscr{V}}_j$ and thus it follows from the definition of $Q_j$ that, for any $v \in L^2(\Gamma)$,

$$\begin{aligned}\big\langle (Q_j - Q_{j-1})v, (Q_i - Q_{i-1})v\big\rangle_{L^2(\Gamma)} &= \big\langle Q_j v - v, (Q_i - Q_{i-1})v\big\rangle_{L^2(\Gamma)} \\ &\quad - \big\langle Q_{j-1} v - v, (Q_i - Q_{i-1})v\big\rangle_{L^2(\Gamma)} \\ &= 0.\end{aligned}$$

This orthogonality and the limiting property above imply

$$\sum_{i=j+1}^{\infty} \|(Q_i - Q_{i-1})v\|_{L^2(\Gamma)}^2 = \lim_{J\to\infty} \Big\|\sum_{i=j+1}^{J} (Q_i - Q_{i-1})v\Big\|_{L^2(\Gamma)}^2$$

$$
\begin{aligned}
&= \lim_{J\to\infty} \|Q_J v - Q_j v\|^2_{L^2(\Gamma)} \\
&= \lim_{J\to\infty} \|\big(Q_J v - v\big) + \big(v - Q_j v\big)\|^2_{L^2(\Gamma)} \\
&= \|v - Q_j v\|^2_{L^2(\Gamma)}.
\end{aligned}
$$

By using the above identity and by interchanging the order of summation we deduce

$$
\begin{aligned}
\sum_{j=0}^{\infty} \widehat{h}_j^{-1} \|v - Q_j v\|^2_{L^2(\Gamma)} &= \sum_{j=0}^{\infty} \widehat{h}_j^{-1} \sum_{i=j+1}^{\infty} \|\big(Q_i - Q_{i-1}\big)v\|^2_{L^2(\Gamma)} \\
&= \sum_{i=1}^{\infty} \|\big(Q_i - Q_{i-1}\big)v\|^2_{L^2(\Gamma)} \sum_{j=0}^{i-1} \widehat{h}_j^{-1}.
\end{aligned}
$$

The second sum on the right-hand side satisfies

$$
\sum_{j=0}^{i-1} \widehat{h}_j^{-1} = \widehat{h}_0^{-1} \sum_{j=0}^{i-1} 2^j < \widehat{h}_0^{-1} 2^i = \widehat{h}_i^{-1}
$$

so that

$$
\sum_{j=0}^{\infty} \widehat{h}_j^{-1} \|v - Q_j v\|^2_{L^2(\Gamma)} < \sum_{i=1}^{\infty} \widehat{h}_i^{-1} \|\big(Q_i - Q_{i-1}\big)v\|^2_{L^2(\Gamma)} \tag{C.62}
$$

On the other hand, Theorem 4 in [167] (see also [4, Theorem 5]) states that for $s \in [0,1]$ and for all $v \in \widetilde{H}^s(\Gamma)$

$$
\|v\|^2_{\widetilde{H}^s_{\mathrm{I}}(\Gamma)} \simeq \|Q_0 v\|^2_{\widetilde{H}^s_{\mathrm{I}}(\Gamma)} + \sum_{i=1}^{\infty} \widehat{h}_i^{-2s} \|\big(Q_i - Q_{i-1}\big)v\|^2_{L^2(\Gamma)} \tag{C.63}
$$

where the constants depend only on $\Gamma$, the initial triangulation $\widehat{\mathscr{T}_0}$, and $s$. In particular, for $s = 1/2$ we deduce from (C.62) and (C.63)

$$
\sum_{j=0}^{\infty} \widehat{h}_j^{-1} \|v - Q_j v\|^2_{L^2(\Gamma)} \lesssim \|v\|^2_{\widetilde{H}^s_{\mathrm{I}}(\Gamma)},
$$

proving the lemma. □

## *C.5.2 The standard interpolation operator*

In this subsection, we provide some properties of the standard interpolation operator. We note that on the one hand, this operator is a projection, a property which is useful in some proofs. On the other hand, it is not bounded as a function on $H^1(\Omega)$. In fact the operator is not even well-defined in $H^1(\Omega)$. The following lemma is recalled here for convenience.

**Lemma C.36.** *Assume that $\mathscr{T}_h$ and $\mathscr{T}_H$ are two shape-regular (not necessarily nested) meshes in a domain $\Omega$ in $\mathbb{R}^d$, $d = 1,2,3$. Assume that the mesh size $h$ of $\mathscr{T}_h$ and $H$ of $\mathscr{T}_H$ satisfy $0 < h < H$. Let $\mathscr{V}_h$ and $\mathscr{V}_H$ be two corresponding finite element spaces consisting of continuous piecewise polynomials defined on $\Omega$, with respect to the two meshes. Let $\Pi_h : C(\overline{\Omega}) \to \mathscr{V}_h$ be the standard interpolation operator. Then*

$$h^{-1}\|\Pi_h v - v\|_{L^2(\Omega)} + |\Pi_h v|_{H^1(\Omega)} \lesssim |v|_{H^1(\Omega)} \quad \forall v \in \mathscr{V}_H.$$

*Consequently,*

$$\|\Pi_h v\|_{H_{\mathrm{I}}^{1/2}(\Omega)} \lesssim \|v\|_{H_{\mathrm{I}}^{1/2}(\Omega)} \quad \forall v \in \mathscr{V}_H$$

*and*

$$\|\Pi_h v\|_{\widetilde{H}_{\mathrm{I}}^{1/2}(\Omega)} \lesssim \|v\|_{\widetilde{H}_{\mathrm{I}}^{1/2}(\Omega)} \quad \forall v \in \mathscr{V}_H \cap \widetilde{H}^{1/2}(\Omega).$$

*The constants are independent of $v$, $h$, and $H$.*

*Proof.* We note that since the meshes $\mathscr{T}_h$ and $\mathscr{T}_H$ are not necessarily nested, in general $\mathscr{V}_H \not\subset \mathscr{V}_h$. For alternative proofs of the first estimate, the reader is referred to [45, Lemma 2.1] and [56, Lemma 1]. By using the triangle inequality and the inverse property, it is also clear from the first estimate that

$$\|\Pi_h v\|_{L^2(\Omega)} \lesssim \|v\|_{L^2(\Omega)} \quad \forall v \in \mathscr{V}_H. \tag{C.64}$$

Hence, the next two estimates follow by interpolation. □

The following lemma is proved in [229, Lemma 4.3 and Lemma 4.4].

**Lemma C.37.** *Assume that $I$ is an interval of length $H$ and that there is a quasi-uniform mesh of size $h$ on $I$. Let $v$ be a continuous function on $I$ and $\Pi_H v$ be the affine function which interpolates $v$ at the endpoints of $I$.*

(i) *If $v$ is a piecewise polynomial of degree $p \ge 1$ on this mesh, then*

$$\|\Pi_H v\|^2_{H_{\mathrm{w}}^{1/2}(I)} \lesssim \left(1 + \log\frac{H}{h}\right)(1 + \log p)\|v\|^2_{H_{\mathrm{w}}^{1/2}(I)}.$$

(ii) *If $v$ is a polynomial of degree $p \ge 1$ on $I$, then*

$$\|\Pi_H v\|^2_{H_{\mathrm{w}}^{1/2}(I)} \lesssim (1 + \log p)\|v\|^2_{H_{\mathrm{w}}^{1/2}(I)}.$$

*The constants are independent of $v$, $H$, $h$, and $p$.*

*Proof.* First we recall the definition of the weighted norm defined in Section A.2.6.1)

$$\|\Pi_H v\|^2_{H_{\mathrm{w}}^{1/2}(I)} = \frac{1}{H}\int_I |\Pi_H v(x)|^2\,dx + \iint_{I\times I} \frac{|\Pi_H v(x) - \Pi_H v(y)|^2}{|x-y|^2}\,dx\,dy.$$

Let $I = (a,b)$ so that $b - a = H$. Then since $\Pi_H v$ is an affine function on $I$, we have for all $x, y \in I$

$$\frac{|\Pi_H v(x) - \Pi_H v(y)|^2}{|x-y|^2} = \frac{|\Pi_H v(a) - \Pi_H v(b)|^2}{|a-b|^2} \lesssim \frac{\|\Pi_H v\|^2_{L^\infty(I)}}{H^2}.$$

Hence

$$\|\Pi_H v\|^2_{H^{1/2}_{\mathrm{w}}(I)} \lesssim \frac{\|\Pi_H v\|^2_{L^\infty(I)}}{H} \int_I \mathrm{d}x + \frac{\|\Pi_H v\|^2_{L^\infty(I)}}{H^2} \iint_{I\times I} \mathrm{d}x\,\mathrm{d}y$$
$$= 2\,\|\Pi_H v\|^2_{L^\infty(I)} \lesssim \|v\|^2_{L^\infty(I)}.$$

The first required result is now a consequent of Lemma C.27. The second required result is obtained by taking $H = h$. □

For the next lemma we define the following two-level mesh for an interval $\Gamma$.

*The coarse mesh:* We first divide $\Gamma$ into disjoint subdomains $\Gamma_i$ with length $H_i$, $i = 1,\dots,J$, so that $\overline{\Gamma} = \cup_{i=1}^J \overline{\Gamma}_i$, and denote by $H$ the maximum value of $H_i$.

*The fine mesh:* Each $\Gamma_i$ is further divided into disjoint subintervals $\Gamma_{ij}$, $j = 1,\dots,N_i$, so that $\overline{\Gamma}_i = \cup_{j=1}^{N_i} \overline{\Gamma}_{ij}$. The maximum length of the subintervals $\Gamma_{ij}$ in $\Gamma_i$ is denoted by $h_i$, and the maximum value of $h_i$ is denoted by $h$. We require that the fine mesh is locally quasi-uniform, i.e., it is quasi-uniform in each subdomain $\Gamma_i$.

The following lemma gives similar results as the previous lemma, in the $\widetilde{H}^{1/2}(\Gamma)$-norm; see [229, Lemmas 4.5–4.7].

**Lemma C.38.** *Assume that a two-level mesh is defined on an interval $\Gamma$ as described above. Assume that $v$ is a continuous function vanishing at the endpoints of $\Gamma$, whose restriction on each subinterval $\Gamma_{ij}$, $i = 1,\dots,J$, $j = 1,\dots,N_i$, is a polynomial of degree $p \geq 1$.*

(i) *If $\Pi_H v$ is the piecewise linear function which interpolates $v$ at the coarse mesh points and if $w_i := (v - \Pi_H v)|_{\Gamma_i}$, then*

$$\|\Pi_H v\|^2_{\widetilde{H}^{1/2}_{\mathrm{I}}(\Gamma)} + \sum_{i=1}^J \|w_i\|^2_{\widetilde{H}^{1/2}_{\mathrm{I}}(\Gamma_i)} \lesssim (1+\log p)^2 \max_{1\le i\le J} \left(1 + \log\frac{H_i}{h_i}\right) \|v\|^2_{\widetilde{H}^{1/2}_{\mathrm{I}}(\Gamma)}.$$

(ii) *If $\Pi_h v$ is the piecewise linear function which interpolates $v$ at the fine mesh points and if $w_{ij} := (v - \Pi_h v)|_{\Gamma_{ij}}$, then*

$$\|\Pi_h v\|^2_{\widetilde{H}^{1/2}_{\mathrm{I}}(\Gamma)} + \sum_{i=1}^J \sum_{j=1}^{N_i} \|w_{ij}\|^2_{\widetilde{H}^{1/2}_{\mathrm{I}}(\Gamma_{ij})} \lesssim (1+\log p)^2 \|v\|^2_{\widetilde{H}^{1/2}_{\mathrm{I}}(\Gamma)}.$$

*The constants are independent of $v$, $H$, $h$, and $p$.*

*Proof.* It suffices to prove the first statement and then set $H_i = h_i$ to obtain the second statement. For each $i = 1,\dots,J$, since $\Pi_H v|_{\Gamma_i}$ is the affine function which interpolates $v$ at the endpoints of $\Gamma_i$, Lemma C.37 gives

$$\|\Pi_H v\|^2_{H^{1/2}_{\mathrm{w}}(\Gamma_i)} \lesssim (1+\log p)\left(1+\log\frac{H_i}{h_i}\right)\|v\|^2_{H^{1/2}_{\mathrm{w}}(\Gamma_i)},$$

which implies

$$|\Pi_H v|^2_{H^{1/2}(\Gamma_i)} \lesssim (1+\log p)\left(1+\log\frac{H_i}{h_i}\right)\|v\|^2_{H^{1/2}_{\mathrm{w}}(\Gamma_i)}.$$

Consequently, for any $\alpha \in \mathbb{R}$ we have

$$\begin{aligned}
|\Pi_H v|^2_{H^{1/2}(\Gamma_i)} &= |\Pi_H v+\alpha|^2_{H^{1/2}(\Gamma_i)}\\
&= |\Pi_H (v+\alpha)|^2_{H^{1/2}(\Gamma_i)}\\
&\lesssim (1+\log p)\left(1+\log\frac{H_i}{h_i}\right)\|v+\alpha\|^2_{H^{1/2}_{\mathrm{w}}(\Gamma_i)}.
\end{aligned}$$

Taking the infimum over $\alpha \in \mathbb{R}$ and using Lemma A.15 we deduce

$$|\Pi_H v|^2_{H^{1/2}(\Gamma_i)} \lesssim (1+\log p)\left(1+\log\frac{H_i}{h_i}\right)|v|^2_{H^{1/2}(\Gamma_i)}$$

where the constant is independent of $v$, $p$, $h$, and $H$. Let $w := v - \Pi_H v$ and let $\widehat{w}_i$ be the affine image of $w_i = w|_{\Gamma_i}$ onto the reference element $\widehat{I}$ when $\Gamma_i$ is mapped onto $\widehat{I}$. It follows from Lemma A.4, the triangle inequality, and the above inequality that

$$|\widehat{w}_i|^2_{H^{1/2}(\widehat{I})} = |w_i|^2_{H^{1/2}(\Gamma_i)} \lesssim (1+\log p)\left(1+\log\frac{H_i}{h_i}\right)|v|^2_{H^{1/2}(\Gamma_i)}.$$

On the other hand, since $w$ vanishes at the fine mesh points which include the coarse mesh points, we have $w_i \in \widetilde{H}^{1/2}(\Gamma_i)$ and thus $\widehat{w}_i \in \widetilde{H}^{1/2}(\widehat{I})$. The equivalence of norms in $\widetilde{H}^{1/2}(\widehat{I})$ (with constants independent of all the parameters), inequality (C.50) and the above inequality yield

$$\begin{aligned}
\|\widehat{w}_i\|^2_{\widetilde{H}^{1/2}_{\mathrm{I}}(\widehat{I})} &\simeq \|\widehat{w}_i\|^2_{\widetilde{H}^{1/2}_{\mathrm{w}}(\widehat{I})}\\
&\lesssim |\widehat{w}_i|^2_{H^{1/2}(\widehat{I})} + (1+\log p)\left(1+\log\frac{H_i}{h_i}\right)\|\widehat{w}_i\|^2_{L^\infty(\widehat{I})}\\
&\lesssim (1+\log p)\left(1+\log\frac{H_i}{h_i}\right)\left(|v|^2_{H^{1/2}(\Gamma_i)} + \|\widehat{w}_i\|^2_{L^\infty(\widehat{I})}\right).
\end{aligned}$$

Note that, for any arbitrary point $x_i$ in $\Gamma_i$,

$$\begin{aligned}
\|\widehat{w}_i\|^2_{L^\infty(\widehat{I})} &= \|w\|^2_{L^\infty(\Gamma_i)}\\
&\lesssim \|v - v(x_i)\|^2_{L^\infty(\Gamma_i)} + \|\Pi_H(v - v(x_i))\|^2_{L^\infty(\Gamma_i)}\\
&\lesssim \|v - v(x_i)\|^2_{L^\infty(\Gamma_i)}.
\end{aligned}$$

There exists $j \in \{1,\dots,N_i\}$ such that

$$\|v - v(x_i)\|^2_{L^\infty(\Gamma_i)} = \|v - v(x_i)\|^2_{L^\infty(\Gamma_{ij})}.$$

Since $v - v(x_i)$ is a polynomial of degree at most $p$ in $\Gamma_{ij}$, if we choose $x_i = x_{ij}$ to be a point in $\Gamma_{ij}$ so that $v - v(x_{ij})$ vanishes at this point, then we can invoke Lemma C.26 to obtain

$$\begin{aligned}
\|\widehat{w}_i\|^2_{L^\infty(\widehat{I})} &\lesssim (1+\log p)|v - v(x_{ij})|^2_{H^{1/2}(\Gamma_{ij})} \\
&= (1+\log p)|v|^2_{H^{1/2}(\Gamma_{ij})} \\
&\le (1+\log p)|v|^2_{H^{1/2}(\Gamma_i)}.
\end{aligned}$$

Altogether, we obtain

$$\|\widehat{w}_i\|^2_{\widetilde{H}_{\mathrm{I}}^{1/2}(\widehat{I})} \lesssim (1+\log p)^2 \left(1 + \log \frac{H_i}{h_i}\right) |v|^2_{H^{1/2}(\Gamma_i)}.$$

Lemma A.6 yields

$$\begin{aligned}
\sum_{i=1}^{J} \|w_i\|^2_{\widetilde{H}_{\mathrm{I}}^{1/2}(\Gamma_i)} &= \sum_{i=1}^{J} \|\widehat{w}_i\|^2_{\widetilde{H}_{\mathrm{I}}^{1/2}(\widehat{I})} \\
&\lesssim (1+\log p)^2 \max_{1\le i\le J} \left(1 + \log \frac{H_i}{h_i}\right) \sum_{i=1}^{J} |v|^2_{H^{1/2}(\Gamma_i)} \\
&\le (1+\log p)^2 \max_{1\le i\le J} \left(1 + \log \frac{H_i}{h_i}\right) |v|^2_{H^{1/2}(\Gamma)} \\
&\le (1+\log p)^2 \max_{1\le i\le J} \left(1 + \log \frac{H_i}{h_i}\right) \|v\|^2_{\widetilde{H}_{\mathrm{S}}^{1/2}(\Gamma)} \\
&\simeq (1+\log p)^2 \max_{1\le i\le J} \left(1 + \log \frac{H_i}{h_i}\right) \|v\|^2_{\widetilde{H}_{\mathrm{I}}^{1/2}(\Gamma)},
\end{aligned}$$

proving one part of the required inequality. Theorem A.10 then gives

$$\|w\|^2_{\widetilde{H}_{\mathrm{I}}^{1/2}(\Gamma)} \lesssim (1+\log p)^2 \max_{1\le i\le J} \left(1 + \log \frac{H_i}{h_i}\right) \|v\|^2_{\widetilde{H}_{\mathrm{I}}^{1/2}(\Gamma)}.$$

The remaining part of the required inequality follows from the triangle inequality.
□

The following result is first proved in [228, Lemma 3.5].

**Lemma C.39.** *Let $\Gamma$ be a bounded interval in $\mathbb{R}$ which is uniformly partitioned by points $x_j$, $j = 0,\dots,N$, into subintervals $\Gamma_j = (x_{j-1}, x_j)$, $j = 1,\dots,N$, of length $h$. We define*

$$\overline{\mathscr{V}} := \left\{u \in C(\Gamma) : u|_{\Gamma_j} \in \mathbb{P}_p, \quad j = 1,\dots,N\right\}$$

*where $\mathbb{P}_p$ denotes the space of polynomials of degree at most p. For any $u \in \overline{\mathcal{V}}$, if $\Pi_h u$ is the piecewise linear function which interpolates u at the mesh points $x_j$, then*

$$\sum_{j=1}^{N} |\Pi_h u|^2_{H^{1/2}(\Gamma_j)} \lesssim (1+\log p) \sum_{j=1}^{N} |u|^2_{H^{1/2}(\Gamma_j)}.$$

*The constant is independent of u, p, and the mesh size h.*

*Proof.* We denote by $u_0$ the interpolant $\Pi_h u$ and generate from $\Gamma$ a rectangular region

$$\Omega = \Gamma \times (0,h),$$

which is partitioned into sub-rectangular regions

$$\Omega_j = \Gamma_j \times (0,h), \quad j = 1,\dots,N;$$

see Figure C.2 for $N = 4$. By [18, Theorem 7.8], there exist $U_j \in \mathbb{P}_p(\Omega_j)$ such that

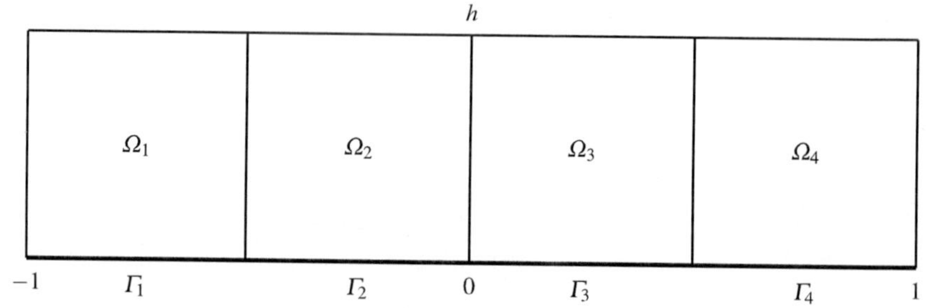


**Fig. C.2** Generating $\Omega$ from $\Gamma = \bigcup_{i=1}^{4} \Gamma_i$

$$U_j = \begin{cases} u & \text{on } \Gamma_j \times \{0\}, \\ 0 & \text{on } \Gamma_j \times \{h\}. \end{cases}$$

Let $U_{0,j}$ be the bilinear interpolant of $U_j$ at the vertices of $\Omega_j$ and let

$$U_0 := \sum_{j=1}^{N} U_{0,j}.$$

Then $U_0|_\Gamma = u_0$. Indeed, both $U_0|_\Gamma$ and $u_0$ are piecewise-linear functions on $\Gamma$ which take the same values $u(x_i)$ at the nodal points $x_i$. Using the trace theorem and Lemma A.6, we deduce

$$|u_0|_{H^{1/2}(\Gamma_j)} \le \|U_0\|_{H^1_I(\Omega_j)} = \|U_{0,j}\|_{H^1_I(\Omega_j)}$$

$$\leq \|\widehat{U_{0,j}}\|_{H^1_{\mathrm{I}}(\widehat{\Omega}_j)} = {}_*\|\widehat{U_{0,j}}\|_{H^1_{\mathrm{w}}(\widehat{\Omega}_j)}$$

where $\widehat{U_{0,j}}$ is the affine image of $U_{0,j}$ onto scaled element $\widehat{\Omega}_j$ of side length 1. Moreover, $\widehat{U_{0,j}}$ is the bilinear interpolant of $\widehat{U}_j$, the affine image of $U_j$ onto $\widehat{\Omega}_j$ under scaling. Hence, invoking Lemma C.40 gives

$$|u_0|_{H^{1/2}(\Gamma_j)} \lesssim (1+\log p)^{1/2}{}_*\|\widehat{U}_j\|_{H^1_{\mathrm{w}}(\widehat{\Omega}_j)} \simeq (1+\log p)^{1/2}\|\widehat{U}_j\|_{H^1_{\mathrm{I}}(\widehat{\Omega}_j)},$$

where we used the equivalence of norms in $H^1(\widehat{\Omega}_j)$ with constants depending only on the diameter of $\widehat{\Omega}_j$ which is a constant. Theorem 7.8 in [18] gives

$$|u_0|_{H^{1/2}(\Gamma_j)} \lesssim (1+\log p)^{1/2}\|\widehat{u}\|_{H^{1/2}_{\mathrm{I}}(\widehat{\Gamma}_j)} \simeq (1+\log p)^{1/2}\|\widehat{u}\|_{H^{1/2}_{\mathrm{w}}(\widehat{\Gamma}_j)}.$$

Using Lemma A.5 we deduce

$$|u_0|_{H^{1/2}(\Gamma_j)} \lesssim (1+\log p)^{1/2}\|u\|_{H^{1/2}_{\mathrm{w}}(\Gamma_j)}.$$

For any $\alpha \in \mathbb{R}$ we write the above inequality for $u+\alpha \in \overline{\mathscr{V}}$ instead of $u$ to have

$$|u_0|^2_{H^{1/2}(\Gamma_j)} = |u_0+\alpha|^2_{H^{1/2}(\Gamma_j)} \lesssim (1+\log p)^{1/2}\|u+\alpha\|_{H^{1/2}_{\mathrm{w}}(\Gamma_j)}.$$

Taking the infimum over $\alpha$ and using Lemma A.15 we then deduce

$$|u_0|^2_{H^{1/2}(\Gamma_j)} \leq c(1+\log p)|u|^2_{H^{1/2}(\Gamma_j)}.$$

By summing over $j$ we obtain

$$\sum_{j=1}^{N} |u_0|^2_{H^{1/2}(\Gamma_j)} \leq c(1+\log p)\sum_{j=1}^{N} |u|^2_{H^{1/2}(\Gamma_j)},$$

proving the lemma. □

**Lemma C.40.** *Let $R$ be a rectangular domain in $\mathbb{R}^2$ of side length $h<1$. If $u_p$ belongs to $\mathbb{P}_p(R)$, the space of polynomials of degree at most $p$ in each variable, and $u_0$ is the bilinear interpolant of $u_p$ at the vertices of $R$, then*

$${}_*\|u_0\|_{H^1_{\mathrm{w}}(R)} \lesssim (1+\log p)^{1/2}{}_*\|u_p\|_{H^1_{\mathrm{w}}(R)},$$

*where the constant is independent of $u_p$, $p$, and $h$.*

*Proof.* Due to Lemma A.5, it suffices to prove

$${}_*\|\widehat{u}_0\|_{H^1_{\mathrm{w}}(\widehat{R})} \lesssim (1+\log p)^{1/2}{}_*\|\widehat{u}_p\|_{H^1_{\mathrm{w}}(\widehat{R})} \tag{C.65}$$

where $\widehat{R}$ has side length 1. A simple calculation reveals

$$ {}_*\|\widehat{u}_0\|_{H^1_w(\widehat{R})} = \|\widehat{u}_0\|_{H^1_S(\widehat{R})} \lesssim \max_{1\le i\le 4} |\widehat{u}_0(\widehat{\mathbf{x}}_i)| $$

where $\widehat{\mathbf{x}}_i$, $i = 1,\dots,4$, are vertices of $\widehat{R}$. Since $\widehat{u}_0$ is bilinear on $\widehat{R}$, it attains its maximum value at one of its vertices. Moreover, since $\widehat{u}_0$ interpolates $\widehat{u}_p$ at these vertices, this maximum value is not greater than the maximum value of $\widehat{u}_p$. Therefore,

$$ {}_*\|\widehat{u}_0\|_{H^1_w(\widehat{R})} \lesssim \max_{1\le i\le 4} |\widehat{u}_0(\widehat{\mathbf{x}}_i)| = \|\widehat{u}_0\|_{L^\infty(\widehat{R})} \le \|\widehat{u}_p\|_{L^\infty(\widehat{R})}. \tag{C.66} $$

Lemma C.22 gives

$$ \|\widehat{u}_p\|_{L^\infty(\widehat{R})} \lesssim |\log\varepsilon|^{1/2} \|\widehat{u}_p\|_{H^1_S(\widehat{R})} + \varepsilon \|\widehat{u}_p\|_{W^{1,\infty}(\widehat{R})}, \quad \varepsilon \in (0,1), $$

where the constant is independent of $p$, $u_p$, and $h$. By Markov's inequality (C.33)

$$ \max_x \left|\frac{\partial \widehat{u}_p}{\partial x}\right| \le p^2 \max_x |\widehat{u}_p(x,y)| \quad \text{and} \quad \max_y \left|\frac{\partial \widehat{u}_p}{\partial y}\right| \le p^2 \max_y |\widehat{u}_p(x,y)|, $$

implying

$$ \|\widehat{u}_p\|_{W^{1,\infty}(\widehat{R})} \lesssim p^2 \|\widehat{u}_p\|_{L^\infty(\widehat{R})}. $$

By choosing $\varepsilon = 1/(2p^2)$ we deduce

$$ \|\widehat{u}_p\|_{L^\infty(\widehat{R})} \lesssim (1+\log p)^{1/2} \|\widehat{u}_p\|_{H^1_S(\widehat{R})} = (1+\log p)^{1/2} {}_*\|\widehat{u}_p\|_{H^1_w(\widehat{R})}. $$

This together with (C.66) gives (C.65), completing the proof of the lemma. □

We finish this subsection by proving the boundedness in the $H^{1/2}$-seminorm of the interpolation operator interpolating continuous functions at the Gauss–Lobatto points $z_1$, $z_2$, …, $z_{p+1}$ which are zeros of $\mathscr{L}_{p+1}$ defined in (C.4). To assure ourselves that $\mathscr{L}_{p+1}$ has $p+1$ distinct zeros in $[-1,1]$, we first recall from [216, Theorem 4.2.1] that the Legendre polynomial $L_p$ satisfies the differential equation

$$ (1-x^2)L_p''(x) - 2xL_p'(x) + p(p+1)L_p(x) = 0. $$

As a consequence, $\mathscr{L}_{p+1}$ satisfies

$$ (1-x^2)L_p'(x) + p(p+1)\mathscr{L}_{p+1}(x) = 0. $$

This implies that $x = \pm 1$ are zeros of $\mathscr{L}_{p+1}$. Moreover, as is well known, $L_p$ has $p$ distinct zeros in $(-1,1)$. Therefore, the above identity implies that $\mathscr{L}_{p+1}$ has $p+1$ distinct zeros in $[-1,1]$, the other zeros besides $\pm 1$ are zeros of $L_p'$ which interleave those of $L_p$. We denote the zeros of $\mathscr{L}_{p+1}$ by

$$ -1 = z_1 < z_2 < \cdots < z_p < z_{p+1} = 1. \tag{C.67} $$

**Lemma C.41.** *Let $J = [-1,1]$, $\mathbb{P}_p(J)$ be the space of polynomials of degree at most $p$ on $J$, and $\widehat{\mathbb{P}}_p(J) := \mathbb{P}_p(J)/\mathbb{R}$. For any continuous function $v$ on $J$, we denote* *by $T_p v$ the polynomial of degree at most p that interpolates v at $z_1, \ldots, z_{p+1}$ defined in (C.67). Then $T_p : \widehat{\mathbb{P}}_{p+1}(J) \to \widehat{\mathbb{P}}_p(J)$ is uniformly bounded in the $|\cdot|_{H^{1/2}(J)}$-norm, i.e.,*

$$|T_p f|_{H^{1/2}(J)} \lesssim |f|_{H^{1/2}(J)} \quad \forall f \in \widehat{\mathbb{P}}_{p+1}(J). \tag{C.68}$$

*The constant is independent of f and p.*

*Proof.* For any $f \in \widehat{\mathbb{P}}_{p+1}(J)$, let $F \in \widehat{\mathbb{P}}_{p+1}(J \times J)$ be the extension of $f$ onto $J \times J$ given by [18, Theorem 7.8] so that

$$|F|_{H^1(J\times J)} \lesssim |f|_{H^{1/2}(J)}.$$

Let $G \in \widehat{\mathbb{P}}_p(J \times J)$ be the interpolant of $F$ at $(z_i, z_j) \in J \times J$, $i, j = 1, \ldots, p+1$. It is shown in [168, Lemma 2] that

$$|G|_{H^1(J\times J)} \lesssim |F|_{H^1(J\times J)}.$$

On the other hand, since $G|_J = T_p(f)$, it follows from the trace theorem that

$$|T_p(f)|_{H^{1/2}(J)} \lesssim |G|_{H^1(J\times J)}.$$

Here we note that on the quotient spaces, the seminorms are norms which are equivalent to the usual norms. Altogether, we obtain (C.68). □

### *C.5.3 Some useful operators*

In this subsection, we define some useful transformations and prove their important properties. It is known that a polynomial on the boundary of a rectangular element can be extended into a polynomial in the rectangle, with appropriate bounds of norms; see e.g. [18]. The following lemma which extends this result to piecewise polynomial functions is first proved in [222].

The next two lemmas require the following sets depicted in Figure C.3.

$$\begin{aligned}
\Omega_1^- &:= [-1,0]\times[0,1], & \Omega_1^+ &:= [0,1]\times[0,1], & \Omega_1 &:= \Omega_1^- \cup \Omega_1^+,\\
\Omega_2^- &:= [-1,0]\times[-1,0], & \Omega_2^+ &:= [0,1]\times[-1,0], & \Omega_2 &:= \Omega_2^- \cup \Omega_2^+,\\
\gamma_1^- &:= [-1,0]\times\{0\}, & \gamma_1^+ &:= [0,1]\times\{0\}, & \gamma_1 &:= \gamma_1^- \cup \gamma_1^+,\\
\gamma_2^- &:= \{0\}\times[0,1], & \gamma_3^- &:= [-1,0]\times\{1\}, & \gamma_4^- &:= \{-1\}\times[0,1],\\
\gamma_2^+ &:= \{0\}\times[0,1], & \gamma_3^+ &:= [0,1]\times\{1\}, & \gamma_4^+ &:= \{1\}\times[0,1],\\
\gamma^- &:= \gamma_1^- \cup\cdots\cup \gamma_4^-, & \gamma^+ &:= \gamma_1^+ \cup\cdots\cup \gamma_4^+, & \Omega &:= \Omega_1 \cup \Omega_2.
\end{aligned} \tag{C.69}$$

We also identify $\gamma_1$ and $\gamma_1^\pm$ with the corresponding one-dimensional intervals.

**Lemma C.42.** *Let $f$ be a continuous function defined on $\gamma_1$ such that $f(\pm 1) = 0$ and $f^\pm := f|_{\gamma_1^\pm}$ are polynomials of degree p. Then there exists $F \in H_0^1(\Omega)$ such that*

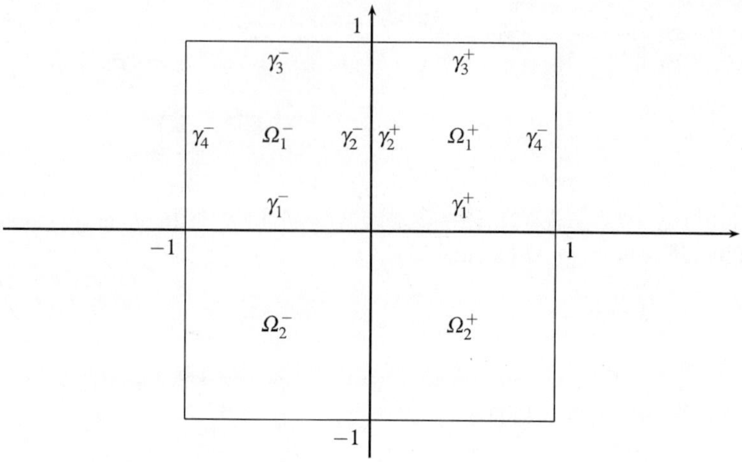


**Fig. C.3** Sets defined in (C.69), $\Omega_i = \Omega_i^+ \cup \Omega_i^-$, $i = 1,2$, $\Omega = \Omega_1 \cup \Omega_2$, $\gamma^\pm = \gamma_1^\pm \cup \ldots \cup \gamma_4^\pm$, $\gamma_1 = \gamma_1^- \cup \gamma_1^+$

$$F|_{\Omega_i^\pm} \in \mathbb{P}_p(\Omega_i^\pm),\ i = 1,2, \quad F|_{\partial\Omega} = 0, \quad F|_{\gamma_1} = f,$$

*and*

$$\|F\|_{H_\mathrm{I}^1(\Omega_i)} \lesssim \|f\|_{\widetilde{H}_\mathrm{I}^{1/2}(\gamma_1)}, \quad i = 1,2, \tag{C.70}$$

*where the constant is independent of $f$ and $p$.*

*Proof.* Let $\tilde{f}^\pm$ be functions defined on $\gamma^\pm$ as

$$\tilde{f}^\pm := \begin{cases} f^\pm & \text{on } \gamma_1^\pm, \\ f^+ \circ \eta & \text{on } \gamma_2^\pm, \\ 0 & \text{on } \gamma_3^\pm \cup \gamma_4^\pm, \end{cases}$$

where $\eta$ is the rotation that maps $\gamma_2^+$ onto $\gamma_1^+$. We note that $\tilde{f}^+$ and $\tilde{f}^-$ take the same values on $\gamma_2^\pm$ and are continuous on $\gamma^\pm$, respectively. By [18, Theorem 7.5] there exist $F_1^\pm \in \mathbb{P}_p(\Omega_1^\pm)$ such that $F_1^\pm|_{\gamma^\pm} = \tilde{f}^\pm$ and

$$\|F_1^\pm\|_{H_\mathrm{I}^1(\Omega_1^\pm)} \simeq \|F_1^\pm\|_{H_\mathrm{S}^1(\Omega_1^\pm)} \lesssim \|\tilde{f}^\pm\|_{H_\mathrm{S}^{1/2}(\gamma^\pm)} \simeq \|\tilde{f}^\pm\|_{H_\mathrm{I}^{1/2}(\gamma^\pm)}, \tag{C.71}$$

where the constant is independent of $f$ and $p$. Since $F_1^+|_{\gamma_2^+} = F_1^-|_{\gamma_2^-}$, if we define

$$F_1 := \begin{cases} F_1^+ & \text{on } \Omega_1^+, \\ F_1^- & \text{on } \Omega_1^-, \end{cases}$$

then $F_1$ is continuous on $\Omega_1$; thus $F_1 \in H^1(\Omega_1)$. Similarly, we can define $F_2 \in H^1(\Omega_2)$ with similar properties as $F_1$, and then define $F \in H^1(\Omega)$ such that $F|_{\Omega_i} =$

$F_i, i = 1,2$. Clearly, $F$ vanishes on the boundary of $\Omega$. In view of (C.71), the required inequality (C.70) will follow if

$$\|\tilde{f}^{\pm}\|_{H_{\mathrm{I}}^{1/2}(\gamma^{\pm})} \lesssim \|f\|_{\widetilde{H}_{\mathrm{I}}^{1/2}(\gamma_1)}. \tag{C.72}$$

It is clear from the definition of $\tilde{f}^{\pm}$ that

$$\|\tilde{f}^{-}\|_{H_{\mathrm{I}}^{1/2}(\gamma^{-})} \simeq \|\tilde{f}^{-}\|_{H^{1/2}(\gamma^{-})} \simeq \|f\|_{\widetilde{H}^{1/2}(\gamma_1)} \simeq \|f\|_{\widetilde{H}_{\mathrm{I}}^{1/2}(\gamma_1)}$$

and

$$\|\tilde{f}^{+}\|_{H_{\mathrm{I}}^{1/2}(\gamma^{+})} \simeq \|\tilde{f}^{+}\|_{H^{1/2}(\gamma^{+})} \simeq \|f^{+}\|_{H^{1/2}(\gamma_1^{+})} \lesssim \|f\|_{\widetilde{H}_{\mathrm{I}}^{1/2}(\gamma_1)}.$$

Thus (C.72) holds and the lemma is proved. □

In the sequel, if $v$ is a function defined on an interval, then we denote by $\hat{v}$ its affine image on the reference interval $[-1,1]$.

**Lemma C.43.** *Let $f$ be a continuous function on $[-1,1]$ vanishing at $\pm 1$ such that $f^{-} := f|_{[-1,0]}$ and $f^{+} := f|_{[0,1]}$ are polynomials of degree $p+1$. Let $T_p^*(f)$ be defined as*

$$T_p^*(f) := \begin{cases} g^{-} & \text{on } [-1,0), \\ g^{+} & \text{on } [0,1], \end{cases}$$

*where $g^{\pm}$ are polynomials of degree $p$, the affine images $\hat{g}^{\pm}$ of which are defined by $\hat{g}^{\pm} := T_p(\hat{f}^{\pm})$, respectively. (Here $T_p$ is the interpolation operator defined in Lemma C.41.) Then $T_p^*(f)$ is continuous on $[-1,1]$ and*

$$\|T_p^*(f)\|_{\widetilde{H}_{\mathrm{I}}^{1/2}(-1,1)} \lesssim \|f\|_{\widetilde{H}_{\mathrm{I}}^{1/2}(-1,1)} \tag{C.73}$$

*where the constant is independent of $f$ and $p$.*

*Proof.* We note that the functions $g^{\pm}$ so defined are polynomials of degree $p$. Moreover, since $f(\pm 1) = 0$ and $f$ is continuous at 0, we have $g^{-}(-1) = g^{+}(1) = 0$ and $g^{-}(0) = g^{+}(0)$. Hence $T_p^*(f)$ is continuous and belongs to $\widetilde{H}^{1/2}(-1,1)$. We now prove (C.73). Recall the sets defined by (C.69). By Lemma C.42, there exists a function $F \in H_0^1(\Omega)$ such that

$$F|_{\Omega_i^{\pm}} \in \mathbb{P}_{p+1}(\Omega_i^{\pm}), \quad i = 1,2, \qquad F|_{\gamma_1} = f,$$

and

$$\|F\|_{H_{\mathrm{I}}^1(\Omega)} \lesssim \|f\|_{\widetilde{H}_{\mathrm{I}}^{1/2}(\gamma_1)} \tag{C.74}$$

where the constant is independent of $f$ and $p$. By using the two-dimensional version of the interpolation operator $T_p$ discussed in [168], we can define $G_i^{\pm} \in \mathbb{P}_p(\Omega_i^{\pm})$ such that

$$|G_i^{\pm}|_{H^1(\Omega_i^{\pm})} \lesssim |F|_{\Omega_i^{\pm}}|_{H^1(\Omega_i^{\pm})}.$$

Let $G$ denote the function defined on $\Omega$ such that $G|_{\Omega_i^\pm} = G_i^\pm$. Then $G \in H_0^1(\Omega)$ (because $F \in H_0^1(\Omega)$) and

$$|G|_{H^1(\Omega)} \lesssim |F|_{H^1(\Omega)}. \tag{C.75}$$

Now let $\partial\Omega_1$ be the boundary of $\Omega_1$ and let $\Gamma := \partial\Omega_1 \setminus \gamma_1$. We denote by $\tilde{g}$ the extension of $T_p^*(f)$ by 0 onto $\Gamma$. On each interval $[-1,0]$ and $[0,1]$, the functions $G$ and $T_p^*(f)$ are polynomials of degree $p$ which coincide at the interpolation points $z_1$, $\ldots$, $z_{p+1}$ defined in (C.67). Therefore, $G|_{\gamma_1} = T_p^*(f) = \tilde{g}|_{\gamma_1}$. On the other hand, $G|_\Gamma \equiv 0 \equiv \tilde{g}|_\Gamma$. Thus $G|_{\partial\Omega_2} = \tilde{g}$, which implies

$$\|T_p^*(f)\|_{\widetilde{H}_\mathrm{I}^{1/2}(-1,1)} = \|\tilde{g}\|_{H^{1/2}(\partial\Omega_2)} \lesssim \|G\|_{H^1(\Omega)} \simeq |G|_{H^1(\Omega)}. \tag{C.76}$$

The estimate (C.73) hence follows from (C.74), (C.75), and (C.76). □

We finish this section by a result concerning the invariance of the $\widetilde{H}^{1/2}$-norm under a special transformation which will be used in the analysis of the $p$-version. First we recall that the norm $\|v\|_{\widetilde{H}_\mathrm{w}^{1/2}(\alpha,\beta)}$ is defined in Section A.2.6.1 by

$$\|v\|^2_{\widetilde{H}_\mathrm{w}^{1/2}(\alpha,\beta)} = |v|^2_{H^{1/2}(\alpha,\beta)} + \int_\alpha^\beta \frac{v^2(x)}{\beta - x}\,\mathrm{d}x + \int_\alpha^\beta \frac{v^2(x)}{x-\alpha}\,\mathrm{d}x \tag{C.77}$$

with $|v|_{H^{1/2}(\alpha,\beta)}$ defined in (A.25).

**Lemma C.44.** *Let $a < b < c$ and let $A : u \mapsto \tilde{u}$ be a mapping from $\widetilde{H}^{1/2}(a,c)$ onto $\widetilde{H}^{1/2}(-1,1)$ defined as*

$$\tilde{u} := \begin{cases} \tilde{u}_1 & \text{on } [-1,0], \\ \tilde{u}_2 & \text{on } [0,1], \end{cases}$$

*where $\tilde{u}_1$ and $\tilde{u}_2$ are the affine images of $u_1 := u|_{[a,b]}$ and $u_2 := u|_{[b,c]}$ on $[-1,0]$ and $[0,1]$, respectively. Then*

$$\frac{1}{\mu}\|u\|^2_{\widetilde{H}_\mathrm{w}^{1/2}(a,c)} \le \|Au\|^2_{\widetilde{H}_\mathrm{w}^{1/2}(-1,1)} \le \mu\|u\|^2_{\widetilde{H}_\mathrm{w}^{1/2}(a,c)}, \tag{C.78}$$

*where*

$$\mu := \max\{\frac{c-b}{b-a}, \frac{b-a}{c-b}\}.$$

*Proof.* Assume without loss of generality that $\mu = (c-b)/(b-a)$. For any $t \in [a,b]$ and $\tau \in [b,c]$, if

$$s = \frac{t-b}{b-a} \in [-1,0] \qquad \text{and} \qquad \sigma = \frac{\tau - b}{c-b} \in [0,1], \tag{C.79}$$

then $\tilde{u}_1(s) = u_1(t)$ and $\tilde{u}_2(\sigma) = u_2(\tau)$. To prove (C.78), in view of (C.77) we consider three terms

$$T_1 := \int_{-1}^{1}\int_{-1}^{1} \frac{|\tilde{u}(x)-\tilde{u}(y)|^2}{|x-y|^2}\,\mathrm{d}x\,\mathrm{d}y,$$

$$T_2 := \int_{-1}^{1} \frac{|\tilde{u}(x)|^2}{1-x}\,\mathrm{d}x, \quad \text{and} \quad T_3 := \int_{-1}^{1} \frac{|\tilde{u}(x)|^2}{1+x}\,\mathrm{d}x.$$

By symmetry we have

$$T_1 = T_{11} + 2T_{12} + T_{13}$$

where

$$T_{11} := \int_0^1\int_0^1 \frac{|\tilde{u}_2(\sigma)-\tilde{u}_2(\sigma')|^2}{|\sigma-\sigma'|^2}\,\mathrm{d}\sigma\,\mathrm{d}\sigma',$$

$$T_{12} := \int_{-1}^0\int_{-1}^0 \frac{|\tilde{u}_1(s)-\tilde{u}_1(s')|^2}{|s-s'|^2}\,\mathrm{d}s\,\mathrm{d}s',$$

$$T_{13} := \int_0^1\int_{-1}^0 \frac{|\tilde{u}_1(s)-\tilde{u}_2(\sigma)|^2}{|s-\sigma|^2}\,\mathrm{d}s\,\mathrm{d}\sigma.$$

Noting (C.79) we have

$$T_{11} = \int_b^c\int_b^c \frac{|u_2(\tau)-u_2(\tau')|^2}{|\tau-\tau'|^2}\,\mathrm{d}\tau\,\mathrm{d}\tau',$$

$$T_{12} = \int_a^b\int_a^b \frac{|u_1(t)-u_1(t')|^2}{|t-t'|^2}\,\mathrm{d}t\,\mathrm{d}t',$$

$$T_{13} = \mu\int_b^c\int_a^b \frac{|u_1(t)-u_2(\tau)|^2}{|\tau-\mu t+(\mu-1)b|^2}\,\mathrm{d}t\,\mathrm{d}\tau.$$

Since $a \le t \le b \le \tau \le c$ and $\mu \ge 1$, we have

$$|\tau-\mu t+(\mu-1)b| = \tau - t + (\mu-1)(b-t),$$

which implies

$$\tau - t \le |\tau-\mu t+(\mu-1)b| \le \mu(\tau-t).$$

Therefore,

$$\frac{1}{\mu}\int_a^b\int_b^c \frac{|u_2(\tau)-u_1(t)|^2}{|\tau-t|^2}\,\mathrm{d}\tau\,\mathrm{d}t \le T_{13} \le \mu\int_a^b\int_b^c \frac{|u_2(\tau)-u_1(t)|^2}{|\tau-t|^2}\,\mathrm{d}\tau\,\mathrm{d}t.$$

Altogether we obtain

$$\frac{1}{\mu}|u|^2_{H^{1/2}(a,c)} \le T_1 \le \mu |u|^2_{H^{1/2}(a,c)}.$$

Similarly, we have

$$T_2 = \int_a^b \frac{|u_1(t)|^2}{2b-a-t}\,\mathrm{d}t + \int_b^c \frac{|u_2(\tau)|^2}{c-\tau}\,\mathrm{d}\tau.$$

Under the assumption $b-a \le c-b$, for $t \in [a,b]$ we have

$$\frac{c-t}{\mu} \le 2b-a-t \le c-t.$$

Hence

$$\int_a^c \frac{|u(z)|^2}{c-z}\,\mathrm{d}z \le T_2 \le \mu \int_a^c \frac{|u(z)|^2}{c-z}\,\mathrm{d}z.$$

Finally, a similar argument yields

$$\frac{1}{\mu}\int_a^c \frac{|u(z)|^2}{z-a}\,\mathrm{d}z \le T_3 \le \int_a^c \frac{|u(z)|^2}{z-a}\,\mathrm{d}z,$$

completing the proof of the lemma. □

## C.5.4 Clément's interpolation

As has been seen in Subsection C.5.2, the standard interpolation is not bounded in $H^1(\Gamma)$ nor $L^2(\Gamma)$. In fact, it is not even well defined in these spaces. It is sometimes necessary, in the analysis, to employ other types of interpolation operators which possess boundedness in these spaces (and thus in $H^{1/2}(\Gamma)$ and $\widetilde{H}^{1/2}(\Gamma)$). In this subsection we recall some properties of Clément's interpolation operator introduced in [61] by using local averaging.

Let $\mathscr{V}_h$ be the space of continuous piecewise linear functions defined on a mesh of $\Gamma$. We define $\Pi_h^0 : L_2(\Gamma) \longrightarrow \mathscr{V}_h$ by

$$\Pi_h^0 v = \sum_{\mathbf{x}_k \in \mathscr{N}_h} (P_k^0 v)\phi_k^h$$

where $P_k^0 v$ being the constant associated with $S_k = \operatorname{supp}\phi_k^h$ (the patch of all elements $K$ sharing the vertex $\mathbf{x}_k$) satisfies

$$\int_{S_k} (v - P_k^0 v) w \, d\mathbf{x} = 0$$

for all $w \in \mathbb{P}_0(S_k)$. The operator $\Pi_h^0$ is a type of local averaging interpolation operators. We note that it is not a projection.

The following results are well known and are recalled here for convenience. See [61]; see also [3, 56, 90].

**Lemma C.45.** *For any $v \in H_0^1(\Gamma)$ the following estimates hold*

$$\|\Pi_h^0 v\|_{L_2(\Gamma)} \lesssim \|v\|_{L_2(\Gamma)}, \quad |\Pi_h^0 v|_{H_0^1(\Gamma)} \lesssim |v|_{H_0^1(\Gamma)},$$

*and*

$$\sum_K h_K^{-2} \|v - \Pi_h^0 v\|_{L_2(K)}^2 + |v - \Pi_h^0 v|_{H^1(\Gamma)}^2 \lesssim |v|_{H^1(\Gamma)}^2.$$

*The constants are independent of v and h.*

Analogous bounds in $\widetilde{H}^{1/2}(\Gamma)$ and $H^{1/2}(\Gamma)$ can be drawn.

**Lemma C.46.** *For any $v \in \widetilde{H}^{1/2}(\Gamma)$ there hold*

$$\|\Pi_h^0 v\|_{\widetilde{H}_{\mathrm{I}}^{1/2}(\Gamma)} \lesssim \|v\|_{\widetilde{H}_{\mathrm{I}}^{1/2}(\Gamma)} \tag{C.80}$$

*and*

$$\sum_K h_K^{-1} \|v - \Pi_h^0 v\|_{L_2(K)}^2 + |v - \Pi_h^0 v|_{H^{1/2}(\Gamma)}^2 \lesssim |v|_{H^{1/2}(\Gamma)}^2. \tag{C.81}$$

*Proof.* The results for quasi-uniform meshes are proved in [3, Lemma 8]. We prove the results when the mesh is only locally quasi-uniform. It follows from Lemma C.45 that

$$\|\Pi_h^0 v\|_{L_2(\Gamma)} \lesssim \|v\|_{L_2(\Gamma)} \quad \text{and} \quad |\Pi_h^0 v|_{H^1(\Gamma)} \lesssim |v|_{H^1(\Gamma)}.$$

So (C.80) is obtained by interpolation.

To prove (C.81) we first note that the inequality

$$|v - \Pi_h^0 v|_{H^{1/2}(\Gamma)}^2 \lesssim |v|_{H^{1/2}(\Gamma)}^2$$

can be obtained by following the argument of the proof of [3, Lemma 8]. We now prove

$$\sum_K h_K^{-1} \|v - \Pi_h^0 v\|_{L_2(K)}^2 \lesssim |v|_{H^{1/2}(\Gamma)}^2.$$

For each element $K$, let $X_K = L_2(K)$ equipped with norm $\|\cdot\|_{X_K} = \|\cdot\|_{L_2(K)}$ and $Y_K = L_2(K)$ equipped with norm $\|\cdot\|_{Y_K} = h_K^{-1}\|\cdot\|_{L_2(K)}$. Let $Z_K$ be defined by Hilbert space interpolation as

$$Z_K := [X_K, Y_K]_{1/2},$$

with norm

$$\|v\|_{Z_K}^2 = \int_0^\infty K(t,v)^2 \frac{\mathrm{d}t}{t^2},$$

where the $K$-functional is defined, for $v \in X_K + Y_K$, by

$$K(t,v)^2 = \inf_{v=v_0+v_1} \left( \|v_0\|_{X_K}^2 + t^2 \|v_1\|_{Y_K}^2 \right).$$

Then

$$\|v\|_{Z_K}^2 \simeq \|v\|_{X_K} \|v\|_{Y_K} = h_K^{-1} \|v\|_{L_2(K)}^2. \tag{C.82}$$

Now if we define

$$\mathscr{X} = \prod_K X_K, \quad \mathscr{Y} = \prod_K Y_K, \quad \text{and} \quad \mathscr{Z} = \prod_K Z_K,$$

with norms

$$\|\underline{v}\|_{\mathscr{X}}^2 = \sum_K \|v_K\|_{X_K}^2, \quad \|\underline{v}\|_{\mathscr{Y}}^2 = \sum_K \|v_K\|_{Y_K}^2, \quad \text{and} \quad \|\underline{v}\|_{\mathscr{Z}}^2 = \sum_K \|v_K\|_{Z_K}^2,$$

where $\underline{v} = (v_K)_K$, then $\mathscr{Z} = [\mathscr{X}, \mathscr{Y}]_{1/2}$ with the $K$-functional satisfying

$$K(t,\underline{v})^2 = \sum_K K(t, v_K)^2.$$

Let $T$ be an operator defined by $Tv = v - \Pi_h^0 v$. Then it is given by Lemma C.45 that

$$\|Tv\|_{\mathscr{X}}^2 = \sum_K \|v - \Pi_h^0 v\|_{L_2(K)}^2 \lesssim \|v\|_{L_2(\Gamma)}^2$$

and

$$\|Tv\|_{\mathscr{Y}}^2 = \sum_K h_K^{-2} \|v - \Pi_h^0 v\|_{L_2(K)}^2 \lesssim |v|_{H^1(\Gamma)}^2.$$

By interpolation and noting (C.82), we infer

$$\sum_K h_K^{-1} \|v - \Pi_h^0 v\|_{L_2(K)}^2 \simeq \|Tv\|_{\mathscr{Z}}^2 \lesssim \|v\|_{H^{1/2}(\Gamma)}^2.$$

Since $T\alpha = 0$ for all $\alpha \in \mathbb{R}$, the above inequality implies, with the aid of Lemma A.6 and Lemma A.5,

$$\sum_K h_K^{-1} \|v - \Pi_h^0 v\|_{L_2(K)}^2 \simeq \|T(v+\alpha)\|_{\mathscr{Z}}^2 \lesssim \|v+\alpha\|_{H_w^{1/2}(\Gamma)}^2.$$

Taking the infimum over $\alpha$ and applying [3, Lemma 5] yields

$$\sum_K h_K^{-1} \|v - \Pi_h^0 v\|_{L_2(K)}^2 \lesssim |v|_{H^{1/2}(\Gamma)}^2,$$

proving the desired result. □

### *C.5.5 The Scott–Zhang interpolation*

The Scott–Zhang interpolation is defined in [191] with a variant considered in [14] for the Sobolev space $\widetilde{H}^s(\Gamma)$, $0 \le s \le 1$. This interpolation, also a local averaging interpolation operator as Clément's interpolation defined Subsection C.5.4, differs from the latter one in that it naturally preserves homogenous boundary conditions.

Let $\mathscr{T}_h$ be a triangulation of $\Gamma$, not necessarily quasi-uniform but shape regular. Let $\mathscr{V}_h$ be the boundary element space defined on this mesh $\mathscr{T}_h$ by

$$\mathscr{V}_h := \{v \in C(\Gamma) : v|_T \in \mathbb{P}_1 \ \forall T \in \mathscr{T}_h,\ v = 0 \text{ on } \partial\Gamma\}.$$

Let $\mathscr{N}_h := \{z_j : j = 1, \dots, N_h\}$ be the set of nodal points in $\mathscr{T}_h$. For any $j = 1, \dots, N_h$, let $\eta_j \in \mathscr{V}_h$ be the hat function which takes the value 1 at $z_j$ and zero at the other nodal points. These functions $\eta_j$, $j = 1, \dots, N_h$, form a basis for $\mathscr{V}_h$.

For each $z_j \in \mathscr{N}_h$, let $T_{z_j} \in \mathscr{T}_h$ be a triangle containing $z_j$ as a vertex. We consider the $L^2$-dual basis for $\mathscr{V}_h$ consisting of functions $\psi_j$ which satisfy

$$\int\limits_{T_{z_j}} \psi_j(x)\eta_i(x)\,\mathrm{d}s(x) = \delta_{ij}, \quad i = 1, \dots, N_h.$$

It is shown in [191, Lemma 3.1] that

$$\|\psi_j\|_{L^\infty(T_{z_j})} \lesssim |T_{z_j}|^{-1}, \quad j = 1, \dots, N_h. \tag{C.83}$$

The Scott–Zhang interpolation operator $J_h : L^2(\Gamma) \longrightarrow \mathscr{V}_h$ is defined by

$$J_h v = \sum_{j=1}^{N_h} \eta_j \int\limits_{T_{z_j}} \psi_j(x)v(x)\,\mathrm{d}s(x) \quad \forall v \in L^2(\Gamma). \tag{C.84}$$

Different choices of the triangle $T_{z_j}$ result in different operators $J_h$. The following properties of $J_h$ are proved in [191, Theorem 3.1 and Theorem 4.1]; see also [14].

**Lemma C.47.** *The operator $J_h : L^2(\Gamma) \longrightarrow \mathscr{V}_h$ has the following properties*

$$\begin{aligned}
J_h v &= v \quad \forall v \in \mathscr{V}_h, \\
\|J_h v\|_{\widetilde{H}^s_{\mathrm{I}}(\Gamma)} &\lesssim \|v\|_{\widetilde{H}^s_{\mathrm{I}}(\Gamma)} \quad \forall v \in \widetilde{H}^s(\Gamma), \quad 0 \le s \le 1, \\
\|v - J_h v\|_{L^2(T)} &\lesssim \mathrm{diam}(T)\|\nabla v\|_{L^2(\omega_h(T))} \quad \forall v \in \widetilde{H}^1(\Gamma), \\
\|\nabla(v - J_h v)\|_{L^2(T)} &\lesssim \|\nabla v\|_{L^2(\omega_h(T))} \quad \forall v \in \widetilde{H}^1(\Gamma),
\end{aligned}$$

*where $\omega_h(T) := \bigcup\{T' \in \mathscr{T}_h : T \cap T' \neq \varnothing\}$. The constant depends only on the shape regularity of $\mathscr{T}_h$.*

## C.6 Gauss–Lobatto Quadrature

Let $L_p$ be the Legendre polynomial normalised so that $L_p(1) = 1$. For each $p \in \mathbb{N}$, let $\tau_{p,j}$, $j = 0, \dots, p$, be the solutions of $(1-t^2)L_p'(t) = 0$, ordered such that

$$-1 = \tau_{p,0} < \tau_{p,1} < \cdots < \tau_{p,p} = 1. \tag{C.85}$$

These points $\tau_{p,j}$ are called the Gauss–Lobatto–Legendre (GLL) points. The Lagrange polynomial associated with $\tau_{p,j}$ is the polynomial of degree $p$ satisfying

$$\ell_{p,i}(\tau_{p,j}) = \delta_{ij}, \qquad i,j = 0,\dots,p. \tag{C.86}$$

It is well-known that the GLL points defined in (C.85) can be used to define the $(p+1)$-point Gauss-Lobatto quadrature rule

$$\int_{-1}^{1} f(t)\,\mathrm{d}t \approx \sum_{k=0}^{p} \omega_{p,k} f(\tau_{p,k}),$$

where the weights $\omega_{p,k}$ are given by

$$\omega_{p,k} = \frac{2}{p(p+1)L_p^2(\tau_{p,k})}, \qquad k = 0,\dots,p, \tag{C.87}$$

so that in particular

$$\omega_{p,0} = \omega_{p,p} = \frac{2}{p(p+1)}. \tag{C.88}$$

The quadrature rule is exact when $f$ is a polynomial of degree at most $2p-1$. Moreover, if $f$ is a polynomial of degree at most $p$, there holds [48]

$$\|f\|^2_{L_2(-1,1)} \le \sum_{k=0}^{p} \omega_{p,k}[f(\tau_{p,k})]^2 \le (2+1/p)\|f\|^2_{L_2(-1,1)}. \tag{C.89}$$

It follows from (C.89) that

$$\|\ell_{p,i}\|^2_{L_2(-1,1)} \le \omega_{p,i} \le (2+1/p)\|\ell_{p,i}\|^2_{L_2(-1,1)}, \tag{C.90}$$

where $\ell_{p,i}$ is defined by (C.86). When there is no confusion present, we write $\omega_k$ for $\omega_{p,k}$.

Let $\gamma \subset \partial K$ be a side of an element $K$ with associated polynomial order $p$. A node $\mathbf{q}_k$ on $\overline{\gamma}$ (the union of $\gamma$ and its two endpoints) is defined from a node $\tau_{p,k} \in [-1,1]$ by $\mathscr{F}_K$. Associated with this node is a weight $\omega_k$ defined by (C.87). Therefore, if $f$ is a function defined on $\gamma$, we can approximate the integral of $f$ on $\gamma$ as

$$\int_{\gamma} f(\mathbf{x})d\gamma_{\mathbf{x}} \approx \frac{h_\gamma}{2} \sum_{\mathbf{q}_k \in \mathscr{W}_p \cap \overline{\gamma}} \omega_k f(\mathbf{q}_k),$$

where $h_\gamma = \text{meas}(\gamma)$.

## C.7 Additional Technical Lemmas

**Lemma C.48.** *Let*

$$F(x) = \begin{cases} x^2 \log|x| & \text{if } x \neq 0, \\ 0 & \text{if } x = 0. \end{cases}$$

*and let*

$$f(x) = F(x+1) + F(x-1) - 2F(x) \quad \forall x \in \mathbb{R}.$$

*Then*

(i) $f''(x) < 0$ *if and only if* $|x| > 1/\sqrt{2}$ *and* $|x| \neq 1$;
(ii) $f'(x) > 0$ *for any* $x > 0$ *and* $f(x) > 0$ *for any* $x \neq 0$;
(iii) $f(2x)/2 - f(x) < 0$ *for* $x > 1/\sqrt{2}$.

*Proof.* We first note that

$$F'(x) = 2x\log|x| + x \quad \text{and} \quad F''(x) = 2\log|x| + 3 \quad \text{for all } x \neq 0.$$

Hence, for $x \neq 0$ and $x \neq \pm 1$,

$$f'(x) = 2(x+1)\log|x+1| + 2(x-1)\log|x-1| - 4x\log|x|$$

and

$$f''(x) = 2\log\frac{|x^2-1|}{x^2}.$$

Therefore (i) holds.

We now prove (ii). Consider a fixed $x > 0$ and define

$$h(t) := 2(t+x)\log|t+x| - 2(t-x)\log|t-x| \quad \forall t \in [0,1].$$

Then

$$h'(t) = 2\log\frac{|t+x|}{|t-x|}.$$

For $x \geq 1$, there exists $\theta \in (0,1)$ such that

$$f'(x) = h(1) - h(0) = h'(\theta) = 2\log\frac{|\theta+x|}{|\theta-x|} > 0.$$

For $0 < x < 1$, there exist $\theta_1 \in (x,1)$ and $\theta_2 \in (0,x)$ such that

$$f'(x) = h(1) - h(x) + h(x) - h(0) = (1-x)h'(\theta_1) + xh'(\theta_2) > 0.$$

Hence $f'(x) > 0$ for all $x > 0$. The remainder of the proof of (ii) is straightforward if one notes that $f(0) = 0$ and that $f$ is an even function.

To prove (iii) we note that if

$$k(x) := \frac{f(2x)}{2} - f(x) \quad \forall x > 1/\sqrt{2},$$

then, since $f''(x) < 0$ for $x > 1/\sqrt{2}$ and $x \neq 1$ we have

$$k'(x) = f'(2x) - f'(x) < 0,$$

and therefore $k(x) < k(0) = 0$. □

In the following lemma, we prove a special version of the mean value theorem, which will be used in the next result.

**Lemma C.49.** *Let $\varphi : \mathbb{R} \to \mathbb{R}$ be a differentiable function. For any $x \in \mathbb{R}$, there exists $\theta \in (0,1)$ satisfying*

$$\varphi(x+1) + \varphi(x-1) - 2\varphi(x) = \varphi'(x+\theta) - \varphi'(x-\theta).$$

*Proof.* Fixing $x \in \mathbb{R}$, we define

$$\psi : [0,1] \to \mathbb{R}, \quad t \mapsto \psi(t) := \varphi(x+t) + \varphi(x-t) - 2\varphi(x).$$

Then, there exists $\theta \in (0,1)$ such that

$$\varphi(x+1) + \varphi(x-1) - 2\varphi(x) = \psi(1) - \psi(0) = \psi'(\theta) = \varphi'(x+\theta) - \varphi'(x-\theta),$$

proving the lemma. □

**Lemma C.50.** *Let $f$ be defined as in Lemma C.48. If*

$$G(x) = f(x+1) + f(x-1) - 2f(x),$$

*then*

(i) $G(x) < 0$ *and* $G'(x) > 0$ *for* $x \geq 1 + 1/\sqrt{2}$;
(ii) *The following statement holds*

$$\sum_{m=4}^{\infty} |G(m)| \leq \frac{\pi^2}{3}.$$

*Proof.* (i) Due to Lemma C.48, $f$ is concave on $(1/\sqrt{2}, \infty)$. Therefore, $G(x) < 0$ when $x - 1 \geq 1/\sqrt{2}$. It follows from Lemma C.49 that there exists $\theta \in (0,1)$ satisfying

$$\begin{aligned} G'(x) &= f'(x+1) + f'(x-1) - 2f'(x) \\ &= f''(x+\theta) - f''(x-\theta) \end{aligned}$$

$$= 2\log\left(\frac{(x+\theta)^2-1}{(x+\theta)^2}\frac{(x-\theta)^2}{(x-\theta)^2-1}\right) > 0,$$

proving (i).

(ii) From (i) and Lemma C.49, we deduce, for $m \geq 2$,

$$\begin{aligned} |G(m)| &= 2f(m) - f(m+1) - f(m-1) \\ &= f'(m-\theta) - f'(m+\theta) \\ &= -2\theta\Big(f''(m-a) + f''(m+a)\Big) \\ &= 2\theta \log \frac{\left(m^2-a^2\right)^2}{\left((m-1)^2-a^2\right)\left((m+1)^2-a^2\right)} \end{aligned} \tag{C.91}$$

where $\theta, \theta' \in (0,1)$ and $a = \theta\theta'$. Since

$$\frac{\left(m^2-a^2\right)^2}{\left((m-1)^2-a^2\right)\left((m+1)^2-a^2\right)} = \frac{1}{\left(1-\frac{2m-1}{m^2-a^2}\right)\left(1+\frac{2m+1}{m^2-a^2}\right)} \leq \frac{1}{1-\left(\frac{2m-1}{m^2-a^2}\right)^2}$$

and

$$\frac{2m-1}{m^2-a^2} \leq \frac{2}{m} < 1 \quad \text{for } m \geq 3,$$

it follows from (C.91) that

$$|G(m)| \leq -2\log(1-\frac{4}{m^2}) = 2\sum_{i=1}^{\infty}\frac{1}{i}\left(\frac{2}{m}\right)^{2i}, \quad m \geq 3.$$

Therefore

$$\begin{aligned} \sum_{m=4}^{\infty}|G(m)| &\leq \sum_{m=4}^{\infty}\sum_{i=1}^{\infty}\frac{1}{i}\left(\frac{2}{m}\right)^{2i} \leq \int_3^{\infty}\sum_{i=1}^{\infty}\frac{1}{i}\left(\frac{2}{x}\right)^{2i} dx \\ &= \sum_{i=1}^{\infty}\frac{1}{i}\int_3^{\infty}\left(\frac{2}{x}\right)^{2i} dx = \sum_{i=1}^{\infty}\frac{3}{i(2i-1)}\left(\frac{2}{3}\right)^{2i} \\ &\leq \sum_{i=1}^{\infty}\frac{2}{i^2} = \frac{\pi^2}{3}. \end{aligned}$$

The lemma is proved. □

**Lemma C.51.** *Let $f$ be defined as in Lemma C.48. For a fixed $n \in \mathbb{N}$, if*

$$\widetilde{G}(x) := f(x) - f(x-n), \quad x \in \mathbb{R},$$

*then*

$$\lim_{n\to\infty} \frac{1}{n} \sum_{j=1}^{n-1} |\widetilde{G}(j)| = 4\log 2.$$

*Proof.* It follows from the parity of $f$ that

$$\widetilde{G}(n/2+x) = -\widetilde{G}(n/2-x) \quad \forall x \in [0, n/2].$$

Therefore,

$$\begin{cases} \displaystyle\sum_{j=1}^{[n/2]-1} |\widetilde{G}(j)| = \sum_{j=[n/2]+1}^{n-1} |\widetilde{G}(j)|, & n \text{ is even,} \\ \displaystyle\sum_{j=1}^{[n/2]} |\widetilde{G}(j)| = \sum_{j=[n/2]+1}^{n-1} |\widetilde{G}(j)|, & n \text{ is odd,} \end{cases}$$

where $[n/2]$ denotes the integral part of $n/2$. Noting that $\widetilde{G}([n/2]) = \widetilde{G}(n/2) = 0$ when $n$ is even, we deduce, for all $n \geq 2$,

$$\sum_{j=1}^{n-1} |\widetilde{G}(j)| = 2 \sum_{j=n^*}^{n-1} |\widetilde{G}(j)|$$

where $n^* := [n/2]+1$. On the other hand, the parity of $f'$ and Lemma C.48 imply

$$\widetilde{G}'(x) = f'(x) + f'(n-x) > 0 \quad \forall x \in [0, n],$$

so that

$$\widetilde{G}(x) \geq \widetilde{G}(n/2) = 0 \quad \forall x \in [n/2, n].$$

Therefore, by using the definition of $\widetilde{G}$, we deduce

$$\begin{aligned} \sum_{j=1}^{n-1} |\widetilde{G}(j)| &= 2 \sum_{j=n^*}^{n-1} \widetilde{G}(j) = 2 \sum_{j=n^*}^{n-1} \Big(f(j) - f(j-n)\Big) \\ &= 2 \sum_{j=n^*}^{n-1} \Big(F(j+1) - F(j)\Big) - 2 \sum_{j=n^*}^{n-1} \Big(F(j) - F(j-1)\Big) \\ &\quad - 2 \sum_{j=n^*}^{n-1} \Big(F(j-n+1) - F(j-n)\Big) + 2 \sum_{j=n^*}^{n-1} \Big(F(j-n) - F(j-n-1)\Big). \end{aligned}$$

Telescoping sums give, noting that $n^* = (n+1)/2$ when $n$ is odd,

$$\begin{aligned} \sum_{j=1}^{n-1} |\widetilde{G}(j)| &= 2\Big(F(n) - F(n^*) - F(n-1) + F(n^*-1) + F(n^*-n) - F(n^*-n-1)\Big) \\ &= \begin{cases} 2\Big(F(n) - F(n/2+1) - F(n-1) + F(n/2-1)\Big), & n \text{ is even,} \\ 2\Big(F(n) - 2F((n+1)/2) - F(n-1) + 2F((n-1)/2)\Big), & n \text{ is odd.} \end{cases} \end{aligned}$$

Consider first the case of even $n$. We have, from the definition of $F$,

$$\begin{aligned}
&\frac{1}{n}\Big(F(n)-F(n/2+1)-F(n-1)+F(n/2-1)\Big)\\
&\quad=\frac{1}{n}\Big[n^2\log n-\Big(\frac{n^2}{4}+n+1\Big)\log\frac{n+2}{2}\\
&\qquad-(n^2-2n+1)\log(n-1)+\Big(\frac{n^2}{4}-n+1\Big)\log\frac{n-2}{2}\Big]\\
&\quad=n\log\frac{n}{n-1}-\frac{n}{4}\log\frac{n+2}{n-2}-\log\frac{(n+2)(n-2)}{(n-1)^2}-\frac{1}{n}\log\frac{(n+2)(n-1)}{n-2}\\
&\qquad+2\log 2\\
&\quad=n\log\Big(1+\frac{1}{n-1}\Big)-\frac{n}{4}\log\Big(1+\frac{4}{n-2}\Big)-\log\frac{(n+2)(n-2)}{(n-1)^2}\\
&\qquad-\frac{1}{n}\log\frac{n(n+2)(n-1)}{n(n-2)}+2\log 2\\
&\quad=n\log\Big(1+\frac{1}{n-1}\Big)-\frac{n}{4}\log\Big(1+\frac{4}{n-2}\Big)-\log\frac{(n+2)(n-2)}{(n-1)^2}\\
&\qquad-\frac{1}{n}\log\frac{(n+2)(n-1)}{n(n-2)}-\frac{1}{n}\log n+2\log 2.
\end{aligned}$$

Letting $n\to\infty$ and noting that the first two terms on the right-hand side tend to $1-1=0$ whereas each of the next three terms tends to 0, we obtain

$$\lim_{n\to\infty}\frac{1}{n}\Big(F(n)-F(n/2+1)-F(n-1)+F(n/2-1)\Big)=2\log 2.$$

Similarly, for the case of odd $n$, we have

$$\begin{aligned}
&\frac{1}{n}\Big(F(n)-2F((n+1)/2)-F(n-1)+2F((n-1)/2)\Big)\\
&\quad=n\log\Big(1+\frac{1}{n-1}\Big)-\frac{n}{2}\log\Big(1+\frac{2}{n-1}\Big)-\log\frac{n+1}{n-1}--\frac{1}{2n}\log(n^2-1)\\
&\qquad+2\log 2.
\end{aligned}$$

Again, when $n\to\infty$ the first two terms on the righ-hand side tend to $1-1=0$ and each of the next two terms tends to 0. Therefore

$$\lim_{n\to\infty}\frac{1}{n}\Big(F(n)-2F((n+1)/2)-F(n-1)+2F((n-1)/2)\Big)=2\log 2.$$

Altogether we deduce

$$\lim_{n\to\infty}\frac{1}{n}\sum_{j=1}^{n-1}|\widetilde{G}(j)|=4\log 2,$$

completing the proof of the lemma. □

**Lemma C.52.** *The following statements hold true.*

(i) *If $p > 1$, $r \neq 1$, and $F(x)$ is defined by*

$$F(x) = \begin{cases} \int_0^x f(y)\,\mathrm{d}y, & r > 1, \\ \int_x^\infty f(y)\,\mathrm{d}y, & r < 1, \end{cases}$$

*then*

$$\int_0^\infty x^{-r} F^p(x)\,\mathrm{d}x < \left(\frac{p}{|r-1|}\right)^p \int_0^\infty x^{-r} (xf(x))^p\,\mathrm{d}x,$$

*unless $f \equiv 0$. The constant is the best possible. When $p = 1$, the two sides are equal.*

(ii) *If $0 \leq x \leq 1$ and $f \in L^2(x,1)$, then*

$$\int_0^{1-x} \frac{1}{y^2} \left( \int_x^{x+y} f(t)\,\mathrm{d}t \right)^2 \mathrm{d}y \leq 4 \int_x^1 f^2(t)\,\mathrm{d}t.$$

*Proof.* This result, which is a consequence of Hardy's inequality [103, Theorem 330, page 245], is proved in [110, Lemma 3.1]. □

# References


1. Adams, R.A.: Sobolev Spaces. Academic Press [A subsidiary of Harcourt Brace Jovanovich, Publishers], New York-London (1975). Pure and Applied Mathematics, Vol. 65
2. Ainsworth, M., Guo, B.: An additive Schwarz preconditioner for $p$-version boundary element approximation of the hypersingular operator in three dimensions. Numer. Math. **85**, 343–366 (2000)
3. Ainsworth, M., Guo, B.: Analysis of iterative sub-structuring techniques for boundary element approximation of the hypersingular operator in three dimensions. Appl. Anal. **81**, 241–280 (2002)
4. Ainsworth, M., McLean, W.: Multilevel diagonal scaling preconditioners for boundary element equations on locally refined meshes. Numer. Math. **93**(3), 387–413 (2003)
5. Ainsworth, M., McLean, W., Tran, T.: The conditioning of boundary element equations on locally refined meshes and preconditioning by diagonal scaling. SIAM J. Numer. Anal. **36**, 1901–1932 (1999)
6. Alfeld, P., Neamtu, M., Schumaker, L.L.: Bernstein-Bézier polynomials on spheres and sphere-like surfaces. Comput. Aided Geom. Design **13**, 333–349 (1996)
7. Alfeld, P., Neamtu, M., Schumaker, L.L.: Dimension and local bases of homogeneous spline spaces. SIAM J. Math. Anal. **27**, 1482–1501 (1996)
8. Alfeld, P., Neamtu, M., Schumaker, L.L.: Fitting scattered data on sphere-like surfaces using spherical splines. J. Comput. Appl. Math. **73**, 5–43 (1996)
9. Amini, S., Profit, A.T.J.: Multi-level fast multipole Galerkin method for the boundary integral solution of the exterior Helmholtz equation. In: Current trends in scientific computing (Xi'an, 2002), *Contemp. Math.*, vol. 329, pp. 13–19. Amer. Math. Soc., Providence, RI (2003)
10. Aronszajn, N., Smith, K.T.: Theory of Bessel potentials. I. Ann. Inst. Fourier (Grenoble) **11**, 385–475 (1961)
11. Ashby, S.F., Manteuffel, T.A., Saylor, P.E.: A taxonomy for conjugate gradient methods. SIAM J. Numer. Anal. **27**(6), 1542–1568 (1990)
12. Aurada, M., Feischl, M., Führer, T., Karkulik, M., Melenk, J.M., Praetorius, D.: Local inverse estimates for non-local boundary integral operators. Math. Comp. **86**(308), 2651–2686 (2017)
13. Aurada, M., Feischl, M., Führer, T., Karkulik, M., Praetorius, D.: Efficiency and optimality of some weighted-residual error estimator for adaptive 2D boundary element methods. Comput. Methods Appl. Math. **13**(3), 305–332 (2013)
14. Aurada, M., Feischl, M., Führer, T., Karkulik, M., Praetorius, D.: Energy norm based error estimators for adaptive BEM for hypersingular integral equations. Appl. Numer. Math. **95**, 15–35 (2015)
15. Aurada, M., Feischl, M., Karkulik, M., Praetorius, D.: A posteriori error estimates for the Johnson-Nédélec FEM-BEM coupling. Eng. Anal. Bound. Elem. **36**(2), 255–266 (2012)
16. Axelsson, O.: Iterative Solution Methods. Cambridge University Press, Cambridge (1994)
17. Babuška, I., Guo, B.Q., Stephan, E.P.: On the exponential convergence of the $h$-$p$ version for boundary element Galerkin methods on polygons. Math. Methods Appl. Sci. **12**, 413–427 (1990)
18. Babuška, I., Craig, A., Mandel, J., Pitkäranta, J.: Efficient preconditioning for the $p$ version finite element method in two dimensions. SIAM J. Numer. Anal. **28**, 624–661 (1991)
19. Babuška, I., Suri, M.: The $h$-$p$ version of the finite element method with quasi-uniform meshes. RAIRO Modél. Math. Anal. Numér. **21**(2), 199–238 (1987)
20. Bänsch, E.: Local mesh refinement in 2 and 3 dimensions. Impact Comput. Sci. Engrg. **3**(3), 181–191 (1991)
21. Baramidze, V., Lai, M.J.: Error bounds for minimal energy interpolatory spherical splines. In: Approximation theory XI: Gatlinburg 2004, Mod. Methods Math., pp. 25–50. Nashboro Press, Brentwood, TN (2005)
22. Bellman, R.: A note on an inequality of E. Schmidt. Bull. Amer. Math. Soc. **50**, 734–737 (1944)






E. P. Stephan and T. Tran, *Schwarz Methods and Multilevel Preconditioners for Boundary Element Methods*, https://doi.org/10.1007/978-3-030-79283-1

23. Ben Belgacem, F.: Polynomial extensions of compatible polynomial traces in three dimensions. Comput. Methods Appl. Mech. Engrg. **116**(1-4), 235–241 (1994). ICOSAHOM'92 (Montpellier, 1992)
24. Benzi, M.: Preconditioning techniques for large linear systems: a survey. J. Comput. Phys. **182**(2), 418–477 (2002)
25. Bergh, J., Löfström, J.: Interpolation Spaces: An Introduction. Springer-Verlag, Berlin (1976)
26. Bernardi, C., Maday, Y.: Approximations Spectrales de Problèmes aux Limites Elliptiques, *Mathématiques & Applications (Berlin) [Mathematics & Applications]*, vol. 10. Springer-Verlag, Paris (1992)
27. Bernardi, C., Maday, Y.: Spectral methods. In: Handbook of numerical analysis, Vol. V, Handb. Numer. Anal., V, pp. 209–485. North-Holland, Amsterdam (1997)
28. Bernardi, C., Maday, Y., Rapetti, F.: Discrétisations Variationnelles de Problèmes aux Limites Elliptiques, *Mathématiques & Applications (Berlin) [Mathematics & Applications]*, vol. 45. Springer-Verlag, Berlin (2004)
29. Bespalov, A., Heuer, N.: The $p$-version of the boundary element method for hypersingular operators on piecewise plane open surfaces. Numer. Math. **100**(2), 185–209 (2005)
30. Bespalov, A., Heuer, N.: The $p$-version of the boundary element method for weakly singular operators on piecewise plane open surfaces. Numer. Math. **106**, 69–97 (2007)
31. Bespalov, A., Heuer, N.: The $hp$-version of the boundary element method with quasi-uniform meshes in three dimensions. M2AN Math. Model. Numer. Anal. **42**, 821–849 (2008)
32. Bică, I.: Iterative substructuring algorithms for the $p$-version finite element method for elliptic problems. Ph.D. thesis, Courant Institute of Mathematical Sciences, New York University, New York (1997)
33. Bielak, J., MacCamy, R.C.: An exterior interface problem in two-dimensional elastodynamics. Quart. Appl. Math. **41**(1), 143–159 (1983/84)
34. Bollobás, B.: Modern Graph Theory. Springer Verlag, New York (1998)
35. Bramble, J., Leyk, Z., Pasciak, J.: The analysis of multigrid algorithms for pseudodifferential operators of order minus one. Math. Comp. **63**, 461–478 (1994)
36. Bramble, J., Pasciak, J., Schatz, A.: The construction of preconditioners for elliptic problems by substructuring. I. Math. Comp. **47**, 103–134 (1986)
37. Bramble, J.H.: A second order finite difference analog of the first biharmonic boundary value problem. Numer. Math. **9**, 236–249 (1966)
38. Bramble, J.H., Pasciak, J.E.: New estimates for multilevel algorithms including the $V$-cycle. Math. Comp. **60**(202), 447–471 (1993)
39. Bramble, J.H., Pasciak, J.E., Wang, J.P., Xu, J.: Convergence estimates for product iterative methods with applications to domain decomposition. Math. Comp. **57**(195), 1–21 (1991)
40. Bramble, J.H., Pasciak, J.E., Xu, J.: Parallel multilevel preconditioners. Math. Comp. **55**(191), 1–22 (1990)
41. Bramble, J.H., Xu, J.: Some estimates for a weighted $L^2$ projection. Math. Comp. **56**(194), 463–476 (1991)
42. Brenner, S.C., Scott, L.R.: The Mathematical Theory of Finite Element Methods. Springer, Berlin (2002)
43. Brezis, H.: Functional Analysis, Sobolev Spaces and Partial Differential Equations. Universitext. Springer, New York (2011)
44. Brezzi, F., Johnson, C.: On the coupling of boundary integral and finite element methods. Calcolo **16**(2), 189–201 (1979)
45. Cai, X.C.: The use of pointwise interpolation in domain decomposition methods with nonnested meshes. SIAM J. Sci. Comput. **16**(1), 250–256 (1995)
46. Cai, X.C., Zou, J.: Some observations on the $l^2$ convergence of the additive Schwarz preconditioned GMRES method. Numer. Linear Algebra Appl. **9**(5), 379–397 (2002)
47. Calderón, A.P.: Lebesgue spaces of differentiable functions and distributions. In: Proc. Sympos. Pure Math., Vol. IV, pp. 33–49. American Mathematical Society, Providence, R.I. (1961)
48. Canuto, C., Quarteroni, A.: Approximation results for orthogonal polynomials in Sobolev spaces. Math. Comp. **38**(157), 67–86 (1982)

49. Carstensen, C.: Nonlinear interface problems in solid mechanics–finite element and boundary element coupling. Habilitation thesis, Leibniz Universität Hannover, Hannover (1992)
50. Carstensen, C., Feischl, M., Page, M., Praetorius, D.: Axioms of adaptivity. Comput. Math. Appl. **67**(6), 1195–1253 (2014)
51. Carstensen, C., Kuhn, M., Langer, U.: Fast parallel solvers for symmetric boundary element domain decomposition equations. Numer. Math. **79**(3), 321–347 (1998)
52. Carstensen, C., Maischak, M., Praetorius, D., Stephan, E.P.: Residual-based a posteriori error estimate for hypersingular equation on surfaces. Numer. Math. **97**(3), 397–425 (2004)
53. Carstensen, C., Stephan, E.P.: Adaptive coupling of boundary elements and finite elements. RAIRO Modél. Math. Anal. Numér. **29**(7), 779–817 (1995)
54. Carstensen, C., Stephan, E.P.: A posteriori error estimates for boundary element methods. Math. Comp. **64**(210), 483–500 (1995)
55. Carstensen, C., Stephan, E.P.: Adaptive boundary element methods for some first kind integral equations. SIAM J. Numer. Anal. **33**(6), 2166–2183 (1996)
56. Chan, T.F., Smith, B.F., Zou, J.: Overlapping Schwarz methods on unstructured meshes using non-matching coarse grids. Numer. Math. **73**(2), 149–167 (1996)
57. Chandler-Wilde, S.N., Hewett, D.P., Moiola, A.: Interpolation of Hilbert and Sobolev spaces: quantitative estimates and counterexamples. Mathematika **61**(2), 414–443 (2015)
58. Chandra, R., Eisenstat, S., Schultz, M.: The modified conjugate residual method for partial differential equations, pp. 13–19. Defense Technical Information Center, New Brunswick (1977)
59. Chen, D., Menegatto, V.A., Sun, X.: A necessary and sufficient condition for strictly positive definite functions on spheres. Proc. Amer. Math. Soc. **131**, 2733–2740 (2003)
60. Ciarlet, P.G.: The Finite Element Method for Elliptic Problems. North-Holland, Amsterdam (1978)
61. Clément, P.: Approximation by finite element functions using local regularization. RAIRO Anal. Numer. **9**, 77–84 (1975)
62. Costabel, M.: Symmetric methods for the coupling of finite elements and boundary elements (invited contribution). In: Boundary elements IX, Vol. 1 (Stuttgart, 1987), pp. 411–420. Comput. Mech., Southampton (1987)
63. Costabel, M.: Boundary integral operators on Lipschitz domains: Elementary results. SIAM J. Math. Anal. **19**, 613–626 (1988)
64. Costabel, M.: On the coupling of finite and boundary element methods. In: Proceedings of the International Symposium on Numerical Analysis (Ankara, 1987), pp. 1–14. Middle East Tech. Univ., Ankara (1989)
65. Costabel, M., Stephan, E.: Boundary integral equations for mixed boundary value problems in polygonal domains and Galerkin approximation. In: Mathematical Models and Methods in Mechanics, *Banach Center Publ.*, vol. 15, pp. 175–251. PWN, Warsaw (1985)
66. Costabel, M., Stephan, E.: A direct boundary integral equation method for transmission problems. J. Math. Anal. Appl. **106**(2), 367–413 (1985)
67. Costabel, M., Stephan, E.P.: An improved boundary element Galerkin method for three-dimensional crack problems. Integral Equations Operator Theory **10**(4), 467–504 (1987)
68. Costabel, M., Stephan, E.P.: On the convergence of collocation methods for boundary integral equations on polygons. Math. Comp. **49**, 461–478 (1987)
69. Costabel, M., Stephan, E.P.: Coupling of finite and boundary element methods for an elasto-plastic interface problem. SIAM J. Numer. Anal. **27**(5), 1212–1226 (1990)
70. Dryja, M.: A method of domain decomposition for three-dimensional finite element elliptic problems. In: R. Glowinski, G. Golub, G. Meurant, J. Périaux (eds.) First International Symposium on Domain Decomposition Methods for Partial Differential Equations, pp. 43–61. SIAM, Philadelphia, PA (1988)
71. Dryja, M., Widlund, O.B.: Multilevel additive methods for elliptic finite element problems. In: W. Hackbusch (ed.) Parallel Algorithms for Partial Differential Equations (Proc. of the Sixth GAMM-Seminar, Kiel, Germany, January, 1990), pp. 58–69. Vieweg, Braunschweig (1991)

72. Dryja, M., Widlund, O.B.: Domain decomposition algorithms with small overlap. SIAM J. Sci. Comput. **15**, 604–620 (1994)
73. Eisenstat, S.C., Elman, H.C., Schultz, M.H.: Variational iterative methods for non-symmetric systems of linear equations. SIAM J. Numer. Anal. **20**, 345–357 (1983)
74. Feischl, M., Führer, T., Heuer, N., Karkulik, M., Praetorius, D.: Adaptive boundary element methods. Arch. Comput. Methods Eng. **22**(3), 309–389 (2015)
75. Feischl, M., Führer, T., Praetorius, D., Stephan, E.P.: Optimal additive Schwarz preconditioning for hypersingular integral equations on locally refined triangulations. Calcolo **54**(1), 367–399 (2017)
76. Feischl, M., Führer, T., Praetorius, D., Stephan, E.P.: Optimal preconditioning for the symmetric and nonsymmetric coupling of adaptive finite elements and boundary elements. Numer. Methods Partial Differential Equations **33**(3), 603–632 (2017)
77. Freeden, W., Gervens, T., Schreiner, M.: Constructive Approximation on the Sphere with Applications to Geomathematics. Oxford University Press, Oxford (1998)
78. Führer, T.: Zur Kopplung von Finiten Elementen und Randelementen. Ph.D. thesis, Vienna University of Technology, Vienna (2014)
79. Führer, T., Haberl, A., Praetorius, D., Schimanko, S.: Adaptive BEM with inexact PCG solver yields almost optimal computational costs. Numer. Math. **141**(4), 967–1008 (2019)
80. Funken, S.A., Stephan, E.P.: The BPX preconditioner for the single layer potential operator. Appl. Anal. **67**(3-4), 327–340 (1997)
81. Funken, S.A., Stephan, E.P.: Fast solvers with block-diagonal preconditioners for linear FEM-BEM coupling. Numer. Linear Algebra Appl. **16**(5), 365–395 (2009)
82. Gallistl, D., Schedensack, M., Stevenson, R.P.: A remark on newest vertex bisection in any space dimension. Comput. Methods Appl. Math. **14**(3), 317–320 (2014)
83. Golub, G.H., Van Loan, C.F.: Matrix Computations, fourth edn. Johns Hopkins Studies in the Mathematical Sciences. Johns Hopkins University Press, Baltimore, MD (2013)
84. Grafarend, E.W., Krumm, F.W., Schwarze, V.S. (eds.): Geodesy: the Challenge of the 3rd Millennium. Springer, Berlin (2003)
85. Graham, I.G., Hackbusch, W., Sauter, S.A.: Finite elements on degenerate meshes: inverse-type inequalities and applications. IMA J. Numer. Anal. **25**(2), 379–407 (2005)
86. Graham, I.G., McLean, W.: Anisotropic mesh refinement: the conditioning of Galerkin boundary element matrices and simple preconditioners. SIAM J. Numer. Anal. **44**(4), 1487–1513 (2006)
87. Greenbaum, A.: Iterative Methods for Solving Linear Systems, *Frontiers in Applied Mathematics*, vol. 17. Society for Industrial and Applied Mathematics (SIAM), Philadelphia, PA (1997)
88. Grisvard, P.: Elliptic Problems in Nonsmooth Domains. Pitman, Boston (1985)
89. Guo, B., Cao, W.: Additive Schwarz methods for the $h$-$p$ version of the finite element method in two dimensions. SIAM J. Sci. Comput. **18**(5), 1267–1288 (1997)
90. Guo, B., Cao, W.: An additive Schwarz method for the $h$-$p$ version of the finite element method in three dimensions. SIAM J. Numer. Anal. **35**(2), 632–654 (1998)
91. Guo, B., Heuer, N.: The optimal rate of convergence of the $p$-version of the boundary element method in two dimensions. Numer. Math. **98**, 499–538 (2004)
92. Gwinner, J., Stephan, E.P.: Advanced Boundary Element Methods, *Springer Series in Computational Mathematics*, vol. 52. Springer, Cham (2018). Treatment of boundary value, transmission and contact problems
93. Hackbusch, W.: The panel clustering technique for the boundary element method (invited contribution). In: Boundary elements IX, Vol. 1 (Stuttgart, 1987), pp. 463–474. Comput. Mech., Southampton (1987)
94. Hackbusch, W.: Integral Equations, *International Series of Numerical Mathematics*, vol. 120. Birkhäuser Verlag, Basel (1995). Theory and numerical treatment, Translated and revised by the author from the 1989 German original
95. Hackbusch, W.: A sparse matrix arithmetic based on $\mathscr{H}$-matrices. I. Introduction to $\mathscr{H}$-matrices. Computing **62**(2), 89–108 (1999)

96. Hackbusch, W.: Iterative Solution of Large Sparse Systems of Equations, *Applied Mathematical Sciences*, vol. 95, second edn. Springer, [Cham] (2016)
97. Hackbusch, W., Börm, S.: Data-sparse approximation by adaptive $\mathscr{H}^2$-matrices. Computing **69**(1), 1–35 (2002)
98. Hackbusch, W., Khoromskij, B., Sauter, S.A.: On $\mathscr{H}^2$-matrices. In: Lectures on applied mathematics (Munich, 1999), pp. 9–29. Springer, Berlin (2000)
99. Hackbusch, W., Khoromskij, B.N., Sauter, S.: Adaptive Galerkin boundary element methods with panel clustering. Numer. Math. **105**(4), 603–631 (2007)
100. Hackbusch, W., Nowak, Z.P.: On the fast matrix multiplication in the boundary element method by panel clustering. Numer. Math. **54**(4), 463–491 (1989)
101. Hahne, M., Stephan, E.P.: Schwarz iterations for the efficient solution of screen problems with boundary elements. Computing **56**(1), 61–85 (1996)
102. Han, H.D.: A new class of variational formulations for the coupling of finite and boundary element methods. J. Comput. Math. **8**(3), 223–232 (1990)
103. Hardy, G.H., Littlewood, J.E., Pólya, G.: Inequalities. Cambridge, at the University Press (1952). 2d ed
104. Hestenes, M.R., Stiefel, E.: Methods of conjugate gradients for solving linear systems. J. Research Nat. Bur. Standards **49**, 409–436 (1953) (1952)
105. Heuer, N.: Additive Schwarz methods for weakly singular integral equations in $\mathbb{R}^3$ – the $p$ version. In: W. Hackbusch, G. Wittum (eds.) Boundary Elements: Implementation and Analysis of Advanced Algorithms (Proc. of the 12th GAMM–Seminar, Kiel, Germany, January, 1996), pp. 126–135. Vieweg-Verlag, Braunschweig (1996)
106. Heuer, N.: Efficient algorithms for the $p$ version of the boundary element method. J. Integral Equations Appl. **8**, 337–361 (1996)
107. Heuer, N.: Preconditioners for the $p$-version of the boundary element Galerkin method in $\mathbb{R}^3$. Habilitation thesis, Leibniz Universität Hannover, Hannover (2000)
108. Heuer, N.: Additive Schwarz method for the $p$-version of the boundary element method for the single layer potential operator on a plane screen. Numer. Math. **88**(3), 485–511 (2001)
109. Heuer, N.: On the equivalence of fractional-order Sobolev semi-norms. J. Math. Anal. Appl. **417**(2), 505–518 (2014)
110. Heuer, N., Leydecker, F.: An extension theorem for polynomials on triangles. Calcolo **45**(2), 69–85 (2008)
111. Heuer, N., Leydecker, F., Stephan, E.P.: An iterative substructuring method for the $hp$-version of the BEM on quasi-uniform triangular meshes. Numer. Methods Partial Differential Equations **23**(4), 879–903 (2007)
112. Heuer, N., Maischak, M., Stephan, E.P.: Exponential convergence of the $hp$-version for the boundary element method on open surfaces. Numer. Math. **83**(4), 641–666 (1999)
113. Heuer, N., Maischak, M., Stephan, E.P.: Preconditioned minimum residual iteration for the $h$-$p$ version of the coupled FEM/BEM with quasi-uniform meshes. Numer. Linear Algebra Appl. **6**(6), 435–456 (1999). Iterative solution methods for the elasticity equations in mechanics and biomechanics, IMMB'98, Part 1 (Nijmegen)
114. Heuer, N., Stephan, E.P.: The $hp$-version of the boundary element method on polygons. J. Integral Equations Appl. **8**(2), 173–212 (1996)
115. Heuer, N., Stephan, E.P.: Boundary integral operators in countably normed spaces. Math. Nachr. **191**, 123–151 (1998)
116. Heuer, N., Stephan, E.P.: Iterative substructuring for hypersingular integral equations in $\mathbf{R}^3$. SIAM J. Sci. Comput. **20**(2), 739–749 (1998)
117. Heuer, N., Stephan, E.P.: Preconditioners for the $p$-version of the Galerkin method for a coupled finite element/boundary element system. Numer. Methods Partial Differential Equations **14**(1), 47–61 (1998)
118. Heuer, N., Stephan, E.P.: An additive Schwarz method for the $h$-$p$ version of the boundary element method for hypersingular integral equations in $\mathbb{R}^3$. IMA J. Numer. Anal. **21**(1), 265–283 (2001)

119. Heuer, N., Stephan, E.P.: An overlapping domain decomposition preconditioner for high order BEM with anisotropic elements. Adv. Comput. Math. **19**(1-3), 211–230 (2003). Challenges in computational mathematics (Pohang, 2001)
120. Heuer, N., Stephan, E.P., Tran, T.: Multilevel additive Schwarz method for the *h-p* version of the Galerkin boundary element method. Math. Comp. **67**(222), 501–518 (1998)
121. Hille, E.: Analytic Function Theory, vol. II. Ginn, Boston (1962)
122. Hille, E., Szegö, G., Tamarkin, J.: On some generalizations of a theorem of A. Markoff. Duke Math. J. **3**, 729–739 (1937)
123. Hiptmair, R.: Operator preconditioning. Comput. Math. Appl. **52**(5), 699–706 (2006)
124. Holm, H., Maischak, M., Stephan, E.P.: The *hp*-version of the boundary element method for Helmholtz screen problems. Computing **57**(2), 105–134 (1996)
125. Holm, H., Maischak, M., Stephan, E.P.: Exponential convergence of the *h-p* version BEM for mixed boundary value problems on polyhedrons. Math. Methods Appl. Sci. **31**, 2069–2093 (2008)
126. Hörmander, L.: Pseudodifferential operators. Comm. Pure Appl. Math. **18**, 501–517 (1965)
127. Hsiao, G.C., Wendland, W.L.: A finite element method for some integral equations of the first kind. J. Math. Anal. Appl. **58**, 449–481 (1977)
128. Hsiao, G.C., Wendland, W.L.: Boundary Integral Equations, *Applied Mathematical Sciences*, vol. 164. Springer-Verlag, Berlin (2008)
129. Johnson, C., Nédélec, J.C.: On the coupling of boundary integral and finite element methods. Math. Comp. **35**(152), 1063–1079 (1980)
130. Karkulik, M., Pavlicek, D., Praetorius, D.: On 2D newest vertex bisection: optimality of mesh-closure and $H^1$-stability of $L_2$-projection. Constr. Approx. **38**(2), 213–234 (2013)
131. Kohn, J., Nirenberg, L.: On the algebra of pseudodifferential operators. Comm. Pure Appl. Math. **18**, 269–305 (1965)
132. Langer, U., Of, G., Steinbach, O., Zulehner, W.: Inexact data-sparse boundary element tearing and interconnecting methods. SIAM J. Sci. Comput. **29**(1), 290–314 (2007)
133. Langer, U., Of, G., Steinbach, O., Zulehner, W.: Inexact fast multipole boundary element tearing and interconnecting methods. In: Domain decomposition methods in science and engineering XVI, *Lect. Notes Comput. Sci. Eng.*, vol. 55, pp. 405–412. Springer, Berlin (2007)
134. Langer, U., Steinbach, O.: Boundary element tearing and interconnecting methods. Computing **71**(3), 205–228 (2003)
135. Langer, U., Steinbach, O.: Coupled boundary and finite element tearing and interconnecting methods. In: Domain decomposition methods in science and engineering, *Lect. Notes Comput. Sci. Eng.*, vol. 40, pp. 83–97. Springer, Berlin (2005)
136. Le Gia, Q.T., Narcowich, F.J., Ward, J.D., Wendland, H.: Continuous and discrete least-square approximation by radial basis functions on spheres. J. Approx. Theory **143**, 124–133 (2006)
137. Le Gia, Q.T., Sloan, I.H., Tran, T.: Overlapping additive Schwarz preconditioners for elliptic PDEs on the unit sphere. Math. Comp. **78**(265), 79–101 (2009)
138. Lions, J.L., Magenes, E.: Non-Homogeneous Boundary Value Problems and Applications I. Springer-Verlag, New York (1972)
139. Maday, Y.: Relèvements de traces polynomiales et interpolations hilbertiennes entre espaces de polynômes. C. R. Acad. Sci. Paris Sér. I Math. **309**(7), 463–468 (1989)
140. Maday, Y.: Relèvements de traces polynomiales et interpolations hilbertiennes entre espaces de polynômes. C. R. Acad. Sci. Paris Sér. I Math. **309**(7), 463–468 (1989)
141. Maischak, M.: The analytical computation of the Galerkin elements for the Laplace, Lamé, and Helmholtz equations in 3d-BEM. Tech. Rep. 95-16, Institut für Angewandte Mathematik, Universität Hannover (1995)
142. Maischak, M., Stephan, E.P.: The *hp*-version of the boundary element method in $\mathbb{R}^3$: the basic approximation results. Math. Methods Appl. Sci. **20**, 461–476 (1997)
143. Maischak, M., Stephan, E.P., Tran, T.: Domain decomposition methods for boundary integral equations of the first kind: numerical results. Appl. Anal. **63**(1-2), 111–132 (1996)
144. Maischak, M., Stephan, E.P., Tran, T.: Multiplicative Schwarz algorithms for the Galerkin boundary element method. SIAM J. Numer. Anal. **38**(4), 1243–1268 (2000)

145. Maischak, M., Tran, T.: A block preconditioner for an electromagnetic FEM-BEM coupling problem in $\mathbb{R}^3$. In: Z.C.S. Wenbin Liu Michael Ng (ed.) Recent Progress in Scientific Computing, pp. 302–318. Science Press, Beijing (2007). Proceeding of the $2^{nd}$ International Conference on Scientific Computing and Partial Differential Equations, 2 December, 2005
146. Maitre, J.F., Pourquier, O.: Condition number and diagonal preconditioning: comparison of the $p$-version and the spectral element methods. Numer. Math. **74**(1), 69–84 (1996)
147. McLean, W.: Strongly Elliptic Systems and Boundary Integral Equations. CUP, Cambridge (2000)
148. McLean, W., Sloan, I.H.: A fully discrete and symmetric boundary element method. IMA J. Numer. Anal. **14**, 311–345 (1994)
149. McLean, W., Tran, T.: A preconditioning strategy for boundary element Galerkin methods. Numer. Methods Partial Differential Equations **13**(3), 283–301 (1997)
150. Meddahi, S.: An optimal iterative process for the Johnson-Nedelec method of coupling boundary and finite elements. SIAM J. Numer. Anal. **35**(4), 1393–1415 (1998)
151. Melenk, J.M.: On condition numbers in $hp$-FEM with Gauss-Lobatto-based shape functions. J. Comput. Appl. Math. **139**(1), 21–48 (2002)
152. Mitchell, W.F.: Unified multilevel adaptive finite element methods for elliptic problems. Ph.D. thesis, University of Illinois at Urbana-Champaign, Illinois, USA (1988)
153. Mitchell, W.F.: Optimal multilevel iterative methods for adaptive grids. SIAM J. Sci. Statist. Comput. **13**(1), 146–167 (1992)
154. Morton, T.M., Neamtu, M.: Error bounds for solving pseudodifferential equations on spheres. J. Approx. Theory **114**, 242–268 (2002)
155. Müller, C.: Spherical Harmonics, *Lecture Notes in Mathematics*, vol. 17. Springer-Verlag, Berlin (1966)
156. Mund, P., Stephan, E.P.: The preconditioned GMRES method for systems of coupled FEM-BEM equations. Adv. Comput. Math. **9**(1-2), 131–144 (1998). Numerical treatment of boundary integral equations
157. Narcowich, F.J., Ward, J.D.: Norms of inverses and condition numbers for matrices associated with scattered data. J. Approx. Theory **64**, 69–94 (1991)
158. Narcowich, F.J., Ward, J.D.: Scattered data interpolation on spheres: error estimates and locally supported basis functions. SIAM J. Math. Anal. **33**, 1393–1410 (2002)
159. Neamtu, M., Schumaker, L.L.: On the approximation order of splines on spherical triangulations. Advances in Comput. Math. **21**, 3–20 (2004)
160. Nédélec, J.C.: Integral equations with nonintegrable kernels. Integral Equations Operator Theory **5**, 562–572 (1982)
161. Nédélec, J.C.: Acoustic and Electromagnetic Equations. Springer-Verlag, New York (2000)
162. Nédélec, J.C., Planchard, J.: Une méthode variationnelle d' éléments finis pour la résolution numérique d'un problème extérieur dans $R^3$. Rev. Française Automat. Informat. Recherche Opérationnelle Sér. Rouge **7**, 105–129 (1973)
163. Nečas, J.: Les Méthodes Directes en Théorie des Equations Elliptiques. Masson et Cie, Paris (1967)
164. Of, G., Steinbach, O.: Is the one-equation coupling of finite and boundary element methods always stable? ZAMM Z. Angew. Math. Mech. **93**(6-7), 476–484 (2013)
165. Of, G., Steinbach, O., Wendland, W.L.: The fast multipole method for the symmetric boundary integral formulation. IMA J. Numer. Anal. **26**(2), 272–296 (2006)
166. Olsen, E.T., Douglas Jr., J.: Bounds on spectral condition numbers of matrices arising in the $p$-version of the finite element method. Numer. Math. **69**(3), 333–352 (1995)
167. Oswald, P.: Multilevel norms for $H^{-1/2}$. Computing **61**(3), 235–255 (1998)
168. Pavarino, L.F.: Additive Schwarz methods for the $p$ version finite element method. Numer. Math. **66**, 493–515 (1994)
169. Pavarino, L.F., Widlund, O.B.: A polylogarithmic bound for an iterative substructuring method for spectral elements in three dimensions. SIAM J. Numer. Anal. **33**(4), 1303–1335 (1996)

170. von Petersdorff, T.: Randwertprobleme der Elastizitätstheorie für Polyeder–Singularitäten und Approximation mit Randelementmethoden. Ph.D. thesis, Technische Hochschule Darmstadt, Darmstadt (1989)
171. von Petersdorff, T., Stephan, E.P.: On the convergence of the multigrid method for a hypersingular integral equation of the first kind. Numer. Math. **57**(4), 379–391 (1990)
172. von Petersdorff, T., Stephan, E.P.: Multigrid solvers and preconditioners for first kind integral equations. Numer. Methods Partial Differential Equations **8**(5), 443–450 (1992)
173. Petersen, B.E.: Introduction to the Fourier Transform & Pseudodifferential Operators, *Monographs and Studies in Mathematics*, vol. 19. Pitman (Advanced Publishing Program), Boston, MA (1983)
174. Pham, D., Tran, T.: A domain decomposition method for solving the hypersingular integral equation on the sphere with spherical splines. Numer. Math. **120**(1), 117–151 (2012)
175. Pham, D., Tran, T., Crothers, S.: An overlapping additive Schwarz preconditioner for the Laplace-Beltrami equation using spherical splines. Adv. Comput. Math. **37**(1), 93–121 (2012)
176. Pham, D.T., Tran, T.: Solving non-strongly elliptic pseudodifferential equations on a sphere using radial basis functions. Comput. Math. Appl. **70**(8), 1970–1983 (2015)
177. Pham, T.D.: Pseudodifferential equations on spheres with spherical radial basis functions and spherical splines. Ph.D. thesis, UNSW Sydney, Sydney (2011)
178. Pham, T.D., Tran, T.: Strongly elliptic pseudodifferential equations on the sphere with radial basis functions. Numer. Math. **128**(3), 589–614 (2014)
179. Pham, T.D., Tran, T., Chernov, A.: Pseudodifferential equations on the sphere with spherical splines. Math. Models Methods Appl. Sci. **21**(9), 1933–1959 (2011)
180. Ratcliffe, J.G.: Foundations of Hyperbolic Manifolds. Springer, New York (1994)
181. Raviart, P.A., Thomas, J.M.: Introduction à l'analyse numérique des équations aux dérivées partielles. Collection Mathématiques Appliquées pour la Maîtrise. [Collection of Applied Mathematics for the Master's Degree]. Masson, Paris (1989)
182. Rychkov, V.S.: On restrictions and extensions of the Besov and Triebel-Lizorkin spaces with respect to Lipschitz domains. J. London Math. Soc. (2) **60**(1), 237–257 (1999)
183. Saad, Y.: Iterative Methods for Sparse Linear Systems, second edn. Society for Industrial and Applied Mathematics, Philadelphia, PA (2003)
184. Saad, Y., Schultz, M.H.: GMRES: a generalized minimal residual algorithm for solving nonsymmetric linear systems. SIAM J. Sci. Statist. Comput. **7**(3), 856–869 (1986)
185. Sarkis, M., Szyld, D.B.: Optimal left and right additive Schwarz preconditioning for minimal residual methods with Euclidean and energy norms. Comput. Methods Appl. Mech. Engrg. **196**(8), 1612–1621 (2007)
186. Sauter, S.A., Schwab, C.: Boundary Element Methods, *Springer Series in Computational Mathematics*, vol. 39. Springer-Verlag, Berlin (2011). Translated and expanded from the 2004 German original
187. Sayas, F.J.: The validity of Johnson-Nédélec's BEM-FEM coupling on polygonal interfaces. SIAM J. Numer. Anal. **47**(5), 3451–3463 (2009)
188. Sayas, F.J.: The validity of Johnson-Nédélec's BEM-FEM coupling on polygonal interfaces. SIAM Rev. **55**(1), 131–146 (2013)
189. Schoenberg, I.J.: Positive definite functions on spheres. Duke Math. J. **9**, 96–108 (1942)
190. Schwab, C., Suri, M.: The optimal $p$-version approximation of singularities on polyhedra in the boundary element method. SIAM J. Numer. Anal. **33**, 729–759 (1996)
191. Scott, L.R., Zhang, S.: Finite element interpolation of nonsmooth functions satisfying boundary conditions. Math. Comp. **54**(190), 483–493 (1990)
192. Seeley, R.T.: Extension of $C^\infty$ functions defined in a half space. Proc. Amer. Math. Soc. **15**, 625–626 (1964)
193. Silvester, D., Wathen, A.: Fast iterative solution of stabilised Stokes systems. II. Using general block preconditioners. SIAM J. Numer. Anal. **31**(5), 1352–1367 (1994)
194. Simpson, R.N., Liu, Z.: Acceleration of isogeometric boundary element analysis through a black-box fast multipole method. Eng. Anal. Bound. Elem. **66**, 168–182 (2016)

195. Smith, K.T., Solmon, D.C., Wagner, S.L.: Practical and mathematical aspects of the problem of reconstructing objects from radiographs. Bull. Amer. Math. Soc. **83**, 1227–1270 (1977)
196. Muñoz Sola, R.: Polynomial liftings on a tetrahedron and applications to the *h-p* version of the finite element method in three dimensions. SIAM J. Numer. Anal. **34**(1), 282–314 (1997)
197. Stein, E.M.: Singular Integrals and Differentiability Properties of Functions. Princeton Mathematical Series, No. 30. Princeton University Press, Princeton, N.J. (1970)
198. Steinbach, O.: Fast solvers for the symmetric boundary element method. In: Boundary elements: implementation and analysis of advanced algorithms (Kiel, 1996), *Notes Numer. Fluid Mech.*, vol. 54, pp. 232–242. Friedr. Vieweg, Braunschweig (1996)
199. Steinbach, O.: Numerical Approximation Methods for Elliptic Boundary Value Problems. Springer, New York (2008). Finite and boundary elements, Translated from the 2003 German original
200. Steinbach, O., Wendland, W.L.: On C. Neumann's method for second-order elliptic systems in domains with non-smooth boundaries. J. Math. Anal. Appl. **262**(2), 733–748 (2001)
201. Stephan, E.P.: A boundary integral equation method for three-dimensional crack problems in elasticity. Math. Meth. Appl. Sci. **8**, 609–623 (1986)
202. Stephan, E.P.: Boundary integral equations for screen problems in $\mathbb{R}^3$. Integral Equations Operator Theory **10**, 236–257 (1987)
203. Stephan, E.P.: The *h-p* boundary element method for solving 2- and 3-dimensional problems. Comput. Methods Appl. Mech. Engrg. **133**(3-4), 183–208 (1996)
204. Stephan, E.P.: Multilevel methods for the *h*-, *p*-, and *hp*-versions of the boundary element method. J. Comput. Appl. Math. **125**(1-2), 503–519 (2000). Numerical analysis 2000, Vol. VI, Ordinary differential equations and integral equations
205. Stephan, E.P.: Coupling of boundary element methods and finite element methods. In: Encyclopedia of Computational Mechanics Second Edition, pp. 1–40. John Wiley & Sons (2017)
206. Stephan, E.P., Maischak, M., Leydecker, F.: Some Schwarz methods for integral equations on surfaces—*h* and *p* versions. Comput. Vis. Sci. **8**(3-4), 211–216 (2005)
207. Stephan, E.P., Maischak, M., Tran, T.: Domain decomposition algorithms for an indefinite hypersingular integral equation in three dimensions. In: Domain decomposition methods in science and engineering XVII, *Lect. Notes Comput. Sci. Eng.*, vol. 60, pp. 647–655. Springer, Berlin (2008)
208. Stephan, E.P., Suri, M.: On the convergence of the *p* version of the boundary element Galerkin method. Math. Comp. **52**, 31–48 (1989)
209. Stephan, E.P., Suri, M.: The *h-p* version of the boundary element method on polygonal domains with quasiuniform meshes. Math. Modelling Numer. Anal. (RAIRO) **25**, 783–807 (1991)
210. Stephan, E.P., Tran, T.: Domain decomposition algorithms for indefinite hypersingular integral equations: the *h* and *p* versions. SIAM J. Sci. Comput. **19**(4), 1139–1153 (1998)
211. Stephan, E.P., Tran, T.: Domain decomposition algorithms for indefinite weakly singular integral equations: the *h* and *p* versions. IMA J. Numer. Anal. **20**, 1–24 (2000)
212. Stephan, E.P., Wendland, W.L.: An augmented Galerkin procedure for the boundary integral method applied to two-dimensional screen and crack problems. Appl. Anal. **18**, 183–219 (1984)
213. Strang, G., Fix, G.: An Analysis of the Finite Element Method. Prentice-Hall, Englewood Cliffs, New Jersey (1973)
214. Svensson, S.: Pseudodifferential operators – a new approach to the boundary problems of physical geodesy. Manuscr. Geod. **8**, 1–40 (1983)
215. Szabó, B., Babuška, I.: Finite Element Analysis. Wiley-Interscience Publication, New York (1991)
216. Szegö, G.: Orthogonal Polynomials. Amer. Math. Soc. Colloquium Publication,23, Providence, R.I. (1959)
217. Tausch, J.: The variable order fast multipole method for boundary integral equations of the second kind. Computing **72**(3-4), 267–291 (2004)
218. Thomée, V.: Galerkin Finite Element Methods for Parabolic Problems. Springer, Berlin (1997)

219. Toselli, A., Widlund, O.: Domain Decomposition Methods—Algorithms and Theory, *Springer Series in Computational Mathematics*, vol. 34. Springer-Verlag, Berlin (2005)
220. Tran, T.: Additive Schwarz algorithms and the Galerkin boundary element method. In: Computational techniques and applications: CTAC97 (Adelaide), pp. 703–710. World Sci. Publ., River Edge, NJ (1998)
221. Tran, T.: Additive Schwarz preconditioners for a fully-discrete and symmetric boundary element method. ANZIAM J. **42**((C)), C1420–C1442 (2000)
222. Tran, T.: Overlapping additive Schwarz preconditioners for boundary element methods. J. Integral Eqns Appl. **12**, 177–207 (2000)
223. Tran, T.: Additive Schwarz preconditioners for the $h$-$p$ version boundary-element approximation to the hypersingular operator in three dimensions. Int. J. Comput. Math. **84**(10), 1417–1437 (2007)
224. Tran, T.: Remarks on Sobolev norms of fractional orders. J. Math. Anal. Appl. **498**(2), 124,960, 19 (2021)
225. Tran, T., Le Gia, Q.T., Sloan, I.H., Stephan, E.P.: Boundary integral equations on the sphere with radial basis functions: Error analysis. Appl. Numer. Math. **59**, 2857–2871 (2009)
226. Tran, T., Le Gia, Q.T., Sloan, I.H., Stephan, E.P.: Preconditioners for pseudodifferential equations on the sphere with radial basis functions. Numer. Math. **115**(1), 141–163 (2010)
227. Tran, T., Stephan, E.P.: Additive Schwarz methods for the $h$ version boundary element method. Appl. Anal. **60**, 63–84 (1996)
228. Tran, T., Stephan, E.P.: Additive Schwarz algorithms for the $p$ version of the Galerkin boundary element method. Numer. Math. **85**, 433–468 (2000)
229. Tran, T., Stephan, E.P.: Two-level additive Schwarz preconditioners for the $h$-$p$ version of the Galerkin boundary element method for 2-d problems. Computing **67**(1), 57–82 (2001)
230. Tran, T., Stephan, E.P.: An overlapping additive Schwarz preconditioner for boundary element approximations to the Laplace screen and Lamé crack problems. J. Numer. Math. **12**(4), 311–330 (2004)
231. Tran, T., Stephan, E.P., Mund, P.: Hierarchical basis preconditioners for first kind integral equations. Appl. Anal. **65**, 353–372 (1997)
232. Wathen, A., Fischer, B., Silvester, D.: The convergence rate of the minimal residual method for the Stokes problem. Numer. Math. **71**(1), 121–134 (1995)
233. Wathen, A.J.: Preconditioning. Acta Numer. **24**, 329–376 (2015)
234. Wendland, H.: Scattered Data Approximation. Cambridge University Press, Cambridge (2005)
235. Wendland, W.L., Stephan, E.P.: A hypersingular boundary integral method for two-dimensional screen and crack problems. Arch. Rational Mech. Anal. **112**, 363–390 (1990)
236. Widlund, O.B.: Iterative substructuring methods: algorithms and theory for elliptic problems in the plane. In: R. Glowinski, G.H. Golub, G.A. Meurant, J. Périaux (eds.) Proc. of the First International Symposium on Domain Decomposition Methods for Partial Differential Equations, pp. 113–128. SIAM, Philadelphia, PA (1988)
237. Widlund, O.B.: Optimal iterative refinement methods. In: T.F. Chan, R. Glowinski, J. Périaux, O.B. Widlund (eds.) Proc. of the Second International Symposium on Domain Decomposition Methods for Partial Differential Equations, pp. 114–125. SIAM, Philadelphia, PA (1989)
238. Wilkinson, J.H.: The Algebraic Eigenvalue Problem. Clarendon Press, Oxford (1965)
239. Wu, H., Chen, Z.: Uniform convergence of multigrid V-cycle on adaptively refined finite element meshes for second order elliptic problems. Sci. China Ser. A **49**(10), 1405–1429 (2006)
240. Xu, J.: Theory of multilevel methods. Ph.D. thesis, Cornell University, Ithaca, New York (1989)
241. Xu, J.: Iterative methods by space decomposition and subspace correction. SIAM Rev. **34**(4), 581–613 (1992)
242. Xu, Y., Cheney, E.W.: Strictly positive definite functions on spheres. Proc. Amer. Math. Soc. **116**, 977–981 (1992)

243. Yserentant, H.: On the multilevel splitting of finite element spaces. Numer. Math. **49**(4), 379–412 (1986)
244. Zhang, X.: Multilevel Schwarz methods. Numer. Math. **63**, 521–539 (1992)
245. Zienkiewicz, O.C., Kelly, D.W., Bettess, P.: Marriage à la mode—the best of both worlds (finite elements and boundary integrals). In: Energy methods in finite element analysis, pp. 81–107. Wiley, Chichester (1979)

# Index




E. P. Stephan and T. Tran, *Schwarz Methods and Multilevel Preconditioners for Boundary Element Methods*, https://doi.org/10.1007/978-3-030-79283-1

# Index of Notation: Function Spaces

# Index of Notation: Norms

# Index of Notation: Operators and Other Symbols